Handbook of Research on Mobile Marketing Management

Key Pousttchi
University of Augsburg, Germany

Dietmar G. Wiedemann
University of Augsburg, Germany

BUSINESS SCIENCE REFERENCE
Hershey · New York

Director of Editorial Content: Kristin Klinger
Senior Managing Editor: Jamie Snavely
Assistant Managing Editor: Michael Brehm
Publishing Assistant: Sean Woznicki
Typesetter: Mike Killian, Carole Coulson
Cover Design: Lisa Tosheff
Printed at: Yurchak Printing Inc.

Published in the United States of America by
Business Science Reference (an imprint of IGI Global)
701 E. Chocolate Avenue
Hershey PA 17033
Tel: 717-533-8845
Fax: 717-533-8661
E-mail: cust@igi-global.com
Web site: http://www.igi-global.com/reference

Library of Congress Cataloging-in-Publication Data

Handbook of research on mobile marketing management / Key Pousttchi and Dietmar G. Wiedemann, editors.
p. cm.

Includes bibliographical references and index.
Summary: "This book provides a compelling collection of innovative mobile marketing thoughts and practices"--Provided by publisher.

ISBN 978-1-60566-074-5 (hardcover) -- ISBN 978-1-60566-075-2 (ebook) 1. Telemarketing. 2. Text messages (Telephone systems) 3. Internet marketing. 4. Internet advertising. 5. Cellular telephones. I. Pousttchi, Key, 1970- II. Wiedemann, Dietmar G., 1977-
HF5415.1265.H357 2010
658.8'72--dc22

2009006922

British Cataloguing in Publication Data
A Cataloguing in Publication record for this book is available from the British Library.

Advances in E-Business Research Series (AEBR)

ISSN: 1935-2700

Editor-in-Chief: In Lee, PhD, Western Illinois University, USA

E-Business Innovation and Process Management

CyberTech Publishing • copyright 2007 • 384 pp • H/C (ISBN: 1-59904-277-0)

E-business research is currently one of the most active research areas. With the rapid advancement in information technologies, e-business is growing in significance and is having a direct impact upon ways of doing business. As e-business becomes one of the most important areas in organizations, researchers and practitioners need to understand the implications of many technological and organizational changes taking place. Advances in E-Business Research: E-Business Innovation and Process Management provides researchers and practitioners with valuable information on recent advances and developments in emerging e-business models and technologies. This book covers a variety of topics, such as e-business models, e-business strategies, online consumer behavior, e-business process modeling and practices, electronic communication adoption and service provider strategies, privacy policies, and implementation issues.

E-Business Models, Services and Communications

Information Science Reference • copyright 2008 • 300 pp • H/C (ISBN: 978-1-59904-831-4)

With the rapid advancement in information technologies, e-business is rapidly growing in significance and is having a direct impact upon business applications and technologies. E-Business Models, Services and Communications provides researchers and practitioners with valuable information on recent advances and developments in emerging e-business models and technologies. This book covers a variety of topics such as e-business models, telecommunication network utilization, online consumer behavior, electronic communication adoption and service provider strategies, and privacy policies and implementation issues.

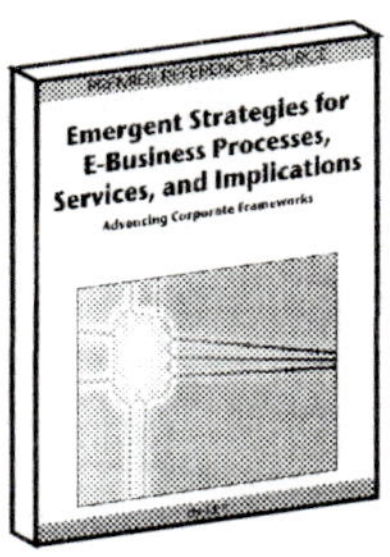

Emergent Strategies for E-Business Processes, Services, and Implications: Advancing Corporate Frameworks

Information Science Reference • copyright 2009 • 374 pp • H/C (ISBN: 978-1-60566-154-4)

Recently, e-business applications have evolved beyond business transactions and services to include customer relationship management (CRM), mobile computing, Web mining, e-healthcare, social networking, and Web 2.0. As e-business continues to create new business models and technologies, researchers, higher education faculty, and practitioners are in great need of appropriate reference resources to enhance their understanding of all aspects of e-business technologies and management.

E-business is broadly defined as a business process that includes not just the buying and selling of goods and services, but also servicing customers, collaborating with business partners, and conducting electronic transactions within an organization via telecommunications networks. E-business technologies and applications continue to evolve in many different directions and are now key strategic assets in business organizations. New e-business applications that have significant implications for the corporate strategies are being developed constantly. Current e-business research comes from diversified disciplines ranging from marketing, psychology, information systems, accounting, economics to computer science. The Advances in E-Business Research (AEBR) Book Series plans to serve as balanced interdisciplinary references for researchers and practitioners in this area.

IGI GLOBAL
DISSEMINATOR OF KNOWLEDGE

Hershey • New York

Order online at www.igi-global.com or call 717-533-8845 x100 – Mon-Fri 8:30 am - 5:00 pm (est) or fax 24 hours a day 717-533-8661

Jörg Link, *University of Kassel, Germany*
Juliane Chudalla, *University of Augsburg, Germany*
Katja Westphal, *University of Augsburg, Germany*
Key Pousttchi, *University of Augsburg, Germany*
Marco Garito, *Digital business, Italy*
Marc-Oliver Reeh, *Leibniz Universität Hannover, Germany*
Matti Leppäniemi, *University of Oulu, Finland*
Niklas Nohlen, *SHS VIVEON AG, Germany*
Nils Urbach, *International University Schloss Reichartshausen, Germany*
Ramin Vataparast, *Nokia Interactive Advertising, USA*
Richard Mannix, *Regent's College, UK*
Rob Kauffman, *Arizona State University, USA*
Seng-Cho T. Chou, *National Taiwan University, Taiwan*
Stefan Smolnik, *International University Schloss Reichartshausen, Germany*
Stuart J. Barnes, *University of East Anglia, UK*
Süphan Nasir, *Istanbul Universitesi, Turkey*
Tobias Haunstetter, *University of Augsburg, Germany*
Tobias Straub, *Baden-Württemberg Cooperative State University, Germany*
Wen-Kuo Chen, *National Taiwan University, Taiwan*
Wolfgang Palka, *University of Augsburg, Germany*

List of Contributors

Table of Contents

Section 3
Research Issues in Mobile Marketing

Detailed Table of Contents

Mobile marketing is a new form of marketing communication using mobile communication techniques to promote goods, services and ideas. The marketing instrument provides anytime and anywhere interaction, location and situation dependency, targeted addressing of consumers, and inherent measurement of campaign effectiveness. Based on case study research and empirical studies this chapter examines marketing objectives, types and implementation techniques of mobile marketing. Moreover, two instruments for mobile marketing are provided: The campaign type implementation toolbox supports the development of campaigns on a tactical level and the campaign type selection toolbox enables a purposive planning of mobile marketing campaigns on a strategic level.

The purpose of this chapter is to illustrate and reflect upon the extant literature on mobile marketing with a view of identifying contributions, gaps and avenues for future research. The review is based on more than two hundred articles published in leading journals and conference proceedings. The authors first

environment supporting device independent, anytime, anywhere publishing, i.e. a Ubiquitous Media Environment (UME). In this paper the authors explore how ubiquitous advertising challenges the role of media houses. Ubiquitous advertising refers to advertising in a UME, anytime, anywhere and in any device. Based on results from interviews and workshops with advertisers and publishers the authors discuss how media houses traditional role is challenged. We contribute with an understanding of how media houses can develop their role to succeed with ubiquitous advertising and to m-advertising literature with insights into the dynamics of balancing consumer and advertiser value of ubiquitous advertising.

Mobile technologies have posed new challenges for marketers as well as opportunities in an increasingly global market with high human mobility within and across borders. To understand and address customer needs more effectively, this chapter proposes to add new variables to market segmentations formula such as change of residence, movement and commuting distance/hours. A discussion of 'mobile nationals', and 'transnationals', as segments, is undertaken. The transnationals segment includes immigrants, refugees, tourists as well as businesspeople and professionals whose careers span into several countries. These groups create multiple reference points, which are likely to determine their characteristics and behaviour. This is an emerging and promising customer segment particularly for mobile marketing and mobile services. To explore the viability of such a segment, the authors have made use of several existing theoretical frameworks and concepts of segmentation. Future research should focus on the identification of transnational and national mobile segments while also developing and fine tuning the new variables –movement, change of residence and commuting for segmentation theory.

The success of Mobile Marketing is to a large extent dependent on two factors: (1) the situation appropriateness, where demands are made on the ability of the supplier to process all the central characteristics of a customer situation and turn them into an offer that is appropriate to the situation; (2) the economic success of the supplier depends on the economic potential of the situation and his accurate appraisal of it. The supplier therefore has to model both the specific features and the profit potential of the various customer situations at the right time and in the right way. This chapter is going to demonstrate that the success of Mobile Marketing measures can be guaranteed by a synergic interplay of market and result orientation. On the one hand customer situations are systemized and placed in the context of an integrated Customer Relationship Management system; on the other hand the evaluation of situations, reference object hierarchies, and the various levels of situation profit and loss statements are shown.

This chapter introduces an integrated way to look into the world of mobile offerings, referred often as m-commerce,- through a cross-country analysis of multiple applications in various countries in Europe and Asia. The authors investigated the role of four dominant factors - technology, government, financial incentives and culture - that influence mobile value offerings and connected them to Methlie and Pederson's (2005) model of extrinsic and intrinsic attributes in mobile space. The authors found that there are many applications and services available that provide added value to customers, but an ill-defined business model prevents their successful realization. An unclear working relationship among the value players and their relative power position in the market also hinder the roll out of many services. The key findings from our research consist of-: i. the role of the four factors in development and deployment of m-commerce across multiple countries; ii. The level of influence of each of the four factors has on various applications; iii. the importance of value delivery, not merely the promotion of the "bells and whistles" of new applications, with a clear focus on a viable business model; iv. the role of socio-cultural differences in influencing mobile value applications and acceptance. In addition, the authors found that competition in the early phase of market growth is counter-productive and cooperative relationship between all value players is absolutely crucial for healthy growth of the mobile market and ultimately leads to a win-win situation for everyone.

The concept of mobile communication has been rapidly changing all over the world and the Global System for Mobile communication (GSM) business has been evolving from voice services to value added services (VAS). The rapid growth in mobile technology enables Mobile Network Operators (MNOs) to provide a wide range of VAS to their corporate customers and these VAS change and transform companies' way of doing business. The purpose of this chapter is to analyze the corporate VAS of a MNO, which has gained global acceptance and recognition, as a case study in order to understand how the mobile communication changes the companies' way of doing business. The key supporting mobile technologies, services and solutions, which are provided by the case company to its corporate customers, are reviewed in order to comprehend the mobile technologies, practices, and devices that support mobile business (m-business) and mobile marketing (m-marketing) applications.

Customer Relationship Management is conducive to providing customer information and the correct way to deal with customers. After the advent of the world-wide-web (www), which brought an increase in customer data and intensified CRM discussions in its wake, the authors are now entering the age of computerized mobile communication with diverse possibilities to enlist the services of mobile people. This chapter outlines the new and latest technological developments and the resultant capabilities of mobile CRM. Looking to a future in which mobile marketing efforts may become increasingly important, the erosion of personal freedom and the threat to this freedom becomes all the more significant. The findings from reactance research do not only specify the reactance conditions but also the reactance consequences. Previous research has sometimes pointed to self-determination as an acceptance factor yet failed to analyze the reactance risk more closely. This chapter will attempt to close this gap. In this respect, the initial findings of an empirical study will be presented, revealing that the acceptance of mobile services is surprisingly low and the reactance risk for mobile services is not to be underestimated, and that reactance involves the corresponding behavior on the part of the user. The plea is to consider the acceptance risk in the mobile CRM of the future and select a customer-friendly permission policy.

Near Field Communication (NFC) is a short-distance wireless technology which allows user friendly networking of mobile terminals, e.g., cellular phones and PDAs, as it does with stationary units like Automated Teller Machines (ATM). Hardware producers, network operators and service providers are willing to implement NFC technologies in order to offer new services to their customers. The main goal is to increase the attraction of the underdeveloped areas of mobile commerce and make them more interesting for owners of mobile devices. If and how much real potential there is in NFC as a basic technology to really provide new impulses in mobile commerce and mobile marketing so that a higher acceptance can be reached are remaining open questions. This chapter tries to approach these questions and to provide possible answers by outlining relevant applicative examples and further configuration options of NFC technology as well as by discussing their acceptability.

The deployment of SMS-based marketing campaigns has been noted as a very successful mean of reaching consumers. With this mind, this chapter reports on three different studies that the authors have conducted aimed at better understanding what makes SMS marketing campaigns successful. The studies were conducted in the past four years and they explore business as well as consumer perspectives: (i) a consumer scenario perspective, which examines the importance of three factors in SMS advertising acceptance; (ii) a consumer decision perspective, which analyses the opt-in decision for an SMS campaign; and (iii) a business, perspective examining managers' perceptions of the critical success factors in advertising campaigns. The chapter rounds off with conclusions and recommendation for future research and practice in the area of mobile advertising.

This chapter investigates the impact of gender and age on mobile marketing responsiveness among regular customers of Finnish night club chain. The study develops sixteen research propositions which are tested with a sample of 8578 members of the company's permission- based, opt-in mobile advertising list . The results contribute to mobile marketing and technology acceptance literature in various ways. First, the authors find support for four out of eight gender related propositions. Second, five out of the eight age-related hypotheses are supported. The results further show that among opt-in customers, gender and age explain various antecedents of intention to engage in mobile marketing. The study provides several theoretical and managerial contributions and outlines vital avenues for further investigation in the field.

This chapter investigates the changes in customers' responses to mobile direct mail (DM) coupons on the shop visit probability (SVP) of a beauty parlor. Two experiments were carried out to examine the promotional effects of mobile DM coupons. The first experiment, conducted in 2004, compared mobile DM coupons with postcard DM coupons. The mobile DM coupons were found to have no effect on SVP, although positive effects were observed for postcard DM coupons. The second experiment, conducted in 2005 with three types of mobile DM coupons, compared the responses of new customers with those of repeat customers. The results varied according to the settings but demonstrated stability in the same settings. For repeat customers, mail with shop's telephone number for reservation had positive effects on SVP; however, ordinary mail had no effect, while hyperlink mail had negative effects. For new customers, all three types of mobile DM coupons exhibited positive effects on SVP.

Chapter 18
Mobile Store Environment Dynamics: An Interdisciplinary Approach 346
Adam P. Vrechopoulos, Athens University of Economics and Business, Greece
Emmanouela E. Manganari, Athens University of Economics and Business, Greece
George J. Siomkos, Athens University of Economics and Business, Greece

The development and rapid diffusion of m-commerce has attracted lately a great deal of research interest. Researchers from many different disciplines and domains (e.g. Strategic Marketing, Human Computer Interaction, Consumer Behavior, Environmental Psychology, Information Technology, Retailing, E-Commerce, etc.) attempt to examine and better understand this new medium, following different scientific paths. The current chapter constitutes an interdisciplinary research effort on that field placing particular emphasis on the design qualities of the mobile store environment-atmosphere and its effects on users'/consumers' behavior. To that end, the concept "m-atmospherics" along with a corresponding conceptual model are introduced as the theoretical vehicles that can well support the initiation of future research attempts measuring m-atmospherics effects on consumer behavior.

Chapter 19
Mobile Loyalty Programs: Relevance for Relationship Management and Consumer Acceptance 362
Andreas Mann, SVI-Stiftungslehrstuhl für Dialogmarketing, Germany
Jan Prein, University of Kassel, Germany

Mobile loyalty programs hold various advantages to companies compared to traditional card-based programs. Mobile customer cards enable companies to precisely assign advertising re-actions and purchase data to individual customers. For this reason customer retention activities can be adjusted to individual preferences and the situational context. However, any successful implementation of mobile technologies in consumer contexts depends primarily on consumer acceptance. In this chapter, the authors begin by illustrating characteristics of mobile loyalty programs, particularly as it relates to relationship management. Following that the authors identify and analyse drivers affecting the acceptance of mobile loyalty programs and present the results of an exploratory survey on consumer acceptance by using a fictional mobile customer card offer.

As mobile devices are personal communication tools, they are platforms for word-of-mouth marketing. Mobile viral marketing is tremendously attractive for marketers but neglected by academic research. Surprisingly, relatively few studies are directed at its basic elements, i.e., directed at willingness to forward different mobile viral content, understanding characters of those who forward mobile viral content frequently, and characters of the recipients of this content. This chapter presents the findings of an online survey conducted to empirically investigate the consumers' intention to forward different kinds of mobile viral content, to identify the primary target groups for the mobile viral marketing in terms of their forwarding behavior, and to analyze to whom mobile viral content is forwarded.

Research shows that the use of mobile phone services is surprisingly low because users would not like to pay money without first using services. The authors consider that a customer is not only a service user, but also the partner of a firm. Therefore, this chapter investigates how the users recommend mobile phone services to other potential customers based on their experiences. The authors focus on two determinants of customer recommendation behavior, namely, overall satisfaction and flow experience. The authors also propose a number of hypotheses related to the constructs of overall satisfaction, and test them using data provided by 3G mobile service subscribers. The results contribute to our understanding of how customers are willing to make recommendations to others. Besides increasing customer satisfaction, the authors should consider a mediating factor—users' flow experiences. The findings provide empirical support for most of the hypotheses. The theoretical and practical implications of the findings are also discussed, and suggestions for further research are offered.

With the increasing growth and sophistication of mobile communications, companies integrate the mobile channel into their marketing communication and campaigns. Companies are able to execute very different and creative mobile marketing campaigns by using a wide range of mobile platforms such as Wireless Application Protocol (WAP), Short Message Service (SMS), and Multimedia Message Service (MMS). The purpose of this chapter is to review and exemplify the concepts, objectives, strategies and promotional tactics associated with mobile marketing campaigns; so that the review of relevant literature and exemplification of mobile marketing campaigns provide an insight into the nature of mobile marketing campaigns.

Mobile Commerce is of growing interest for vendors and customers and with that its importance within the mix of marketing and distributing channels increases. This is of particular significance when customer retention and improved service are essential success factors. Because of their focus on customer integration the instruments of Web 2.0 offer new interactive possibilities for customer retention management. With a systematization of the new web-based applications linked to an evaluation of existing possibilities of customer retention this chapter offers a reasonable frame of reference for the utilization of Web 2.0 within the success chain of customer retention management in Mobile Commerce. Combined with existing studies of the acceptance of Web 2.0 recommendations for a successful timely order of introduction are offered to reach optimal diffusion and retention rates.

The increasing availability of affordable mobile business services (or "m-services") across different telecommunication carriers and platform is now a reality that is changing the way customers perceive products and solutions: mobile TV, one of the latest m-services, can be now seen either as a stand-alone channel or in conjunction with other media, thus providing several levels of engagement and interaction. Marketeers are now talking about "experiential marketing" to explain how customers and businesses are now reacting to market stimuli: Philip Kotler talked about "sense and respond" in his book "Marketing moves". Technology development is now taking a step forward enabling customers to be users and creators of products, services and brand image. This chapter provides and overview about traditional mobile marketing and mobile TV, describing how those interact and how business can be affected in some crucial areas: platform interoperability and revenue models. Some real examples of mobile initiatives are also provided.

Despite the fact that the mobile phone is a perfectly suited advertising medium also for brand campaigns a creative perspective is not only desirable but also necessary. Because: Without creativity, the latest mass medium would be swiftly rendered desolate. In the creative conceptual design of brand campaigns it is important above all to carry out advertising not only on the mobile phone but with it. On the basis of five theses it is illustrated how the mobile phone becomes the essential building block within commercial multi-channel concepts. One key issue being the telling of 'brand stories', told across various media or 'stages'. If the stories being told are compelling enough one can convert brand supporters into brand addicts, who in the best case scenario wouldn't want to miss a single performance of their favorite brand.

Foreword

Marketing aims at creating, communicating, delivering, and exchanging offerings that (hopefully) have value for customers, clients, partners, or other relationships. Much of its success depends on reaching the right people with the right message in the right situation. Whatever can be of help in this endeavour is to be used or at least tried out. Trying different media is part of these experiments and has lead to many experiences - more or less successful ones among them.

In today's media-rich society with many messages (more or less desperately) searching for receivers successful marketing means especially to reach people when they are open to receive a marketing message. This may be one reason, why mobile marketing – or more precisely – marketing making use of mobile communication is considered as such a promising field. The vast coverage that mobile communication networks (e.g., GSM) offer allows to reach more and more people at almost every location and consequently at almost every time. In addition mobile communication subscribers are often identified to at least some degree via their mobile phone number. And last but not least the infrastructure of cell-based networks together with ever-improving navigation based on satellites can tell quite a bit about customers and their situation. Thus, marketers can come closer to consumers than ever before.

Coming close and not being annoying requires special skills. While not all of them can be acquired by research, this situation clearly asks for a Handbook of Research on Mobile Marketing Management. Key Pousttchi and Dietmar G. Wiedemann can therefore be applauded for having chosen the right topic at the right time demonstrating good marketing skills. Will a handbook be the right medium in an age of content overflow? It can be, if it is structured well, and the structure of this book does make sense: Starting with an "Introduction to Mobile Marketing" and then following up with "Technology Issues in Mobile Marketing" avoids the mistake to believe that technology would solve every problem, even if it is not even explained, while demonstrating at the same time that mobile marketing cannot be successful without understanding the technology basics. Both sections give a good basis for describing "Research Issues in Mobile Marketing", as done in Section 3, before then "Managerial Issues in Mobile Marketing" can be discussed in Section 4.

This handbook, collected by editors, who for a long time have fostered the communication and exchange between researchers and practitioners, should help both to understand current issues as well as open questions and last not least each other.

May it receive the success it deserves.

Kai Rannenberg

Kai Rannenberg *(www.m-chair.net) holds the T-Mobile Chair for Mobile Business and Multilateral Security at Goethe University Frankfurt since September 2002. Before he was with the System Security Group at Microsoft Research Cambridge, UK focussing on "Personal Security Devices and Privacy Technologies". 1993-1999 Kai worked at Freiburg University and coordinated the interdisciplinary "Kolleg Security in Communication Technology". Kai is active in the international (ISO/IEC) standardization of IT Security and Criteria (JTC 1/SC 27/WG 3 "Security evaluation criteria"), is serving as Convenor of the SC 27 Working Group 5 on "Identity management and privacy technologies", and is leading coordinator of the Network of Excellence "Future of Identity in the Information Society" (FIDIS) funded by the European Union. His current research interests include mobile applications and Multilateral Security in M-Business and M-Commerce; privacy and identity management, communication infrastructures and devices; and IT security evaluation and certification.*

Preface

There is little doubt about the profound effect wireless and mobile networks have on everyday life. With the development of wireless and mobile networks, mobile marketing is creating significant benefits and new business opportunities for companies. Thus, mobile marketing – a new form of marketing communication using mobile communication techniques to promote goods, services and ideas – will be the subject of interest for many years to come. Mobile marketing revenues, as indicated by many researchers, make up a large share of m-commerce revenues. The high penetration rate of cell phones has resulted in the increasing use of handheld devices to deliver advertisements of products and services. What is more compelling is that mobile marketing provides for interaction anytime and anywhere, location and situation dependency, targeted addressing of consumers, and inherent measurement of campaign effectiveness.

In general, the success of mobile marketing largely depends on whether it supports advertisers in coping with their major challenge today: getting time and attention from consumers. However, constraints in using both mobile networks and devices impose significant negative influences on the operational performance of mobile marketing campaigns. For example, the small screens of the devices allow the user to view only limited pages of information. In addition, these constraints range from issues of minimal battery power on wireless devices, limited and error-pronewireless links, limited input and output capacity, the diversity of devices, and the myriad differences in user profiles.

TARGETS AND OVERALL OBJECTIVES OF THE HANDBOOK

Our understanding of some of the basics as well as the technology, research, and managerial issues surrounding mobile marketing continues to increase but information on these issues remain sparse. Because the mobile market is still nascent, mobile marketing campaigns still have to be developed by "artists" who rely solely on intuition and personal experience. In order for a professional worker to better perform this task, he should be provided with a handbook, which provides current thinking in designing, managing, and evaluating mobile marketing campaigns as well as best practices. This Handbook of Research on Mobile Marketing Management aims to be just such a book. It will be most helpful as it will provide the current state-of-the-art developments in this area and includes case studies, challenges and trends. Contributions to this important publication will bring together the expertise of authors from different parts of the world and, as a result, the best in thoughts and practices of this fascinating and innovative marketing instrument. Rather than being the work of one author, the book is a collection of edited chapters written by experts and practitioners in the field of business informatics, information systems reseacrh, marekting, software engineering, economics, sociology, etc.

The editors received fourty chapters initially, but after an extensive peer-review process, only twenty-fife chapters were selected for inclusion in this handbook. The diversity in thoughts and experiences of the authors represents a joint work and cooperative action approach to mobile marketing. The subject area is wide and entails many facets that the book attempts to cover:

- Mobile Marketing Campaign Types
- Mobile Technologies Supporting Mobile Marketing
- Managing Interactions among Mobile Devices, Mobile Applications, and Users
- Location Based Services and Mobile Marketing
- Conceptualization of Mobile Marketing
- Theory Building in Mobile Marketing
- Consumers' and Advertisers' Perception and Attitudes Toward Mobile Marketing
- Cross-Country Comparison of Consumer Attitudes
- Success Factors of Mobile Marketing
- Planning and Management of Mobile Marketing Campaigns
- Mobile Marketing Case Studies

The handbook chapters combine techniques with practice to establish a synergic view of mobile marketing so that it will benefit researchers, marketers, managers in advertising agencies, IT-managers, developers, and students. We hope it is useful to anyone involved in mobile marketing.

ORGANIZATION OF THE BOOK

The book is organized into twenty-five chapters. A brief description of each of the chapters follows:

Section 1: Introduction to Mobile Marketing

Chapter 1, *Mobile Marketing Management: Marketing Objectives, Types and Implementation Techniques*, by Key Pousttchi and Dietmar G. Wiedemann. Mobile marketing is a new form of marketing communication using mobile communication techniques to promote goods, services and ideas. The marketing instrument provides anytime and anywhere interaction, location and situation dependency, targeted addressing of consumers, and inherent measurement of campaign effectiveness. Based on case study research and empirical studies this chapter examines marketing objectives, types and implementation techniques of mobile marketing. Moreover, two instruments for mobile marketing are provided: The campaign type implementation toolbox supports the development of campaigns on a tactical level and the campaign type selection toolbox enables a purposive planning of mobile marketing campaigns on a strategic level.

Chapter 2, *Reviewing Mobile Marketing Research to Date: Towards Ubiquitous Marketing*, by Dimitris Drossos and George M. Giaglis. The purpose of this chapter is to illustrate and reflect upon the extant literature on mobile marketing with a view of identifying contributions, gaps and avenues for future research. The review is based on more than two hundred articles published in leading journals and conference proceedings. The authors first discuss key areas that have already attracted the attention of researchers, such as consumer acceptance in m-marketing and location-sensitive mobile marketing. The authors then focus on the emerging area of ubiquitous marketing and illustrate how mobile and wireless devices and technologies can become the enablers of meaningful dialogues between customers and marketers on an omnipresent basis. Ubiquitous marketing could be the next frontier in electronic commerce and customer relationship management.

Chapter 3, *Framework for Mobile Marketing: The Locales Framework*, by Dennis Lee, Ralf Muhlberger, Mark Brown. This chapter suggests that mobile marketing research take an Interaction Design approach through the adoption of the Locales Framework as an alternative perspective to investigate the context of mobile marketing. The research shows how such an approach can integrate marketing and new technology research, highlighting the issues and opportunities brought about by the technology within a domain context, i.e. mobile marketing. Given that there are not many theoretically based frameworks available to support mobile marketing research, the application of this framework provides a roadmap for future research in mobile marketing and helps build the body of knowledge in this field.

Chapter 4, *Factors Affecting Mobile Advertising*, by Ramin Vatanparast. Mobile advertising holds strong promises to become the most highly targeted advertising medium offering new ways to target messages to users. By utilizing mobile advertising, companies can run marketing campaigns targeted to tens of thousands of people with a fraction of the costs and time compared to other direct marketing mediums. However, as mobile advertising is a novel approach, many aspects of it still need further investigation. Little is known regarding the effectiveness of mobile advertising campaigns and the factors contributing to their success. This chapter aims to provide a comprehensive understanding of the advertising space and its influencing factors. First, the study investigates factors that influence mobile advertising from both the industry's and consumer's point of view. Second, based on a review of previous studies in the field, the author proposes a conceptual model for mobile advertising, which categorizes the factors in different groups and provides a holistic view of their impact in the mobile advertising space.

Chapter 5, *Advertising Challenges in Ubiquitous Media Environments*, by Maria Åkesson and Carina Ihlström Eriksson. Mobile advertising opportunities have attracted interest from industry and academics as a response to the trend of diffusion of new mobile technology. One industry particularly interested in new advertising opportunities is the newspaper industry. Many media houses envision a future multi-channel media environment supporting device independent, anytime, anywhere publishing, i.e. a Ubiquitous Media Environment (UME). In this paper the authors explore how ubiquitous advertising challenges the role of media houses. Ubiquitous advertising refers to advertising in a UME, anytime, anywhere and in any device. Based on results from interviews and workshops with advertisers and publishers the authors discuss how media houses traditional role is challenged. We contribute with an understanding of how media houses can develop their role to succeed with ubiquitous advertising and to m-advertising literature with insights into the dynamics of balancing consumer and advertiser value of ubiquitous advertising.

Chapter 6, *Segmentation Challenges Posed by 'Transnationals' in Mobile Marketing*, by Ibrahim Sirkeci and Richard Mannix. Mobile technologies have posed new challenges for marketers as well as opportunities in an increasingly global market with high human mobility within and across borders. To understand and address customer needs more effectively, this chapter proposes to add new variables to market segmentations formula such as change of residence, movement and commuting distance/hours. A discussion of 'mobile nationals', and 'transnationals', as segments, is undertaken. The transnationals segment includes immigrants, refugees, tourists as well as businesspeople and professionals whose careers span into several countries. These groups create multiple reference points, which are likely to determine their characteristics and behaviour. This is an emerging and promising customer segment particularly for mobile marketing and mobile services. To explore the viability of such a segment, the authors have made use of several existing theoretical frameworks and concepts of segmentation. Future research should focus on the identification of transnational and national mobile segments while also developing and fine tuning the new variables –movement, change of residence and commuting for segmentation theory.

Chapter 7, *Situation Approach as Success Factor of Mobile Marketing*, by Jörg Link and Franziska Seidl. The success of Mobile Marketing is to a large extent dependent on two factors: (1) the situation appropriateness, where demands are made on the ability of the supplier to process all the central characteristics of a customer situation and turn them into an offer that is appropriate to the situation; (2) the economic success of the supplier depends on the economic potential of the situation and his accurate appraisal of it. The supplier therefore has to model both the specific features and the profit potential of the various customer situations at the right time and in the right way. This chapter is going to demonstrate that the success of Mobile Marketing measures can be guaranteed by a synergic interplay of market and result orientation. On the one hand customer situations are systemized and placed in the context of an integrated Customer Relationship Management system; on the other hand the evaluation of situations, reference object hierarchies, and the various levels of situation profit and loss statements are shown.

Chapter 8, *A Framework for Understanding Mobile Value Offering through Multi-Country Studies,* by Kala Chand Seal. This chapter introduces an integrated way to look into the world of mobile offerings, referred often as m-commerce,- through a cross-country analysis of multiple applications in various countries in Europe and Asia. The authors investigated the role of four dominant factors - technology, government, financial incentives and culture - that influence mobile value offerings and connected them to Methlie and Pederson's (2005) model of extrinsic and intrinsic attributes in mobile space. The authors found that there are many applications and services available that provide added value to customers, but an ill-defined business model prevents their successful realization. An unclear working relationship among the value players and their relative power position in the market also hinder the roll out of many services. The key findings from our research consist of-: i. the role of the four factors in development and deployment of m-commerce across multiple countries; ii. The level of influence of each of the four factors has on various applications; iii. the importance of value delivery, not merely the promotion of the "bells and whistles" of new applications, with a clear focus on a viable business model; iv. the role of socio-cultural differences in influencing mobile value applications and acceptance. In addition, the authors found that competition in the early phase of market growth is counter-productive and cooperative relationship between all value players is absolutely crucial for healthy growth of the mobile market and ultimately leads to a win-win situation for everyone.

Section 2: Technology Issues in Mobile Marketing

Chapter 9, *Supporting Marketing Practices: Mobile Network Operators' Value Added Services Changing the Way of Doing Business*, by Süphan Nasir. The concept of mobile communication has been rapidly changing all over the world and the Global System for Mobile communication (GSM) business has been evolving from voice services to value added services (VAS). The rapid growth in mobile technology enables Mobile Network Operators (MNOs) to provide a wide range of VAS to their corporate customers and these VAS change and transform companies' way of doing business. The purpose of this chapter is to analyze the corporate VAS of a MNO, which has gained global acceptance and recognition, as a case study in order to understand how the mobile communication changes the companies' way of doing business. The key supporting mobile technologies, services and solutions, which are provided by the case company to its corporate customers, are reviewed in order to comprehend the mobile technologies, practices, and devices that support mobile business (m-business) and mobile marketing (m-marketing) applications.

Chapter 10, *Mobile Customer Relationship Management (mCRM): Constraints and Challenges*, by Günter Silberer and Sebastian Schulz. Customer Relationship Management is conducive to providing customer information and the correct way to deal with customers. After the advent of the world-wide-web (www), which brought an increase in customer data and intensified CRM discussions in its wake, the authors are now entering the age of computerized mobile communication with diverse possibilities to enlist the services of mobile people. This chapter outlines the new and latest technological developments and the resultant capabilities of mobile CRM. Looking to a future in which mobile marketing efforts may become increasingly important, the erosion of personal freedom and the threat to this freedom becomes all the more significant. The findings from reactance research do not only specify the reactance conditions but also the reactance consequences. Previous research has sometimes pointed to self-determination as an acceptance factor yet failed to analyze the reactance risk more closely. This chapter will attempt to close this gap. In this respect, the initial findings of an empirical study will be presented, revealing that the acceptance of mobile services is surprisingly low and the reactance risk for mobile services is not to be underestimated, and that reactance involves the corresponding behavior on the part of the user. The plea is to consider the acceptance risk in the mobile CRM of the future and select a customer-friendly permission policy.

Chapter 11, *Employment and Acceptance of Near Field Communication in Mobile Marketing*, by Klaus-Peter Wiedmann, Marc-Oliver Reeh, and Henrik Schumacher. Near Field Communication (NFC) is a short-distance wireless technology which allows user friendly networking of mobile terminals, e.g., cellular phones and PDAs, as it does with stationary units like Automated Teller Machines (ATM). Hardware producers, network operators and service providers are willing to implement NFC technologies in order to offer new services to their customers. The main goal is to increase the attraction of the underdeveloped areas of mobile commerce and make them more interesting for owners of mobile devices. If and how much real potential there is in NFC as a basic technology to really provide new impulses in mobile commerce and mobile marketing so that a higher acceptance can be reached are remaining open questions. This chapter tries to approach these questions and to provide possible answers by outlining relevant applicative examples and further configuration options of NFC technology as well as by discussing their acceptability.

Chapter 12, *Mobile Customer Acquisition in the Swiss Health Care Industry: An Empirical Study*, by Niklas Nohlen, Stefan Smolnik, and Gerold Riempp. Applications based on mobile technologies are predicted to achieve increasing distribution as well as high potential for business processes especially with regard to customer interfaces. This chapter is going to analyze and present perspectives and approaches for initiation efforts by means of mobile technologies of Switzerland's health care industry. On the one hand, acceptance and success factors of such mobile applications are analyzed by means of empirical studies in-volving customers and companies, on the other hand a prototype of mobile processes for initiation efforts is introduced, which was developed in the course of a case study.

Chapter 13, *Opportunistic Networks as an Enabling Technology for Mobile Word-of-Mouth Advertising*, by Andreas Heinemann and Tobias Straub. The growing share of people using mobile devices, that support wireless peer-to-peer interaction, offers the opportunity to build a ubiquitous infrastructure for electronic word-of-mouth messaging and advertising. This chapter introduces Opportunistic Networks as a layer for one-hop communication that opens up electronic word-of-mouth messaging for mobile devices. The reader will learn about adPASS (short for advertisement passing), a system build on top of Opportunistic Networks for digital advertisement distribution stimulated by an anonymous and tamper-resistant bonus point model. A simulation based on empirical movement patterns indicates the feasibility of our approach. This chapter concludes with a summary and provides an outlook on further research paths starting from the findings.

Section 3: Research Issues in Mobile Marketing

Chapter 14, *Theories behind Mobile Marketing Research*, by Ramin Vatanparast. The future of mobile communication is expected to rely on mobile services and revenue generated through mobile marketing.

Marketing activities supported by mobile devices allow companies to directly communicate with their consumers without location or time barriers. It is becoming vital for today's marketers to understand the processes behind the factors affecting consumers' intention to use and adopt mobile marketing. One can often argue that the mobile marketing adoption is difficult to understand due to a lack of relevant research. However, much research has already been conducted on the adoption of mobile services and technology acceptance that is likely can support mobile marketing research. Thus it is essential for mobile marketing researchers to get a through understanding of the theory behind mobile service adoption and technology acceptance. Theories in this area have been developed gradually and built up on each another. This chapter covers some of the theories and related models and show how those could be used in mobile services and mobile marketing research.

Chapter 15, *In Search of Successful Mobile Advertising: Consumer and Business Perspectives*, by Stuart J. Barnes, Eusebio Scornavacca, Jonathon McKenzie, and Amy Carroll. The deployment of SMS-based marketing campaigns has been noted as a very successful mean of reaching consumers. With this mind, this chapter reports on three different studies that the authors have conducted aimed at better understanding what makes SMS marketing campaigns successful. The studies were conducted in the past four years and they explore business as well as consumer perspectives: (i) a consumer scenario perspective, which examines the importance of three factors in SMS advertising acceptance; (ii) a consumer decision perspective, which analyses the opt-in decision for an SMS campaign; and (iii) a business, perspective examining managers' perceptions of the critical success factors in advertising campaigns. The chapter rounds off with conclusions and recommendation for future research and practice in the area of mobile advertising.

Chapter 16, *The Impact of Gender and Age on Consumer Responsiveness to Permission-Based Mobile Advertising*, by Heikki Karjaluoto, Heikki Lehto, and Matti Leppäniemi. This chapter investigates the impact of gender and age on mobile marketing responsiveness among regular customers of Finnish night club chain. The study develops sixteen research propositions which are tested with a sample of 8578 members of the company's permission- based, opt-in mobile advertising list . The results contribute to mobile marketing and technology acceptance literature in various ways. First, the authors find support for four out of eight gender related propositions. Second, five out of the eight age-related hypotheses are supported. The results further show that among opt-in customers, gender and age explain various antecedents of intention to engage in mobile marketing. The study provides several theoretical and managerial contributions and outlines vital avenues for further investigation in the field.

Chapter 17, *Mobile DM Coupon Promotion in Japan: A Case Study on Response Behavior Changes in Services Consumption*, by Fumiyo N.Kondo, Yasuhiro Uwadaira, Mariko Nakahara, and Shahriar Akter. This chapter investigates the changes in customers' responses to mobile direct mail (DM) coupons on the shop visit probability (SVP) of a beauty parlor. Two experiments were carried out to examine the promotional effects of mobile DM coupons. The first experiment, conducted in 2004, compared mobile DM coupons with postcard DM coupons. The mobile DM coupons were found to have no effect on SVP, although positive effects were observed for postcard DM coupons. The second experiment, conducted in 2005 with three types of mobile DM coupons, compared the responses of new customers with those of repeat customers. The results varied according to the settings but demonstrated stability in the same settings. For repeat customers, mail with shop's telephone number for reservation had positive effects on SVP; however, ordinary mail had no effect, while hyperlink mail had negative effects. For new customers, all three types of mobile DM coupons exhibited positive effects on SVP.

Chapter 18, *Mobile Store Environment Dynamics: An Interdisciplinary Approach*, by Adam P. Vrechopoulos, Emmanouela E. Manganari, and George J. Siomkos. The development and rapid diffusion of m-commerce has attracted lately a great deal of research interest. Researchers from many different disciplines and domains (e.g. Strategic Marketing, Human Computer Interaction, Consumer Behavior, Environmental Psychology, Information Technology, Retailing, E-Commerce, etc.) attempt to examine and better understand this new medium, following different scientific paths. The current chapter constitutes an interdisciplinary research effort on that field placing particular emphasis on the design qualities of the mobile store environment-atmosphere and its effects on users'/consumers' behavior. To that end, the concept "m-atmospherics" along with a corresponding conceptual model are introduced as the theoretical vehicles that can well support the initiation of future research attempts measuring m-atmospherics effects on consumer behavior.

Chapter 19, *Mobile Loyalty Programs: Relevance for the Relationship Management and Consumer Acceptance*, by Andreas Mann and Jan Prein. Mobile loyalty programs hold various advantages to companies compared to traditional card-based programs. Mobile customer cards enable companies to precisely assign advertising re-actions and purchase data to individual customers. For this reason customer retention activities can be adjusted to individual preferences and the situational context. However, any successful implementation of mobile technologies in consumer contexts depends primarily on consumer acceptance. In this chapter, the authors begin by illustrating characteristics of mobile loyalty programs, particularly as it relates to relationship management. Following that the authors identify and analyse drivers affecting the acceptance of mobile loyalty programs and present the results of an exploratory survey on consumer acceptance by using a fictional mobile customer card offer.

Chapter 20, *Analyzing the Forwarding Behavior in Mobile Viral Marketing: An Empirical Study*, by Dietmar G. Wiedemann, Tobias Haunstetter, and Key Pousttchi. As mobile devices are personal communication tools, they are platforms for word-of-mouth marketing. Mobile viral marketing is tremendously attractive for marketers but neglected by academic research. Surprisingly, relatively few studies are directed at its basic elements, i.e., directed at willingness to forward different mobile viral content, understanding characters of those who forward mobile viral content frequently, and characters of the recipients of this content. This chapter presents the findings of an online survey conducted to empirically investigate the consumers' intention to forward different kinds of mobile viral content, to identify the primary target groups for the mobile viral marketing in terms of their forwarding behavior, and to analyze to whom mobile viral content is forwarded.

Chapter 21, *Understanding Consumer Recommendation Behavior*, by Wen-Kuo Chen, Heng-Chiang Huang, and Seng-Cho T. Chou. Research shows that the use of mobile phone services is surprisingly low because users would not like to pay money without first using services. The authors consider that a customer is not only a service user, but also the partner of a firm. Therefore, this chapter investigates how the users recommend mobile phone services to other potential customers based on their experiences. The authors focus on two determinants of customer recommendation behavior, namely, overall satisfaction and flow experience. The authors also propose a number of hypotheses related to the constructs of overall satisfaction, and test them using data provided by 3G mobile service subscribers. The results contribute to our understanding of how customers are willing to make recommendations to others. Besides increasing customer satisfaction, the authors should consider a mediating factor—users' flow experiences. The findings provide empirical support for most of the hypotheses. The theoretical and practical implications of the findings are also discussed, and suggestions for further research are offered.

Section 4: Managerial Issues in Mobile Marketing

Chapter 22, *Integrating Mobile Marketing into the Marketing Communication: Exemplification of Mobile Marketing Campaigns*, by Süphan Nasir. With the increasing growth and sophistication of mobile communications, companies integrate the mobile channel into their marketing communication and campaigns. Companies are able to execute very different and creative mobile marketing campaigns by using a wide range of mobile platforms such as Wireless Application Protocol (WAP), Short Message Service (SMS), and Multimedia Message Service (MMS). The purpose of this chapter is to review and exemplify the concepts, objectives, strategies and promotional tactics associated with mobile marketing campaigns; so that the review of relevant literature and exemplification of mobile marketing campaigns provide an insight into the nature of mobile marketing campaigns.

Chapter 23, *Interactive Customer Retention Management for Mobile Commerce*, by Dirk Möhlenbruch, Steffen Dölling, and Falk Ritschel. Mobile Commerce is of growing interest for vendors and customers and with that its importance within the mix of marketing and distributing channels increases. This is of particular significance when customer retention and improved service are essential success factors. Because of their focus on customer integration the instruments of Web 2.0 offer new interactive possibilities for customer retention management. With a systematization of the new web-based applications linked to an evaluation of existing possibilities of customer retention this chapter offers a reasonable frame of reference for the utilization of Web 2.0 within the success chain of customer retention management in Mobile Commerce. Combined with existing studies of the acceptance of Web 2.0 recommendations for a successful timely order of introduction are offered to reach optimal diffusion and retention rates.

Chapter 24, *Mobile Business and Mobile TV: Available Technologies, Future Opportunities and New Marketing Trends*, by *Marco Garito*. The increasing availability of affordable mobile business services (or "m-services") across different telecommunication carriers and platform is now a reality that is changing the way customers perceive products and solutions: mobile TV, one of the latest m-services, can be now seen either as a stand-alone channel or in conjunction with other media, thus providing several levels of engagement and interaction. Marketeers are now talking about "experiential marketing" to explain how customers and businesses are now reacting to market stimuli: Philip Kotler talked about "sense and respond" in his book "Marketing moves". Technology development is now taking a step forward enabling customers to be users and creators of products, services and brand image. This chapter provides and overview about traditional mobile marketing and mobile TV, describing how those interact and how business can be affected in some crucial areas: platform interoperability and revenue models. Some real examples of mobile initiatives are also provided.

Chapter 25, *Brand Driven Mobile Marketing: 5 Theses for Today and Tomorrow*, by Christopher Quente. Despite the fact that the mobile phone is a perfectly suited advertising medium also for brand campaigns a creative perspective is not only desirable but also necessary. Because: Without creativity, the latest mass medium would be swiftly rendered desolate. In the creative conceptual design of brand campaigns it is important above all to carry out advertising not only on the mobile phone but with it. On the basis of five theses it is illustrated how the mobile phone becomes the essential building block within commercial multi-channel concepts. One key issue being the telling of 'brand stories', told across various media or 'stages'. If the stories being told are compelling enough one can convert brand supporters into brand addicts, who in the best case scenario wouldn't want to miss a single performance of their favorite brand.

Key Pousttchi and Dietmar G. Wiedemann
Editors
wi-mobile Research Group
Chair of Business Informatics and Systems Engineering
University of Augsburg
2009

Section 1
Introduction to Mobile Marketing

Chapter 1
Mobile Marketing Management:
Marketing Objectives, Types and Implementation Techniques

Key Pousttchi
University of Augsburg, Germany

Dietmar G. Wiedemann
University of Augsburg, Germany

ABSTRACT

Mobile marketing is a new form of marketing communication using mobile communication techniques to promote goods, services, and ideas. The marketing instrument provides anytime and anywhere interaction, location and situation dependency, targeted addressing of consumers, and inherent measurement of campaign effectiveness. Based on case study research and empirical studies this chapter examines marketing objectives, types and implementation techniques of mobile marketing. Moreover two instruments for mobile marketing are provided: The campaign type implementation toolbox supports the development of campaigns on a tactical level and the campaign type selection toolbox enables a purposive planning of mobile marketing campaigns on a strategic level.

INTRODUCTION

Mobile communication techniques provide the opportunity to revolutionize marketing as they support advertisers in coping with their major challenge today: getting time and attention from consumers. For years, advertisers have created communication strategies based on print, radio, TV, and Internet to broadcast their messages, but the advertising effectiveness of these media has significantly decreased in recent years. However, the rapid growth in mobile phone ownership throughout the world (Fife & Pereira, 2005) has opened a new mass marketing channel that enables advertisers to come closer to consumers than ever before. The groundbreaking characteristics of mobile communication techniques (ubiquity – context sensitivity – identifying functions – command and control functions) enable potentials not reachable with other media types. (Pousttchi & Wiedemann, 2006)

Mobile marketing – a new tactical marketing instrument of marketing communication using mobile communication techniques to promote goods, services and ideas, providing anytime and

DOI: 10.4018/978-1-60566-074-5.ch001

anywhere interaction, location and situation dependency, targeted addressing of consumers, and inherent measurement of campaign effectiveness. With these options, the mobile channel is ideal for building a real one-to-one relationship with the individual customer (Scharl et al., 2005). This is reflected by response rates far higher than those for traditional advertising: Up to 20 percent responses within certain mobile marketing campaigns are not uncommon (Kavassalis et al., 2003). M:Metrics (2006) found that a sizable percentage of mobile subscribers are responding to short codes placed in advertisements or in other media, with Spain topping the list at 29.1 percent, followed by the UK at 18.5 percent, France at 10.1 percent, the United States at 7 percent and Germany at 3.4 percent. According eMarketer (2005), it is predicted that advertisers will spend from $254.8 million to $760.4 million on mobile marketing in the U.S. by 2009. Moreover, it is already apparent that mobile marketing will in the long run become an important source of revenue for mobile network operators (Sharma & Wildmann, 2009).

Despite these optimistic predictions of current and future success of mobile marketing our understanding of mobile marketing strategy formulation and implementation remains deficient. Leppäniemi and Karjaluoto (2008) argue that mobile marketing is too often implemented at ad hoc basis. Thus, in this early stage a mobile marketing campaign still has to be developed by an "artist" who relies solely on intuition and personal experience. In order for a professional worker to perform this task, he must be provided with well-defined toolboxes that distinguish the available elements of a mobile marketing campaign and link them to the different marketing objectives. To develop these toolboxes, a four-step approach is used: identify and classify major mobile marketing campaign characteristics, derive four mobile marketing standard types, identify and discuss the reachable mobile marketing objectives, and finally link standard types and objectives in a conceptual framework, complete with typical instances of the respective combinations.

The findings reported here draw upon the results of three empirical studies conducted in 2005 and 2007, including a case study research of 71 mobile marketing campaigns[1] and, based on the results, a web-based survey with 44 invited experts (both part of the MM1 study as described by Pousttchi & Wiedemann (2006)), as well as in-depth interviews with 10 senior managers of advertising agencies specialized in mobile marketing (the MM2 study).

CAMPAIGN TYPE IMPLEMENTATION TOOLBOX

Case study research indicated that a broad variety of characteristics is necessary to specify a single mobile marketing campaign. Basically, two main categories can be distinguished: push and pull. In push campaigns, advertisers send relevant but not explicitly requested mobile ads. According to the MM2 study, the number of push messages should be limited to two per month; more are typically considered mobile spam. In pull campaigns, other supporting media serve as triggers for customer requests and result in higher customer involvement. Therefore, pull campaigns are clearly preferred by the MM2 informants. The high suitability of on-pack advertising as a supporting medium (Figure 1) reveals the potential of mobile marketing for the food industry, as campaign instructions adequately describing the campaign process and the added value can be printed on the packaging at low additional cost. Providing this latter type of material is the primary challenge in the use of mobile marketing campaigns. Mobile marketing is effective only when it creates a win-win situation for both parties – consumer and advertiser. Bauer et al. (2005) identified added value through entertainment and information as the most important drivers of mobile marketing acceptance. If the added value is high enough, advertisers will even have the opportunity to charge a premium rate and refinance costs for sent messages. The MM2 study revealed that fluctuations in response rates are minimal between premium rates of €0.29 and €0.49, but that higher premium rates should be avoided due to possible image-damaging effects. Moreover, the premium rate model seems to be flawed due to the current unfavorable revenue share model in view of service providers (Zmijewska & Lawrence, 2006; Au & Kauffman, 2008) or, in our case, advertisers. Another crucial success factor is the permission of the consumers (opt-in) to send mobile ads. The characteristic *none* indicates that the opt-out principle is used instead – unsolicited messages are sent to the cell phone. This is allowed in some markets, provided that consumers are given the possibility to refuse easily to receive subsequent messages. However, for countries where such permission is regulated by law, e.g., in the European Union, we refer to the instance *none* as mobile spam. Unlike the email spam problem, the issue in mobile marketing campaigns is not one of scale (because SMS is not free), but of sensitivity due to the intimate nature of cell phones. 75% of the MM1 experts stated that mobile spam is a serious problem and may endanger their whole industry. Once an opt-in is received through one of the various channels available, the ideal type of further communication is dialogue. In contrast to this multi-bidirectional communication, message-reaction refers to one mutual bidirectional message exchange between advertisers and consumers, whereas one-time message refers to one single message only. Besides the type of mobile network used and the enabling technique for sending, replying, and forwarding mobile ads, positioning of the consumer is another technical issue. One of the particularities of mobile technology is that a consumer's location can be used to deliver relevant, targeted, and timely information exploiting this physical context. According to the MM1 study, location-based advertising is even more fruitful in the pull scenario, as in a push scenario, most consumers feel controlled and refuse outright to give permission for using their positioning data.

Table 1 aggregates the main characteristics of a mobile marketing campaign, representing the major implementation options of a campaign in the form of a morphological box. This will be referred to as the *campaign type implementation toolbox*.

Figure 1. Suitability of supporting media for mobile marketing (MM2 study)

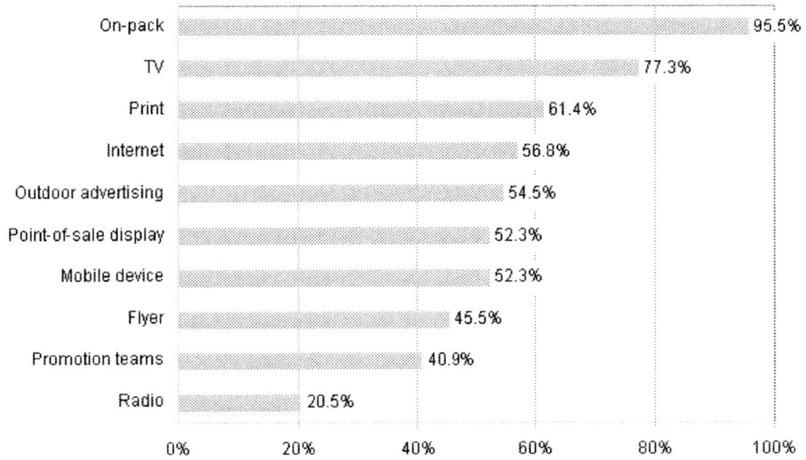

Table 1. Campaign type implementation toolbox

Characteristic	Instances							
Initiation	Push	Pull						
Supporting medium	Print	Outdoor advertising	Radio	TV	Internet	On-pack	Mobile device	Other
Added value	Information	Entertainment	Raffle	Coupon				
Cost	Premium rate	Transmission costs	None					
Opt-in	Mobile	Online	Offline	None				
Communication type	Dialogue	Message-reaction	One time message					
Mobile network type	Cellular phone network	Wireless local area network	Bluetooth	Radio frequency identification	Infrared	Near field communication		
Enabling technology	High-level language	Wireless Application Protocol	Multimedia Messaging Service	Short Message Service	Interactive Voice Response			
Positioning	Mobile network dependent technology	Specialized positioning system	Manual	None				

MOBILE MARKETING STANDARD TYPES

Using case study research, Pousttchi and Wiedemann (2006) established a typology of mobile marketing based on the characteristic of *added value* (Table 1). The respective instances turned out to be representative of a major category of mobile marketing campaigns. These categories are referred to here as *mobile marketing standard types*. Whereas the standard types represent different major concepts, they are not totally disjoint: any of the current mobile marketing campaigns can be classified into (at least) one standard type, but a single mobile marketing campaign may use more than one standard type, thus accumulating added value for the consumers.

Typical examples of the *information standard type* include mobile newsletters or special information about the brand, e.g., the nearest point of sale. However, mobile information services like news, weather, traffic, market rates, or horoscopes are also possible. Sophisticated campaigns use custom-built Java clients bringing a wealth of information in the form of a multi-faceted portal. The information must be relevant for customers; otherwise it is construed as spam. Higher relevance will be achieved if consumers receive context-sensitive information based on time of day, location, or profiles previously given. In this way, the information fulfills the particular needs of the customers in their current situation.

The starting point for the *entertainment standard type* is often amusing or humorous elements. Typical examples are mobile marketing campaigns providing games, videos, music, or customization options for cell phones like logos, wallpapers, themes, ring tones, or answering machine announcements. If the entertainment is amusing and emotional enough, the advertising message will be communicated subliminally. The MM2 experts stated that the provision of entertaining elements causes high response rates. The motives for this are consumers' natural playfulness and desire to express their uniqueness through mobile entertainment services; this is interestingly not significantly conditioned by age (Hong et al., 2006).

Although the prizes vary widely, the *raffle standard type* follows the same process as known from the offline or stationary Internet world. However, the main advantage of mobile raffles is the ubiquity of the medium, resulting in anywhere anytime participation. In general, the more attractive the prize is to the target group, the higher the chance of success for the campaign. Campaigns are most effective if they either have one very valuable first prize or if they supply prizes that each consumer wins.

The latter represents an intersection with the *coupon standard type* defined by the instance monetary incentive. With this type, advertisers create real value via mobile advertising and do not restrict themselves to sending general promotions. According to Drosos et al. (2007), the use of incentives in mobile marketing leads to more positive attitudes and purchase intentions. The most common mobile coupon is a straightforward SMS that has to be shown at a point of sale. To ensure against double spending, means of fraud prevention such as validation codes or the use of special scanners for 2D barcodes at the cash desk may be incorporated. Typical coupons include discounts or trial packages. In addition, free SMS or Multimedia Message Service (MMS) available on a brand website, or discount tickets collected in a virtual discount ticket book on the cell phone, belong to this standard type.

OBJECTIVES OF MOBILE MARKETING

When developing a mobile marketing campaign, advertisers must always start by identifying the target group and setting campaign objectives. In the MM1 study, the following primary campaign mobile marketing objectives were identified. *Building brand awareness* aims to reinforce the consumer's ability to recognize and recall the brand in purchase and consumption situations. *Changing brand image* aims to change the consumer's perception of the brand. Brand image refers to the perceptions about a brand as reflected by the compound held in consumer memory. These connotations may include product-related or non-product-related attributes, functional, experiential, or symbolic benefits, and overall brand attitudes. Basically, high levels of brand awareness and a positive brand image should increase the probability of brand choice and produce greater consumer loyalty. The objective of *increasing sales* aims to stimulate quicker or greater purchase of the brand and is mostly short-term focused. The objective of *establishing brand loyalty* aims to strengthen the consumer's psychological commitment to repurchase the brand, which requires larger and longer-term investments in relationship building and developing consumer intimacy with the brand.

Besides these primary objectives, two secondary objectives can be achieved with mobile marketing campaigns. Permission marketing (Krishnamurthy, 2001) has as a precondition the activity *build consumer database*. Such databases contain the Mobile Subscriber Numbers (MSISDN) of customers, together with their opt-in information or even detailed profiles for using the information during subsequent campaigns. Building profiles with detailed information to segment, target, and position individual consumers requires motivating them to visit a website where the opportunity for (self-)profiling is provided. *Mobile viral marketing* (also called mobile word-of-mouth) aims at passing on advertising messages from consumer to consumer via mobile communication techniques. Viral elements help to expand significantly the number of recipients and increase the impact of the campaign [ScDM05]. Respondents stated that personal messages sent by friends gain more credibility than those coming directly from the self-interested advertiser. In addition, consumers who got the initial message from a familiar recommender participate more frequently in a campaign as initial contacts.

CAMPAIGN TYPE SELECTION TOOLBOX

The *campaign type implementation toolbox* helps to design and customize a single mobile marketing campaign of a certain type. To support in the same way the higher-level decision to select the campaign type, the *campaign type selection toolbox* is now provided as the final step of the bottom-up approach presented in this paper.

After the mobile marketing standard types and the mobile marketing objectives have been identified, they are now linked in a matrix-type conceptual framework, complete with typical examples for any combination (Table 2). With this toolbox, advertisers have the ability to select a mobile marketing campaign concept according to the basic elements specified. Afterwards, the concept has to be specified using the campaign type implementation toolbox (Table 1).

Building Brand Awareness

Repeated exposures to a brand lead to increased consumer ability to recognize and recall the brand (Keller, 1993). Therefore, repeated information like mobile newsletters as well as branded logos, wallpapers, or themes presented permanently on the cell phone are appropriate to build brand awareness. A raffle in which all consumers win a prize (e.g., free download of a mobile game) contributes to this objective as well, since consumers are engaged with the brand for a longer time and memorize the feeling of success. An opportunity to send free SMS or MMS from the brand website not only involves the sender with the brand, but also the recipient, as the brand will be named in the message ("free SMS service by…").

Changing Brand Image

A brand image is a complex construct (Keller, 1993). Depending on the specific association to be changed, advertisers need to select a certain added value that conveys the desired brand image. For instance, BMW demonstrated product-related attributes by a mobile racing game, conveying the experience not only of a sporty car, but also of a technically perfect car. Games are especially suited to this purpose, as consumers play them several times and thus repeatedly encounter the embedded advertising message.

Increasing Sales

Mobile marketing provides many avenues for increasing sales. If advertisers have a rough knowledge of needs, e.g., with seasonal products

Table 2. Campaign type selection toolbox

	Information standard type	Entertainment standard type	Raffle standard type	Coupon standard type
Building brand awareness	Mobile newsletter	Logo, wallpaper or theme	Raffle in which all customers win	Free SMS or MMS contingent on the brand website
Changing brand image	Mobile portal informing about attributes	Mobile game	Placing image through the prize of the raffle	Coupon for a promotional gift that places the desired image
Increasing sales	Information about season products	Ring tone provided via an on-pack campaign	Raffle provided via an on-pack campaign	Cash, quantity, or seasonal discounts provided via SMS
Establishing brand loyalty	Question and answer application	WAP download-center	Provision of a raffle for existing customers	Mobile discount ticket book
Building customer database	Personalized information on condition of registration	Provision of a logo on condition of registration	Participation in a raffle on condition of registration	Provision of a coupon on condition of registration
Motivating mobile viral marketing	Targeted recommendation	Mobile greeting card	Participation in a raffle requires participation of other people	Discount in return of a recommendation

like fuel oil, they will be able to send information about special offers and evoke the need. Discounts as well as on-pack campaigns can facilitate the purchase decision and also increase short-term sales. This last is especially true for homogeneous products that become more interesting and valuable through the added value provided.

Establishing Brand Loyalty

Brand loyalty requires developing a relationship with consumers that is more than product-based. Using the information standard type, advertisers can make available a question and answer application that creates a great experience of an anytime-anywhere interaction with the brand. For example, the Austrian Raiffeisen bank in Tirol developed a virtual character, "Lynn". This SMS application acts as a "contact person" for financial questions and is based on an artificial intelligence application including about 7000 rules. A similar relationship can be established by a special Java-based download center application that is downloaded to the cell phone and provides various entertainment elements. Since content can be updated dynamically over the mobile Internet, advertisers have a direct relationship tool to access consumers. Another way to establish brand loyalty is the translation of the loyalty card idea into the cell phone environment in the form of a mobile discount ticket book (ideally in conjunction with a mobile payment procedure).

Building a Consumer Database

The consumer's interest in joining a permission marketing program is positively affected by message relevance and tangible benefit (Krishnamurthy, 2001). Personalization is an important antecedent of message relevance. Hence, if standard information is used, advertisers should provide the opportunity to enter a personal profile that specifies the type of information desired as well as the date and frequency of dispatch (Lippert, 2004). Tangible benefits can be offered by the standard types of entertainment, raffles, or coupons on condition of registration.

Motivating Mobile Viral Marketing

Targeted recommendation (Subramani & Rajagopalan, 2003) applies the "send to a friend" option in information portals or simply adds this request under the mobile ad. Typically, the recommender can roughly predict the recipient's interests and preferences based on private information. Hence, targeted recommendation shows great promise for spreading specific product information within a target group. Current viral mobile marketing campaigns typically adopt MMS-based greeting cards that fulfill the desire to communicate with peers in a humorous way and provide added value for both recommender and receiver. For mobile viral marketing purposes, the standard raffle will be suitable if the raffle requires the participation of other people and the prize is conducive to the desired corporate impression, e.g., a group cinema visit. As is already evident from Internet commerce, advertisers can offer discounts for recommending a brand. For instance, consumers may have to fill in a web form with five mobile subscriber numbers to get a 10% discount. However, privacy concerns of consumers must be addressed.

Typically, a single mobile marketing campaign combines different standard types to achieve several objectives, as seen in the following example. To build brand awareness, the mobile marketing campaign of a company producing dairy products starts with an on-pack raffle in which all consumers win. The consumer is informed about the prize by using a WAP Push that leads to a Java-based download-center. Besides product information, the consumer can now access a mobile game that establishes associations with freshness and youthfulness to change the brand image. On the packaging, the consumer is also informed about the possibility of obtaining more mobile gimmicks if he buys another dairy product and enters a special code into the Java application. Sales may increase as a result. In addition, the Java application will provide a coupon for the newest product if the consumer subscribes to the monthly mobile newsletter about events in her immediate vicinity. In order to motivate mobile viral marketing, the possibility of recommendation is also offered.

CONCLUSION

Mobile marketing provides an unparalleled opportunity to build personal one-to-one relationships with consumers. However, mobile marketing is not just a new mechanism for advertising, but an entirely new medium that requires non-traditional strategies; mobile customers have a very special set of needs and expectations. Moreover, marketers must respect the personal nature of the cell phone and create a win-win situation where the customer provides the marketer with real added value. Sole use of the new channel, however, is not enough; advertisers must integrate mobile marketing campaigns into their total promotion mix to take full advantage of the medium.

REFERENCES

Au, Y. A., & Kauffman, R. J. (2008). The economics of mobile payments: Understanding stakeholder issues for an emerging financial technology application. *Electronic Commerce Research and Applications*, *7*(2), 141–164. doi:10.1016/j.elerap.2006.12.004

Bauer, H. H., Barnes, S., Reichardt, T., & Neumann, M. M. (2005). Driving consumer acceptance of mobile marketing: A theoretical framework and empirical study. *Journal of Electronic Commerce Research*, *6*(3), 181–192.

Drossos, D., Giaglis, G. M., Lekakos, G., Kokkinaki, F., & Stavraki, M. G. (2007). Determinants of effective SMS advertising: An experimental study. *Journal of Interactive Advertising, 7*(2). Retrieved January 30, 2009, from http://www.jiad.org/article90

eMarketer. (2005). Mobile marketing and m-commerce: Global spending and trends. Retrieved January 15, 2006, from http://www.emarketer.com/ Report.aspx?mobile_mktg_feb05

Fife, E., & Pereira, F. (2005). *Global acceptance of technology (GAT) and demand for mobile data services*. Paper presented at the Hong Kong Global Mobility Roundtable, Hong Kong.

Hong, S., Tam, K. Y., & Kim, J. (2006). Mobile data service fuels the desire for uniqueness. *Communications of the ACM, 49*(9), 89–94. doi:10.1145/1151030.1151034

Kavassalis, P., Spyropoulou, N., Drossos, D., Mitrokostas, E., Gikas, G., & Hatzistamatiou, A. (2003). Mobile permission marketing: Framing the market inquiry. *International Journal of Electronic Commerce*, *8*(1), 55–79.

Keller, K. L. (1993). Conceptualizing, measuring, and managing customer-based brand equity. *Journal of Marketing*, *57*(1), 1–22. doi:10.2307/1252054

Krishnamurthy, S. (2001). A comprehensive analysis of permission marketing. *Journal of Computer-Mediated Communication, 6*(2). Retrieved January 30, 2009, from http://jcmc.indiana.edu/vol6/issue2/krishnamurthy.html

Leppäniemi, M., & Karjaluoto, H. (2008). Mobile marketing: From marketing strategy to mobile marketing campaign implementation. *International Journal of Mobile Marketing*, *3*(1), 50–61.

Lippert, I. (2004). Mobile marketing. In R. Gerstner, G. Hunke & H. Sabel (Eds.), *Innovatives marketing* (pp. 109-123). Frankfurt am Main: Fritz Knapp.

M:Metrics. (2006). M:Metrics unveils industrys first definitive mobile marketing metrics. Retrieved January 30, 2009, http://www.mmetrics.com/press/articles/20061003-sms-shorttext.pdf, 2008-08-20.

Pousttchi, K., & Wiedemann, D. G. (2006). *A contribution to theory building for mobile marketing: Categorizing mobile marketing campaigns through case study research.* Paper presented at the 5th International Conference on Mobile Business, Copenhagen.

Scharl, A., Dickinger, A., & Murphy, J. (2005). Diffusion and success factors of mobile marketing. *Electronic Commerce Research and Applications*, *4*(2), 159–173. doi:10.1016/j.elerap.2004.10.006

Sharma, R. S., & Wildman, S. S. (2009). The economics of delivering digital content over mobile networks. *Journal of Media Business Studies*, *6*(2).

Subramani, M. R., & Rajagopalan, B. (2003). Knowledge-sharing and influence in online social networks via viral marketing. *Communications of the ACM*, *46*(12), 300–307. doi:10.1145/953460.953514

Zmijewska, A., & Lawrence, E. (2006). *Implementation models in mobile payments*. Paper presented at the International Conference Advances in Computer Science and Technology, Puerto Vallarta, Mexico.

KEY TERMS AND DEFINITIONS

Mobile Marketing: Mobile marketing is a new form of marketing communication using mobile communication techniques to promote goods, services and ideas.

Push Campaigns: In push campaigns, advertisers send relevant but not explicitly requested mobile ads to users.

Pull Campaigns: In pull campaigns, other supporting media serve as triggers for customer requests of mobile ads.

Mobile Marketing Standard Types: Mobile marketing standard types are categories of mobile viral marketing campaigns.

Mobile Marketing Objectives: Mobile marketing objectives are goals that advertisers seeking to achieve with mobile marketing campaigns, such as building brand awareness, changing brand image, increasing sales, establishing brand loyalty, building customer database, or motivating mobile viral marketing.

Campaign Type Implementation Toolbox: The campaign type implementation toolbox supports the development of campaigns on a tactical level and helps to design and customize a single mobile marketing campaign.

Campaign Type Selection Toolbox.: The campaign type selection toolbox enables a purposive planning of mobile marketing campaigns on a strategic level and links mobile marketing standard types and the mobile marketing objectives in a matrix-type conceptual framework including typical examples for any combination.

ENDNOTE

1 Descriptions of the case studies are available from the authors upon request.

Chapter 2
Reviewing Mobile Marketing Research to Date:
Towards Ubiquitous Marketing

Dimitris Drossos
University of the Aegean; Athens University of Economics and Business, Greece

George M. Giaglis
University of the Aegean; Athens University of Economics and Business, Greece

ABSTRACT

The purpose of this chapter is to illustrate and reflect upon the extant literature on mobile marketing with a view of identifying contributions, gaps and avenues for future research. The review is based on more than two hundred articles published in leading journals and conference proceedings. We first discuss key areas that have already attracted the attention of researchers, such as consumer acceptance in m-marketing and location-sensitive mobile marketing. We then focus on the emerging area of ubiquitous marketing and illustrate how mobile and wireless devices and technologies can become the enablers of meaningful dialogues between customers and marketers on an omnipresent basis. Ubiquitous marketing could be the next frontier in electronic commerce and customer relationship management.

INTRODUCTION

Research in mobile marketing has proliferated steadily, fueled by the unprecedented penetration of mobile telephony, coupled with the overwhelming acceptance of text messaging and the recent uptake of mobile data services. Acknowledging this surge in research output, academic journals, in the disciplines of both e-commerce and marketing, have published numerous papers related to mobile marketing. The majority of these papers investigate user attitudes against m-marketing and the factors that may inhibit or boost the success of mobile marketing campaigns. A sizeable portion of this research is theoretical and argumentative in nature, while empirical explorations have only recently started to emerge. This may reflect that mobile marketing research is still in its infancy; however, it also serves as an indicator of a need to launch new explanatory and conclusive empirical research endeavors that will shed more light in this emerging research area. Such studies should arguably build on a concrete synthesis of research results to date and a carefully drafted set of questions about the

DOI: 10.4018/978-1-60566-074-5.ch002

future. The purpose of this chapter is to provide the basis for this: we aim to synthesize findings from m-marketing research studies to date and then use this analysis to identify research gaps that may motivate future research.

We start by a systematic review of extant research in the area of mobile marketing. Since such research has already produced a significant contribution to the field of mobile marketing, the review helps in understanding factors contributing to successful mobile marketing campaigns as well as identifying limitations of extant knowledge in the area. This analysis is then used to identify gaps that emerge from reviewing the literature. Several theoretical and practical issues are discussed in order to encourage further research in the field. We identify and present a need to move research beyond the confines of m-marketing into the wider encompassing realm of *ubiquitous marketing* and present both technological and consumer related research challenges in this new area of scientific investigation.

CURRENT M-MARKETING RESEARCH

We have obtained our pool of literature on which the review was based by querying *Business Source Premier*, one of the most well-known academic electronic text databases and *M-lit*, the first online bibliographical database dedicated to mobile business literature. Additional research from other resources was added to the EBSCO and M-lit results if a topic was deemed to be under-studied based solely on these references.

Our queries have produced over 400 publications from scholarly journals and conference proceedings, which, after initial analysis and screening, have produced a database of two hundred and nine articles related to m-marketing. In what follows, we use the review of this literature as a basis for understanding the current status and future prospects of m-marketing research. It must be noted that, for reasons of brevity, we could not of course include a review of all 207 papers in this chapter. Instead, we have chosen to focus on the most representative *research areas* and include the complete reference list in the Bibliography on Mobile Marketing that is contained at the Appendix of this chapter. The interested reader can further consult the freely available academic database of mobile literature at www.m-lit.org with over 1,000 references can help the reader to elaborate more on specific topics of m-commerce.

It is interesting to note that m-marketing research seems to be gaining interest and popularity, since papers per year are increasing (Figure 1),

Figure 1. m-Marketing research per year

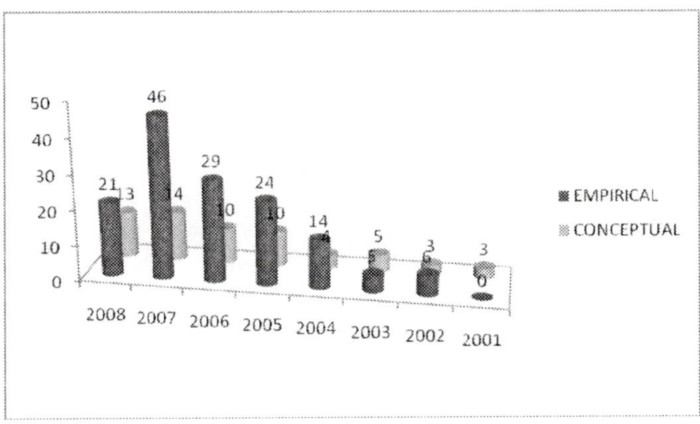

while empirical papers dominate the research agenda in recent years. Please note that 2008 papers were quantified since October 1st.

M-advertising, mobile services and applications diffusion and acceptance are the main topics that researchers have focused upon. However, other topics, such as viral marketing and conducting research via wireless devices, are still relatively unexplored (Figure 2).

Early Research: Establishing the Potential of m-Marketing

Some of the pioneering publications in mobile marketing research were the works of Balasubramanian et al. (2002) and Mort and Drennan (2002). While being a good starting point for investigating mobile marketing, such early works were mostly argumentative in nature, identifying concepts related to m-marketing rather than exploring and explaining its potential on an empirical basis. This is perhaps unsurprising, given the early stages of m-marketing research at that time. One of the most notable exceptions of an early publication that shed some empirical light into consumer mindsets against mobile marketing was the work of Barwise and Strong (2002). The authors report the results of a study of *permission-based advertising* via mobile phones exploring the effectiveness of *text messaging* as an advertising medium. The results suggest that the vast majority of participants (81%) did not delete any of the text messages before reading them. Even more interestingly, more than two out of three (69%) were able to recall eleven or more of the twenty-one advertised brands, including fourteen percent who recalled sixteen or more. In brand attitude effectiveness terms, the experimental group had higher recall than the control group and a full sixty-three percent of participants claimed to have either replied or taken action as a result of receiving a text advert. This study was one of the first to provide marketers with hands-on data for various advertising effectiveness measures, such as recall, brand attitude and direct response, thus establishing the potential of permission-based mobile advertising.

Shortly after, De Reyck and Degraeve (2003) reported the development of a *broadcast scheduling system* developed for a precision marketing firm specialized in *location-sensitive, permission-based, text messaging mobile advertising*. Text ads were sent to registered customers when they were shopping in one of two shopping centers in the vicinity of London. The ads typically contained a limited-time promotional offer. The system assisted the company in deciding which ads to send to which customers at what time, given a

Figure 2. Topics researched under the umbrella of m-marketing

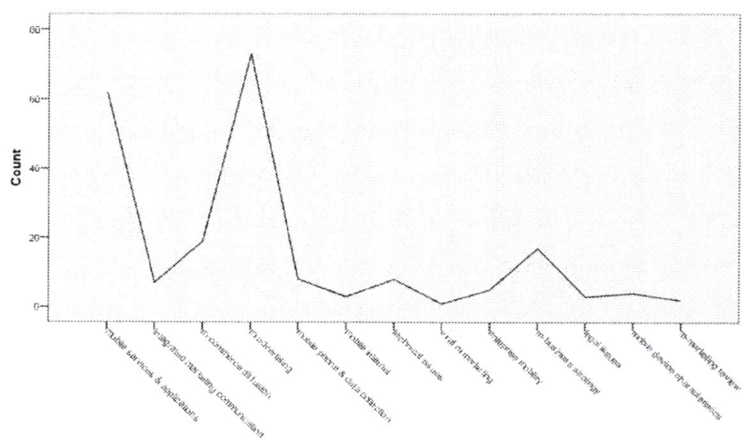

limited capacity of broadcast time slots, while maximizing customer response and revenues from retailers paying for each ad broadcast. The proposed system significantly reduced the time required to schedule the broadcasts, and resulted both in increased customer response and revenues, thus demonstrating the potential of automating the procedure of broadcasting mobile ads in a practically efficient manner.

Early Theorization in Mobile Marketing

Kleijnen, De Ruyter and Wetzels (2004) were amongst the first to adopt a well-established theory to explain the dynamics of a specific mobile service. Specifically, they modified the *innovation adoption theory* of Rogers (1995) to explain the adoption patterns of mobile gaming. Their empirical findings suggest that *perceived risk* (particularly functional risk relating to performance) is a key factor in the adoption process, followed by *complexity* and *compatibility*. They have also postulated that *security* and *privacy* issues are of crucial importance and recommended that companies need to invest in systems that assure reliable and secure data transmission, prohibiting bad system performance as well as misuse/abuse of personal information.

A different mobile service that attracted the attention of researchers was the use of multimedia messaging. Pagani (2004) has reviewed the information technology acceptance literature and formulated a model of consumer adoption of third-generation mobile multimedia services, which was then validated through a qualitative exploratory study comprising twenty-four focus groups in six markets. The model proposed was tested on the Italian market. The results show that *perceived usefulness*, *ease of use*, *price*, and *speed of use* are the most important determinants of the adoption of multimedia mobile services, in that order.

In the context of information and communication technology (ICT) adoption process, Pedersen (2005) applied a modified version of the decomposed theory of planned behavior to the adoption behavior of early adopters of mobile commerce services. The model provides a good basis for industry players developing a service evaluation framework to determine the adoption potential of new services, since it explains forty-nine percent of the early adopters' intentions to use mobile commerce services.

The Importance of Segmentation

Kleijnen, Wetzels, and De Ruyter (2004) provide another study focusing on consumer acceptance of wireless finance. One of their most important findings in this study is the significant moderating effects that indicate that *segmentation* and *targeting* are essential in the mobile context. Particularly, *age*, *computer skills* and *mobile technology readiness* can help companies establish effective customer profile segments.

In the same line of thought, the results from an exploratory study by Haque (2004) depict that *gender* and *occupation* can serve as additional segmentation variables. Namely, male recipients show greater interest than female ones in acquiring information while on the move. At the same time, students perceive access to entertainment as the most important service feature, while professionals show more interest on banking, financial, news and travel information.

Kleijnen, De Ruyter, and Andreassen (2005) then go beyond demographic segmentation and propose that marketers should also consider segmenting the market in terms of *consumer image*. Their results indicate the importance of fit between a consumer's image and the image of wireless services. More specifically, consumers who perceive high image congruence will experience innovative, wireless services more positively than consumers who experience low image congruence in terms of attitude and intention to adopt.

Another quantitative survey by Suoranta and Mattila (2004) sheds even more light on the issue of segmentation. The authors comment on the influence of certain demographic characteristics (such as *age* and *household income*) and provide useful managerial and practical implications on the adoption of mobile banking services based on 1,253 survey responses.

Finally, Mort and Drennan (2005) establish patterns of relationships between mobile user characteristics and m-services. *Innovators* have strong associations with location/information services, communication services, mobile chat, and financial services, while *shopping lovers* prefer value-added shopping services like receiving coupons, searching for and receiving product information, and booking cinema/theatre tickets online.

Consumer Attitudes against m-Marketing and Consumer Acceptance

A major theme of mobile marketing research that is steadily emerging recently is consumer attitudes toward mobile marketing and the acceptance of m-marketing practices by the public. Empirical results indicate that consumers have generally *negative attitudes* toward mobile advertising *unless they have specifically consented to it* (Carroll, Barnes, Scornavacca, & Fletcher, 2007; Tsang, Shu-Chun, & Ting-Peng, 2004). Results also indicate that *entertainment* is the most significant of the factors affecting recipient attitudes, followed by *credibility* and *irritation*. Bauer, Reichardt, Barnes, and Neumann (2005) investigated the factors that induce consumers to accept the mobile phone as a means of communicating promotional content. Their empirical results (n = 1,028) based on a Theory of Reasoned Action framework, identify *entertainment* value as well as *information value* as the strongest drivers of the m-marketing acceptance. In other studies, *credibility* and *trust* have also been found to affect recipient attitudes toward mobile ads (Chowdhury, Parvin, Weitenberner, & Becker, 2006; Park & SuJin, 2006). Maneesoonthorn and Fortin (2006) have also observed a fairly *lukewarm attitude* toward the use of SMS advertising and support the notion of permission marketing. Their results were based on a sample of 295 students in New Zealand. Hanley, Becker, and Martinsen (2006) have also tested college students and their reactions to receiving wireless advertising as well as the factors that may influence their willingness to accept ads on their cell phones. Some important findings emerged: while twenty-nine percent of those surveyed said they would accept ads on their cell phone if they got something free, one out of two (51%) said they are not willing to accept advertisements on their cell phones, even if they get something free. Moreover, sixty-six percent will accept cell phone ads if they are paid to accept them. Finally, adult youth, especially college students, are more willing to consider accepting ads on their cell phone than cell phone users in general.

Pihlström and Brush (2008) examined differences between entertainment and information mobile content service users in how their value perceptions influence intentions to repurchase, intentions to spread positive word-of-mouth, and willingness to pay a price premium. Using a sample of 579 mobile service users, this study found empirical support for the role of conditional and epistemic value as antecedents that intensify the need to use a service in certain situations and consequently enhance emotional, social, monetary, and convenience value derived from the use of mobile services.

Based on a review of consumer innovation studies, consumer behavior theory, and advertising research, Muk and Babin (2006) tested a model in an attempt to distinguish relationships between *consumer attitude*, *normative influence*, and *intention* corresponding to SMS advertising. Responses from 171 American university students show that consumer attitudes and *social influences* significantly impact a consumer's propensity to adopt services like SMS advertising. By employing variables adapted from the Extended Technology Acceptance Model, Yang (2007) has demonstrated that social influence processes significantly influence consumer attitudes toward using mobile commerce. Findings from an Australian study also reveal that the *youth* segment utilizes mobile services for both utilitarian and hedonic needs, albeit significantly more so for the latter, while *peer influence* affects their behavior in regards to the choice of mobile service providers and mobile services usage (Lee & Murphy, 2006).

Cross-Cultural Studies

Arguably, the formulation of perceptions against m-marketing is culturally influenced. It is then somewhat surprising that there is a scarcity of studies exploring cross-cultural differences. Two of the few cross-market studies that we have found are the ones by Rohm and Sultan (2006) and Muk (2007). Despite the small sample that cannot ensure generalizabilty of the findings, the authors of the first study report that some factors, such as *age* and *sex*, were common across markets. However, they also found that *regional differences* exist. For instance, emerging markets are more likely to accept mobile marketing: respondents in Pakistan indicated greater willingness to give permission to marketers to engage them in the mobile space, yet they also exhibited greater vulnerability to privacy abuse. The second study reports that cultural orientations embedded in consumer belief systems may play an influential role in decision making processes. The findings of this study reveal that the decisions of American consumers on accepting SMS ads via their mobile phones are primarily based on *attitudinal considerations* (for example, privacy concerns), whereas Taiwanese consumer intentions to act are more influenced by *social norms*.

Effects of m-Marketing on Consumer Behavior

While most of the empirical studies reviewed so far seem to have focused on the antecedents for use of mobile services, a few studies explore the effects of mobile marketing on consumer cognition, affect and behavior. However, these are admittedly few: empirical studies focusing on the effects of mobile services on the relationship between a brand and its customers seem to be missing in the literature. Nysveen et al. (2005) were amongst the first to study SMS and MMS effects on consumer-brand relationship dimensions. The authors emphasize on these effects and have found that the SMS channel is primarily perceived as a complement to the brands' main advertisement channels. Thus, they suggest that companies build an emotional relationship with their customers by the use of SMS/MMS. At the same time, Rettie, Grandcolas, and Deakins (2005) reported results from twenty-six text messaging marketing campaigns involving 5,401 respondents. SMS advertising has been shown to be effective both as a *branding vehicle* and in *stimulating response*. Prompted brand recall was forty three per cent on average, while most messages were read (89%), and five per cent were even forwarded to friends. The vast majority (85.7%) of those who felt more positively toward the brand claimed they were more likely to make a purchase.

Hsu, Wang, and Wen (2006) employed the decomposed theory of planned behavior to reveal that the decisive factors influencing the behavior and intention of consumers in using m-coupons are *attitude* and *perceived behavioral control*, while subjective norms are not evident. Kondo, Uwadaira, and Nakahara (2007) have proceeded to study the effect of sending of direct mail coupons via mobile phone to the probability of visiting a store. Their research confirmed that mobile coupons increased shop visits; at the same time, the authors point to the need to let the customer take some action (e.g. via hyperlink mails) in order to avoid forgetting the mere fact of mobile coupon receipt.

In addition, Bojin and Spencer (2006) studied how the advertising content and the product profile are related to the opt-in process or purchase intention. They found that the key point for content and transmission personalization is *offering credible and trustworthy information*. In addition, they contend that the mobile channel is more useful for the promotion of *famous brands* and *low-value products*, at least in China. Okazaki's (2004) findings from a large-scale mobile survey suggest that a mobile campaign's *recall* largely depends on perceptions of both the medium and the advertised content, and that (a) *mobile advertising trust* directly and positively affects attitudes toward mobile advertising, (b) *brand trust* directly and positively affects attitudes toward the brand, (c) *attitude toward mobile advertising* directly and positively affects attitudes toward the brand, and (d) *attitude toward brand* directly and positively affects mobile campaign recall.

Wang (2007) has examined the effect of cross-media integration between an advertiser's text messages and a website that featured relevant advertising messages on consumer media engagement, message strength, and brand attitudes. The results revealed that the cross-media effect of mobile and Internet advertising was evident on all dependent variables. Drossos et al. (2007) investigated the significance of a number of factors associated with SMS advertising effectiveness through an experimental study. The findings indicate that *incentive*, *interactivity*, *advertisement appeal*, *product involvement*, and *attitude toward SMS advertising in general* directly influence attitudes toward the advertisement and toward the brand, as well as purchase intention. The results of the study suggest that a stronger focus on these factors is necessary to improve the effectiveness of SMS advertising campaigns.

Jingjun Xu (2006) has identified *entertainment*, *credibility* and *personalization* as the three most important factors that will affect consumer attitudes toward mobile advertising. Furthermore, through a correlation analysis, he has concluded that the attitude toward mobile advertising is a good indicator of intention to use mobile advertising to consume.

Mobile Services and Applications

Mobile and wireless services have been visualized as the next step of electronic commerce. Due to their significance, many researchers have shifted their attention toward understanding how consumers will embrace mobile services and which will be the '*killer*' application. The study of Kleijnen et al. (2004) were among the first that investigated critical factors influencing mobile services adoption. The findings indicate the large influence of perceived risk on the mobile service adoption process. As a result, focus on risk reduction and m-trust formation should be of major concern for mobile operators and application providers for successful establishment of mobile services. Mort and Drennan (2005) argue that while there is a vast range of m-services, both existing and emerging, consumers are seeking specific bundles of usage benefits from m-services. Additionally, the importance of affiliation and belonging in the use of m-services affirms that the social network, peer bonding and celebrity endorsement strategies that were effective in building the market for the mobile phone handset are also highly likely to be effective in building the product extension market for m-services. Furthermore, youth adopts the services for utilitarian and hedonic reasons and reference group influence (Lee & Murphy, 2006). Moreover, Weitenberner et al. (2006) reveal that mobile phone usage and interest in mobile services are more advanced in Asian countries including Japan, South Korea and Singapore than the U.S. The differences of using mobile services between Asian countries is affected by factors including government support, consumer culture and infrastructure investment.

Smartphone service evolution is another interesting topic studied by Verkasalo (2007). A typical Finnish spends a daily average of 33 minutes with a smartphone. Most of this time is allocated on voice calls (33 per cent) and messaging (24 per cent). Younger people and men are still the most active users of smartphones. People still consider pricing and technical implementation of the mobile web and email as bottlenecks for wider mobile internet usage.

The literature also provides some insights concerning *mobile financial services*. M-banking and consumer behavior have been explored by Suoranta and Mattila (2004). In the light of the findings of this study the next customers willing to adopt mobile banking are those over 50 years old. Moreover, in this survey there is a clear indication that wealthier respondents were less willing to adopt the new mobile banking services. Findings illustrated in this study do not align with Internet banking research, meaning that the typology of electronic service users is changing depending on the medium. In this context, Hourahine and Howard (2004) highlight that the most popular financial applications are (a) checking account balances (b) viewing recent account transactions and (c) paying a bill while on the move.

Mobile TV is described as the next big step, as mobile phones get more multimedia features. According to a survey of 1,361 cellular telephone users, more than 50 percent were either not at all or not very interested in watching television with mobile telephones or other handheld devices. Only 32 percent display some interest and a mere 9 percent are very interested in mobile television technology (Zid, 2005). Bayartsaikhan et al. (2007) mention that in the case of mobile TV "content is king." To be successful, compelling content must be developed or else consumers will not buy the service.

Sharma and Ojha (2004) studied *service performance* in mobile communications. Their findings indicate that service performance in mobile communication service industry has three distinct constituents, network-based service performance, retailer-related process performance and network operator-related process performance. Consequently with this multi-item measure of service performance, managers can spot their strengths and weaknesses and improve service provision.

While the areas mentioned above have been extensively studied, other mobile applications and services remains less researched. Indeed, we have relatively less knowledge on the level of readiness to adopt *mobile learning* (Venkatesh, Nargundkar, Sayed, & Shahaida, 2006), credibility and effectiveness of *mobile political communications* (Prete, 2007), *mobile phone photography* and its role in the visitor experience and motivation to learn and interact with visitor groups (Leighton, 2007). The m-revolution will impact numerous facets of our daily lives and the way business is conducted. However, several important research areas remain unexplored.

Location-Sensitive Mobile Marketing

Even if location-based marketing has long been proposed as a fundamental driver of mobile marketing, relatively few studies demonstrate its significance in an empirical fashion. Unni and Harmon (2007) used an experimental setting to test the effects of location-based advertising (LBA) on privacy concerns about location tracking, perceived benefits, value, and intentions to try LBA. Results indicate that *privacy concerns* are high, while *perceived benefits and value* of LBA are low; users prefer LBA when it is based on opt-in.

However, other studies indicate a higher potential for *context-aware marketing*. Merisavo et al. (2007) argue that mobile advertising would benefit from being location-, time- and consumer profile-specific. Gopal and Tripathi (2006) argue that the distance between the retail location and point of ad delivery has a significant impact on recipients' recall of ads and thus reaching consumers at the right location can have significant monetary implications for sellers. In addition, context-aware marketing depends on the ability to provide the consumer with ads not only at the right place but also at the right time. Sophonthummapharn and Tesar (2007) report that 18:01 to 21:00 is slightly preferable than other time periods of the day, while the most preferred receiving frequency of mobile ads is one to three times per day. However, more research is arguably needed since in-depth examination of the topic has just begun. Researchers should start elaborating on experimental designs and scales [i.e. *attitude toward location-based advertising*, as proposed by Bruner II and Kumar (2007)] that could shed light on the effects of location-based marketing.

FUTURE RESEARCH ISSUES

A careful scrutiny of the extant literature on mobile marketing reveals some interesting conclusions. In line with expectations, research has gradually progressed from initial exploratory work to more robust empirical investigations that provide new knowledge in specific dimensions of advertising through a mobile phone. However, the road is far from traveled yet. Little empirical research has been performed in issues like *mobile marketing systems* (Gewei, 2007), *integrated marketing communications* in which the mobile channel is but one ingredient (Wang, 2007), methods for *mobile data collection* (Balabanis, Mitchell, & Heinonen-Mavrovouniotis, 2007), *campaign management* (Kavassalis et al., 2003), the *legal and regulatory landscape* surrounding mobile business-to-customer communication, and others. It is therefore essential to codify and discuss several issues of future research to assist researchers in augmenting extant knowledge and draw agendas for future investigation.

A few papers have already addressed issues related to the current state of m-marketing research and possible unexplored subjects (at least during the time those studies were published). Leppaniemi, Sinisalo, and Karjaluoto (2006) identify and evaluate research conducted in the field of mobile marketing between January 2000 and February 2006. They assessed academic as well as industry sources focusing on mobile or wireless applications aimed at marketing, business and management or information systems domains. From this process they found that a substantial number of studies have focused on user behavior and attitudes toward mobile marketing and that mobile marketing effectiveness is gaining more research interest. They found fewer studies focusing on the effect of mobile marketing in branding, the m-marketing value chain and business models. In addition, Okazaki (2005) mentioned that important progress has been made in mobile Internet and SMS-based mobile advertising adoption, but other important topics, such as e-commerce and m-commerce comparison, mobile-based word-of-mouth, and methodological issues, have rarely been addressed.

Thinking m-marketing as the next step of e-marketing will lead to several themes that are still unexplored, for example *efficient m-campaign targeting*. Are women more likely than men to have a positive attitude toward m-advertising? Does age have a significant impact on attitudes and behaviors? Both education and income have been historically found to have significant relationships with advertising attitudes. Moreover, are less educated and lower income consumers more likely to enjoy advertising and rely on advertising in making purchasing decisions compared to their more educated and wealthier counterparts (Shavitt, Lowrey, & Haefner, 1998)?

What is more, although demographic variables have received some exposure in the context of attitude toward SMS advertising, the literature review reveals a relative lack of research on the *effects of psychographic variables*. Factors that are related to activities, interests, and opinions of consumers are typically clustered together under the title of psychographic factors. Are activities, interests, and opinions more effective than demographics in understanding m-consumer behavior? Researchers should examine a number of other personality characteristics. For example, impulse buying, situational involvement, commitment to the brand, variety seeking, information seeking, and price interest are just a few factors that may affect the mindsets of the consumer.

In this line of thinking, researchers should examine the *level of fit between the mobile channel and different categories of products/services*. It is essential to identify the mobile marketing effects for high-, medium-, and low-involvement product categories, durables and non-durables, and industrial versus consumer products. Researchers can also explore the effects of the stage of the product life cycle and investigate whether m-marketing effects are different for new products than those for mature or post-mature products.

The real challenge, however, is to think m-marketing as a radical innovation due to forthcoming ubiquitous and pervasive networks. M-marketing should sense and respond not only to who the customer is, but where she is and what she is doing with transparent and undisruptive approaches. The user does not have to think about how to use the mobile device; rather, the device automatically helps the user carry out a task. Mort and Drennan (2002) propose a shift from the PC-based website to mobile applications following a change in consumer behavior from '*sit and search*' to '*roam and receive*'. Under the strong effects of ubiquitous network development (i.e. use of RFID on all products, diversified but transparent relationships between users, mobile networks and content), high-bandwidth channels, ownership of '*smart*' mobile devices, corporate m-marketing and integrated marketing communications will undergo considerable changes. All these developments and changes are forcing m-marketing toward ubiquitous marketing (Figure 3).

Within this context, research could be technology or consumer driven. Some technological issues are:

- The diversity of standards, devices and software (e.g. micro-browsers, operating

Figure 3. Challenges for ubiquitous marketing

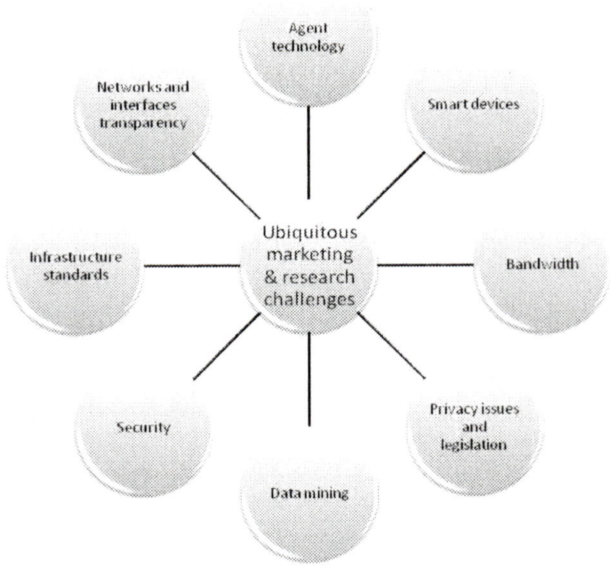

systems for mobile-client) can confuse a company planning to implement ubiquitous campaigns. Therefore, there is a strong need to conclude with few acceptable standards.

- Integrating the mobile device into the company's supply chain. With this integration it would be possible to exchange product information and easily conduct transactions while on the move. Mechanisms for assisting with secure transactions and billing are top priorities.
- Instant conversion of content to be transparently presented on various end user devices (PCs, mobile phones, palmtops, etc.)
- Innovative interfaces that will allow people to control their ubiquitous services with simple commands (i.e. voice commands)

Consumer driven research could emphasize on:

- Understanding the user's interactions within a context, to efficiently recognize what products or services consumers might possibly be interested in at specific space and time boundaries.
- The provision of ground-breaking content and services optimized for mobility.
- Pull vs. push content. Will permission-based marketing continue to be the baseline for conducting m-marketing?
- Retail pricing in a mobile competitive environment. If consumers can use their mobile device to gain information about a product or service at the time that they are making a buying decision then isn't it probable that they would buy from the closest and cheapest retailer?
- Marketing efficiency and effectiveness. Before companies commit to this market they want evidence of consumer acceptance and effectiveness. Thus, new metrics, different from those developed for traditional and fixed Internet marketing, need to be developed in order to measure the effectiveness of ubiquitous marketing campaigns.

In summary, this chapter represents an effort to provide a platform for capturing m-marketing issues that deserve the attention of researchers and future investigations. As such, we hope that it will contribute to an ongoing debate on the future of m-marketing research, stimulate theory building, and generate constructive managerial insights on a phenomenon likely to alter current marketing paradigms.

REFERENCES

Balabanis, G., Mitchell, V.-W., & Heinonen-Mavrovouniotis, S. (2007). SMS-based surveys: Strategies to improve participation. *International Journal of Advertising*, *26*(3), 369–385.

Balasubramanian, S., Peterson, R. A., & Jarvenpaa, S. L. (2002). Exploring the implications of m-commerce for markets and marketing. *Journal of the Academy of Marketing Science*, *30*(4), 348–361. doi:10.1177/009207002236910

Barwise, P., & Strong, C. (2002). Permission-based mobile advertising. *Journal of Interactive Marketing*, *16*(1), 14. doi:10.1002/dir.10000

Bauer, H., Reichardt, T., Barnes, S., & Neumann, M. (2005). Driving consumer acceptance of mobile marketing: A theoretical framework and empirical study. *Journal of Electronic Commerce Research*, *6*(3), 181–192.

Bayartsaikhan, K., Danielak, P., Dunst, K., Guibert, J., Luxford, L., & Romanossian, R. (2007). Market for third screen: a study of market potential of mobile TV and video across the U.S. and selected European countries. *International Journal of Mobile Marketing*, *2*(1), 12–27.

Bojin, P., & Spencer, I. (2006). Mobile marketing-the Chinese perspective. *International Journal of Mobile Marketing, 1*(2), 50–59.

Bruner, G. C. II, & Kumar, A. (2007). Attitude toward location-based advertising. *Journal of Interactive Advertising, 7*(2), 1.

Carroll, A., Barnes, S. J., Scornavacca, E., & Fletcher, K. (2007). Consumer perceptions and attitudes towards SMS advertising: Recent evidence from New Zealand. *International Journal of Advertising, 26*(1), 79–98.

Chowdhury, H. K., Parvin, N., Weitenberner, C., & Becker, M. (2006). Consumer attitude toward mobile advertising in an emerging market: An empirical study. *International Journal of Mobile Marketing, 1*(2), 33–42.

De Reyck, B., & Degraeve, Z. (2003). Broadcast scheduling for mobile advertising. *Operations Research, 51*(4), 509–517. doi:10.1287/opre.51.4.509.16104

Drossos, D., Giaglis, G., Lekakos, G., Kokkinaki, F., & Stavraki, M. (2007). Determinants of effective sms advertising: An experimental study. *Journal of Interactive Advertising, 7*(2).

Gewei, Y. (2007). Mobile marketing systems: Framework and technology enabler. *International Journal of Mobile Marketing, 2*(1), 43–49.

Gopal, R. D., & Tripathi, A. K. (2006). Advertising via wireless networks. *International Journal of Mobile Communications, 4*(1), 1–16. doi:10.1504/IJMC.2006.008194

Hanley, M., Becker, M., & Martinsen, J. (2006). Factors influencing mobile advertising acceptance: Will incentives motivate college students to accept mobile advertisements? *International Journal of Mobile Marketing, 1*(1), 50–58.

Haque, A. (2004). Mobile commerce: Customer perception and It's prospect on business operation in Malaysia. *Journal of American Academy of Business, Cambridge, 4*(1/2), 257–262.

Hourahine, B., & Howard, M. (2004). Money on the move: Opportunities for financial service providers in the 'third space'. *Journal of Financial Services Marketing, 9*(1), 57–67. doi:10.1057/palgrave.fsm.4770141

Hsu, T.-h., Wang, Y.-s., & Wen, S.-c. (2006). Using the decomposed theory of planning behavioural to analyse consumer behavioural intention towards mobile text message coupons. *Journal of Targeting. Measurement & Analysis for Marketing, 14*(4), 309–324. doi:10.1057/palgrave.jt.5740191

Jingjun Xu, D. (2006). The influence of personalization in affecting consumer attitudes toward mobile advertising in China. *Journal of Computer Information Systems, 47*(2), 9–19.

Kavassalis, P., Spyropoulou, N., Drossos, D., Mitrokostas, E., Gikas, G., & Hatzistamatiou, A. (2003). Mobile permission marketing: Framing the market inquiry. *International Journal of Electronic Commerce, 8*(1), 55–79.

Kleijnen, M., De Ruyter, K., & Andreassen, T. W. (2005). Image congruence and the adoption of service innovations. *Journal of Service Research, 7*(4), 343–359. doi:10.1177/1094670504273978

Kleijnen, M., De Ruyter, K., & Wetzels, M. (2004). Consumer adoption of wireless services: discovering the rules, while playing the game. *Journal of Interactive Marketing, 18*(2), 51–61. doi:10.1002/dir.20002

Kleijnen, M., Wetzels, M., & De Ruyter, K. (2004). Consumer acceptance of wireless finance. *Journal of Financial Services Marketing, 8*(3), 206–217. doi:10.1057/palgrave.fsm.4770120

Kondo, F. N., Uwadaira, Y., & Nakahara, M. (2007). Stimulating customer response to promotions: The case of mobile phone coupons. *Journal of Targeting . Measurement & Analysis for Marketing, 16*(1), 57–67. doi:10.1057/palgrave.jt.5750064

Lee, R., & Murphy, J. (2006). The consumption of mobile services by Australian university students. *International Journal of Mobile Marketing, 1*(1), 13–20.

Leighton, D. (2007). In the frame: investigating the use of mobile phone photography in museums. *International Journal of Nonprofit and Voluntary Sector Marketing, 12*(4), 308–319. doi:10.1002/nvsm.320

Leppaniemi, M., Sinisalo, J., & Karjaluoto, H. (2006). A review of mobile marketing research. *International Journal of Mobile Marketing, 1*(1), 30–40.

Maneesoonthorn, C., & Fortin, D. (2006). Texting behaviour and attitudes toward permission mobile advertising: An empirical study of mobile users' acceptance of SMS for marketing purposes. *International Journal of Mobile Marketing, 1*(1), 66–72.

Merisavo, M., Kajalo, S., Karjaluoto, H., Virtanen, V., Salmenkivi, S., & Raulas, M. (2007). An empirical study of the drivers of consumer acceptance of mobile advertising. *Journal of Interactive Advertising, 7*(2), 1.

Mort, G. S., & Drennan, J. (2002). Mobile digital technology: Emerging issues for marketing. *Journal of Database Marketing, 10*(1), 9. doi:10.1057/palgrave.jdm.3240090

Mort, G. S., & Drennan, J. (2005). Marketing m-services: Establishing a usage benefit typology related to mobile user characteristics. *Journal of Database Marketing & Customer Strategy Management, 12*(4), 327–341. doi:10.1057/palgrave.dbm.3240269

Muk, A. (2007). Cultural influences on adoption of SMS advertising: A study of American and Taiwanese consumers. *Journal of Targeting . Measurement & Analysis for Marketing, 16*(1), 39–47. doi:10.1057/palgrave.jt.5750062

Muk, A., & Babin, J. (2006). U.S. consumers' adoption -- nonadoption of mobile SMS advertising. *International Journal of Mobile Marketing, 1*(1), 21–29.

Nysveen, H., Pedersen, P. E., Thorbjornsen, H., & Berthon, P. (2005). Mobilizing the brand: The effects of mobile services on brand relationships and main channel use. *Journal of Service Research, 7*(3), 257–276. doi:10.1177/1094670504271151

Okazaki, S. (2004). How do Japanese consumers perceive wireless ads? A multivariate analysis. *International Journal of Advertising, 23*(4), 429.

Okazaki. (2005). New perspectives on m-commerce research. *Journal of Electronic Commerce Research, 6*(3), 160 - 164.

Pagani, M. (2004). Determinants of adoption of third generation mobile multimedia services. *Journal of Interactive Marketing, 18*(3), 46–59. doi:10.1002/dir.20011

Park, J., & SuJin, Y. (2006). The moderating role of consumer trust and experiences: Value driven usage of mobile technology. *International Journal of Mobile Marketing, 1*(2), 24–32.

Pedersen, P. E. (2005). Adoption of mobile Internet services: An exploratory study of mobile commerce early adopters. *Journal of Organizational Computing and Electronic Commerce, 15*(3), 203–222. doi:10.1207/s15327744joce1503_2

Pihlström, M., & Brush, G. J. (2008). Comparing the perceived value of information and entertainment mobile services. *Psychology and Marketing, 25*(8), 732–755. doi:10.1002/mar.20236

Prete, M. I. (2007). M-politics: Credibility and effectiveness of mobile political communications. *Journal of Targeting . Measurement & Analysis for Marketing, 16*(1), 48–56. doi:10.1057/palgrave.jt.5750063

Rettie, R., Grandcolas, U., & Deakins, B. (2005). Text message advertising: Response rates and branding effects. *Journal of Targeting . Measurement & Analysis for Marketing, 13*(4), 304–312. doi:10.1057/palgrave.jt.5740158

Rogers, E. M. (1995). *Diffusion of innovations* (4 ed.). New York, NY: The Free Press.

Rohm, A. J., & Sultan, F. (2006). An exploratory cross-market study of mobile marketing acceptance. *International Journal of Mobile Marketing, 1*(1), 4–12.

Sharma, N., & Ojha, S. (2004). Measuring service performance in mobile communications. *Service Industries Journal, 24*(6), 109–128. doi:10.1080/0264206042000299202

Shavitt, S., Lowrey, P., & Haefner, J. (1998). Public attitudes toward advertising: More favorable than you might think. *Journal of Advertising Research, 38*(4), 7–22.

Sophonthummapharn, K., & Tesar, G. (2007). Technology readiness and propensity of cell phone users to subscribe to commercial messaging services. *Marketing Management Journal, 17*(2), 81–95.

Suoranta, M., & Mattila, M. (2004). Mobile banking and consumer behaviour: New insights into the diffusion pattern. *Journal of Financial Services Marketing, 8*(4), 354–366. doi:10.1057/palgrave.fsm.4770132

Tsang, M. M., Shu-Chun, H., & Ting-Peng, L. (2004). Consumer Attitudes Toward Mobile Advertising: An Empirical Study. *International Journal of Electronic Commerce, 8*(3), 65–78.

Unni, R., & Harmon, R. (2007). Perceived Effectiveness of Push vs. Pull Mobile Location-Based Advertising. *Journal of Interactive Advertising, 7*(2), 1.

Venkatesh, B., Nargundkar, R., Sayed, F. K., & Shahaida, P. (2006). Assessing Indian students' perceptions towards m-learning - some initial conclusions. *International Journal of Mobile Marketing, 1*(2), 75–79.

Verkasalo, H. (2007). Handset-based measurement of smartphone service evolution in Finland. *Journal of Targeting . Measurement & Analysis for Marketing, 16*(1), 7–25. doi:10.1057/palgrave.jt.5750060

Wang, A. (2007). Branding over mobile and Internet advertising: The cross-media effect. *International Journal of Mobile Marketing, 2*(1), 34–42.

Weitenberner, C., Chapman, M., Miranda, G., Tobar, S., Wagoner, E., & Akasaki, D. (2006). United States and South East Asian mobile markets: a comparative analysis of infrastructure and cultural differences. *International Journal of Mobile Marketing, 1*(1), 73–82.

Yang, K. C. C. (2007). Exploring factors affecting consumer intention to use mobile advertising in Taiwan. *Journal of International Consumer Marketing, 20*(1), 33–49. doi:10.1300/J046v20n01_04

Zid, L. A.-S. (2005). TV to GO. *Marketing Management, 14*(1), 5–5.

ADDITIONAL READING

Abraham, R. (2007). Mobile phones and economic development: Evidence from the fishing industry in India. *Information Technologies & International Development, 4*(1), 5–17. doi:10.1162/itid.2007.4.1.5

Anckar, B. (2002). *Value-added services in mobile commerce: An analytical framework and empirical findings from a national consumer survey.* Paper presented at the 35th Hawaii International Conference on System Sciences, Maui, Hawaii.

Anckar, B. (2002). Value creation in mobile commerce: Findings from a consumer survey. *Journal of Information Technology Theory and Application*, *4*(1), 43–64.

Anderson, J., & Kupp, M. (2008). Serving the poor: Drivers of business model innovation in mobile. *Info*, *10*(1), 5–12. doi:10.1108/14636690810850120

Aydin, S., & Ozer, G. (2006). How switching costs affect subscriber loyalty in the Turkish mobile phone market: An exploratory study. *Journal of Targeting . Measurement & Analysis for Marketing*, *14*(2), 141–155. doi:10.1057/palgrave.jt.5740176

Bailey, T. (2007). Mobile enhancing the performing arts center: a case study and conceptual view. *International Journal of Mobile Marketing*, *2*(2), 86–89.

Balabanis, G., Mitchell, V.-W., & Heinonen-Mavrovouniotis, S. (2007). SMS-based surveys: strategies to improve participation. *International Journal of Advertising*, *26*(3), 369–385.

Barnes, S. J. (2002). Wireless digital advertising: nature and implications. *International Journal of Advertising*, *21*(3), 399–420.

Barnes, S. J., & Scornavacca, E. (2004). Mobile marketing: the role of permission and acceptance. *International Journal of Mobile Communications*, *2*(2), 128–139. doi:10.1504/IJMC.2004.004663

Barutcu, S. (2007). Attitudes towards mobile marketing tools: A study of Turkish consumers. *Journal of Targeting . Measurement & Analysis for Marketing*, *16*(1), 26–38. doi:10.1057/palgrave.jt.5750061

Barwise, P., & Strong, C. (2002). Permission-based mobile advertising. *Journal of Interactive Marketing*, *16*(1), 14–24. doi:10.1002/dir.10000

Bayartsaikhan, K., Danielak, P., Dunst, K., Guibert, J., Luxford, L., & Romanossian, R. (2007). Market for third screen: a study of market potential of mobile TV and video across the U.S. and selected European countries. *International Journal of Mobile Marketing*, *2*(1), 12–27.

Becker, D. (2007). Participation TV: premium SMS vs. toll-free IVR. *International Journal of Mobile Marketing*, *2*(2), 50–52.

Benni, E., Hjartar, K., & Laartz, J. (2003). The IT factor in mobile services. *McKinsey Quarterly*(3), 87-93.

Birdsong, T., Brauner, K., Crowe, S., Harmon, G., Johnson, T., & Lucas, A. (2008). A theater of self and family: mobile device use, cultural idiom and appropriation. *International Journal of Mobile Marketing*, *3*(1), 41–49.

Bojin, P., & Spencer, I. (2006). Mobile marketing - the Chinese perspective. *International Journal of Mobile Marketing*, *1*(2), 50–59.

Bouwman, H., Haaker, T., & de Vos, H. (2007). Mobile service bundles: The example of navigation services. *Electronic Markets*, *17*(1), 20–28. doi:10.1080/10196780601136757

Bragge, J., Tuunanen, T., & Virtanen, V. (2005). *A repeatable collaboration process for developing a road map for emerging new technology business: Case mobile marketing.* Paper presented at the Eleventh Americas Conference on Information Systems, Omaha.

Braiterman, J., & Savio, N. (2007). Design sketch: the context of mobile interaction. *International Journal of Mobile Marketing*, *2*(1), 66–68.

Bruner Li, G. C., & Kumar, A. (2007). Attitude toward location-based advertising. *Journal of Interactive Advertising*, *7*(2), 1.

Buerkler, E. (2005). Dramatically different price developments for mobile services in India and Switzerland and their underlying drivers. *Info*, *7*(2), 72–75. doi:10.1108/14636690510587225

Busk, D. (2007). Election 2008: the mobile tipping point. *International Journal of Mobile Marketing*, *2*(2), 5–9.

Butt, M. M., & De Run, E. C. (2008). Measuring Pakistani mobile cellular customer satisfaction. *ICFAI Journal of Services Marketing*, *6*(1), 40–50.

Callahan, T. (2007). Needs-based calling: staying connected in a disconnected world. *International Journal of Mobile Marketing*, *2*(1), 69–75.

Callegaro, M., Steeh, C., Buskirk, T. D., Vehovar, V., Kuusela, V., & Piekarski, L. (2007). Fitting disposition codes to mobile phone surveys: experiences from studies in Finland, Slovenia and the USA. *Journal of the Royal Statistical Society. Series A, (Statistics in Society)*, *170*(3), 647–670. doi:10.1111/j.1467-985X.2006.00461.x

Carroll, A., Barnes, S. J., & Scornavacca, E. (2005). *Consumers perceptions and attitudes towards SMS mobile marketing in New Zealand.* Paper presented at the Fourth International Conference on Mobile Business, Sydney, Australia.

Carroll, A., Barnes, S. J., Scornavacca, E., & Fletcher, K. (2007). Consumer perceptions and attitudes towards SMS advertising: recent evidence from New Zealand. *International Journal of Advertising*, *26*(1), 79–98.

Carter, E. (2008). Mobile marketing and generation Y African-American mobile consumers: the issues and opportunities. *International Journal of Mobile Marketing*, *3*(1), 62–66.

Caruana, A. (2004). The impact of switching costs on customer loyalty: A study among corporate customers of mobile telephony. *Journal of Targeting . Measurement & Analysis for Marketing*, *12*(3), 256–268. doi:10.1057/palgrave.jt.5740113

Casini, A. (2007). If you ask me: Establishing a connection. *Engineering & Technology*, *2*(9), 24–24. doi:10.1049/et:20070901

Chowdhury, H. K., Parvin, N., Weitenberner, C., & Becker, M. (2006). Consumer attitude toward mobile advertising in an emerging market: an empirical study. *International Journal of Mobile Marketing*, *1*(2), 33–42.

Cleff, E. B. (2007). Privacy issues in mobile advertising. *International Review of Law Computers & Technology*, *21*(3), 225–236. doi:10.1080/13600860701701421

Constantiou, I. D., Damsgaard, J., & Knutsen, L. (2007). The four incremental steps toward advanced mobile service adoption. *Communications of the ACM*, *50*(6), 51–55. doi:10.1145/1247001.1247005

Curwen, P., & Whalley, J. (2007). Tele2 and the strategic role of virtual operations. *Info*, *9*(4), 55–69. doi:10.1108/14636690710762138

Das, B., & Mohanty, S. (2007). Service usability and users' satisfaction in India: an exploratory study on mobile phone users. *ICFAI Journal of Services Marketing*, *5*(4), 53–66.

Davis, R., & Sajtos, L. (2008). Measuring consumer interactivity in response to campaigns coupling mobile and Television media. *Journal of Advertising Research*, *48*(3), 375–391. doi:10.2501/S0021849908080409

Davis, R., & Yung, D. (2005). Understanding the interactivity between television and mobile commerce. *Communications of the ACM*, *48*(7), 103–105. doi:10.1145/1070838.1070869

De Los Salmones, M. M. G., Crespo, A. H., & del Bosque, I. R. g. (2005). Influence of corporate social responsibility on loyalty and valuation of services. *Journal of Business Ethics*, *61*(4), 369–385. doi:10.1007/s10551-005-5841-2

De Marez, L., & Verleye, G. (2004). Innovation diffusion: The need for more accurate consumer insight. Illustration of the PSAP scale as a segmentation instrument. *Journal of Targeting . Measurement & Analysis for Marketing*, *13*(1), 32–49. doi:10.1057/palgrave.jt.5740130

De Marez, L., Vyncke, P., Berte, K., Schuurman, D., & De Moor, K. (2007). Adopter segments, adoption determinants and mobile marketing. *Journal of Targeting . Measurement & Analysis for Marketing*, *16*(1), 78–95. doi:10.1057/palgrave.jt.5750057

De Reyck, B., & Degraeve, Z. (2003). Broadcast scheduling for mobile advertising. *Operations Research*, *51*(4), 509–517. doi:10.1287/opre.51.4.509.16104

Dickinger, A., & Haghirian, P. (2004). *An investigation and conceptual model of SMS marketing.* Paper presented at the 37th Hawaii International Conference on System Sciences, Big Island, Hawaii.

Dolian, B. (2008). Text-to-screen emerges: a conceptual approach to a powerful interactive marketing tool. *International Journal of Mobile Marketing*, *3*(1), 81–85.

Facchetti, A., Rangone, A., Renga, F. M., & Savoldelli, A. (2005). Mobile marketing: an analysis of key success factors and the European value chain. *International Journal of Management and Decision Making*, *6*(1), 65–80. doi:10.1504/IJMDM.2005.005966

Ferris, M. (2007). Insights on mobile advertising, promotion, and research. *Journal of Advertising Research*, *47*(1), 28–37. doi:10.2501/S0021849907070043

Folinas, D., Vlachopoulou, M., Manthou, V., & Zogopoulos, D. (2002). *The value system of m-business.* Paper presented at the First International Conference on Mobile Business, Athens, Greece.

Frank, C., & Wuersch, M. (2006). Implicit opt-in mobile advertising via the Ipointer selection model. *International Journal of Mobile Marketing*, *1*(2), 68–74.

Friel, A. L. (2005). The spam spat. *Marketing Management*, *14*(3), 48–50.

Fu, G. L., Yang, C., & Tzeng, G. H. (2007). A multicriteria analysis on the strategies to open Taiwan's mobile virtual network operators services. *International Journal of Information Technology & Decision Making*, *6*(1), 85–112. doi:10.1142/S0219622007002320

Funk, J. L. (2003). *Key technological trajectories and the expansion of mobile Internet applications.* Paper presented at the Stockholm Mobility Roundtable, Stockholm, Sweden.

Funk, J. L. (2004). Key technological trajectories and the expansion of mobile Internet applications. *Info*, *6*(3), 208–215. doi:10.1108/14636690410549543

Garner, P., Edwards, R., & Coulton, P. (2006). *Card-based macro-payment for mobile phones.* Paper presented at the International Conference on Mobile Business, Copenhagen, Denmark.

Gerpott, T. J., & Jakopin, N. M. (2005). International marketing standardization and financial performance of mobile network operators - an empirical analysis. [SBR]. *Schmalenbach Business Review*, *57*(3), 189–228.

Gewei, Y. (2007). Mobile marketing systems: framework and technology enabler. *International Journal of Mobile Marketing*, *2*(1), 43–49.

Gilbert, L. A. (2003). *A marketing model for mobile wireless services.* Paper presented at the 6th Hawaii International Conference on System Sciences, Big Island, Hawaii.

Gill, T. (2008). Convergent products: What functionalities add more value to the base? *Journal of Marketing, 72*(2), 46–62. doi:10.1509/jmkg.72.2.46

Goode, M. M. H., Davies, F., Moutinho, L., & Jamal, A. (2005). Determining customer satisfaction from mobile phones: A neural network approach. *Journal of Marketing Management, 21*(7/8), 755–778. doi:10.1362/026725705774538381

Grant, I., & O'Donohoe, S. (2007). Why young consumers are not open to mobile marketing communication. *International Journal of Advertising, 26*(2), 223–246.

Grant, I., & O'Donohoe, S. (2007). Brand in the hand or sand in the hand? A contextualized account of adolescents' mobile phone consumption. *Advances in Consumer Research - North American Conference Proceedings, 34*, 280-285.

Gururajan, R. (2001). *Wireless applications: Influences and risks of location identification technologies.* Paper presented at the Twelfth Australian Conference on Information Systems, Coffs Harbour, Australia.

Haghirian, P., & Dickinger, A. (2005). *Identifying success factors of mobile marketing.* Paper presented at the Advances in Consumer Research - Asia-Pacific Conference Proceedings.

Haghirian, P., & Madlberger, M. (2006). *Consumer attitude toward advertising via mobile devices - An empirical investigation among Austrian users.* Paper presented at the 14th European Conference on Information Systems (ECIS), Goeteborg, Sweden.

Haghirian, P., Madlberger, M., & Tanuskova, A. (2005). *Increasing advertising value of mobile marketing - An empirical study of antecedents.* Paper presented at the 38th Hawaii International Conference on System Sciences, Big Island, Hawaii.

Hairong, L., & Stoller, B. (2007). Parameteres of mobile advertising: a field experiment. *International Journal of Mobile Marketing, 2*(1), 4–11.

Hairong, L., & Townsend, L. (2008). Mobile research in marketing: Design and implementation issues. *International Journal of Mobile Marketing, 3*(1), 32–40.

Han, S.-Y., Cho, M.-K., & Choi, M.-K. (2005). *Ubitem: A framework for interactive marketing in location-based gaming environment.* Paper presented at the Fourth International Conference on Mobile Business, Sydney, Australia.

Hanley, M., & Becker, M. (2008). Cell phone usage and advertising acceptance among College students: a four-year analysis. *International Journal of Mobile Marketing, 3*(1), 67–80.

Hanley, M., Becker, M., & Martinsen, J. (2006). Factors influencing mobile advertising acceptance: will incentives motivate College students to accept mobile advertisements? *International Journal of Mobile Marketing, 1*(1), 50–58.

Hao, L., Gao, C., & Liu, Z. a. (2007). Customer-based brand equity and improvement strategy for mobile phone brands: Foreign versus local in the Chinese market. *International Management Review, 3*(3), 76–83.

Haque, A. (2004). Mobile commerce: Customer perception and It's prospect on business operation in Malaysia. *Journal of American Academy of Business, Cambridge, 4*(1/2), 257–262.

Harris, P., Rettie, R., & Kwan, C. C. (2005). Adoption and usage of m-commerce: a cross-cultural comparison of Hong Kong and the United Kingdom. *Journal of Electronic Commerce Research, 6*(3), 210–224.

Hawkins, R., & Ballon, P. (2007). When standards become business models: reinterpreting 'failure' in the standardization paradigm. *Info, 9*(5), 20–30. doi:10.1108/14636690710816426

Haydon, J. M. (2004). Capitalizing on content in Australia. *I-Ways, 27*(3/4), 159–160.

Heinonen, K., & Pura, M. (2006). *Developing a conceptual framework for mobile services.* Paper presented at the Helsinki Mobility Roundtable

Heinonen, K., & Strandvik, T. (2003). *Consumer responsiveness to mobile marketing.* Paper presented at the Stockholm Mobility Roundtable, Stockholm, Sweden.

Heitmann, M., Prykop, C., & Aschmoneit, P. (2004). *Using means-end chains to build mobile brand communities.* Paper presented at the 37th Hawaii International Conference on System Sciences, Big Island, Hawaii.

Heller, N. E. (2006). Growth of mobile multimedia advertising. *International Journal of Mobile Marketing, 1*(1), 41–49.

Henten, A., Olesen, H., Saugstrup, D., & Sw-En, T. (2004). Mobile communications: Europe, Japan and South Korea in a comparative perspective. *Info, 6*(3), 197–207. doi:10.1108/14636690410549534

Hosbond, J. H., & Skov, M. B. (2007). Micro mobility marketing: Two cases on location-based supermarket shopping trolleys. *Journal of Targeting . Measurement & Analysis for Marketing, 16*(1), 68–77. doi:10.1057/palgrave.jt.5750058

Hourahine, B., & Howard, M. (2004). Money on the move: Opportunities for financial service providers in the 'third space'. *Journal of Financial Services Marketing, 9*(1), 57–67. doi:10.1057/palgrave.fsm.4770141

Hsu, T., Wang, Y., & Wen, S. (2006). Using the decomposed theory of planning behavioural to analyse consumer behavioural intention towards mobile text message coupons. *Journal of Targeting. Measurement & Analysis for Marketing, 14*(4), 309–324. doi:10.1057/palgrave.jt.5740191

Jaeki, S., Chulmo, K., & Yongjin, K. (2007). Investigating Antecedents of Behavioral Intentions in Mobile Commerce. *Journal of Internet Commerce, 6*(1), 13–34.

Jha, S. (2008). Understanding mobile phone usage pattern among college-goers. *ICFAI Journal of Services Marketing, 6*(1), 51–61.

Jingjun Xu, D. (2006). The influence of personalization in affecting consumer attitudes toward mobile advertising in China. *Journal of Computer Information Systems, 47*(2), 9–19.

Jong Woo, J., & Sangmi, L. (2007). Mobile media use and its impact on consumer attitudes toward mobile advertising. *International Journal of Mobile Marketing, 2*(1), 50–58.

Kannan, P. K., & Whinston, A. B. (2001). *Wireless commerce: Marketing issues and possibilities.* Paper presented at the 34th Hawaii International Conference on System Sciences, Maui, Hawaii.

Kapostasy, J. L. (2008). Adopting mobile payment channels: Key challenges for US financial institutions. *Journal of Payments Strategy & Systems, 2*(2), 167–174.

Karjaluoto, H., Karvonen, J., Kesti, M., Koivumaki, T., Manninen, M., & Pakola, J. (2005). Factors affecting consumer choice of mobile phones: Two studies from Finland. *Journal of Euromarketing, 14*(3), 59–82. doi:10.1300/J037v14n03_04

Karjaluoto, H., Lehto, H., Leppaniemi, M., & Mustonen, T. (2007). Insights into the implementation of mobile marketing campaigns. *International Journal of Mobile Marketing, 2*(2), 10–20.

Karjaluoto, H., Leppaniemi, M., Standing, C., Kajalo, S., Merisavo, M., & Virtanen, V. (2006). Individual differences in the use of mobile services among Finnish consumers. *International Journal of Mobile Marketing, 1*(2), 4–10.

Karp, G. (2007). Mobile marketing and interactive promotions on mobile devices: navigating legal hurdles. *International Journal of Mobile Marketing, 2*(2), 78–85.

Kauffman, R. J., & Techatassanasoontorn, A. A. (2004). *Does one standard promote faster growth? An econometric analysis of the international diffusion of wireless technology.* Paper presented at the 37th Hawaii International Conference on System Sciences, Big Island, Hawaii.

Kavassalis, P., Spyropoulou, N., Drossos, D., Mitrokostas, E., Gikas, G., & Hatzistamatiou, A. (2003). Mobile permission marketing: Framing the market inquiry. *International Journal of Electronic Commerce, 8*(1), 55–79.

Khalifa, M., & Ning Shen, K. (2008). Drivers for transactional B2C m-commerce adoption: extended theory of planned behavior. *Journal of Computer Information Systems, 48*(3), 111–117.

Kiseol, Y., & Jolly, L. D. (2006). Value-added mobile data services: the antecedent effects of consumer value on using mobile data services. *International Journal of Mobile Marketing, 1*(2), 11–17.

Kleijnen, M., De Ruyter, K., & Wetzels, M. (2004). Consumer adoption of wireless services: discovering the rules, while playing the game. *Journal of Interactive Marketing, 18*(2), 51–61. doi:10.1002/dir.20002

Kleijnen, M., Wetzels, M., & De Ruyter, K. (2004). Consumer acceptance of wireless finance. *Journal of Financial Services Marketing, 8*(3), 206–217. doi:10.1057/palgrave.fsm.4770120

Kondo, F. N., & Nakahara, M. (2007). Differences in customers' responsiveness to mobile direct mail coupon promotions. *International Journal of Mobile Marketing, 2*(2), 68–74.

Kondo, F. N., Uwadaira, Y., & Nakahara, M. (2007). Stimulating customer response to promotions: The case of mobile phone coupons. *Journal of Targeting . Measurement & Analysis for Marketing, 16*(1), 57–67. doi:10.1057/palgrave.jt.5750064

Koski, H., & Kretschmer, T. (2005). Entry, standards and competition: Firm strategies and the diffusion of mobile telephony. *Review of Industrial Organization, 26*(1), 89–113. doi:10.1007/s11151-004-4085-0

Krishnan, R., & Kothari, M. (2008). Antecedents of customer relationships in the telecommunication sector: An empirical study. *ICFAI Journal of Services Marketing, 6*(3), 38–59.

Laukkanen, T., & Lauronen, J. (2005). Consumer value creation in mobile banking services. *International Journal of Mobile Communications, 3*(4), 325–338. doi:10.1504/IJMC.2005.007021

Lee, C. (2008). Review on the implementation of mobile commerce in Malaysia. *Journal of Internet Banking & Commerce, 13*(2), 1–10.

Lee, D., & Muhlberger, R. (2006). Mobile marketing as technology intervention: a locales framework analysis. *International Journal of Mobile Marketing, 1*(1), 59–65.

Lee, R., & Murphy, J. (2006). The consumption of mobile services by Australian university students. *International Journal of Mobile Marketing, 1*(1), 13–20.

Leighton, D. (2007). In the frame: investigating the use of mobile phone photography in museums. *International Journal of Nonprofit and Voluntary Sector Marketing, 12*(4), 308–319. doi:10.1002/nvsm.320

Leppaniemi, M., & Karjaluoto, H. (2005). Factors influencing consumers' willingness to accept mobile advertising: a conceptual model. *International Journal of Mobile Communications, 3*(3), 197–213. doi:10.1504/IJMC.2005.006580

Leppaniemi, M., & Karjaluoto, H. (2008). Mobile marketing: from marketing strategy to mobile marketing campaign implementation. *International Journal of Mobile Marketing, 3*(1), 50–61.

Leppaniemi, M., Sinisalo, J., & Karjaluoto, H. (2006). A review of mobile marketing research. *International Journal of Mobile Marketing, 1*(1), 30–40.

Lin, H.-H., & Wang, Y.-S. (2006). An examination of the determinants of customer loyalty in mobile commerce contexts. *Information & Management, 43*(3), 271–282. doi:10.1016/j.im.2005.08.001

Lipner, W. E. (2007). The future of online market research. *Journal of Advertising Research, 47*(2), 142–146. doi:10.2501/S002184990707016X

Loh, C. C., Stadlen, A. D., Basole, R. C., Moses, J. D., & Tuohy, C. (2008). Enterprise mobility and support outsourcing: A research model and initial findings. *Information Knowledge Systems Management, 7*(1/2), 183–210.

Mackay, M. M., & Weidlich, O. (2007). Case study: conducting a web-based survey of mobile content usage by Australian mobile phone customers. *International Journal of Mobile Marketing, 2*(1), 28–33.

Maneesoonthorn, C., & Fortin, D. (2006). Texting behaviour and attitudes toward permission mobile advertising: an empirical study of mobile users' acceptance of SMS for marketing purposes. *International Journal of Mobile Marketing, 1*(1), 66–72.

Mattila, M. (2002). Introducing existing financial services over new electronic channels. *International Journal of Innovation Management, 6*(4), 431–447. doi:10.1142/S1363919602000689

McManus, P., & Scornavacca, E. (2005). *Mobile marketing: Killer application or new hype?* Paper presented at the Fourth International Conference on Mobile Business, Sydney, Australia.

Merisavo, M., Kajalo, S., Karjaluoto, H., Virtanen, V., Salmenkivi, S., & Raulas, M. (2007). An empirical study of the drivers of consumer acceptance of mobile advertising. *Journal of Interactive Advertising, 7*(2), 1.

Mi Jung, K., & Jong Woo, J. (2008). A case study of mobile advertising in South Korea: Personalisation and digital multimedia broadcasting (DMB). *Journal of Targeting . Measurement & Analysis for Marketing, 16*(2), 129–138. doi:10.1057/jt.2008.5

Mijung, K., & Jun, J. (2008). *A case study of mobile advertising in South Korea: personalization and digital multimedia broadcasting (DMB).* Paper presented at the American Academy of Advertising Conference Proceedings.

Milne, G. R., & Rohm, A. J. (2003). The 411 on mobile privacy. *Marketing Management, 12*(4), 40–45.

Moldrup, C. (2007). Individualised health marketing using SMS - A smoking cessation case. *Journal of Medical Marketing, 7*(3), 255–259. doi:10.1057/palgrave.jmm.5050087

Mort, G. S., & Drennan, J. (2002). Mobile digital technology: Emerging issues for marketing. *Journal of Database Marketing*, *10*(1), 9–23. doi:10.1057/palgrave.jdm.3240090

Mort, G. S., & Drennan, J. (2005). Marketing m-services: Establishing a usage benefit typology related to mobile user characteristics. *Journal of Database Marketing & Customer Strategy Management*, *12*(4), 327–341. doi:10.1057/palgrave.dbm.3240269

Mort, G. S., & Drennan, J. (2007). Mobile communications: A study of factors influencing consumer use of m-services. *Journal of Advertising Research*, *47*(3), 302–312. doi:10.2501/S0021849907070328

Muk, A. (2007). Consumers' intentions to opt in to SMS advertising. *International Journal of Advertising*, *26*(2), 177–198.

Muk, A. (2007). Cultural influences on adoption of SMS advertising: A study of American and Taiwanese consumers. *Journal of Targeting . Measurement & Analysis for Marketing*, *16*(1), 39–47. doi:10.1057/palgrave.jt.5750062

Muk, A., & Babin, J. (2006). U.S. consumers' adoption -- nonadoption of mobile SMS advertising. *International Journal of Mobile Marketing*, *1*(1), 21–29.

Muneer, S., & Sharma, C. (2008). Enterprise mobile product strategy using scenario planning. *Information Knowledge Systems Management*, *7*(1/2), 211–224.

Neeru, S., & Ojha, S. (2004). Measuring service performance in mobile communications. *Service Industries Journal*, *24*(6), 109–128. doi:10.1080/0264206042000299202

Nerger, P. (2008). Breaking free from 'dotcom thinking' in a mobile world. *International Journal of Mobile Marketing*, *3*(1), 19–22.

Newell, J., Hanley, M., & Meier, M. (2008). Consumer justifications for accepting cell phone advertising: a field study. *50th American Academy of Advertising Conference Proceedings*, 140-140.

Newell, J., & Meier, M. (2007). Desperately seeking opt-in: a field report from a student-led mobile marketing initiative. *International Journal of Mobile Marketing*, *2*(2), 53–57.

Newton, T. (2008). Reaching those on the go: Mobile Web users present next major challenge for public relations. *Public Relations Tactics*, *15*(5), 11–11.

Niemela, J. (2006). Mobile semantics: Defining concepts and their interrelationships. *Electronic Markets*, *16*(4), 329–336. doi:10.1080/10196780600999759

Nohria, N., & Leestma, M. (2001). A moving target: The mobile-commerce customer. *MIT Sloan Management Review*, *42*(3), 104–104.

Nysveen, H. r., Pedersen, P. E., Thorbjornsen, H., & Berthon, P. (2005). Mobilizing the brand: The effects of mobile services on brand relationships and main channel use. *Journal of Service Research*, *7*(3), 257–276. doi:10.1177/1094670504271151

Okazaki, S. (2007). Assessing mobile-based online surveys. *International Journal of Market Research*, *49*(5), 651–675.

Okazaki, S. (2008). Exploring experiential value in online mobile gaming adoption. *Cyberpsychology & Behavior*, *11*(5), 619–622. doi:10.1089/cpb.2007.0202

Okazaki, S. (2008). Browse your mobile and spread the word to your friends: a brand promotion campaign. *50th American Academy of Advertising Conference Proceedings*, 141-143.

Okazaki, S., Katsukura, A., & Nishiyama, M. (2007). How mobile advertising works: The role of trust in improving attitudes and recall. *Journal of Advertising Research, 47*(2), 165–178. doi:10.2501/S0021849907070195

Pagani, M. (2004). Determinants of adoption of third generation mobile multimedia services. *Journal of Interactive Marketing, 18*(3), 46–59. doi:10.1002/dir.20011

Pagani, M. (2008). A value-choice model to forecast market consequences of 3G mobile service design decisions. *International Journal of Mobile Marketing, 3*(1), 23–31.

Pagiavlas, N., Stratmann, M., Marburger, P., & Young, S. (2005). Mobile business - comprehensive marketing strategies or merely IT expenses. *Journal of Electronic Commerce Research, 6*(3), 251–261.

Park, J., & SuJin, Y. (2006). The moderating role of consumer trust and experiences: value driven usage of mobile technology. *International Journal of Mobile Marketing, 1*(2), 24–32.

Pedersen, P. E. (2005). Adoption of mobile Internet services: An exploratory study of mobile commerce early adopters. *Journal of Organizational Computing and Electronic Commerce, 15*(3), 203–222. doi:10.1207/s15327744joce1503_2

Pihlström, M., & Brush, G. J. (2008). Comparing the perceived value of information and entertainment mobile services. *Psychology and Marketing, 25*(8), 732–755. doi:10.1002/mar.20236

Pousttchi, K., & Wiedemann, D. G. (2006). *A contribution to theory building for mobile marketing: Categorizing mobile marketing campaigns through case study research.* Paper presented at the International Conference on Mobile Business, Copenhagen, Denmark.

Prete, M. I. (2007). M-politics: Credibility and effectiveness of mobile political communications. *Journal of Targeting. Measurement & Analysis for Marketing, 16*(1), 48–56. doi:10.1057/palgrave.jt.5750063

Prykop, C., & Heitmann, M. (2006). Designing mobile brand communities: Concept and empirical illustration. *Journal of Organizational Computing and Electronic Commerce, 16*(3/4), 301–323. doi:10.1207/s15327744joce1603&4_8

Ramkumar, G. D. (2007). Image recognition as a method for opt-in and applications for mobile marketing. *International Journal of Mobile Marketing, 2*(2), 42–49.

Ranchhod, A. (2007). Developing mobile marketing strategies. *International Journal of Mobile Marketing, 2*(1), 76–83.

Ranganathan, C., Seo, D., & Babad, Y. (2006). Switching behavior of mobile users: do users' relational investments and demographics matter. *European Journal of Information Systems, 15*(3), 269–276. doi:10.1057/palgrave.ejis.3000616

Rangone, A., & Renga, F. M. (2006). *Mobile advertising: A framework for the appraisal of the campaigns.* Paper presented at the International Conference on Mobile Business, Copenhagen, Denmark.

Reday, P. A., & Na, W. (2005). *Information technology and the sales function.* Paper presented at the Advances in Consumer Research - Asia-Pacific Conference Proceedings.

Repo, P., & Pantzar, M. (2004). *Users inventing ways to enjoy new mobile services.* Paper presented at the 37th Hawaii International Conference on System Sciences, Big Island, Hawaii.

Rettie, R., Grandcolas, U., & Deakins, B. (2005). Text message advertising: Response rates and branding effects. *Journal of Targeting . Measurement & Analysis for Marketing, 13*(4), 304–312. doi:10.1057/palgrave.jt.5740158

Ridgeon, E. (2007). Creating a center of communication expertise at Orange. *Strategic Communication Management, 11*(3), 20–23.

Riivari, J. (2005). Mobile banking: A powerful new marketing and CRM tool for financial services companies all over Europe. *Journal of Financial Services Marketing, 10*(1), 11–20. doi:10.1057/palgrave.fsm.4770170

Rohm, A. J., & Sultan, F. (2006). An exploratory cross-market study of mobile marketing acceptance. *International Journal of Mobile Marketing, 1*(1), 4–12.

Rondeau, D. B. (2005). For mobile applications, branding is experience. *Communications of the ACM, 48*(7), 61–66. doi:10.1145/1070838.1070867

Sabat, H. K. (2008). The business economics of growing a next generation mobile wireless business. *IIMB Management Review, 20*(1), 47–68.

Salaberry, R., & Upton, L. (2007). Language assistance products support mobile advertising in underprivileged countries. *International Journal of Mobile Marketing, 2*(1), 59–65.

Sangwon, L., & Chan-Olmsted, S. M. (2006). The development of mobile television: examining the convergence of mobile and broadcasting services in Korea. *International Journal of Mobile Marketing, 1*(2), 60–67.

Scharl, A., Dickinger, A., & Murphy, J. (2005). Diffusion and success factors of mobile marketing. *Electronic Commerce Research and Applications, 4*(2), 159–173. doi:10.1016/j.elerap.2004.10.006

Schierholz, R., Glissmann, S., Kolbe, L. M., Brenner, W., & Ostrowski, A. (2006). Don't call us, we'll call you -- Performance measurement in multi-channel environments. *Journal of Information Science & Technology, 3*(2), 45–63.

Schierholz, R., Ostrowski, A., Glissmann, S., & Kolbe, L. M. (2005). *Performance measurement of mobile marketing in multi-channeleEnvironments.* Paper presented at the Fourth International Conference on Mobile Business, Sydney, Australia.

Scornavacca, E., & McKenzie, J. (2006). *Perceived critical success factors of SMS based campaigns: A managerial perspective.* Paper presented at the 14th European Conference on Information Systems (ECIS), Goeteborg, Sweden.

Sharma, C., Herzog, J., & Melfi, V. (2008). A five-points measurement framework for mobile advertising. *International Journal of Mobile Marketing, 3*(1), 4–11.

Sharma, N., & Ojha, S. (2004). Measuring service performance in mobile communications. *Service Industries Journal, 24*(6), 109–128. doi:10.1080/0264206042000299202

Sophonthummapharn, K., & Tesar, G. (2007). Technology readiness and propensity of cell phone users to subscribe to commercial messaging services. *Marketing Management Journal, 17*(2), 81–95.

Sorensen, C., Al-Taitoon, A., Kietzmann, J., Pica, D., Wiredu, G., & Elaluf-Calderwood, S. (2008). Exploring enterprise mobility: Lessons from the field. *Information Knowledge Systems Management, 7*(1/2), 243–271.

Spurgeon, C. (2005). Losers and lovers: Mobile phone services advertising and the new media consumer/producer. *Journal of Interactive Advertising, 5*(2).

Srivastava, R. K. (2007). Determination of brand loyalty factors age group-18-24. *Academy of Marketing Studies Journal, 11*(1), 1–16.

Steinbock, D. (2006). The missing link why mobile marketing is different. *International Journal of Mobile Marketing, 1*(1), 83–94.

Stoica, M., Miller, D. W., & Stotlar, D. (2005). New technology adoption, business strategy and government involvement: The case of mobile commerce. *Journal of Nonprofit & Public Sector Marketing, 13*(1/2), 213–232. doi:10.1300/J054v13n01_12

Suoranta, M., & Mattila, M. (2004). Mobile banking and consumer behaviour: New insights into the diffusion pattern. *Journal of Financial Services Marketing, 8*(4), 354–366. doi:10.1057/palgrave.fsm.4770132

Swilley, E., & Hofacker, C. F. (2006). Defining mobile commerce in a marketing context. *International Journal of Mobile Marketing, 1*(2), 18–23.

Tanakinjal, G. H., Amin, H., Lajuni, N., & Bolongkikit, J. (2007). Mobile devices and communication: An analysis. *Journal of Internet Banking & Commerce, 12*(3), 1–12.

Thurner, R. (2008). Adding mobile to the marketing mix. *Journal of Telecommunications Management,* pp. 122-124.

Treiblmaier, H., & Dickinger, A. (2006). *The importance of previous experience for the trial of mobile self-service technologies.* Paper presented at the 27th International Conference on Information Systems, Milwaukee, U.S.A.

Tsang, M. M., Shu-Chun, H., & Ting-Peng, L. (2004). Consumer attitudes toward mobile advertising: An empirical study. *International Journal of Electronic Commerce, 8*(3), 65–78.

Unhelkar, B., & Agrawal, S. (2008). Globalization wave and managing mobile enterprise transitions: MET framework. *ICFAI Journal of International Business, 3*(1), 7–15.

Unni, R., & Harmon, R. (2007). Perceived effectiveness of push vs. pull mobile location-based advertising. *Journal of Interactive Advertising, 7*(2), 1.

Van den Dam, R., & Nelson, E. (2008). How telcos will change advertising. *Journal of Telecommunications Management, 1*(3), 237–246.

Vatanparast, R., & Asil, M. (2007). Factors affecting the use of mobile advertising. *International Journal of Mobile Marketing, 2*(2), 21–34.

Venkatesh, B., Nargundkar, R., Sayed, F. K., & Shahaida, P. (2006). Assessing Indian students' perceptions towards m-learning - some initial conclusions. *International Journal of Mobile Marketing, 1*(2), 75–79.

Verkasalo, H. (2007). Handset-based measurement of smartphone service evolution in Finland. *Journal of Targeting . Measurement & Analysis for Marketing, 16*(1), 7–25. doi:10.1057/palgrave.jt.5750060

Verkasalo, H. (2008). Handset-based measurement of mobile service demand and value. *Info, 10*(3), 51–69. doi:10.1108/14636690810874395

Verkasalo, H., & Hammaminen, H. (2007). A handset-based platform for measuring mobile service usage. *Info, 9*(1), 80–96. doi:10.1108/14636690710725085

Vlachos, P., & Vrechopoulos, A. (2003). *Predictors of consumer attitudes towards mobile music services: A cross-cultural research approach.* Paper presented at the 2nd International Conference on Mobile Business, Vienna, Austria.

Vrechopoulos, A., Constantiou, I., Sideris, I., Doukidis, G., & Mylonopoulos, N. (2003). The critical role of consumer behaviour research in mobile commerce. *International Journal of Mobile Communications*, *1*(3), 329–340. doi:10.1504/IJMC.2003.003498

Vrehopoulos, A., Constantiou, I., & Sideris, I. (2002). *Strategic marketing planning for mobile commerce diffusion and consumer adoption.* Paper presented at the 1st International Conference on Mobile Business, Athens, Greece.

Wais, J. S., & Clemons, E. K. (2008). Understanding and implementing mobile social advertising. *International Journal of Mobile Marketing*, *3*(1), 12–18.

Wang, A. (2007). Branding over mobile and Internet advertising: the cross-media effect. *International Journal of Mobile Marketing*, *2*(1), 34–42.

Wang, A. (2007). How consumers perceive free offers: implications for mobile marketing. *International Journal of Mobile Marketing*, *2*(2), 35–41.

Wang, A., & Acar, S. A. (2006). Information search and mobile promotions. *International Journal of Mobile Marketing*, *1*(2), 80–87.

Wang, W., & Cheong, F. (2004). *Using system dynamics to model mobile commerce diffusion.* Paper presented at the 4th International Conference on Electronic Business, Beijing, China.

Wehmeyer, K. (2005). *Mobile couponing - measuring consumers. Acceptance and preferences with a limit conjoint approach.* Paper presented at the 18th Bled Conference, Bled, Slovenia.

Weitenberner, C., Chapman, M., Miranda, G., Tobar, S., Wagoner, E., & Akasaki, D. (2006). United States and South East Asian mobile markets: a comparative analysis of infrastructure and cultural differences. *International Journal of Mobile Marketing*, *1*(1), 73–82.

Wyner, G. A. (2006). Space and Time. *Marketing Research*, *18*(4), 6–8.

Xiaoyun, H., Kwortnik, R. J. Jr, & Chunxiao, W. (2008). Service loyalty: An integrative model and examination across service contexts. *Journal of Service Research*, *11*(1), 22–42. doi:10.1177/1094670508319094

Yang, K. C. C. (2007). Exploring factors affecting consumer intention to use mobile advertising in Taiwan. *Journal of International Consumer Marketing*, *20*(1), 33–49. doi:10.1300/J046v20n01_04

Yeo, J., & Huang, W. (2003). Mobile e-commerce outlook. *International Journal of Information Technology & Decision Making*, *2*(2), 313. doi:10.1142/S0219622003000641

Yol, S., Serenko, A., & Turel, O. (2006). *Moderating roles of user demographics in the American customer satisfaction model within the context of mobile services.* Paper presented at the 12th Americas Conference on Information Systems, Acapulco, Mexico.

Young-Hyuck, J., Jong-Kun, J., & Byung-Do, K. (2002). Encouraging customers to pay less for mobile telecommunication services. *Journal of Database Marketing*, *9*(4), 350. doi:10.1057/palgrave.jdm.3240083

Zid, L. A.-S. (2005). TV to go. *Marketing Management*, *14*(1), 5–5.

KEY TERMS AND DEFINITIONS

Mobile Advertising: A paid form of context-sensitive personal communication from an identifiable source, designed to persuade the receiver to take some action, either in real time via the mobile device or in the future.

Permission-Based Marketing: Prospect-initiated communication followed by the possibility of an active two-way exchange

Consumer Attitude: A lasting general evaluation of something; a state of mind or a feeling

Purchase Intention: An individual's conscious plan to make an effort to purchase a brand

Mobile Services: In this context m-services are perceived as all non-voice services afforded through mobile networks, except for interpersonal SMS exchanges, that the end users can employ whilst mobile

Context-Aware Marketing: Marketing that is location-, time- and consumer profile-specific.

Ubiquitous Marketing: This type of marketing is sensing and responding not only to who the customer is, but where she is and what she is doing with transparent and undisruptive approaches

Chapter 3
Framework for Mobile Marketing:
The Locales Framework

Dennis Lee
American University in Dubai, UAE

Ralf Muhlberger
The University of Queensland, Australia

Mark Brown
The University of Queensland, Australia

ABSTRACT

This chapter suggests that mobile marketing research take an Interaction Design approach through the adoption of the Locales Framework as an alternative perspective to investigate the context of mobile marketing. The research shows how such an approach can integrate marketing and new technology research, highlighting the issues and opportunities brought about by the technology within a domain context, i.e. mobile marketing. Given that there are not many theoretically based frameworks available to support mobile marketing research, the application of this framework provides a roadmap for future research in mobile marketing and helps build the body of knowledge in this field.

BACKGROUND

Many people incorrectly perceive the term "marketing" as the same thing as advertising or promotion (Kerin, Hartley & Rudelius, 2007). This definition only highlights the meaning of "marketing" as a form of activity or tool for delivering marketing contents. In a similar manner, many people also find it difficult to distinguish the term "mobile marketing" (Leppaniemi, Sinisalo & Karjaluotoaniemi, 2006). Currently, no clear definition of mobile marketing exists in academic or industry publications (Leppaniemi, Karjaluoto & Salo, 2004). Numerous definitions of mobile marketing have been proposed by marketing practitioners and research scholars. Facchetti, Rangone, Renga and Savoldelli (2005) define mobile marketing as "any paid form of impersonal presentation and promotions of goods, services, ideas by well-identified promoter (Kotler, Jain, & Maesincee, 2002) using 'the wireless' as delivery channel". Tsang, Ho, &

DOI: 10.4018/978-1-60566-074-5.ch003

Liang (2004) refer to the phenomenon as sending advertising to mobile devices such as mobile phones or Personal Digital Assistants (PDAs) through wireless network, and the delivery of mobile advertising as involving digital texts, images and voice with interactive, immediate, personalized, and responsive capabilities. A much simpler explanation of mobile marketing is offered by Bauer, Barnes, Reichardt and Neumann (2005) where they describe it as "using the mobile phone as a means of conveying commercial content to customers". Mobile marketing practitioners like Becker (2005) offer another definition. Becker (2005) states that "the mobile channel is a multi-faceted interactive, interoperable, network composed of various mobile delivery technologies used by companies to market to an individual, and deliver to the individual communication, personalization, information and entertainment products and services".

Building on these insights into mobile marketing, we define mobile marketing as leveraging the usage of mobile phones for conducting marketing activities over a telecommunication network, to transmit advertising and promotional contents to consumers in the form of time- and location-sensitive, personalized information with the intention of promoting goods and services, thereby generating value for all stakeholders involved. This paper also builds on the traditional definition of marketing which involves advertising and promotion as channels to conduct marketing activities, such that mobile marketing is viewed as a 'pre-launch' problem that occurs during the planning of a marketing campaign. Mobile marketing is regarded as a form of technology intervention in the context of a marketing campaign, looking at either advertising or promotion, or both.

INTRODUCTION

According to the marketing literature, the perspective of mobile marketing is often related to consumers' traits and behaviors such as acceptance, attitude, experiences, responsiveness and perception. These have been have been intensively investigated over the past few years by Barwise and Strong (2002), Bauer et al. (2005), Haghirian and Madlberger (2005), Okazaki (2004), Rohm and Sultan (2006), and Tsang et al. (2004). Much of this research was conducted using a single research approach such as interview, experiment or survey. This form of research mainly focuses on the characteristics of individuals in terms of how they view mobile marketing. The outcomes from the analyses may provide justifications for the phenomenon studied but they are not thorough enough to comprehensively detail the full mobile marketing concept (Gopal & Tripathi, 2006).

LITERATURE REVIEW

The means of determining whether marketing communications are effective have been well documented; however such research has largely been predicated on traditional, and more recently Internet, based platforms. There is a distinct lack of literature regarding the mobile medium, particularly in terms of studying the effectiveness of this new marketing communication tool in facilitating marketing campaigns for companies (Perlado & Barwise, 2004). For many companies, mobile marketing still represents a mysterious and challenging new component of a company's communication mix. There are still challenges that could hinder the effective implementation of advertising through mobile phones. For example, the screen sizes of mobile phones are small and therefore it is unlikely to be possible to exhibit the entire same advertisement as in print.

Furthermore, no studies to date have compared the effectiveness of this medium in delivering advertising or promotions with other more established media such as print (Lee & Muhlberger, 2006). Some of the fundamental issues that marketers are concerned with in evaluating the mobile channels for advertising and promotional purposes are: "What is the difference between mobile advertising and traditional advertising?"; "Will this mobile advertising be more effective than traditional advertising?"; "How will consumers respond to mobile advertising at different times in different locations?", and "What forms of privacy are there to protect consumers when they are presented with mobile advertising?"

The adoption of the mobile phone as a major medium to market products and services by companies has just begun. In the United States of America, companies like Coca-Cola, BMW, ESPN, McDonald's, MTV, Nike, Visa USA, and Volvo have begun to experiment with SMS and WAP-based marketing campaigns to influence consumers' purchasing behaviors (Richtel, 2006; Rohm & Sultan 2005; Varanparast & Asil, 2007). Similarly in Australia, sales promotional materials can be delivered to mobile phones via a SMS when one signs up with companies such as www.mobile-dockets.com.au. In countries like Japan, China, Thailand and Singapore, mobile phone users are also beginning to receive mobile marketing contents on their mobile phones in exchange for free services (Sgriccia et al., 2007).

In spite of the popularity of adopting the mobile phone as a tool to engage in marketing activities, work on what makes mobile marketing effective, and how to utilize this new channel for marketing efforts, is an open area for research (Bauer et al., 2005; Becker, 2006; Karjaluoto, Lehto, Leppäniemi & Mustonen, 2007; Okazaki, 2005). Research in mobile marketing has sparked many diverse interests in different disciplines and perspectives over the past few years. The research area of mobile marketing is still inconsistent and highly fragmented (Bauer et al., 2005; Leppaniemi & Karjaluoto, 2005; Okazaki, 2005). These have caused concern about the need to build a common conceptualization of a theoretical framework to examine the mobile marketing phenomenon (Leppaniemi, Sinisalo & Karjaluotoaniemi, 2006). A common conceptualization serves not only as a ground to build knowledge but also allows further exploration and enhancement to the knowledge domain, with researchers in related fields having a common language in which to share and contribute (Heflin, 2004).

Particularly, research frameworks to fully comprehend the mechanism of running successful mobile marketing campaigns are still lacking. These theoretical frameworks are considered as underlying foundations in which they can help further explore and evaluate the mobile marketing research to their potential (Karjaluoto et al., 2007). Frameworks, research methods and theories are essential ingredients in academic research because these needed elements are contributors to the building of knowledge to the specific domain (Babbie, 2005).

THE LOCALES FRAMEWORK

In this paper we put forth the introduction of a research framework known as Locales Framework (LF) to study a mobile marketing context, i.e. a marketing campaign. Frameworks like the LF are able to provide principles, rules and guidelines to analyze phenomena (Rogers, 1997). The usage of a framework like LF also allows the context of mobile marketing to be viewed from a broader holistic perspective that overcomes the narrowness of findings from the traditional marketing research methods. The LF is able to incorporate multiple research methods to provide a richer and better basis for assessing mobile marketing research. Moreover, such frameworks permit a better understanding of the different stakeholders involved within the studied phenomenon (Blackwell & Green, 2003; Fitzpatrick, 2003). Most frameworks aim to provide pragmatic solutions and applications to bridge the gaps between research design and the social world (Rogers & Muller, 2006). The use of the LF can also achieve this objective as it serves as an analytical tool in understanding the different aspects of mobile marketing (Lee & Muhlberger, 2006).

Usages of the Locales Framework

The Locales Framework was primarily developed as a design framework to improve issues related to the context of interaction design, Human-Computer Interaction (HCI) and Computer Supported Collaborative Work (CSCW) (Fitzpatrick, 2003). It had been used in the creation of a groupware tool to support informal awareness and casual interactions for small communities (McEwan & Greenberg, 2005), in facilitating and improving information and communication flows within a government department (Fitzpatrick, 2003), and in improving the medical pathology services (Green, 2005). The LF is an evolvable framework which can be made relevant for both understanding the social world (Fitzpatrick, 1998) as well as for improving the applications and designs in mobile marketing (Lee & Muhlberger, 2007). The LF uses Anselm Strauss' (1993) *Theory of Action* as its underlying theoretical foundation. According to Strauss (1993), this theory is grounded in and abstracted from a lifetime of social observations of how people work and is able to provide a rich multifaceted way of understanding the structure and dynamics of collaborative group interactions within the social world (Strauss, 1993).

To conduct a mobile marketing campaign, the medium itself involves many stakeholders and entities. The components of the various stakeholders and entities include people, mobile technologies, socio and economic environments, software and hardware, telecommunication standards, legislations and geographical locations. Thus, a broader and general overview to mobile marketing research is required as it can provide a better overall understanding of this phenomenon not bounded by one discipline's research domain (Becker, 2006; Leppaniemi, Sinisalo & Karjaluoto, 2006; Pousttchi & Wiedemann, 2006). The main objective of this paper is then to demonstrate that the LF can be used as a general approach to study a mobile marketing campaign with a holistic and multidisciplinary lens.

Applying a framework approach to facilitate technology interventions in social contexts seems to be a stronger approach than treating mobile technology just as a different medium for carrying the marketing material (Lee & Muhlberger, 2007). Besides, many of the business models in mobile marketing have been developed without any conceptual foundation in their descriptions (see e.g. Camponovo & Pigneur, 2003; Komulainen, Mainela, Sinisalo, Tähtinen, & Ulkuniemi, 2004; Yuan & Zhang, 2003). Frameworks of theoretical underpinnings are important to aid any existing business models in their explanation of the empirical phenomenon found (Amit & Zott, 2001). Thus, a framework approach such as the LF provides empirical answers to support mobile marketing if companies are considering implementing marketing strategies (Yunos, Gao, & Shim, 2003).

ANALYSIS OF THE LOCALES FRAMEWORK

The approach we have developed in using the LF to analyze mobile marketing depends on a certain context and involves three phases. Further phases can be added for validation, and some iteration of phases is recommended to refine the work, however the core is threefold.

First Phase

The first phase is established by data collection. This will facilitate the building of a better understanding of the current locale of interest. The traditional method to collect data under the LF is based on ethnographic study practices. Ethnography involves the study of people performing activities and interacting in complex social settings in order to obtain a qualitative understanding of these interactions (Flick, 2006). It also allows development of detailed descriptions of the lived experience since it focuses on what people actually do in natural settings (Punch, 2005).Conducting an ethnographical study purely based on observing people may take time and effort (Punch, 2005) that may result in months or years of intensive fieldwork. Hence, ethnography research is rarely acceptable to or adopted by companies as a research method for understanding their markets (Ireland, 2003). A shorter market research time frame seems to be favored by most companies as most research objectives have to be met typically within a few weeks or even days (Plowman, 2003). However, we need to understand all the stakeholders involved in the phenomenon and we need a much faster time frame for the data collection process to understand the marketing campaign. Therefore, the use of common marketing research techniques like experiments, interviews and surveys are adopted for data collection purposes and applied according to the need of the researcher. A small amount of general observation by the researcher may also be needed for the second phase.

Second Phase – General LF

The second phase is to frame the attributes and findings from the data collection by associating them to a general template of the LF, based on their close association to the aspects' descriptions. Attributes represent either the context in which a mobile marketing campaign is being carried out or any general observations made during the context where stakeholders are involved. Findings are results of the analysis of any given marketing research methods used. The general template for the LF consists of five aspects. These five aspects are able to accept interdependent and overlapping attributes and findings from the output of data collected. Duplicated entries of the attributes and findings can be allocated into two or more aspects of the LF.

The five aspects of the LF are the locale foundation, the civic structure, the views of individuals, the trajectory interaction and the mutuality factor. Figure 1 highlights the 5 aspects of the Locales Framework.

The first aspect, the locale foundation, is a site or the means that a social world uses in its pursuit of interactions (Fitzpatrick, 1998). For example, the locale foundation in the context of mobile marketing would be any given locations where the campaign was carried out by the marketers and technologies involved.

The second aspect of the LF is civic structures. It concerns the facilitation of interaction with the wider community within and beyond a person's

Figure 1. The Five Aspects of the Locales Framework

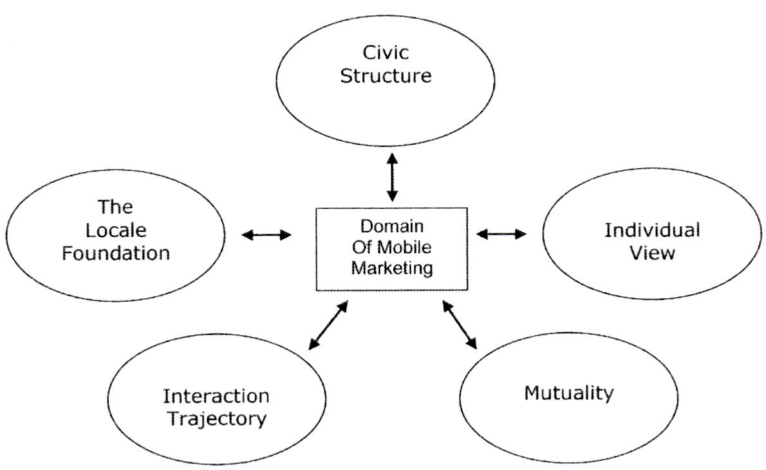

known social worlds and locales. The interaction with a wider community can possibly relate to an environment that is physical, spatial, geographical, organizational, informational, professional, or legislative and so on. (Fitzpatrick, 1998). This aspect highlights the various relationships of the stakeholders when the mobile marketing campaign is carried out.

The third aspect of the LF's analysis describes an individual's perspective(s) on his or her social world(s). Individuals have unique sets of perspectives and opinions. They come from different backgrounds; they are of different ages, genders, and cultures. Individuals personalize their views to suit their tasks according to their current level of engagement.

Interaction trajectory is the fourth aspect of the LF. According to Fitzpatrick (1998), it identifies the dynamic, temporal aspects of the social world in action. An important consideration in understanding this aspect is to look at what perspective or point of view is applied to any particular domain. Companies need to understand that the social world is always changing and evolving. Therefore companies need to involve the interaction trajectory aspect which identifies actual interactions individuals have over time within a given context setting and with each other. In the context of mobile marketing, this aspect identifies the actual interactions between the company whose product or service is being marketed and the participants during the marketing campaign.

The last aspect of the LF is known as mutuality. The mutuality aspect is regarded as the heart of collaborative activity that draws specific attention to how the locale supports presence and how awareness of that presence is supported for the achievement of shared activity (Fitzpatrick, 1998). In other words, mutuality concerns the support for interactions within locales for the purposes of maintaining a sense of shared place. For instance, for marketing, awareness and presence are of the product from the consumer's point of view, but also of the consumer from the company's perspective.

A general list of questions has been developed to highlight the objectives of each marketing research method used to study the context of mobile marketing. These questions are mostly associated to a marketing campaign. They are not comprehensive, but they do provide a general overview for associating the attributes and findings to the aspects of LF.

- Locales Foundation
  1. Where is the locale for mobile marketing?
  2. What is used to carry out mobile marketing?
  3. Who are the stakeholders involved in the context of mobile marketing?
  4. What are the existing marketing communications tools?
- Civic Structure
  1. What are the internal and external environments for the context of mobile marketing?
  2. How do we generate customers' interest in visiting the locale?
  3. How can the locale support the implementation of mobile marketing?
  4. What would be an appropriate time to deliver marketing materials?
- Individual View
  1. What are the views of customers?
  2. Are there means to determine the purpose of consumers coming into the locale?
  3. What are the cultural, social, technological and ethical characteristics of the potential customer?
  4. What are the experiences of the consumers with mobile phones?
- Interaction Trajectory
  1. How to conduct mobile marketing?

2. What types of marketing materials are used and how are they used?
3. How will stakeholders affect the mobile marketing?
4. When will be an appropriate time to send out mobile marketing materials?
5. What are the potential issues or barriers that will arise over time when conducting mobile marketing?

- Mutuality
  1. How do we facilitate mobile marketing?
  2. What are the available technologies to support mobile marketing?
  3. Why do we use the particular type of mobile marketing?
  4. How can mobile marketing be complemented by existing marketing tools?

Final Phase – Specific LF

The third and final phase is to evolve new locales, informed by the data that has been collated into the general LF. This is where the specific LF will evolve based on a certain context. Specifically, there are a number of possible questions that will help to drive this phase:

- How can the existing locale be enhanced to support the five aspects of the Locales Framework?
- What are the newer technologies to be applied to the locale?
- Will newer social worlds(e.g., new communities, new consumers) evolve if the resources are used in these new locales?
- How can mismatches in similar aspects of the domain, when viewed from different social perspectives, be reconciled?
- How can matches in similar aspects of the domain, when viewed from different social perspectives, be leveraged?

APPLICATION OF THE LF FOR A MARKETING CAMPAIGN

We illustrate the use of the LF for mobile marketing analysis with a case study. A marketing campaign was carried out by two cafés located in Brisbane, Australia. The two cafes are managed by the same organization. The objective of this marketing campaign was to access the effectiveness of mobile coupons as a channel to attract consumers. The cafés adopted the Multimedia Messaging Service (MMS) as a mode of marketing communication when sending out their marketing contents to their consumers. In particular, mobile coupons were sent to all of their registered customers as a form of enticement to attract customers to visit the cafés during the marketing campaign. The mobile coupons provided a special deal for a cup of coffee if consumers present them at the café during the suggested period of time. Each mobile coupon was also embedded with additional text information that served as free reminders to the general public for several upcoming major events to be held in the city. All consumers were contacted at least 3 weeks prior the start of this marketing campaign. This campaign attracted 133 individual customers to the two cafés. All participants had used mobile phones for at least one year.

First Phase of the LF

For this marketing campaign, an experiment was carried out. The main aim of this experiment was to undercover a better understanding of what is mobile marketing, how it can be used and what are the associated challenges an organization faces when the marketing campaign is based on coupons to promote their business. Basically, this experiment involved examining all the operational aspects of the cafe when the marketing campaign took place. A survey was incorporated into this experiment to further determine consumers' behavior toward mobile marketing. Each consumer is tasked to complete this survey a week after the marketing campaign had ended regardless if they have redeemed the mobile coupon. The questions for this survey had two objectives. One was to understand the general shopping behavior of people, and the other was to examine people's perceptions of a marketing campaign when the advertisement includes a promotional tool like a coupon. The questions adopted for the survey were formulated based on the existing scales found in the marketing literature. All scales adopted had a Cronbach's alpha of more than 0.7 which is deemed as reliable and valid (Tabachnick & Fidell, 2001). The survey tool was also pre-tested before all the selected scales were used. This was to ensure that all responses to the questions were not misunderstood, ambiguous or incomprehensible. Data collected from this experiment were based on the findings of the survey and observations made during the experiment. The direct observation of consumers' behaviors in the context of the cafés allowed for a better interpretation of those involved and also provided additional useful description for the context (Atkinson, Coffey & Delamont, 2003). Particularly, the researchers hoped to fulfill a number of observational tasks like understanding the different body language demonstrated by participants, the context of the café, mobile phones usage patterns, participants' buying behavior at the café, and a number of situational factors like time and day. The researchers were also reflexive throughout the entire study, being careful not to impose any personal biasness and perspectives on what was described.

Second Phase of the LF

The five aspects of the LF are associated with the attributes and findings from phase one.

Locale Foundation

- The staffs were not well informed on mobile promotion. At the start of the campaign, some participants tried to redeem the coupons but failed. Due to a poor communication link between the staffs and store managers, these participants were turned away. They may have made their purchases from competitors.
- No computerised system to keep track of purchases existed. The organisation did not use a computerised system to track purchasing patterns, instead a manual system was used.
- Two specific locations were used for the mobile campaign. The two cafés were operating at different locations but they were operated by the same management.
- Participants had previous experience with promotions. Their experiences had accumulated from their past patronage of promotions with other companies as well as with one of the two cafés.
- Other customers come to the two cafés for their needs and wants. The cafés attract all sorts of people; this was not limited to those participants who acted as consumers for this study.
- Promotions and advertisements are common marketing tools used for the respective cafés.
- Physical layouts in the two cafés support the running of the operations. At both locations, there were adequate chairs and tables to maintain the daily crowds at the respective café. The staffs are capable of handling the respective types of equipment found in the cafés.

Civic Structure

- Past experiences of promotion can affect participants' behavior. It can be assumed that most participants have engaged previously with other types of promotional events.
- The attitudes of 133 persons towards mobile promotion were considered. Most of

the participants considered the mobile phone as one possible promotional tool for companies to adopt as a marketing channel. Participants also perceived the mobile channel as a more effective means to communicate marketing materials than the traditional print media.

- Mobile phones are able to operate within the premises of the cafés. Participants' mobile phones were fully functioning because they were all maintained and covered by their related mobile phone network operators in these two areas.

Individual View

- No attitude difference between print and mobile coupon exists. Most consumers showed no difference in their attitudes to paper based or mobile based coupons. This suggests that neither type of coupon has any outstanding advantages over the other.
- Novelty has no influence for mobile coupons. The novelty factor of a mobile coupon had no positive influence on an individual when compared to a paper-based coupon.
- Going to the store can be unrelated to the mobile promotion. An individual goes into the café for different objectives; an individual can visit the store to meet an acquaintance or to attend a group meeting.
- Individuals choose the respective cafés based on their own choices and preferences.

Interaction Trajectory

- Regardless of how positively or negatively the individual viewed promotional tools, such as mobile and paper coupons, these prior experiences had no impact on the individual's attitude toward coupons in general.
- Mobile coupons are viewed the same as paper-based coupons. In this campaign, the mobile coupon was not seen as a technologically superior product when compared with a paper-based coupon. Some participants highlighted that mobile coupons have no novelty factor. They may view mobile coupons as sharing functions and features similar to those of paper-based coupons.
- Different activities are carried out within the premises of each café daily. Although the main business for both cafés is selling food and beverages, there are different activities conducted within the cafés themselves. For example, at one cafe, there was a special meal offered for a number of days where the other did not have this special offer.
- A once-off setting to activate MMS. In this campaign, it was found that some difficulties arose when individuals had to configure their mobile phones for MMS to function before their mobile phones were ready to accept MMS messages.
- There are people who come into the stores for different purposes such as meeting someone or having a group discussion, without consuming any type of food or beverages from the cafés. Some people may sit at a table to observe the activities that are carried out in the cafés.

Mutuality

- Mobile service providers facilitate the delivery of MMS. Most participants subscribe to their telephony services under a certain mobile network provider. They are able to send SMS or MMS as well as receiving these sorts of message. Likewise, the participants were able to obtain marketing materials from the cafes.
- Daily sales due to the mobile coupons were not tracked. Without any tracking system,

the cafés were not able to check on the daily transactions related to the marketing campaign. Only manual calculations were conducted to check for the coupon redemption.

- Participants were able to interact with each other while the campaign was conducted. Some participants knew each other and therefore there may have been views and ideas exchanged about the campaign. Participants may influence their friends and cause a change of mindset as a result.
- Mobile phones are used for awareness and presence in various capabilities. Most mobile phones are used for communication purposes and most participants appeared to be fairly comfortable in using their mobile phones. Most participants are between 16 and 25 years old and people in this age group are considered to be heavy users of SMS for mobile communication purposes.

In summary for the second phase, the final findings from the survey showed that there were no differences between an individual's attitude toward a paper-based coupon and a mobile coupon; neither type of coupon influenced purchase intention; both types of coupon were viewed as socially acceptable at the store; the novelty factor for the mobile coupon did not influence the choice factor between a paper or a mobile based coupon; and the credibility for both types of coupon appeared to be similar. Additional issues discovered through observation included the lack of a system to track buying behavior of the consumers; the lack of communication between the café management and staffs; the lack of a system to maintain customer relationships; and the need for customers to have their mobile phones configured so that MMS is activated.

Final Phase – The Specific LF

The specific LF is based on the context of the mobile marketing conducted by the cafés. This specific LF provides the results and the findings based on the marketing campaign of the cafes. This specific involve the analysis of the contextual mobile marketing based on the locale foundation, the civic structure, the views of individuals, the trajectory interaction and the mutuality factor. Table 1 highlights this specific LF.

Attributes and Findings on the Aspect of Locale Foundation

The relevant attributes and findings for the locale foundation showed that the cafés were able to carry out the marketing campaign via MMS within its premise. About 70% of the participants believed that the redemption of a mobile coupon is more convenient as compared to a paper-based coupon. There was no effort required to cut out the coupon as was the case for a paper-based one. But the café management was not aware that for some mobile users, it is necessary to have their mobile phones activated by their service providers in order to receive an MMS. The specific LF also highlighted that the cafés could consider using Bluetooth, GPRS or Wifi to monitor the redemption process within the locale. The cafés were able to prevent duplication of the redemption since each mobile coupon was assigned a different mobile code which the staff at the café needed to take note of when registering the coupon into their cash register. However, the analysis does suggest that providing a barcode such as a QR code or a Data Matrix standard (see e.g., Semacode, 2006) on the mobile coupon can be a possible solution to simplify the redemption process.

Attributes and Findings on the Aspect of Civic Structure

The relevant attributes and findings of the aspect on civic structure showed that there is a need to understand thoroughly the relationship between entities (or stakeholders) such as the cafés and the consumers, the marketing campaign and the

Table 1. Specific LF Based on the Mobile Marketing Conducted by the Cafés

The Aspects of LF ---------------- **Characteristics of the Context**	**Locale Foundation**	**Civic Structure**	**Individual View**	**Trajectory Interaction**	**Mutuality**
Time & Location	Located in a national region. The cafés are located in premises with good reception to any mobile phones signal. Both cafés have ongoing promotion.	Customers can join the loyalty program to become registered members of the cafés.	The cafés invite different types of people. Customers come to the cafés for different needs.	Both cafés are opened from 8am to 5.30pm during weekdays. Customers visit the café for food and beverages. They may also come for different purposes. Marketing campaign lasts only two weeks.	Customers come into the café for coupon redemption.
Stakeholders	The setting involves Consumers, the Staff at the café, The Management of the Cafés, and Telecommunication Service Providers.	Relationship is established between • Customer & Cafés • Cafés & Promotion • Promotion & Customer • Cafés and Telco. Service Providers.	Varies from person to person; company to company. Customers have many different types of attitude toward mobile marketing. The main objective of both cafes is to serve people's needs. Telecommunica-tion service providers ensure mobile phones subscribers have stable connections.		
Type of Telecommunication Standard Available	• 3G • 2.5G • Wifi • Bluetooth • MMS/SMS	The management of the cafés is not willing to invest in any solutions to track redemption. Use existing systems based on customers' handsets.	Phones are found to be 2.5G- and 3G-enabled.		Choose the most common telecommunication standard.
Mobile Medium	MMS and Paper-based.	The cafés could have adopted Bluetooth, Wifi, RFID to track coupon redemption.	Some customers were not able to receive MMS. They had to activate system through their service providers. Customers decide the types of coupon to be used at café: mobile or paper coupon.	The staffs at the cafés accept either type of coupon by registering at their cash register.	No available solutions to detect customers who did not receive the mobile coupon. Only MMS-enabled phones would accept mobile coupons.

Continued on following page

Table 1. continued

The Aspects of LF ---------------- **Characteristics of the Context**	**Locale Foundation**	**Civic Structure**	**Individual View**	**Trajectory Interaction**	**Mutuality**
Marketing Content	Both cafés offer coupons for their marketing campaign.	Society may validate the use of mobile coupons.	Customers view coupons as savings. Consumers see mobile coupon as convenient Customers view privacy threaten. Customers have different views on marketing content.	Expiry date was set to inform the validity of the campaign.	
Target audience		Relationship established between targeted consumers.	Interested consumers were informed.	Promotion for the campaign started three weeks beforehand.	Permission was obtained from consumers Opt-out required for customers to cease participation in the campaign.

consumers, the marketing campaign and the cafés, the cafés and the telecommunication service providers, the consumers and the telecommunication service providers. If the relationship between the pair of stakeholders is positive, it will show that mobile marketing can be implemented in an efficient and perhaps effective manner. But if there is a negative relationship, some additional measures are required to rectify the problem. For instance, in order to understand if mobile marketing can be feasible, the cafés need to understand whether their customers have MMS-enabled phones. Failure to do so may result in an ineffective launch of their marketing campaign through MMS. Moreover, the interaction between the entity of the café and the telecommunication service provider may affect mobile marketing as it is difficult to consider the type of services or solutions offered by the telecommunication service providers. The cafés thus have to choose the most appropriate method to deliver its marketing content. At the present moment, SMS and MMS are the most feasible choices but in Japan the use of the QR code in mobile content has gained much popularity and has improved the sales figures in a number of companies (Funk, 2004).

Attributes and Findings on the Aspect of Individual Views

The relevant attributes and findings on individual views show that individuals who participated in the marketing campaign had various perspectives that related to his or her social world(s). Many different types of consumer come into the two cafés each day. They all hold different views and opinions toward the concept of mobile marketing. Their background — such as education, culture and traditions — can also be factors affecting how they view mobile marketing. For the consumers, the mobile coupon was a kind of savings, enabling them to buy a cup of coffee at a discounted price. At least 65% of the participants claimed that special offers like the mobile coupon should appear on mobile phones if companies intend to advertise and promote their products and services. In the analysis of LF, the five most wanted types of content that participants liked to see on their mobile phones included food and beverages, upcoming concerts and events, fashion news, travel discounts and electronics discounts. These preferences toward marketing promotional materials were largely in line with those of the participants who have at least 4 to 6 years of mobile phone experience. The establishment of trust between the cafés and the consumers also signify an important factor that dictates individuals' views on mobile marketing. This is supported by the fact that most participants registered their interest in receiving mobile marketing materials from the cafés that they patronized. In this way, the participants opted-in to ensure that their mobile numbers were used appropriately by cafés that they trusted.

Attributes and Findings on the Aspect of Interaction Trajectory

The related attributes and findings associated with the aspect of interaction trajectory enabled the identification of the various forms of interaction that go on in the context of the cafés. It can be seen that there was interaction between the cafés and the participants during the marketing campaign. It can also be seen that other types of consumers came into the cafés as well. The daily operation of the cafés did not undergo any major changes when the campaign was conducted. However, on the first two days, when participants tried to redeem the mobile coupon, there were some teething issues. First, due to poor communication from the marketing manager, the staff was not informed of the marketing campaign. Some participants were thus not able to redeem the mobile coupon. These participants did however come back on another day to make the redemption. Second, the staffs were not informed on how to carry out the redemption of the mobile coupon. But these issues were sorted out fairly quickly when the management provided guidelines on what to do for the campaign.

One important point to consider is the difficulty in evaluating why participants came to the cafés and the affect of the mobile coupon on the participants' attitudes toward the mobile marketing. A certain participant can come in to the café just to redeem the voucher and then be off to somewhere else, but that not does warrant that he or she will come back to the same café in the future to make a purchase. The cafés can consider inviting their customers to sign up for a membership program at no cost, and customers can then be required to give their mobile number as a form of membership requirement. In return, they may get a QR code from the café to certify that they are members. As part of joining the membership program, a special privilege is that all members wiil get a cup of beverage for free after purchasing ten cups of beverage from the cafés. The cafés can perhaps invest in some types of solutions like Bluetooth, Wifi, RFID to detect the 'presence' of their members if they are near the vicinity of the cafés. Updates of the latest promotions or happenings could then be sent to all members provided they opt-in. Having a database of opted-in consumers for the purposes of delivering marketing materials to them can help to prolong and enhance the relationship between the cafés and the consumers.

Consumers opted-in to receive the marketing materials prior the start of the marketing campaign. Therefore it is important follow up with them if possible whether they will be willing to continue to receive any form of marketing materials from the café. One way is to allow these participants to have some forms of interaction with the system on which the café operated. For every marketing item sent by the café, there should be an opt-in or opt-out mechanism embedded in the materials to allow interaction between the customers and the cafés. In this way, customers may not easily get irritated by any marketing messages delivered to their mobile phones. During the campaign, the cafés did not maintain any systems to keep track of the profile of the participants, each of whom may become a potential consumer later. The other possible solution is to allow the customers to dictate the number of marketing materials they want and the amount of material to be received. The cafés need to collect these requirements manually in order to know what their customers deem acceptable to marketing materials. These marketing materials may involve monthly promotions from the cafés and advertising the introduction of any new food or beverage. In this way, customers who want to become members of the cafés, can decide the types of material they want to receive.

Attributes and Findings on the Aspect of Mutuality

Finally, the relevant attributes associated with mutuality showed that the cafés could consider creating further awareness of their mobile marketing campaign. This could be further explored by the cafes by requesting of their consumers if they would want upcoming marketing materials. Further consideration involves understanding the environment that the two cafés operate in, thereby providing insights into how to develop better marketing plans. For this aspect to be effective, the cafés need to further evaluate the type of technology that they believe is best to deliver marketing materials for the given context. The investment in appropriate technology will help to deliver quality marketing materials to their consumers at the right cost, time and location. Additional issues to consider involve the appropriate time to deliver marketing materials, the types of customer that the cafés want to target, the types of promotions to run, and the time period for running campaigns.

The management that operates both of the cafés can determine the preferences of all their registered consumers if they invest and build some types of CRM systems. Setting up these systems can allow consumers to register their preferences based on the cafés' menu, so that, whenever consumers walk into the cafés, their presence may trigger the system to send a list of their regular purchases to the cash register. This can even allow the staff to infer what the particular consumer wants to order. In an ideal mobile marketing scenario (Sinisalo, Salo, Karjaluoto, & Leppaniemi, 2006), this type of system allows the organization to understand the needs and wants of their consumers. Particularly, the interactivity and personalized nature of mobile phones can allow the cafés to engage the customers in a two-way dialogue which traditional media like print are not able to do.

Additional Findings Based on the Specific LF

Building on the analysis of the specific LF, the following factors should also be considered when the management of the two cafes attempts another marketing campaign via the mobile channel.

Firstly, the marketing department at the firm needs to have a thorough understanding of the nature of mobile commerce and its related technology. The marketing department in particular needs to have a good overview of what mobile marketing is. To understand this domain, the marketing department needs to know the characteristics of mobile commerce, the types of network operators and network spectrums involved, the different forms of delivery technique for their marketing materials, the types of cost involved for sending out mobile marketing materials through MMS, and also the types of mobile phone and mobile phone operating systems. This information can be gathered simply whenever a new member attempts to join the café for its loyalty program. The registration form of the loyalty program should request the new member for detailed information about his or her mobile phone.

Secondly, the marketing department may provide all their customers with a choice of possible time slots to receive mobile marketing materials for upcoming events. If their consumers can nominate the best time to receive any mobile marketing materials, this may allow a higher chance of acceptance of the mobile marketing material.

Thirdly, to entice the customers to pay attention to the mobile marketing materials, monetary incentives should be provided, for instance, a special deal that allows two cups of coffee to be purchased at the price of one, or a discount offer of 50% on the price of any main meal at the cafés. Non-monetary incentives — like that the first 100 consumers to respond to the mobile marketing material may receive a free coffee mug or a free coffee coaster, and having a monthly contest to win a week's supply of free beverage — may help.

Fifthly, the management needs to offer their consumers a choice in deciding the types of marketing materials they want to receive. Currently, the use of printed promotional pamphlets and the online website are used for advertising and promotions to draw consumers to the cafes. The mobile phone serves as a third alternative channel to deliver marketing materials to their registered users. If the registered users are able to customize in what they would like to see as marketing material, they are more likely to be favorable in receiving a marketing content via the mobile phones as compared to a print or online promotional advertisement.

Lastly, the marketing officer needs to ensure that the details of the next mobile marketing campaign are disseminated to all staff involved. This is to enable a smooth run on the mobile marketing campaign. To ensure that there is proper communication established among the staff, the management may consider establishing a 'system' to facilitate the running of the mobile marketing campaign. This system may provide details of when the mobile marketing campaign is conducted and what types of mobile marketing are being used, such as mobile coupons for food and beverages. Moreover, this system may allow the tracking of all mobile marketing materials such as mobile coupons used during the period of marketing campaign.

MANAGERIAL AND PRACTICAL IMPLICATIONS DERIVED FROM THE SPECIFIC LF

One of the biggest issues identified from the use of the LF analysis was privacy. The question of "How will consumers know that they will not be 'tracked down' when a certain retailer beams marketing material to their mobile phones via MMS, or even Bluetooth, or GPS technology?" Consumers want their personal details to be protected. Organizations then need to provide some form assurance to their consumers, e.g. by means of a simple text message which highlights the terms and conditions that govern the protection of consumers' personal details. This "code of mobile marketing ethics" may provide the consumers with peace of mind when they interact and exchange personal details with organizations.

Another issue is the personalization of the marketing content for the consumers. Marketers need to understand that different mobile phone users have different attitudes and opinions toward marketing. Getting the permission of mobile phone users to receive is highly recommended before sending to individuals the required marketing contents. To personalize the right type of message for the consumers, much effort is required to be carried out by organizations if they intend to deliver their marketing contents via the mobile channel. These organizations would need constant diligence to communicate with their value consumers in building up a relevant up-to-date database of their preferences and likes/dislikes. Besides, the option for consumers to opt-out of this material is required to build a lasting bond with the consumers. The opt-out option allows consumers to perceive that the companies are responsible for their marketing effort. However, this does not guarantee a "buy-in" for the marketing content when the mobile users view it.

There is also the problem of the screen size of all mobile phones. The difference of the various sizes as produced by the different manufacturers might impact the appeal of the marketing content as well as the acceptance of the marketing materials themselves. The marketing materials may appear reasonable in size for one type of mobile phone, yet on another phone may be viewed as too small. Hence, the size of mobile marketing materials needs considerable effort to be considered. Companies may consider the design of such materials by following the guidelines as outlined by Mobile Marketing Association (MMA) Global for their advertising and promotional contents.

Yet another issue that surfaced from this framework describes the importance of offering good benefits like considerable savings for consumers in order for them to read the marketing content on their mobile phones. The value and benefit of mobile promotion can be associated with monetary incentives, information and intangible benefits like satisfaction. As all consumers view benefits such as discounts, savings, and special deals differently, the use of mobile marketing materials to entice consumers to engage in the act of purchase may then require various "conviction" methods. Hence, organizations need to be creative and vigilant in designing their marketing materials to convince their consumers into the buying mood.

All related stakeholders involved in any locales need to be aware of the ongoing marketing events. For example, if there is a campaign that takes place for two weeks, all relevant parties involved need to be adequately informed. Failure to do so may result in a loss of business such as due to a loss of consumers' confidence in the store for failing to honor a mobile coupon or even a heated argument between potential consumers and the company for not accepting the discount redemption based on a text message. Companies may even require additional assistance from telecommunication providers to "guarantee" that mobile marketing materials that need to be delivered to potential consumers should be successfully delivered during the period of the marketing campaign.

The usage of MMS as a tool to deliver the marketing materials may not always be the most appropriate tool. For instance in the above campaign, consumers who are on pre-paid subscriptions had to activate their mobile phones with their service providers before that can effortlessly receive MMS contents. This form of teething incident could be corrected if all organizations are willing to take precautionary and extra measurements by questioning their local mobile telecommunication service providers on the pros and cons of various contents delivery methods. Only then, organizations can practice mobile marketing with success if they are well-acquainted with the local mobile network's specifications.

CONCLUSION

To ensure the success of a good marketing plan, researchers are often encouraged to conduct adequate market research. The market research should consist of a framework that involves adopting mixed research techniques in order to derive effective and efficient opportunities and dimensions for the research findings (Onwuegbuzie & Leech, 2005). This is where the approaches to understanding people and technology interventions through a holistic perspective (Blackwell & Green, 2003; Fitzpatrick, 2003) can assist the study of mobile marketing.

The implementation of the LF provides a structured approach to examining mobile marketing contexts. The use of such a framework allows the context of mobile marketing to be viewed from a broader holistic perspective that overcomes the narrowness of findings from the traditional marketing research methods. The LF by its own is a methodology that explores and analyses research problems in a general approach. It can also incorporate multiple research methods to provide a richer and better basis for assessing mobile marketing research. This framework further helps by contributing and building a theoretical foundation to support future mobile marketing research. In doing so, this framework, and the "three plus" process described in this paper, can equip marketers to implement their marketing campaigns via mobile phones with effective and efficient marketing contents.

FUTURE WORK ON USING THE LOCALES FRAMEWORK

One future work is to generalize the usage of the LF to different mobile applications and the environments that mobile marketing operates. This will helps to strengthen and add considerable value to the establishment of this approach. One possible direction of future research would be in the domain of learning whereby we can examine the LF to understand how learning can be enhanced or improved through the use of the mobile phones.

Another project to be considered is to replicate the same type of marketing campaign as mentioned earlier in this chapter for another context. These contexts might involve places like a bookstore, a cinema, or even a clothing store. This is to see if the five aspects of the LF can provide additional new insights and implications. This could also allow the LF to be explored in another domain — say a global context or at a national level.

REFERENCES

Amit, R., & Zott, C. (2001). Value creation in e-business. *Strategic Management Journal, 22*(6/7), 493–520. doi:10.1002/smj.187

Atkinson, P., Coffey, A., & Delamont, S. (2003). *Key themes in qualitative research: Continuities and change*. Walnut Creek: Altamira Press.

Babbie, E. (2005). *The basics of social research* (3rd ed.). Belmont, CA: Thomson/Wadsworth.

Barwise, P., & Strong, C. (2002). Permission-based mobile advertising. *Journal of Interactive Marketing, 16*(1), 14–24. doi:10.1002/dir.10000

Bauer, H. H., Barnes, S. J., Reichardt, T., & Neumann, M. M. (2005). Driving consumer acceptance of mobile marketing: A theoretical framework and empirical study. *Journal of Electronic Commerce Research, 6*(3), 181–192.

Becker, M. (2005). *Effectiveness of mobile channel additions and a conceptual model detailing the interaction of influential variables*. MMA Global. Retrieved December 22, 2008, from http://mmaglobal.com/modules/article/view.article.php/131

Becker, M. (2006). *Academic review: Consumer acceptance of mobile marketing*. MMA Global. Retrieved July 15, 2008, from http://mmaglobal.com/modules/wfsection/article.php?articleid=289

Blackwell, A., & Green, T. (2003). Notational systems—the cognitive dimensions of notations framework. In J. Carroll (Ed.) *HCI Models, Theories and Frameworks*. Los Altos, CA: Morgan Kaufmann.

Camponovo, G., & Pigneur, Y. (2003). *Business model analysis applied to mobile business*. Paper presented at the 5th International Conference on Enterprise Information Systems (ICEIS 2003) Angers, France.

Facchetti, A., Rangone, A., Renga, F. M., & Savoldelli, A. (2005). Mobile marketing: An analysis of key success factors and the European value chain. *International Journal of Management and Decision Making, 6*(1), 65–80. doi:10.1504/IJMDM.2005.005966

Fitzpatrick, G. (1998). *The locales framework: Understanding and designing for cooperative work*. Unpublished doctoral dissertation, The University of Queensland, Australia.

Fitzpatrick, G. (2003). *The locales framework: Understanding and designing for wicked problems.* Kluwer Academic Publishers.

Flick, U. (2006). *An introduction to qualitative research* (3rd ed.). London: Sage Publications.

Gopal, R. D., & Tripathi, A. K. (2006). Advertising via wireless networks. *International Journal of Mobile Communications, 4*(1), 1–16. doi:10.1504/IJMC.2006.008194

Green, B. (2005). *Reflections on the circle of life: Interactions in a pathology environment (medical): A 'wicked problem'*. Unpublished master's thesis, The University of Queensland, Brisbane, Australia.

Haghirian, P., & Madlberger, M. (2005). *Consumer attitude toward advertising via mobile devices-an empirical investigation among Austrian users.* Paper presented at the 13th European Conference on Information Systems, Regensburg, Germany.

Heflin, J. (2004). *OWL Web ontology language use cases and requirements. W3C Recommendation.* Retrieved December 22, 2008, from http://www.aaai.org/AITopics/html/ontol.html#what

Ireland, C. (2003). Qualitative methods: From boring to brillant. In B. Laurel (Ed.), *Design research: Methods and perspectives.* Cambridge, Massachusetts: The MIT Press.

Karjaluoto, H., Lehto, H., Leppäniemi, M., & Mustonen, T. (2007)... *International Journal of Mobile Marketing, 2*(2), 10–20.

Kerin, R. A., Hartley, S. W., & Rudelius, W. (2007). *Marketing: The core.* (2nd ed.). McGraw-Hill:Irwin.

Komulainen, H., Mainela, T., Sinisalo, J., Tähtinen, J., & Ulkuniemi, P. (2004). *Business models in the emerging context of mobile advertising.* Paper presented at the Frontiers of e-Business Research 2004, FeBR 2004, Tampere, Finland.

Kotler, P., Jain, D., & Maesincee, S. (2002). *Marketing moves: A new approach to profits, growth and renewal.* London: McGraw-Hill.

Lee, D., & Muhlberger, R. (2006). Mobile marketing as technology intervention: A locales framework analysis. *International Journal of Mobile Marketing, 1*(1), 59–65.

Lee, D., & Muhlberger, R. (2007). A technology introduction view on mobile marketing. In D. Taniar (Ed.), *Encyclopedia of mobile computing & commerce* (Vol. 1). USA: Idea Group Reference.

Leppaniemi, M., & Karjaluoto, H. (2005). Factors influencing consumers' willingness to accept mobile advertising: A conceptual model. *International Journal of Mobile Communications, 3*(3), 197–213. doi:10.1504/IJMC.2005.006580

Leppaniemi, M., Karjaluoto, H., & Salo, J. (2004). The success factors of mobile advertising value chain. *The E-business Review, 14*(1), 93–97.

Leppaniemi, M., Sinisalo, J., & Karjaluoto, H. (2006). A review of mobile marketing research. *International Journal of Mobile Marketing, 1*(1), 30–40.

McEwan, G., & Greenberg, S. (2005). *Supporting social worlds with the community bar.* Paper presented at the 2005 international ACM SIGGROUP conference on supporting group work, Sanibel Island, FL.

Okazaki, S. (2004). How do Japanese consumers perceive wireless ad? A multivariate analysis. *International Journal of Advertising, 23*, 429–454.

Okazaki, S. (2005). New perspectives on m-commerce research. *Journal of Electronic Commerce Research, 6*(3), 160–164.

Onwuegbuzie, A. J., & Leech, N. L. (2005). On becoming a pragmatic researcher: The importance of combining quantitative and qualitative research methodologies. *International Journal of Social Research Methodology, 8*(5), 375–387. doi:10.1080/13645570500402447

Perlado, V. R., & Barwise, P. (2004). Mobile advertising: A research agenda. In M. Stafford & R. J. Faber (Eds.), *Advertising and the New Media.* Armonk, NY: M.E. Sharpe.

Plowman, T. (2003). Ethnography and critical design practice. In B. Laurel (Ed.), *Design research: Methods and perspectives*. Cambridge, MA: The MIT Press.

Pousttchi, K., & Wiedemann, D. G. (2006). *A contribution to theory building for mobile marketing: Categorizing mobile marketing campaigns through case study research.* Paper presented at the International Conference on Mobile Business (ICMB2006), Copenhagen, Denmark.

Punch, K. F. (2005). *Introduction to social research.* Thousand Oaks, California: Sage Publications.

Richtel, M. (2006). *Marketers interested in small screen.* Retrieved December 22, 2008, from http://www.smh.com.au/news/breaking/small-screen-bonanza/2006/01/17/1137466994899.html

Rogers, Y. (1997). Reconfiguring the social scientist: Shifting from prescription to proactive research. In G. C. Bowker, S. L. Star, W. Turner & L. Gasser (Eds.), *Bridging the great divide: Social science, technical systems and cooperative work* (pp. 55-77). Mahwah, NJ: L. Erlbaum.

Rogers, Y., & Muller, H. (2006). A framework for designing sensor-based interactions to promote exploration and reflection in play. *International Journal of Human-Computer Studies, 64*, 1–14. doi:10.1016/j.ijhcs.2005.05.004

Rohm, A. J., & Sultan, F. (2005). Brand in the hand: An exploratory study of mobile marketing communications. *Proceeding of EURAM 2005*, April, Munich.

Rohm, A. J., & Sultan, F. (2006). An exploratory cross-market study of mobile marketing acceptance. *International Journal of Mobile Marketing, 1*(1), 4–12.

Rossiter, J. R., & Bellman, S. (2005). *Marketing communications: Theory and applications.* Frenchs Forest, NSW: Prentice Hall.

Sgriccia, M., Nguyen, H., Edra, R., Alworth, A., Brandeis, O., & Escandon, R. (2007). Drivers of mobile business models: Lessons from four Asian countries. *International Journal of Mobile Marketing, 2*(2), 58–67.

Shimp, T. A. (2003). *Advertising, promotion and supplemental aspect to integrated marketing communications* (6th ed.). Mason, Ohio: Thomson South-Western.

Sinisalo, J., Salo, J., Karjaluoto, H., & Leppäniemi, M. (2006). *Managing customer relationships through mobile medium-underlying issues and opportunities.* Paper presented at the 39th Hawaii International Conference on System Sciences, Hawaii.

Tabachnick, B. G., & Fidell, L. S. (2001). *Using Multivariate Statistics* (4th ed.). Needham Heights, MA: Allyn & Bacon.

Tsang, M. M., Ho, S. H., & Liang, T. P. (2004). Consumer attitudes toward mobile advertising: An empirical study. *International Journal of Electronic Commerce, 8*(3), 65–79.

Varanparast, R., & Asil, M. (2007). Factors affecting the use of mobile advertising. *International Journal of Mobile Marketing, 2*(2), 21–34.

Yuan, Y., & Zhang, J. J. (2003). Towards an appropriate business model for m-commerce. *International Journal of Mobile Communications*, *1*(1-2), 35–56. doi:10.1504/IJMC.2003.002459

Yunos, H. M., Gao, J. Z., & Shim, S. (2003). Wireless advertising's challenges and opportunities. *Computer*, *36*(5), 30–37. doi:10.1109/MC.2003.1198234

KEY TERMS AND DEFINITIONS

Civic Structure: Aspect 2 of the LF, civic structure show how locales involve and interaction with each other. It concerns the facilitation of interaction with the wider community within and beyond a person's known social worlds and locales. In addition, gaining understanding of locales requires prior knowledge of events and or actions, along with a reasonable expectation of what events are to follow.

Individual Views: Aspect 3 of the LF, individual views define the perspectives, concerns and roles of a particular locale, as well as of the multiple social worlds they are in. A single perspective is how an individual sees one social world (the people and the locales), and it is dependent on the level of engagement with the center of that world, whereas multiple viewsets incorporate the individual's views of all the social worlds with which he or she is engaged. Individuals personalize their views to suit their tasks according to their current level of engagement.

Interaction Design (ID): ID has a firm foundation in the theory, practice, and methodology of traditional user interface design, its focus is on defining the complex dialogues that occur between people and interactive devices (Preece, Rogers, & Sharp, 2002) such as computers, mobile communications devices and portable digital assistants. ID can also be considered as the design of the behaviour of, and determination of goals for, artefacts, systems, and environments (Palino, 2004). ID does not simply aim at providing the end users with a positive usage experience achieved through traditional industrial design. Rather, it goes beyond that, providing people with positive experiences both from a technological and non-technological perspective (Danzico, 2006). In the context of mobile marketing, ID serves not only as an approach for user-centered design practice but also as a solution to help overcome the limitations of the marketing research perspective. Applying ID know-how that aims to facilitate technology interventions in social contexts seems to be a stronger approach than treating mobile technology as just a different type of marketing channel (Lee & Muhlberger, 2007). In other words, ID serves as a methodological space where informed research methods enhance a framework to provide interesting and valuable results for a mobile marketing context. The ID approach is applied with the use of a framework known as the Locales Framework (LF). This framework enables an holistic overview of what mobile marketing means and the solution offered by ID is enhanced due to the populated mobile marketing research methods which overcome the individual limits of a single mobile marketing research method.

Interaction Trajectory: Aspect 4 of the LF, interaction trajectories serve as a means for understanding how a social world develops, how people interact, and how the activities in which they engage contribute to the various courses of action over time that can be past, present and future.

Locales Foundation: Aspect 1 of the LF, the locales foundation involves a group of individuals, related either physically or remotely, who are brought together by their needs, who require cooperative actions, and the sites and means that support these groups. The elements of foundation include a basic understanding of how communication will take place in order to direct the group towards achieving their goals.

Mobile Coupon: It is generally a form of discounts or entitlements embedded as either on a text message via Short Message Service (SMS) or a graphical message Multimedia Messaging Service (MMS) with expiration dates to entice consumers to make purchases for suggested items.

Mobile Marketing: In this chapter, it is regarded as a form of technology intervention in the context of a marketing campaign, looking at either advertising or promotion, or both.

Mutuality: Aspect 5 of the LF, mutuality concerns how group members make their presence known to other in order to maintain both their actions and their interactions with the social world.

The Locales Framework (LF): The LF offers an alternative approach for investigating the context of mobile marketing. The research shows how such an approach can integrate marketing and new technology research, highlighting the issues and opportunities brought about by the technology within a domain context, i.e. mobile marketing. Given that there are not many theoretical based frameworks available to support mobile marketing research, the application of this framework provides a roadmap for future research in mobile marketing and helps build the body of knowledge in this field.

Chapter 4
Factors Affecting Mobile Advertising

Ramin Vatanparast
Nokia Interactive Advertising, USA

ABSTRACT

Mobile advertising holds strong promises to become the most highly targeted advertising medium offering new ways to target messages to users. By utilizing mobile advertising, companies can run marketing campaigns targeted to tens of thousands of people with a fraction of the costs and time in comparison to other direct marketing mediums. However, as mobile advertising is a novel approach, many aspects of it still need further investigation. Little is known regarding the effectiveness of mobile advertising campaigns and the factors contributing to their success. This chapter aims to provide a comprehensive understanding of the advertising space and its influencing factors. First, the study investigates factors that influence mobile advertising from both the industry's and consumer's point of view. Second, based on a review of previous studies in the field, the author proposes a conceptual model for mobile advertising, which categorizes the factors in different groups and provides a holistic view of their impact in the mobile advertising space.

INTRODUCTION

The heightened interest towards mobile-advertising recently observed among the academic and industry circles has led to the need for the formulation of a simple yet comprehensive conceptual framework to ascertain the key factors effecting the use and potency of current as well as future endeavors in the realm of mobile advertising. A number of attempts have been made by the academia as well as industry practitioners to lay out an intuitive conceptual model to analyze this exciting and innovative field of technology based marketing. The underlying chapter, at first, this chapter reviews the factors that influence mobile advertising from the industry's and consumer's point of view (Vatanparast and Asil 2007). Second, based on this literature, the author proposes a preliminary conceptual model

DOI: 10.4018/978-1-60566-074-5.ch004

for factors effecting mobile advertising. At last, a summary result of a quantitative evaluation of the model will be presented. This proposed conceptual model/framework can serve as a solid base for the evaluation of the critical success factors for a market-worthy mobile advertising strategy. The proposed conceptual model may also help researchers and managers to better understand the critical components of mobile advertising and provide ideas for further research in this emerging field.

BACKGROUND

Advances in mobile technology have changed the business environment significantly. Devices and systems based on mobile technologies have become a common place in our everyday lives (Balasubramanian et al., 2002), increasing the accessibility, frequency and speed of communication. As a result, mobile technologies have the potential to create new markets, change the competitive landscape of business, create new opportunities, and change existing community and market structures (Stewart & Pavlou, 2002).

Today's development in information technology helps marketers to keep track of customers and provide new communication venues for reaching smaller customer segments more cost effectively and with more personalized messages. This resulted in two major changes of permission marketing and targeted marketing in marketing communications environments (Kotler et al., 2005). Developments in information technology are speeding up the shift towards permission marketing (Godin, 1999), and at the same time marketers are shifting to targeted marketing. As a result, modern advertisers are increasingly relying on various modes of interactive technology to advertise and promote their products and services (Pavlou & Stewart, 2000). Gradually, many companies are redirecting their marketing spending to interactive marketing, which can be focused more effectively on targeted individual consumer and trade segments. Despite this, potential customers so far did not have the opportunity to signal their likes and dislikes with marketing activities via mobile devices. This puts marketers at a high risk (Robins, 2003) as they are unsure whether their marketing activities cause positive or negative reactions by the customer.

There are a couple of reasons why many experts consider mobile advertising as an encouraging branch of mobile business (Bulander et al., 2005): high penetration rate of mobile terminals, mobile terminals as personal communication devices, individually addressable, multimedia capabilities and interactivity. However, there are also some serious challenges when talking about mobile advertising, such as spam, limited user interface, privacy concerns, and the expense of mobile data communication. The mentioned features and challenges of mobile advertising show an area of divergence: personalized advertising requires sensitive information about the end user, his fields of interest, latest activities or his current location. On the other side, there are privacy concerns when providing this information for a mobile advertising application.

Although the global mobile advertising industry is in its formative years (Ovum Forecast, 2002), forecasts concerning growth of mobile advertising have been quite enthusiastic (Future Mobile Handsets, 2006). Mobile advertising holds strong promises to become the best targeted, one-to-one, and the most powerful digital advertising medium, offering new ways to aim messages to users that existing advertising channels can never do. The mobile advertising market is expected to grow to over $600m by 2007 (Taylor, 2006) and will jump to $11.35bn in 2011 (Future Mobile Handsets, 2006). Europe will lead the mobile advertising market, whereas in the USA, the fragmented nature of its cellular markets (Nelson, 2000) and users' fear of mobile spam have slowed the market growth (Fuller, 2003). By utilizing mobile advertising, companies can run marketing campaigns targeted to tens of thousands of people with a fragment of the costs in just few seconds of time (Enpocket Research Report, 2002).Top international brands such as Nike, BMW, Virgin, and McDonald's have by now launched mobile advertising campaigns to convey commercial content to customers.

FACTORS AFFECTING MOBILE ADVERTISING

Factors affecting mobile marketing and advertising has been researched from established information technology and application point of views. Based on Rogers' innovation and diffusion theory (Rogers, 1995), Barnes and Huff (2003) look at the diffusion of Internet access via mobile phones. Attitude of consumer has been studied in the field of marketing and advertising using a form of theory of reasoned action by Fishbein and Ajzen (1975). Tsang et al., (2004) studied the link between attitude, intention and behavior in mobile marketing. Their study validates that incentives, such as free calls, have a big impact on attitude and intention to receive mobile marketing and also consumers have a negative attitude towards mobile advertising, unless they have given permission to it. They suggest that entertainment, informativeness, irritation and credibility also affect attitudes of consumer. Within the application stream of research, the typical features of the application such as consumer and industry technology, location awareness, personalization, benefits of mobile advertising and fears in relation to mobile advertising are gathered together in different research models (Leppäniemi & Karjaluoto, 2005; Bauer et al., 2005). Krishnamurthy (2001) identifies the key factors impacting on mobile marketing adoption as message relevance, personalization, privacy costs, message processing costs and monetary benefits.

Venkatesh et al., (2003) suggest that intention is a predictor of use behavior and performance expectancy, effort expectancy, social influence and facilitating conditions are four factors impact on intention and usage. He also mentions that the key moderators are gender, age, experience and voluntariness of use and attitude was considered to be in overlap with performance and effort expectancies. Mobile marketing acceptance and adoption are on the grow (Becker, 2005), but without a clear picture about the elements affecting mobile advertising, marketers and advertisers will have a modest aptitude to implement and also have a positive return on their campaigns.

Consumer

The achievement of mobile advertising depends largely on its acceptance by consumers. Potential customers so far did not have the opportunity to signal their likes and dislikes with marketing activities via mobile devices. This puts marketers at a high risk (Robins, 2003) as they are unsure whether their marketing activities cause positive or negative reactions by the customer. If marketers want to use the mobile media communication channels in an efficient way, they need to understand how mobile consumers perceive, value and evaluate mobile devices as a source of advertising.

Consumer's Privacy

Privacy refers to the degree to which personal information is not known by others (Rust et al., 2002). Customer privacy has always been a critical issue in marketing, but with the rise of Internet-based commercial transactions have brought a greater meaning in recent years (Rust et al., 2002). As mobile communication is becoming more and more ubiquitous and the amount of information traveling wirelessly increases, the analysis, planning, and implementation of a security architecture become extremely important (UMTS Forum Report, 2003).

Consumers regard their mobile phone as a very private item (Fortunati, 2001). Mobile consumers are very sensitive about receiving messages from unknown persons or organizations. Most consumers are still quite uncomfortable and skeptical about mobile business and whether these businesses are feasible and secure (Siau & Shen, 2003). From the users' point of view, invasion of privacy and general security concerns relating to the mobile medium have been identified as one of the main obstacles to the success of mobile advertising (Gohring, 2002). Users who turn to new media services tend to have concerns about unauthorized data access, data manipulation, and unwanted tracking of usage patterns. Godin (2001) described that majority of anonymous mass advertising is despised by consumers leading them to reject the messages. However, it is possible for marketers and advertisers to reach consumers anytime and anywhere using the mobile medium. This means accounts for consumer's fear of privacy violations in one hand but also high-potential, personalized mobile advertising on other hands.

According to studies by Forrester Research, Inc. (2001) and Cahners In-Stat Group (Saunders, 2003), consumer concerns over privacy and invasiveness regarding mobile advertising could threaten the entire mobile advertising market, at least in the short term. In accordance with Forrester Research, Inc. (2001), 80% of consumers fear invasion of their privacy in SMS campaigns. Ackerman, Darrel and Weitzner (2001) concluded that, privacy concerns cannot be dismissed, but consumers will accept a certain degree of privacy loss if the benefit is perceived as being sufficient and satisfying. This was confirmed by another study (eMarketer Research Report, 2003) which found that roughly 65% of US consumers are willing to give their personal information to marketers in exchange for relevant mobile marketing information. Quoting a review of several consultancy studies conducted by Lewis (2001), US consumers seem to be willing to accept advertising to subsidies the cost of other services such as e-mail and news services. It is critical to remember that customers, eventually, judge the right ways, time, and place for what is perceived as convenient communication.

Consumer's Purpose

According to the uses-and-gratification-approach (Katz et al., 1973 & 1974), consumers consciously select and use certain media and content to gratify a specific purpose. Based on this approach, mobile marketing will only be accepted by consumers if perceived as an opportunity to gratify the needs for information, knowledge and social acceptance. Clearly mobile advertising is a unique form of advertising, which brings new way of reaching consumer and also new ways that consumer can interact with advertisements.

An individual's purpose is often referred to as a person's acknowledgment of what he is pursuing in a particular situation and to an associated inner state of arousal (Eysenck, 1982). Thus an individual's mobile media purpose is his acknowledgment of the processing goal he is pursuing when attending to his mobile device, which in this case is the medium for mobile advertisement. Depending on what type of purpose the receiver is trying to satisfy by using a mobile device also affects his processing of the advertisements. For example, if the customer's main intention is gathering information, the customer will be more interested in advertisements that provide him relevant information on products, services or companies.

On the other hand, if the customer's intention is garnered more towards entertainment, he will enjoy advertisements that are entertaining and provide experiential satisfaction through aesthetic pleasure, emotional stimulation, or social experience (Barwise & Strong, 2002). A consumer may wish to satisfy different kinds of purposes at the same time, and the relative importance of the types may change according to the situation that he is in. Katz et al. (1973) identify the following three categories of needs as being the most important 1) information, knowledge and understanding,

2) aesthetic, pleasurable and emotional experience, and 3) contact with family, friends, and the world.

Consumer's Performance

The indignity people feel towards advertisements greatly influences their attitude towards advertising (Shavitt et al., 1998). Consumers are likely to perceive advertisements as an unwanted and irritating influence when advertising employs techniques that annoy, distract, offend, insult, or are excessively manipulative (Ducoffe, 1996). Mobile advertising can provide an array of information that distracts, confuses, and overwhelms the consumer (Stewart & Pavlou, 2002). Consumers may therefore, feel confused about them and react negatively.

Irritation does negatively influence the value of mobile advertising and irritation is not as strong as the influence of the variables entertainment and informativeness (Haghirian et al., 2005). Irritation is a phenomenon that is similar to reactance, with consumers feeling that the advertisements are too intrusive and tend to refuse advertisements. Rodgers and Thorson (2000) provide an extensive summary of responses to online advertising measures of effectiveness (Wells, 1997) in traditional media. A critical measure is attention. Gaining attention, often measured after the fact as recall, is a precondition for influencing consumer action. Compared with other media, the recall rate for mobile advertisements is high (Windwire Report, 2000).

Viral marketing, or word-of-mouth (Helm, 2003) defined as an advertising message is spread indirectly by consumers among other consumers, not by the advertiser or an agency. Viral effects (Godin, 2001) multiply the impact of appealing mobile advertising messages and consumer attention by enlarging a campaign's reach. A viral effect expands if recipients of advertising messages forward the message to additional recipients who are not part of the initial target group of the advertising campaign. Through viral effects, it is thus not only possible to enlarge the reach of mobile advertising campaigns, but also to enlarge their effectiveness and respond to rates. By forwarding messages to their friends, recipients create a strong peer influence. Other intended consumer reactions to mobile advertising messages include following a link to a Web page, e-mailing the advertiser, purchasing a product, and placing a telephone call (Barnes, 2002). In camparision to click-through rates of less than 1% for Web-based advertising, average click and call through rates for wireless devices are 19% and 12%, respectively (Windwire Report, 2000). Well targeted campaigns in Japan achieved click-through rates up to 33% (Natsuno, 2003).

Message

Today's development in information technology helps advertisers to keep track of customers and provide new communication venues for reaching smaller customer segments more cost effectively and with more personalized messages. This resulted in three major changes of messages toward permission marketing, targeted marketing and interactive marketing in marketing communications environments (Kotler et al., 2005). Modern advertisers are increasingly relying on various modes of interactive through messages targeted to consumers. to advertise and promote their products and services (Pavlou & Stewart, 2000).

Message Content

Advertising messages refer to communication exchanges between advertisers and consumers (Ducoffe, 1996). Mobile advertising activities strongly depend on message characteristics, which need to be developed carefully. Similar to attracting attention in other media, the content of a mobile advertisement is crucial. Content is a key factor in creating a service that attracts the users and keeps them coming back (Paavilainen,

2002). We should bear in mind that the golden rule to all advertising, 'content is king', is also not complete. Here I can argue that if the content is not relevant then the advertisement does not have much value. As result I have to add the relevancy to the equation, 'relevant content is king' in mobile advertising.

Reading from mobile devices may take more time and effort than reading from a desktop computer (Duchnicky & Kolers, 1983) due to space limitations and constrained user interfaces. Therefore, the message should be kept short and the use of graphics or photos should be encouraged (Edens & McCormick, 2000). Kotler (2000) highlights that formulating the content of message, logically and symbolically, is critical in mobile advertising. Mobile advertising should contain an attractive idea, convey this idea concisely, employ language or image understood by the target group, and utilize the available space or characters effectively (Barwise & Strong, 2002).

Although the receiver pays little or nothing for information services, advertisements must complement consumer interests. Advertising messages should also disclose how to stop receiving further messages. Humor and surprises in the design of the advertisement create positive feelings toward the advertisement and may lead to viral marketing, especially among the younger receivers (Barwise & Strong, 2002). Entertainment has turned out to increase advertising value in different empirical investigations (Ferrazzi et al., 2003). The higher the entertainment factor of mobile advertising messages, the higher the perceived advertising value of the consumer (Haghirian et al., 2005).

Message Credibility

Advertising credibility can be defined as the "consumers' perception of the truthfulness and believability of advertising in general" (McKenzie & Lutz, 1989) or "predictability and fulfillment of implicit and explicit requirements of an agreement" (Pavlou & Stewart, 2000). Building customer trust is a complex process that involves technology and business practices, but it is crucial for the growth and success of mobile commerce (Siau & Shen, 2003). Since communication with consumers via their mobile devices is a very young phenomenon, marketers are requested to build and breed trust. Therefore advisable to build awareness via other media as well (Enpocket Research Report, 2002).

Credibility of an advertisement is influenced by different factors, especially by the company's credibility and the bearer of the message (Goldsmith et al., 2000), but it is also influenced by the advertising medium. Credibility of a mobile advertising message has a positive influence on the perceived advertising value of the consumer (Haghirian et al., 2005). Viral marketing is very beneficial for the advertiser, as the customer forwarding the advertisement becomes the sender of the message and therefore, the message gains in credibility. In addition, it is assumed that advertisements received from relatives rather than marketers have a higher credibility.

Message Customization

Most mobile users maintain a very personal and intimate relationship with their mobile phone. The mobile phone is rarely used by any other person than its owner, allowing for highly personalized marketing measures. To communicate effectively, marketers need to understand the critical elements underlying the advertising. Various studies have found that customization (DeZoysa, 2002) is the key element of successful services, especially advertising (Yunos et al., 2003). Customization in general means understanding different kinds of individual preferences, needs, mindsets, lifestyles, and cultural as well as geographical differences, to build a customer loyalty and meaningful one-to-one relationships (Riecken, 2000). In other words, customization is about mapping and satisfying the customer's goal in a specific context with a business' goal in its corresponding context (Riecken,

2000). Mobile advertising provides potential for customization, because mobile devices usually carry the user's assigned identity (Lee & Benbasat, 2003) and also can utilize time, activity and location awareness as customization variables.

Similar to traditional media, a customized mobile advertising campaign relies upon databases with enough active and potential clients to reach the target group affectively. Mobile marketers can customize the mobile messages based on the consumer's profile, local time, location, and preferences (Richards & Curran, 2002; Rao & Minakakis, 2003). The aspect of customization turns mobile advertising into a very important mobile commerce application, as it allows the usage of demographic information collected by wireless service providers and information on the current location of a mobile user. Thus, advertising can be carried out very precisely and with a clear focus on the target group (Varshney & Vetter, 2002). Users can be localized through operators to adapt marketing messages by using Geo-location technologies such as the Global Positioning System (GPS) or Cell of Origin (COO) (Barnes, 2003). As the result, consumers increasingly expect tailored and location-based services, thereby underlining the importance of customized mobile marketing (Watson, 2000). Consumer satisfaction and willingness will rise, as becomes easier for the customer to pick the best service in particular location (Rao & Minakakis, 2003). If it properly applied, location-based services with customizing messages can create or reinforce virtual communities (Geser, 2004). However, currently these kinds of location based services have hardly been realized due to limitation in needed infrastructures.

Device

Many experts consider mobile advertising as an encouraging branch of mobile business (Bulander et al., 2005) because mobile devices, as personal communication devices which could be addressed individually, have high penetration rate with multimedia capabilities and interactivity. However, there are also some serious challenges in mobile advertising area, such as spam, limited user interface, privacy concerns, and the expense of mobile data communication.

Device Interface

Mobile terminals have a very limited user interface: the display; which does not allow presenting substantial graphic elements or significant amounts of text, and lack a full keyboard, for most mobile devices (Naccarato & Neuendorf, 1998), which makes it difficult for users to enter a lot of data (Bulander et al., 2005). There are also some differences between different mobile devices, such as display format and color display, that cause limited scope of design. Unless the message is tailored to the terminal, the receiver will face problems in receiving and understanding the message. Smart phones and wireless PDAs are better suited for mobile advertising purposes as they have larger screens, higher storage capacity and many have touch screens or keyboards. Smart phones and wireless PDAs will, however, still have more complicated user interfaces, because of the many features that are integrated into them (Gartners Forecast, 2006).

In any event, if marketers want to use efficiently the communication channels that mobile media provide, they need to understand how mobile consumers perceive and evaluate mobile devices as a source of advertising. For a long period of time, it has been clear that perceptions of the advertising medium affect attitudes toward individual advertising (Ducoffe, 1996). A poor user experience in accessing content in a mobile device, discourage consumers and operators (Blaber et al., 2006). A reliable and user friendly user interface will shorten learning curves and increase demand for multiple services. These issues seem to be the most difficult obstacles to overcome when applying advertising to mobile devices. Despite these limitations of mobile devices, an optimal interface is usually one of the main criteria that determine how well an application is received by its users, and must be designed to support the limited attention of its user (Kurkovsky, 2005).

Device Interactivity

Interactive media demonstrate a bi-directional mode of communication enabling the beneficiary to affect the communication process actively. There is little agreement on the definition of interactivity, as it is a multidimensional and complex concept (Liu & Shrum, 2002). However, while practitioners and academics would agree that the true value of electronic media lies in its interactivity, the way to proceed to make this reality is much more ambiguous (Peltier et al., 2003). Thus, the concept of a completely interactive marketing approach is more of a challenging idea than a practiced reality (Davenport et al., 2001). Interactive marketing is the iterative process taken by a firm to uncover, meet, modify and satisfy customer's needs and desires (Bezjian-Avery et al., 1998). Interactive marketing can also be defined as an approach to use captured customer data to create messages that can affect the behavior of at least one customer (Haeckel, 1998).

At the device level, interactivity is generated through facilitation of one-way and two-way communication through applications (Duncan & Moriaty, 1998). Thus the mobile phone lends itself to effective mobile marketing with its application's interactivity capability. The mobile phone enables the beneficiary of a message to reply to it instantly, which makes this medium extremely interactive. Therefore, mobile phone displays all characteristics essential to enable a direct communication between consumer and advertiser. Based on the above definitions, electronic interactive media can be defined as an electronic medium that can establish a two-way communication system between buyers and sellers (Peltier et al., 2003). So far, most companies have failed to take the advantage of the full potential of interactive marketing (Deighton & Glazer, 1998), in particular, mobile advertising (Okazaki, 2005). Challenges for having full interactive mobile advertising are the creation of interactive messages and applications and also having devices, which facilitate and enable such interaction.

Device Intelligence

Mobile technology provides greater flexibility in communication, collaboration, and information sharing (Sheng et al., 2005). Communication through the mobile medium has several limitations, such as limited bandwidth, device diversity, lack of standardization, and heterogeneous screen sizes and displays, which make certain fonts and picture formats, look awkward (Yunos et al., 2003). Moreover, communication through the mobile medium is often considered more complex than through other media since the content has to be optimized for a numerous of different devices (Nysveen et al., 2005).

Advances in mobile technology will enhance communication through the mobile medium step by step, from basic SMS messaging to a more intelligent means of communication, such as MMS and Java-based applications, enabling previously unfeasibly richer forms (Geysken et al., 2002). Intelligent devices are able to collect, processes, integrate, analysis, evaluate, and interpret available information to facilitate certain services. Intelligence devices are able to understand user's habits; data mine the content itself and determining the consumer's likes, dislike, context or environment that the consumer is in. Third-generation mobile technology and devices enable services with an ability to transfer simultaneously both voice data and non-voice data.

Mobile advertising will be a more attractive marketing medium for both advertisers and consumers by developing technology, such as intelligent properties at the handsets, higher connection speed, and more accurate location awareness, context (Leppäniemi & Karjaluoto, 2005). In addition, strong investments by the software industry to lay the foundation for Web services in the mobile environment should fuel the diffusion of mobile technology (Scharl et al., 2005). Converged devices combine more than one form of wireless communication. Initially, this was limited to combinations of mobile technology, such as GSM, and CDMA, to allow for roaming. Today there is an increasing number of handsets with features such as WLAN, EDGE, WCDMA and so on. This multi-band and multi-mode capability can enable their use almost anywhere in the world. While initially these devices had lower data capabilities than the business segment, this is not the case for higher tier devices.

Media

The history of media industry and related advertising is long enough for marketers to be familiar with the technicalities of this medium. Advertising has changed radically over the years due to new techniques and technologies (Richards & Curran, 2002). The mobile phone has also recently emerged as a promising advertising channel (Barwise & Strong, 2002).

Media Price

Based on information economical model of communication, consumers will only accept mobile advertising if they recognize a benefit in receiving advertising messages on their mobile phone (Kavassalis et al., 2003). Based on this model, the consumer acts as an active participant in the decision making process, by not only trading off the perception of advertising against other possible activities, which can be done in same time, but also has to select from different advertising sources. Financial barriers emerge as most important determinants of an individual's decision to use the mobile advertisements. The cost considerations shift attention from the value creating capabilities and reflect consumers' desires for reasonably priced services over anything else. It could be suggested that instant and recognizable benefits such as discounts or special offers convince people to consider mobile advertisements.

Incentive based advertising is a move toward providing specific financial rewards to consumers who agree to receive ads into their mobile phones, such as free connection time to lower their monthly subscription bill or information about bargains and new products (Haghirian, & Madlberger, 2006). Varshney (2003) describes that recipients react very positively to advertisement that transfers incentives. Incentive based services considered to influence consumer attitudes towards mobile advertising. Several trials have verified that mobile phone users are more positive to receive incentive based advertising messages into their phones (Varshney, 2003). According to Varshney (2003) consumer attitudes towards mobile advertising will be positively affected when receiving services of financial value for accepting an advertisement. In regard to campaign participation, financial incentives can substantially improve the level of participation (Standing et al., 2005).

Media Process

Advertising experts see mobile advertising as a four-stage process (Drossos & Giaglis, 2006): deciding on the campaign strategy, designing based on brand, accessing data to find targeted customers, and making the message interactive, informative and entertaining. Assael (1984) describes that the advertising procedure involves, essentially, four core segments: the brand's overall marketing strategy, the advertisement itself, the medium through which the stimuli are delivered, and the prospect who receives the communication message.

Mobile advertisement should be able to provide relevant information and rewards, be delivered by trusted organization, and also give the viewer control over the message. A determinant for the success of the mobile advertising campaign (Baltas, 2003), as a form of permission based marketing, is an adequate market research resource. Agencies should be competent in creating opt-in lists through traditional marketing media, such as websites, television, or flyers. The advertiser must deliver a consistent message for an identified consumer, based on the information about consumer needs.

Subsequently, agencies need to create meaningful opportunities for consumers to interact with advertisers with significant incentives (Kavassalis et al., 2003). In case of low response rates, contingency plans should permit the modification of the campaign strategy by employing alternative types of campaigns and service modules. Since a reputable brand will probably initiate more responses and overcome the clutter created from competitive mobile advertising, the advertised brand is the important factor for the effectiveness of the campaign.

Media Policy

Norris (2003) has collected ten legal considerations for today's mobile marketers; 1) data protection, 2) unsolicited commercial communication, 3) location data, 4) online contracting, 5) information requirements, 6) distance selling, 7) industry code, 8) contractual commitments, 9) specific activities, market or products, and 10) roaming and cross-border risk. As mobile phones cannot distinguish between spam and genuine communication automatically, unsolicited messages, commonly known as spam (Hinde, 2003), overwhelm user acceptance. Unwanted messages are illegal in some countries and annoy consumers regardless of the medium. A list of critical factors that helps marketers to distance themselves from spam, collected by Fuller (2003), includes: frequency, relevance, control, confidentiality and unsolicited. The Mobile Marketing Association (Mobile Marketing Industry Glossary, 2007) has also developed a code of conduct for the players in the mobile advertising industry that regulates spam. The code of conduct includes six principles that the Mobile Marketing Association, has labeled as the six C's of privacy: choice, control, constraint, customization, consideration, and confidentiality.

An extensive survey of mobile advertising in the UK (Enpocket Research Report, 2002) reveals that consumers fear that spamming might occur during mobile advertising, thus, they prefer to give their permission only to organizations they trust. Corporate policies must consider legalities such as electronic signatures, electronic contracts, and conditions for delivering mobile advertisements. The European Union (Official Journal at OJ L201/37, 2002, Vol. 31) approved a new directive (Directive/58/EC) in 2002 for establishing standards for the processing of personal data and the protection of privacy in electronic communications sector. Mobile advertising in the EU area is regulated by law and involves asking end users' permission to send unsolicited marketing messages via all electronic communications for marketing purposes. This is called opt-in mobile advertising. Furthermore, the directive adopts an opt-in rule for the use of traffic data. Finally, the legislation sets rules for the collection and procession of location-based data. Location data cannot be collected or used without user's explicit prior permission. Regulatory guarantees privacy for customer in market practices. New regulations in the United States that allow people to keep their phone numbers when switching cellular carriers (CNN, Wireless Society Report, 2003) may reinforce fears of unwanted messages and misuse of personal data, thereby keeping consumers from registering for SMS based information services.

FUTURE RESEARCH

This chapter targeted to explore mobile advertising's challenges and future directions by evaluating factors that seem to influence mobile advertising. The materials for the literature review are gathered from different sources including research reports from consulting and marketing research firms, publications, as well as academic studies. This chapter is put together from a previously written article that author published in a journal. The theoretical contribution of this work lies in outlining the key factors of mobile advertising. Even though the work is trying to cover the field, but surely there are some other factors, which may have an impact on mobile advertising. For example, in order to include customer heterogeneity, environmental or social factor such as religion or literacy may have an impact on consumer, privacy, policy or media. Author hopes that this chapter opens new dialogues on multiple unexamined issues concerning mobile advertising industry. A potential extension of this research work is to validate the factors through a survey from mobile phone consumers across different demographics in different countries to assess how the consumers' perception stack-rank against the expert's opinion.

CONCLUSION

Mobile marketing adoption and acceptance are on the rise (Becker, 2005), but without a clear understanding of the elements affecting mobile advertising value for consumer, marketers will have a little ability to constantly generate positive returns from their programs. Bauer et al. (2005), identified the influential variables affecting consumer behavior. Based on this review, the main factors that are affecting consumer behavior are 1) Privacy; the degree to which personal information is not known by others, 2) Purpose; the type of purpose the receiver is trying to satisfy by using a mobile device, and 3) Performance; respond and attitude towards mobile advertising.

According to Kotler (2000), formulating the message is critical in mobile advertising and will require solving four problems: what to say (message content), how to say it logically (message structure), how to say it symbolically (message format), and who should say it (message source). Based on the review, the following factors are critical regarding messaging in mobile advertising; 1) Content; 'relevant content is king' in mobile advertising, content being informative and entertaining, 2) Credibility; consumer's perception of the truthfulness and believability of advertising in general, 3) Customization; mapping and satisfying of customer's goal in a specific context with a business's goal in its respective context.

Mobile handsets are seen as very powerful communication devices by advertisers, due to the personal, immediate, and interactive nature of them (Koranteng, 2001). The mobile device may be an attention getter, but an attention getting device that is unrelated to the message will not attract consumers interested in the message or the product (Ogilvy, 1963). The device should have the following characteristics to maximize user experience in mobile advertising: 1) Interface; a good and suitable user interface maximizes user experience, 2) Interactivity; enable interactive solutions for communication between mobile marketer and consumer, 3) Intelligent; include the latest telecommunications platforms and also location based technologies.

For maximizing mobile advertising effect, business models that can capture the synergy of two existing industries, advertising and telecommunications, must be conceived. In any future sustainable business model, all players will have to reach a consensus on the structure of the system and on the importance of each player in that system (Leppäniemi et al., 2004). The factors affecting the mobile advertising media are: 1) Price; instant and recognizable benefits such as discounts or special offers convince people to consider mobile advertisements, 2) Process; relationship and cross media working in the mobile advertising value chains, 3) Policies; legislation and regulation which place sets of rules for the collection and procession of location-based data.

Figure 1. A conceptual model of factors affecting mobile advertising

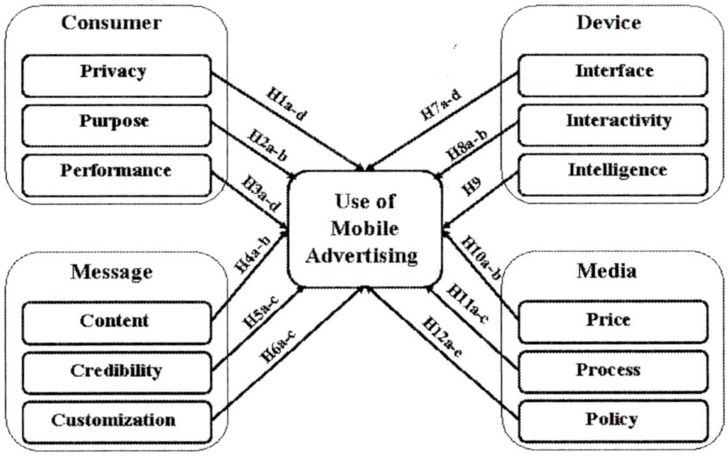

To summarize, the above mentioned information lead to the development of a conceptual model of factors effecting mobile advertising, Figure 1.

Based on above literature study, a set of hypotheses was proposed that lay the foundation for the conceptual model of factors affecting mobile advertising, Figure 1. The model is tested by evaluating the hypotheses against a five point Likert scale (1 strongly agree to 5 strongly disagree). The data was gathered through a survey study during spring 2008, to help evaluate the hypotheses and statistical hypothesis testing techniques. The survey population comprised of 115 mobile advertising experts working across 19 different countries in North America, Europe and Asia. The experts were selected in this early stage of model evaluation with the intention of first proofing the model through contemporary thought leaders in the area before getting it tested by the consumers. A contradictory case could be made that our research methodology of validation through experts would tend to replicate what we already know and hence tend to validate models more easily but the survey results show that still there is a not strong consensus over some of the tenants of our model (Vatanparast and Butt, 2009)

A macro-level analysis of the basic statistical metrics of the survey data reveals that the panel of mobile advertising experts agrees with the twelve important factors proposed in our previous paper and the only factor which demands a look closer look and some debate is Device Intelligence. A closer look at the top-5 rated 'sub-factors' reveals that all the four core tenants/factors of our model are seen as critical to the success of mobile advertising. The range of mean rating values is not very wide [3.38-4.17] and most of the ratings fall fairly close to each other, with Device Intelligence being the only outlier. The plot below (Figure 2) presents the average scores for the factor on the X-axis and the Standard Deviation on the Y axis. This plot can be seen as a simple representation of what factors are more important than others and to what degree the respondents agreed with each other in their responses. It is evident to this graph that the strongest consensus was found on the 'Message Content' factor of our model, which had the lowest Standard Deviation and a 4+ average rating. The detail evaluation of the model can be found in up coming article of (Vatanparast and Butt, 2009).

The graph also gives clear evidence that the expert panelists believe that the core marketing and advertising principles of targeted and relevant messaging will hold true for this emergent advertising domain as well. The higher mean rating

Figure 2. Simple representation of what factors are more important based on average scores for the factor on the X-axis and the Standard Deviation on the Y axis

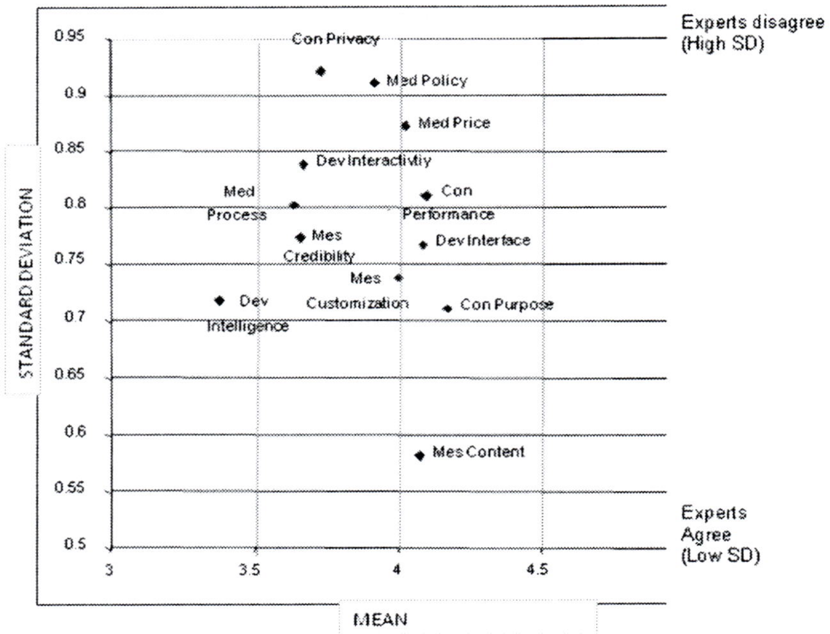

for Message Content, Consumer Purpose and Consumer Performance all reveal that the 'softer' aspects of the message are going to be critical for mass-market adoption of this new advertising medium by the consumer audience. Furthermore, the high mean rating on Media Pricing also shows that pricing is going to be an important consideration not just for the consumers, who would expect rebates or discounts for receiving mobile advertising but would also be important for other players in the mobile advertising value chain. To prove the viability of this promising but yet-to-be proven advertising avenue, mobile advertisers will have to offer attractive pricing propositions to the advertisers (Vatanparast and Butt, 2009). The results of the survey have proven the conceptual model in the respective areas of Consumer (Privacy, Purpose & Performance), Message (Content, Credibility and Customization) and Media (Price, Process & Policy) at a statistically significant level. This proposed Conceptual Model/Framework can serve as a solid base for the evaluation of the critical success factors for a market-worthy mobile advertising strategy.

REFERENCES

Ackerman, M., Darrel, T., & Weitzner, D. (2001). Privacy in context. *Human-Computer Interaction*, *16*, 167–176. doi:10.1207/S15327051HCI16234_03

Assael, H. (1984). *Consumer behavior and marketing action.* 2nd edition. Boston: Kent Publishing Company.

Balasubramanian, S., Peterson, R. A., & Järvenpää, S. L. (2002). Exploring the implications of m-commerce for markets and marketing. *Journal of the Academy of Marketing Science*, *30*(4), 348–361. doi:10.1177/009207002236910

Baltas, G. (2003). Determinants of internet advertising effectiveness: An empirical study. *International Journal of Market Research*, *45*(4), 505–513.

Barnes, S., & Huff, S. (2003). Rising sun: iMode and the wireless Internet. *Communications of the ACM*, *46*(11), 79–84. doi:10.1145/948383.948384

Barnes, S. J. (2002). Wireless digital advertising: Nature and implications. *International Journal of Advertising*, *21*, 399–420.

Barnes, S. J. (2003). Location-based services: The state-of-the-art. *e-Service Journal*, *2*(3), 59-70.

Barwise, P., & Strong, C. (2002). Permission based mobile advertising. *Journal of Interactive Marketing*, *16*(1), 14–24. doi:10.1002/dir.10000

Bauer, H. H., Barnes, S. J., Reichardt, T., & Neumann, M. M. (2005). Driving consumer acceptance of mobile marketing: A theoretical framework and empirical study. *International Journal of Electronic Commerce*, *16*(3), 181–192.

Becker, M. (2005). Effectiveness of mobile channel additions and a conceptual model detailing the interaction of influential variables. Access January 2007, from http://www.iloopmobile.com/news/mb_research_111705.htm

Bezjian-Avery, A., Calder, B., & Iacobucci, D. (1998). Interactive advertising vs. traditional advertising. *Journal of Advertising Research*, *38*(4), 23–32.

Blaber, G., Bouchard, J. P., & Brown, A. (2006). IDC Western European market analysis, mobile phone forecast and analysis 2006-2010.

Bulander, R., Decker, M., Schiefer, G., & Kolmel, B. (2005). Comparison of different approaches for mobile advertising. Proceedings of the 2nd IEEE International Workshop on Mobile Commerce and Services (WMCS '05), IEEE Computer Society, Munich, Germany, July, 174-182.

CNN, Wireless Society Report. (2003). New switching rules may trash millions of cell phones, cable network news. Access January 2007, from http://www.cnn.com/2003/TECH/ptech/10/31/sprj.ws.cellphone.garbage.ap/

Davenport, T. H., Harris, J. G., & Kohli, A. (2001). How do they know their customers so well? *Sloan Management Review*, *42*(2), 63–73.

Deighton, J., & Glazer, R. (1998). From the editors. *Journal of Interactive Marketing*, *12*(1), 2–4. doi:10.1002/(SICI)1520-6653(199824)12:1<2::AID-DIR1>3.0.CO;2-P

DeZoysa, S. (2002). Mobile advertising needs to get personal. *Telecommunications International Ed., 36(2),* 8.

Drossos, D., & Giaglis, G. M. (2006). Mobile advertising effectiveness: An exploratory study. *Proceedings of the International Conference on Mobile Business (ICMB'06)*

Duchnicky, R. L., & Kolers, P. A. (1983). Readability of text scrolled on visual display terminals as a function of window size. *Human Factors*, *25*, 683–692.

Ducoffe, R. H. (1996). Advertising value and advertising on the Web. *Journal of Advertising Research*, *36*(September/October), 21–36.

Duncan, T., & Moriaty, S. E. (1998). A communication-based marketing model for managing relationships. *Journal of Marketing*, *62*(2), 1-13. doi:10.2307/1252157

Edens, K. M., & McCormick, C. B. (2000). How do adolescents process advertisements? The influence of ad characteristics, processing objective, and gender. *Contemporary Educational Psychology*, *25*, 450–463. doi:10.1006/ceps.1999.1031

eMarketer. (2003). SMS marketing yields strong results. *eMarketer Research Report, UK.*

Enpocket Research Report 1. (2002). Consumer preferences for SMS marketing in the UK, August–November 12. Access January 2007, from http://www.enpocket.com/news/press-releases/consumers-find-sms-marketing-as-acceptable-as-tv-or-radio-advertising

Eysenck, M. (1982). *Attention and arousal, cognition and performance*. New York: Springer-Verlag.

Ferrazzi, K., Chen, J., & Li, Z. (2003). Idea: Playing games with customers. *Harvard Business Review, 81*(4), 21.

Fishbein, M., & Ajzen, I. (1975). *Belief, attitude, intention, and behaviour: An introduction to theory and research.* Reading, MA: Addison-Wesley.

Gartners Forecast: GPS in Mobile Devices, Worldwide, 2004-2010, Dataquest. (2006, November 29). ID Number: G00144746

Forrester Research, Inc. (2001). The marketer's guide to use SMS. *Forrester Research Report*, European Research Center, Amsterdam.

Fortunati, L. (2001). The mobile phone: An identity on the move. *Personal and Ubiquitous Computing, 5*, 85–98. doi:10.1007/PL00000017

Forum, U. M. T. S. (2003). UMTS security awareness. *UMTS Forum Report*, November 5, 30. Access January 2007, from http://www.umts-forum.org/servlet/dycon/ztumts/umts/Live/en/umts/Resources_Reports_30_index

Fuller, P. (2003). Why spam doesn't have to happen on mobile devices. Mobile Marketing Association, November 2. Access January 2007, from http://mmaglobal.com/modules/wfsection/article.php?articleid=44.

Future Mobile Handsets, 8th edition. (2006, October). Informa

Geser, H. (2004). Towards a sociological theory of the mobile phone. Access January 2007, from www.socio.ch/mobile/t_geser1.htm

Geysken, I., Gielens, K., & Dekimpe, M. G. (2002). The market valuation of Internet channel addition. *Journal of Marketing, 66*(April), 102–119. doi:10.1509/jmkg.66.2.102.18478

Godin, S. (1999). *Permission marketing-turning strangers into friends, and friends into customers.* New York: Simon & Schuster.

Godin, S. (2001). *Permission marketing.* FinanzBuch-Verlag, Munich.

Gohring, N. (2002). And now a word from our sponsors… . *America's Network, 106*(3), 17.

Goldsmith, R. E., Lafferty, B. A., & Newell, S. J. (2000). The impact of corporate credibility and celebrity credibility on consumer reaction to advertisements and brands. *Journal of Advertising, 29*(3), 43–54.

Haeckel, S. H. (1998). About the nature and future of interactive marketing. *Journal of Interactive Marketing, 2*(1), 63–71. doi:10.1002/(SICI)1520-6653(199824)12:1<63::AID-DIR8>3.0.CO;2-C

Haghirian, P., & Madlberger, M. (2005). Consumer attitude toward advertising via mobile devices-an empirical investigation among Austrian users. *Proceedings of ECIS (European Conference of Information Systems),* Regensburg, Germany.

Haghirian, P., & Madlberger, M. (2006). Cross-cultural consumer perceptions of advertising via mobile devices-some evidence from Europa and Japan. *Proceedings of the ECIS (European Conference on Information Systems),* Goeteborg, Sweden.

Haghirian, P., Madlberger, M., & Tanuskova, A. (2005). Increasing advertising value of mobile marketing–an empirical study of antecedents. *Proceedings of the 38th Hawaii International Conference on System Sciences,* HI.

Helm, S. (2003). Viral marketing–establishing customer relationships by word-of-mouse. *Electronic Markets*, *10*(3), 158–161. doi:10.1080/10196780050177053

Hinde, S. (2003). Spam: The evolution of a nuisance. *Computers & Security*, *22*(6), 474–478. doi:10.1016/S0167-4048(03)00603-5

Katz, E., Blummer, J. G., & Gurevitch, M. (1974). Uses and gratifications research. *Public Opinion Quarterly*, *23*, 389–398.

Katz, E., Gurevitch, M., & Haas, H. (1973). On the use of mass media for important things. *American Sociological Review.*

Kavassalis, P., Spyropoulou, N., Drossos, D., Mitrokostas, E., Gikas, G., & Hatzistamatiou, A. (2003). Mobile permission marketing: Framing the market inquiry. *International Journal of Electronic Commerce*, *8*(1), 55–79.

Koranteng, J. (2001). ZAP! There's no escaping the mobile advertisement. *Ad Age Global, January, 1(5),* 9.

Kotler, P. (2000). *Marketing management: The millennium edition*. Upper Saddle River, NJ: Prentice Hall.

Kotler, P., Wong, V., Saundersa, J., & Armstrong, G. (2005). *Principles of marketing* (fourth European edition). Pearson Education Limited.

Krishnamurthy, S. (2001). A comprehensive analysis of permission marketing. *Journal of Computer Mediated Communication, 6(2)*. Retrieved from http://www.ascusc.org/jcmc/vol6/issue2/krishnamurthy.html

Kurkovsky, S. (2005). Using principles of pervasive computing to design m-commerce applications. *Proceedings of the International Conference on Information Technology: Coding and Computing (ITCC'05),* Volume II, 59-64.

Lee, Y. E., & Benbasat, I. (2003). Interface design for mobile commerce. *Communications of the ACM*, *46*(12), 49–52. doi:10.1145/606272.606300

Leppäniemi, M., & Karjaluoto, H. (2005). Factors influencing consumers' willingness to accept mobile advertising: A conceptual model. [IJMC]. *International Journal of Mobile Communications*, *3*(3), 197–213. doi:10.1504/IJMC.2005.006580

Leppäniemi, M., Karjaluoto, H., & Salo, J. (2004). The success factors of mobile advertising value chain. *eBusiness Review IV,* 93-97.

Lewis, S. (2001). M-commerce: Ads in the ether. *Asian Business, January, 37(1),* 31.

Liu, Y., & Shrum, L. J. (2002). What is interactivity and is it always such a good thing? Implications of Definition, Person, and Situation for the Influence of Interactivity on Advertising Effectiveness. *Journal of Marketing*, *31*(4), 53–64.

McKenzie, S. B., & Lutz, R. L. (1989). An empirical examination of the structural antecedents of attitude toward the ad in an advertising pretesting context. *Journal of Marketing*, *53*(April), 48–65. doi:10.2307/1251413

Mobile Marketing Association. Mobile Marketing Industry Glossary. (2007). Access January 2007, from http://www.mmaglobal.com/modules/content/index.php?id=6

Naccarato, J., & Neuendorf, K. (1998). Content analysis as a predictive methodology: Recall, readership, and evaluations of business-to-business print advertising. *Journal of Advertising Research*, *38*(3), 19–33.

Natsuno, T. (2003). *The i-mode wireless ecosystem*. Wiley, Yokohama.

Nelson, R. (2000). Wireless advertising: What's in it for telcos? *Global Telephony, July, 8(7),* 12.

Norris, S. (2003). Make the mobile connection. *Revolution Magazine*. Access Jan 2007, from http://www.brandrepublic.com/login/News/182912/

Nysveen, H., Pedersen, P. E., Thorbjornsen, H., & Berthon, P. (2005). Mobilizing the brand: The effects of mobile services on brand relationships and main channel use. *Journal of Service Research*, *7*(3), 257–276. doi:10.1177/1094670504271151

Ogilvy, D. (1963). *Confessions of an advertising man*. NY: Ballantine Books.

Okazaki, S. (2005). Mobile advertising adoption by multinationals. Senior executives' initial responses. *Internet Research*, *15*(2), 160–180. doi:10.1108/10662240510590342

Ovum. (2002). Ovum forecasts: Global wireless markets 2002–2006. Research Report.

Paavilainen, J. (2002). Mobile business strategies. Understanding the technologies and opportunities. London: Addison Wesley.

Pavlou, P. A., & Stewart, D. W. (2000). Measuring the effects and effectiveness of interactive advertising: A research agenda. *Journal of Interactive Advertising, 1(1)*. Access January 2007, from http://jiad.org/vol1/no1/pavlou/

Peltier, J. W., Schibrowsky, J. A., & Schultz, D. E. (2003). Interactive integrated marketing communication: Combining the power of IMC. New Media and Database Marketing. *International Journal of Advertising*, *22*, 93–115.

Rao, B., & Minakakis, L. (2003). Evolution of mobile location-based services. *Communications of the ACM*, *46*(12), 61–65. doi:10.1145/953460.953490

Report, W. (2000). First-to-wireless: Capabilities and benefits of wireless marketing and advertising based on the first national mobile marketing trial. Access January 2007, from www.imap-project.org/imapproject/downloadroot/public3/ftw_report.pdf

Richards, J., & Curran, C. (2002). Oracles on 'advertising': Searching for a definition. *Journal of Advertising*, *XXXI*(2), 63–77.

Riecken, D. (2000). Personalized views of personalization. *Communications of the ACM, August, 43(8),* 27-28.

Robins, F. (2003). The marketing of 3G. *Marketing Intelligence & Planning*, *21*(6), 370–378. doi:10.1108/02634500310499239

Rodgers, S., & Thorson, E. (2000). The interactive advertising model: How users perceive and process online ads. *Journal of Interactive Advertising, 1(1)*. Access January 2007, from www.jiad.org/vol1/no1/rodgers/

Rogers, E. M. (1995). *Diffusion of innovations*. New York: Free Press.

Rust, R. T., Kannan, P. K., & Peng, N. (2002). The customer economics of Internet privacy. *Journal of the Academy of Marketing Science*, *30*(4), 455–464. doi:10.1177/009207002236917

Saunders, C. (2003). Studies: Mobile advertisement market to grow, amid risks. *Internet Advertising Report,* November 1. Access January 2007, from http://www.clickz.com/showPage.html?page=915121

Scharl, A., Dickinger, A., & Murphy, J. (2005). Diffusion and success factors of mobile marketing. *Electronic Commerce Research and Applications*, *4*, 159–173. doi:10.1016/j.elerap.2004.10.006

Shavitt, S., Lowrey, P., & Haefner, J. (1998). Public attitudes towards advertising: More favourable than you might think. *Journal of Advertising Research*, *38*(4), 7–22.

Sheng, H., Nah, F. F.-H., & Siau, K. (2005). Strategic implications of mobile technology: A case study using value-focused thinking. *The Journal of Strategic Information Systems*, *14*(3), 269–290. doi:10.1016/j.jsis.2005.07.004

Siau, K., & Shen, Z. (2003). Building customer trust in mobile commerce. *Communications of the ACM*, *46*(4), 91–94. doi:10.1145/641205.641211

Standing, C., Benson, S., & Karjaluoto, H. (2005). Consumer perspectives on mobile Advertising and marketing. *Proceedings of the Australian & New Zealand Marketing Academy Conference (ANZMAC)*. Perth, Australia, December 5-7, 135-141.

Stewart, D. W., & Pavlou, P. A. (2002). From consumer response to active consumer: Measuring the effectiveness of interactive media. *Journal of the Academy of Marketing Science*, *30*(4), 376–396. doi:10.1177/009207002236912

Taylor, P. (2006). Mobile consumer applications outlook 2007. *Wireless Internet Applications Service, Strategy Analytics,* 24-26

The European Union. (2002). 2002/58/EC. *Official Journal at OJ*, *L201*(37), 31.

Tsang, M. M., Ho, S., & Liang, T. (2004). Consumer attitudes toward mobile advertising: An empirical study. *International Journal of Electronic Commerce*, *8*(3), 65–78.

Varshney, U. (2003). Location management for mobile commerce applications in wireless Internet environment. *ACM Transactions on Internet Technology*, *3*(3), 236–255. doi:10.1145/857166.857169

Varshney, U., & Vetter, R. (2002). Mobile commerce: Framework, applications, and networking support. *Mobile Networks and Applications*, *7*, 185–198. doi:10.1023/A:1014570512129

Vatanparast, R. (2007). Piercing the fog of mobile advertising. *The Sixth International Conference on Mobile Business ICMB 2007 July 9-11*, 19. Toronto, Ontario, Canada

Vatanparast, R., & Asil, M. (2007). Factors affecting the use of mobile advertising. *International Journal of Mobile Marketing*.

Venkatesh, V., Morris, M. G., Davis, G. B., & Davis, F. D. (2003). User acceptance of information technology: Toward a unified view. *MISQ*, *27*(3), 425–479.

Watson, R. T. (2000). Electronic commerce: The strategic perspective. Fort Worth, TX: Dryden.

Wells, W. D. (1997). *Measuring advertising effectiveness*. Mahwah, NJ: Lawrence Erlbaum.

Yunos, H. M., Gao, J. Z., & Shim, S. (2003). Wireless advertising's challenges and opportunities. *IEEE Computer*, *36*(5), 30–37.

KEY TERMS AND DEFINITIONS

Consumer Performance: the degree to consumer's reactions or respond.

Consumer Privacy: the degree to which consumer's personal information is not known by others.

Consumer Purpose: the degree to consumer's acknowledgment of what he is pursuing in a particular situation to satisfy.

Device Intelligence: mobile technologies which enhance communication through the mobile medium to a more intelligent and richer means of communication.

Device Interactivity: facilitation of one-way and two-way communication through applications.

Device Interface: enabling features in device hardware or software which allows consumer to interact.

Media Policies: rules and regulations around media.

Media Price: the cost associated to interact with media.

Media Process: the steps required to enable delivery and collection of interaction between consumers and media.

Message Content: communication exchanges between advertisers and consumers.

Message Credibility: consumers' perception of the truthfulness and believability of message in general.

Message Customization: mapping and satisfying the customer's goal in a specific message.

Chapter 5
Advertising Challenges in Ubiquitous Media Environments

Maria Åkesson
Halmstad University, Sweden

Carina Ihlström Eriksson
Halmstad University, Sweden

ABSTRACT

Mobile advertising opportunities have attracted interest from industry and academics as a response to the trend of diffusion of new mobile technology. One industry particularly interested in new advertising opportunities is the newspaper industry. Many media houses envision a future multi-channel media environment supporting device independent, anytime, anywhere publishing, i.e. a Ubiquitous Media Environment (UME). In this chapter we explore how ubiquitous advertising challenges the role of media houses. Ubiquitous advertising refers to advertising in a UME, anytime, anywhere and in any device. Based on results from interviews and workshops with advertisers and publishers we discuss how media houses traditional role is challenged. We contribute with an understanding of how media houses can develop their role to succeed with ubiquitous advertising and to m-advertising literature with insights into the dynamics of balancing consumer and advertiser value of ubiquitous advertising.

INTRODUCTION

The diffusion of new mobile technology, rapid growth of mobile phone penetration and the improvements of mobile device capabilities are trends that have led to an increased interest in mobile advertising (m-advertising). The unique features of mobile devices such as their mobility, personalization, interactivity, and location awareness enable advertisers to reach consumers anytime and anywhere with targeted advertising adapted to consumers preferences and location. Indeed, this phenomenon has attracted the interest of academia as well as industry.

The interest of academic scholars has led to numerous studies of mobile marketing and advertising. Commonly, studies have been focused on the effects of m-advertising and consumer attitudes and behavior (see e.g. Bauer *et al.*, 2005; Heinonen & Strandvik, 2007; Läppäniemi, 2007; Merisavo *et al.*,

DOI: 10.4018/978-1-60566-074-5.ch005

2007; Okazaki, 2004; Scharl *et al.*, 2004; Tsang, 2004). Research representing the advertiser point of view is not as common, examples are studies on m-advertising success factors (Vatanparast & Butt, 2009), mobile marketing strategy (Okazaki, 2005), challenges for ubiquitous advertising (Ihlström Eriksson & Åkesson, 2008) and factors influencing m-advertising value from an advertiser as well as a consumer point of view (Komulainen *et al.*, 2006; Vatanparast, 2007).

One industry that has shown special interest in m-advertising is the newspaper industry. The digitization of news media has led to the emergence of new opportunities but also to disruptive implications for traditional media (Christensen & Davis, 2006). Lately, the newspaper industry has experienced declining circulation and advertising revenues of their printed editions forcing them to seek new opportunities in digital media (The Economist, 2006). Many newspaper organizations have met this changing media landscape by organizing themselves as media houses publishing in multiple channels such as print, online, PDAs, and mobile phones. The vision is a future media environment supporting device independent, anytime, anywhere publishing, i.e. a Ubiquitous Media Environment (UME) (Åkesson & Ihlström Eriksson, 2008).

However, the digitalization of news publishing has not been trouble-free to the newspaper industry. It has been very difficult to identify business models and value networks that enable profitable online business. The same difficulty is now experienced in publishing mobile services. Now yet another mobile device, the e-reader with e-paper displays (for example the iRex iLiad or the Amazon Kindle) is introduced as a digital publishing channel. Indeed, mobile devices and wireless access to content do not only offer new opportunities but also challenges the core business of the newspaper industry. Since the revenue from advertising is the backbone of newspaper economy, there is a need to innovate advertising models in order to offer new values to advertisers and maintain a strong position on the advertising market.

The traditional business model in the newspaper industry has long been to sell content to media consumers and exposure to advertisers. In digital media, the advertising opportunities are built on the possibility of getting closer to individual consumers. Thus, there are inherent risks of violating consumers' integrity, risk of *SPAM* effects and irritating the consumer (Vatanparast & Butt, 2009; MMA, 2008a). Balancing these risks with advertiser values will demand that media houses develop their role in the value network. In this chapter we will address the question of how ubiquitous advertising challenges the role of media houses. We focus on the relationships between media houses, consumers and advertisers. The aim is to provide an understanding of how media houses can develop their role to succeed with ubiquitous advertising and to contribute to m-advertising literature with insights into the dynamics of balancing consumer and advertiser value of ubiquitous advertising.

Background to Newspaper Industry and Research Setting

In this section we present a short background to the newspaper industry interest in m-advertising and some background statistics about the advertising market. Thereafter a presentation of the setting in which the research reported in this chapter was conducted is given. Finally, we give a brief description of the method for analyzing the material on advertising.

As described in the introduction, newspaper industry has a long tradition of selling content to an audience and selling access to this audience to advertisers. This model was not challenged until the introduction of the internet. Since then, the newspaper industry is undergoing disruptive change. Constant introduction of new digital technology, increased mobility, changing media consumption and advertising patterns, as well as digital convergence are radically changing the value networks in newspaper industry (Åkesson & Ihlström Eriksson, 2008).

This has led to a situation where the younger audience is losing interest in traditional media such as printed newspapers and is going for digital media (The Economist, 2006). Adding to this, the advertisers are following. According to the Newspaper Association of America, revenues of print advertising fell 9.4% in 2007 while online advertising revenues grew by 18.8% (NNA, 2008). Recent statistics show that newspaper advertising revenues in the US declined 16.4% in 2008 and eMarketer estimates that the online advertising revenues dropped as well with about 0.4% (eMarketer, 2008). Further eMarketer (2008) predicts that newspaper advertising in print will decline in 2009 more than any other medium, as a consequence of the economic crisis that started late 2008.

On the advertising market in digital media, newspaper industry does not have as strong position as in print. Other actors such as Google and Yahoo are strong competitors owning about 60% of the online advertising market (eMarketer, 2008). Advertisers are also communicating directly with their market with for example advanced websites, e-mail offers and in social media. The online advertising grew with 26% in the US and 40% in Europe in 2007. The total global online advertising market was 45 billion US dollars which is 7.4% of the total advertising market. The newspaper industry's share of the global market is expected to decrease from 26.5% in 2008 to 25.0% in 2011, while internet's share is expected to increase from 10.8% to 13.8% during the same period (WAN, 2008). The m-advertising market is growing but still it is only 0.3% of the total advertising market (ZenithOptimedia, 2008). However, Gartner forecast that the mobile advertising market will grow with more than 100% per year (ZenithOptimedia, 2008).

An example of a new mobile media interesting for m-advertising recently introduced on the market is e the e-reader. The e-reader is enabled by a digital display technology called e-paper technology. E-paper technology is reflecting, giving the same reading experience as on paper (such as high contrast and the possibility to read in sunlight), and is thin, flexible and non-sensitive. The resolution is 160 dpi, which is the same resolution as on printed newspapers, compared to the 72 dpi of a LCD screen. The electronically made print also enables e-paper the same viewing angles as printed material since it is a reflecting display technology in contrast to projecting technologies such as LCD screens. Another advantage of the technology is the low power consumption; power is only needed when the display is updated. These properties makes e-paper technology especially suitable for content traditionally offered in print such as newspapers, books, magazines, making it an interesting addition to the existing channels.

Today there exist several e-readers on the market, e.g. Sony Reader, iRex iLiad, Bookeen Cybook, STAReBOOK and the Amazon Kindle. Several international newspapers such as New York Times, Wall Street Journal, Le Monde, and The Times are currently available as text versions with limited use of pictures, in some of these devices. However, the value chain in this context is in favor of the manufacturers, they own the distribution and customer base. There is a lot of effort put into developing e-readers further. Manufacturers like Philips, Plastic Logic, Polymer Vision and Bridgestone announces flexible, bendable and color e-paper to enter the market.

The research presented in this chapter has been conducted in collaboration within the newspaper industry. It started in the DigiNews project, a two-year collaborative research project, including several major technology firms, media houses and universities across Europe, which ended mid-year 2006. The overall goal of the project was to explore research and development issues for an electronic newspaper of the future, including m-advertising.

Table 1. Data sources

	No. of interviews	No. of workshops	Total no. of respondents
Publishers	13	4	34
Advertisers	2	5	20

After the DigiNews project ended in mid-year 2006, the research continued within the project Designing Ubiquitous Media Services (UbiMedia), which was a Swedish project with partners from 9 Swedish newspapers, the Swedish Newspaper Publishers´ Association and Stampen (a parent company for several newspapers, printing houses and distribution companies). This two-year project targeted the challenge of designing ubiquitous media services for a multitude of devices and contexts to be consumed anytime and anywhere. This is referred to as a Ubiquitous Media Environment (UME), a media environment with integrated infrastructures, globally supported mobility and capable of distributing services seamlessly accessible independently of device.

This means that a future UME is characterized by a multitude of available media channels with different properties and capacities (see Åkesson & Ihlström Eriksson, 2008 for a more extensive description). Depending on the resources a user has available in a given context, services will be consumed in different devices and in different ways. In turn, this means that the same service will be distributed in different formats and adapted to different devices and most importantly adapted to user´s profile and context. Indeed, this creates new opportunities for advertising.

Within these two projects we have done interviews with newspaper management, workshops with newspaper staff working with advertising as well as with advertisers. In table 1, an overview of data sources used in this chapter is presented.

The interview study included a total of 13 respondents from media houses and to 2 advertisers on the topic of visions of future e-newspapers, visions of UME and advertising in the future. All interviews were recorded and later transcribed. In the workshops we discussed different scenarios of future ubiquitous advertising and these scenarios were illustrated in concept movies. The data was analyzed using the themes from literature presented in the next section. Excerpts in the transcribed material were marked with assigned colors, facilitating data categorization according to corresponding themes.

RELATED LITERATURE

We start this section with a brief introduction to advertising in digital media. Thereafter we present a literature review of m-advertising regarding the value of m-advertising to advertisers and consumers and the associated tradeoffs when adapting m-advertising to consumers context. We will then proceed to an introduction of how the concept of context is discussed in ubiquitous computing literature. We will argue that there are several risks of context adaptation of advertising in UME. This section is concluded by a summary of the concepts from literature employed in the analysis of the empirical material.

Advertising in Digital Media

Since the internet started to be used for commercial web sites, it has become an important advertising medium with banner advertising, search advertising, online video advertising and email marketing. The values discussed for advertisers compared to advertising in traditional media are for example the interactive features, the unlimited reach in time and space, and the opportunity to reach individuals (Wind & Mahajan, 2000). Another value is the possibility of providing advertisers with information and measurements of advertising effects such as reach, frequency, media impact, information that is useful for advertisers to follow up on for example sales effects (Wind & Mahajan, 2000). Already in 1996 this targeted internet advertising was labeled ubiquitous advertising by Kohda and Endo (1996).

Today, advertising in digital media has expanded to include mobile and handheld devices. The penetration of mobile phones has grown rapidly and penetration has now reached more than 60% global penetration and many countries have passed the 100% barrier according to the International Telecommunications Union (ITU, 2008). This high penetration has taken these advertising possibilities further, offering opportunities of approaching consumers even closer, most often referred to as m-advertising.

What is m-advertising and how does it differ from traditional advertising? Well, there are several concepts in literature referring to advertising in mobile digital media, for example mobile marketing, wireless advertising and m-advertising. With the ambition to clarify the meaning of m-advertising Tähtinen (2006) performed a conceptual analysis of the terms related to mobile commercial communication in practice and research. Drawing on the general definition of advertising by Richards and Curran (2002)*a paid, mediated form of communication from an identifiable source that is designed to persuade the receiver to take some action, now or in the future* (p. 74), an analysis on existing terms used to describe m-advertising and m-marketing was performed. In this analysis it was found that these terms are interrelated and not distinctly used (Tähtinen, 2006). In this chapter we refer to literature using all of these terms. As a result of the analysis Tähtinen (2006) suggest that the m-advertising features that differs from traditional advertising are interactivity of the media, personal nature of mobile devices, potential customization, and its ubiquity and fastness. Thus m-advertising enables sending unique, personalized and customized advertisements engaging consumers to dialogue and transactions with the advertiser.

The Mobile Marketing Association (MMA, 2008b) has published an industry glossary with definitions of terms related to mobile marketing targeted at both industry and academia. In this glossary the definition of the term mobile advertising is: *A form of advertising that is communicated to the consumer/target via a handset. This type of advertising is most commonly seen as a Mobile Web Banner (top of page), Mobile Web Poster (bottom of page banner), and full screen interstitial, which appear while a requested mobile web page is "loading". Other forms of this type of advertising are SMS and MMS, mobile gaming ads, mobile video ads (pre, mid and post roll.* (MMA, 2008a, p.21)

This definition refers to the targeted devices and the various forms of advertising rather than the potential associated with m-advertising. The potential value of m-advertising has been widely discussed in academic literature; one of the pioneers was Barnes (2002) who recognized the potential of customizing and personalizing market communication in mobile media. To advertisers examples of such values are the ability to target consumers (Salo & Tähtinen, 2005), the possibility to tailoring advertising (Komulainen *et al.*, 2006), interactivity with consumers (Salo & Tähtinen, 2005; Bauer *et al.*, 2005; Barnes, 2002) and increased reachability (Komulainen *et al.*, 2006). These values build on mobile devices being a personal, always on medium, uniquely identifiable and identification of location.

Thus, to leverage these values adaptations based on context data is required. Context awareness is an important part of m-advertising (Salo & Tähtinen, 2005; Komulainen *et al.*, 2006). However, understanding mobile context is challenging (Tamminen *et al.*, 2004). Research addressing context information as the source of adapting m-advertising in ubiquitous environments has addressed challenges and opportunities (see e.g. Ranganathan & Campbell, 2002; Yunos *et al.*, 2003) and technical solutions (see e.g. Bulander *et al.*, 2007; Kurkovsky & Harihar, 2006). In recent research conducted among experts of m-advertising, it was revealed that leveraging the potential advertiser values of m-advertising is very challenging due to consumers' divergent contexts (Vatanparast & Butt, 2009). Adapted advertising requires context data including personal information about interests, current location, movements etc.

On the other side of these values there are a number of concerns related to using context information to adapt advertising, for example privacy concerns (Vatanparast & Butt, 2009). With the ambition to personalize advertising messages there is a risk of intrusion in consumers' private spheres. As emphasized by Scharl *et al.* (2005) there is a tradeoff between personalization and privacy, and respecting consumers privacy is one key success factor for m-advertising (Vatanparast & Butt, 2009). As m-advertising targets individuals, it is perceived as personal, and could be apprehended as an intrusion of privacy. It is therefore recommended that a service provider collect consumers' permissions (Salo & Tähtinen, 2005), which is required by law in some countries (Scharl *et al.*, 2005). Another risk acknowledged by the Mobile Marketing Association is that m-advertising turns into SPAM, thus perceived as disturbing and unwanted by consumers. Disturbing and interruptive effects of advertising are also a risk with online advertising, these risks apply to all digital media. There are applications with different degrees of privacy on the internet as well as in mobile media. The personal e-mail inbox and the personal *SMS* inbox are far more private applications than internet applications, weather on PC or on the mobile.

What values are there then for consumers? Despite the concerns described above, there are also values perceived by consumers. The possibility of receiving targeted and personalized advertising (Salo & Tähtinen, 2005; Bauer *et al.*, 2005), at the time and location when needed (Haghirian *et al.*, 2005; Tsang *et al.*, 2004; Bauer *et al.*, 2005), brings value to consumers. According to MMA's Commitment to Consumer Satisfaction there are six fundamental elements to a positive consumer experience of m-advertising. The first is that m-advertising should only be sent on consumers' *choice*. Second, consumers should have the *control* over when and how they receive messages. Third, m-advertising should be *customized* to consumers' interests to make content as relevant as possible. Fourth, *consideration*, consumers should receive or be offered something in return for accepting the m-advertising such as a discount, entry to a competition etc. Fifth, m-advertising should be *constraint* with responsibility, and finally *confidentiality* which refers to the commitment of not sharing consumer information with non-affiliated third parties (MMA, 2008b). This is in line with the findings by the online market research company Harris Interactive who found that consumers accept m-advertising if there is a distinct value proposition, if the ads are relevant, and if the consumer is in control of what they get and how they are profiled (Harris Interactive, 2008).

There is some academic research regarding consumers' perceptions of m-advertising. In one study it was shown that the value of m-advertising is strongly related to the content and the frequency of the advertising (Haghirian *et al.*, 2005). In another, entertainment and information value were identified as the central acceptance drivers of mobile marketing (Bauer *et al.*, 2005). In a study investigating the more personal spaces of mobile media such as *SMS*, *MMS* and email, showed that these mobile channels are perceived to be highly personal creating high expectations for the relevance of marketing communication messages. In addition to the importance of relevant content, the consumers' context proved to be important, for example how and when the message is delivered (Heinonen & Strandvik, 2007).

Context Adaptation

A fundamental aspect of advertising in UME is adaptation to consumers' context. In the following, we derive themes for analyzing the challenges of balancing context adaptation of m-advertising leveraging advertiser and consumer value. The conceptualization of context draws on ubiquitous computing literature and advertiser and consumer value primarily on literature on m-advertising. The aim is an analytical basis that can provide a comprehensive understanding of challenges related to adaptation of advertising in future UME. This analysis will provide insights into the dynamics of balancing consumer and advertiser value of ubiquitous advertising.

In ubiquitous computing literature, context is a central concept that can be understood as the background and specific circumstances of a subject. Context information is any information that can be used to characterize a situation such as location, identity, state of people, groups, and computing resources (Dey, 2001). In this view, context is regarded as information related to the situation in which interaction occurs. Context can also be regarded as a relational property between objects and activities (Dourish, 2004). This means that context is dynamically shaped in action rather that pre-defined and stabile. Requiring and representing context information is very complex. The most commonly applied conceptualization in ubiquitous computing by Abowd and Mynatt (2000) suggest a minimum of five interrogatives necessary to comprehensively describe context; who, what, where, when and why. '*Who*' refers to the user's identity and possibly to the identity the people in the user's surrounding. '*What*' refers to the activity the user is engaged in. '*Where*' is about location and movements, '*When*' about time of use, and '*Why*' regards the use purposes. This is the conceptualization we have adopted in this chapter.

In this chapter we are discussing ubiquitous advertising, that means advertising anytime, anywhere, in any media, and adapted to users' context. We refer to this as ubiquitous advertising as it is not limited to mobile hand held devices but include all possible media recourses in the users' environment, stationary and mobile, in other words advertising in a UME. A motivation for this more inclusive view is that the distinctions between mobile devices and for example personal computers are blurring. With the advent of new mini PCs using 3G connections, large smart phones using wifi, the same type of services and applications available on all types of devices, and users expectations of being able to use stationary as well as mobile devices for the same purposes. We believe that making distinctions based on types of devices will not be as relevant when discussing the future ubiquitous advertising. Therefore, in this framework we primarily draw on literature on m-advertising as described above, but we adopt a more inclusive view of advertising not limited to mobile handheld devices. In addition we draw on values of m-adverting as discussed in m-marketing industry. In the following we summarize values discussed in literature. This is the basis for the analysis of how ubiquitous advertising can be adapted to context balancing consumer and advertiser values.

In table 2, a summary of values for advertisers and consumers are presented in relation to the concepts of context. This will serve as guidance in the analysis of the empirical material from the projects in newspaper industry. The analysis aims at an understanding of context adaptation of advertising content balancing user and advertiser values.

ADVERTISING IN UBIQUITOUS MEDIA ENVIRONMENTS

In this section we provide the publishers and advertisers reflections of adapting ubiquitous advertising to consumer context. We relate these thoughts to the themes derived in the literature

Table 2. M-advertising values, balancing risks and context concepts

Consumer value of m-advertising	Context adaptation and balancing risks	Advertiser value of m-advertising
Personalized offers Demanded advertising Control over personal profile	Who - is interested in and susceptible to the message/offer? Risk: Violation of personal integrity, security risks related to context data management	Targeting Willing receivers
Relevant message content Distinct value proposition Customized to individual interest Consumer control Consumer reward	What – advertising content is relevant to the targeted user? Risk: Irritation over irrelevant offers	Message exposure Interactivity Tracing and tracking information
Timely Consumer control Balanced frequency	When – is the targeted user susceptible and free enough to act on the advertisement Risk: over frequency perceived as SPAM	Reachability any-time Target at the right time
Suitable to media Relevant to location Consumer control	Where - is the targeted user susceptible? Risk: wrong location perceived as intrusive	Reachability any-where Location based Target at the right location
Fit with individual media goals Entertainment and information value	Why – does the advertising content match the consumer's intent? Risk: advertising perceived as non useful	Tailoring and customizing advertising

review aiming at an understanding of challenges associated with context adaptations of ubiquitous advertising balancing user and advertiser values.

WHO: The Identity

The most fundamental aspect of ubiquitous advertising is the opportunities to reach individuals and thereby come much closer to consumers (see e.g. Barnes, 2002; Salo & Tähtinen, 2005). However, to truly target individuals it is necessary to identify individuals rather than devices. When advertising in for example mobile phone, this is done by identifying a device and a mobile phone subscription. However, to leverage the potential values of ubiquitous advertising individuals need to be identifiable independently of media and device as illustrated with this quote:

You cannot focus on a device to reach your target group. If you want to reach teenagers during the Idol show, you must advertise online in their social community. As soon as the TV commercial starts they go online to discuss the show with their friends. During the TV commercial break you might reach their parents. (Advertiser)

Consumers would then be targeted cross-media and throughout their media day anytime and anywhere. Being targeted with personalized advertising that is relevant is one of the fundamental values from the consumer point of view (Salo & Tähtinen, 2005; Bauer *et al.*, 2005). On the other hand this means that consumers are targetable all the time and everywhere and it is important to respect the integrity of consumers and to make sure that the consumer is in control of the flow of advertising (Vatanparat & Butt, 2009; Scharl *et al.*, 2005; MMA, 2008a). According to this quote this is challenging to advertisers:

The most irritating advertisements are those that are not targeted, that is mass marketing and SPAM. Of course, it's not fun to receive ads from Via and Pampers if you aren't interested. We as advertis-

ers are so trained at thinking in mass-media, I mean that if we reach as many as possible some of them will be interested an most of them not. We will not be able to restrict the targeting enough. We are too inhibited when it comes to advertising. (Advertiser)

This implies that there really is a need for a balancing role if the consumers in a UME are to be willing receivers of advertising (MMA, 2008b) and thus more receptive to the advertising messages. The publishers recognize that this can be accomplished if the consumer is in control of their personal profile as shown by Harris Interactive (2008).

We have to develop our ability to target. This is what we can sell to advertisers in the future. If we can define relevant target groups based on knowledge about the consumers and as a result reach the right people with the right advertising and provide the means for interaction between consumers and advertisers, we have really added new value to advertising. (Publisher)

The communication with consumers would then be far more personal and interactive than in traditional advertising.

WHAT: The Activity

Information about consumer activities could be used to adapt advertising content. In traditional advertising the message is general to a wide audience, whereas in ubiquitous advertising the relevance of messages could be increased by adapting advertising to consumer actions (Vatanparat & Butt, 2009). Advertising content and format could be personalized to add value according to activities. For example, while cooking, recipes could be offered together with ads about food, kitchen utilities, kitchens, dining furniture, but not about drilling machines. Another example is illustrated by this quote form a newspaper publisher:

Your washing machine is broken. I start searching for washing machines, get some consumer tests together with advertising offers. I connect directly with the advertiser with the most attractive offer, place an order and the next day I have the washing machine delivered. (Publisher)

It is all about reaching people at the right moment and in the right mood. This means that the advertising would be adapted to consumers' actual situation. Targeting on a situation, like when you are looking for a house to buy is a well working model. Everybody wanting to buy or sell a house is interested. However, even if consumers try to avoid the advertising not perceived as relevant, some advertisers will try to push through anyway.

Nobody can avoid advertising. Then we only say: how are we going to reach this customer in another way? The company will talk to the consumers anyway. We will have to find other ways. Sure, we can try to sort and control as a consumer, but Pampers will not give up on parents of small children just because they have rejected advertising. (Advertiser)

Traditional advertising models are not selective enough since it also reaches people that are not interested. This could be an important part of the media house role in a UME.

WHERE: The Position

One of the most common context concepts discussed is making use of position to customize advertising, for example location based advertising (see e.g. Haghirian *et al.*, 2005; Tsang *et al.*, 2004; Bauer *et al.*, 2005). This might be even more interesting to consumers while on travel as this advertiser is visualizing.

Think to get a last minute ticket offer to a musical when you are in New York. That would be cool.

Yes, if I have given my permission. But I don't want them to know that I'm in New York. (Advertiser)

Another aspect of consumer position is increased reachability for advertising, to reach the consumers wherever they are. This fact also brings a risk of over exposing consumers with advertising and thereby causes irritation.

It is a delicate balance between advertising and disturbance, between surprise and irritation. If people are irritated they think SPAM – that does not work. (Publisher)

A mediating role coordinating advertising could help overcome this negative effect. Based on consumers individual acceptance level advertising could be channeled to consumers at the right location. The advertising format could be adapted to fit the location, for example the advertising message might be silent on the bus but maybe audio in the car. In traditional advertising in for example print newspapers, the audience in a specific area such as a town or a region is reached. In ubiquitous advertising it would be possible to reach the consumers independently of location.

WHEN: The Time

Based on context information, the best possible time to deliver advertising can be determined. Such context information may be based on time of day the consumers schedule etc (see e.g. Haghirian *et al.*, 2005; Salo & Tähtinen, 2005). One aspect of timing is adapting advertising to time of day, so called day-parting. For example, ads about lunch menus are not very interesting after 2 pm. Relevance is not only to target a specific time of day but also about holidays and vacation times, so called seasoning. Relevant information about the consumer could be used to customize offers according to for example Christmas, Ramadan, Mother's Day etc. However the right time can also be related to a situation, for example when the consumer is searching for something as this quote shows.

That is the Google model, when the user is there, when they are searching for something, and not when we think they want it. That is to say it is on the user's terms. That is really timing! (Publisher)

The back side of these opportunities is if consumers perceive the frequency of advertising to be interruptive as demonstrated by this quote.

If it is the right time too often consumers will not accept advertising. We must learn from TV viewers' behavior. (Advertiser)

In traditional advertising messages are exposed to an audience during a specific time-slot, such as a specific day in a morning paper or at a specific time in TV. Ubiquitous advertising could offer opportunities to present advertising messages in real time when users are interested.

WHY: The Purpose

Adapting advertising to fit with consumer use purposes and needs is probably one of the most challenging (Vatanparast & Butt, 2009). Discover consumer intent and to match the message with the consumer use purpose is very complex. Search based advertising is the model that comes the closest to this adaptation, but still how do you know *why* a consumer is searching for something?

Search based advertising is very interesting but how do you really know why somebody is searching for something based on one word. If I type the word sun in the search engine, I could be looking for tomorrows weather, for a charter trip to the sun or maybe I am searching for Sun Microsystems. (Advertiser)

To manage the fit between consumer goals and advertising offers, detailed information of

Figure 1. Ubiquitous advertising related to traditional advertising

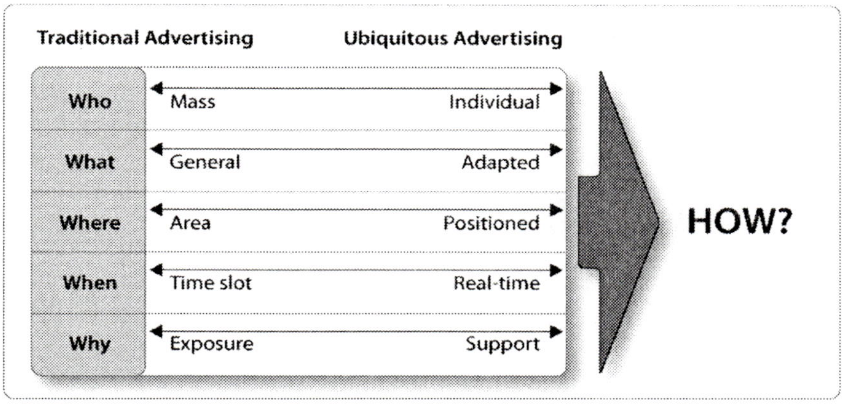

consumer preferences is required to support advanced interpretations of consumer media behavior. Furthermore, there must be a process of requesting or prioritizing advertising without predefined criteria. This is very challenging for media houses to achieve.

Maybe this could be a possibility if we can coordinate our efforts between the newspapers. In the future we will see more behavioral advertising. What you do decides what ads you get. (Publisher)

Thus, ubiquitous advertising could add value by supporting consumers use purposes, what the individual is trying to achieve, in contrast to the traditional model of exposure.

Summary

As shown true targeting of individuals with ubiquitous advertising rely on information about the consumer about all five context concepts. In figure 1, a summary of how ubiquitous advertising differs from traditional advertising is summarized.

These differences will have implications for media house strategies. While the advertising and pricing models of today largely build on traditional advertising as described above, media houses have to rethink and reorganize in order to be able to expand their value offers to advertisers with ubiquitous advertising integrating all their channels as recognized by this respondent.

We are very bad at using the strengths from each channel the best way and combine them for best effect. We should be able to use the fact that we have many channels as a huge advantage. But we do not do that today.

In addition, to be able to attract advertisers, new models for tracing and tracking advertising impact are needed as illustrated by this quote.

We really need to develop our methods for how to measure to produce targeting and tracing information. We really need to develop such techniques for the future. This is really a huge challenge and maybe the most important one because then we can communicate what we are selling to the advertisers.

The publishers and advertisers that participated in interviews and workshops reported in this chapter recognize that for ubiquitous advertising to be successful, content, format and distribution of advertising need to be adapted to be accepted and appreciated by consumers. For media houses to take a role in accomplishing consumer value of advertis-

Figure 2. Consumer and advertiser values, and media house role in UME

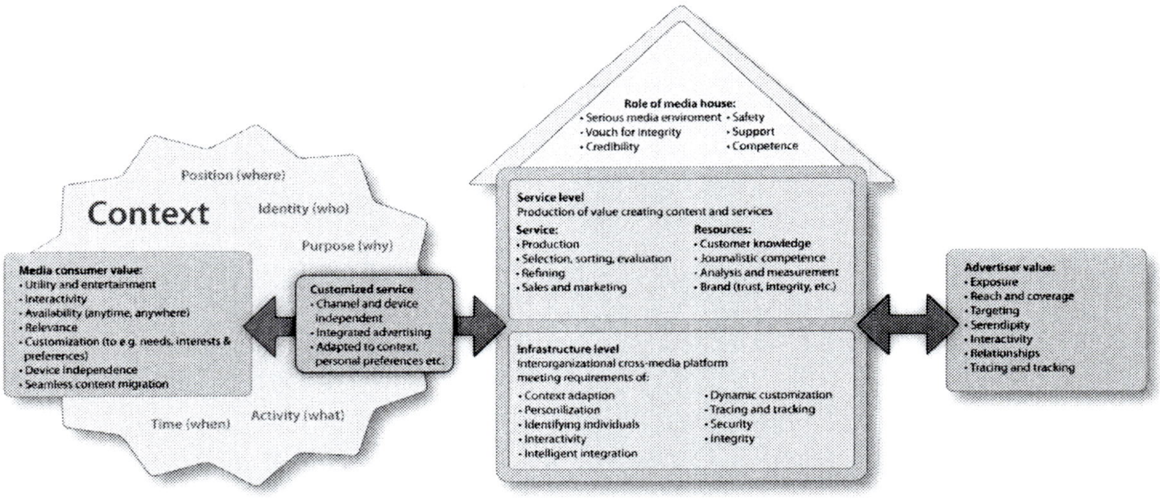

ing with context adaptations they have to rethink their relationships to advertisers and consumers. Their business would be much more relational and closer to individuals, in other words, a shift from mass-media to relational media. In the following we will discuss a scenario of the media house role in a UME and the associated challenges.

Media House Role in Ubiquitous Media Environments

To develop a role in UME that can balance advertiser and consumer value, media houses must be able to communicate the capability of handling context data. The media house role must vouch for individual integrity. Towards advertisers media houses need to offer a serious and trustworthy media environment. To be trustworthy, the supporting platform needs to be stable, secure and capable of intelligent and dynamic service and advertisement distribution.

Further, the opportunities of ubiquitous advertising adapted to consumer context also means that new values can be offered to advertiser customers. Apart from exposure, reach, coverage, targeting, and interactivity close relationship building can be offered in a serious and trustworthy media environment. In addition, a UME platform could offer new possibilities for detailed feedback to advertiser customers based on tracking and tracing data on how users react and act on advertising. Further, a UME offers possibilities of integrating relevant advertising in services, adding to service values. In other words, new possibilities for advertisers to contribute to and co-create consumer service value by supporting their use purposes are opened. In figure 2, the future role of media houses balancing consumer and advertiser values in UME is illustrated.

However, to develop the role as outlined in figure 2, a UME needs to be built on an interorganizational cross-media platform. A platform supporting a UME would apart from supporting a wide variety of mobile devices such as mobile phones, iPods, e-readers and other resources such as displays, wireless data access etc, be highly dependent on context information as described earlier. In our scenario of the media houses role in UME, this context data would be collected and managed by media houses. Context data together with the ability to identify individuals independently of device are hence the core resources in a UME. In turn, this is dependent on the ability to identify individuals. In light of the above we can establish that there are a number of challenges that media houses need to address in order to develop their role in UME.

Challenge 1 - Context adaptation: In what ways can advertising be integrated with services/content adjusted to context adding to consumer relevance?

Managing context data challenges the role of media houses dramatically compared to the role they have in traditional media. Adapting services and advertising to context information is of course very complex. It requires integration between systems such as publishing systems and advertising systems in and between organizations, and integrated technologies for intelligent and dynamic context adaptations. It would also require a very sensitive approach to consumer reactions to advertising adjusted to context.

Challenge 2 - Real-time advertising adjustment (and channeling) to consumer behavior and preferences: How to adjust advertising to consumer preferences, i.e. advertising on consumer terms?

Media houses would have to balance the interests of advertisers with user's integrity and preferences. Not only are there different acceptance levels for advertising in different media channels and different advertising formats, but also different acceptance levels between different situations and between individuals. These aspects are also part of the adaptations that need to be made of advertising. When advertising is integrated with editorial content in new ways this challenges the relationship between users, advertisers and media houses. For media houses to keep the respectability in their relations it is of utmost importance to develop models to balance advertising on user terms.

Challenge 3 - Dynamic data exploitation: How to develop models to produce and make use of consumer information dynamically over time in a ubiquitous media environment vouching for integrity and security.

Advertising in a UME requires huge amounts of data need to be managed in real-time. In addition, data need to be stored and managed to customize services as well as to analyze user behavior. In turn, this requires advanced integrity and security capabilities. This means that media houses would have to strengthen their relationships and build stronger value networks among themselves.

Challenge 4 - Consumer-advertiser relations: How to create an arena for building relations between users, users and advertisers as well as between advertisers?

A UME needs to support a very high level of interactivity. When advertising becomes more interactive and relationship building, there is also a need for structures to uphold relationships over time. Relationships might also be between consumers and between advertisers. If consumers can build communities as consumers within the media environment they can also communicate with advertisers in more powerful and influencing ways. Further, advertisers can be more adaptive to consumers and build relationships to create joint offers. Further, media houses can host interactive consumer forums with many different but related firms offering support to consumer needs. Thus media houses role is far more relational and would require new competences.

Challenge 5 - Advertising business models: How to develop business models that leverage advertiser value in ubiquitous media environments.

Ubiquitous advertising would demand totally new business models. In traditional advertising, the revenue models are built on selling a space and place or a time-slot to expose an advertising message to users. In a UME this would be impossible. New revenue models might instead be based on actual targeting effects, user reactions and actions or even on actual impact of for example sales. In addition, the business offers would include production of advertising content to a larger extent than today. Further, detailed feed-back and analysis of advertising and market analysis are new business opportunities for media houses to offer advertising customers.

Concluding Remarks

In this chapter we have addressed the question of how ubiquitous advertising challenges the role of media houses in UME. We did this by deriving themes from a literature review as guidance for interpreting publisher and advertiser reflections and ideas about future advertising. By this analysis we have established that the media house needs to be extensively developed to leverage the values of ubiquitous advertising.

When new mobile media enter the media market this does not only challenge media houses to innovate services but definitely also compel a need to innovate advertising offers. As we have shown, the need for innovating advertising is as important as exploring new services to media consumers. This is not limited to the efforts of publishing an e-newspaper, the need for innovating advertising is related to all channels integrated in a UME. Making use of the opportunities offered in a UME and pursuing a development towards ubiquitous advertising might be one path to innovate traditional newspaper media.

However, this development is not only challenging to media houses in general, it also challenges the existing value networks in the newspaper industry. Leveraging the values of ubiquitous advertising would expand the relationships and exchanges between media houses and advertisers. Further, to be able to leverage the advertiser values, the relationships between media houses would change to a more open network, exchanging knowledge, technical solutions and customer relationships. A single media house is not capable of building the platforms required, an interorganizational effort is needed. Whether the newspaper industry with its long tradition is capable of this change is still to be seen.

The development towards UME is not limited to the newspaper industry. The concepts described in this chapter are applicable to other industries, content providers and distributors of services. One example is the mobile service market in telecom infrastructure offering services to mobile phones. In this chapter we have discussed ubiquitous advertising challenges from a media house perspective, suggesting the media house as a suitable and trustworthy host for the context information needed in a UME. However, similar visions have been produced by for example DoCoMo and Nokia. There is no question about the digitalization towards ubiquitous advertising, it will become a reality in due time, in one form or another. The question is who is capable of taking the role of the spider in the value networks of UME.

REFERENCES

Abowd, G. D., & Mynatt, E. D. (2000). Charting past, present, and future research in ubiquitous computing. *ACM Transactions on Computer-Human Interaction*, *7*(1), 29–58. doi:10.1145/344949.344988

Åkesson, M., & Ihlström Eriksson, C. (2008). From multi channel publishing towards ubiquitous media environment. *Journal of Graphic Technology*, *10*, 126–148.

Barnes, S. J. (2002). Wireless digital advertising: Nature and implications. *International Journal of Advertising*, *21*, 399–419.

Bauer, H. H., Barnes, S. J., Reichardt, T., & Neumann, M. M. (2005). Driving consumer acceptance of mobile marketing: A theoretical framework and empirical study. *Journal of Electronic Commerce Research*, *6*(3), 181–192.

Bulander, R., Decker, M., Schiefer, G., & Kölmel, B. (2005). Comparison of different approaches for mobile advertising. In Proceedings of the International Workshop on Mobile Commerce and Services (WMCS'05).

Christensen, C. M., & Davis, A. B. (2006). The future of newspapers. Retrieved September 29, 2008, from http://www.forbes.com/home/leadership/2006/10/10/leadership-newspapers-media-lead-innovation-cx_cc_1011clayton.html

Dey, A. K. (2001). Understanding and using context. *Personal and Ubiquitous Computing*, *5*, 4–7. doi:10.1007/s007790170019

Dourish, P. (2004). What we talk about when we talk about context. *Personal and Ubiquitous Computing*, *8*(1), 19–30. doi:10.1007/s00779-003-0253-8

eMarketer. (2008). E-marketer's key predictions for 2009. Available at http://www.emarketer.com/Article.aspx?id=1006818

Haghirian, P., Madlberger, M., & Tanuskova, A. (2005). Increasing advertising value of mobile marketing–an empirical study of antecedents. In Proceedings of the 38th Annual Hawaii International Conference on System Sciences, Hawaii.

Harris Interactive. (2008). Available at http://www.harrisinteractive.com/news/newsletters.asp

Heinonen, K., & Strandvik, T. (2007). Consumer responsiveness to mobile marketing. International Journal of Mobile Communications archive, 5(6), 603-617.

Ihlström Eriksson, C., & Åkesson, M. (2008). Ubiquitous advertising Challenges. In Proceedings of 7th International Conference on Mobile Business, (ICMB´08), Barcelona, Spain.

ITU. (2008). International Telecommunications Union statistics. Available at http://www.itu.int/net/home/index.aspx

Kohda, Y., & Endo, S. (1996). Ubiquitous advertising on the WWW: Merging advertisement on the browser. *Computer Networks and ISDN Systems*, *28*, 1493–1499. doi:10.1016/0169-7552(96)00070-0

Komulainen, H., Ristola, A., & Still, J. (2006). Mobile advertising in the eyes of retailers and consumers–empirical evidence from a real life experience. In Proceedings of the 5th International Conference on Mobile Business, (ICMB'06), Copenhagen, Denmark.

Kurkovsky, S., & Harihar, K. (2006). Using ubiquitous computing in interactive mobile marketing. *Personal and Ubiquitous Computing*, *10*, 227–240. doi:10.1007/s00779-005-0044-5

Merisavo, M., Kajalo, S., Karjaluoto, H., Virtanen, V., Salmenkivi, S., Raulas, M., & Leppäniemi, M. (2007). An empirical study of the drivers of consumer acceptance of mobile advertising. Journal of Interactive Advertising, 7(2). Retrieved on September 29, 2008, from http://www.jiad.org/article92

MMA. (2008a). MMA's commitment to consumer satisfaction. Available at http://www.mmaglobal.com/consumer

MMA. (2008b). Mobile marketing glossary, version 2008.12. Available at http://mmaglobal.com/uploads/glossary.pdf

NNA. (2008). Newspaper association of America-advertising expenditure. Available at http://www.naa.org/TrendsandNumbers/Advertising-Expenditures.aspx

Okazaki, S. (2004). How do Japanese consumers perceive wireless ads? A multivariate analysis. *International Journal of Advertising, 23*(4), 429–454.

Okazaki, S. (2005). Mobile advertising adoption by multinationals. Senior executives′ initial responses. *Internet Research, 15*(2), 160–180. doi:10.1108/10662240510590342

Ranganathan, A., & Campbell, R. H. (2002). Advertising in a pervasive computing environment. In Proceedings of the 2nd International Workshop on Mobile Commerce. ACM Press.

Richards, J. I., & Curran, C. M. (2002). Oracles on "advertising": Searching for a definition. *Journal of Advertising, XXXI*(2), 63–77.

Salo, J., & Tähtinen, J. (2005). Retailer use of permission-based mobile advertising. In Clarke III I & Flaherty TB (Eds.), Advances in electronic marketing (pp.139-156). Hershey, PA: Idea Publishing Group.

Scharl, A., Dickinger, A., & Murphy, J. (2005). Diffusion and success factors of mobile marketing. *Electronic Commerce Research and Applications, 4*, 159–173. doi:10.1016/j.elerap.2004.10.006

Tähtinen, J. (2006). Mobile advertising or mobile marketing. A need for a new concept? Frontiers of e-business Research, 1, 152-164.

Tamminen, S., Oulasvirta, A., Toiskallio, K., & Kankainen, A. (2004). Understanding mobile contexts. *Personal and Ubiquitous Computing, 8*, 35–143. doi:10.1007/s00779-004-0263-1

The Economist. (2006). Who killed the newspaper? Retrieved September 29, 2008, from http://www.economist.com/opinion/displaystory.cfm?story_id=7830218

Tsang, M., Ho, S., & Liang, T. (2004). Consumer attitudes towards mobile advertising: An empirical study. *International Journal of Electronic Commerce, 8*(3), 65–78.

Vatanparast, R. (2007). Piercing the fog of mobile advertising. In Proceedings of the 6th International Conference on Mobile Business (ICMB07′), Toronto, Canada.

Vatanparast, R., & Butt, A. H. (2009). Factors affecting use of mobile advertising: A quantitative study. In Proceedings of the 42nd Hawaii International Conference on System Sciences (HICSS42), Hawaii, USA.

WAN. (2008). World press trends. Available at http://www.wan-press.org/worldpresstrends/home.php

Wind, J. a&Mahajan, V. (2000). Digital marketing: Global strategies from the world's leading experts. John Wiley & Sons.

Yunos, H. M., Gao, J. Z., & Shim, S. (2003). Wireless advertising's challenges and opportunities. *Computer, 36*(5), 30–37. doi:10.1109/MC.2003.1198234

ZenithOptimedia. (2008). Advertising expenditure forecast December 2008. Available at http://www.zenithoptimedia.com/gff/pdf/Adspend%20forecasts%20March%202008.pdf

KEY TERMS AND DEFINITIONS

Context Adaptation: Adapting advertising to consumer context

Context: A central concept in ubiquitous computing literature that can be understood as the background and specific circumstances of a subject

Mobile Advertising Market: The market for mobile advertising

Mobile Advertising (m-Advertising): A form of advertising that is communicated to the consumer/target via a mobile channel

Ubiquitous Advertising: Advertising anytime, anywhere, in any media, and adapted to users' context.

Ubiquitous Media Environment: A future media environment supporting device independent, anytime, anywhere publishing

Chapter 6
Segmentation Challenges Posed by 'Transnationals' in Mobile Marketing

Ibrahim Sirkeci
European Business School London, Regent's College, UK

Richard Mannix
European Business School London, Regent's College, UK

ABSTRACT

Mobile technologies have posed new challenges for marketers as well as opportunities in an increasingly global market with high human mobility within and across borders. To understand and address customer needs more effectively, this chapter proposes to add new variables to market segmentations formula such as change of residence, movement and commuting distance/hours. A discussion of 'mobile nationals', and 'transnationals', as segments, is undertaken. The transnationals segment includes immigrants, refugees, tourists as well as businesspeople and professionals whose careers span into several countries. These groups create multiple reference points, which are likely to determine their characteristics and behaviour. This is an emerging and promising customer segment particularly for mobile marketing and mobile services. To explore the viability of such a segment, we have made use of several existing theoretical frameworks and concepts of segmentation. Future research should focus on the identification of transnational and national mobile segments while also developing and fine tuning the new variables –movement, change of residence and commuting for segmentation theory.

INTRODUCTION: MOBILE MARKETING IN RESPONSE TO HUMAN MOBILITY AND MOBILE TECHNOLOGIES

Mobile marketing is defined as "the use of wireless media as an integrated content delivery and direct response vehicle within a cross-media or stand-alone marketing communications program" by Mobile Marketing Association (MMA, 2008, p.24). Writing in what is arguably claimed to be the first book exclusively devoted to exploring the potential of mobile marketing, Haig underlined the future potential of this form of marketing and discussed the possible availability of mobile broadband while SMS itself (the short message service)

DOI: 10.4018/978-1-60566-074-5.ch006

was still seen as a revolutionary marketing idea (Haig, 2002, p.7). Haig outlined the rapid development in this area and made predictions about the pace of change by arguing that text messaging was still in its early infancy as a marketing tool in 2000 but had developed significantly between then and 2002 (Haig, 2002, p.12). Barnes pointed out the importance and the rich potential of location based services in the development of mobile marketing (Barnes, 2003). The key areas for marketers to consider while this relatively new marketing tool is being established are obtaining permission from the recipient and acceptance as well as attributes inherent to mobile marketing such as personalisation, location based, time based, interactivity, and ubiquity (Bauer et al., 2005; Barnes and Scornavacca, 2004). This area is rapidly developing in response to changes in global communication behaviour, mobile needs and ever-improving mobile technologies. In this paper, following a brief background on mobile marketing, we are going to elaborate on segmentation theory in relation to mobile populations; and delineate on the "mobile nationals" concept and develop it to introduce [mobile] "transnationals" along with the challenges for segmentation that this emerging segment poses.

Mobile marketing, encounters the same fundamental challenges as other forms of marketing namely targeting and segmentation. Customers and clients using mobile technologies but still settled and identifiable by their relatively fixed and stable geographical location would, perhaps, present few difficulties for marketers to overcome. A more complex challenge comes from increasing human mobility: locally, nationally, and internationally. Although the total number of people who live outside their country of birth (i.e. immigrants) is around 200 million, people are increasingly mobile (e.g. overseas students, transnational professionals, business people, and tourists). This on the one hand contributing to the need for mobile services, and on the other posing new challenges for marketers regarding segmentation and targeting as *the movement* needs to be included as a segmentation variable.

From the first mobile phones to the rich array of portable multimedia devices available today, we have reached to the era of ubiquitous computing, which is characterised by "changing relationship (one user–many computers), pervasiveness, invisibility and mobility of computing nodes" (Valavanis et al. 2003, p.2). Thus, notebook computers with wireless internet connectivity enabling access to web sites as well as handheld devices such as PDAs or mobile phones, which are often equipped with GPS services, are rapidly becoming mainstream items. GPS technologies are providing "location information and maps, but more importantly used in smart rooms and buildings to offer internet connectivity and location-specific services to their users" while posing "research challenges, such as ubiquitous connectivity, heterogeneity of sources, effective and precise resource discovery" (Valavanis et al. 2003, p.2).

"Consumers and businesspeople no longer need to be near a computer to send and receive information." (Kotler and Keller 2008, p.477) Devices such as mobile phones and personal digital assistants (PDA) enable increasingly interactive and quick methods of communication, thus greatly improving marketing opportunities. Mobile marketing is one of the fastest growing areas of the marketing function, where new challenges and opportunities are constantly emerging. Mobile commerce (*m-commerce*) is predicted to have an ever larger role in our business dealings (Kotler and Keller 2008, p.477). The use of various mobile devices has changed the daily routines of hundreds of millions of people in an example of what consumer behaviour theorists would call discontinuous innovation (Schiffman, Kanuk and Hansen, 2008, p.438).

Bauer and colleagues (2005) list "personalization, ubiquity, interactivity and localization" as the key characteristics of mobile marketing. *Localization* "The development of positioning technologies has led many experts to forecast a promising future for mobile commerce applications. Location based services (LBS) are predicted to become the "killer application" of mobile commerce" (Kölmel 2003, p. 88 cited in Bauer et al. 2005). Barnes (2003) earlier underlined LBS as an emerging marketing tool and today with smart phones and GPS services we have seen a significant expansion of them. Knowing the position of the user enables the provider to pre-select services, or customise them to satisfy customers and clients. Using geo-location technologies marketers can adapt the marketing offer to the customer's position (Barnes 2003). Barnes (2002) categorizes these as push- and pull-models. In the former, the marketer can inform the customer about relevant product offers, for example, at the point-of-sale to tempt them for impulse purchases. The latter has developed to be relatively wide spread. It requires a demand from the customer but yet can lead to customised offers from the current position.

As Barnes and Scornavacca (2004) explain, mobile marketing is based on obtaining permission from the message recipient. Thus "permission marketing" addresses one of the ills of marketing, spam in new media communication, as it demands explicit consent from the receiver. We can relate this to the *interactivity* attribute of mobile marketing. After push- and pull-models, a third type is defined based on long-term and more intense interaction with the customer: *mobile dialogue* (Jelassi and Enders, 2005:88).

Personalization is aimed at convincing the receiver of marketing communications to perceive them as valuable information services (Barnes and Scornavacca 2004) as many consumers despise mass advertising as bothersome "interrupt marketing" and reject the messages (Godin, 2001 cited in Bauer et al. 2005). This kind of customisation is likely to reduce negative reaction among customers (Barnes 2002). As personal objects, mobile phones allow marketers to address specifically the needs of the person targeted while recognising the context and personal characteristics as well as location (Barnes, 2002, Barnes and Scornavacca, 2004, Jelassi and Enders, 2005).

Ubiquity relates to this personalisation and location as individuals with their personal mobile devices can be reached virtually *anywhere* and *everywhere*. We give a brief account of the high global penetration of mobile communication devices, which is an indicator of viability of this type of reach for mobile marketing. A mobile device such as mobile phone is almost always attributable to one single person allowing for highly personalized marketing measures. As Bauer and colleagues (2005, p.182) put "using the mobile medium for communication also enables the advertiser to contact potential customers anytime and anywhere. Mobile phone users typically have their device with them at all times. Furthermore the SIM (Subscriber Identity Module) card allows for the exact identification of each mobile phone and its user."

The number of mobile device users and service subscribers has rapidly grown over the last decade. This coincides with increasing human mobility globally. Wide spread availability and the use of the internet have expanded the tools available to marketers without much changing the fundamental principles of the marketing function. Thus we still have to deal with the basic marketing problem of identifying and addressing human needs and wants appropriately.

As seen in Table 1, according to International Telecommunications Union, a United Nations agency, the total number of mobile phone subscribers has increased from 1.57 billion in 2002 to 3.35 billion by the end of 2007 making mobile communication accessible to half the world's population (ITU, 2007). Nevertheless, in advanced countries usage of mobile services expanded and varied to include more and more picture and video messaging as well as increased use of the internet from mobile devices. The European Union exhibits the highest mobile penetration rates (Table 1). In the UK, for example, it is not only SMS but MMS (video and picture messaging) that has rapidly

Table 1. Mobile subscription rates, 2002-2007

	Number of Subscribers			
	2002 (000s)	*2007 (000s)*	*Change (%)*	*Per 100 inhabitants*
Africa	36 923.8	274 067.8	49.3	28.44
Americas	255 451.3	656 918.1	20.8	72.21
of USA	*141 800.0*	*255 395.6*	*12.5*	*83.51*
Asia	443 937.4	1 497 253.5	27.5	37.64
Europe	405 447.7	897 515.7	17.2	111.26
of EU	*341 209.9*	*569 521*	*66.9*	*117.04*
Oceania	15 458.9	27 011.4	11.8	79.17
World	1 159 221.1	3 354 773.5	23.7	50.10

Source: ITU World Telecommunication/ICT Indicators Database (ITU, 2007).

expanded. According to recent figures from the UK Mobile Data Association, the number of MMS messages sent per month reached 45 million in addition to the 6.5 billion text messages sent per month in 2008, while mobile internet usage has grown 25% to 16.4 million in the last two years (MDA, 2008).

Covey (2008) also provides a summary of mobile internet usage among mobile phone users. The USA is leading this market with about 40 million (15.6%) subscribers using the service alongside with 96 million others who have subscribed but do not necessarily use it (Covey, 2008). This represents over a 70% increase from 2006. Regarding significantly high internet penetration rates in some other large countries (e.g. over 50% in most European countries, thus larger than the US in total), one can recognise the enormous size of the current global mobile internet users market segment which is very likely to include some of the 'transnationals' group which we are going to discuss later in the paper. The differences between different parts of the world may offer distinct opportunities. For example, the African continent, which may have been seen as an initially unappealing market has been a target for mobile companies offering mobile banking to nearly a billion Africans who suffer from lack of bank branches in most parts of the continent (Wray, 2008). Due to rapid and constant technological improvements, various expansion opportunities and expectations in mobile services (Shiels, 2008), we appear to be entering a golden age for mobile marketing.

In this chapter, we will focus on one broad, yet particular customer segment that is heavily dependent on mobile communication technologies: internationally mobile people. This chapter will be divided into three sections. The first part will review segmentation theory and relate it to mobile marketing. In the second section, an understudied concept, "mobile nationals", will be discussed to pave the way for mobile transnationals, an emerging segment to be explored. Transnationals will be described in tandem with the framework set out in the first section earlier. Finally, we pose some questions regarding the segmentation of mobile customers and indicate possible future directions for research in this field.

AN EXAMINATION OF THE SEGMENTATION VARIABLES AND SUCCESSFUL SEGMENTATION CRITERIA IN MOBILE MARKETING

Segmentation as a concept was introduced in the 1950s and with the evolution of market-oriented thought in firms, it became a central concept in both marketing theory and practice (Wedel and Kamakura, 2000). Smith (quoted in Wedel and Kamakura, 2000, p.3) defined the concept in 1956 as a process that "involves viewing a heterogeneous market as a number of smaller homogeneous markets, in response to differing preferences, attributable to the desires of consumers for more precise satisfaction of their varying wants." His definition still retains its value and this principle is extremely important from the firm's perspective. As Kotler et al. (2008) argue companies have varying resources and competencies which lead to differing abilities in the potential to satisfy customer needs and wants. As they state, 'a company must identify the parts of the market that it can serve best and most profitably. It must design customer-driven marketing strategies that build the *right* relationships with the *right* customers' (Kotler et al, 2008, p.410). New technologies provide new channels and richer information which can be directly accessible by marketers enabling micro-marketing. At the same time, the increasing globalisation of both markets and firms is leading to a global-oriented approach to marketing which targets consumers that cut across geographic and cultural boundaries. These issues have led to a questioning of the appropriateness of traditional approaches to segmentation in this new era marked by the rise in the importance of the internet (Day and Montgomery, 1999).

With the decline of traditional mass marketing, many organisations have begun to further explore niche markets, local areas and even marketing to individual consumers. This focus on targeting customers within a very narrow segment is termed *micromarketing* (Blythe, 2008, p.358). Identifying and understanding segments in this way is of increasing importance. Common variables for segmenting markets are outlined in many standard texts on the subject of marketing (Brassington and Pettitt, 2006, Kotler et al 2008, Schiffman and Kanuk, 2007). These include; demographic (gender, age, income education and occupation) geographic (nations, regions, states, countries, cities or neighbourhoods) psychographic (personality and lifestyle choices) behavioural (use-related, use-situation and benefits sought) and hybrid systems (examples being geodemographic systems such as ACORN). Using a single segmentation base to explore the segments for mobile marketing, such as a traditional demographic approach focusing on age, gender, income, education and occupation will certainly yield some useful results but choosing a more encompassing view of the customer would provide better insights into this particular market.

When considering segmentation and mobile marketing, researchers might approach the issue by examining a number of different segmentation base concepts in conjunction with the established criteria for successful segmentation. Hill and O'Sullivan (2004, p.165) rate segments or segmentation schemes according to four key criteria; whether the segments size, purchasing power, and other characteristics are measurable, large and profitable enough, possible to reach and serve the segment effectively, segments are distinguishable in terms of their responses to the marketing mix and whether it is possible to create appropriate marketing programmes to attract and serve the segments. This follows the framework developed by Wedel and Kamakura (2000, Ch. 2). They argue that the foundations of good segmentation lie in the ability to successfully apply certain viability, or feasibility, criteria to the potential market segments that the marketer has identified.

The six criteria are as follows. *Identifiability,* or the extent to which a particular segment can be distinguished from other groups in the market place and accurately measured. *Substantiality,* or size of the market segment and its ability to generate profits for the enterprise. This is contextual to the market and the objectives of the firm, so a relatively small group of high spending customers could generate enough revenue to keep a business viable and indeed profitable. *Accessibility,* or the ease by which a marketer can

Table 2. Segmentation variables matrix

	General	**Product-Specific**
Observable	Cultural, geographic, demographic and socio-economic variables	User status, usage frequency, store loyalty and patronage, situations
Unobservable	Psychographics, values, personality and life-style	Psychographics, benefits, perceptions, elasticities, attributes, preferences, intention

Adapted from Wedel and Kamakura, 2000, p. 7.

communicate with the customer and distribute products or services to them. *Stability,* or the extent to which the market remains constant and sure while a marketing strategy and tactics are being planned and implemented. *Responsiveness,* or the degree to which segments respond uniquely to the marketing mix. If two segments respond in exactly the same way to the firm's marketing efforts then it should be argued that they should not be categorised into more than one segment. The final criterion is *Actionabilty*, or the extent to which the segment provides guidance for the firm on the marketing mix it should use and the ability of the firm to effectively satisfy the needs of the target segment profitably.

Wedel and Kamakura take the conventional segmentation variables (demographic, geographic, behavioural, psychographic and hybrid) and reclassify them in a matrix system as illustrated in Table 2.

The authors explain that *General* bases of segmentation focus on the characteristics of the customer, regardless of the product category. Age as a factor in demographic segmentation would be an example of this. In contrast, *Product-Specific* segmentation looks at the relationship between the customer and a specific product. An example of this would be the behavioural variable of usage frequency, or how often a consumer uses a product in any given time period. *Observable* variables are those that can be easily and directly measured such as the age of customers or their geographical location. *Unobservable* variables are those which are less easy to measure in a direct, quantitative way and may well need to be inferred. These would include psychographic issues such as lifestyle choices and personality characteristics as defined by Freudian, Neo Freudian or trait theory (Schiffman and Kanuk, 2008). They would also include behavioural issues such as consumer perception and benefits sought.

When initially segmenting markets and then subsequently choosing which segments to target, Wedel and Kamakura suggest that marketers may combine one or more of the elements listed in the segmentation variables matrix (Table 2) and examine their viability through the lens of successful segmentation criteria previously mentioned. When considering potential groups that may be open to the marketing of goods and services through mobile devices one might certainly suggest that product-specific variables be examined as the mobile device is a focal point here, whether simply as a portal for the communication of messages about other products and services, or as an integral aspect of the service being provided, for example, Location Based Services (LBS) such as navigational tracking and aid (Barnes, 2003). Both observable variables (such as usage frequency and usage situation) and unobservable characteristics (such as perception and benefits sought) might be employed. The question is to build effective and useful segmentation schemes, which can be applied to today's context of mobile marketing, perhaps within a *needs-based segmentation approach*, where marketers combine several variables to identify smaller, better-defined target groups (Greengrove, 2002).

In terms of the criteria for successful segmentation, there are a number of issues to consider.

Firstly, in terms of *identifiability* marketers should be able to both identify and measure segments of customers open to mobile marketing from an observable aspect as they have electronic records of usage and device movement and location. Unobservable factors are harder to infer here although caller destinations might provide some clues. This does of course raise ethical issues however, as the even the unobtrusive tracking of mobile device users by service providers raises major ethical concerns and meets harsh criticisms from users (BBC, 2007). In terms of *substantiality*, the market should prove a lucrative prospect for firms as shown by the figures (see Table 1). *Accessibility* should also prove a fairly minor obstacle to marketers although legislative and/or self regulatory action are needed to prevent saturation and customer rebellion (DeZoysa, 2002, Barnes and Scornavacca, 2004). *Stability* may well be a major issue for marketers as some groups of customers may change their contact details (both phone numbers and web addresses) relatively frequently and this could present a problem in frequently updating databases. *Responsiveness* may also be an issue as marketers may find a trial and error process occurring in how segments respond in similar or different ways. A certain amount of counter segmentation, or the combining of a number of smaller segments into fewer, larger ones may take place in the early stages of segmentation. Finally, the segments being examined should be *actionable* as marketers could well track user preferences in order to help them identify products and services they might also wish to purchase, or purchase more of for example. Only segmentation schemes that satisfy the key criteria explained earlier should be implemented. Careful consideration of the alternatives is vital before any decision to commit resources is made. Mobile populations present an added challenge as due to their possibly different routines which marketers may not be familiar with as it is the case for 'settled' customers.

Digital technology greatly improves the ability of marketers to construct and update customer databases. They can more readily track the behaviour of customers through the use of direct marketing techniques. Businesses are much more aware of how effective promotional messages have been as 'log on' and 'click through' of customers have been recorded. This allows for a much greater level of relationship building with mobile customers and also enables marketers to target smaller, more focussed groups of customers that are likely to respond to the business's message. This delivery of more customised messages to smaller segments has been termed "narrowcasting" by some authors (Waterman, 1992; Smith-Shomade 2002). Applying psychographic analysis including activities, interests and opinions would also improve our knowledge of the various consumer types.

The rise of niche marketing and so the need for improved segmentation has led to creative thinking. For example, Chris Anderson (2006) defined *the Long Tail* marketing era as the one where blockbuster marketing ended as markets fragmented into countless niches, a proposal virtually underlining the importance of detailed and complex segmentation in today's marketing environment. The Long tail denotes that success is not coming from the core product/benefit but from the numerous niche offerings based on the core benefit/product.

The bases for segmenting markets classify the major variables we already use in recognizing customer differences. Increasing adoption of mobile marketing by organisations and firms influences the use of these categories. Geographic variables within the mobile marketing context become more fluid, for example. Customers need to be defined not only by their location but also by their travels, spatial movements, or by their mobile existence in space (e.g. commuting distance or commuting hours depending on the context whether it is local, national or transnational). Such fluidity influences other variables too. For example, *cosmopolitans* are defined as immigrants and descendants of multicultural backgrounds with big families in multiethnic neighbourhoods are such a segment in

US (Claritas, 2008). When immigrants or mobile populations became part of a segment, mobility or movement becomes an important variable. The difficulty here is the fact that mobile people (migrants and commuters alike) are likely to have multiple reference points guiding their behaviour.

According to the United Nations, migrants are defined as those who have changed their place of residence from one country to another for a period longer than twelve months. However, internationally the mobile population is much larger than that. For example, professionals and business people working in multiple locations across borders, as well as people living in borderlands and earning their lives on the 'foreign' side of the border while living on the 'home' side would fall into this grouping. A broader definition still would add tourists and travellers to this segment. The demographics of these groups are likely to be significantly different than the more sedentary populations. For example, the family structures, relationships, gender roles, income, and life stages all come with multiple references: country of origin, country in which the person works, occupational class, place of residence all become more difficult to measure and analyse with the certainties of sedentary populations.

Cultural and national differences have been frequently referred to in the literature as well as being used in practice. A rich array of bases benefiting from a wealth of variables (age, class, ethnicity, occupation, lifestyles, opinions, values, attitudes, and so on) is used in many studies with different focuses (Usunier and Lee, 2005, p.232). However, our argument here is that these differences may not be of much help when the target population is on the move. The main dilemma lies in understanding the characteristics of the customers' multiple reference points: in the country of origin, in the country of transit and destination(s).

In 2001, Louvieris and Driver claimed that marketing success in cyberspace depends on understanding IP address in the same sense that marketers can interpret postcodes or zip codes. They hoped to find geo-demographics and related lifestyle and behavioural variables through the fixed IP code that give the virtual address details of the internet user. Their Classification of Internet People (CLIP) can be defined as an adaptation of geo-demographics to the internet which could be further enhanced by additional life style variables while also improving the emerging cyber-demographics as a new basis for market segmentation. They propose an alternative conceptualisation of cyber segmentation taking loyalty level and decision type as key parameters. However, there are various problems with this. Firstly, this claim precedes the recent improvements in wireless services which do not need a fixed residence. Secondly, IP addresses are not able to identify the individuals within a household or other entity using the same internet connection.

Nevertheless, challenges and novelties, as proposed by Louvieris and Driver (2001) should also be taken into account. One should not forget the fact that successful segmentation also needs creativity, intuition and educated guesswork as well as robust scientific methods. Transnationally mobile customers (from here onwards, we will call them 'transnationals'), perhaps, pose such a challenge for marketers.

RELATING MOBILE MARKETING SEGMENTS TO MOBILE CUSTOMERS

Case: Axis Mobile

Mainly focusing on behavioural and psychographic variables, Axis Mobile, a leading global firm in consumer mobile email solutions, refers to three segments for their consumer mobile email service:

Professional Super Parents

In developed countries there are a group of well educated, affluent, 30-45 year old parents with very busy lifestyles. With both parents in the workforce or with one parent absent (in a single parent scenario), after school hours are complex and hectic; thus the coordination of extracurricular activities is often done via email. Many parents even have a dedicated email account for child– related messages. At the same time, these parents are dedicated professionals, and often need to get work-related email after hours, even when they can't access their laptop (for example, at a child's birthday party.)

SOHO

Small and home offices are more and more prevalent, as professionals such as accountants, lawyers, and consultants leave large firms and "set up shop" as independent contractors. These offices are characterized by a high level of professional demands and mobility, but a low level of willingness or means to invest in infrastructure. Thus, SOHO subscribers may need access to email anywhere but lack the technical resources or know-how to implement complex solutions.

Golden Age

Some retirees benefiting from good pension and savings schemes are also very active and very mobile. They are highly tech-aware, but less tech-savvy. Grandparents are likely to use email on a daily basis when at home, but they also travel for several months a year. While on the road, they miss the frequent correspondence with children and grandchildren, and would like to share their photos and travel experiences with friends and relatives. Source: AxisMobile (2007)

Progress in the mobile and wireless communications drive also occurs in the development of services to handle needs in this area, for example, responding to wireless needs and addressing those needs while on holiday or away. *AxisMobile* in the vignette above is segmenting their market for mobile service offerings. They segment their markets to serve them more effectively. The customer segments they have identified are good indicators of potential segments for other companies, products and services, and thus can be implemented with revisions. This should not be much different than strategic organisation of, for instance, TV advertisements. Hence, marketers can see how complex the unique needs and user requirements could be including attitudes towards mobile marketing, cost concerns, differences in lifestyle complexities, and so forth; thus raising the possibility of a number of segments.

With increasing access rates, mobile services are becoming more main stream and therefore will be attracting more and more businesses as a means for direct marketing. The internet delivered through wireless services are almost on a par with the quality of immobile PC based internet services and are increasingly widely used, as mentioned previously. SMS based marketing is also well-established. However, these two channels need to be treated differently, not only because of technical differences but also for the different types of customers that the marketer can reach. Location based mobile marketing through wireless GPS services are used by a customer segment that may be very different from, for example, mothers using their mobile phones occasionally when their children are away on a school trip.

It is very possible that different customer segments may need to be reached by either telephone-based or computer based internet access. The rise of virtual communities on the web in terms of review sites, blogs, web forums and so forth would lead to another distinct group of people who are likely to be much more demanding and critical of product offerings and much more knowledgeable about product alternatives. Schiffman, Kanuk and Hansen (2008, p.52) suggest a psychographic category 'Techno Road Warrior' who is also different in needs and wants. They are the businesspeople who spend a high percentage of their working week travelling, equipped with laptop computers, PDAs and mobile phones.

These new categories arising from different (technical or usage related) factors need to be integrated with the conventional segmentation criteria. Gender differences are evident in internet use, mobile phone use, and choices of wireless services. The type of business and leisure surfing done is remarkably different between men and women. Geographic segmentation is important as the mobile customer is converging in tastes and buying habits with his or her counterparts around the world, which enables a standardised approach to marketing. On the other hand, existing differences between geographies (regions, countries and districts) can still be catered to and the rise of more focused micro marketing through digital technology makes this more viable. Regional integration projects such as EU are creating larger markets with opportunities for standardised marketing. However, socio-cultural differences between joining units still may need attention. Similarly demographic segmentation is still important for many businesses. Mobile technologies and innovation are conducive to customised messages and products to be delivered to individual customers.

Within such a framework, the likely groups which mobile marketing may target can include everybody who spends a substantial amount of time away from home and/or office. This includes business people on the move, some holiday makers, children, students, and some immigrant categories. A systematic classification of these segments can be made for each country, region, city and zone using demographic and attitudinal profiling. Okazaki (2006) identified four distinct clusters of mobile internet adopters in Japan: *affluent single youth* (this group had the most negative perception of mobile internet adoption), *clerical office workers* (the most positive perception), *married housewives* and *company executives* (also positive perceptions). These adopter clusters, if we have had the data, could have been linked to some kind of mobile living patterns. Company executives travelling between company sites and customers, house wives picking up children from the school which may involve long waiting times, affluent single youth is the difficult group with the most negative attitude. One reason for that could be that these people are working with computers and internet almost constantly and want to avoid using them a great deal if outside working hours. However, these segments are settled and relatively easy to identify, reach, and serve as they, to a great extent, are location-bound. The difficulty arises when they become transnationally mobile.

The contemporary international business environment is increasingly well-connected, so much so that it has been termed a *Global Village*. Improved telecommunications, further economic and political integration have brought globalisation of markets and thus consumers. We are now confidently referring to global markets, global brands, global products, and yes, global consumer segments. People living in different parts of the world today share some tastes, needs and wants and therefore are served similarly. This can be seen as a benefit for marketers and managers as it facilitates standardisation helping to build economies of scale and increase efficiency. People see, know and share experiences through mediums such as the internet, satellite television and increasingly frequent travel abroad.

Transnational migration involves only about three percent of the world population which varies among countries and continents. In Europe, it comprises almost 9 percent of the total population while it is between 13 and 15 percent in North America and Oceania (UNDP, 2008). Travel statistics often indicate much larger figures. For example, the UK statistical office reported 30 million foreign citizens entered and 64 million British citizens left the country in 2005 (ONS, 2006). These people travelling around the world purchase products and services while they are on the move and at their destinations. A large portion of them are expected to be tourists on package holidays and so on, but sizeable portions are composed of business people, highly skilled professionals, and immigrants of various kinds (seasonal workers, students, high skilled migrants, asylum seekers, refugees, etc). Their needs and wants are expected to be relatively different from their fellow citizens left behind in the origin country and in their destination(s). There are no exact statistics of those who travel frequently across borders. However, it is clear that we are looking at a very large customer segment that includes about 200 million migrants along side with those who are internally mobile in their own countries. The latter group is much larger than the former. To this one should also add holiday makers and professionals working in multiple or far away locations.

We term both the groups mentioned above as transnational migrants because of their links, networks and involving reference points in more than two countries. As an illustration of this, students going abroad for exchange programs and for postgraduate degrees often build and maintain networks involving friends and teachers acquired in these visits. Professionals in certain industries have to work in different locations for various durations across the world. Similarly, other migrants groups establish links with their friends, families, partners in Diasporas (e.g. Asians, Turks, Moroccans, and Iraqis in Western European countries). Some of these groups can constitute a distinct market segment for some products and services, allowing for a more standardised offering. However, in some other groups' variations would be inevitable. This raises concerns in terms of reference criteria. Some of these people are heavily dependent on mobile internet services while some others are heavy users by choice. Yet some may prefer it because of practical concerns such as foreign students who does not want to (or cannot) have a land line. Thus, mobile marketers have to deal with the issue of the 'address'; as these people are likely not to have a stable address. Address here can be expanded to include the reference point in terms of segmentation bases too. How these customers will be identified and clustered into our geographic classifications? What will be the reference point if in case we need to take into account contextual characteristics: those of the origin or those of the destination?

MOBILES AND MOBILE NATIONALS: AN OVERLOOKED SEGMENT?

National mobiles are those migrating within any country as an established category in segmentation studies and are considered to be a large and lucrative segment to target as Andreasen (1966) pointed to in his US study. These 'mobiles' are likely to constitute about one fifth of the population, are often aged around 18-35, and have ethnic minorities overrepresented amongst them. The relevance of this group for mobile marketing would come from not resettled populations within any national market but those who commute, and commute relatively long distances for work. Again, it is not easy to estimate the exact size of such populations but major cities, for example London in the UK, would attract a large number of such people from nearby and distant places commuting for work and would benefit from and be likely to use mobile communication services.

Despite being inherently considered in marketing, 'movement' has received little attention as a segmentation variable. As Bell (1969, p.37) suggested nearly four decades ago, "geographic location or residence has long been a recognised factor" but "change in residence has received very little attention". After Bell's paper, there are only a handful of studies partially considering this issue. There are few studies on immigrant consumers (e.g. Askegaard et al., 2005, Chung, 1999 and 20002000), some studies on transit (or public transport) marketing utilising commuting distance to work as a variable (e.g. Frumkin-Rosengaus, 1987), and also some indirectly dealing with immigrant minority consumers (e.g. Sirkeci, 2008, Burton, 2002). Bell's work (1969) thus remains a pioneering work. Clearly, when he was using the term "mobile consumers" there was no m-marketing, but his arguments concerning the significance of a considerable number of people changing their place of residence in the United States seem strong. In the mid-20th Century, he quotes that about 20 per cent of "American consumers changing addresses each year and each year, two-thirds of those making local or intracounty moves" (Bell, 1969, p.37). He defined this unique segment as "mobiles"; in other words "inter-county mobile families".

Bell's study answered the three basic questions of segmentation: a) defined their socio-economic and life cycle characteristics, b) explained how their shopping behaviour is affected by being 'mobile', c) whether these mobiles need special marketing programs. Bell's mobiles were relatively better off in terms of income, often with families, white-collar professionals, and mostly educated at the college level at the time (Bell, 1969, p.39). Rebuilding consumption patterns in the new location was one of the key challenges for the segment. His verdict on sustainability of specialised marketing programs targeting 'mobiles' was positive. Mobiles, he found, tend to spend more on various goods and services compared to non-mobiles which makes the segment attractive to even for small businesses (Bell, 1969, pp.43-44). One can find similar figures for national mobiles (or 'internal migrants' in demographic terminology) in other countries. For example, in England and Wales, over 1.3 million (nearly 3 percent of the total) people changed their place of residence permanently in 2006 according to UK Office of National Statistics.

Bell's work and mobility statistics are urging us to look into new variables for segmentation that will include movement, spatial mobility, and commuting distance/hours to identify these mobile segments and possibly sub-segments within for designing specialised marketing programs. Thereby, internal migrants, long-distance commuters, and transnational mobiles appear as key segments to look into.

Transnationals are likely to share many characteristics with the 'mobiles' of Bell, but apparent differences will be noted. Nevertheless, the same questions regarding segment characteristics and attractiveness need to be asked about the 'transnationals' we propose here.

TRANSNATIONALS: AN EMERGING SEGMENT FOR MOBILE MARKETING

Case: YOC AG

YOC AG is one of the leading mobile marketing providers in Europe. Their clients include Germany's Bild newspaper, Coca-Cola, Ford and H&M. YOC provides tailored solutions to companies that integrate mobile phone technologies with other forms of marketing communications. For example, when working with Coca-Cola in the UK they designed an SMS based campaign to coincide with the company's sponsorship of the British football league. The 'Buy a Player' campaign offered some winners a cash prize of £10,000 for themselves and a further £250,000 for a football team of their choice to spend on signing new players. The competition worked by having codes on the inside of Coca-Cola bottle caps which the customers could text to the firm to see if they had won a prize. The football fans travel across the World for various competitive games (e.g. World Cup, UEFA Cup, etc.) and they are part of the temporarily mobile customer segment. Their needs can be categorised more by psychographic segmentation variables. Source: Promny, M. (2008)

The above vignette provides a good example of (perhaps a sub-segment of) transnationals. These football fans are likely to share most other characteristics with non-travelling fans. They are however, likely to have a higher income and could well need a range of services and products that non travelling fans wouldn't. Today's world is increasingly mobile and more and more people are living outside the countries where they born; more and more living in transnational settings without any sedentary attachment to a single location; more and more studying abroad; and more and more taking holidays abroad. These groups can be segmented based on their causes for international mobility, lifestyles, work styles relating to the particular needs and wants of each group. However, one could argue that they are dependent on mobile communication services more than others who pursue more sedentary lives. Therefore, the 'transnationals' segment is likely to use mobile services more than others and in perhaps different ways (for example, international or foreign service providers, multiple uses and so on) compared to the natives of their destination (or temporary residence) countries.

Forty years ago, arguing "mobile customers are a relatively neglected market segment", James Bell (1969) proposed there was 'the change of residence' as a variable in market segmentation. The same concern is shared here with particular reference to the transnationally mobile populations of today's rapidly globalising world. We would like to expand the scope of it by adding 'the movement' as a variable rather than change of residence; because for many, it is travelling across borders and spending significant periods in other towns, cities, countries, and continents. Also we need to clarify here the differences between mobile nationals and transnationals. We can deploy simple environmental screening to show these differences: The former is likely to move shorter distances compared to the latter while the latter is also passing across into (a) a new jurisdiction, a new country with separate rules and regulations, a new political environment, (b) a new economic environment and (c) possibly a different technological environment too. Perhaps more importantly, transnationals have to face (d) a new socio-cultural environment. Examples for these could be: (a) legality regarding smoking in different countries, (b) differences in inflation and interest rates which is likely to affect buying and saving behaviour, (c) different telecommunications infrastructure (e.g. availability of 3 G), and (d) roles, customs, values and traditions (e.g. gender roles in different countries).

One major similarity between mobile nationals and transnationals is perhaps the multiple reference points for these two groups. For example, they are likely to share characteristics with those at the current place of residence as well as with those at the former place of residence. In the case of transnationally mobile professionals, there could be more than these two reference points.

A systematic classification of the 'transnationals' may generate the following categories which can be considered within a multivariate segmentation model as sub-segments of this broader group:

1. Mobile professionals employed by transnationals or organisations with international operations and links. Executives, consultants, IT consultants, engineers, high rank government officials, and professors could be considered amongst this group. These people live abroad for relatively short periods up to six months or a year; often working on projects or with clients, providing training, giving lectures and consulting. It is expected many of them travel alone or with colleagues but not with families. In fact many of them might be single. One should note the increasing importance of these market segments. For example, according to the latest Barclaycard Business Travel Survey (2006), in the last decade, the number of business travel miles of leading executives

increased 32 per cent, which has implications for the mobile technologies supporting these activities while creating opportunities for marketers (Faulconbridge and Beaverstock, forthcoming). Another recent study explores the mobility of highly skilled staff within transnational corporations (Millar and Salt, 2008a). These corporations are keen to source within their internal labour markets and thus move staff around the world. This indicates the significance and size of such mobile populations.

The UK case could be representative for advanced markets, at least in Europe. Thus, business travel trends are rapidly growing. According to ONS (2006) *Travel Trends*, the numbers of overseas residents visiting the UK and of the UK residents going abroad for business purposes have almost doubled in the last two decades. On average the length of stay is around 5 nights amongst the 8.5 million UK residents going abroad and the 8 million overseas residents visiting the UK. One can, again, we can see there a regional pattern as ONS (2006) *Travel Trends* reports; for example a quarter of these visits take place between UK and France and Germany while Europe constitutes three quarters of the total visits. The size of this market is significantly large and expected to be lucrative.

2. Students spending a term or a year abroad enrolled in programmes abroad including language courses and degree programmes will be the second group. As Millar and Salt (2008b) report, there were over 300,000 international students in higher education in the UK benefiting the economy in excess of £5 billion. Perhaps, university students and young researchers share characteristics (hence needs and wants) with mobile professionals. For instance, with some variation by disciplinary field, those transnationals constitute between 24 to 59 percent of research staff in the UK (Smetherham et al., 2008). Despite the fact that some of these are migrating for settlement, many are temporarily working abroad for career development. However, in both cases we are looking at mobile populations. For young people this often means living in shared houses and also not having land lines, private TVs and so on. One may expect then, that they will be more inclined to use the internet, with mobile broadband on laptops and other mobile devices such as phones and PDAs.
3. Immigrants are composed of various subgroups based on their motivations, patterns and mechanisms of migration which are heavily determined by the admission regulations of the destination countries. High skilled migrants are largely covered above in terms of business travellers and transnational corporations' internal sourcing. However, there are also many individuals who resettle in other countries for longer or shorter periods. Medical doctors and IT specialists are examples in this category. In many advanced countries, for example, overseas doctors constitute a large portion of those in the profession. As an indication of that, in New Zealand, more than a third of doctors have overseas qualifications (MOHnz, 2008). Similar figures found for the UK: about a quarter of consultants have overseas qualifications while one third is from immigrant ethnic minority backgrounds (Goldacre et al. 2004).
4. A sub-category of immigrants are refugees. This group is often smaller than others and can be counted among settled immigrants as they legally have equal rights to citizens. However, like other immigrant groups they do maintain a transnational life style where connections with the country of origin and other Diaspora communities are maintained. Their consumption patterns are very likely to be similar to other immigrant groups.

Thus, as a politically and administratively defined category of immigrants, they would not require any special attention in terms of segmentation.

5. Another sub-category similar to refugees is asylum seekers. Asylum seekers are heavily restricted by government regulations in most countries. Their entitlements are limited in terms of employment and settlement. They often live in purpose built centres and/or are subject to close monitoring. This bars them choosing a 'normal' life style where they can get land lines, permanent addresses and have access to a range of communication services. One may expect them to be dependent on mobile communication. They are likely to have closer connections with their countries of origin and as their uncertain resident status implies, there is a chance of immediate and sudden return to home (Gibney, 2008). While being smaller than the other groups mentioned they are still a sizeable group in many advanced countries reaching into millions Europe wide, for example. One important aspect of this group is the long waiting times, and relative high rejection rates in recent years. This makes it a temporal group.
6. Another similarly temporal but distinct sub-category is the 'undocumented', 'illegal', or 'clandestine' migrants category. By nature and by definition, this group is unidentifiable and difficult to estimate in terms of market size. It may not be a lucrative group but still worthy of consideration, we believe. One distinction here could be made between working holiday-makers, visa-over-stayers, and illegal entrants. These three kinds of people may be able to obtain legal resident status but this is often very restricted and not always. Geographically, they are likely to be found in larger cities and towns, particularly where an established sizeable immigrant minority population exists. London, New York, Paris, Berlin and Rome would be example cities. We can expect these customers to be on low incomes, and not to possess bank accounts, credit cards and the like. These factors make them difficult to track and target. Nevertheless, they are still using some way of communication, which is very likely to be some kind of mobile service. Is it a significant market segment? We hope so, because estimations, again in most advanced countries and emerging markets indicate that they make up to 4 percent of the total population (Sirkeci, 2006). Thus, it should be a significant concern for segmentation in mobile marketing. Of course there are also legal and ethical implications involved in the active targeting of these groups, as the firms involved may be accused of encouraging law breaking.

Internationally mobile groups are a challenging market segment. There are not many organisations tailoring marketing programmes to these particular groups. Many industrialised countries, so advanced markets, have become increasingly multicultural as telecommunication and transportation links have turned the Earth into a global village. Acknowledging the existence of cultural differences and significantly large groups with different cultural backgrounds has been a difficult political issue in many countries. Afraid of a backlash, perhaps, many organisations and firms have avoided special marketing programmes targeting those newcomers, immigrants, travellers and so forth. In the UK, only after more than a million Polish immigrants' arrived, did some companies consider tapping into this market, while similar moves took more than ten years to capture the interest of Asian customers in the country (Gray, 2007). HSBC and NatWest were among the first to realise the opportunities in banking. Regarding the special nature or dependency on international or transnational communication needs we have mentioned above, some UK mobile phone service providers have begun to specifically target these markets. Orange, O2 and Vodafone, for example, moved in advance of their rivals to seize this market of strangers, or as we call them 'transnationals', and particularly those of Polish origin (Gray, 2007, pp.30-31).

A major marketing challenge stems from the difficulty of reaching these customers through mainstream media. A UK study, for example, found that less than 1 in 6 Polish immigrants read UK newspapers (Gray, 2007, p.31). This definitely has a clear implication for marketers; new and appropriate alternative channels of communication with these customers. Issues of reach and availability of channels can be tackled by spatial research. Geographical concentration of transnationals (cross border mobiles) as well as nationals (internal mobiles) and patterns identifying the routes of movements will help to understand potential avenues of marketing communication directed at the peculiar segments of these mobile populations. Marketers thus will need the demographic expertise looking at spatial analysis of internal and international movements of natives and foreign-born. Availability of some services or lack of them (e.g. local bank branches, ATM points, etc.) may reveal some opportunities for mobile services and thus for mobile marketing. Identification of such areas would enhance marketing planning. Future research should focus on these areas of crucial interest to marketing managers.

According to industry reports, advertising is changing from broadcast to "uni" cast, where advanced advertising technology allows sending localized information to consumers while advertising on mobile phones gains increasing market share at the expense of TV and traditional media. Increasing cooperation between mobile service provider companies across nations is likely to offer economies of scale opportunities as well as enabling service providers to reach their customers while they are abroad and also design specific products and services to appeal the needs and wants of the 'transnational' customers.

Future research should focus on the profiling of mobile populations, their differentiated needs and the variety amongst them. Their usage rates of mobile devices would be of certain interest to the industry as well as customer loyalty patterns regarding mobile service providers. 'Transnationals' as a peculiar segment need to be studied in relation to the availability of services at transnational level as well as capability of firms to serve across this relatively unsettled terrain, as several countries jurisdictions will impact upon operations and a good deal of international cooperation and coordination effort may need to go into such businesses.

CONCLUSION

This chapter attempted to point out an important gap in our marketing knowledge which can be addressed by research on movement as a segmentation variable. This variable is very likely to be useful in mobile marketing. Reviewing the current segmentation literature along with growing mobile marketing literature, we have provided some elaboration and direction on this gap hoping marketing researchers and students will take up and further the knowledge in this area.

A significant avenue to pursue in addressing the difficulties in segmenting mobile markets is to follow the segmentation strategies of mobile services markets (mobile phones, mobile email clients, service providers). These would give a good indication of current and potential segments and could be easily understood, as these segments would be closely linked to the four segmentation bases already often used in marketing. However, this may not help a great deal in segmenting the 'transnationals'. For these new ways, methods and techniques might be needed and innovations and creativity should be channelled towards this, if organisations wish to take these particular groups seriously.

The main issue for segmenting the 'transnationals' arises from the difficulty of identifying the reference points. This is why country of origin, country of destination or countries in transit are becoming relevant. Perhaps due to this difficulty, we should ask the question whether marketers should focus on behavioural and psychographic variables with less emphasis on geographic characteristics. The 'transnationals' are a significantly large and enormously diverse range of groups incorporating all four bases of the segmentation theory. The mobile marketing effort is also growing rapidly. More and more companies are moving into mobile advertising. In four years time, the total value of mobile advertising is expected to climb above $14 billion (Glanz, 2008).

Migration is one of the key areas in demographics along with births, deaths, and marriage. Hence, reformulating the question, a further avenue for investigation would be to ask whether the transnationally mobile is a unique realm for undertaking segmentation or simply another demographic variable to take into account. Thus, 'spatial mobility', 'the movement', and "commuting hours/distance" are three variables that need to be taken into account in segmentation and targeting decisions.

Segmentation is an important area in marketing and the segmentation of 'mobiles' and 'transnationals' is particularly challenging. Conventional segmentation schemes (for example, VALS, ACORN and their like) need to be revisited. Researchers and practitioners need to think about ways in which those multiple reference points (cultural and geographical) can be integrated into new segmentation approaches in order to create successful segments. There might be many new niches to identify among 'transnationals' but in any event the size of this particular group of customers is certainly large enough to accommodate multiple segments that require to be treated separately. Mobile marketers also need to think about the ways in which these 'transnationals' can be reached.

It has also become evident that the concept of 'mobile nationals' was introduced nearly three decades ago and relatively little work has been devoted to the idea since. With increasingly mobile populations, crowded metropolitan areas generating large commuter populations all relate to this concept. We believe there is some potential for marketing and that marketing researchers and students would find this stream of research important and interesting. Another avenue for research to be exploited here is to expand this 'mobiles' category towards covering long distance commuters, which may constitute a sub-segment.

REFERENCES

Anderson, C. (2006). *The long tail: Why the future of business is selling less of more.* New York: Hyperion.

Andreasen, A. R. (1966). Geographic mobility and market segmentation. *JMR, Journal of Marketing Research*, *3*(4), 341–348. doi:10.2307/3149849

Askegaard, S., Arnould, E. J., & Kjeldgaard, D. (2005). Postassimilationist ethnic consumer research: Qualifications and extensions. *The Journal of Consumer Research*, *32*(1), 160–170. doi:10.1086/426625

AxisMobile. (2007). AxisMobile consumer mobile email marketing segmentation White Paper. Retrieved May 5, 2008, from http://www.axismobile.com/ files/ AxisMobile_ Consumer_ Mobile_ Email_ Marketing_ Segmentation_ White_ Paper.pdf

Barnes, S. J. (2002). Wireless digital advertising: Nature and implications. *International Journal of Advertising*, *21*(3), 399–420.

Barnes, S. J. (2003). Location-based services: The state of the art. *E-Service Journal*, *2*(3), 59–70. doi:10.2979/ESJ.2003.2.3.59

Barnes, S. J., & Scornavacca, E. (2004). Mobile marketing: The role of permission and acceptance. *International Journal of Mobile Communications*, *2*(2), 128–139. doi:10.1504/IJMC.2004.004663

BBC. (2007). Ask rolls out search privacy tool. Retrieved December 11, 2007, from http://news.bbc.co.uk/1/hi/technology/7138260.stm

Bell, J. E. (1969). Mobiles-a neglected market segment. *Journal of Marketing*, *33*(2), 37–44. doi:10.2307/1249400

Bird, D. (1989). *Commonsense direct marketing*. London: Kogan Page.

Blythe, J. (2008). *Consumer behaviour*. London: Thomson.

Brassington, F., & Pettitt, S. (2006). *Principles of marketing fourth edition*. Hemel Hempstead: Prentice Hall.

Burton, D. (2002). Incorporating ethnicity into marketing intelligence and planning. *Marketing Intelligence & Planning*, *20*(7), 442–451. doi:10.1108/02634500210450873

Chung, E. (2000). Navigating the primordial soup: Charting the lived worlds of the migrant consumer. *Journal of Consumer Marketing*, *17*(1), 36–54. doi:10.1108/07363760010309537

Claritas. (2008). 66 Prizm marketing segments. Claritas Prizm NE. Retrieved June 3, 2008, from http://www.claritas.com/claritas/ Default.jsp?ci= 3&si= 4&pn= prizmne_ segments

Covey, N. (2008). Critical mass, the worldwide state of the mobile Web. Nielsen Mobile. July 2008. Retrieved July 13, 2008, from http://www.nextgenweb.org/ wp-content/ uploads/ 2008/07/ nielsen-critical-mass-the-worldwide-state-of-the-mobile-web.pdf

Day, G. S., & Montgomery, D. B. (1999). Charting new directions for marketing. *Journal of Marketing*, *63*(special issue), 3–13. doi:10.2307/1252096

DeZoysa, S. (2002). Mobile ads need to get personal. *Telecommunications International*, *36*(2), 8.

Faulconbridge, J. R., & Beaverstock, J. V. (forthcoming). Geographies of international business travel in the professional service economy. In D. Hislop (Ed.), *Mobile work/technology: Changing patterns of spatial mobility and mobile technology use in work*. London: Routledge.

Frumkin-Rosengaus, M. (1987). *Increasing transit ridership through a targeted transit marketing approach*. Unpublished doctoral dissertation, State University of New Jersey, New Brunswick, NJ.

Gibney, M. J. (2008). Asylum and the expansion of deportation in the United Kingdom. *Government and Opposition*, *43*(2), 146–167. doi:10.1111/j.1477-7053.2007.00249.x

Glanz, O. (2008). How to get ahead in mobile advertising. *Mobile Marketing Magazine*. Retrieved May 23, 2008, from http://www. mobilemarketingmagazine. co.uk/ 2008/05/how-to-get-ahea.html

Goldacre, M. J., Davidson, J. M., & Lambert, T. W. (2004). Country of training and ethnic origin of UK doctors: Database and survey studies. *British Medical Journal*, *329*(7466), 597–600. doi:10.1136/bmj.38202.364271.BE

Gray, R. (2007). Immigration, world of opportunity. *The Marketer*, November, 28-31.

Greengrove, K. (2002). Needs-based segmentation: Principles and practice. *International Journal of Market Research*, *44*(4), 405–422.

Haig, M. (2002). *Mobile marketing: The message revolution*. London: Kogan Page.

Hill, L., & O'Sullivan, T. (2004). *Foundation marketing*. London: Pearson, Prentice Hall.

ITU (International Telecommunications Union). (2007). ITU world communication/ICT indicators database. Retrieved January 3, 2009, from http://www.itu.int/ITU-D/ict/statistics/

Kotler, A. Saunders, & Wong. (2008). *Principles of marketing, fifth European Edition.* Harlow, Pearson Education Limited

Kotler, P., & Keller, K. L. (2008). *Marketing management, 13th edition.* London: Pearson, Prentice Hall.

Louvieris, P., & Driver, J. (2001). New frontiers in cybersegmentation: Marketing success in cyberspace depends on IP address. *Qualitative Market Research: An International Journal, 4*(3), 169–181. doi:10.1108/13522750110393071

MDA (Mobile Data Association). (2008). The Q3 2008 UK mobile trends report. Retrieved December 15, 2008, from http://www.themda.org/mda-press-releases/the-q3-2008-uk-mobile-trends-report.php

Millar, J., & Salt, J. (2008a). Portfolios of mobility: The movement of expertise in transnational corporations in two sectors–aerospace and extractive industries. *Global Networks, 8*(1), 25–50.

Millar, J., & Salt, J. (2008b). *International students and the labour market: Experience in the UK.* Paper presented at Leverhulme Programme Conference on Mobility in International Labour Markets, University College London, UK.

MMA (Mobile Marketing Association). (2008). *Mobile marketing industry glossary.* Version 2008.12. Retrieved December 15, 2008, from http://mmaglobal.com.

MOHnz (Ministry of Health New Zealand). (2008). Doctors in New Zealand. Retrieved June 10, 2008, from http://www.moh.govt.nz/ moh.nsf/wpg_Index/ About-Statistics+ about+ doctors

Okazaki, S. (2006). What do we know about mobile Internet adopters? A cluster analysis. *Information & Management, 43*(2), 127–141. doi:10.1016/j.im.2005.05.001

ONS (Office of National Statistics). (2006). Travel trends. Retrieved May 5, 2008, from http://www.statistics.gov.uk

Promny, M. (2008) YOC AG, integrating the mobile phone into the marketing mix. INSEAD, ECCH case study, Cranfield, UK: ECCH. Retrieved June 15, 2008, from http://www.ecch.com

Schiffman, L., Kanuk, L., & Hansen, H. (2008). *Consumer behaviour: A European outlook.* London: Pearson, Prentice Hall.

Shiels, M. (2008). The mobile future is calling. BBC News. Retrieved July 14, 2008, from http://news.bbc.co.uk/ go/pr/ fr/-/1/hi/ technology/7366403.stm

Sirkeci, I. (2006). *The environment of insecurity in Turkey and the emigration of Turkish Kurds to Germany.* Lewiston, NY: E. Mellen Press.

Sirkeci, I. (2008). Ethnic marketing potential in England and Wales: New evidence from the 2001 UK census. *Asian Journal of Marketing, 2*(1), 1–9.

Smetherham, C., Fenton, S., & Modood, T. (2008). *How global is higher education in the UK?* Paper presented at Leverhulme Programme Conference on Mobility in International Labour Markets, University College London, UK.

Smith-Shomade, B. E. (2002). Narrowcasting in the new world information order, a space for the audience? *Television & New Media, 5*(1), 69–81. doi:10.1177/1527476403259746

UNDP (Population Division of the Department of Economic and Social Affairs of the United Nations Secretariat). (2008). Trends in total migrant stock: The 2005 Revision. Retrieved May 5, 2008 from http://esa.un.org/migration

Usunier, J.-C., & Lee, J. A. (2005). *Marketing across cultures, 4th edition.* London: Pearson, Prentice Hall.

Valavanis, E., Ververidis, C., Vazirgianis, M., & Polyzos, G. C. (2003). MobiShare: Sharing context-dependent data and services among mobile devices. Retrieved January 3, 2009, from http://www.db-net.aueb.gr

Waterman, D. (1992). "Narrowcasting" and "broadcasting" on nonbroadcast media, a program choice model. *Communication Research, 19*(1), 3–28. doi:10.1177/009365092019001001

Wedel, M., & Kamakura, W. A. (2000). Market segmentation, conceptual and methodological foundations, 2nd edition. ISQM8. Boston, MA: Kluwer Academic Publishers.

Wray, R. (2008). Cash in hand: Why Africans are banking on the mobile phone. The Guardian, June 17, 2008.

KEY TERMS AND DEFINITIONS

Behavioural: The sub-categorisation of a market based upon the relationship that the customer has with the product or service. For example, the frequency of purchase, occasion of purchase, and the specific benefits sought the awareness status of the customer and the brand loyalty levels. In the context of mobile marketing segmentation this might include, for example, potential customer awareness of a variety of new mobile services such as m-banking that would revolutionise the way many people transfer funds and pay bills.

Demographics: The separating of a total market for particular goods and services into smaller groupings based on observable, objective characteristics such as age, sex, marital status, income, education and occupation as often identified in official statistics and surveys.

Direct Marketing: Any activity which creates and exploits a direct relationship between you and your customer as an individual (Bird, 1989). The central aim is to produce a direct response from the prospect. This response may take a variety of forms ranging from an immediate purchase through to a request for further information.

Geographics: The division of a market into smaller groupings based on location specific characteristics such as region, country, city, post-code or zip code and climate type. This type of segmentation approach is frequently combined with demographic data to form a rich hybrid base known as geodemographics. An example of this is the British ACORN system (A Classification Of Residential Neighbourhoods).

International Migration: changing one's place of residence for short or long term from one country to another. People migrate for various reasons and often with mixed motivations including escaping from fear of persecution, from wars, conflicts, and natural hazards, for job opportunities, for joining families, for education, to avoid regulations (e.g. compulsory military service), due to job relocation within international companies, to start up businesses, and also for adventure. Depending on the available admission rules and routes, international migration may occur through legal or undocumented channels and mechanisms.

Mobile Nationals: As defined by Bell (1969), these are customers who have changed their place of residence within a country.

Psychographic: This approach to segmentation is based on the psychological profiling of the potential customers. This often focuses upon the customers' activities interests and opinions. In the mobile marketing context this is an important approach as it would help marketers to understand how potential customers feel about the use of new technologies and innovative processes and the role that mobile devices already play in their working and social lives.

Segmentation Bases/Variables: categories of individuals and organisations used for grouping them for targeting and positioning purposes. These include demographic, geographic, behavioural, psychographic, and hybrid sets of variables. These categories are often used in complementing each other to obtain fine segments. Application of these variables in business to business markets slightly differs from its uses in consumer markets.

Transnationals: Transnationally mobile customers, who either relocated to other countries from their countries of origin or follow a circulatory pattern of migration, or professionals working in different divisions of transnational companies for short/long periods, or students studying abroad, or holiday makers who spend substantial periods abroad.

Chapter 7
Situation Approach as Success Factor of Mobile Marketing

Jörg Link
University of Kassel, Germany

Franziska Seidl
University of Kassel, Germany

ABSTRACT

The success of Mobile Marketing is to a large extent dependent on two factors: (1) the situation appropriateness, where demands are made on the ability of the supplier to process all the central characteristics of a customer situation and turn them into an offer that is appropriate to the situation; (2) the economic success of the supplier depends on the economic potential of the situation and his accurate appraisal of it. The supplier therefore has to model both the specific features and the profit potential of the various customer situations at the right time and in the right way. This chapter is going to demonstrate that the success of Mobile Marketing measures can be guaranteed by a synergic interplay of market and result orientation. On the one hand customer situations are systemized and placed in the context of an integrated Customer Relationship Management system; on the other hand the evaluation of situations, reference object hierarchies, and the various levels of situation profit and loss statements are shown.

SITUATION APPROPRIATENESS AND POTENTIAL AS SUCCESS FACTORS OF MOBILE MARKETING

In view of the growing importance of flexibility and mobility together with an extremely high penetration success rate of mobile terminals – e.g. mobile phone, personal digital assistants (PDA), palm tops, smart phones or notebooks – companies are beginning to show more interest in the use of these media for marketing purposes.

If marketing is defined as market-oriented leadership or market-derived leadership (Meffert, 2000, p. 3), Mobile Marketing can be understood as market-derived leadership using mobile terminals.

The mobile channel offers increasing possibilities of individualization in terms of a mobile one-to-one call (direct marketing). While individualization in the classic one-to-one call means simply attending to the special needs of the individual customer,

DOI: 10.4018/978-1-60566-074-5.ch007

there is an additional possibility of attending to the special situation. This situation is marked by a different macrostructure (idle times situation, search situation, emergency situation, quasi-stationary situation) and microstructure (current location, time etc.). Individualization possibilities and requirements are no longer limited to the data profiles of the customer database, but follow an extended concept of individualization including the situation data (Kriewald, 2007). The situation appropriateness thus complements the customer appropriateness (on many of the following aspects cf. Link & Seidl, 2008).

Situation appropriateness accordingly represents, on the effect level, a central success factor in Mobile Marketing. Situation-appropriate offers are more promising, i.e. they offer companies the promise of higher profits.

The height of these profits is determined by the situation potential. This is the (economic) potential revealed by a situation, and this is a further significant success factor in Mobile Marketing. It must be gauged whether it is profitable to make a particular customer in a particular situation an offer appropriate to the situation. This aspect reflects both the situation and customer evaluation.

While the *success factor "situation appropriateness"* refers principally to the marketing perspective, the *success factor "(economic) situation potential"* is related more to the controlling perspective. This makes it clear that a close connection between marketing – market-derived leadership – and controlling – success-derived leadership – is indispensable. This is the starting point for this paper. The overall goal of this chapter is to show that the success of situation-specific offers is determined by the use of the synergy potential of Mobile Marketing and controlling respectively situation appropriateness and situation potential to a large extent.

First, (customer) situations are rendered concrete within the framework of a systematic consideration. The customer and situation data underlying the (customer) situations are then placed in the context of an integrated Customer Relationship Management (CRM) system. Afterwards the evaluation of situations, reference object hierarchies and the various levels of situation profit and loss statements are shown.

SYSTEMATIZATION OF SITUATIONS

As already shown, a special situation relationship marks the use of mobile terminals as information, communication, and sales channels. People normally use mobile terminals when they are on the move and confronted by changing situations. As their situations change, so do their requirements.

The following will analyse and systemize situations in fundamental detail. A basic distinction is to be made between macrostructure and microstructure.

In Figure 1 the connections are shown in outline (on microstructure cf. Kriewald, 2007). It is clear from this that both the particular basic situation – *macrostructure* – and the specific situation factors – *microstructure* – determine the customer situation to be considered for the situation-specific offer. The supplier, after analysing the customer situation, in the ideal case can make an offer tailored to the specialized individual situation. The formula for targeted direct marketing thus becomes considerably more effective as a result of the mobile factor.

Macrostructure: Basic Situations

In terms of macro-perspective the following need and/or starting situations for the use of mobile terminals can be distinguished (Link, 2003):

- *"idle times situation"*,
- *"search situation"*,
- *"emergency situation"*,
- *"quasi-stationary situation"*.

Figure 1. Systematics of situations

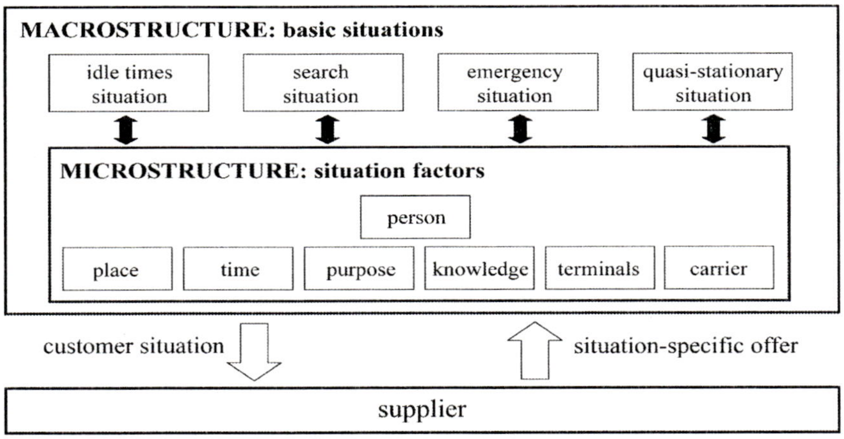

The first three situations are marked by the fact that a communication or transaction relationship between customer and supplier is able to take place in them exclusively via the mobile channel. If a supplier does not want to eliminate possible customer or business relationships from the start, he has to include the mobile channel in his multi-channel marketing (cf. the front office area in Figure 3).

The *idle times situation* describes the unproductive time taken up by transport. In management theory the concept of idle times – as distinct from productive times – is used for times that are not used or usable. Idle times in turn mean corresponding idle time costs. Those travelling privately or on business are confronted by the problem of possibly being unable to employ their personal potential satisfactorily. In the case of someone on a business trip, who could really perform certain tasks and thereby derive some productive benefit, the impossibility of achieving this goal, imposed by the transport time, represents opportunity costs. In the case of non-business travel the opportunity costs are located more in the immaterial area, such as for example the loss of enjoyment generated by leisure activities.

The use of mobile terminals offers numerous opportunities to employ inherently unproductive travel time productively and thus avoid idle times and costs. The various opportunities available for purpose include, among other things, the use of consumer to consumer (c2c)-, business to consumer (b2c)-, business to business (b2b)- and business to employee (b2e)- communication (e.g. phone conversations, Short Message Service (SMS), Multimedia Message Service (MMS), email; mobile office).

In the b2b and b2e areas the main purpose is to incorporate the mobile services into the total inner- and inter-operational value added process, which is subsumed under the heading of mobile Supply Chain Management. A typical field of application in this connection is mobile E-procurement. Further mobile applications include remote control/telemetry, vehicle fleet management, job dispatch and external staff support (mobile CRM) (Bennemann & Möllenberg, 2001).

In addition the traveller has an abundance of options open to him/her in the field of mobile entertainment. For example, it is possible via mobile terminals with Internet access to retrieve information from search machines and databases. As with newsletter services mobile information services are also conceivable. Music, computer games or films/videos can be downloaded in the same way as books, newspapers, magazines. Furthermore, Internet radio and TV offer a broad spectrum of information and entertainment. An additional form of entertainment is also available wih participation in gaming activities, such as lotteries or (sports) gambling.

It is of course also possible, when one is travelling with the appropriate mobile terminal, to activate all other types of transaction that are feasible on the Internet. This includes, among other things, mobile shopping and booking (e.g. purchase of rail and air tickets, booking of hotel rooms and hired cars, reservations) or even mobile banking (electronic wallet; payments; financial transactions; share dealing at the most current share price).

The second basic type of possible occasion for the use of mobile terminals, the *search situation*, is marked by a short-term and situation-dependent information or service need, that involves an appropriate search behaviour of the person concerned. A typical example is the traveller, who in an unfamiliar environment is reliant on electronic support and prepared to pay for situation-dependent value added services. Examples include directions/navigation (e.g. to hotels, restaurants, banks, authorities, gas stations, hospitals etc.) and information about prices, services, capacity utilization (e.g. of hotels, restaurants, parking areas, businesses, theatres) in a defined radius of the relevant place in addition to transport connections (e.g. underground stations, trams, buses, long-distance trains), local traffic information (e.g. diversions, traffic jams, traffic reports) and local weather news. With all these information and service requirements the functionality of an automated geographical positioning system takes on a key role, where different localization technology (Location Based Services), as is well known, is available (like, for example, GPS, Enhanced Observed Time Difference, Cell of Origin (COO) and Cell-ID, Bluetooth; cf. Kriewald, 2007 for a detailed exposition).

In an *emergency situation* there is an involuntary and unforeseen need for information and services. This need is articulated by the customer either via a push-button system or via the evaluation of measurement data fully automatically relayed to an appropriate central control. With respect to situations such as attacks, medical emergencies, burglary/theft, car accident or breakdown, emergency situations in the mountains or forests, both alternatives are conceivable and have to some extent been made possible by existing technology, so that with mobile terminals an improvement in the direction of better services or lower costs can be aimed at.

Already available on the market are cell phone-based terminals that can activate an alarm signal in an emergency call centre which will automatically locate the cell phone user. Work has been going on for many years on cell phone-based systems that record and transmit the measurement values of the body (e.g. heart frequency, blood pressure) to medical centres. Several car manufacturers also offer systems that can initiate an automatic emergency call after an accident or breakdown and locate or deactivate stolen vehicles.

In the basic type of *quasi-stationary situation* the customer finds himself in a situation where there are two alternative forms of Internet access: the mobile and the stationary. The decision depends ultimately on certain context variables, such personal, product and situation characteristics. But the two alternatives have to be seen as competitors with other communication possibilities in the multi-channel concept (cf. Figure 3).

Figure 2 represents in overview the four basic situations, including the mobile possibilities of each one.

Microstructure: Situation Factors

An analysis of the microstructure within the basic situations reveals what situation factors have to be considered in detail if the provider wants to approach the customer with an individual and situation-appropriate offer.

The concept of individualization in the narrower sense, which is aimed exclusively at the

Figure 2. Mobile possibilities in the basic situations

	idle times situation	search situation	emergency situation	quasi-stationary situation
characterization of the situation	unproductive time taken up by transport typical situation: travelling privately or on business by train or car	short-term and situation-dependent information or service need, that involves an appropriate search behaviour typical situation: traveller in an unfamiliar environment	involuntary and unforeseen need for information and services typical situation: medical emergencies, attacks, burglary/theft, car accident or breakdown, emergency situations in the mountains or forests	situation where there are two alternative forms of Internet access: the mobile and the stationary typical situation: private person at home, businessman at the office
selected mobile possibilities	▪ c2c-, b2c, b2b-, b2e-communication: phone, SMS, MMS, email ▪ mobile office ▪ mobile Internet (downloads, Internet TV and radio) ▪ mobile shopping/ booking ▪ mobile gaming ▪ mobile banking ▪ mobile Supply Chain Management (e.g. mobile E-procurement) ▪ external staff support (mobile CRM) ▪ remote control/ telemetry ▪ job dispatch ▪ vehicle fleet management	▪ directions/navigation (e.g. to hotels, banks, restaurants, hospitals, authorities, theatres, gas stations) ▪ information about prices, services, capacity utilization (e.g. of parking areas, hotels, restaurants, theatres, businesses) ▪ transport connections (e.g. underground stations, trams, buses, long-distance trains) ▪ local traffic information (e.g. diversions, traffic jams, traffic reports) ▪ local weather news	▪ activating an alarm signal in an emergency call centre and automatically locating the cell phone user ▪ record and transmit the measurement values of the body (e.g. heart frequency, blood pressure) to medical centres ▪ systems that can initiate an automatic emergency call after an accident or breakdown and locate or deactivate stolen vehicles	evaluation of both alternatives (pros and cons); If the decision is in favour of the mobile channel, in principle all shown mobile possibilities are thinkable and realizable.

person, is too narrowly conceived for the context of Mobile Marketing and is thus inadequate. In order to replicate the (mobile) customer situation comprehensively and then be able to draw up a situation-appropriate offer, an *extended concept of invidualization* has to be introduced, which in addition to the person includes other situation factors – place, time, purpose, knowledge, terminals, carrier (Kriewald, 2007). Individualization possibilities and requirements are thus no longer limited to the customer data (as in Database Marketing), but rather follow the extended concept of individualization including the situation data (cf. also the combined customer and situation database in Figure 3). The individual situation factors that principally determine the customer situation will now be explained in more detail (Figge, 2007; Grandjot, 2006; Kriewald, 2007).

The *person* as situation factor has an outstanding importance, because the personal context has a major influence on the customer situation. In the first place, social-demographic characteristics, such as gender, age, family status, household size or job, saved in the customer database as basic data (Figure 3), are important. In the second place, psychographic characteristics are part of the personal profile. What counts here above all are the personal characteristics (e.g. non-smoker), interests (e.g. sport, music) and individual preferences, attitudes and needs. Finally, potential data are important, because they affect the economic success of a customer relationship.

The situation factor *place* describes the local context of a person. From the place situation different needs can arise (e.g. in another city: navigation service; at home: pizza delivery service).

Among the possible ways of locating mobile customers the static residence and the dynamic whereabouts can be enlisted to provide place-related services and information.

The residence can be adduced for the macro- and micro-geographical planning of sales measures. While the macro-geographical segmentation is based on administrative and postal regions (e.g. state, city, district postal code/area), the micro-geographical information of the address provides information about consumer behaviour. The micro-geographical market segmentation is based on the behavioural science supposition that persons with homogeneous living and consumer habits occupy contiguous space (neighbourhood effect/segregation), which leads to the emergence of homogeneous population districts (e.g. students, artists and so on). In this way homogeneous, segment-specific customer groups can be filtered for the purposes of Mobile Marketing.

The use of the dynamic whereabouts of the customer can provide, via geographical proximity, consumer reference to shops, restaurants etc. Customers are prepared to pay for situation-dependent value added services that would satisfy their existing needs (e.g. navigation to selected shops or restaurants in a defined area of the place where they happen to be).

The situation factor *time* refers to particular points of time (year, month, day, hour etc.) or time periods/intervals (e.g. work, vacation, weekend), when customers experience certain needs. These needs can appear individually (e.g. in an emergency) or regularly (e.g. daily retrieval of stock exchange prices, weekly purchase of food items). From the time context of the mobile customer an action context (e.g. on the way to work, leisure time) can be derived, which can in turn be used as the basis for sales policy measures within the framework of Mobile Marketing.

The customer situation is then influenced by the situation factor *purpose* to the extent that the goal and/or intention of an action is affected. For example, the goal of trip in the direction of the city may be to go shopping, get cash or visit the theatre.

The situation factor *knowledge* embraces the knowledge status or foreknowledge of the customer with regard to the problem solving possibilities in the concrete situation. The less extensive this knowledge, the more urgent and comprehensive the need of the customer for offers of help. By contrast a high degree of knowledge or foreknowledge with reference to the possibilities offered by the mobile terminal will be necessary for handling situations.

With respect to the situation factor *terminals* the main question is what terminals will be used (e.g. stationary or mobile terminal; mobile phone, PDA or notebook) and how they will be equipped (e.g. UMTS). Here technological compatibility, bandwidth and the net configuration, among other things, are important, and this is also to be taken into consideration when making a situation-specific offer.

The situation factor *carrier* means the choice of transport means (e.g. car, plane, train, ship, bicycle, walking), and each one imposes its own requirements. For example, car drivers are interested in navigation services and traffic information, while rail and air travellers want to know about current rail and flight schedules.

The customer situation is determined by the interplay of all previously mentioned situation factors, and it is this situation that is targeted by the situation-specific offer. The longer and more intensive the relationship with the customer, the more situation factor-specific information can be gathered by the provider. The individual profiles – person, place, time, purpose, knowledge, terminal, and carrier – become more distinct with time and can gradually be compacted into an informative total profile (Link & Weiser, 2006). Thus, a holistic, situation-specific customer model is able to emerge.

When profiles are created, data protection laws and regulations have, of course, to be taken into account, and this is known as Permission Marketing (Kaeding, 2002; Möhlenbruch & Schmieder, 2002). To get the agreement of customers on data collection of this sort it is also important for companies to be able to show benefits for the customer. One of these benefits might, for example, be that the customer saves on transaction costs in the search situation. Another might be that certain movement and use profiles might generate knowledge and future recommendations for the customer in terms of savings or increased use (Link & Weiser, 2006).

INTEGRATED SITUATION-ORIENTED CUSTOMER RELATIONSHIP MANAGEMENT SYSTEM

The two important characteristics of Mobile Marketing – the mobile channel and the situation reference of sales-oriented measures – are now presented in a total context. Figure 3 shows an *integrated situation-oriented Customer Relationship Management system* (Link & Weiser, 2006).

It is becoming more and more important for an entrepreneur to get an exact and comprehensive customer and situation profile in the sense of "one face of the customer" and at the same time to integrate and harmonize, holistically and interrelatedly, all customer-related business processes in the sense of "one face to the customer".

The complete integration of front and back office area is essential for a smooth development of customer contact within the framework

Figure 3. Integrated situation-oriented Customer Relationship Management system

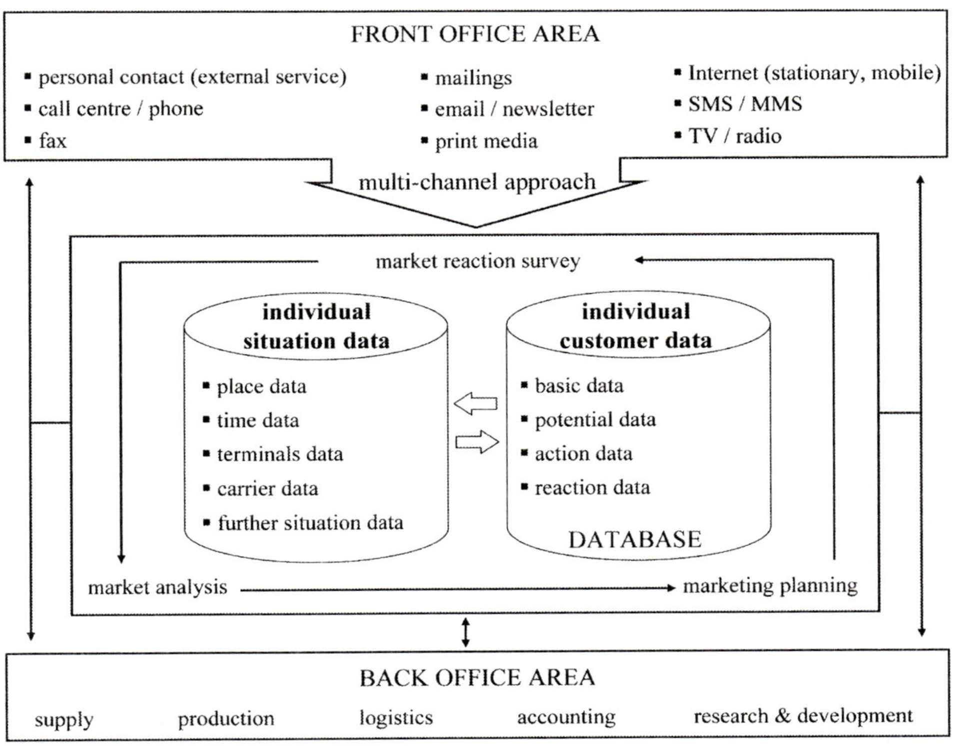

of customer-oriented business processes. With integrated data processing the customer can be guaranteed a high quality long-term information service. Moreover, for reasons of speed, data security and cost lowering in the transaction processes and the overall economic optimization possibilities, integration of the front and back office, as well as within the back office, should be the target (Link, 2003).

In the front office area, i.e. the interface with the customer, the aim is to make it possible for the customer, via the integrative design of all communication and marketing channels, the multi-channel approach, to contact the company via any channel and at any time.

Mobile Commerce, in Customer Relationship Management, represents only one of several touchpoints of the multi-channel concept and offers particular opportunities to build up competitive advantages, such as for example individualization, convenience, omnipresence, and multimediality (Link, 2003).

In terms of transaction suitability care must be taken that the dialogue with the customer can always be resumed and continued precisely at that point where it ended last time. This means that all dialogues, both during and after the contact, are stored on the customer and/or situation database.

The precondition here is the availability of a customer database and – precisely in the context of Mobile Marketing – a situation database that is constantly updated as a central integration platform for all customer-oriented information systems and the rest of the touchpoints in the front office area.

With respect to individual customer data stored in the database, four data categories – basic, potential, action, and reaction – can be distinguished (Link & Weiser, 2006). The continuous recording, storing, and analysis of these data helps to "model" the customer and create a customer profile that is suitable for realistically mapping current customer behaviour and forecasting future behaviour as accurately as possible. At the same time the continuous recording, storing and analysis of individual situation data can be understood as an attempt to model the situation or create a situation profile. Both of these types of profile represent the basis for decisions in Mobile Marketing.

If one adds the customer data available in the database to the context-sensitive information from the situation database, personalized, situation-specific offers can be made to the customer via push-mechanisms (Link & Weiser, 2006). For example, the combination of the situation data with the action and reaction data is useful, because certain success patterns of Mobile Marketing can possibly be identified from this data. By means of the customer and situation profile gained, the ideal goal of individual or direct marketing in Mobile Marketing can be achieved, viz. to make a tailor-made offer of information or a service to the "right" customer in the "right" place at "right" time with the "right" arguments.

Whether and how such offers can "pay" for the provider has to await specialist analysis, and this forms the subject of the next sections.

EVALUATION OF SITUATIONS

Reference Object Hierarchies in Marketing Profit and Loss Statements

Decisions in general and calculations for decision-making of marketing accounting in particular depend on whether the decision-maker has a suitable image of the situation, including the existing action alternatives (Link & Weiser, 2006). This leads, for example, to the concept of the model as image of reality. The more this image models the real structure of the decision situation and the action alternatives, the less danger there is of a wrong decision. In contrast, the precision of the image is limited both in its user-friendliness and the profitability of the model design.

The precision of the image in decision making, in concrete terms, means that all relevant values are contained in the decision model and that all irrelevant values remain outside (Link & Weiser, 2006, p. 211, with references cited there). Relevant for the decision about an alternative in a particular decision situation are those values that accrue if the planned alternative is carried out and that cease to apply or do not happen if the alternative is not carried out. In this sense decision-relevant variables have to be future-related, expectation-driven and susceptible of being influenced and individually calculable.

The following contains a more detailed account of the image precision of decision situations taking attribution aspects into account. The attribution of variables happens in the case of items that are the subject of operational decisions, which is why they can be designated as decision objects or reference objects.

Reference objects in marketing accounting can be the most varied characteristics of service packages (e.g. products, brands, orders), service recipients (e.g. customers, markets, regions), service providers (e.g. departments, systems, distribution channels) or actions (individual measures, campaigns, strategies). A part of these reference objects is subsumed under the heading of sales segments, and so the term sales segment accounting is used. Here, we wish to use the more comprehensive term marketing profit and loss statements, which can be used for all reference objects.

A systematic treatment of selected reference objects in marketing accounting is shown in Figure 4 (Link & Weiser, 2006). Unlike the "established" reference objects the situation represents a wholly new reference object for marketing profit and loss statements. This depends, among other things, on the rapid advance of Mobile Commerce. The following section contains a detailed treatment of the individual levels of situation profit and loss statements.

A decision problem, in terms of attribution aspect, is correctly modelled if the values are as-

Figure 4. Reference object hierarchies in marketing profit and loss statements (can refer in principle to every super or subordinate unit and combinations of units)*

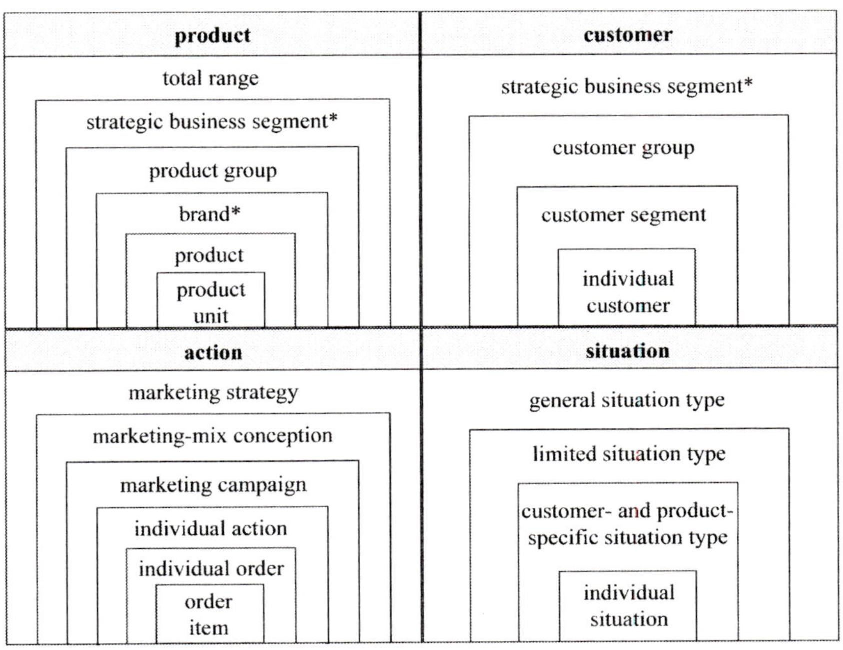

signed respectively to lowest levels from Figure 4, at which this is possible directly, causation-appropriate and coding-free (according to Riebel, 1994).

Because the same value often affects reference objects simultaneously, several descriptors are adduced in such cases to indicate the values: for example, a particular order position can be assigned to a particular product and a particular total order, which in turn can be assigned to a particular customer and distribution channel. One should try to keep open as many evaluation possibilities according to different dimensions and reference objects as one can, and at the same time pay attention constantly to the profitability of data capture, storage and evaluation.

Situation Profit and Loss Statements

The tasks of the marketing profit and loss statement are above all to be seen against the background of a selective sales policy (Link & Weiser, 2006, p. 218, with references cited there). Selective sales policy describes the approach, sales and the sales efforts to concentrate on or restrict the selected products, customers, orders (type, size), sales areas ways and methods as also situations, which in the long-term can yield the greatest success for the company and lead to competitive advantages in the sales market. There, elimination decisions also play a role. In this regard it is to be noted that rationalization processes determined by the exclusion of concrete reference objects regularly last for long periods, with the result that information about the availability of (capacity) costs becomes especially important.

With regard to the maturity of the results that can be achieved with a profit and loss statement one must differentiate whether only variable costs have been apportioned on one level or also fixed costs and capacity costs. Provided the declared level amount of coverage has been calculated only via the variable costs, it is relevant for short-term decisions, because in determining it the starting point has been a given operational stand-by and/or an already activated investment volume.

The time decision horizon against which a step contribution margin for a reference object is to be seen depends moreover on the degradability or earmarking length of fixed costs and/or stand-by costs. If long-term effective decisions for reference objects are made, which, for example, affect investments in investment funds, personnel, marketing and sales measures, investment calculations should be carried out as a basic principle.

With the focus on the situation as new reference object for marketing profit and loss statements – *situation profit and loss statements* – it may be useful at this point to provide a brief excursus on the fundamental situation reference in business dealings (Link & Weiser, 2006).

Both the theory and practice of business dealings are inherently strongly characterized by the image of the resourceful entrepreneur who is constantly on the lookout for new business opportunities. Opportunities can be interpreted from an entrepreneurial point of view generally as situations in which a profit can be made. This is typically the case when there is an unsatisfied need that one can meet as either the first or the best. In this sense the entrepreneur is always situation-oriented. The situation-orientation of the entrepreneur ranges from detecting and analysing a concrete individual situation in a concrete current market situation to the recognition and analysis of future business and general situations of a macro-economic, management, technological or other nature.

This is reflected also clearly in several descriptions of the planning and control process (management process) in companies and also especially in marketing. The entire process of marketing planning and control begins significantly with situation analysis and prognosis with regard to surrounding systems and the business venture (Link & Weiser 2006).

The entrepreneur, in the individual and entrepreneurial situation, weighs his input against the prospective profit offered by the situation. This means that a *situation evaluation* is a necessary component of the management process and business transaction. In the case of the analysis of the entrepreneur situation the result is the company evaluation; in the case of the concrete individual situation in the concrete current market situation the result is the action evaluation as comparison between costs and profits of a particular action.

If an action is no more than the economic exploitation of a concrete situation – do not action and situation profit and loss statements in principle amount to the same thing? Why is a situation profit and loss statement in Figure 4 also proposed in addition to an action profit and loss statement?

Situation profit and loss statements reveal the entire profit potential associated with a particular situation and not only the profit share "cut out" from the potential by a provider through his actions. The entire potential of a market situation is made up in the extreme case of the following components:

1. Sales potentital of an individual customer for an individual product group (in Database Marketing commonly to be found under the heading of potential data; the sales share, "cut out" by provider for himself, is the customer share), minus costs;
2. Cross-selling sales potential of this invidual customer with reference to other product groups, minus costs;
3. Corresponding sales potential of other customers, who are likely to be in the same situation at some time in the future, minus costs;
4. Corresponding cross-selling potential of other customers, who are likely to be in the same situation at some time in the future, minus costs.

If one takes the travel situation as an example, the various fields in Figure 5 correspond to the above-mentioned.

This example has been chosen deliberately. Situation profit and loss statements might be of special interest in the area of Mobile Commerce, where travel situations in the sense of mobile situ-

Figure 5. The various levels of situation profit and loss statements

	1 product	**several / all products**
1 customer	**concrete customer- and product-specific individual situation** (e.g. passenger XY with a certain need for catering), concrete place, time, carrier	**limited, customer-specific situation type** (e.g. passenger XY with a certain need for catering, accommodation, entertainment and wellness), typical place, time, carrier
several / all customers	**limited, product-specific situation type** (e.g. standard passenger with need for catering)	**general situation type** (e.g. standard passenger with need for catering, accommodation, entertainment and wellness)

ations are typical. Here the situation profit and loss statement is employed in two variants:

- Calculation of profit potential for different types of travel situation, in which profit- and location-dependent, automated and in real-time customer-specific offers and the appropriate navigation aids can be shown on the display (e.g. midday break situation of the traveller during a motorway journey from the point of view of service areas and gas station operators; evening situation of the traveller on the periphery of a city from the point of view of hotel and restaurant operators).
- New calculation of all profit potential of customers by inclusion of customer-individual situation data (cf. the situation database as supplement to customer database in Figure 3).

So far, the reference objects, product and customer, are seen in particular as triggering offers in Customer Relationship Management, especially in Database Marketing. Then there is the situation where in Mobile Marketing a real-time reaction of the provider is necessary if he wants to make profitable use of the spontaneous, to some extent "transient" need situation. For this he needs access to all the information about the profitability of such situations, possibly calculated in advance on the basis of Figure 5.

In future therefore a basic distinction is to be made in mobile Customer Relationship Management between

- *Product-triggered*,
- *Customer-triggered*, and
- S*ituation-triggered* offers.

But also for numerous other situations in Mobile Commerce there are similar analysis opportunities (on the details of the forms and possibilities of situation-oriented considerations in Mobile Commerce cf. Kriewald, 2007; Grandjot, 2006).

Outside Mobile Commerce the example given in Figure 5 in the same form would be relevant for the basic profit potential of tourism customers; also for tourist providers the art is to cut as large a profit share as possible from the fields of Figure 5.

A situation profit and loss statement of this sort can in the short-term (for an individual period) or in the long-term (over all relevant periods) be drawn up. In the first case, it is a matter of an operative profit and loss statement corresponding to the contribution accounting and in the second case there is a strategic profit and loss statement according to the model of an investment appraisal.

SUMMARY

Mobile Marketing is becoming the "marketing instrument of the future" (Bauer, Reichardt, & Neumann, 2004, p. 33). Increasing numbers of companies are integrating the mobile channel into their multi-channel marketing and using it in a targeted way as a sales policy measure.

As has been demonstrated, situation appropriateness is to be regarded as a major success factor of Mobile Marketing. Only by taking the relevant customer situation into consideration the supplier can make a situation-specific offer via the mobile channel. Thus, the customer situation is determined both by the basic situation – idle times, search, need or quasi-stationary – and by the interplay of individual situation factors – person, place, time, purpose, kowledge, terminals, carrier. Possibilities of individualization and requirements are thus limited no longer to customer data, but include situation data in the sense of the extended concept of individualization.

The combined customer and situation database represents the central integration platform in an integrated situation-oriented CRM system. The customer and situation profiles generated as a result, i.e. customer/situation models, form the basis for decisions in Mobile Marketing.

Decisions of this sort should be based on analyses as well as profit and loss statements, which deal with the evaluation of the (economic) situation potential – customer value and situation value consideration –, to be able to estimate whether and how situation-specific offers "work out" for the supplier.

Finally, it can be established that a situation-specific offer via mobile terminals are promising for the entrepreneur when the two central success factors of Mobile Marketing – situation appropriateness and (economic) situation potential – have been clarified, i.e. when the offer is context-sensitive and at the same "works out". The success of mobile sales policy measures can accordingly be guaranteed by a synergic interplay of market and result orientation and/or Mobile Marketing and controlling.

REFERENCES

Bauer, H. H., Reichardt, T., & Neumann, M. M. (2004). Zu jeder Zeit an jedem Ort – oder über die Potenziale des Mobile Marketing. *Direktmarketing*, *40*(10), 32–37.

Bennemann, S., & Möllenberg, A. (2001). *M-Commerce aus der Marketing-Perspektive. Technische Voraussetzungen, State of the Art, Künftige Möglichkeiten*. Norderstedt, Books on Demand.

Figge, S. (2007). *Innovatives Mobile Marketing. Kontextabhängige Kundenansprache mit Hilfe mobiler Portale*. Hamburg: Verlag Dr. Kovač.

Grandjot, T. (2006). *Mobile Scorecard. Entwicklung einer Balanced Scorecard für das mobile Customer Relationship Management*. Hamburg: Rainer Hampp Verlag.

Kaeding, N. (2002). Mobile Business und Datenschutz. In R. Reichwald (Ed.), *Mobile Kommunikation – Wertschöpfung, Technologien, neue Dienste* (pp. 193-204). Wiesbaden, Gabler Verlag.

Kriewald, M. (2007). *Situationsabhängiges mobiles Customer Relationship Management. Analysen – Wettbewerbsvorteile – Beispiele*. Hamburg: Verlag Dr. Kovač.

Link, J. (2003). M-Commerce: Die stille Revolution hin zum Electronic Aided Acting. In J. Link (Ed.), *Mobile Commerce. Gewinnpotenziale einer stillen Revolution* (pp. 1-39). Berlin/Heidelberg: Springer Verlag.

Link, J., & Seidl, F. (2008). Der Situationsansatz als Erfolgsfaktor des Mobile Marketing. In H.H. Bauer, T. Dirks, & M.D. Bryant (Ed.), *Erfolgsfaktoren des Mobile Marketing – Strategien, Konzepte und Instrumente* (pp. 51-70). Berlin/Heidelberg: Springer Verlag.

Link, J., & Weiser, C. (2006). *Marketing-Controlling. Systeme für mehr Markt- und Unternehmenserfolg*. München: Vahlen Verlag.

Meffert, H. (2000). *Marketing – Grundlagen marktorientierter Unternehmensführung*. Wiesbaden, Gabler Verlag.

Möhlenbruch, D., & Schmieder, U.-M. (2002). Mobile Marketing als Schlüsselgröße für Multichannel-Commerce. In G. Silberer, J. Wohlfahrt, & T. Wilhelm (Ed.), *Mobile Commerce – Grundlagen, Geschäftsmodelle, Erfolgsfaktoren* (pp. 67-89). Wiesbaden, Gabler Verlag.

Riebel, P. (1994). *Einzelkosten- und Deckungsbeitragsrechnung: Grundfragen einer markt- und entscheidungsorientierten Unternehmensrechnung*. Wiesbaden, Gabler Verlag.

KEY TERMS AND DEFINTIONS

Extended Concept of Invidualization: The extended concept of individualization includes the situation data (situation factors) in addition to the customer data. Therefore the supplier is able to model the (mobile) customer situation comprehensively and then to draw up a situation-appropriate offer.

Macrostructure of Situations: The macrostructure of situations contains the basic situations: idle times situation, search situation, emergency situation and quasi-stationary situation.

Microstructure of Situations: The microstructure of situations contains the situation factors: person, place, time, purpose, knowledge, terminals and carrier.

Mobile Marketing: Mobile Marketing is market-derived leadership using mobile terminals.

Situation Appropriateness: The *adequacy* of an offer made by a supplier in a specific customer situation (considering the situation factors), central success factor of Mobile Marketing.

Situation Potential: The economic potential revealed by a situation, central success factor of Mobile Marketing.

Situation Profit and Loss Statement: Situation profit and loss statements imply the evaluation of the reference object "situation" in order to concentrate on situations which can yield the greatest success for the company and lead to competitive advantages in the sales market. Situation profit and loss statements reveal the entire profit potential associated with a particular situation. Reference object hierarchy in situation profit and loss statements: individual situation, customer- and product-specific situation type, limited situation type, general situation type.

Chapter 8
A Framework for Understanding Mobile Value Offering through Multi-Country Studies

Kala Chand Seal
Loyola Marymount University, USA

ABSTRACT

Our chapter introduces an integrated way to look into the world of mobile offerings, referred often as m-commerce,- through a cross-country analysis of multiple applications in various countries in Europe and Asia. We investigated the role of four dominant factors - technology, government, financial incentives and culture - that influence mobile value offerings and connected them to Methlie and Pederson's (2005) model of extrinsic and intrinsic attributes in mobile space. We found that there are many applications and services available that provide added value to customers, but an ill-defined business model prevents their successful realization. An unclear working relationship among the value players and their relative power position in the market also hinder the roll out of many services. The key findings from our research consist of-: i. the role of the four factors in development and deployment of m-commerce across multiple countries; ii. The level of influence of each of the four factors has on various applications; iii. the importance of value delivery, not merely the promotion of the "bells and whistles" of new applications, with a clear focus on a viable business model; iv. the role of socio-cultural differences in influencing mobile value applications and acceptance. In addition, we found that competition in the early phase of market growth is counter-productive and cooperative relationship between all value players is absolutely crucial for healthy growth of the mobile market and ultimately leads to a win-win situation for everyone.

INTRODUCTION

Mobile commerce (m-commerce) is touted as the next revolution for businesses by the technology, business, and academic communities. Within the last few years, the mobile phone market has seen extraordinary growth and cell phones now have become prominent in the lives of consumers worldwide and present a huge opportunity to businesses

DOI: 10.4018/978-1-60566-074-5.ch008

in offering new goods and services as well as in delivering their existing goods and services in many innovative ways to the customers. Technological innovations continue to add to the utility of these devices so that they are viewed as much more than simple telephones. Businesses are now eager to take advantage of the anytime-anywhere promises of m-commerce, thereby opening up another channel to reach existing and potential customers in order to grow their revenues. However, the business community is unsure which application will be the driver of mass adoption. Prior research postulates that this "killer application" does not exist (Yuan & Zhang, 2003; Seal et. al., 2003; Cooke et. al., 2004), and applications must be developed and tailored for each market where deployment is taking place.

This chapter is intended to provide a framework to understand the values in various types of mobile offerings and the market characteristics that govern the uptake of such offerings with the consumers. The framework is based on the findings from multiple studies conducted over a period of time across several countries. With four aspects including (1) technology, (2) government, (3) market characteristics, financial and value incentive, and (4) social culture supported by previous researches on mobile phones and mobile services, this chapter demonstrates the mobile phone and mobile services subscribers and the operation of telecom industries in Europe, America, and Asia. For both practitioners and academicians, this chapter provides a reference tool for analysis of various mobile markets for making significant business decisions in mobile commerce.

The studies in this paper consisted of firsthand interviews with senior managers from various mobile value players in the U.S.A as well as in Belgium, Iceland, Ireland, Netherlands, United Kingdom, China, Japan, Singapore, South Korea and Thailand. All the studies were conducted with small groups of MBA students who took part in it for completion of their degrees. The MBA program at the University of the author drove the selection of the countries and the companies with very little room for deviation. Thus it can be considered as a convenient sampling with qualitative data collection through interviews that lasted from 1 to 4 hours. While such a data collection does not provide the rigorous strength of random sampling, surveys and subsequent statistical analyses, it does provide very information rich findings and overarching macro views of the markets. Such views are helpful for strategic positioning and marketing of new mobile value offerings and identifying opportunities.

The paper is organized to provide the reader with an overall definition of mobile commerce and the value players in the mobile value network followed by the actual mobile value offerings and their attributes. Next section presents the central theme of the paper:-the various business models that are in existence in multiple countries and how they are affected by the four influencing factors described above. This analysis provides the framework that would allow the reader to connect a value proposition in mobile space to a concrete and viable execution plan and evaluate existing mobile business models in a particular market. The paper concludes with a summary of the findings and conclusion.

M-COMMERCE: DEFINITION AND VALUE NETWORK

There are multiple viewpoints on what constitutes m-commerce (see e.g. Barnes, 2002; Sabat, 2002). However, for the purpose of this research, we define m-commerce as the transaction of goods and services through a mobile hand held device. These goods and services include text messaging (SMS), multimedia messages (MMS), ringtones, downloadable music, mobile-banking, entertainment services through streaming video, ticket purchases, micro-payments, and many others. The deliveries of such goods and services require coordination and interaction between multiple stakeholders.

Figure 1. Mobile Value Network

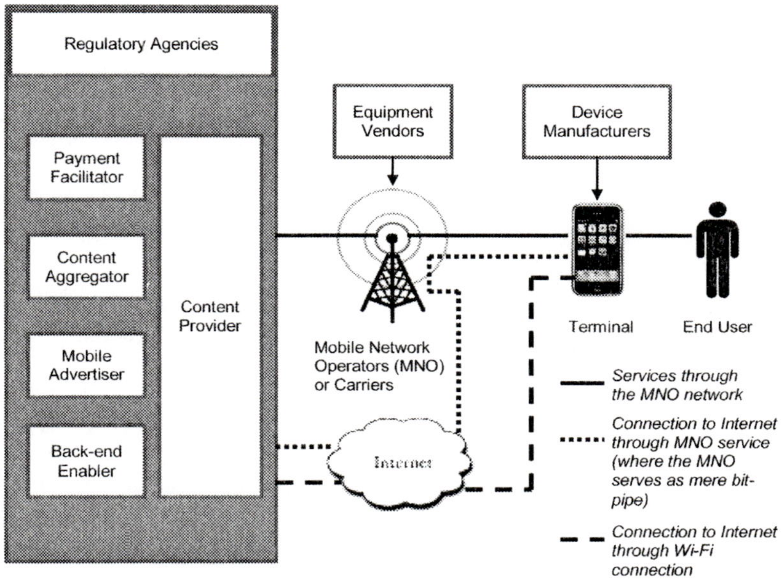

The mechanism and process through which the stakeholders in the mobile commerce transfer value while creating, providing, delivering, and maintaining products and services is called the mobile value chain (Lu et. al., 2007). Sometime it is also referred to as the mobile eco-system.

However, value chain is probably a misnomer because some of the value activities can take place simultaneously. Stabelll and Fjelstad (1998) termed such a configuration as value network. Based on that, Molberg (2006) presented a value network for the mobile commerce. Figure 1 shows a slightly modified version of the network where the terminals are shown to be connected with the end users directly and serve the purpose of the final destination of the mobile content and values before they reach the end-users. Such a modification is necessary to understand the power of the terminal in the mobile market. While the mobile ecosystem is typically Mobile Network Operators (MNO) dominated, an innovative terminal vendor can completely change the power distribution in the mobile eco-system. For example, i-phone in the U.S. changed the dominance of the MNOs and made the device to be a very powerful player in the value network. Customers' demand for i-phone forced the MNOs to offer very competitive services for the i-phone.

Relationships between various players in the value network dictate revenue flow and comparative advantages, directly affecting the nature of products and services and their corresponding price levels. Members of the mobile value network include, but are not limited to, mobile network operators or carriers, content creators, content aggregators, application and back-end systems developers, equipment vendors, and handset or terminal manufacturers. A brief overview of the roles of each player is given below.

Content Creator: As the name implies, this is the source of the actual material that will be viewed or consumed by the end users. The content creator (or owner) controls the rights to the material, and negotiates deals with the entities that will distribute the content. Some examples of the content creators will be the various TV stations and studios like NBC, CBS, Sony, and games and other types of content developers such as Electronic Arts, MForma (www.mforma.com), and iLoopMobile (www.iloopmobile.com).

Carrier (MNO): Often a major telecom company, the carrier or the mobile network operator (MNO) owns the transmission equipment, brands and controls the software on the handsets, operates retail stores to sell plans and subsidized handsets, and does end-user billing for basic services. However, depending on their position in the value chain, MNOs can serve as a simple network provider or a "bit pipe" to an intermediary gateway, portal, or trusted third party (Tsalgatidou and Pitoura, 2001).

Aggregator: Content aggregators stand in-between carriers and content owners and provide several key value-added services. For a small content company, it may be very difficult to gain access to the large carrier networks. By going through an aggregator, they now have access to ALL the networks. Conversely, the carriers only need to deal with one company, the aggregator, in order to offer all of the content of those small individual providers to their customers. Aggregators typically act as a one-stop shop for both content creators and carriers and can provide additional services such as intellectual property protection and management, payment clearing and billing of the end users. JAMDAT (www.eamobile.com) and Bango (www.bango.com) are examples of content aggregators.

Applications and Back-End Enabler/Content Management Systems: Unlike e-commerce that is based on the common standard of TCP/IP and html (later XML) protocol, m-commerce is rife with proprietary technologies that need to work together. Each content must be tailored for the type of network (e.g. GSM or CDMA), the application environment (brew or java), the operating systems and the types of handset. Various content providers, aggregators, and carriers somehow have to keep track of all of these so that the content can be encoded in the proper format to deliver over a particular network to a particular handset. This process is called "porting". On top of porting, there should be some way of keeping track of the user activities, billings and collection of payments. It has to be managed in terms of status of rights clearance, price for billing, and billing processing. Furthermore, the content has to be backed up in case of equipment failure, it has to be easy for the provider to add content, and easy for the consumer to access content. All of these functions are the role of the application providers and the back-end enablers. In some cases it is simply a software package that a carrier either develops in-house or purchases from a third party. Or, it can be a service business that handles all of these issues and charges a fee to the providers and carriers that use the service. Back-end enablers allow service or content providers to focus on the core of their business to better serve customers. They also have shared risk-reward arrangements with the carriers with various kinds of revenue sharing models, usually tied to the number of subscribers. The main value proposition for this group is having a proprietary or enabling technology to bring people together. This group provides more value to business entities in the value network as opposed to consumers since they often create store fronts, assist with marketing on the retail side, pool resources together (i.e., video and SMS aggregators), and provide one-stop shop services to the carriers.

Payment Facilitator. These companies provide solution for both making and collecting payment for mobile services. They typically develop the technological solution required to make and accept payment from a mobile device and then partner with a bank for the final clearing and settlement of the payment.

Equipment Vendors. These companies provide the equipment for the infrastructure. They are typically the telecommunication vendors that supply equipment for the network of the carriers.

Device Manufacturers: It goes without saying that none of the mobile offerings would be possible without the device in the consumers' hand. Device manufacturers manufacture the handsets.

Sometime MNOs will contract the handset makers to create their own brand of handsets, but the handset makers are still present behind the scene. For example, Sony and Panasonic are two major vendors who create NTT DoCoMo branded handsets. Nokia, Motorola, Samsung, and many other vendors (including Sony Erickson), on the other hand, create a majority of the handsets under their own brand.

Regulatory Agencies: The regulatory agencies in the value network has an over encompassing effect and that is why they are shown to stretch across the various value providing items in the mobile value network.

Mobile Value Offerings

To clearly understand the values of the offerings in mobile domain, we have to understand the values created by the participants in the value network. Pederson et. al. (2005) suggested a simple model for this purpose. His model connects the actors in the supply side of the mobile business with the end-users who are on the demand or consumption side of the mobile business through delivery of attributes. The content and service level expressions of these attributes are typically termed as mobile values. Some examples of such expressions are time saving, mobility, entertainment, spontaneous needs and localization (Anckar and D'Incau, 2002). Cooke et. al. (2004) further postulate these to be the derivative of values provided by their house of values consisting of ubiquity, convenience, location and personalization capabilities of the mobile devices and the associated network. Faber et. al. (2003) also provides a framework to understand the value of the service offerings in mobile domain. However, the framework by Pederson et. al. (2005) probably provides the simplest form of abstraction of the values along with the connection between the providers and users of such values. An expanded version of his model along with the details of the attributes as explained in Methlie and Pederson (2005) is shown in Figure 2 followed by the explanations for the various parts of the model.

Actors: Actors are on the supply side of the mobile domain and consist of all the players described above. They are in charge of delivering the value to the end-users and collect revenue in return. In the value network, the revenue generated may be shared amongst all the actors involved in delivering the attributes.

Attributes: The attributes are the mobile value propositions for the end users and various expressions of these attributes through business models deliver the value to the customers. Methlie and Pederson (2005) divide these attributes into extrinsic and intrinsic attributes. Extrinsic attributes are related to the values provided by the underlying mobile network while the intrinsic attributes such as the nature of the content, the quality of the handset, etc., are more personal to customers. A detailed discussion of the various attributes is given below and is necessary to understand the values provided by various mobile offerings in the current market.

Extrinsic Attributes

Direct Network Effect: Direct network effect is created by the presence and property of the mobile network as the mediating technology between users and the supply side of the value network. It is also a mediating technology between the users in creating peer to peer (P2P) connections. Direct network effects are related to increasing value of a service as the size of the network increases (Leibowitz and Margolis, 1999). This is also known as network externalities which has been one of the dominant reasons behind the success of the Internet and e-commerce (Affuah and Tucci, 2002, Chapter 3). As the mobile network grows, it should provide the similar values to the end users and due to its personalization capabilities, convenience and location based services, should be able to take it to the next level. We have already started witnessing such an effect in many

Figure 2. Actor-Attributes-User model of the Mobile Eco System

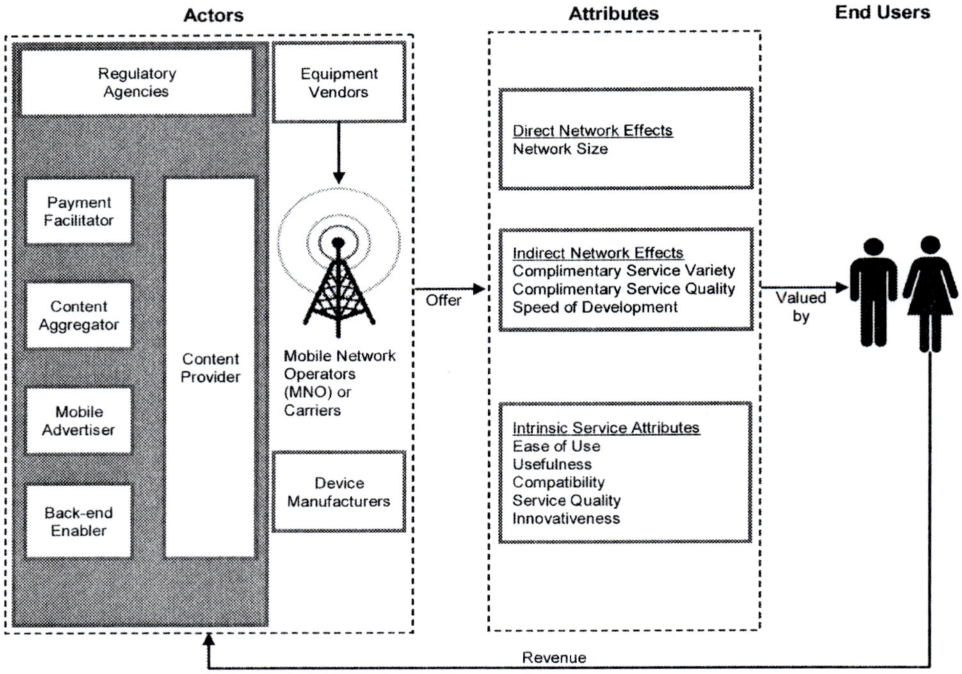

interpersonal communications taking place in the mobile space. SMS, MMS, multiplayer games, e-mails, and mobile advertising are all examples of such communications and services and all of them are dependent on the numbers of end-users attached to and supporting the network (Molberg, 2006).

Indirect Network Effects: Indirect network effects attributes are the auxiliary values delivered to the end-users due to the presence of the network providing a platform for complementary services and products (Gupta et. al, 1999; Methlie and Pederson, 2005; and Molberg, 2006). The indirect network effects in the mobile services can be divided into three categories: complementary service variety and quality, speed of development, and common reference (Methlie and Pederson, 2005).

Complementary service results from creation of opportunities or demand of a service based on the presence or usage of another service. For example, usage of SMS may give rise to the use of e-mail on the mobile phone as users are already comfortable in typing through the phone keys and thus are not as affected by the form factor. Similarly, presence of MMS capabilities make the phone a perfect channel for offering video based entertainment services such as TV and movie downloads. In South Korea, the development of Internet and subsequent on-line gaming created a huge demand for multi-player games in the mobile space, a feat that is unmatched by any other countries (Weitenberner et. al., 2006). The demand for western content in Thailand and the challenge of maintaining quality of multimedia mobile entertainment on a 2G network precipitated back-enablers as a viable and profitable business model (Sgriccia et. al., 2007).

Speed of development can accelerate or hinder the uptake of a particular mobile service. When mobile services beyond voice started to come out in the market, development took place at a rapid pace and many companies wanted to join the "band wagon" to avoid being left behind on the technology curve. However, later we witnessed a reverse trend in 3G where multiple failure efforts, upsets and glitches (e.g. see Savio, 2001; McGinty and Da Bona, 2004) made everyone in the value network skeptical about the promise of the 3G network and impeded 3G based content development.

Common reference relates to the educational effect of a network that allows various types of users to understand an offering by experiencing the similar network presence in the mobile services. For example, in the United States, the i-phone sets the standard for the touch terminal and display quality and thus educates the users about the possibilities of receiving multimedia entertainment on their phone. The prevalent presence of Blackberry amongst the business users made the e-mail in the mobile media as an accepted service. Japan's Felica and Suica chip used in the beginning for the station entry for Japan Railway gave rise to the near field communication (NFC) based systems and possibility of expanding the mobile phone as the integrated life style device (Weitenberner et. al., 2006; Sgriccia et. al., 2007).

Intrinsic Attributes

Intrinsic attributes are the value offerings that are easily understood by the current end-users of the mobile contents. Methlie and Pederson (2005) as well as Molberg (2006) divide them into *service specific intrinsic attributes*, *network specific intrinsic attributes*, and *terminal specific intrinsic attributes*. Out of the three groups, *service and some of the terminal specific attributes* bring value to the end-users while the others provide the invisible but necessary support functions to deliver the service attributes. While the presence and perfect execution of the *network and terminal specific attributes* will not be noticed that much by the users (with some notable exceptions such as i-phone), absence or the lack of quality in these attributes will affect the users' view of the service level attributes negatively.

Service specific attributes are mainly the content and the content services provided to the users along with some convenience services. It can be informative, entertaining and convenience enhancing and all of them aim to improve the quality of life or experience for the consumers. Table 1 below shows the various content and content services that are available in the mobile domain and are usually viewed as value offerings by the consumers. Note that the membership to the category is not mutually exclusive, one value offering can belong to multiple areas. The integrated life style device is the concept that is becoming popular with the consumers (Weitenberner et. al, 2006) and it encompasses many of the services in the other categories thus making the cell phone as the one-stop shop for everything.

Network specific intrinsic attributes are decided by the quality of the network such as the data rates, network security, and network coverage. The main players here are the infrastructure vendors and the MNOs.

Terminal specific intrinsic attributes are decided by the influence of the quality and design of the terminals used by the end-users to interact with the mobile value network. They consist of the ease of use, the design (size and display), resolution, connecting support such as generation of the network as well as Bluetooth and other near field communications technology, battery life and the quality of integration of the various software and the operating platforms. As one can see, Apple's phenomenal success with i-phone came from the perfect execution of the terminal level intrinsic attributes that surpassed the value perception for the end-users thus making the company a dominating player in the value network.

Mobile Commerce Business Models and Influencing Factors

To understand the current mobile commerce environments, we have to understand the various business models that are in existence for delivering

Table 1. Service Specific Attributes in Mobile Domain

	Content Services		
Content	Interpersonal Communications	Non-interpersonal Communications	Convenience Services
Text	SMS (Peer to Peer or P2P)	SMS (Machine to Person or M2P and P2M as in the case of voting for American Idol show)	Phone concierge providing various location based services
Voice Messages	MMS	TV	Mobile payment
Music	E-mail	Ringback tones	Friend locator
Video	Multiplayer games	Internet connection and browsing	Mobile banking
Ring tones and Ringback tones	Gambling	Mobile advertising	Mobile offers such as instant coupons, discounts, sale announcements, etc.
Pictures including wallpapers	Instant Messaging		Mobile shops, marketplace, and information portals
Games	Video Telephony		Integrated life style support device

the values discussed in the sections above to the customers. However, the term *business model* in electronic and mobile commerce is defined in various fashions and there is no consensus for the exact definition (see for example, Lambert, 2005, for an excellent discussion on the various taxonomies of business models). For our purpose in this paper we will use a modified version proposed by Timmers (1999) and Tsalgatidou and Pitoura (2001) and define business model as an architecture for execution of the delivery of the value propositions including a description of the business actors and their roles and revenue sources. These architectures result in labels that sometime describe the final form of the business models (such as E-Shop, Content Portal, etc.) or the activities that are being carried out (such as ASP, Value Chain Integrator, etc.). Success of a mobile business model is dependent on various environmental factors such as the market needs, the presence of competing and complimentary services, and the various technological and regulatory constraints within which the business model operates. Based on the work by Henten et. al. (2003) we can classify these factors into four general categories: *Technology Infrastructure, Government* that includes policy and regulation and a part of the market development incentive, *market characteristics that include financial incentives, power of the various players in the value network, and the pricing effect*, and the *socio-cultural characteristics of the market and the end-users*. The attributes discussed in the previous sections are all affected by these factors and the players in the value network need to take them into account to deliver the attributes in a way so that their values are apparent to the end-users.

In the rest of the paper, we will be using the above factors as the principal backbone for our framework to describe the development path and efforts taken in the mobile commerce area by different countries in Europe and Asia. The information in this section is based on selected extract from our published research carried out in those countries (Cooke et. al., 2004; Weitenberner et. al., 2006; Sgriccia et. al., 2007). We would also provide characterizations of the various types of mobile market based on the dominant roles of the various key players. We hope that our analysis of the effect of each factor in various countries will provide an insight for understanding the overall nature of the mobile commerce development and also to identify business models more likely to succeed in a given context.

Technology

Technology defines the access to and the capabilities of the network itself and existence of a dominant or multiple competing standards can affect the network attributes. For example, the high bandwidth based entertainment content cannot be delivered without the 3G technology and the availability of the corresponding "smart phones" to the end-users.

In the U.S., Mobile Network Operators (MNO) and application developers have technology and network infrastructure hurdles to overcome due to the highly fragmented networks. The limitation of the infrastructure presents itself through problems such as dropped calls while traversing certain areas of a city and complete sections of the country not being served by any operator. There is no single standard, and principal MNOs use different technologies resulting in lack of true interoperability and disparate user experiences. This decreases the network specific intrinsic value available from the m-commerce. 3G has been introduced very recently in the U.S. and the overall user experience has not been very positive due to the spotty availability and poor support.

In Europe and Asia, the technology is lot more uniform and stable. It creates opportunities for interoperability and encourages more value added services. A number of the mobile companies in Europe and Asia are offshoot from the government based telecommunications giant that used to operate in those markets. In Iceland, the three wireless carriers, OG Vodafone, Landssiminn (Iceland Telecomm) and IMC Island ehf (Viking Wireless) are transformation of former government telecommunications companies and they offer coverage to areas that encompass 96% of the population through GSM network. By standardizing the network structure under GSM, interoperability is less of a concern. The popularity of Short Messaging Services (SMS) across all three networks is a testament to Iceland's interoperability. Similar trend can be found in the rest of Europe. For example, GSM is a dominant standard in the UK, Ireland, Netherlands, and Belgium. Most of these countries are now operating on a 2.5 G to 3G networks and offer a high level of ubiquity in their network presence (Cooke et. al., 2004). Such ubiquities allow them to offer interpretabilities and new values. In the UK, Vodafone wants to standardize how the transaction is done by mandating all new applications to use an applet (application object) framework. This would allow them to offer a service similar to the i-mode service offered by NTT DoCoMo in Japan on multitude of complex end-user devices and extend the similar services in other countries (personal interview with Vodafone representative, 2004).

Asia is a mixed bag when it comes to mobile technology with Japan and South Korea at the forefront of development. Though Japanese network operators have had difficulty introducing 3G services using W-CDMA (NTT DoCoMo and Vodafone K.K.) and CDMA2000 (KDDI) technologies (Bradley et. al., 2006) they have overcome those issues and both Japanese and South Korean markets now boasts some of the most advanced 3G technologies and are looking ahead for the 4G implementation (Dholkia et.al., 2004; Onoe et. al., 2007)

3G mobile networks and services have been available to consumers in Singapore since 2005 (Myers, 2007). However, while mobile phones are widely used throughout the country, consumers still do not demand high functionality on their phones, preferring to use personal computers for value added services (Myers, 2007). Indeed, Singapore's high home PC and broadband penetration rate encourages using the Internet on the PC, which consequently limits mobile use to basic voice and SMS functions (Myers, 2007). Further, the high mobile phone penetration rate has caused the market in Singapore to become overly saturated, making it very difficult for mobile phone companies to be profitable. Subsequently, mobile companies are now looking for other avenues, such as mobile advertising (e.g., ACME Mobile's AdWrap) in order to maintain profits.

Unlike other advanced nations such as Japan and Singapore, China profits from outdated technologies. Currently, value added services are limited to 2 and 2.5G platforms involving GSM transport technologies, which may explain the popularity of SMS services. Additionally, SMS was initially priced low - approximately one-fourth of the price of one minute of airtime (Aditham, 2007). Subsequently, SMS volumes are astronomical, with the Chinese sending an average of 1 billion SMS messages a day (China Mobile Press Release, 2007). However, Chinese government is pushing the 3G platform to foster growth and be a leader in the mobile market. Analysts predict a growth in mobile entertainment including video, games, and music within next 1-2 years (Ning, 2007).

Like China, Thailand is focused on 2G applications and services. The market is largely centered on "phone candy," a term describing mobile wallpaper, themes, ring-tones, icons and menu display (Chapin, 2007). Content providers maintain leverage by offering unique packages to satisfy this demand. MovieSeer.com has had great success in the Thai marketplace by offering premium Western phone candy content born of strategic relationships with major media companies like Disney and Sony Pictures Entertainment (Chapin, 2007). Also, like China, consumers can go to any second-hand store and purchase an "unlocked" phone across a wide variety of options (Chapin, 2007). The majority of Thai consumers purchase handsets that are relatively simple and cheap. Given the primary use of low-data content in the Thai market, businesses aren't expressing an urgent need to introduce 3G applications to the marketplace. Network operators are holding off on 3G services until technology becomes cheaper, 3G phones become widely available, and users start showing more interest in advanced 3G applications.

Government

The government has a huge say in defining a mobile marketplace because unlike Internet, which is very minimally controlled by any regulatory bodies, mobile communications network in each country is dependent on the frequency spectrum that is typically controlled by the government or government affiliated regulatory bodies. The government also defines the nature of competition that can exist in the market, the method through which the mobile customers can be reached and the tax structure on the revenue stream.

The U.S. government directly influences the wireless industry through the establishment and enforcement of telecommunications regulations. For example, the E-911 initiative requires wireless service providers to track the locations of mobile phones. This requires an investment in various global positioning technologies. It would force a huge investment and thus a longer recovery of capital layouts for mobile operators. Similarly, mobile gambling is greatly impacted by federal laws that prohibit gambling over any telecommunications lines and yet this is a very lucrative industry in the mobile commerce area.

The telecommunications industry in Iceland has been encouraged to grow by their government for the last decade. Beginning with complete deregulation in 1998, the government has been investing heavily in telecommunications (Carmel, 2004). The fiber optic networks have been mostly state-funded, and the government envisions having the entire population technically connected. The government is pushing the mobile phone sector to be at the forefront of all new technology and is encouraging foreign investment and market penetration. However, due to the small size of the country, partnerships must be forged with larger foreign corporations that have the capital to propel these small companies forward.

Ireland has already invested $5 billion in the telecommunications infrastructure over the last 10 years and offers a reliable and fully digital telecommunications system. Still, Ireland is described as "a first world economy with third world infrastructure". The Irish government is aware of these problems and has instituted an aggressive plan to transform the overall infrastructure including communications through federal funding. Of particular importance is the widespread availability of broadband access at competitive prices. In 2003, the Department of Communications has provided over EUR 250,000 in funding for five wireless Internet pilots in Dublin, Cork, Mayo, Sligo, Galway, Limerick and Louth to provide wireless connectivity in a range of locations such as hotels, specific towns, ferry terminals, universities and conference centers (personal communication with representative from Meteor, Ireland, 2004).

In the United Kingdom, in order to offer wireless services, a carrier must have both a Telecommunications Act license and a Wireless Telecommunications Act license. In addition, the VAT (Value-Added Tax) affects m-commerce strategies. This tax needs to be considered when planning, or companies stand to lose a great deal of money. The VAT tax is added on to the price of services and is a fairly hefty percentage. This adds to the cost of operations and creates a layer of bureaucracy for MNOs in the U.K.

As a member state of the European Union, Belgium participates in the action plan e-Europe, "An Information Society for All," adopted at the European Council of Feira in June 2000 (Information Society Directorate-General). The plan promises to create an integrated, competitive and knowledge-based society that remains at the forefront of electronic communications. According to a EU directive, the Belgian telecommunications market has been deregulated since 1998. The regulatory body responsible for telecom is the Belgian Institute for Post and Telecommunications, which reports to the Minister of Telecommunications and Public Enterprises.

The government of the Netherlands has multiple initiatives in place to keep them on the edge of emerging technologies – in particular electronic and mobile commerce. Specifically, fund is set aside to promote research & development and maintain an effective, high-quality knowledge infrastructure (Cooke et. al., 2004).

In addition to specific government regulations, most countries in Europe depend on the EU to establish m-commerce standards for all member nations. In many situations, the EU will provide an advisory role until there is a market failure such as lack of competition or price fixing. In an interview with Christer Hammarlund, Principal Administrator from the European Commission's Information Society Directorate-General, it was revealed that the EU sets the course for the future of mobile services. There are three policy areas studied within mobile telephony: call origination fees, call termination fees and access for other operators-creation of Mobile Virtual Network Operators (MVNOs). Their goal is to keep the regulatory framework in line with the rapid progression of mobile technology. They are continually meeting with manufacturers, MNOs and application developers to come up with a regulatory framework that serves to protect consumers without hindering economic progress. For example, when MMS was introduced in Ireland in 2004, a nude photo of a teenager was transmitted across Ireland's mobile network using the protocol resulting in a consumer backlash. EU, along with the Irish government, ordered the carriers to develop a solution that would filter out and prevent such files from being viewed (personal communications with representatives from O2 and Meteor in Dublin, Ireland, 2004) and created an indirect regulation on the MMS content.

The Japanese government has played a large part in the sculpting of the current Japanese mobile market. New entrants in the mobile realm are administered by the Japanese Ministry of Internal Affairs and Communications, who have only granted licenses to three network operators from 1993-2005: NTT DoCoMo, KDDI and Vodafone K.K (Towns, 2005). These companies thrived in the wake of cellular phone deregulation, announced by the government in April 1994 to spur cellular phone purchases and mass adoption. Until that point, all cellular phones in Japan were leased. This bred a few effects resonant in today's mobile arena, similar to a "first to market" advantage: an initial sharp reduction of handset prices to the consumer level, lowered subscription fees, and heightened usage rates. This transcendence, mitigated by the Japanese government to limit foreign investment and nullify would-be competitors, allowed the Japanese mobile market to mature sooner than most neighboring markets. It was only in late 2005 that the Japanese government began sanctioning competition, when Softbank Corporation, eAccess and IP Mobile entered the already mature market with aims to exploit broadband Internet services and build on the existing mobile communications infrastructure (Towns, 2005).

South Korea's rapid evolution can be directly attributable to government sponsorship and regulation. The high speed Internet approach and the broadband drive through the fixed line in South Korea created a high technology and network savvy consumer base that hungers for content over the mobile network as well. The telecom industry in South Korea is still regulated by the government's ministry of information and communication (MIC), although its role at this stage is to provide financial support and fair competition along the value chain. In order to stabilize the development of the telecom industry regardless of global economic recessions, all major carriers have agreed to make significant investments into small and medium sized technology companies in 2003 totaling approximately US$2.3 billion.

The government in Singapore plays an active role in the telecom sector and is working to improve the industry by launching a "next generation information communications infrastructure" in order to create high speed national broadband and wireless networks throughout the country (Singapore Telecom Climate Report, 2007). In April 2000, Singapore began to liberalize the telecom industry by removing all barriers to entry for foreign investors. Additionally, the U.S. Singapore Free Trade Agreement ("FTA") enacted in 2004 allows U.S. telecom service providers to interconnect with Singapore networks at competitive rates and under transparent terms and conditions.[1]

The Chinese mobile market is defined to a large extent by direct guidance from government policies and actions. They have stimulated the mobile market by lowering prices of value-added services such as SMS and WAP, and concurrently limiting the entry of new service and content providers to control competition and quality of service (Ning, 2007). Service is dominated by two network operators, and only 1,400 (7%) of some 20,000 Chinese content providers claim cross-province licenses allowing them to conduct business in different regions (Sun, 2007). Further, the government heavily regulates foreign companies so that domestic companies do not have to worry about international competition. The government does not allow foreign shares to exceed 49% of total ownership, and requires joint ventures to obtain specific licenses before operating (Ning, 2007).

Chinese government regulations protect consumers through policies such as free one-month trial periods, billing reminders, and double confirmation (Sun, 2007). For example, the government has required content providers to offer a free trial in the first month of all subscription offerings. It has also mandated billing reminders to SMS, MMS, and WAP subscribers for fees charged. Next, the government has required two positive confirmations from the user before being billed for such services. If no confirmation is sent, the subscription is cancelled. Overall, these policies negatively impact mobile companies, leading to slower revenue growth, increased customer acquisition costs, and lower profit margins.

In Thailand, network operators must apply to the Telephone Organization of Thailand ("TOT") and/or to the Community Authority of Thailand ("CAT") to provide mobile phone services (Ure and Vivorakij, 1997). Additionally, legal ownership transfers from the network operator to the TOT or CAT once the entity installs and activates its phone services. Further, under government provisions, network operators are required to pay the TOT or CAT a large percentage of revenues (20-30%) derived from both their postpaid and prepaid billing systems as excise tax (AIS, 2007).

Market Characteristics and Financial and Value Incentives

Success of mobile offerings in a market is dependent on a complex interaction of many factors such as the economic condition, the relative power of the players in the value network (e.g. network operators, equipment manufacturers, content providers, etc.), installed bases of alternate technologies, and the business models adopted. Economic condition of a market plays a significant role in deciding the sustainability of many product and services. In developing economy with low disposable income per capita, penetration of high end devices and services will be limited as it will be not affordable by most of the population. Economy also determines the availability of capitals that in turn can decide the level of competition and the variety of offerings in the marketplace. Clearly, the pricing structure and the revenue sharing model between the various participants in the value network will determine the attractiveness of a value added service in the mobile space and in turn will determine the profitability and the viability of the services. In general, end users are not willing to pay a lot for the mobile value added services specially when such value added services are available through other channels. In fact, existence of a developed non-mobile based infrastructure for communications and high penetration of Internet have been found to be a deterrent in the uptake of mobile value offerings in many developed market with the notable exception of South Korea (Weitenberner et. al., 2006). The relative power of the various players in the market also has profound effect on the dynamics of mobile marketplace. From our research of the various markets, we observed that there are four basic classifications that can be attributed to mobile markets based on the dominance of the value players in that market. These characterizations are:- carrier dominant, content dominant, device dominant, and user dominant. It is possible to have a mix of multiple kinds in a market and that, in fact will be the ideal situation as it would make the competition healthy and proliferate innovation. However, the current markets around the globe are mainly carrier dominant with a few exceptions where the device and content play a significant role in defining the market (e.g. U.S. with introduction of i-phone, Singapore, and Thailand). This is due to the control that the carriers exert over the main communications infrastructure and the fact that primary relationship is built first between the carrier and the end-users for the voice services. This has a profound impact on the acceptance and growth of mobile value offerings. Carriers either can be a *Dominating Authority* that was described by Moseberg as a *Soviet Ministry* approach where "like the old bureaucracies of communism, they sit athwart the market, breaking the link between the producers of goods and services and the people who use them" (Moseberg, 2007, Wall Street Journal, Oct. 22, p R3), or they can be a *Benevolent Dictator* like NTT DoCoMo in Japan thus encouraging creation of a mobile market through growth of various mobile value added offerings to the end-users.

In the US, a growing number of consumers are becoming accustomed to fast Internet connections at work and at home, and over half of the population has a mobile phone. Moreover, almost all Internet access in the U.S. is paid for with one flat monthly fee – including broadband. The flat fee nature of Internet and mobile access makes all premium features seem relatively more expensive than voice over cellular or Internet over landlines.

Iceland's economy has seen a boom in the past few years before the current global economic crisis. This can be both a blessing and a problem for m-commerce ventures. According to Landssiminn, services offered by the company may actually become more popular during economic downturns as merchants look to reduce the cost of transactions. On the other hand, since the Icelandic people are relatively well off, they are very willing to pay for premium services offered by carriers. According to Landssiminn, three important factors in getting the service off the ground are: the low consumer education required due to the use of familiar SMS technology, getting all players to agree to the SMS solution by having a centralized computer center with all bank account numbers and the focus on merchants who have difficulty selling their content due to a lack of point of sale technology thereby offering the mobile device as an alternative (personal interview with representatives from Landssiminn/FSG, Iceland, May, 2004).

O2 in Ireland echoed Landssiminn's view that wireless carriers are more careful in their investments in new offerings given the economic downturn caused by the events of the dot-com and telecom busts. Moreover, companies without viable business plans were weeded out, allowing only the fittest business models to survive.

In the UK, as in most of Europe, there is a financial incentive away from voice. Since most of the wireless users are on a prepaid model, they can see the comparable price of premium services in contrast to voice. SMS has become very popular in the UK because of its lower cost as opposed to landline and voice telephony. According to Vodafone, consumers are willing to pay a premium for services that use the SMS system. In addition, a low PC penetration makes m-commerce more interesting to consumers since the mobile phone penetration is much higher.

In Belgium, voice still accounts for a majority of revenue, but SMS is the most used application. Belgium's other key application is M-Payment, which allows the cellular device to be used like a wallet to conduct transactions. M-Payment was a logical outgrowth of the extensive financial infrastructure already in place. With only one major clearinghouse, Belgium has been able to eliminate the use of checks altogether. The next step is to eliminate cash by placing micro-payments on cards and phones. This will greatly provide users the added convenience of not carrying checks or cash, and the clearance fee expenses for banks will be greatly reduced.

Within Asia, Japan's mobile environment continues to be the most mature, highly saturated, and competitive market. As of April 2007, there are over 96 million cellular and personal phone subscribers across three primary Japanese network operators: NTT DoCoMo (54.4% market share); KDDI (29.1% market share); and Softbank Mobile (16.4% market share) (KDDI Annual Report, 2007). Mobile phone culture has become so pervasive and embedded within Japanese society that it has developed its own moniker, or "*keitai*" culture (Matsuda, 2004), seamlessly integrating the mobile device into the end user's command and enabling a "lifestyle infrastructure" industry (MacDonald, 2003).

In Japan, the network operator is the most dominant player in the mobile value chain. NTT DoCoMo, for example, oversaw the entire value chain in presenting the i-mode service, and worked closely with handset manufacturers, content providers, server vendors, and other third party solution vendors in delivering the end product (Matsunaga, 2000). Ultimately, NTT DoCoMo decided each specific detail of the product offering, from the design and functionality of the mobile device to exactly what type of content could be accessed through its mobile portal to create a seamless user experience (Matsunaga, 2000). The company has systematically addressed issues regarding consumer transaction security, with attributes ranging from password protection mechanisms on enabled handsets to operator-based locking services (Yokokuchi, 2007). Handset theft and mal-use is prevented by preemptive online registration (i.e., blacklisting) and transaction-based PIN authorization at the point of sale (Yokokuchi, 2007). NTT DoCoMo has also addressed merchant concerns through device installment subsidies, which are particularly important for smaller vendors reliant on word-of-mouth advertising. In cases of fraud, liability is directed at the credit card issuer through its insurance (Yokokuchi, 2007).

Even though MNOs control and monitor the access to content and services, the revenue sharing percentage as well as the willingness of the MNOs to act as the trusted third party solution providers has made them *Benevolent Dictator* and helped in phenomenal growth of the market. For example, NTT DoCoMo keeps a commission of 9 percent only from a content charge paid by the user and forwards the rest onto the content provider. In most of the other carrier dominated market, the carrier keeps, on the average about 20% of the content charge, and it can be as high as 40%! The favorable revenue split coupled with NTT DoCoMo's strategy of adopting compact HTML (cHTML) standard thus enabling service developers to implement new services in an inexpensive and efficient way created a harmonious environment for both content providers and aggregators and has provided tremendous incentives for the development of compelling content.

Japanese mobile companies continually look for new areas for generating revenue. Mobile payment and Mobile Virtual Network Operators are two such avenues being pursued by the Japanese Service Providers (Weitenberner et. al., 2006). Market research suggests that credit services are an untapped revenue source that may spark new growth in a mature mobile environment (Yokokuchi, 2007). NTT DoCoMo deployed it's e-wallet service, *iD*, in November 2005 to explore the mobile payment niche (Yokokuchi, 2007). As of March 2007, 2.3 million of NTT DoCoMo's customers have enlisted the *iD* service (Sharma, 2007).

South Korea exhibits a mobile environment very similar to that of Japan. Korean Consumers have been nurtured by the mobile industry into a unique buying culture that readily adapts to new product offerings and services. The MNO's in conjunction with the Korean government have created a dynamic market where innovation outweighs utility. Korean consumers have bought into this rapidly changing market by purchasing the latest technologies and services even though some of which become obsolete or out of fashion within a short time period (3 months – 1 year). Korean consumers are also willing to pay a premium for the latest phones and services as these products are considered a cultural and peer status symbol (Rhee, 2005). Furthermore, the multiplayer gaming, a highly successful phenomenon in the Internet, has quickly found its way as one of the top services in the mobile arena.

Singapore is one of Asia's leading banking and financial centers with a highly educated workforce and a government that promotes the development of industries in modern technology (Gannon, 2003). Singapore's thriving economic environment has gradually created a healthy atmosphere where people feel compelled to adopt the latest technologies. In fact, the mobile phone penetration rate has reached 111.2% as of June 2007.[2] The most dominant members of the value chain in Singapore are content providers and companies that allow users to personalize their phones. Singapore was a connected society before the mobile revolution and the carriers therefore, do not wield their clout as they do on users in Japan. In such a market, the value of the mobile offering has to surpass existing applications to gain notice, which can also be construed as an obstacle for content providers (Miles, 2007).

Because content providers maintain leverage in the mobile value chain in Singapore, off-deck content delivery is popular. This model allows a content provider to work with a network operator to provide the consumer with value added services such as ring-tones and wallpapers, since many network operators do not provide an internal deck. Further, the network operators benefit from this agreement because they are able to provide their customers with value added services and not have to manage the administration and billing of this content. Simultaneously, the content providers benefit from this arrangement because they get instant visibility, credibility and a large user base that they would not have been able to attain without the account. The off-deck business model has become an extremely important revenue-generating avenue in Singapore because of both the high mobile penetration rate and the influence of the content provider in the value chain. It has become one of the few sustainable ways left for businesses to profit in the mobile industry in Singapore.

China's mobile market is relatively young, yet it hosts the world's largest mobile subscriber base with an estimated 500 million subscribers (Robinson, 2007). In 2007, there were two major network operators in China: China Mobile (67.9% market share) and China Unicom (33.1% market share), both of which were majority state-owned (Chan, 2007). China is a carrier-dominated market and the MNO's not only offer mobile networks, but may also function as mobile portals, WAP gateways, intermediaries, and content providers (Lu et. al., 2007). China Mobile's Monternet and China Unicom's Uni-Info services have adopted billing and revenue sharing models somewhat similar to NTT DoCoMo's i-mode service (Kallio, 2007). However, in China content providers collect the full price for services offered, and then compensate the service provider for using its network. High transaction fees reinforce the dominance of the service provider in the Chinese mobile market, but fail to prime the market by hindering the profitability of new ventures (Robinson, 2007).

Off-deck traffic in China is significantly higher than on network operator sites. This is the resulting combination of higher MNO revenue sharing percentages and the glut of over 20,000 Chinese content providers (Sun, 2007). From that perspective, China is actually a content dominated market, with providers facing the challenge of differentiating themselves from their competitors. For instance, KongZhong ("KZ"), a leading content provider, offers a bundle of unique services that include exclusive motion picture studio content and customized social networking (Sun, 2007). KZ has also developed strategic alliances with handset manufacturers to create phones with pre-loaded buttons or links that automatically direct the user to the KZ website (Sun, 2007). Further, KZ employs approximately 100 in-house game developers, enabling the company to offer cutting-edge gaming to consumers while diversifying their service offerings (Sun, 2007). Overall, KZ's strategy is to develop alliances with major partners that will help leverage its brand relationship with users.

Simultaneously, China Mobile has started decreasing its service fee for WAP Internet access, which encourages the use of mobile Internet. Current WAP offerings include readable news, downloadable pictures, downloadable ring tones, downloadable ring-back tones, and wallpaper (Ning, 2007). So far, 100 million in China have used WAP, with 39 million active WAP users (Robinson, 2007).

A fragmented handset market has hindered technological advancement. In China, users go the mom-and-pop shop around the corner to purchase the device of their choice, usually sold at a low price (Tan, 2007). With over 1,000 handset models on the market, each phone differs in size, weight, memory, operating systems, and applications (Aditham, 2007). This lack of consistency has made it difficult to impose widespread technology and security improvements, and to adapt new business models and advertising campaigns to different types of devices

In Thailand, the mobile market can be classified as a mix of user dominated and content dominated market. As of July 2007, Thailand has 46 million mobile subscribers, and a mobile penetration rate of 69%, up from 18% in 2002 (Sangophan, 2004; Phoosuphanusorn, 2007). There are three dominant network operators in Thailand: Advanced Information Services ("AIS") with 50% market share; Total Access Communication ("DTAC") with 32% market share; and True Move with 14% market share. Thailand's low-GDP market is saturated with pre-paid customers, accounting for 90% of national mobile commerce revenues (Chapin, 2007). Thus, there are no contracts between a user and a network operator, and customers have the ability to switch from one network operator to another with incredible ease, making Thailand a price sensitive mobile market. As a result, the key for the network operators in sustaining users is familiarity and the quality of on-deck content offered. AIS (an industry veteran) recently signed a deal in June 2007 with Celestial Pictures to offer a new mobile television channel, Kung Fu TV (Ho, 2007). Here, consumers will have the ability to get thirty second to three minute clips, video ring tones, animated wallpapers, movie trailers and music videos of the famous Shaw Brothers Kung Fu movies.

Alternatively, True Move is shifting towards a "convergence" business model. Expanding lifestyle shops will cater to consumers' various appetites, from cuisine and coffee to Internet browsing, mobile payments and "True" mobile TV Programming (Banomyoung, 2007). This type of unique customer experience is something that no other network operator is providing, and True Move is banking on a competitive advantage. True Move will also provide exclusive content throughout all platforms offered under the "True" umbrella, by offering television, cable, Internet and e-wallet services in conjunction with its mobile phone services (Banomyoung, 2007).

Socio-Cultural Factors

The user culture plays a significant role as well in the acceptance of mobile values and such culture is determined by a complex interaction of various socio economic conditions including housing, daily commute, the collectivistic versus individualistic nature of the user groups, and users view towards new technology. Societal culture of a country plays an important role in determining the potential success of m-commerce ventures. Differences in the cultures of the various countries require custom-tailored offerings, as well as adaptations to the business climate.

The cost of landline telephone service in the U.S. is cheap and very accessible. Ninety five percent of US homes and businesses have a landline telephone, although this is expected to decline to 40% by 2009, as cellular phones gain in popularity in the U.S. (USA Telecoms and Technology Forecast from Economic Intelligence Unit, June 9, 2005). Home computers and Internet services are very inexpensive in the US as well. These factors helped make the US the largest market for electronic commerce, mainly in the B2C and B2B segments. US companies have been reluctant to invest in m-commerce, because Americans appear to prefer purchasing items on a PC, which is more user-friendly than a mobile phone and allows for quick transactions at work or home. Because of the high PC penetration rate in the U.S., more functionality is desired to entice the mass market to embrace an application on small handsets. This observation is supported by the tremendous success of the i-phone in the U.S. With the exception of a few metropolises in the East Coast, the car is the dominant form of transportation in the US, which precludes the use of mobile commerce services during commuting hours—an important success factor for mobile commerce services in Asia and other regions. Furthermore, unlike Asia, lack of availability and reliability of cellular networks in the U.S. is a definite barrier for use of m-services. Moreover, use of credit cards is prevalent and there is no social stigma attached to the use of cards to pay for goods and services. On the other hand, the individualistic nature of the U.S. society values privacy rather highly and the personal privacy provided by mobile phones is driving their use amongst young teens as a tool to express their individuality by purchasing ring tones, message backgrounds and other entertainment content.

Iceland is a relatively young country and its relatively small geographic size and population provide an ideal test bed for innovative uses of mobile technologies. They are very comfortable using electronic media to make purchases. Over 85% of the population has access to the Internet, giving it one of the highest penetration rates in the world. More than one-third of all Icelanders have purchased something over the Internet (Carmel, 2004). With a fraud rate of 0.0003%, Icelanders use electronic means of payment for over 90% of their transactions. Icelanders also pride themselves in keeping up with the newest technologies and the country has positioned itself at the forefront of m-commerce and most telecom technologies.

Although Ireland's Internet penetration rate at the end of 2002 was 40%, the mobile phone penetration rates are at a current high of 79%. O2 also believes that consumers greatly desire an audit trail of their purchases. They want to be able to view and verify their transaction on one bill. According to O2, although the Irish are open minded, the presentation and education of the mobile phone potential beyond voice/text is lacking. It is mainly a pre-paid market with a majority SMS users for various reasons, one reason being that people view text messaging as non-intrusive. However, location based services must jump the cultural hurdle of users "opting in" to allow such services to become a part of society (personal interviews with representatives from O2 and Meteor, 2004).

The UK, as a whole, likes to keep their identity anonymous when it comes to businesses and purchases and from our discussions with Vodafone in the UK, we learned that one of the major forces towards prepaid phone bills is the attractiveness of anonymity by the users (personal interview with representative from Vodafone, 2004). This anonymity also results in the UK being a high cash culture where credit cards are not as popular as cash and debit cards.

Belgian mobile users are still reluctant to consistently use m-commerce applications. Changing consumers' habits still proves to be an obstacle, even in the E-commerce arena. However, there is a likelihood that these sentiments may change as consumers move along the learning curve and become more comfortable with technology. Security continues to be a concern to Belgians. Because of this concern, Belgium remains a debit-oriented society. Currently 65% of users are prepaid. They like the security of debit cards as opposed to credit cards. Belgians are still nervous about online purchases, as 60% of the population is not comfortable buying with credit cards online.

Due to the costs of upgraded bandwidths, the early adopters of m-commerce in Belgium are most likely to be business users. Many of the services now available online are mainly useful for a businessperson who is a frequent traveler. Network operators that have a greater proportion of business customers have the first-user advantage of the system. Another important target market is the under-25 age group. These are the customers of the future and capturing this market will require marketing savvy on youth-oriented brands.

The population of the Netherlands is highly proficient in technology, with 74% of the population having a computer in the household and 69% of all households have an Internet connection. Wireless access points and Internet cafés are popular. With this acceptance of technology, the people of the Netherlands, whose daily average travel has increased through each of the past three years, have adopted mobile communication technology as a standard part of daily life. Market penetration of mobile phones has stabilized at 74%. As the density of mobile telephone traffic has rapidly increased, the people are comfortable and welcome further modernization of the mobile system. This modernization, however, must follow the mobile attributes (both intrinsic and network level) discussed earlier.

Overall, there is a heavy reliance on public transportation in practically every European city and the population is heavily concentrated in the cities. Furthermore, in most populated areas of Europe, mobile telecommunication is highly developed and it is possible to have an uninterrupted signal in the high-speed underground train systems. Commuters in Europe therefore, have greater amounts of idle time either waiting for their train or during the actual commute thus presenting an opportunity to engage in m-activities beyond voice. European users, in general seem to prefer pre-paid model.

Japan is an interesting case study because there are over 200 million credit cards circulated in Japan, placing them as the second largest country after the U.S. in numbers of credit cards in use among the population, yet the use of credit is not viewed very positively in the Japanese culture and the actual usage of credit cards is very low. Approximately 5-7% of consumer spending is made by the credit card. In a society that sees borrowing (debt or buying on credit) as a sign of weakness, there is a tendency to be conservative with the use of debt (Shimokawa, 2005) However the younger generation is much more open with credit card usage. Micro-payments provide Japanese consumers with an excellent alternative, as they can prepay a certain amount and then have the convenience of using their handset just as they would a credit card.

Transportation mode in Japan also plays an important role in adoption of mobile phone for communication beyond voice. Japan has a very efficient public transportation system like Europe and the majority of the population utilizes the system. Japan is also one of the densest countries in the world in terms of population per area. With limited room, the people have learned to adapt to certain etiquettes, one of which is not intruding into other people's personal space. Cell phones have filled a need for communication and entertainment within certain bounds. Text messaging is a very popular application for this quiet and considerate culture. With windows of time open during the commute (average of 70 minutes per day), application providers have the perfect opportunity to create attractive options.

Japanese citizens, specifically in urban areas, spend a lot of time outside their homes. It is not uncommon for a majority of working people to come back home late even on workdays thereby presenting an opportunity to deliver mobile entertainment since mobile phone based entertainment content and services, unlike TV and PC, can be delivered without being held to a fixed location. The influx of women in the Japanese workforce has created a significant boost in dual family incomes thus increasing the amount of household disposable income and this increase in disposable income is seen as a source of growth for additional mobile revenue (Koike, 2005)

The South Korean lifestyle is somewhat similar to Japan, i.e. the predominance of public transportation; the desire for new products and services, etc.—but there is one large difference. The South Koreans have a high adoption rate for e-commerce and personal computers, which has subsequently *encouraged* the adoption of mobile commerce. As South Koreans became acclimated to the idea of purchasing items online, they became quickly comfortable with purchasing goods and services through m-commerce (Woolley, 2005)

South Korean consumers have a very liberal attitude towards credit card usage. On average each consumer holds approximately 4 credit cards and has default rates of approximately 8% (BBC News, February, 2003). High adoption of credit cards was partly fueled by a government policy to spark consumer spending in 2002 (Gluck, 2003). This accepting attitude towards credit appears to extend towards mobile commerce as consumers have eagerly purchased products and services through their phones. SK Telecom announced that digital services grew 39% between 2004 and 2005, accounting for approximately 23% of total ARPU (Kim, 2005).

Mobile companies in Singapore primarily operate on a post-paid subscription model. The post-paid billing model represents about 65% of billing operations in Singapore and is very unique to Southeast Asia, given the low income levels of neighboring countries such as Thailand, Malaysia, and Indonesia (Yap, 2007). Consumers with higher income are more likely to subscribe to a post-paid billing method while consumers with lesser income are more likely to use pre-paid. Furthermore, Singapore has an established system of credit from which carriers could evaluate whether they will grant a customer a post-paid account.

Singaporeans are not afraid to use their credit cards for most purchases. The country holds both a high usage and penetration rate for credit cards (Lorenz, 2005). Each person owns about three or more credit cards. As mentioned before, people still find it easier and more convenient to use their credit card and other established ways of payment rather than to try something new like pay with their mobile phones. Mobile phones are seen more as a "cultural accessory" in Singapore, highly correlated with social status (Miles, 2007). Though the ideal target demographic is 18-28 year olds with disposable income, the 14-18 year old demographic is gaining the attention of content providers with their inclination towards mobile gaming (Miles, 2007). Being able to influence customers through brand relationships, therefore, is an integral part to understanding the mobile phone consumer behavior in Singapore. High handset turnover is also a byproduct of social status: new mobile phones are introduced into the market every two months and consumers replace them as often as every six to eight months (Miles, 2007). Further, studies have shown that when people leave their house they take their mobile phone, keys, and wallet, prioritizing them in that order (Myers, 2007).

Similar to Japan, Singaporeans rely heavily on public transportation. Nearly 50% of the countries' citizens use Mass Rapid Transit (MRT) rail, public buses, or taxis, and average a 42-minute workplace commute.[3] The Singaporean government has been very successful at promoting and increasing the use of public transportation, which helps drive the mobile phone traffic and usage as commuters seek to entertain themselves during long commutes.

"Mobile Expression" is the catch-phrase cultural phenomenon driving the Chinese mobile market (Robinson, 2007). The Chinese, particularly the younger generation, are using mobile devices as an extension of their personality, character, and image. For instance, the Chinese are interested in downloading content, such as ring-back tones, that will make their phone stand out from other mobile phones and users. Gaining rapid popularity, ring-back tones are the tones the person dialing hears on the line while connecting. This is the epitome of "mobile expression," given that the user is paying for a service that he does not personally enjoy, but rather paying a price for the way he expresses himself to others. Additionally, with the wide availability and selection of handsets, the Chinese quickly replace mobile devices every one to two years (Tan, 2007). The high turnover rate is also in part due to the mobile phone as a strong status symbol, and its external appearance as another form of "mobile expression." Currently, social networking is becoming another mobile expression trend. Users are consistently logging into WAP sites such as the Chinese version of myspace.com or linkedin.com, as well as using SMS and IM, to stay connected with friends, family and co-workers (Robinson, 2007).

The traditional m-commerce activity of buying goods with the phone is a difficult proposition in China because most Chinese require seeing and touching products before they are willing to buy them. This user-culture along with a credit card penetration rate well below 30%, (China Steps into Credit Card Era, 2004) has kept the Internet shopping, on both a computer as well as a mobile phone, limited to small items such as books and CDs (Wu, 2005). Most mobile users in China commute using public transportation, where they can utilize their mobile phones over long travel distances. For instance, many rural workers migrate long distances to big cities, and need to entertain themselves on daily commutes. In fact, China Mobile is heavily targeting rural areas, and has already penetrated 60% of the rural market (Ning, 2007). In 2006, rural subscribers accounted for more than half of China Mobile's 53 million new subscribers (Chandler, 2007). The vast rural market also explains why approximately 70% of China's billing services are pre-paid. In rural areas, users can simply head to a newsstand, and purchase a pre-paid SIM card to insert into the low-cost device they obtained from the corner shop and immediately start using services – without having the hassle to sign up or input registration information (Tan, 2007).

One major obstacle for Chinese mobile companies is piracy. Chinese, particularly in rural areas, enjoy having mobile content such as games, ringback tones, and wallpaper, but do not wish to pay for such content (Tan, 2007). Many companies fear that in the future, users will no longer be willing to pay for downloaded content. Recently, off-deck content providers have been trying to address this issue by attempting to monetize content through other means, such as mobile advertising.

Unlike the majority of the other countries in this study, only half of Thai mobile users utilize SMS data services (Banomyoung, 2007). A contributing factor is that the Thai language characters are not easy to type and it takes much longer to send an SMS as opposed to communicating by voice (Banomyoung, 2007). Price also plays a factor in low SMS usage since its cost is high compared to voice. This is in direct contrast to China, where SMS is priced substantially lower than voice.

Piracy is also a challenge in Thai mobile market. Currently there is no government regulation to effectively fight piracy. Content providers often hire enforcement agents to pursue piracy offenders. Government must step in and introduce strict laws to help this content driven market by reducing the worry of the content providers in entering and operating in this market.

Thailand's primary method of transportation may also explain why it is a voice-dominated market. Because most people own cars, it is easier to use voice functions than SMS functions (which require more attention) while driving. However, the Skytrain public transportation service is already in place in Bangkok and we hope that it would foster much faster adoption of mobile services by the Thai population.

Summary of the Findings

From the discussions of the market characteristics and other factors in multiple mobile commerce markets around Europe, Asia and the U.S., we can get an idea about the key drivers for the mobile offerings and their corresponding value attributes as described in Figure 2. Figure 3 provides a classification of the various countries according to the level of values provided by the extrinsic and intrinsic attributes and we present an organization of the various business models that we feel are likely to succeed in various types of market characterized by the dominance of the players of the mobile value network in Table 2. However, this is not an exhaustive and mutually exclusive list, many of the models in one type of market can also be successful in the other types of markets depending on influence and interplay of the four influential factors discussed earlier in the paper with examples from multiple countries.

CONCLUSION

This paper presents a framework for understanding the influencing factors for mobile value offerings and their impacts through analysis of the various mobile markets around the globe along with market characterization based on the relative power of the players in the mobile value network and the mobile offerings. We found that government support, technology infrastructure, certain market characteristics and consumer culture from a socio-cultural perspective are the leading influences in these markets in delivering the extrinsic

Figure 3. Relative Approximate Positions of the Mobile Markets of the Countries Studied according to the Levels of Extrinsic and Intrisic Values offered to the Consumers

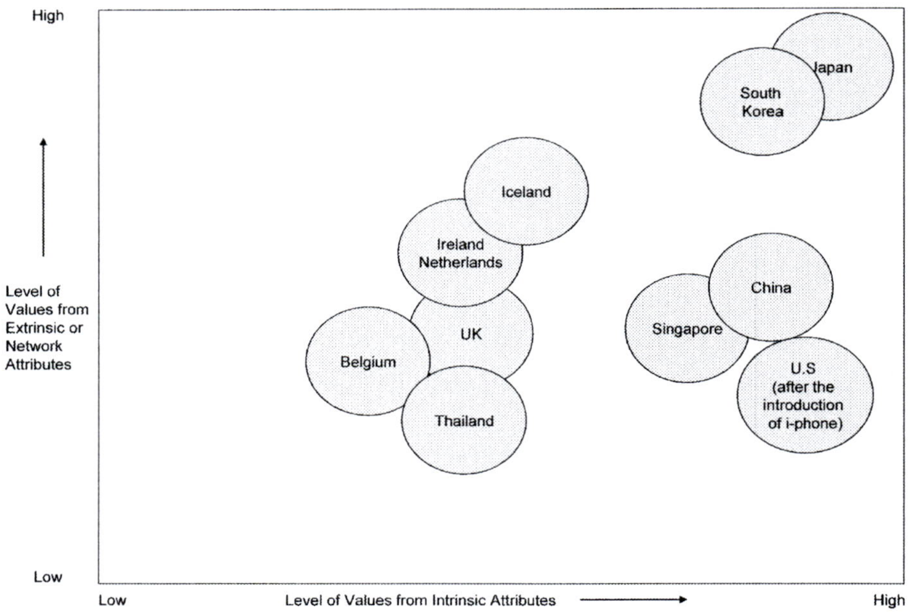

and intrinsic attributes identified by Methlie and Pederson (2005). For example, South Korea's rapid evolution can be directly attributable to government sponsorship and regulation, whereas China's overprotection of the consumers and the dominant power of the MNO is hurting the development of the mobile market.

We further observe that the users can understand the values of the intrinsic attributes of a mobile offering rather easily compared to the values delivered by the extrinsic attributes, especially at the early stage of the development of the m-commerce (Methlie and Pederson, 2005). This is still a push market from the value players' perspective and the customers do not know what offerings to demand and expect from the m-commerce space. Thus the extrinsic values are not well understood and customers mainly focus on intrinsic values. However, as the market matures, the extrinsic values will start coming to the forefront and anyone who is in charge of delivering those values will tend to gain. MNOs must understand this as they are in control of the extrinsic attributes and thus they have to leverage them to create long term values for the customers. This will be beneficial for them as well as they can then enjoy the position of a facilitator (of course, they have to take into consideration of the other market defining factors as discussed in the paper) for anyone who wishes to join the value network. One can clearly see this being the case with Japan where NTT DoCoMo served that exact role and made Japan the poster child for mobile commerce.

In a carrier dominant market, the carrier cannot afford to behave like anything other than a benevolent dictator. They have to understand that by delivering the network related services and making it easy for the other to connect to the network and demanding a fair share of the revenue will grow the market and create a bigger pie to share. Domineering Authority or Soviet Ministry Model as depicted by Moseberg (2007), will restrict the market growth. This will create a losing market situation for all the players in the value network, or will force the other participants of the value network to seek alternate channels

Table 2. Market Characteristics and possible Mobile Business Models

Market Characteristics	Suggested Focus of the Dominant Player to facilitate growth in mobile offerings	Revenue Source	Business Models likely to succeed
Carrier Dominant	• Benevolent Dictator (or Trusted Third Party Host)	Fair revenue sharing with other players, data charges, and gain from the market growth.	• Intellectual property management solution • Marketplace portals • Premium content developers
Content Dominant	• Innovation in content development • Availability of Repurposed and Premium Content	Sale of content and services.	• Search engines (information brokerage) • Ad-wrapped content • Off-deck content delivery • Premium content provider • Content aggregator • Application Service Providers • Mobile Value Network Operators (MVNO)
Device Dominant	• Innovation on terminal level intrinsic attributes • Seamless User experience through tight integration • Cool factor	Sales of the device itself and also cross selling of co-branded content and services (e.g. i-phone and Apple).	• Integrated mobile portal • Ad supported smart phones • Support and services for mobile phone as the life style device
User Dominant	• Innovative use of phones • Networking opportunity • Use for personal productivity and life style • Innovative and exciting content	User generated content, P2P marketplace, advertisement.	• Mobile auction sites • P2P services and social networking • All content based models including MVNO • Support and services for mobile phone as the life style device

of delivery of their products and services to the customers.

MNOs in Japan and Korea (NTT DoCoMo, KDDI, SK Telecom, etc.) and Vodafone in Europe are prime examples of the benevolent dictator models. On the other hand, the MNOs in China and other places create barriers to the growth by creating restrictive policies and demanding too much of the revenue share from the content or service providers. Unless someone is in possession of premier content (e.g. Electronic Arts with their games), the bargaining power of the participants of the value network with the MNOs are limited and sometime it makes it difficult for them to create a sustaining business model.

i-Phone is an interesting case where the power in the mobile eco-system is redefined by the terminal vendor. The innovative and high quality nature of the i-phone as a terminal with very high intrinsic values to the customers forced the other members in the value network, including the MNOs to cooperate with Apple to deliver an integrated service. Once again, such integrated services provide very high value for the customers and create a long term sustainability of the various value offerings and an impetus for market growth. It turned the U.S. into somewhat of a device dominant market especially with the applications (or Apps) available on the handsets through the "Apps Stores"

Overall we saw that Japan and South Korea remain the worldwide leaders among these mobile markets, and platforms like gaming, video-on-demand, and i-mode have already become popular standards among its users. However, they are currently encountering the problems of saturated markets and must rely on price and product differentiation for competitive advantage. Japanese and South Korean MNOs are adjusting their business model in anticipation of off-deck delivery and supporting them through an advertising business model. They are opening up new revenue streams through various revenue share services such as mobile payment, marketing and advertising. They are also encouraging growth of Mobile Value Network Operators, threat to the MNOs in many countries, to employ new commercial revenue streams. European mobile players are somewhat in the middle and are slowly making progress with various mobile offerings beyond the voice. SMS appears to have led consumer awareness of mobile commerce offerings and is considered a gateway to the 3G services. China is attempting to overcome its infrastructure, bureaucratic, and economic hurdles to become a leader in the worldwide mobile commerce market. Singapore has taken a more conservative evolution compared to it South East Asian neighbors. In the United States, there is an enormous opportunity but the market is still predominantly voice driven, though that is changing with the development of the i-phone culture. The burden is on the MNO's to convince consumers that data services are desirable and reliable.

Finally, we present an organization of the various business models that we feel are likely to succeed in various types of market characterized by the dominance of the players of the mobile value network with an understanding that many of the models in one type of market can also be successful in the other types of markets depending on influence and interplay of the four influential factors discussed earlier in the paper. A complete classification and development of an integrated framework that would connect Pederson's model with the market characteristics to derive attributes of successful business models for mobile offerings is open to future research.

REFERENCES

Aditham, K. (2007). China's Mobile Charge: Madhouse CEO Joshua Maa Reveals the Framework and Future of the Chinese Mobile Beat. Retrieved August 12, 2007, from Adotas.com

Afuah, A., & Tucci, C. (2002). *Internet Business Models and Strategies: Text and Cases 2/e*. McGraw-Hill/Irwin

AIS-About US Company Profile. (2007). Retrieved Oct 31, 2008, from http://www.ais.co.th.

Anckar, B., & D'Incau, D. (2002). Value-Added Services in Mobile Commerce: An Analytical Framework and Empirical Findings from a National Consumer Survey. *Proceedings of the 35th Hawaii International Conference on System Sciences.*

Banomyoung, A. (2007). Personal Interview at TrueMove. May 2007.

Barnes, S. (2002). The mobile commerce value chain: analysis and future developments. *International Journal of Information Management*, *22*, 91–108. doi:10.1016/S0268-4012(01)00047-0

Basu, A., Mazumdar, T., & Raj, S. P. (2003). Indirect network effects on product attributes. *Marketing Science*, *22*(2), 209–221. doi:10.1287/mksc.22.2.209.16037

Chan, I. (2007). China's Big, Really Big, Mobile Market. *BusinessWeek*, Retrieved August 1, 2007, from Technology Section of BusinessWeek Online located at http://www.businessweek.com/globalbiz /content/jul2007

Chandler, C. (2007). *China's Mobile Maestro*. Retrieved August 12, 2007, from CNNMoney web site, located at http://money.cnn.com/magazines/fortune/fortune_archive/2007/08/06/100156748/index2.htm

Chapin, R (2007), MovieSeer.com interview, May 2007.

China Mobile Press Release. (2007). Retrieved August 12, 2007, from China Mobilewebsite located at http://www.chinamobileltd.com/doc/pr/2007/20070321.htm

China Steps into Credit Card Era. CISCA. http://www.cisca.org.hk/english/industry-page.asp?news_id=791, October 8, 2004

Cooke, D., Flores, G., Garcia, J., Guzman, E., Lopez, E., Muselli, C., & Seal, K. C. (2004). Influencing Factors for Mobile Values: A Comparative Analysis Between the United States and Iceland, Ireland, the United Kingdom, Belgium and the Netherlands. In the *Proceedings of The Second International Conference of AIMS*, India, 2004.

Dholkia, N., Dholkia, R., Lehrer, M., & Kshetri, K. (2004). Global Heterogeneity in the Emerging M-Commerce Landscape. Un Wireless communications and mobile commerce, 2004, accessed at http://cybertech-pub.com/downloads/excerpts/1591401844E.pdf, last accessed on Oct 31, 2008.

Faber, E., Ballon, P., Bouwman, H., Haaker, T., Rietkerk, O., & Stern, M. (2003). Designing business models for mobile ICT services. Paper presented at the *16th Bled Electronic Commerce Conference, Bled, June 9-11*, available from http://www.mozaiek.dds.nl/docs/PanelBled03.pdf, last accessed Oct 31, 2008

Gannon, M. (2003). *Understanding Global Cultures: Metaphorical Journeys Through 28 Nations, Clusters of Nations, and Continents*. California: Sage Publications.

Gluck, C. (2003), South Korea's credit card headache. *BBC News*.

Gupta, S., Jain, D. C., & Sawhney, M. B. (1999). Modeling the evolution of markets with indirect network externalities: An application to digital television. *Marketing Science, 18*(3), 396–416. doi:10.1287/mksc.18.3.396

Henten, A., Olesen, H., Saugstrup, D., & Tan, S. E. (2003). New mobile systems and services in Europe, Japan and South-Korea. Paper presented at the *Stockholm Mobility Roundtable*, Stockholm, May 22-23.

Hitoshi, K. (2005). Personal Interview, VISA, Tokyo Japan, May 2005.

Ho, V. (2007). Celestial Launches Mobile Channel. Retrieved on August 20, 2007 from http://www.varietyasiaonline.com/content/view/1600/

ITU. (March 2003). Broadband Korea: Internet Case Study.

Kallio, J. (2007). Personal Interview, China, June 2007.

KDDI. Annual Report. (2007). Retrieved August 12, 2007, from http://www.kddi.com/english/corporate/ir/library/annual_report/2007/pdf/kddi_ar2007_e.pdf

Kodama, M. (2005). Innovation through networked strategic communities: case study on NTT DoCoMo. *Journal of Management Development, 24*, 169–187. doi:10.1108/02621710510579527

Koike, K. (2005). Personal Interview, NMS Communications, Tokyo, Japan, May 2005.

Lambert, S. C. (2005). Making Sense of Business Models. *The Magnus Journal of Management*, 1.

Liebowitz, S., & Margolis, S. (1998). Network externality. *The New Palgraves Dictionary of Economics and the Law*, Macmillan, London.

Lorenz, R. (2005). Personal Interview, NETS, Singapore, May 2005.

Lu, Y., Dong, Y., & Wang, B. (2007). The Mobile Business Value Chain in China. *Sixth International Conference on the Management of Mobile Business*.

MacDonald, D. (2003). NTT DoCoMo's i-mode: Developing Win-Win Relationships for Mobile Commerce. In B. Mennecke & T. Strader, Mobile Commerce: Technology, Theory, and Applications (1-25). Hershey, PA: IGI Publishing.

Matsuda, M. (2004). International Symposium: The Social Use of Anthropology in the Contemporary World, Session 1: Children and Media. Retrieved August 2, 2007, from http://www10.plala.or.jp/misamatsuda /Mead_Symposium.html

Matsunaga, M. (2000). [*-mode: An Analogue Account of the Mobile Internet.* Singapore: Chuang-Yi Publishing Pte Ltd.]. *Birth (Berkeley, Calif.)*, i.

McGinty, A., & Da Bona, D. (2004). 3G Licensing in China: A Waiting Game. *Computer Law & Security Report*, *20*, 480–481. doi:10.1016/S0267-3649(04)00093-7

Methlie, L. B., & Pedersen, P. E. (2005). Network effects and intrinsic attributes of mobile services: Mapping perceived service values on business model choices. Presented at the *Hong Kong Mobility Roundtable, 2005,* Hong Kong, June, 1-3.draft pdf version http://ikt.hia.no/perep/hk_mobility.pdf accessed last on Oct 31, 2008

Miles, C. (2007). iPOP interview. May 2007.

Molberg, M. (2006). Success Criteria for Future Mobile Products. Masters thesis, available at http://www.dtu.dk/upload/centre/cict/master_thesis/martin_molberg_success_criteria_for_future_mobile_products.pdf. last accessed, Oct 31, 2008

Mossberg, W. (2007). Free My Phone. *Wall Street Journal,* Oct 22, R1-R3.

Myers, P. (2007). ACME Mobile interview, May 2007.

BBC News. (2005). S Korean consumers spending again. February 2005

Ning, L. (2007). Personal Interview, China, 2007.

Onoe, S., Nakamura, T., & Higuchi, K. (2007). 3G evolution scenario toward 4G: Super 3G concept. *Wireless Communications and Mobile Computing*, *7*, 1013–1019. doi:10.1002/wcm.511

Pedersen, P. E., Methlie, L. B., Gressgård, L., Nysveen, H., & Thorbjörnsen, H. (2005). An exploratory study of the relationships between mobile data services business models and customer value. SNF report no 13/2005, Bergen, March 2005.

Phoosuphanusorn, S. (2007). Thailand: Mobile users growing on discounts. Retrieved on August 14, 2007 from http://www.asiamedia.ucla.edu/article.asp?parentid=74884

Rhee, S. (2005). Personal Interview, Clemensen Capital Company, LLC, May 2005

Robinson, R. (2007). Dada Asia interview, May 2007.

Sabat, H. (2002). The evolving mobile wireless value chain and market structure. *Telecommunications Policy*, *26*, 505–535. doi:10.1016/S0308-5961(02)00029-0

Sanongphan, S. (2004). Thailand - Telecom and Broadcasting Market Brief. Retrieved July 1, 2007, from http://strategis.ic.gc.ca/epic/site/imr-ri.nsf/en/gr124774e.html

Savio, D. C. (2004). Usability aspects in 3G/4G mobile telephony. At http://userpage.fu-berlin.de/~jmueller/its/conf/Madrid02/abstracts/Savio.pdf, last accessed on Oct 31, 2008.

Seal, K. C., Gomez, A., Kong, D., Hernandez, X., Macias, A., Soohoo, K., & Villa, M. (2003). A Comparative Analysis of M-Commerce issues Between Australia, New Zealand, and the United States. In the *Proceedings of The First International Conference of AIMS*, India, Dec 28-30, 2003.

Sgriccia, M., Nguyen, H., Edra, R., Alworth, A., Brandeis, O., & Escandon, R. (2007). Drivers of Mobile Business Models: Lessons from Four Asian Countries. *International Journal of Mobile Marketing*, *2*, 58–67.

Sharma, A. (2007). What's New in Wireless. Retrieved March 26, 2007, from The Wall Street Journal Online website located at http://online.wsj.com/article/SB117434453204441967.html?mod=todays_us_The_journal_report

Shimokawa, T. (2005). Personal Interview, JCB, Tokyo, Japan. May, 2005.

South Korea and Japan agree to launch 4G Forum and Pilot Network. *Asia Pacific Telecom*, January, 2002, http://findarticles.com/p/articles/mi_hb6530/is_200201/ai_n25884820, last accessed Oct 31, 2008

Stabell, C. B., & Fjeldstad, Ø. D. (1998). Configuring value for competitive advantage: on chains, shops, and networks. *Strategic Management Journal*, *19*, 413–437. doi:10.1002/(SICI)1097-0266(199805)19:5<413::AID-SMJ946>3.0.CO;2-C

Sun, S. (2007). KongZhong interview, May 2007.

Tan, B. (2007). Microsoft interview, May 2007

USA Telecoms and Technology Forecast. (2005). Economist Intelligence Unit.

Timmers, P. (1998). Business models for e-commerce. *Electronic Markets*, *8*, 3–7. doi:10.1080/10196789800000016

Towns, S. (2005). 3 New Entrants Granted Access to Japan's Mobile Phone Market (DCM, NTT, EWJ). http://japan.seekingalpha.com/article/4168, last accessed Oct 31, 2008

Tsalgatidou, A., & Pitoura, E. (2001). Business Models and Transactions in Mobile Electronic Commerce: Requirements and Properties. *Computer Networks*, *37*, 221–236. doi:10.1016/S1389-1286(01)00216-X

Ure, J., & Vivorakij, A. (1997). Privatization of Telecoms in Asia. Retrieved July 1, 2007, from Telecom Research Project Hong Kong website located at http://www.trp.hku.hk/publications.php

Weitenberner, C., Chapman, M., Miranda, G., Tobar, S., Wagoner, E., Akasaki, D., & Seal, K. C. (2006). United States and South East Asian Mobile Markets: A Comparative Analysis of Infrastructure and Cultural Differences. *International Journal of Mobile Marketing*, *1*, 73–82.

Woolley, S. (2005). Best Leaders. *Business Week*, 62.

Wu, P. (2005). Personal Interview, General Manager at Walkgame. May 2005.

Yap, C. (2007). ESPN Mobile Interview, May 2007.

Yokokuchi, S. (2007). Personal Interview, Japan, May 2007.

Yuan, Y., & Zhang, J. (2003). Towards an Appropriate Business Model for m-commerce. *International Journal of Mobile Communications*, *1*(1-2), 35–56. doi:10.1504/IJMC.2003.002459

KEY TERMS AND DEFINITIONS

CDMA: The CDMA, that stands for Code Division Multiple Access is a mobile communications standard developed by Qualcom.

GSM: GSM stands for Global Systems Mobile and is the prevalent standard for the communications in mobile network for most part of the world.

Killer Applications: Applications in a new area with such compelling values that end users adopt it at a very high speed and it becomes a critical part of their everyday use. E-mail, for example, is considered to be the "killer application" for the Internet.

M-Commerce: Refers to Mobile Commerce and is defined as the transaction of goods and services through a mobile hand held device that can include text messaging (SMS), multimedia messages (MMS), ringtones, downloadable music, mobile-banking, entertainment services through streaming video, ticket purchases, micro-payments, and many others.

Mobile Business Models: An architecture for execution of the delivery of the value propositions in mobile domain including a description of the business actors and their roles and revenue sources.

Mobile Network Operator or MNO: Probably the most significant player in the mobile value network. Often a major telecom company, the carrier or the mobile network operator (MNO) owns the transmission equipment, brands and controls the software on the handsets, operates retail stores to sell plans and subsidized handsets, and does end-user billing for basic services.

Mobile Value Network: Collection of all the mobile players who create and deliver value and their corresponding interactions.

Short Messaging Services (SMS): A means of communications in the mobile network through short texts that can be typed using the phone keypads.

ENDNOTES

1 Telecommunication Infrastructure Regulation and Liberalization. Retrieved August 21, 2007 from http://www.american.edu/initeb/at1178a/telecom.htm

2 Statistics on Telecomm Services for 2007 (Jan-June). Retrieved August 21, 2007, from http://www.ida.gov.sg/publications/20070209134246.aspx.

3 Transport. Retrieved August 22, 2007 from, http://www.singstat.gov.sg/pubn/popn/ghrs2/chap1.pdf

Section 2
Technological Issues in Mobile Marketing

Chapter 9
Supporting Marketing Practices:
Mobile Network Operators' Value Added Services Changing the Way of Doing Business

Süphan Nasir
Istanbul University, Turkey

ABSTRACT

The concept of mobile communication has been rapidly changing all over the world and the Global System for Mobile communication (GSM) business has been evolving from voice services to value added services (VAS). The rapid growth in mobile technology enables Mobile Network Operators (MNOs) to provide a wide range of VAS to their corporate customers and these VAS change and transform companies' way of doing business. The purpose of this chapter is to analyze the corporate VAS of a MNO, which has gained global acceptance and recognition, as a case study in order to understand how the mobile communication changes the companies' way of doing business. The key supporting mobile technologies, services and solutions, which are provided by the case company to its corporate customers, are reviewed in order to comprehend the mobile technologies, practices, and devices that support mobile business (m-business) and mobile marketing (m-marketing) applications.

INTRODUCTION

The contemporary era has witnessed the rapid evolution of mobile technology. Global System for Mobile communications (GSM) is an open, digital cellular technology used for transmitting mobile voice and data services. The finding of the wireless intelligence subscriber statistics end Q4 2007 report indicates that GSM is the fastest growing communications technology and there are approximately 2.7 billion GSM subscribers in the world that accounts for 80.5% of the global mobile market (Wireless Intelligence, 2007). According to the ComScore M:Metrics (2008b), there are 226 million mobile users in the United Stated, 49 million mobile users in Germany, 47.5 million mobile user in United Kingdom and 47 million mobile users in Italy. And as an emerging country, there are approximately 62 million mobile users in Turkey and the mobile line penetration rate in Turkey is approximately 88%

DOI: 10.4018/978-1-60566-074-5.ch009

at the end of December, 2007 (Turkish Telecommunications Authority, 2007).

Mobile communication has transformed the way people live and work. Mobile devices allow mobile users to send email, text message, access internet, take pictures and video, listen to music, play games, access data and much more. Azuki's mobile phone user survey details the behavior patterns of mobile phone users in the United States and the finding of the survey indicates that mobile phone users are spending a significant amount of time on their phones in the United States (Azuki Systems, 2008). The Azuki mobile phone user survey findings also indicate that 50 percentage of the respondents access their emails from their mobile phones and 52 percentage of the respondents access the Web via their mobile phones (Azuki Systems, 2008). Moreover, ComScore M:Metrics (2008b) reports that 30.2 percentage of United States, 19.9 percentage of Germany, 28.4 percentage of UK and 25.3 percentage of Italy population use mobile media. In the light of the above statistics, it can be concluded that mobile communication industry has been growing significantly over recent years.

The recent advances in mobile communication technologies and striking adoption of mobile phone and other handheld devices give rise to the evolution of mobile business (m-business) and mobile marketing (m-marketing) activities. As mobile device manufacturers and mobile network operators improve their services and products, mobile communication becomes the mainstream and changes the way of doing business. MNOs play a critical role in the transformation of m-business; because MNOs are the ones who offer technical facilities for the development of m-business models. MNOs play a major role in m-business and m-marketing by being strategically positioned between customers and content providers. The main purpose of this chapter is to review key supporting mobile technologies, services and solutions of a MNO and to comprehend mobile technologies, practices, and devices that support marketing activities during the implementation of m-business and m-marketing applications.

Towards this objective, corporate VAS of a MNO, which has gained global acceptance and recognition, is reviewed as a case study. Case study is particularly welcome in new situations where only a little is known about the phenomenon and in situations where current theories seem inadequate (Eisenhardt, 1989). The aim of the case study is to attain a comprehensive and detailed a description of the phenomena (Eisenhardt and Graebner, 2007). Case studies can be used to provide description, test theory, or generate a theory (Eisenhardt, 1989). Since single-case studies enable to describe the existence of a phenomenon in a detail way (Eisenhardt and Graebner, 2007); in this exploratory study, VAS of a MNO is analyzed. The purpose of analyzing the VAS of MNO as a case is not to make a generalization and attain a generalizable result; but to have a better understanding of the nature of the supporting role of MNOs in m-marketing applications. Moreover, this chapter will provide a better understanding about how the mobile communication and value added services of MNOs change the companies' way of doing business.

LITERATURE OVERVIEW

Mobile communication has a significant impact on the business world. As the interest in the usage of mobile applications in the business environment increases, mobile business related issues become more important for researchers and practitioners in the recent years. Tsai and Gururajan (2007) develop a framework in order to identify the factors that influence the motivations for a successful mobile business transformation and the possible challenges during the process; Leem et.al. (2004) make a mobile business model classification; Varshney and Vetter (2002) identify several important classes of mobile commerce applications and attempt to identify how to successfully define, architect, and implement the necessary infrastructure in support of mobile commerce; Gebauer and Shaw (2005) try to identify success factor and impacts of mobile business applications; Lee and Shim (2006) discuss the issue of satisfaction with mobile business applications; Chen and Nath (2004) also presents a framework for mobile business applications that is designed to give managers a systematic approach to discover mobile business opportunities in their organizations; Nah et al. (2005) focus on the advantages of using mobile business applications and report that mobile and wireless devices are enabling organizations to conduct business more effectively.

Forrester (2007) defines enterprise mobility as the ability for an enterprise to communicate with suppliers, partners, employees, assets, products, and customers irrespective of location. Today's workforce is increasingly mobile. Mobile communication and devices enable mobile employees to stay in constant communication with their company, consumers, and suppliers at anywhere and at any time. With mobile communication, mobile employees can access to business and personal related information and business operations become more flexible. Mobile communication and m-business applications create tremendous business opportunities in achieving greater work flexibility, higher operational efficiency, and higher performance and productivity in organizations; so that m-business applications help to lower operational cost, improve customer service and satisfaction, access to critical information from anywhere at anytime, and so on (Chen and Nath 2004; Gebauer and Shaw 2004; Tsai and Gururajan 2007). Gupta et al. (2005) asses impact of mobile communications on organizations and argue that sales/marketing; operations and maintenance; and general management are the three major functions where m-business applications play a leading role for achieving higher performance. MNOs play a critical role in the transformation of m-business by offering technical facilities. Moreover, as the companies embrace the mobility to increase operational efficiency, MNOs' value added services are leaping from voice calls towards innovative services. MNOs try to supply and deploy a range of innovative value-added services in order to satisfy the needs of business world.

INFORMATION ABOUT THE CASE COMPANY

Mobile phone usage in Turkey is also booming. The number of people using mobile phones in Turkey has grown by over 208% from 2000 to 2006 and in the year 2000 there were only 16.1 million mobile phone users, while in 2006 this figure increased to 49.7 million users (Hofmann, 2007). According to the Turkish Telecommunications Authority (2007) annual report, there are approximately 62 million GSM lines and the mobile line penetration rate in Turkey is approximately 88% at the end of December, 2007 (Turkish Telecommunications Authority, 2007). The traffic volume of mobile operators shows an increase parallel with the number of subscribers. The traffic realised by mobile subscribers is above 57.7 billion minutes in 2007 (Turkish Telecommunications Authority, 2007).

Considering the 70 million population of Turkey, the mobile phone penetration rate is quite high. Turkey as a big emergent market offers lots of opportunities for the evolution of innovative m-business and m-marketing activities. Since Turkey has a very young population, the adaptation of mobile phone and other handheld devices is remarkable. Innovation diffusion research indicates that young individuals are very likely to adopt a new technology because of their tendency to pursue innovativeness (Rogers 1995). ComScore M:Metrics (2008a) reports that 52 percentage of overall mobile users in the United States send text message and 74 percentage of teen population (ages 13-17) in the United States send text messages; so that report concludes that teens are avid text messagers compared to overall mobile population in the United States. According to the same report, teens are more likely to create and share personal multimedia compared overall mobile population (ComScore M: Metrics, 2008a). The Azuki mobile phone user survey findings also show that younger generation in United States spends more time on their mobile phones compared to overall mobile population (Azuki Systems, 2008). As it can be seen from the findings of the above research, young population are more likely to adopt and use mobile technologies and services; therefore, having a young population can be considered as an advantage for Turkey for the wide spread adaptation of mobile technologies and services. This fact forces MNOs to provide wide range of mobile services to their users in Turkey. Hence based on the above findings, it is worthy to review VAS of a Turkish Mobile Network Operator as a case study.

GSM-based mobile communication started in Turkey in February 1994. There are currently three mobile communications operators in Turkey: Turkcell, Vodafone, and Avea. The largest, Turkcell, has over 35.1 million mobile subscribers (Turkcell Communication Services, n.d.). With its 58% market share, Turkcell is the number one player in Turkish GSM market (Turkcell Communication Services, n.d.). Turkcell is not only the leading operator in Turkey, but is also the third biggest GSM operator in Europe in terms of subscriber numbers (Turkcell Communication Services, n.d.). As a large-scale GSM operator in a fast growing market, Turkcell has gained global acceptance and recognition. In July 2000, Turkcell became the first Turkish company to be listed in the New York Stock Exchange. Turkcell provides mobile telecommunications services to Eurasian emerging markets: Azerbaijan, Kazakhstan, Georgia, Moldova and Ukraine. Turkcell, as the leading operator of Turkey and nearby regions, entered the Board of Directors of GSM Association in early 2003. Turkcell won the first prize in the Best New Service category in World Communication Awards 2002, with its "GPRSLand" service and also won the Gartner CRM Excellence Award.

Turkcell focuses on three strategic issues: to cope with competition (by introducing new tariffs for different segments, focusing on localization, and strengthening sales channel); to drive loyalty (by underlining value propositions, improving customer experience, retaining valuable customer and maintaining strong brand); and to grow business (by focusing on postpaid subscriptions, promoting usage, and focusing on VAS) (Turkcell Communication Services, 2008). Turkcell vision of growth focuses on new technologies and new business models because the concept of mobile communication has been rapidly changing all over the world and the GSM business has been evolving from voice services to VAS. In the new era, Turkcell converges mobility, internet and computer technology in order to create VAS to its customer base. The rapid growth in mobile technology enables Turkcell to provide a wide range of VAS to their customers especially to their corporate customers and these VAS change and transform companies' way of doing business. Since Turkcell has gained global acceptance and recognition, it is worthy to review VAS of the Turkcell as a case study.

KEY SUPPORTING VAS FOR M-BUSINESS AND M-MARKETING

Integration of mobile communication into the business life leads the emergence of a contemporary concept that is called m-business. With the introduction of mobile communication, m-business models are evolving. MNOs are the ones who offer technical facilities for the development of mobile business models and VAS of MNOs assist companies to achieve their business objectives and improve their competitive position at the market place. Since MNOs offer an unlimited number of VAS, in this study several important VAS, which supports the m-marketing and m-business

Figure 1. Key Supporting Value Added Services

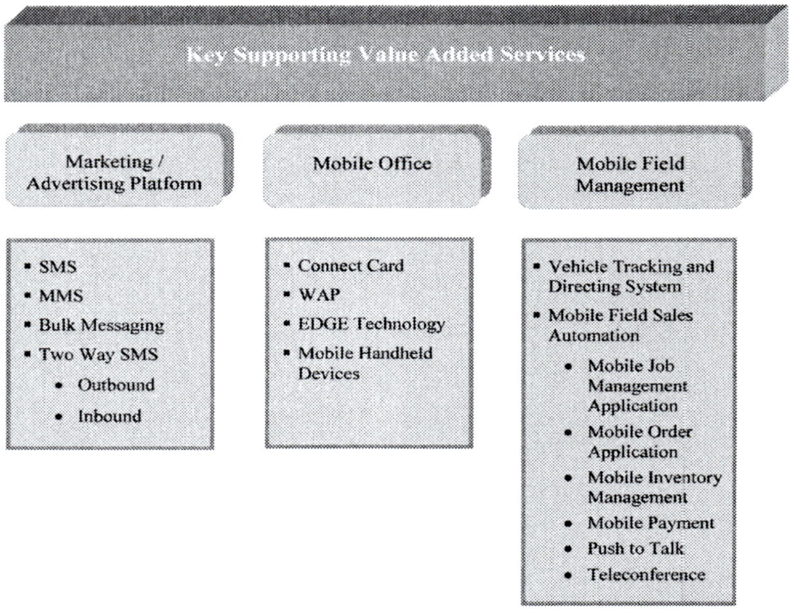

activities, will be reviewed. This chapter takes the Turkcell's Solutions Partners Program, which is developed for corporate customer base, as a case in order to point out what can companies do with the mobile technology. These solutions partner program services are custom designed business models to meet specific needs of corporations during their implementation of m-business and m-marketing applications. So that the readers can get a picture of what can a company do in the mobile age? Key supporting VAS, which are provided by MNOs to support m-marketing and m-business activities of a company, is analyzed under three main categories: mobile marketing/advertising, mobile office technologies, and mobile field management (see Figure 1).

Supporting VAS for Marketing and Advertising

Mobile network operators' value added services support the mobile marketing and mobile advertising activities of a company. Mobile advertising and mobile marketing campaigns are very important elements of m-business and m-marketing applications (Varshney and Vetter, 2002). Marketing and promotional advertising are the areas in which mobile messaging can be useful for the company. MNOs' messaging services are considered as a critical marketing and advertising platform (Kavassalis et al., 2003). Mobile advertising applications deliver promotional information to consumers using mobile phones and other smart handheld devices. Mobile advertising and mobile marketing campaigns can be performed by using messaging services such as SMS (Short Messaging Service) or MMS (Multimedia Messaging Service).

SMS and MMS

With the increasing popularity of m-marketing, companies integrate mobile channel into their communication and marketing campaigns. Today, many companies are effectively using mobile messaging technologies, which are provided by MNOs, during the implementation of marketing campaigns. Mobile messaging campaigns are the most familiar form of m-marketing campaigns and come in two different varieties: SMS-based (text-only messages) and MMS-based. SMS is a messaging infrastructure that allows mobile users to send and receive text messages of up to 160 characters across operator network. MMS is a type of messaging that enables the sending of multimedia objects (images, audio, and video) as opposed to the simple text enabled by SMS messages. SMS is considered as an important tool in mobile marketing campaigns. Companies invite consumers to send text messages to a short code in order to encourage them to attend m-marketing campaigns. To complete an operation of sending, receiving, and processing large volumes of SMS messages in m-marketing, a company requires the use of an SMS Gateway. It means that short messages are not sent directly from the company to consumer, but they are sending through an SMS center, which is an element of the MNO (Kavassalis et al., 2003). Varshney and Vetter (2002) indicate that MNOs' networking support is crucial in realizing mobile commerce applications. Mobile advertising requires multicast, roaming across multiple networks, and location management support (Varshney and Vetter, 2002). Mobile advertising also needs asymmetric multicast to send messages to certain users in certain locations, but such multicast does not have to be real-time (Varshney and Vetter, 2002). Hence, MNOs' technical facilities enable companies to use text messaging services for the purpose of marketing activities.

Bulk Messaging

As the effectiveness of short messaging services has been realized, it has become a crucial element of mobile communications and mobile marketing (Barwise and Strong, 2002; Rettie, Grandcolas, and Deakins, 2005; Merisavo et al., 2006). Bulk messaging can be considered as one of the VAS that is provided by mobile network operators. Bulk messaging offers ideal communication opportunities for the company and the company can reach thousands of customers, employees, or partners at the same time only by one click. Many companies need to reach their customers periodically and/or at special occasions for various reasons such as:

- To send automated warning/notification or reminder messages (e.g., to remind the deadline of the payment, to notify when a certain product is available in stock, to inform the consumer about a new product and price changes, to give notice about delivery status, etc.)
- To announce new marketing campaigns and promotions (e.g., to invite consumers to attend marketing campaigns such as text & win campaigns, to invite consumers to attend events or visit to the store, etc.)
- To send greeting messages

Bulk messaging service, which is supported by MNOs, enables to send messages to all customers at once, to send messages whenever the company wants, to receive SMS failure/success reports, to use alphanumeric (corporate name) in the text heading, and to send message with a personalized content.

Two-Way SMS

Two-way SMS enables companies to communicate with their customers and employees interactively via SMS. Two-way messaging is communication that uses both mobile originated SMS and mobile terminated SMS. Outgoing SMS, which is also called mobile terminated, means SMS traffic from computer systems to mobile phones. On the other hand, SMS traffic from mobile phones to computer systems called incoming SMS, or mobile originated SMS. In mobile originated two way SMS, a mobile user sends a text message to a short code number that the company has announced and the company then reply. In mobile terminated two-way SMS, the company sends the first message (such as sending messages to the customers regarding new product launches, discount offers, promotional activities, price alerts, event announcements, etc.) to the customer's mobile phone and the recipient replies to company's message.

Under the solution partner program, Turkcell's solution partners develop custom design two-way SMS business models, which are created according to the needs of the corporate customers, and Turkcell offers the necessary technological infrastructure for the custom design business solutions. The custom design solution for two way SMS builds a messaging gateway between institutions' database and the users' cellular phones. The two way SMS solution also provides "ask and learn" type interaction, in which the user queries are replied by the institution's information system. Firstly command set for "ask and learn" two way SMS service has to be developed; so that information in the institutions database can be queried by the commands. All commands received from user's mobile phone via SMS are processed by the institution's servers and the results of the queries are sent to user's mobile phones via SMS. Therefore, MNOs' messaging infrastructure enables to build an interactive communication between the company and its consumers.

Below examples are the some of the activities that can be done through two-way SMS:

- Companies ask customers to attend mobile marketing campaigns such as instant text to win campaigns, sweepstakes, mobile contests, mobile voting, mobile survey, mobile quiz etc. These types of SMS called outgoing SMS (mobile terminated SMS). Example 1: An instant win mobile campaign invites consumers to send a text message to short code in order to enter for the chance to instantly win various campaign related prizes and a text message is returned to the consumer from the company; so that consumers instantly learn if they have won a prize. Example 2: As a postpurchase satisfaction measurement, the company can send a SMS and asks if the customer satisfied with the product or service that the company provided. The SMS message asks customer to text yes or no and send a SMS to short code XXXX. Example 3: A bank can launch its new consumer loan product via SMS campaign and in the SMS, consumers were asked to write their citizenship ID and send it to the short code XXXX in order to initiate the loan application process. Then, loan approval is instantly sent to the consumer by Bank via SMS.
- Consumers ask and make queries through SMS and users' SMS queries are replied by the institution's information system via SMS. These are called incoming SMS (mobile originated SMS). Example 1: Patients can take an appointment from the hospital through SMS. The patient texts the hospital code that he wants to go, his citizenship ID, and the department that he wants to go. Then an approval SMS, which informs appointment date, is sent through SMS. Example 2: Students can inquire their academic data such as exam results, registration and tuition fee warnings, exam dates, advisor schedule over the Student Affairs Information System automation by sending SMS. And the related information is received through SMS. A student can text his student ID and the command word of date and send it to short code XXXX and receive the exam dates automatically via SMS.

Supporting Technologies for Mobile Office Applications

Mobile network operators also provide advanced technologies to support the Mobile Office Application (MOA) of a company. Replying an important e-mail immediately, planning the schedule, accessing Internet whenever needed, are the basic needs of the mobile professionals and employers when they are doing business outside of the office. Business and professional customer segment are interested in having access to standard business applications, including standard word processing, e-mail, and other business related information when they are out off office. Turkcell addresses the needs of mobile business and professional customers with its MOA technologies. MOA provide the complete office environment to mobile users any where any time (Varshney and Vetter, 2002). MOA give flexibility to the business and professional customer segment. MOA technologies enable mobile work force to stay connected and interact with their offices from their Personal Digital Assistance (PDA), Blackberry, mobile handsets, or laptops. PDA, Blackberry, and other mobile handheld devices can be considered as a laptop that is reduced to the size of mobile phone.

MOA and its supporting technologies, which are provided by MNOs, enhance the efficiency and productivity of mobile work force. MOAs give mobile work force easy and instant access to essential customer account, lead and product information from their mobile phones, smart mobile handheld devices such as PDA and Blackberry. These smart mobile handheld devices, which are supported by MNO technological infrastructure, enable:

- To receive real time email messages safely with "push technology" (The technology that displays e-mail messages automatically on the screen without the user having to do anything. Push mail refers to the automatic receipt of emails to the handheld device, as soon as they arrive at the destination email server) when the user is out of office
- Wireless push synchronization of all email and calendar activities
- Compose, send, read, forward, reply to, and delete email messages
- Immediate access to documents, spreadsheets and presentations while mobile
- To open, delete and archive attachments
- To access and surf the Internet with EDGE Technology
- To build the user's address book and organize the personal tasks with its calendar
- Real-time wireless access to schedule, contacts and tasks through push technology
- To access company network
- To receive and send SMS

MNOs provide technological infrastructure for the marketing and business activities of the mobile work force. MOA technologies also provide mobile field support and field sales automation. MOA technologies and mobile handheld devices enable companies to mobilize their order management, inventory management, scheduling, invoicing and more. All of which leads to productivity and efficiency for the companies in the mobile era.

Connect Card

MNOs provide connect card as a wireless laptop solution that allows the user mobile access to Internet, or to the corporate network, from the user's laptop via General Packet Radio Service (GPRS). The Connect Card is a credit card sized wireless communication device that plugs into the user's laptop to act as a modem. With the modem card, the user can establish wireless, convenient and fast internet connection without using the cellular line. Connect card is ideal for travelling professionals and mobile sales people. Connect card enables mobile corporate employers and sales people:

- To connect to the Internet from the laptop
- To access the corporate network and emails
- Receive and send short messages from the laptop

Internet AccessThrough Wireless Application Protocol (WAP)

Wireless Application Protocol (WAP), which is one of the MNOs' services, enables the user mobile access to internet just with a few taps from the user's mobile phone. WAP is an open international standard for applications that use wireless communication. Its principal application is to enable access to the Internet from a mobile phone. Internet standards such as HTML are inefficient over mobile networks because standard HTML content can not be effectively displayed on the small-sized screens of mobile phones, mobile terminals and pagers. WAP plays an important role in m-commerce by optimizing Internet standards for the constraints of the wireless environment and thus bridging the gap between Internet and mobile world (Tsalgatidou and Veijalainen, 2000). WAP is a technology platform that helps to create web sites that are easily accessed from handsets (MMA, 2007). A mobile web site, which is also known as WAP sites, is specifically designed and formatted for display on a mobile device. Mobile web sites can be viewable on any phone with a mobile Web browser. WAP sites are the perfect way to reach consumers or give marketers an additional tool for reaching mobile users (MMA, 2007). Mobile web portals can also be incorporated into m-marketing campaigns and in this case WAP page can provide access to images, information, terms and conditions about the promotion.

Internet Access Through Enhanced Data Rates for GSM Evolution (EDGE) Technology

MNOs have multiple choices to fulfil the demand for faster data connection. In the GSM system a GPRS capable network can provide data rate up to 115 kbps. GPRS based services, such as WAP, MMS, content download and others, causes network blocking. So that, MNOs face the need to boost their networks' capacity and data speeds (Nokia, 2003). However, EDGE provides up to three times the data capacity of GPRS. This means that EDGE technology enables theoretical data speeds of up to 384 kbps. EDGE technology is a third-generation and high-speed technology that supports fast data transfer and faster access to Internet from mobile devices such as mobile handsets, PDA, etc. EDGE technology makes possible to achieve higher data rates; so that EDGE technology enables MNOs to provide more sophisticated services to the subscribers. With EDGE, the dial up time is shorter, WAP browsing for the video clip is faster and the MMS video clip is delivered more quickly (Nokia, 2003). Higher data rates and greater network capability of EDGE technology offer many opportunities to the mobile business person such as faster access to data, rapid synchronization of email and other personal information when away from the office (Nokia, 2003). EDGE also brings much higher quality to new services such as Push to Talk over mobile phone. Hence, EDGE technology can be considered as an enabler technology for mobile office applications.

Supporting VAS for Mobile Field Management

Specifically, marketing communications mix consists of five major components of communication: advertising, sales promotion, public relations, personal selling and direct marketing (Kotler, 2000). And an integrated marketing communication mix relies heavily on the employees who have the most interactions with customer (Clow & Baack, 2007). Thus, the quality of the relationship between sales person and the customer are very important. At this point, effective management of field sales force becomes very crucial.

MNOs also provide value added services to support and manage field work force effectively. MNOs' mobile office applications and devices, which are discussed above, enable companies to mobilize their order management, inventory management, scheduling, invoicing and more. Value added services and solutions, which are provided by the MNO and its partners, to support the field force, provide a bundle of benefits to the company's productivity. Some of these benefits are:

- Access to information (i.e., customer, product, and inventory information) anywhere and anytime
- Ability for company executives to access information on inventory, revenues and clients while away from office
- Instant access to customers' account information, invoices, orders, and promissory notes when sales representatives are at customers' office quickly and easily without having to call office
- Ability for field personnel to reach all information about incoming and outgoing orders, sales transactions, inventory levels, price lists and creditor/debtor lists
- Providing the best service to the company's customers by effective supervision of field teams

VAS of mobile network operators enhances the efficiency and productivity of the field sales person; because of that reason, VAS for effective field management can be considered as one of the key supporting activities of MNOs. Value added services and wireless solutions for field management, which are provided by the case MNO, will be reviewed below.

Vehicle Tracking and Directing System (VTDS)

Vehicle tracking system is an electronic device installed in a vehicle to enable to track the vehicle's location through Global Positioning System (GPS). VTDS provides the company to track its sales force vehicles and get information about their positions, statuses and movements. By this way, the company can direct its sales force. VTDS enables tracking of vehicles from any location with internet access. With the assistance of VTDS, the company can access detailed information about sales force vehicles, such as speed, distance, waiting period and location. Moreover, the system can send notification messages when the vehicle exceeds pre-designated speed limits, when it makes a stop at a location for too long, or when it starts moving. Managers can track their mobile workforce's customer visits, and see whether they make the visits or not. Therefore, this service protects the vehicles against misuse, enables the company to be in constant communication with the vehicles, and lowers the cost of operation of vehicles. With MNO's GPS/GPRS substructure, the company tracks, manages and directs its field teams. And also keeps customers up-to-date with the exact location of their package and when it's likely to arrive. As a result, the company gives fast service to its customers.

Mobile Field Sales Automation

MNO's technological infrastructure enables to coordinate the marketing, sales, and distribution functions of the company from a single centre. Mobile field sales automation solutions integrate GSM, GPRS, EDGE and other wireless internet technologies into mobile software applications in order to strengthen the company's ability to manage its mobile work force. The mobile field sales automation solutions, which are offered by MNOs as a VAS, improve management of mobile work force, leading to greater operational efficiency and increased productivity. All these lead to improved customer satisfaction.

Mobile field sales automation allows the enterprise organizations manage their workflow for mobile sales and service professionals. MNOs support the necessary technology for the mobile field sales application platforms. Mobile field automation supports all sales activities from collecting orders (Presell) to invoicing in vehicle (Hotsell). With mobile sales force automation, it is possible to make order, payment, delivery and surveying data entries on the field. And also many other essential customer data that is gathered at the field can be collected and transmitted to the headquarters with hand terminals (such as PDA, BlackBerry, tablet PC, notebook, and smart mobile phones). Mobile sales automation, with GPRS, GSM, GIS, and other wireless technologies, provides mobile work force wireless access to sales information such as customer account information, opportunity alerts, sales order status, inventory availability, etc. Detailed information about the products, prices, vehicle stock, head-office stock, promotion-discount information can be also installed on the software of hand-held terminals. Instant delivery of field data to headquarter, ability to transmit the desired data to the field quickly, real time access to back office information from the field, constant communication between the field teams and headquarter, faster and more effective mobile reporting, and ability to gather all sales data in a single database can be considered as the benefits of mobile sales force applications.

Mobile job management application, which is the one of the important elements of mobile field sales automation, is designed for issuing job orders to mobile field teams, monitoring the process and completion of ordered jobs, receiving feedback from teams and monitoring job planning and team performance from the back office. This mobile application connects over mobile handheld sets and transfers the data through GPRS. There is two way data flow between headquarters and field teams. Real time communication using GPRS technology enables to transmit data instantly, accurately and thoroughly that is obtained on the field to headquarters.

Mobile order application is another crucial element of mobile field sales automation. Mobile order management application lets field sales force to process orders remotely using a PDA, BlackBerry, or any other smart mobile devices. The mobile sales personnel enter orders using smart mobile handheld sets and the application also provides for real time checks of the stock availability. These handheld sets can be synchronized with the company's back office software via GPRS. Mobile work force can invoice the goods in the vehicle and take orders. Moreover, *mobile payment* is also possible with the credit card POS (Point of Sale) terminals that operate over the GSM network. Mobile order application allows sales force to spend less time on manual order processing and more time meeting customer need.

Push to Talk (P2T)

Push to Talk is a feature similar to walkie-talkie that is provided over a mobile phone. P2T service provides the opportunity to reach one or more people at the same time by pressing one single button. Whether in the office or out of office, one can quickly connect with one or multiple team members at the same time with just a push of a button. P2T service is a useful application especially for the coordination of field sales teams and provides the ability to exchange voice messages between company and field sales teams.

Teleconference

Teleconference is a powerful platform, whereby mobile phone conversations with multiple participants can easily be made. It can be utilized easily and instantly for regional meetings, sales force meetings, dealer conferences, technical support meetings, project management conventions and for convening various units that are dispersed domestically or abroad. Teleconference, which enables to instantly exchange and assess views and allows quick decision-making, is especially ideal for informing and coordinating the sales team that operate over a dispersed geography.

CONCLUSION

GSM business has been evolving from voice services to value added services (VAS). Mobile network operators bring the advantages of next generation wireless technology to mobile businesses. The rapid growth in mobile technology enables Mobile Network Operators to provide a wide range of VAS to their customers especially to their corporate customers and these VAS change and transform companies' way of doing business. The review of VAS of a MNO, which has gained global acceptance and recognition in Turkey and as well as in the world, indicates that mobile technology changes the companies' way of doing business.

With the increasing popularity of mobile marketing, companies integrate mobile channel into their communication and marketing campaigns. Many marketing and business activities, such as marketing campaigns, field communications, information services, interactive services etc., are now possible with short messaging technologies that are provided by MNOs. Bulk messaging and two way SMS technologies are the two common applications enable companies to communicate with their customers. Bulk messaging offers ideal communication opportunities for the company to reach thousands of customers, employees, or partners at the same time only by one click. On the other hand, two way SMS enables companies to communicate with their customers and employees interactively via SMS.

MNOs address the needs of mobile business customers with the mobile office applications that enable them to access business related information when they are mobile. Mobile office applications enable to mobile work force to remotely interact with the company's back office applications through mobile hand terminals such as PDA, BlackBerry, tablet PC, notebook, and smart mobile phones. And these smart devices are supported by the wireless technologies such as GSM, WAP, EDGE, GPRS, GPS, and others, which are provided by the MNOs. MNOs offer various services that facilitate mobile workforce to access internet such as connect card, WAP, and EDGE. These technologies enable the companies to mobilize their business and marketing activities. Companies can now mobilize almost all their functions such job management, order management, inventory management, scheduling, invoicing and more. Mobile technologies of MNOs such as vehicle tracking and directing, push to talk, teleconference, etc. make possible to coordinate mobile filed activities. Use of mobile technologies strengthens the companies' ability to manage their mobile work force. The mobile field sales automation solutions, which are offered by MNOs as a VAS, improve management of mobile work force, leading to greater operational efficiency and increased productivity. All these lead to improved customer satisfaction and improve the companies' competitive position at the marketplace.

REFERENCES

Azuki Systems. (2008). *Today's Mobile Phone Users: Current and Emerging Trends,* September, 2008. Retrieved November 1, 2008 from http://www.azukisystems.com/images/uploads/Azuki_User_Survey_Sept_2008_vfinal.pdf

Barwise, P., & Strong, C. (2002). Permission-based Mobile Advertising . *Journal of Interactive Marketing, 16*(1), 14–24. doi:10.1002/dir.10000

Chen, L.-D., & Nath, R. (2004). A Framework for Mobile Business Applications. *International Journal of Mobile Communications*, *2*, 368–381. doi:10.1504/IJMC.2004.005857

Clow, K. E., & Baack, D. (2007). *Integrated Advertising, Promotion, and Marketing Communications*. NJ: Pearson Education Inc.

ComScore M. Metrics. (2008a). *Teens' Mobile Activities by Jen Wu,* August 28, 2008. Retrieved October 25, 2008, from http://www.comscore.com/press/presentations.asp

ComScore M. Metrics. (2008b). *The Mobile Media Authority by Mark Donovan,* August 19, 2008. Retrieved October 25, 2008, from http://www.comscore.com/press/presentations.asp

Eisenhardt, K. M. (1989). Building Theories from Case Study Research. *Academy of Management Review*, *14*(4), 532–550. doi:10.2307/258557

Eisenhardt, K. M., & Graebner, M. E. (2007). Theory Building from Cases. *Academy of Management Journal*, *50*(1), 25–32.

Forrester. (2007). Topic *Overview: Enterprise Mobility. Mobility Becomes an Integral Part of Firms' Network Strategy by Maribel D. Lopez,* July 20, 2007, Retrieved October 29, 2008, from http://www.forrester.com/Research/PDF/0,5110,42404,00.pdf

Gebauer, J., & Shaw, M. J. (2004). Success Factors and Impacts of Mobile Business Applications: Results from a Mobile e-Procurement Study. *International Journal of Electronic Commerce*, *8*(3), 19–41.

Gupta, M. P., Sahu, G. P., & Gauri, S. (2005). Assessing Impact of Mobile Communications on Organizations: A Flexibility Analysis. *Global Journal of Flexible Systems Management*, *6*(3/4), 11–23.

Hofmann, O. (2007). *Young Population Drives Turkey Mobile Market.* Retrieved June 20, 2008, from http://www.euromonitor.com/Young_population_drives_Turkeys_mobile_market

Kavassalis, P., Spyropoulou, N., Drossos, D., Mitrokostas, E., Gikas, G., & Hatzistamatiou, A. (2003). Mobile Permission Marketing: Framing the Market Inquiry. *International Journal of Electronic Commerce*, *8*(1), 55–79.

Kotler, P. (2000). *Marketing Management.* NJ: Prentice-Hall, Inc.

Lee, C., & Shim, J. P. (2006). An Empirical Study on User Satisfaction with Mobile Business Applications Use and Hedonism. [JITTA]. *Journal of Information Technology Theory and Application*, *8*(3), 57–74.

Leem, C. S., Suh, S. H., & Kim, D. S. (2004). A Classification of Mobile Business Models and Its Applications. *Industrial Management & Data Systems*, *104*(1/2), 78–87. doi:10.1108/02635570410514115

M-marketing Association. (2007). Understanding M-marketing: Technology and Reach, May 2007. Retrieved July 25, 2008, from http://www.mmaglobal.com/mobilemarketing102.pdf

Merisavo, M., Vesanen, J., Arponen, A., Kajalo, S., & Raulas, M. (2006). The Effectiveness of Targeted Mobile Advertising in Selling Mobile Services: An Empirical Study. *International Journal of Mobile Communications*, *4*(2), 119–127.

Nah, F. F.-H., Siau, K., & Sheng, H. (2005). The VALUE of Mobile Applications: A Utility Company Study. *Communications of the ACM*, *48*(2), 85–90. doi:10.1145/1042091.1042095

Nokia. (2003). *Make More Profits Now With The Nokia End-To-End EDGE Solution*. Retrieved December 12, 2008, from http://www.nokia.com/BaseProject/Sites/NOKIA_MAIN_18022/CDA/Categories/Operators/MobileOperators&ServiceProviders/MobileNetworks/GSMGPRSEDGE/_Content/_Static_Files/edge_brochure_final_lo_res.pdf

Rettie, R., Grandcolas, U., & Deakins, B. (2005). Text Message Advertising: Response Rates and Branding Effects, *Journal of Targeting. Measurement and Analysis for Marketing*, *13*(4), 304–312. doi:10.1057/palgrave.jt.5740158

Rogers, E. (1995). Diffusion of Innovations, 4th ed. NY: Free Press.

Tsai, H. S., & Gururajan, R. (2007). Motivations and Challenges for M-Business Transformation: A Multiple-Case Study. *Journal of Theoretical and Applied Electronic Commerce Research*, *2*(2), 19–33.

Tsalgatidou, A., & Veijalainen, J. (2000). Mobile Electronic Commerce: Emerging Issues. in the *Proceedings of the International Conference Electronic Commerce and Web Technologies*. Springer Berlin Heidelberg, pp. 477-486.

Turkcell Communication Services. (2008). First Quarter 2008 Results, Retrieved July 25, 2008, from http://www.turkcell.com.tr/c/docs/presentations/presentations2008Q1.pdf

Turkcell Communication Services. (n.d). *A World Company*, Retrieved July 25, 2008, from http://www.turkcell.com.tr/en/AboutTurkcell/corporateInfo/companyHistory

Turkish Telecommunications Authority. (2007). Annual Report 2007. Retrieved July 25, 2008, from http://www.tk.gov.tr/eng/pdf/2008/fr2007eng.pdf

Varshney, U., & Vetter, R. (2002). Mobile Commerce: Framework, Applications and Networking Support. *Mobile Networks and Applications*, *7*, 185–198. doi:10.1023/A:1014570512129

Wireless Intelligence. (2007). *Subscriber statistics end Q4 2007 Report*. Retrieved July 25, 2008, from http://www.gsmworld.com/news/statistics/pdf/gsma_stats_q4_07.pdf

KEY TERMS AND DEFINITIONS

Bulk Messaging: Many companies need to reach their customers periodically and/or at special occasions in order to send automated warning/notification or reminder messages, to announce new marketing campaigns and promotions, and/or to send greetings messages. Bulk messaging enables to reach thousands of customers, employee, or partners at the same time only by one click.

Connect Card: The Connect Card, which is a credit card sized wireless communication device that plugs into the user's laptop, allows the user mobile access to Internet via General Packet Radio Service (GPRS).

Enhanced Data Rates for GSM Evolution (EDGE) Technology: EDGE technology is a third-generation and high-speed technology that supports fast data transfer and faster access to Internet from mobile devices. EDGE technology enables theoretical data speeds of up to 384 kbps.

Mobile Field Sales Automation: Mobile field sales automation solutions integrate GSM, GPRS, GPS and other wireless technologies into mobile software applications in order to strengthen the company's ability to manage its mobile work force. Mobile field sales automation supports all sales activities from collecting orders to invoicing on the field.

Mobile Job Management Application: Mobile job application is designed for issuing job orders to mobile field teams, monitoring the process and completion of ordered jobs, receiving feedback from teams and monitoring job planning and team performance from the back office.

Mobile Office Applications (MOA): MOA technologies enable mobile work force to stay connected and interact with their offices from their PDA, Blackberry, mobile handsets, or laptops. And these smart devices are supported by the wireless technologies such as GSM, WAP, EDGE, GPRS, GPS, and others, which are provided by the MNOs.

Mobile Order Application: Mobile order application is a crucial element of mobile field sales automation. Mobile order management application lets field sales force to process orders remotely by using mobile hand terminals. Order that is gathered at the field can be collected and transmitted to the headquarters with hand terminals (such as PDA, BlackBerry, tablet PC, notebook, and smart mobile phones).

Mobile Originated SMS: SMS traffic from mobile phones to computer systems called incoming SMS, or mobile originated SMS.

Mobile Terminated SMS: Outgoing SMS, which is also called mobile terminated, means SMS traffic from computer systems to mobile phones.

Push to Talk (P2T): Push to Talk is a feature similar to walkie-talkie that is provided over a mobile phone. P2T service provides the opportunity to reach one or more people at the same time by pressing one single button.

Two-Way SMS: Two-way SMS enables companies to communicate with their customers and employees interactively via SMS. Two-way messaging uses both mobile originated (incoming messages) and mobile terminated (outgoing messages) SMS. The two way SMS solution also provides "ask and learn" type interaction, in which the user queries are replied by the institution's information system.

Vehicle Tracking and Directing System (VTDS): Vehicle tracking system is an electronic device installed in a vehicle to enable to track the vehicle's location and movements through Global Positioning System (GPS).

Chapter 10
Mobile Customer Relationship Management (mCRM): Constraints and Challenges

Günter Silberer
University of Mannheim, Germany

Sebastian Schulz
University of Göttingen, Germany

ABSTRACT

Customer Relationship Management is conducive to providing customer information and the correct way to deal with customers. After the advent of the world-wide-web (www), which brought an increase in customer data and intensified CRM discussions in its wake, we are now entering the age of computerized mobile communication with diverse possibilities to enlist the services of mobile people. This chapter outlines the new and latest technological developments and the resultant capabilities of mobile CRM. Looking to a future in which mobile marketing efforts may become increasingly important, the erosion of personal freedom and the threat to this freedom becomes all the more significant. The findings from reactance research do not only specify the reactance conditions but also the reactance consequences. Previous research has sometimes pointed to self-determination as an acceptance factor yet failed to analyze the reactance risk more closely. This chapter will attempt to close this gap. In this respect, the initial findings of an empirical study will be presented, revealing that the acceptance of mobile services is surprisingly low and the reactance risk for mobile services is not to be underestimated, and that reactance involves the corresponding behavior on the part of the user. The plea is to consider the acceptance risk in the mobile CRM of the future and select a customer-friendly permission policy.

CRM AS A FACTOR OF SUCCESS

The relationship between the supplier and his customers is crucial for the success of both the supplier and the customer. Consequently, customer relations have to be nurtured as a symbiotic relationship with mutual benefits for both parties. This transpires through good benefits and acceptable prices, as well as need-based channels of distribution and presentations and pointed communication. The mixture

DOI: 10.4018/978-1-60566-074-5.ch010

of these measures is vital over a longer period of time (for CRM as System cf. (Meyer, 2005; Paulissen et al, 2007)). Anyone who is looking to cultivate customer relations has to know what the customer wants, what he rejects and what he is prepared to spend. Consequently, CRM not only entails an action mix but also an analysis mix (Silberer, 2001, pp. 161-176 (especially on Data quality see (Missi, 2005)). The variety of CRM definitions in different disciplines (e.g. Information System (IS) or Marketing literature) must not belie this (Boulding et al, 2005, pp. 155-166; Ngai, 2005, pp. 582-605; Paulissen et al, 2007; Payne & Frow, 2005, pp. 167-176). Payne und Frow (2005, p. 167) also criticize the many partly contradictory or incomplete definitions. Based on an analysis of various sources (twelve different definitions in total), the authors worked out the relevant aspects and similarities and came up with the following work definition: "CRM is a strategic approach that is concerned with creating improved shareholder value through the development of appropriate [...]. CRM unites the potential of relationship marketing strategies and IT to create profitable, long-term relationships [...]. CRM provides enhanced opportunities to use data and information to both understand customers and cocreate value with them. This requires a cross-functional integration of processes, people, operations, and marketing capabilities that is enabled through information, technology, and applications." (ebda). The definition reflects the current state of knowledge and the essence of CRM particularly well and will also be taken as a basis here.

This chapter focused on the adoption of CRM in mobile communication and the acceptance and reactance of mobile services. Therefore, we outline the role of CRM in the age of mobile communication in the next section. Afterwards, we specify the technical parameters of mobile CRM, before highlighting possibilities of context-aware Services. In the following, we deal with the state of acceptance and reactance within the mCRM and using mobile services. Furthermore, we present the findings of empirical concerning mobile services, acceptance and reactance. The chapter finished with a conclusion.

CRM IN THE AGE OF MOBILE COMMUNICATION

The spread of mobile communication, which has become a global mega-trend, has opened up fresh opportunities to get to know the customer, attract new customers and cultivate customer relations (Silberer, 2004, pp. 453-470). This applies to suppliers of goods and services on the classical market and providers of services on the electronic market alike. Mobile communication means that people with a mobile device can be reached both at home and on the road and are capable of reaching other people and institutions when they are at home or on the road themselves. Access to web pages and other information services can be offered to the customer anytime and anywhere, which generate sales themselves or promote sales for other services (Attwood et al, 2000). Approximately 50% of mobile network users can already access mobile services today. Furthermore, various studies predict that the number of mobile Internet users is set to increase to about one billion by the year 2011 (Schäfer & Toma, 2008, p. 18). As a result, it is worth considering the possibilities of mobile CRM (mCRM) and trying them out (Frolick & Chen, 2004, pp. 53-61; Homburg & Bruhn, 2003, pp. 3-37; Nicolai & Petersmann, 2001; Nysveen, 2005, pp. 330-346; Silberer, 2002; van Ackeren, 2002, pp. 343-360; Zeng, 2003, pp. 302-322). In this context, we defined mCRM as an adoption of CRM in the context of mobile Communication by the use of mobile technologies and offering mobile services to the customer.

This is all the more important today as the technical capabilities have developed so much and reports of successful applications increase; recognizing the new possibilities mobile CRM

Table 1. Technical Aspects of the Mobile Radio Networks

	Local Area Networks	Wireless Local Networks	Wide Area Mobile Networks	
Network Technologies (Physical Transmission Procedures)	Bluetooth	WLAN	GSM (GPRS,HSCDS)	CDMA/ UMTS
Base Protocols for Information Services	Bluetooth specifications	TCP/IP, UPD, HTTP, etc.	TCP/IP, UPD, HTTP, WAE, WSP, WTP, etc.	
Presentation and Programming Standards	HTML, c-HTML, Java, SMS, MMS, WAP			
Terminals	Cell phones (smart phones) / PDAs / mobile PCs (particular hardware necessary for access via WLAN, Bluetooth or CDMA/UMTS)			

can offer is essential in order to keep up with the competition.

TECHNICAL PARAMETERS OF MOBILE CRM TODAY

The range of mobile services in the context of mCRM initially results from the technical possibilities available. In this context, we would like to briefly examine network technology, base protocols, the presentation and programming norms and the terminals (see Table 1 following Kaspar & Hagenhoff, 2005).

Network Technology

Mobile customers can be reached via different networks. Mobile network platforms or wide area mobile network platforms, such as GSM or UMTS networks, have dominated radio technology in recent years (Kaspar, 2006). Mobile network platforms reach theoretical transmission rates from 14.5 Kbit/s (GSM) to 384 kBit/s (CDMA/UMTS). The fully digital networks of the second generation which are nevertheless equipped with low data transmission rates (2G GSM networks, HSCSD & GRPS merely constitute developments of the GSM standard and are based upon the same infrastructure) are widely available in many countries. The faster standard of the third generation (3G, Wideband CDMA/ UMTS networks) are still being upgraded in many countries but are making breakthroughs in an increasing number of them (BITKOM, 2006)[1]. Both kinds of network already play an important role in offering mobile services (Siau & Shen, 2003, pp. 3-14). Moreover, the next development stage, "high speed download packet access" or HSDPA, is already in the pipeline. It will facilitate transmission rates of almost 2 megabits per second, effectively five times higher.

However, in virtue of the low costs (e.g. connection costs), local area network technologies are increasingly given priority (Kaspar & Hagenhoff, 2005, pp. 645-651; Sikora, 2001; Webb, 2001). This includes close-range networks (Bluetooth) or wireless local networks, which differ in terms of both the transmission rates and the range. For example, Bluetooth has a maximum range of 100 meters and a data capacity of 723 kbit/s, whereas WLAN (802.11b,g,a) has a range of 300 meters and a data capacity of 108 Mbit/s (Frolick & Chen, 2004, pp. 53-61).

Local area network technologies have the advantage that the local aspect automatically results from the connection to the respective base station, especially in the context of local-based or context-aware services. Consequently, the disadvantage of the limited radio range of Bluetooth or WLAN networks goes hand in hand with an advantage in supplying local-based services. Locally-oriented subscriber localization is also possible within mobile radio networks as a cell phone or smart phone has to log into a radio cell. This kind of localization incurs additional costs for mobile service providers and is not offered by all network operators. A new path is the integration of a GPS receiver into mobile terminals, which then allows a definition of the locus (Kaspar et al, 2007).

Base Protocols and Presentation Techniques

The next level is a matter of the different protocols on which mobile services can minute. Both known Internet protocols, such as TCP, IP and UPD, and specifications (Bluetooth) and protocols from the mobile radio sector, such as WAE and WSP, play an important role in this (Kaspar & Hagenhoff, 2005, pp. 645-651). They make reaching customers on the road through particular presentation and programming norms within a network per call possible, not to mention the fact that they can access various services per SMS, MMS, WAP or HTML and receive information. The wireless application protocol (WAP) currently plays the most significant role on the mobile service market (Frolick & Chen, 2004, pp. 53-61). In doing so, the corresponding presentation programs, e.g. browser or specific client software, have to be installed on the terminals.

Mobile Terminals

The hardware used on the user page also influences the range of mobile service, i.e. mobile PCs, cell phones and PDAs, as well as multifunctional appliances, e.g. smart phones (cell phone and PDA). The features relevant for the distribution of and access to mobile services are primarily the size of the display, the computing power and the memory capacity (Königstorfer & Gröppel-Klein, 2007, pp. 34-47) (for the usability of mobile appliances cf. (Wohlfahrt, 2004)). The features used to access networks (e.g. Bluetooth or WLAN) are also relevant. A mobile service which is primarily supposed to feature on cell phones has to fulfil different presentation specifications than a service required by the users of a PDA with a much bigger display. However, the disadvantages of many cell phones today are not only the limited display area but also the limited computing power and memory, not to mention the incompatibility between the terminals and the networks due to the numerous different specifications of the manufacturers. Nevertheless, cell phones of the latest generation already have the necessary hardware to access the various networks (WLAN, Bluetooth & CDMA/ UMTS).

The software installed on the terminal is also significant. Software for the presentation of HTML or WAP subject matter is already available on most terminals. However, if particular software is required, problems in the operating system can arise in cell phones and smart phones.

POSSIBILITIES OF MOBILE CRM USING THE EXAMPLE OF CONTEXT-AWARE SERVICES

Initially, the communication possibilities and services for the customer are significant for customer relations (Silberer, 2004, 453-470). In addition, the customer relations can also be improved by providing a provider's field service with helpful services by way of mobile communication and offering the possibility of communicating with the customer in transit. We have singled out a handful of the numerous possibilities of "mobile CRM", which also includes the improvement of customer knowledge through analytical CRM (see Table 2) (Xu & Walton, 2005, pp. 955-971). We primarily consider context-aware offers in the form of communication services, information offers, navigational aids, tracking services,

Table 2. Examples of Mobile Services

	Event Marketing	Regional & Town Marketing	Automobile Industry	Field Work
Tracking Services	Location at events	Location on sightseeing tours	Location of a vehicle	Location of staff
Navigation Services	Way to the event (e.g. stadium)	Route on sightseeing tours	Way to the next garage	Way to the next customer
Information Services	Information on artists	Information on sights	Traffic information	Information on the status of the order processing
Communication Services	Chat amongst user groups	Mobile contact with the town information service	Contact with technical support	Coordination between staff
Entertainment	Distribution of video clips	Mobile Town rally	Offer of an interactive game	Deployment of the latest advertisements
Transaction Services	Purchasing of ring tones	Mobile possibility of paying the parking ticket	Town information with a fee	Mobile acceptance of orders

entertainment offers and order and purchasing possibilities (transaction services) (Bauer et al, 2006, pp. 36-38; Silberer, 2004, 453-470). A service is then context-aware if it considers references to the whereabouts or context of the user (Dev & Abowd, 2000; Kaspar et al, 2007; Russel & Norvig, 2003). A context is any information that is used to characterize relevant situations of people, locations, or objects that are important for the interaction between application and user (Kaspar et al, 2007).

Local-Based Content Distribution via Bluetooth

We primarily have recourse to Bluetooth technology in the field of information and entertainment services. The market leader Qwikker (UK) deploys a technology which consists of two components: first, the stationary network which sends content matter and information to mobile terminals via Bluetooth and, second, a client (as Java software on the cell phone), which can be installed in the first approach and has to be installed for use (www.qwikker.com & www.gsmawards.com). The software facilitates access to content matter via Bluetooth in corresponding local area networks but also access from GSM or CDMA/UMTS networks if one is beyond the range of a local area network. Qwikker currently maintains approx. 1000 Bluetooth content distribution points in the USA and Europe.

Among other things, the Qwikker system was used on the Robbie Williams Summer Tour 2006 and drink manufacturer Pepsi's 2007 Bluetooth marketing campaign. The organizers of the Robbie Williams concerts enabled the visitors to access information about the artist and further tour dates, as well as download a video. Bluetooth hotspots were distributed at the concert and the visitors made aware of the service via signs.

The Pepsi Company integrated Bluetooth into multimedia advertising space at bus stops and kiosks to enable the consumers to download, watch and forward up to five music videos with their cell phones after installing the Qwikker client. This enabled the customers to take the viewed advertisement away with them on their cell phones.

One could imagine similar mobile applications based upon Bluetooth hotspots and client software in the realm of, say, town marketing on an interactive sightseeing tour in which information on sights are accessed or delivered as soon as the user enters the range of the local area network (see Table 2 following Silberer (2004) and Bauer et al. (2006).

Mobile Information & Multimedia Delivery via GSM & CDMA/ UMTS

If a user is inside a mobile radio network (GSM & CDMA/ UMTS), he or she can, for example, access WAP content matter through via browser software. Consequently, the mobile service is in the generation, concentration and distribution of content matter (Kaspar, 2006).

One interesting development of such mobile services is offered by the company ShoZu (UK, USA) (www.ShoZu.com & www.gsmawards.com). After the installation of client software on the terminal, the user indicates in a profile which content he or she is interested in. The aim is to facilitate the exchange of multimedia content matter. Opening a browser and connection-based waiting periods are omitted for the users. The preferred content matter is constantly updated via the mobile radio connection and made available. In the process, continuing connection costs are incurred, only rendering the use of such services useful in the case of a flat rate.

This kind of information delivery presents itself above all in the information, communication and entertainment service sector. Consider a user who subscribes to important content matter shortly before embarking on a long journey in order to be able to access it anytime and anywhere. Companies could use this to supply their sales representative with any important, updated information.

If it were a matter of services which provide access to information on a mobile terminal in the application examples outlined above, the other direction, in which content matter is delivered and, for example, made available to others on the WWW, is also feasible. The company ShoZu offers a the mobile application ShareIt, which can be used to place content matter created on or loaded onto cell phone (e.g. photos) directly onto partner web pages on the WWW easily after installation on the cell phone and the creation of a profile, e.g. MyTube.com. By these means, companies could offer to document product experiences for their customers and make them directly (just in time) available to other users with their mobile terminals.

LIMITS OF MOBILE CRM: REACTANCE IN LIEU OF ACCEPTANCE

In the case of mobile CRM, not only are the technological and functional possibilities of interest but also the limits which draw acceptance in the user. If one assumes that promotional expenditures could increase tenfold from 1.4 billion dollars in 2007 to 14 billion dollars in mobile marketing alone (Strategy Analytics, 2007), these limits will become increasingly important over the coming years. The acceptance and reactance are therefore set to become key factors in the future mobile competition. For example, such limits are mentioned by Boulding et al. (2005, pp. 155-166) when they warn against the loss of the customers' trust in the future CRM (cf. [9, 39]). Incidentally, Boulding et al. (2005, pp. 155-166) mention the "dark side of CRM" in a similar context. Previous studies on the acceptance of mobile services and the acceptance of mobile commerce have primarily focused on user-friendliness, utility and usage willingness but have ignored the topic of acceptance barriers, which involve a restriction of the personal spheres of influence (see e.g. (Bruner & Kumar, 2005, pp. 553-558; Koivumäki, 2006, pp. 418-435; Luarn & Lin, 2005, pp. 873–891; Nysveen, 2005, pp. 330-346; Russel & Norvig, 2003; Wang et al, 2006, pp. 157-179; Wu & Wang, 2005, pp. 719-729) – studies which all proceed from the so-called technology acceptance model, TAM).

Acceptance barriers, which arise from the restriction of personal freedom (Ngai mentions an encroachment on "customer privacy" (Nicolai & Petersmann, 2001)), can primarily be expected in the following situations:

1. Calls and offers which have not been ordered, for which no permission has been obtained or even granted.
2. Calls and offers which in principle have been approved but which gradually exceed the expected, acceptable amount in terms of amplitude and intrusiveness.
3. The gathering and use of user data for which no permission has been obtained or even granted.
4. Use of the user data gathered under approval which gives the user the impression not only of being scrutinized and analyzed but also of being defined and handled as a particular type by the service provider.

When it comes to assessing such risks more accurately than has previously been the case, it seems logical to refer to Brehm's "theory of psychological reactance" (Brehm, 1966; Brehm, 1972). This theory endeavors to provide an answer the question as to when a perceived limitation of freedom occurs and what such a perception can cause. The reactance theory triggered numerous studies after its publication and has been developed considerably so that its empirical content meanwhile can be regarded as highly respectable. State-of-the-art reports on relevant research confirm this assessment – for example, the English reports by Grabitz-Gniech (Grabitz-Gniech, 1971, pp. 188-196), Wicklund (Wicklund, 1974), Clee & Wicklund (Clee & Wicklung, 1980, pp. 389-405), Brehm & Brehm (Brehm & Brehm, 1981), Wright & Brehm (Wright & Brehm, 1982, pp. 608-618), Dickenberger & Gniech (Dickenberger & Gniech, 1982, pp. 311-341) and (Miron & Brehm, 2006, pp. 9-18), and the German-language reports by Grabitz-Gniech & Grabitz (Grabitz-Gniech & Grabitz, 1973a, pp. 19-35; Grabitz-Gniech & Grabitz, 1973b, pp. 361-365), Gniech & Grabitz (Gniech & Grabitz, 1978), Dickenberger (Dickenberger, 1979), Silberer (Silberer, 1980, pp. 386-391) and Gniech & Dickenberger (Gniech & Dickenberger, 1987).

The core assumptions of the reformulated reactance theory and results say (cf. the listed state-of-the-art reports):

1. Limitations of and threats to personal or individual freedom on the part of a third party generate reactance and thus an attempt to re-establish the lost freedom or secure the freedom under threat.
2. The magnitude of the psychological reactance primarily depends upon the significance of the limited or threatened freedom and the probability that the threat will be realized.
3. The possibilities of re-establishing lost freedom and securing threatened freedom include the appreciation of the lost or threatened freedom space and more or less undirected "anger" on a cognitive-affective level and, on a behavioral level, the disregard for freedom limitations and aggression towards the actor with the aim of inducing the latter to cease the restriction or withdraw the threat [14]. In the context of CRM, Boulding et al. (2005, pp. 155-166) point out the danger that the users or customers "circumvent" or "undermine" CRM activities.

The fact that the research reveals additional conditions which have to be satisfied if reactance is to occur also illustrates that reactance cannot be expected for every restriction of freedom. Reactance arises in particular when the freedom restriction is regarded as unjustified or even illegitimate. This is unlikely in a case where the user has granted permission. Furthermore, reactance can only be expected if the freedom restriction cannot be reversed immediately (see (Silberer & Wohlfahrt, 2001, pp. 161-176) on customer retention as a "ligation"). Reactance also assumes that there are not any significant accomplishments connected to the freedom restriction which compensate the loss of freedom (marketing can thus work with incentives or appropriate pledges, as indeed is often the case). Therefore, reactance does not occur if "learned helplessness" dominates and the restriction of freedom is answered with passiveness and resignation (see the aforementioned state-of-the-art reports).

In reality, what is considered strong or weak, legitimate or illegitimate and revisable or irreversible not only depends on personality structure of the person affected but also their cultural sphere – be it more reliant on personal, individual freedom of choice or more on social norms and authority (Miron & Brehm, 2006, pp. 9-18).

What does this research state imply for the reactance risk in mCRM? The findings of the reactance research support the assumption that unsolicited calls or mobile services, violations of the permission granted, a marketing overload and the predefinition of users as a particular type can generate reactance effects. At the same time, the reactance research explains that such effects can only be expected if several conditions are met. These include the importance of freedom, the magnitude of the restriction, its perceived legitimacy, its (feared) sustainability and the absence of advantages which could compensate for the loss of freedom (see the study on "promotion reactance" by Kivetz (2005, pp. 725-736), which highlights the importance of cost-benefit considerations). Anyone who concludes that the risk of reactance is low in mobile CRM after glancing at the list of antecedent conditions should consider that we still know far too little about the existence of these conditions. Furthermore, we should also take into account the fact that, with the spread of mCRM and an increase in mobile marketing, even that which is still perceived as acceptable today can easily become unacceptable in the future.

But how is the reactance and acceptance of mobile services reflected both in general and specifically by users of push / pull services in this day and age? To provide an initial response to this question, a study on the topic of the acceptance and reactance of mobile services will be presented in the following. In terms of mobile services, we will be referring to mobile push services, i.e. services that the provider sends to a mobile terminal without the action of the user, and mobile pull services, i.e. services that the user actively solicits from the provider from or to his mobile terminal.

RESULTS OF AN EMPIRICAL STUDY ON THE ACCEPTANCE AND REACTANCE OF MOBILE SERVICES IN GERMANY

To answer the question posed in the last section, i.e. how the reactance and acceptance of mobile services is reflected both in general and specifically by users of mobile push / pull services today, we conducted a survey with N= 187 students at the University of Göttingen. 50.3% of the participants were male and the average age was 23.4 (SD=2.50). On average, the respondents declared that they had a rather average knowledge of mobile services (M= 3.51; SD= 1.07). The whole questionnaire was computerized and completed by the participants directly on their notebooks. For the operationalization of the acceptance, the attitude acceptance was ascertained through global judgments (two items; Cronbach's α in the study .89) and three trait judgments. The behavioral acceptance comprised the use behavior and use intention (the kind of the services used and the reasons for use). The items for the reactance comprised the conditions of the reactance: the perceived restriction of freedom, the importance of freedom, the legitimacy of the restriction felt and the possibilities of compensating for the restriction of freedom with the benefits mobile services can offer (Cronbach's ; Cronbach's α of the reactance scale in the study: .77; the scales developed by Prof. Dr. G. Silberer for the measurement of acceptance and reactance can be found at the end of this article). Possible reactance effects were illustrated using one item on subjective feeling and behavior for the reactance in each case.

Table 3. Reasons for using mobile services (N= 187; multiple entries possible)

Reason for Using Mobile Services	Entered	Not Entered
Service is tailored to suit personal needs.	83 (44.4%)	104 (55.6%)
The service is accessible everywhere and at any time.	117 (62.6%)	70 (37.4%)
The service provides information in an uncomplicated and convenient fashion.	103 (55.1%)	84 (44.9%)
The service is inexpensive.	66 (35.3%)	121 (64.7%)
The service is reliable.	22 (11.8%)	165 (88.2%)

First of all, however, the reasons why the participants use or basically would use mobile services need to be explained. The results are presented in Table 3. Unsurprisingly, the main reasons were the ubiquity of the corresponding services and comfortable, uncomplicated use.

The analyses reveal, however, that mobile services still have a low overall acceptance. If one considers the acceptance more closely, the attitude acceptance only lies at a medium level of M=2.8 (SD= .87) but is considerably better than the actual behavior acceptance of mobile services for users (M=4.9; SD=1.24). As was to be expected, the acceptance (the overall, attitude and behavior acceptance) for push services was considerably worse than for mobile pull services, where the users become active and solicit the services (both differ significantly at p<.001; see Fig. 1). Accordingly, there was not any permission on the part of the user in the case of push services. The overall acceptance ascertained in this study was only M=3.6 (SD= 1.11).

Above all, the poor acceptance values are astonishing because the target group polled (students) is basically a technically affine target group (M= 2.63; SD= 1.28; scale 1-7) that has already owned a mobile terminal, such as a mobile phone, for M=7.8 (SD= 1.91) years. The consumer-specific variables, age and income, recorded in the study did not have a significant impact on acceptance. In contrast, there was a positive correlation between acceptance and the experience of mobile services (r= .51; p< .001) and the experience of mobile terminals (mobile phone; r= .18; p<.01).

Figure 1. Acceptance of mobile services

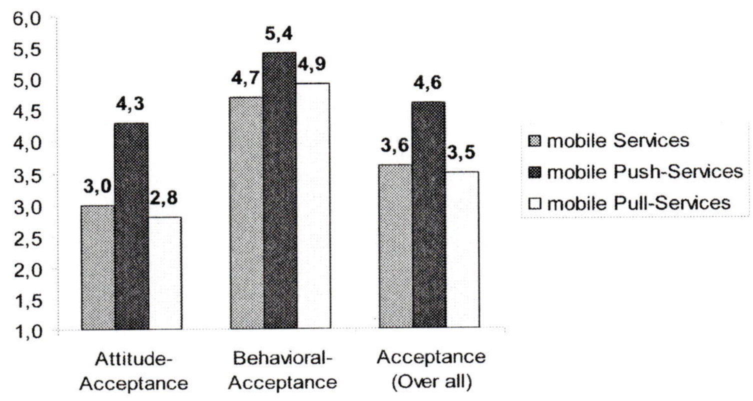

Subject to the respondent's gender, it became apparent that men exhibit a higher acceptance than women (M_M= 3.7; SD_M=0.94 and M_F= 4.2; SD_F= 0.95); p< .001).

In the following, the question arises as to whether this low acceptance is also reflected in the reactance to mobile services. In this, only the reactance risk for mobile push services was considered because it does not involve compliance on the part of the user. According to the understanding taken as a basis here (see section 5), reactance or reactance risk arises if a total of four conditions are fulfilled: restriction of freedom present, importance of the limited freedom present, illegitimacy of the restriction felt and no compensation possibilities for the restriction of freedom through the benefits of mobile services. If one assumes a close understanding of reactance, i.e. if one only considers the respondents that ticked "strongly agree" and "agree" for the corresponding reactance items, there is a reactance risk through mobile services for 40 (21.9%) of the 187 participants in the study. For a further understanding of reactance (i.e. also considering those who ticked "agree a little"), there is even a reactance risk for 111 of the 187 people polled (59.3%). Moreover, the study reveals a strong correlation between reactance and a resulting malaise in the person (r= .60; p< .01), and between reactance and a corresponding reactance behavior in the form of ignoring and deleting the information with r= .49 (p< .01) (as mobile push services restrict my personal freedom as a person, I try to ignore them and delete the corresponding information immediately).

In the scope of this study, the person's general reactance affinity was also recorded, whereby no correlation between reactance affinity and reactance could be ascertained. As far as the consumer-specific control variables are concerned, neither the age, income, frequency of mobile phone use, knowledge of mobile services nor the interest in technology had an influence on the reactance risk. However, it did reveal that reactance differs for mobile services depending on the gender (χ^2 (df= 5)=12.576; p=0.03), men exhibiting a higher reactance level than women. Moreover, possibilities of reducing the reactance risk lie in the content of the services, where both a high information value (r= - .41; p< .01) and a high entertainment value (r= - .54; p< .01) have a negative influence on the reactance risk (i.e. reduce it).

In the survey, the participants were also asked to assess a further four fictitious situations as regards the restriction of freedom they perceived and their consequent reaction. Here, the sending of information to the user's mobile phone served as the example. It became apparent that 63.3% of the users felt their freedom was restricted by a procedure commonly used in practice (see Table 4). The reactions to such a service and the consequent restriction of freedom are multifaceted: only 12.8% of the users would not react at all and only 26.2% would complain to the provider. Such an incident would lead to negative word-of-mouth communication for nearly 58% of the users. Approximately 80% of the participants perceived sending multimedia content matter (e.g. ring tones) as a restriction of freedom. However, when it came to relevant information from the mobile telephone provider (e.g. changes to the contractual data), the figure dropped to a mere 15.5%. In all, it was found that reactance is most frequently reduced through anger and annoyance and thus most commonly results in a subjective behavior effect.

Recapitulating, it is fair to say that the results found reveal a surprisingly low acceptance for mobile services. On the other hand, there is a distinct reactance risk for mobile push services that render a user-friendly permission strategy indispensable. Dispensing with such an approach would lead to numerous negative consequences on the part of the user.

Table 4. Example of the perceived restriction of freedom and reactions to the sending of information to a person's mobile phone (N= 187)

Your provider sends you information on politics and economics on your mobile phone.		
Restriction of freedom?	Yes: 63.6%	No: 35.4%
Reactions to the restriction	**Entered**	**Not entered**
Don't do anything at all.	12.8%	87.2%
Become annoyed with provider.	57.8%	42.2%
Don't use provider any more.	55.1%	44.9%
No longer use any of provider's services.	41.2%	58.8%
Pass on negative experience to third party.	56.7%	43.3%
Complain to provider.	26.2%	73.8%
Fundamentally change my opinion of mobile services for the worse.	35.3%	64.7%

PLEA FOR A CUSTOMER-FRIENDLY PERMISSION STRATEGY

The reactivity risk can be contained by allowing a service provider to have his or her customers decide when, by whom and how often he or she can be contacted. It is therefore a matter of selecting a user-friendly permission policy. What is required can be recorded in detail via the user's entries, even a "fine-tuning" oriented towards time, areas and terminals. He or she would then be called upon to determine when and where which area-specific offers can be offered to him or her without asking. Furthermore, such information could be revisable at any time. The user's right to determination should also include deciding the collection of data which affects his or her activities, its storage, use and its passing on to another party. A provider could also allow his or her customers to decide whether their information should serve as a basis for the customization of offers or services. Such a permission policy may well fail because of the mobile marketing guidelines but it certainly will not fail due to the technical possibilities which nowadays allow the realisation of such systems with minimal effort.

For the technical implementation of a user-friendly permission policy, we need to distinguish between a pull strategy and a push strategy. Should the user access a mobile WAP via a mobile terminal, for example (pull), he or she has already granted initial permission through the accessing process. If the user then be shown the information or service expected, this should not produce reactance. If, for example, the required information is overloaded with advertising or contains the actual service, this does not have to comply with the permission granted. Providers can counteract this risk using suitable software in the form of a personalization function of the content matter. For example, the user can be asked to register for the desired service (log in) and then only receives the information relevant to him or her (on the Internet, e.g. on My Yahoo). The personalization would occur by way of the information provided during the previous visit. A direct registration would also be possible via a cookie so that the entry of the registration data is omitted during access.

As the results of the study conducted also reveal, a permission policy and its technical realization becomes more important in the case of push services, than pull services. In the process, the user may have granted general permission for these services beforehand. If this is not the case, the timing (content, time, place) of the transmission is crucial and whether the user recognizes a use which requires acceptance. If the user grants general permission for a push service, (e.g. current information on the cell phone, MMS subscr. on a city journey) he or she should still have a differentiated possibility to choose. A personalization based upon profiles should be preferred, particularly in the case of push services. The personalization possibility can be realized using the appropriate software (e.g. news service software), which has to be installed on the mobile terminal. The establishment of the profile can then occur during the initial installation. As profiles for registration data already exist in most software applications, these only need to be developed with regard to service-specific information. The form in which this user information is considered in the delivery of mobile services is left to the discretion of the provider.

SUMMARY

Customer Relationship Management (CRM), which serves the purpose of improving customer knowledge and dealing with the customer, has become increasingly significant with the spread of the Internet and, more recently, mobile communication. One outcome of this manifests itself in targeted, interactive and specially tailored marketing. CRM's augmented possibilities primarily arise from the increase in customer data and the countless possibilities of reaching mobile people with services on the move, everywhere and any time and thus serving them in a situation-specific manner. Consequently, we first showed the latest technological developments and the resulting possibilities of mobile CRM. Based on the assumption that we can expect an increased amount of mobile marketing efforts in future, the potential encroachment on personal freedom and its threat in light of the reactance research was considered in more detail in this article. The theses and results of this research direction specify both the reactance conditions and the reactance consequences, which include so-called boomerang effects. The low acceptance and high reactance risk, as well as the corresponding reactance consequences, could not only be deduced in this article but also proven empirically. A low acceptance for mobile services is also evident in such a young and technically affine target group as the students polled here. This low acceptance is also reflected in the reactance risk. There is already such a risk for a close understanding of reactance in 22% of the respondents. Furthermore, the correlation between reactance and negative sentiments and reactance behavior, such as ignoring or deleting the services, could be proven. A high information value and/or entertainment value for a corresponding mobile service can admittedly minimize the reactance risk but never eliminate it entirely. The summation of this article is centered around heeding acceptance in the mCRM of the future and, above all, the reactance risk in the user field, as well as realizing a customer-friendly permission policy. The technical options in the context of such a policy are already available to companies and should be exhausted fully.

Scales for the Measurement of Acceptance and Reactance in the Study Conducted

Acceptance (Attitude-Acceptance: Question 1-5; Behavioral-Acceptance: Question 6 & 7)

1. Do you judge mobile services as rather good or rather bad?

(very good - good quite - good - quite bad - bad - very bad

2. Do you judge mobile services as rather attractive or rather non-attractive?

(very attractive - attractive - quite attractive - quite unattractive - unattractive - very unattractive)

Now, we ask you to rate mobile services by means of individual criteria.

3. Is the information value of mobile services high or low?

(very high - high - quite high - quite low - low - very low)

4. Is the entertainment value of mobile services high or low?

(very high - high - quite high - quite low - low - very low)

5. Is the impact of mobile services on your everyday life high or low?

(very high - high - quite high - quite low - low - very low)

We also want to know, how often you used mobile services in the past and how often you will use mobile services in the future.

6. How often have you used mobile services in the past?

(very often - often - quite often - quite rarely - rarely - very rarely)

7. How often will you use mobile services in the future?

(very often - often - quite often - quite rarely- rarely - very rarely)

Reactance

Introduction: Please think again of a mobile service, which will send you information without your active intervention. In the following you don't need to rate these mobile Services, instead we will concentrate on the question whether or not mobile Services restrict your personal freedom.

Always: strongly agree - agree - agree a little - disagree - strongly disagree

1. I perceive mobile Push-Services as a restriction of my personal freedom as a person.
2. I think that companies shouldn't restrict my personal freedom as a person with mobile Push-Services.
3. My personal freedom as a person, which we just talked about, is important for me.
4. The benefits of mobile Push-Services are more valuable than the restriction of my personal freedom.
5. The restriction of my personal freedom is higher than the benefits of mobile Push-Services.

Sentiment for reactance (6) and behavior for reactance (7):

6. I perceive the restriction of my personal freedom as a person through mobile Push-Services as unpleasant.
7. As mobile Push-Services restrict my personal freedom as a person, I try to ignore them and to delete corresponding information immediately.

REFERENCES

Strategy Analytics. (2007). *Global Mobile Advertising Update: Outlook Bright as Inventory Expands*.

Attwood, R., & Duncan, S. (2000). *Mobile Commerce: Strategies for New Business Paradigm*, London Ft: Telecoms.

Bauer, H. H., Haber, T. E., & Reichhardt, T. (2006). Location Based Services. Große Potenziale für Direktmarketing. *Direkt Marketing*, o. Jg., Heft 8, S. 36-38.

BITKOM. (2006). *UMTS - Fünf Länder haben bei UMTS den Durchbruch geschafft*, from http://www.bitkom.org/de/presse/30739_37663.aspx

Boulding, W., Staelin, R., Ehret, M., & Johnston, W. J. (2005). A Customer Relationship Management Roadmap: What Is Known, Potential Pitfalls, and Where to Go. *Journal of Marketing*, *69*(4), 155–166. doi:10.1509/jmkg.2005.69.4.155

Brehm, J. W. (1966). *A theory of psychological reactance*, New York, NY: Academic Press.

Brehm, J. W. (1972). *Response to loss of freedom. A theory of psychological reactance*, Morristown, NJ: General Learning Press.

Brehm, S. S., & Brehm, J. W. (1981). *Psychological reactance: A theory of freedom and control*, New York, NY: Academic Press.

Bruner, G. C., & Kumar, A. (2005). Explaining consumer acceptance of handheld Internet devices. *Journal of Business Research*, *58*(5), 553–558. doi:10.1016/j.jbusres.2003.08.002

Cannon, D. A. (2002). The ethics of database marketing. *Information Management Journal*, *36*(3), 42–44.

Clee, M. A., & Wicklund, R. A. (1980). Consumer behavior and psychological reactance. *The Journal of Consumer Research*, *6*(4), 389–405. doi:10.1086/208782

Dey, A., & Abowd, G. (2000). *The context toolkit: Aiding the development of context-aware applications*. Proceedings of the Workshop on Software Engineering for Wearable and Pervasive Computing, Limerick, Ireland.

Dickenberger, D. (1979). *Ein neues Konzept von Freiheit: Konsequenzen für die Theorie der psychologischen Reaktanz*, Weilheim: Belz .

Dickenberger, D., & Gniech, G. (1982). The theory of psychological reactance. In Irle, M. (Ed.), *Studies in decision making. Social psychological and socio-economic analyses* (pp. 311-341). Berlin, New York: de Gruyter.

Dickenberger, D., Gniech, G., & Grabitz, H.-J. (1993). Die Theorie der psychologischen Reaktanz, In Frey, D., Irle, M. (Ed.), *Theorien der Sozialpsychologie, Band I: Kognitive Theorien* (pp. 243-273). Bern-Göttingen-Toronto-Seattle: Hans Huber Verlag.

Frolick, M. N., & Chen, L.-d. (2004). Assesing M-Commerce opportunities. *Information Systems Management*, *21*(2), 53–61. doi:10.1201/1078/44118.21.2.20040301/80422.8

Gniech, G., & Dickenberger, D. (1987). Reaktanz. In Frey, D., Greif, S. (Ed.). *Sozialpsychologie (Ein Handbuch in Schlüsselbegriffen)*, 2. Aufl., München-Weinheim: Psychologische Verlagsunion.

Gniech, G., & Grabitz, H.-J. (1978). Freiheitseinengung und psychologische Reaktanz. In Frey, D. (Hg.), *Kognitive Theorien der Sozialpsychologie*. Wien: Huber.

Grabitz-Gniech, G., & Grabitz, H.-J. (1973a). Psychologische Reaktanz: Theoretisches Konzept und experimentelle Untersuchungen. *Zeitschrift für Sozialpsychologie*, *4*, 19–35.

Grabitz-Gniech, G., & Grabitz, H.-J. (1973b). Der Einfluß von Freiheitseinengung und Freiheitswiederherstellung auf den Reaktanz-Effekt. *Zeitschrift für Sozialpsychologie*, *4*, 361–365.

Grabitz,-Gniech, G. (1971). Some restrictive conditions for the occurrence of psychological reactance. *Journal of Personality and Social Psychology*, 19(2), 188-196.

Homburg, C., & Bruhn, M. (2003). Kundenbindungsmanagement – Eine Einführung in die theoretische und praktische Problemstellungen. In Bruhn, M., Homburg, C. (Hg.), *Handbuch Kundenbindungsmanagement – Strategien und Instrumente für ein erfolgreiches CRM* (3-37). 4. Aufl., Wiesbaden: Gabler.

Kaspar, C. (2006). *Individualisierung und mobile Dienste am Beispiel der Medienbranche – Ansätze zum schaffen von Kundenmehrwert*, Göttingen: Universitätsverlag.

Kaspar, C., Diekmann, T., & Hagenhoff, S. (2007). Context-Adaptive Mobile Systems, In Taniar, D. (Ed.), *Encyclopedia of Mobile Computing and Commerce*. Hershey, PA: Idea Group Publishing.

Kaspar, C., & Hagenhoff, S. (2005). An overview of mobile radio technologies. In Pagani, M. (Ed.), *Encyclopedia of Multimedia Technology and Networking* (pp. 645-651). Hershey, PA: Idea Group Publishing.

Kivetz, R. (2005). Promotion reactance: The Role of Effort-Reward Congruity. *The Journal of Consumer Research, 31*(2), 725–736. doi:10.1086/426606

Koivumäki, T., Ristola, A., & Kesti, M. (2006). Predicting consumer acceptance in mobile services: empirical evidence from an experimental end user environment. *International Journal of Mobile Communications, 4*(4), 418–435.

Königstorfer, J., & Gröppel-Klein, A. (2007). Experience of Failure and Anger when Using the Mobile and Wired internet – The Interference of Acceptance- and Resistance-driving Factors. *Marketing JRM, 27*(1), 34–47.

Luarn, P., & Lin, H. H. (2005). Toward an understanding of the behavioral intention to use mobile banking. *Computers in Human Behavior, 21*(6), 873–891. doi:10.1016/j.chb.2004.03.003

Meyer, M. (2005). *Multidisciplinarity of CRM Integration and its Implications*, in: Proceedings of the 38th Hawaii International Conference on System Sciences, (HICCS 2005).

Miron, A. M., & Brehm, J. W. (2006). Reactance Theory – 40 Years Later. *Zeitschrift für Sozialpsychologie, 37*(1), 9–18. doi:10.1024/0044-3514.37.1.9

Missi, F., Alshawi, S., & Fitzgerald, G. (2005). *Why crem efforts fail? A study of the impact of data quality and data integration*, in: Proceedings of the 38th Hawaii International Conference on System Sciences, (HICCS 2005).

Ngai, W. W. T. (2005). Customer relationship management research (1992 – 2002). An academic literature review and classification. *Marketing Intelligence & Planning, 23*(6), 582–605. doi:10.1108/02634500510624147

Nicolai, A. T., & Petersmann, T. (Eds.). (2001). *Strategien im M-Commerce*, Stuttgart: Schäffer-Poeschel.

Nysveen, Pedersen, & Thorbjörnsen. (2005). Intentions to Use Mobile Services: Antecedents and Cross-Service Conparisons. *Journal of the Academy of Marketing Science, 33*(3), 330–346. doi:10.1177/0092070305276149

Paulissen, K., Milis, K., Brengman, M., Fjermestad, J., & Romano, N. C., Jr. (2007). *Voids in the current CRM Literature: Academic literature review and classification (2000-2005)*, in: Proceedings of the 40th Hawaii International Conference on System Sciences.

Payne, A., & Frow, P. (2005). A Strategic Framework for Customer Relationship Management. *Journal of Marketing, 69*(4), 167–176. doi:10.1509/jmkg.2005.69.4.167

Pedersen, P. E. (2005). Adoption of Mobile Internet Services: An Exploratory Study of Mobile Commerce Early Adopters. *Journal of Organizational Computing and Electronic Commerce, 15*(2), 203–222. doi:10.1207/s15327744joce1503_2

Russel, S., & Norvig, P. (2003). *Artificial intelligence, a modern approach*, Upper Saddle River, NJ: Pearson Education.

Schäfer, J., & Thoma, D. (2008). Trends und Strategien im Mobile Marketing. In H. Bauer, T. Dirks & M. D. Bryant (Ed.). *Erfolgsfaktoren des Mobile Marketing*, Heidelberg: Springer-Verlag.

Schoenbachler, D. D., & Gordon, G. L. (2002). Trust and customer willingness to provide information in database-driven relationship marketing. *Journal of Interactive Marketing, 16*(3), 2–16. doi:10.1002/dir.10033

Siau, K., & Shen, Z. (2003). Mobile communications and mobile services. *International Journal of Mobile Communications, 1*(1-2), 3–14. doi:10.1504/IJMC.2003.002457

Sikora, A. (2001). *802.11: Standard für drahtlose Netze*. In tecChannel, from http://www.techChannel.de/hardware/680/index.html

Silberer, G. (1980). Reaktanz bei Konsumenten. In Graf Hoyos, G., Kroeber-Riel, W., v. Rosenstiel, L., Strümpel, B. (Ed.). *Grundbegriffe der Wirtschaftspsychologie* (386-391), München: Kösel.

Silberer, G. (2004.). Grundlagen und Potentiale der mobilfunkbasierten Kundenbeziehungspflege (mobile eCRM). In Hippner, H., Wilde, K. D. (Ed.). *IT-Systeme im CRM. Aufbau und Potenziale* (453-470), Wiesbaden: Gabler.

Silberer, G., & Wohlfahrt, J. (2001). Akzeptanz und Wirkungen des Mobile Banking. In Nicolai, A.T., Petersmann, T. (Ed.). *Strategien im M-Commerce: Grundlagen - Management – Geschäftsmodelle* (161-176), Stuttgart: Schäffer-Poeschel.

Silberer; G., Wohlfahrt, J., & Wilhelm, T. (Hrsg.) (2002). *Mobile Commerce - Grundlagen, Geschäftsmodelle, Erfolgsfaktoren*, Wiesbaden: Gabler.

Van Ackeren, R. (2002). Mit dem Handy auf Kundenfang – Potentiale und Erfolgsfaktoren einer mobilen Kundenkommunikation im Einzelhandel. In Ahlert, D., Becker, J., Knackstadt, R., Wunderlich, M. (Ed.). *Customer Relationship Management im Handel – Strategien – Konzepte – Erfahrungen* (343-360), Berlin, New York etc.: Springer.

Wang, Y.-S., Lin, H.-H., & Luarn, P. (2006). Predicting consumer intention to use mobile service. *Information Systems Journal, 16*(2), 157–179. doi:10.1111/j.1365-2575.2006.00213.x

Webb, W. (2001). *The future of wireless communication*, Boston, MA: Artech House.

Wicklund, R. A. (1974). *Freedom and reactance*, New York, NY: Wiley.

Wohlfahrt, J. (2004). *Akzeptanz und Wirkungen Mobile-Business-Anwendungen*, Hamburg: Dr. Kovac.

Wright, R. A., & Brehm, S. S. (1982). Reactance as impression management: A critical review. *Journal of Personality and Social Psychology, 42*(4), 608–618. doi:10.1037/0022-3514.42.4.608

Wu, J.-H., & Wang, S.-C. (2005). What drives mobile commerce? An empirical evaluation of the revised technology acceptance model. *Information & Management, 42*(5), 719–729. doi:10.1016/j.im.2004.07.001

Xu, M., & Walton, J. (2005). Gaining customer knowledge through analytical CRM. *Industrial Management & Data Systems, 105*(7), 955–971. doi:10.1108/02635570510616139

Zeng, E. Y., Yen, D. C., & Huang, S.-M. (2003). Mobile Commerce: the convergence of e-commerce and wireless technology. *International Journal of Services technology and Management*, 4(3), 302-322.

KEY TERMS AND DEFINITIONS

Acceptance: Acceptance is process by which something new is approved by a customer. It consists of a attitude acceptance and express itself in a behavioral acceptance of something, e.g. mobile communication.

Mobile CRM: Mobile CRM is the implementation of CRM in mobile business. Thereby

CRM is defined as "a strategic approach that is concerned with creating improved shareholder value through the development of appropriate [...]. CRM unites the potential of relationship marketing strategies and IT to create profitable, long-term relationships [...]. CRM provides enhanced opportunities to use data and information to both understand customers and cocreate value with them. This requires a cross-functional integration of processes, people, operations, and marketing capabilities that is enabled through information, technology, and applications. (Payne und Frow 2005, p. 167)".

Mobile Terminals: Mobile PCs, cell phones and PDAs, as well as multifunctional appliances, e.g. smart phones (cell phone and PDA) considered as mobile terminals. The hardware used on the user page influences the range of mobile service. The features relevant for the distribution of and access to mobile services are primarily the size of the display, the computing power and the memory capacity.

Network Technology: Mobile customers can be reached via different networks. Mobile network platforms or wide area mobile network platforms, such as GSM or UMTS networks. Mobile network platforms reach theoretical transmission rates from 14.5 Kbit/s (GSM) to 384 kBit/s (CDMA/UMTS).

Permission Policy: A user-friendly permission policy seems to be a important factor to avoid reactance in mobile CRM. The reactivity risk can be contained by allowing a service provider to have his or her customers decide when, by whom and how often he or she can be contacted. For the technical implementation of a user-friendly permission policy, we need to distinguish between a pull strategy and a push strategy.

Reactance: Reactance occurs if there is a perceived restriction of freedom of a person. The core assumptions of the reformulated reactance theory and results defined three more terms for the state of reactance: the importance of freedom, the legitimacy of the restriction felt and the possibilities of compensating for the restriction of freedom with the benefits mobile services can offer.

Technical Base Protocols: There are different protocols on which mobile services can minute. Both known Internet protocols, such as TCP, IP and UPD, and specifications (Bluetooth) and protocols from the mobile radio sector, such as WAE and WSP, play an important role in this.

ENDNOTE

1 The number of users increased from 16.1 to 47.3 million subscribers worldwide between 2004 and the beginning of 2006. Japan has the largest share with 47%, followed by Italy (19%) (BITKOM, 2006).

Chapter 11
Employment and Acceptance of Near Field Communication in Mobile Marketing

Klaus-Peter Wiedmann
Leibniz Universität Hannover, Germany

Marc-Oliver Reeh
Leibniz Universität Hannover, Germany

Henrik Schumacher
Leibniz Universität Hannover, Germany

ABSTRACT

Near Field Communication (NFC) is a short-distance wireless technology which allows user friendly networking of mobile terminals, e.g., cellular phones and PDAs, as it does with stationary units like Automated Teller Machines (ATM). Hardware producers, network operators and service providers are willing to implement NFC technologies in order to offer new services to their customers. The main goal is to increase the attraction of the underdeveloped areas of mobile commerce and make them more interesting for owners of mobile devices. If and how much real potential there is in NFC as a basic technology to really provide new impulses in mobile commerce and mobile marketing so that a higher acceptance can be reached are remaining open questions. This chapter tries to approach these questions and to provide possible answers by outlining relevant applicative examples and further configuration options of NFC technology as well as by discussing their acceptability.

DEVELOPMENT TRENDS IN MOBILE MARKETING

Nothing seems to be more consistent in the sector of Information and Communication Technology (ICT) than the change of its structure, as market dynamics and development velocity are still growing. The fact that incremental innovations constantly pass off the will, needs and perceptions of customers had to be experienced a few years ago by the mobile telephony industry in its early days. A much-cited example is the Wireless Application Protocol (WAP), which was meant to provide the new economy with "twenty-four-seven" distribution channels, and

DOI: 10.4018/978-1-60566-074-5.ch011

which was supposed to constitute the entrance to the mobile internet era. The problem of WAP was the enormous fissure between the much praised functionality, the high expectations of service providers as well as users on the one hand and hardware capabilities of that time like low duty times, monochrome matrix displays, low Global System for Mobile communications (GSM) data rates etc. on the other hand (Pleil, 2005). Even technophile innovators and early adopters began to associate WAP with expensive downloads, high transfer delays and non-transparent billing models. However, the market introduction of the Short Message Service (SMS) product was unexpectedly successful and everything but an economic flop, which once again shows that services are mostly used if a) they substantially and adequately satisfy a real existent need and b) they are easily accessible and available from a costumer's point of view (Kuhn, Lehner, 2003).

Although until lately mobile phone services and SMS seemed to be the only suitable products for the mass-market, current trends show that mobile commerce - which is characterized by a new kind of delivery of novel location based services using handheld terminals (Dholakia, Dholakia, 2004) - is facing a real renaissance based on a wide range of equipment manufacturers' model offensives, e.g., Apple's iPhone and new, innovative services. In today's communication media, the presence of smartphones, mobile internet, mobile TV and mobile navigation is winning strong public attention. Meanwhile, mobile phones offer computing power that could only be reached by high-performance PCs a couple of years ago. These facts demonstrate that technical obstacles, in mobile commerce as in other areas, tend to lose complexity and their limiting effects are alleviated as time passes (Pils, 2004b).

Considering the fact that technical prerequisites for successful mobile commerce and consequently for effective and mature mobile marketing are in place, the following question arises for the marketing developer: "Why should users abandon the comfort of using their PCs and start purchasing products via comparatively small and "uncomfortable" cell phone displays?" Marketing has to be aware of this open issue and the need to urge the development of adequate convenience technologies (in general regarding the development of mobile commerce see Wiedmann, Buckler, and Buxel 2000a and 2000b2000b). In addition to the necessary increases concerning user-friendliness by latest generations of mobile devices it has to be examined, how mobile marketing can penetrate life- and consumption styles of a constantly growing mobile society, in a way that that real use-benefits can be experienced from a customer's point of view (Wiedmann, 2001). In this case, a very important aspect is the ease of access to ubiquitously available offers of information, entertainment, service and goods. Eventually, mobile internet shall be of use for today's mobile society, if it is easy, intuitively to control and have reasonable handling charges.

An enabler and appropriate technology to develop such user-friendly and secure access to virtual world can be found in NFC (Messmer, 2005). In the near future, NFC will allow the user to connect its mobile device with other mobile phones, all kinds of small electronic devices or customer terminals and ATMs by an easy handshake, which is called "touching", in order to access the services of these terminals (Gonzalez, Organero, Kloos, 2008). Most of the applications related to the use of NFC are technically proven to work. Hence, at this stage it is a matter of acceptance, perception of the offered added value and the actual utilization by and for the customers. Optimizing customer orientation over all parts of the process, companies and NFC-based mobile commerce are able to come into contact with the user fast and straightforward in every condition of life, thus arousing interest in context-relevant contents and services (Kagermann, Österle, 2006).

The purpose of this article is to present the potentials of NFC for mobile marketing and mobile business as well as to introduce and discuss its acceptability by means of application examples. In this context, mobile marketing shall be defined as "any form of marketing, advertising or sales promotion activity aimed at consumers and conducted over a mobile channel" (Mobile Marketing Association, 2003). Mobile business by itself can be distinguished and classified by various items and schemes (Leem, Suh, Kim, 2004). Selected acceptability research approaches serve as a theoretical framework and subsequently help to consider NFC in context of actual cellular market needs for a first understanding of current NFC discussion. The chapter starts with an introduction to the technical properties and functionalities of NFC, the connection to smart card infrastructure and Radio Frequency Identification (RFID), a description of the Unique Selling Proposition (USP) as well as a short presentation of technology alternatives.

A clarification of the relevance of mobile marketing and the mobile phone respectively for direct marketing is carried out as well as a characterization of current customer acceptance problems within the scope of payment services. Therefore, the characteristics of the mobile marketing mix are shown with an exemplary assignment of NFC applications. Furthermore, an explanation of interdependencies in mobile marketing is carried out with reference to starting-points for NFC. In the next section benefits of NFC for advertising, communication and brand management are being pictured. A representation of application areas which are important for network operators, Payment and Ticketing, is then carried out with reference to technical, conceptual and user obtained challenges as well as suggestions concerning market stimulation possibilities.

Finally, an outlook shows the role of NFC in case of "internet of things" in the future. The theoretical part tries to show, how an examination of the use acceptance of NFC could work and why the construct use acceptance has a high relevance in mobile commerce. An estimation of the acceptance chances of NFC as a service enabler follows based on its easy way of use. Afterwards, challenges and limitations in the research process are shown with reference to the various changes in mobile commerce and digital social networks caused by NFC. Partially, a reflection of the NFC performance features is linked to the mentioned functional relationships. A brief analysis of the b2b acceptance constitutes the end.

An analysis of chances and risks follows a look at general market conditions, challenges as well as conflicts and the search for revenue models in the NFC alliance from different views of market participants. Also, selected development forecasts are introduced. The final results with reference to different NFC organisations point once again to the important consumer- and use acceptance in the profitable b2c market and the placement of NFC in future social structures.

TECHNICAL AND CONCEPTUAL NFC BASICS

NFC is a wireless technology that allows short distance data transfer. The development of this technology is the result of a research project by Philips and Sony from 2002 through 2004 (Hanhart, 2008). Its aim was to create a unified communication protocol which could combine the two manufacturers to date competing smart card standards. Smart cards had already being used for years as customer-, credit- or access-cards, for information storage and personal data storage, as well as for electronic authentication and authorization (McElroy, Turban, 2000). The MiFare-card by Philips is the most prevalent system in Europe with a market volume of more than 1 billion cards used worldwide (NXP, 2008). Furthermore, NFC is technically very similar to the well-known RFID systems. Advanced miniaturizing, new production techniques, and accompanying cost degression allow expanding fields of application for RFID tags in commerce and industry.

The basics of NFC systems like MiFare, RFID and other wireless systems are based on the concept of electromagnetic induction. This mechanism allows the exchange of data between transmitter and receiver, without both of them requiring an energy supply. Examples of this are smart cards or RFID tags used by logistics firms in electronic wireless labels which obtain the required energy to work as passive elements from the reading device of the wireless system (Hansen, Neumann, 2005). The technical innovation in NFC systems is that NFC capable devices can act both passively, like a common RFID tag, and actively, like a reading device with an autonomous RF field. In energy saving passive mode, data will be saved on the NFC chip for later request and can only be acquired again by a corresponding reading device. The energy used for this process is obtained from the initializing reading device's RF field. In the active NFC mode, both transmitter and receiver send an electromagnetic field, which can be used to connect or initialize different device functions (Mayes, Markantonakis, 2007).

The adoption of the "correct" operating mode works automatically, which enhances energy management of the NFC devices, raises the bandwidth for possible contact points and broadens the basis for mobile marketing applications. The operating range of NFC is physically limited to 10 - 20 centimeter. Practically, the "touching" or "peering" of two different terminals does not require physical contact. The distance between devices should be kept to a few centimeters from each other. Furthermore, the connection takes place in milliseconds, which is something exceptional in the wireless world (Berndt, 2008). The maximum data transmission velocity is about 424Kbit/s at very short contact distances. Therefore NFC can be used to initialize, configure and establish faster connection types, e.g., Bluetooth or WLAN.

The short range of NFC systems is not a handicap; instead it can be seen as an advantage. On the one hand, it makes the use of NFC very simple, on the other hand, the system is very failure resistant and reliable (Kern, 2006). Compared to other non-contact systems short range quality of NFC is a factor of protection against attacks on data integrity or attempts to manipulate the data transfer, which is a great security advantage. Well known assault scenarios like "man in the middle" attacks are completely impossible. Best utilization scenarios for NFC systems are, in consideration of these security aspects, mobile payment and mobile ticketing, which both requiring comparable security instructions, e.g., closed contact-driven ATM card systems. Basic intrinsic attributes of NFC protocols can be summarized as follows (Wiedmann, Keuling, Reeh, 2005):

- NFC is a peer to peer protocol. This means that only two devices are able to communicate with each other at the same time. The establishment of NFC-based networks is not possible, in contradiction to master-slave based wireless networks.
- NFC uses two different transmission systems. In active mode both transmitter and receiver generate electromagnetic fields. In passive mode only the initiator generates an electromagnetic field. The target device gets the energy needed to accomplish the data transfer within physical distance and coupling to the initiator.
- NFC is a very short range protocol according to International Organization for Standardization (ISO) 14443, 18092 at 13,56MHz. For the initialization and preservation of the wireless communication, the distance between devices should not be greater than a few centimeters.
- NFC can be used for the emulation of smart cards, for reading RFID tags, for accomplishing peering between electronic devices and for the establishment of faster connections like Bluetooth and WLAN.

- NFC is a "secure solution" for mobile payment, ticketing, and authentication uses, compared to other wireless data transfer systems.

Finally, describing the capability characteristics of NFC systems, it is necessary to consider other technologies which represent an alternative to NFC systems or will represent an alternative in the future. For mobile marketing and marketing in general it is usually indifferent how devices physically work, particularly as the inside lifes of cell phones, PCs and other devices show a very high complexity. However, it is absolutely decisive that the devices present the particular attributes and requested characteristics.

Presuming a rigorous delimitation of technologically related systems like RFID, there is no real alternative to NFC. Of course, every wireless system can be deliberately limited in its operating range and device coupling can be replaced by other protocol variants, but the combination of all mentioned characteristics is only offered by NFC. This could change in the future. Although, there are multifaceted developments up till now, the digital photo and image recognition is taking big steps forward. A picture taken by a mobile device, indifferently of which angle, could be interpreted and be related to a real object, which generally make the "detour" over RFID tags or NFC obsolete (Hehl, 2008). A system based on the scanning of two-dimensional barcodes, which is consequently not really comparable, is being tested in New York. The so called "Semapedia project" shows the same conceptual architecture as NFC and is meant to alleviate information transition between the real and the virtual environment. Turning touching into pointing or scanning still requires a lot of time and research, though, because the quantity of possible objects to interpret equals the real object range (Rukzio et al., 2007).

MARKETING STRATEGY PARAMETERS FOR NFC SYSTEMS

Most of the presently discussed tendencies of mobile commerce are widely known. As explained above, a few years ago, technologies were not mature enough to provide the dimension of quality necessary to satisfy users' wishes. Contemporary, mobile phones are now accepted as the constant omnipresent companion and geared up as multifunctional all-rounder with a penetration rate of hundred per cent, making them more interesting than ever for marketing purposes (Hansen, Neumann, 2005). The progress and development of marketing into an interactive direct marketing which satisfies multilayered dimensions of success is accompanied by the idea that advertisement and mobile devices are a necessity of our time. This idea is supported by the fact that the hybrid, multioptional customer requires a more individual and context-specific approach (Silberer, 2004). The mobile phone as an extended part for innovative dialog marketing methods presents a new interface to the customer for manufacturers, distributors and advertisers, which is target-aimed, target audience specific and always accessible. As long as classic web applications in wired electronic commerce base on the growing integration of services and offers, the depth of integration in mobile commerce is increased by NFC, as applications can be ubiquitously initialized and accessed.

For all these reasons NFC can be seen as an initiator for innovative concepts in sales and marketing. From user's point of view, mobile services have the specific additional value to be accessible at any time from any place, which is actually the most important precondition for basic functions in mobile commerce like chargeable services and content. Despite of this, customers have been constantly refusing these services. That is why mobile commerce often shows little self purpose; instead, it acts as a platform for cross-financing other mobile services that would not be able to hold their ground against the "everything-free" mentality of the internet community (Bourreau, Lathiais, 2006). Furthermore, mobile commerce is even more strongly linked to non-material goods and services than electronic commerce. These goods and services are expected to be immediately consumed contemplating the uno actu principle. Based on this, the support capabilities of NFC should be placed exactly at the interface between electronic and mobile commerce in order to facilitate transition to chargeable mobile services (which have very popular web-based free options) by the non-chargeable use of NFC. Immediate consumption and related payment reserves have to be immediately opened by the enabler NFC.

For the following classification of NFC's manifold support capabilities, the definition of the mobile marketing mix involving dynamic service linkages has to be given case-by-case. In a rough classification, the following scopes can be differentiated: a) communication scope: mobile advertising - defined as "sending tailored messages to consumers according to individual personal needs, interests and preferences" (Komulainen et al., 2007, p. 371) - with response channel, b) sales area: preparation and accomplishment of transactions, c) product area: support of physical products by supplementary mobile services, d) price area: allowing discounts for products and services. Furthermore, the individual measures have to be considered in connection with a mobile marketing strategy which has to orientate by the situational context as well as the "capacity" of the receiver more intensely than comparable offline campaigns.

Since communication politics, referring to advertisement, direct marketing etc., constitute the most significant issue of mobile marketing, this will be one of the most important and multi-variant application fields for NFC systems. Examples of NFC applications for each instrument are given in Figure 1 and in the following.

The inherent success factors of mobile marketing present themselves in terms of affinity services - exceptionally target group specific services which are based on existing products, services, or also, e.g., television series - in consequence of their high coverage and low dispersion: a high degree of individualization accompanied by personalization and accordingly a distinctive commitment of the user. This is enabled by the ability of mobile phones to display place-, time-, and context-related contents, which will be further improved by NFC. Context sensitive information, which is mostly based on location data in the mobile world, present a high degree of information acceptance. The huge coverage is in turn a result of place-, and time- independent, ubiquitous accessibility (Wiedmann, Buckler, Buxel, 2000a)

Another very important characteristic of differentiation in mobile marketing is its classification in push and pull services (Homburg, Krohmer, 2006). Push services as a sales promoting strategy are sent without request, pull services, in contrast, only on immediate request. NFC can clearly be identified as a pull service. Compared to push services, pull services show a higher user interest and a higher degree of interaction, whereas reactance risks are lower because the requested information is considered as personally relevant and initialization is preceded by an user action. Rejecting reactions like those experienced when receiving advertisement short messages or any other kind of contacting initiated by the provider are not expected. This mitigates the problems of permission marketing (Kroeber-Riel, Weinberg, 2003).

As an important part of direct marketing, permission marketing has to be an essential component of each mobile marketing campaign (Barwise, Strong, 2002; Schwarz, 2003). This means the delivery of advertisement or information messages that only takes place after explicit and revocable admission by the mobile device user. This matter is of great importance in the process of acceptance of mobile services, since the user perennially carries his switched on mobile device

Figure 1. NFC applications in mobile marketing

Mobile marketing mix	NFC add-on	Customer value
Promotion: • Mobile advertising • Info services • Response channel	Promotion: • Smart poster • Location based services • Deactivator touchpoint	Promotion: • Surveillance • Service quality • Permission
Place: • Sale of digital content • Sale of physical products • Mobile payment	Place: • NFC download • RFID product identification • NFC authorization	Place: • Expeditious • Convenience • Security
Product: • New digital products • Supplemental service • Brand management	Product: • NFC cell phone ticketing • RFID product information • NFC as image medium	Product: • Simplicity • Information • Me Too
Price: • Bonuses • Discounts • Price worthiness	Price: • NFC download for free • Retrograde price fixing • NFC shopping comparision	Price: • Price advantage • Price attractiveness • Transparency

along, giving it a closer personal proximity than his/her PC at home. From the marketing point of view the "accessibility dilemma" should be understood as the desire of not being disturbed at any time and in every place. However, a very interesting implementation of NFC could be presented as an opt-out-procedure, which would permit the user to withdraw or cancel the use of subscribed services by only touching a "deactivation touchpoint". This deactivator tag could be used at the point of sale, with newspaper advertisements or as a NFC dongle at the PC. The implementation of NFC systems could help to reduce inhibitions and fears of otherwise obscure and difficultly comprehensible technical procedures.

NFC prospectively has to fulfill these and other specifics, success variables and acceptability factors of mobile marketing situationally and supportingly (Kollmann, 2007b). NFC has to be understood as an access technology and not as universal remedy, and so it should be integrated into the variety of marketing measures as an enabler or an impact amplifier. This is not only true for the process design, but also for reaction of the customers. The search for the shiny killer application has become extinct, and now we discover that the faultless interaction between different services and their entertaining, partially amusing packaging result in a very interesting and demanded entirety (Yang, Jolly, 2008). If the search for a killer application is supposed to be target-aimed, target groups have to be differentiated.

For younger mobile phone users, for example, sponsored SMS and the download of ringtones from a touchpoint could become attractive applications (Bamba, Barnes, 2007). This way, NFC services will be offered either by the network providers or through a content broker. Adults occupationally use premium services like Blackberry, where NFC could be utilized to accomplish the peering of the terminals. For a large spectrum of age groups, use of the hands-on Web 2.0 on their mobiles would play an important role. Besides the utilization and the potential of viral marketing measures will increase by means of NFC, since peering is simple and quick (Langner, 2006).

NFC APPLICATIONS RELEVANT TO MARKETING

NFC in Advertisement and Communications

The number of NFC applications is nearly unlimited, only taking into account the variety of applications that connected to the emulation of smart cards (Lu, 2007). Hence, it is really difficult to predict which applications will eventually evolve into big sellers. It can be expected that, on the analogy of electronic commerce, applications in the scope of advertisement and communication will quickly gain importance (Sander, 2004). The smart poster constitutes one first simple and inexpensive with immediate relevance to marketing.

A smart poster consists of a write-once RFID tag which will be integrated into a display or an advertising poster. As soon as the user's NFC phone gets close enough to a certain part of the poster marked with the NFC logo, the phone will automatically recognize the RFID tag and start peering and data request (Mattern, 2005). Via the context menu which appears on the phone's display, audible and visual elements can be activated and other processes can be initialized. If the URL of a certain company's website is stored on the RFID tag, for example, it can be adopted into the phone's web browser which opens the website. This simple mechanism constitutes quite a complex matter that costs time and effort with conventional devices which are not capable of NFC. At this point, NFC can simplify the interaction between user and application (Stanoevska-Slabeva, 2004). In conjunction with the browser loading a website, the download of software, games, files, demo mp3 songs and even larger files like video clips can be offered at no charge from the RFID tag or with the support of mobile networking technologies like General Packet Radio Service (GPRS) or Universal Mobile Telecommunications System (UMTS). Since the user playing an active role and being aware of the advertisement, is presenting his identity towards the system through his Subscriber Identity Module (SIM)-Card, adapted offers and links can be carried out according to a location- and customer profile (Bodenstorfer, Hasenauer, 2005).

Assuming that the download of larger data amounts from a smart poster is possible, the creative action range of mobile marketing is considerably extended (Gottwald, 2007). Many mobile phones allow changing the appearance of the user interface by using different design skins. Normally, this does not add any practical benefit, but satisfies the play instincts, esthetic perception and even the desire for individuality. A very innovative configuration for NFC downloads could be the temporary branding of the device's menu according advertisers' ideas in order to even further emotionalize the promotional message (Seiler, 2005). Targeting marketing objectives like brand building and cross selling this represents a very interesting option which offers variety and individuality at the moment of reception of the advertising message. The combination of NFC posters, proactive user participation and the subsequent "I've-got-something-new" impression creates higher sustainable product awareness than conventional SMS advertisement campaigns, particularly as the involvement in advertisement depends more on the way information is transferred as on the product itself.

NFC in Payment & Ticketing

From the point of view of network providers, mobile payment and mobile ticketing strategically represent very important topics (Hendry, 2007). In addition to the stimulation of market demand, thorough coverage and a working billing system can be regarded as prerequisites of the success of mobile commerce services. Hence, a working payment system which is established and accepted by general public can be considered as the most important precondition and as a key factor for any approach of mobile business. In this context, mobile payment can be defined as "payment for goods, services, and bills with a mobile device (such as a mobile phone, smart-phone, or personal digital assistant) by taking advantage of wireless and other communication technologies" (Dahlberg et al., 2008, p. 165). The same applies for mobile marketing, as long as customer approximation, transaction and distribution of the mostly digital media shall be handled without disruption of the medium. The requirements concerning mobile payment systems are manifold (Kreyer, Pousttchi, Turowski, 2003). Aside from several legal regulations and resulting procedural security solutions, an optimal access without cryptical sequences of menus, but with comprehensible wire transfer processes, is necessary for user acceptance.

If examined from user's perspective, portability and ease of use are crucial for success. Portability means that the payment system has to ensure service independence and cross platform operation, since a simple and consistent payment procedure is of great value for the user regarding efficiency as well as service aspects (Pousttchi, Wiedemann, 2005). Paying for utilized services after all represents an obligation and not a pleasure, therefore the disposition to deal with different payment procedures is enormously low. This is closely related to the ease of use of the whole system, which should be easy and intuitive to handle. At this point, a great correlation to the unique characteristics of NFC becomes obvious (Skinner, 2007). Analog premises apply to mobile ticketing - more precisely the use of mobile phones for every kind of mobile pay services in public transport (Mallat et al., 2008).

On the one hand, future usage of NFC as a payment system does not change anything with respect to the billing and accounting processes within the mobile supply chain. NFC also does not remove the hazard of users being initially irritated by higher phone bills caused by mobile payment. On the other hand, the way of accessing mobile payment and mobile ticketing services as well as the variety of supplementary add-on services is changed. Users will only have to hold out their NFC mobile devices close to the touchpoint of the cash register or ticket machine for authentication and authorization of the payment process. The communication range of NFC, which has been purposely kept low, does not only allow the definitive association of application and user, but it also includes an inherent protection against accidental misuse as well as against possible attacks from thirds on the integrity of the connection. Nevertheless, conceptual security is a very important issue for the developers of NFC payment functions and several additional authentication methods are available (Lehner, 2003). For instance, entering a PIN code or prospectively matching with biometrical data could be used for the confirmation of a transaction.

Consequently, the provision of maximum process security in order to reduce retentions against the utilization of mobile payment systems is of substantial relevance for its commercialization and for the acquisition of early adopters. Regardless of the individual risk aversion the perceived risk as a major risk determinant can be influenced by NFC. (Im, Kim, Han, 2008; Kroeber-Riel, Weinberg, 2003). At this point marketing must exactly analyze the innovation related acceptability measures simplicity as well as derived security and has to communicate these as a confidence-building method in a corresponding explicit manner. (Büttner, Schulz, Silberer, 2006). A NFC logo or touch point as first user interface and visible component in mobile commerce can also facilitate the mental access to this new technology.

Meanwhile NFC payment and NFC ticketing gain importance for mobile marketing by the above-mentioned feasibility to implement supplementary add-on services as acquisitive methods in customer-connectivity management of this still young market (Laker, Pohl, Dahldoff, 2004). Examples of this are customized discounts and bonus systems in short-range public transportation. Casual or spontaneous passengers, who have to buy comparatively expensive single tickets without the benefits of a long-term ticket, could be rewarded with a customized bonus based on their check-in and check-out data within the scope of customer-connectivity measure. Depending on the configuration of the conditions, savings could result from the ex post choice of the cheapest ticket alternative. Henceforth, this service can be viewed as marketing for NFC as well as for the essential service itself.

NFC ticketing can also be combined with a smart poster in the form a transport schedule. For this purpose, RFID tags with position data are placed behind the depicted stops. The user only has to hold his NFC device to the destination point, causing the device to automatically obtain the desired ticket equipped with a timestamp. In Germany, NFC ticketing is currently being used in regular operations by the linked transportation system "Rhein-Main-Verkehrsbund". Moreover, the national railway company "Deutsche Bahn" is conducting customer acceptability tests under the name "Touch&Travel" on the rail route between Hannover and Berlin (Langheinrich, 2007). Several other NFC field trials are being carried out in France, the United Kingdom, the United States, and Asia.

NFC in the Age of "Internet of Things"

Next to the emulation of typical smart card functions, the smart poster or the peering of user devices, the above-mentioned technological relationship between NFC and RFID will have considerable and comprehensive influence on our prospective life. In computer science, the concept of "ubiquitous computing" (Welfens et al., 2005) is used to describe this development more precisely. In this context, a futuristic but realistic scenario is outlined, where objects of daily use are equipped with intelligent electronics, enabling them to exchange information about themselves and their environment with each other. The intention of this omnipresent and autonomously operating communication technology is the active and passive support of heterogeneous (background) processes in our professional and private surroundings (Fleisch, Thiesse, Österle, 2002). For instance, in mentioned scenarios the mobile phone acts as key for initialization and supervision of different processes at home. Similar conceptual ideas already exist in the automobile industry, where personal information about the driver are stored on the car key, enabling the car to automatically adjust to the driver's preferences.

The key technology for the "internet of things" is RFID, as new manufacturing processes in the segment of RFID labeling create the necessary cost degression, allowing economically sensible and manifold application on the level of items (Becker, Winkelmann, 2008). The refrigerator which is able to independently perform replenishment orders, or the dressing room in a store which presents the matching blouses to a certain skirt on a display represent examples of this ambient intelligence.

In this connected home scenario, the NFC phone serves as a connecting link and builds a bridge between the user and the virtual world (Fleisch, Thiesse, Österle, 2002). Moreover, assuming that mobile phones will continue to be replaced every two years, a NFC infrastructure will establish more quickly in the scope of mobile devices than in the scope of PCs. It can be stated as a fact that users – no matter if they are young or old, technophile or not – are comparatively familiar with the operation of their mobile phones. Hence, they usually know and control basic menu sequences, which constitute a valuable advantage (Hubschneider, Kölmel, 2002). In the internet of things users will be able to authenticate themselves anywhere to their digital environment. In addition, coincidentally they will be capable to gather information from and about this digital environment by the RFID tags. This offers various starting points for mobile marketing. The consumer in a supermarket could acquire additional information about products before buying, which is especially important for allergy sufferers in case of food, cosmetics, etc. (Wolfram, 2008). Mobile price comparison services, which indicate the cheapest retailer after reading the product ID from the RFID tag, are further examples of consumer empowerment and of the mobile prosumer as a prospective consumer type.

In both cases the existence of databases, which allow the matching of the encoded RFID data and the product information that is kept elsewhere, is a technical requirement. For manufacturers, distributors and advertisers, the sudden increase of market transparency results in a plenitude of chances and challenges. This situation is comparable to the introduction of electronic commerce, particularly as the difference between stationary PC-based services and applications of the mobile internet will increasingly fade from the spotlight due to the conversion of media and devices (Caspers, 2002). Moreover, since the connected home of the future and the employment of NFC in mobile commerce are going to reveal completely new insights concerning consumer behavior and product usage, these scenarios are very interesting for market research.

IMPLEMENTATION ACCEPTANCE AS A FACTOR OF SUCCESS FOR NFC

The economical success of a technological innovation is solely determined by the market. Thus, the main function of marketing research is to analyze the market potentials of new products before market entrance. In accordance with Willrodt, the needs being addressed by a technological innovations – and hence the question of how to satisfy them – are often more than ambiguous (Willrodt, 2005). Even if innovative products are being acquired it is quite uncertain, whether these products are used in the intended form. Findings by Kollmann clearly show, that discrepancies are found between (a) the acceptance of the innovation in general, the repeatedly usage, and a usage acceptance based on the exhaustion of new possibilities of utilization, and (b) commercializing innovative technologies (Kollmann, 2007b).

To stimulate the global demand of mobile commerce applications an increased adoption of new technologies has to be achieved, which lead to an increase of consumption (Baron, Patterson, Harris, 2006). The rate of demand predominantly depends on hardware availability, publicity about offered services and the word of mouth propaganda through users. The consumption, on the other hand, depends on usability, which is represented by a user-friendly and easy menu control, high service quality and a high price-performance-ratio (Yom, 2001). New generations of mobile phones can be considered as a proper example: new configurations play a major role when deciding whether or not to buy a certain cell phone, which will only be used rarely in practice. The best example to picture this matter is the WAP application.

Referring to our former explanations, the abiding market success of NFC is based on the intensive usage of NFC-based systems and related applications. Hence, the concept of user acceptance as a success factor is of special importance for the NFC-based mobile marketing. Knowledge or – at least – valid forecasts of the expected user acceptance of NFC-products, characteristics and functions are crucial preconditions for the development and implementation of a effective NFC-based mobile marketing; empirical studies of the field of acceptance research can be made fruitful for further development of NFC-specific concepts.

One of the most common models of acceptance research is the Technology Acceptance Model (TAM) developed by Davis (1989), which is based on findings of the perceived usefulness and perceived easy-of-use of a particular system, especially. Elementary factors of examination of TAM were the perceived usage as well as user-friendliness of a technology by the user himself. This basic model is being upgraded to more sophisticated ones such as that of Koll-

Figure 2. Acceptance model of Kollmann (1998)

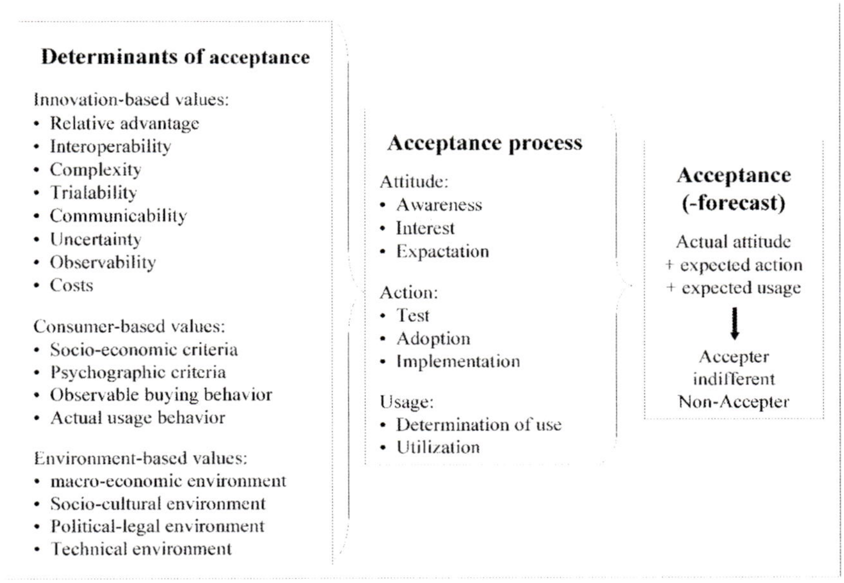

mann (1998 and 2000), who compared different acceptance and usage levels of an innovation, or Silberer, Magerhaus and Wohlfahrt (2002) in which the expectation conformity was included in the prognosis of acceptance and user satisfaction (Fritz, 2008). Last but not least, the Unified Theory of Acceptance and Use of Technology Model (UTAUT) by Venkatesh et al., which takes a closer look at expectancy values, has to be mentioned (Venkatesh et al., 2003).

Kollmann's model (Figure 2), however, is perfect for a case by case, single application analysis, and prognosis of user acceptance in the mobile market. In this model a multitude set of innovation- and environment-based acceptance determinants are put into relation for user-related expectation and acceptance formation. At the same time the three-step, attitude focussed analysis procedure reflects the real decision and assumption processes in the area of mobile services very well and hence, makes a first, basic forecast of the expected use acceptance of NFC possible.

It is very likely, that NFC will be able to meet the acceptance targets mentioned in Figure 2.

The fact that the user is in control of every single NFC-application – he has to initialize every application actively; e.g., by holding his cell phone close to a touchpoint – is the inimitable USP of NFC. This kind of man-machine interaction is comparable to the human communication-mechanism (Eisenhauer, Oppermann, Prinz, 2007). If two people want to talk and change information, they move close to each other and even whisper if the information is confidential. This process is simulated by the NFC-technology and, hence, should result in a high(er) acceptance level of NFC (van Luijt, 2005). Alternative technologies such as SMS or Bluetooth require more individual knowledge, skills, and active action by the users. They are, however, more anonymous and undisclosed to accomplish, what depending on the situation and the contents might be preferred by the users.

The question to be asked independently is, in which way and how frequently NFC-applications will be used by the consumer. Most approaches of the acceptance research are not capable for NFC, since they are either based on ex ante assumptions with respect to benefits of use (and with an ex post verification) or they do not reflect the whole set of usage-relevant determinants of the new product or service (Königstorfer, Gröppel-Klein, 2007). Since mobile commerce applications are obviously different in many respects compared to common applications, however, a valid inquiry regarding user acceptance and actual benefits has to be preceded. Within the range of mobile commerce applications NFC is "only" the enabler for accessing "classic" mobile services, whereby the degree of user acceptance is measured through the increment value of the main application (Kavassalis et al., 2003).

Nevertheless, since usability always depends on additional factors such as easiness, comfort, etc. NFC plays a significant role; the actual complexity of the technology is of less importance for the user (Isomursu et al., 2007). NFC is designed to assist the user to accomplish an easy and secure access through a sign or a brand logotype and thus lead to an acceptance increase – e.g. payment services, which could conclude in a confidence-building step (Esch, 2008). This is of special importance in a rapidly growing digital environment, in which verifiable security is being substituted by trusting the system and providers with limited rationality and lack of objectivity (Petrovic, Fallenböck, Kittl, 2003). Technical and conceptual securities are important preconditions for the user acceptance of NFC-based systems. A very decisive factor at this point is the degree of perceived security and risk aversion of every single user related to each action taken; several categories of risk are given: financial risks, the qualitative impairment of serviceability and the risk of losing personal data.

Modern models of modern acceptance research with special respect of technological innovations do not include social communication factors, social-network structures and viral effects to their investigation. This gab, however, could be filled appropriately with non-technological theories such as adoption- and diffusion research (Khalifa, Ning Shen, 2008; Corsten, Gössinger, Schneider, 2006). Social phenomena like "the wish to belong to something" can cause a certain pulling effect and cause some "I-want-it-effect". These effects mainly occur when seeing other users, who already have the product or service. Since the utilization of NFC is free for everyone to watch and easy to comprehend, NFC fulfills acceptance determinants like observability and connectivity and can cause a "contagious effect".

A differentiation must be done between the influences from a closer social environment and the influences from a farther social environment. The deeply influencing close social-environment includes the group of persons, with whom the consumer stays at closer contact and with whom continuous informal interactions take place. To this group belong family and friends, but also neighbors and coworkers could influence the consumer (Kotler, Armstrong, 2006). The farther social-environment includes all person or groups, to which sporadically the consumer meet in a formal basis and whose background-system is distinguish for a homogeneous set of values and norms. The possibility of a direct influence in the what-to-buy-decisions and the consumer's behavior does not normally exist. Since the visibility of mobile applications is of special importance (vide supra), a promotional campaign for NFC should address the closer social environment primarily or the wishes of the consumer explicitly.

All existing and widely accepted models and theories concerning consumer behavior basically differ in the idea of man. Most of them connect hybrid consumption structures with the fact that consumption of products and services is a way of searching for identity and is interpreted as experience (Foscht, Swoboda, 2005). NFC and mobile commerce respectively, can be interpreted as lifestyle and has to be analyzed and described precisely by modern (neo-behavioristic) research. Thus, it will be one of the most important tasks of future research approaches in marketing to put together controllable stimuli of the NFC macro-environment and personal variables such as motivations, feelings, and preferences.

The automation of intrapersonal actions and processes is not yet viable for a wide market introduction. Hence, the only proper way to gain attention and to show the potential usability of NFC is the utilization of very simple applications to teach new NFC users in a funny, spontaneous way about the new functionality. The future market launch of more sophisticated applications will follow in a next communication phase in the course of "Flow-Experiences" (Fritz, 2008). When using high personalized NFC dialog elements, it is indispensable to ask for the permission of the user before marketing application starts in order not to lose the user confidence in the system (Kotler, Keller, Bliemel, 2007). Contrarily to push campaigns, pull campaigns, like the smart poster utilization or similar NFC applications, are characterized for the conceptual agreement of the user. It can be spoken of a certain degree of involvement of the information seeking user, e.g., movement towards the touchpoint which is a pointer to higher degree of information reception (Pepels, 2005).

The fact that attention, interest in product, awareness of acquisition and even customer retention can be prepared and reached is not only shown in the ambit of product-development, but in the advertising-communication area of mobile commerce. Since the system promotes active participation of the user to reach the information, it is not possible to cause any repulsion from this first one, like it sometimes occur with SMS unwished advertisement campaigns.

It is a possible expectation, to show a close standing point of interest as a location based service using NFC technologies and to reach an accurate geographic localization of the users, which could be rejected by these ones. Similar problem can be expected when using ticketing services which need localization information to be produced. If these services are often used, the production of movement profiles is made very easy. But this is already possible today using triangulation of network infrastructure (Pils, 2004a). Another example is the NFC poster: If the poster shows a product that is a bit disgusting to a group of users standing close to him, the attention of the consumer might be stimulated but he wouldn't approach the touchpoint, in order to prevent to show his intention. Another aspect that could create a rejection is possible data being placed down on the mobile device (same kind of data as the PC cookies). The rejection caused by this, could be so big that a possible total rejection of NFC could be the result.

As final statement should be clarify, both, from the producers' side and the dealers' side, a spectrum of criteria for the acceptance should apply in order to integrate the user in the payment, trade, and promotion services (Komulainen et al., 2007). Acceptance, at the traders' side will only occur, if a big enough quantity of NFC equipment has already penetrated the market at offer-start (Hinrichs, Stroborn, van Ball, 2004). This implies the classical "Hen-Egg-Problem". Studies demonstrate that new payment system experience a huge range of acceptance problems (Silberer, Wohlfahrt, 2001). While costumers are having doubts about unknown providers, cryptic control-menus and unclear payment-mechanisms, the dominating matter at the provider side is in what kind of technology to invest.

FUTURE MARKET DEVELOPMENTS FOR NFC

In the future NFC will have to fight for acceptance as any other newcomer technology, and get so much attention to not drawn in the information- and innovation-sea. For the market entrance, the most important precondition is that technical matureness and uniform standards assure an absolute working and functioning system (Krcmar, 2005). Another important point is the robustness and low error susceptibility of the NFC system

(Rannenberg et al., 2005). Because the more flexible a system is, the more opportunities arise for consumer to misuse the applications. The same applies for technological applications from different producers, which may cause severe communication problems between different platforms. The evaluation of the NFC mechanisms and the development of a process blueprint are very important for the general understanding between network-partners, the "social capital" of strategic marketing management.

The way of saving information on the hardware in mobile phones is of special relevance from a marketing and strategic management perspective: information can be either a) saved on the NFC chip or b) the SIM card (Hendry, 2007). Variable a, would lead in a long term to a change of the structure of the mobile value chain, since, e.g., the phone manufacturers could offer different NFC payment functions independently and merely have to obtain the data transmission without any further services from the network operators, however. Variable b, would keep the network providers longer as only participators of the mobile supply chain who can intelligently combine user- and application- data. To take marketing actions towards specific target groups is this, an essential requisite which can only be productive exposed to the commercial interested public after a given quality form, and has a very high strategic pertinence (Silberer, 2004). A further development and so to speak an expansion solution for cell phones is the implementation of NFC chip and -functionality on a memory card. This solution would extensively make customers and particularly suppliers of firm NFC solutions independent of the result of negotiations of handset maker and Mobile Network Operator (MNO).

The developers of NFC have to integrate their products more efficiently to current and actual existing mobile markets. An essential problem of mobile providers rests on the fact that the UMTS network infrastructure in some countries is not yet carrying the amounts of data they should, and subsequent technologies are already in the pipelines to be implemented (Kollmann, 2007a). The search of information is a factor that will contribute to support the success of online and infotainment offers in the mobile world.

Service providers are afraid of converting into a "bit pipe". Internet users are no longer willing to carry high costs for data transfers. Only content is good honored, and even in this case sell models are based on re- or cross-financing through advertisement, because offer and diversity of free services are simply superior (Link, 2003). If a combination of global players start offering contents through their own web portals and devices´ software (like the new Android open source operating system by Google), and they sell it and charge it through the mobile devices, then the only service left for providers would be the data transfer, what diminish almost any win margins. The low macroeconomic security in mobile market, raising emigration to other systems and the decreasing benefits per customer are the reasons why the start of NFC marketing is being delayed (Rummert, Schiefer, Bulander, 2005). On the other hand, there is no lack of use cases or pilot projects, if it is taken into account that in the last few years there have been done dozens of NFC tests in areas like payment, ticketing, advertising and services with some excellent results.

Seen from the service provider point of view, there is not sufficient information about the properly working of NFC interfaces and shares proportions in business and selling models (Shneyderman, Casati, 2008). The biggest profiteers of a fast implementation of a NFC infrastructure would be the service providers, which would have no business model without NFC and control only little market segments with b2b applications and use NFC only as a middle-term differentiation characteristic. Special b2b applications in which the utilization of mobile NFC data services implies a significant added value already exist. The constitution of long lasting partnership networks based on new technologies is very complicated, because innovations can be rapidly copied and adapted by the market leaders (Kuß, Tomczak, Reinecke, 2007). The durability of the competition advantage for startup companies is very limited and while the standardization of interfaces grow, the network is becoming more open for established market competitors.

For device manufacturers and network providers looking for high profit private customer business, this is not a satisfactory target value, being this one of the reasons of an insufficient market supply of NFC devices. Nokia is one of few producers, which have included NFC ability to their 6131 and 6212 mobile phones. Motorola together with Samsung and in cooperation with service providers offer NFC phones for field trials. This could be an indication to the fact that the device manufacturers are not yet convinced of the market success of NFC-based devices and therefore are not willing to invest some Venture Capital on integrating this technology in their systems (Welfens et al., 2005). If the media perception would be a factor of success for an innovation, then there would not be the best time for NFC right now. On the other hand it would not be smart to establish a new technology without practical use for the user, especially when the consumer is not willing to pay an extra price for it.

For market research institutions it is very difficult to express a consistent opinion about NFC. One of the few accessible NFC market studies made by Abi Research, forecasts a market share of NFC phones of no more than 20% (approx. 350 million devices) by the year 2011. A new analysis by Strategic Analytics expects that amount in the year 2012, which still means an outstanding market development. Analysts mean that the most important criterion for the success of NFC is based on the conviction of the business finance in the security of NFC-based payment systems and the offer of a standardized connection (Abi Research, 2007). In a study made by Juniper Research, the total amount of payment services through NFC systems in the year 2013 is calculated to be 75 billion US-Dollars. The decreasing price of NFC chipsets which is calculated to reduce at the beginning from 5 US$ to 1 US$ should not be a negative factor of growth.

Closely related to this prognosis is the development of the mobile payment sector, but the user groups in areas like commerce and industry are not showing much interest in this matter. If something has to be learned from the past, it is the fact that the adaptation of new technologies and processes happens from day to day slower than it is wished by the developers and suppliers (Bohlin, Burgelman, Casel, 2007). What success factors and multipliers are decisive for bigger market prosperity of NFC is still to watch and doubtful. System-inbuilt- and environmental- factors of success, which have proved in different studies repeated by now, are conclusive shown in Figure 3 (Jaring et al., 2007; Ondrus, Pigneur, 2007; Mayes, Markantonakis, 2008).

CONCLUSION

It can be concluded that technical innovation in the area of mobile human-machine-interactions which directly targets consumer acceptance in can definitely open up new markets and more efficient ways of attracting customers and creating customer-connectivity. NFC opens the chance to establish an easy access to modern mobile data and information services, which is suitable for the mass-market for the first time. The diversity of applications that NFC allows, starting from smart card alternatives, add-ons and smart posters to innovative location based services, offers mobile marketing huge potential for assistance and presents starting points for marketing specialists to participate in a future of networked communication media. The fear of top deciders in the mobile telephony business of becoming a bit pipe is justified and NFC could even accelerate this development. The NFC Forum, which has been founded in 2004, as well as the Pan-European

Figure 3. NFC SWOT analysis

System's inherent strenghts:	System's inherent weaknesses:
Low-priced chip set Standardized protocol Smart card emulation Bluetooth peering Secured authentification Communication mechanism Service enabler	Proprietary system Pioneering stage Missing blueprint
External opportunities:	**External threats:**
Ongoing media convergence Linkage to mobile internet Synergies with MiFare and FeliCa Technical RFID convergence Bidirectional CRM expansion Incitement of UbiCom	Alternative technology Contray MNO strategy No established business model Unclear revenue model User acceptance problem Privacy and legal position

NFC consortium StoLPaN will keep leading the international standardization, distribution and implementation of NFC in companies and bodies of the GSM association as well as ISO, European Computer Manufacturers Association (ECMA), European Telecommunications Standards Institute (ETSI), etc. (NFC Forum, 2008; StoLPaN, 2008).

Finally, the acceptance of customers is crucial for the holistic success of NFC. Only if the NFC usage concept can be brought to the market as really easy and simple, if no technological or procedural barriers constrain the access and if aversions to RFID do not manifest in people's heads, then NFC can become the carthorse of mobile commerce and marketing (Wiedmann, Ludewig, Reeh, 2008). Along with additional services and advantages this development should be supported by monetary incentives and rebate systems. All other application-specific factors of success still have to be conclusively determined by acceptance research, which implies that a lot of research work still has to be dealt with.

Finally from a social point of view the question remains, whether and if applicable to which extent a successful distribution of NFC applications brings along drawbacks and how requirements of society-oriented marketing have to be integrated in the design and early commercialization of innovative NFC solutions (Wiedmann, 1989). Certainly NFC is able in the foreseeable future to contribute to a considerably easier access to the virtual world, to reduce social isolation effects and consequently to allow equal opportunities of participation to all social classes. An issue that has to be reflected skeptically is that this could result in an additional "hypertrophy" of consumption, which might have consequences in everyday life, enforcing problems in other areas of society, e.g., lack of assistance and social integration of disadvantaged groups and deficits regarding arts and culture. On the other hand, new techniques of access to the virtual world could be used for social and cultural purposes, creating a humanization and reinforcement of social commerce (Töpfer, Silbermann, William, 2008). Altogether humanization in a comprehensive manner seems to be essential, not only in view of NFC but as well in the whole sector of mobile marketing.

REFERENCES

Abi Research. (2007). Near Field Communication. Retrieved October 20, 2008, from http://www.abiresearch.com/products/market_research/Near-Field_Communications_(NFC)

Bamba, F., & Barnes, S. J. (2007). SMS advertising, permission and the consumer. *Business Process Management Journal, 13*(6), 815–829. doi:10.1108/14637150710834578

Baron, S., Patterson, A., & Harris, K. (2006). Beyond technology acceptance - understanding consumer practice. *International Journal of Service Industry Management, 17*(2), 111–135. doi:10.1108/09564230610656962

Barwise, P., & Strong, C. (2002). Permission based mobile advertising. *Journal of Interactive Marketing, 16*(1), 14–24. doi:10.1002/dir.10000

Becker, J., & Winkelmann, A. (2008). Handelscontrolling. Berlin: Gabler.

Berndt, H. (2008). Towards 4G Technologies. New York: Wiley.

Bodenstorfer, M., & Hasenauer, R. (2005). Location Based Services. In Holzmüller, H.H., & Schuh, A. (Ed.), *Innovationen im sektoralen Marketing* (pp. 337-350). Heidelberg: Physica.

Bohlin, E., Burgelman, J.-C., & Casal, C. R. (2007). The future of mobile communications in the EU. *Telematics and Informatics, 24*(3), 238–242. doi:10.1016/j.tele.2007.01.011

Bourreau, M., & Lethiais, V. (2006). Pricing information goods: free vs. pay content. In Brousseau, E., & Curien, N. (Ed.), *Internet and Digital Economics* (pp. 345-367). Cambridge: Cambridge University Press.

Büttner, O. B., Schulz, S., & Silberer, G. (2006). Vertrauen, Risiko und Usability bei der Nutzung von Internetapotheken. In Bauer, H. H., Neumann, M.M., & Schüle, A. (Ed.), *Konsumentenvertrauen* (pp. 355-366). München: Vahlen.

Caspers, R. (2002). Neue Geschäftsmodelle der Internet-Ökonomie. In Bieger, T., Bickhoff, N., Caspers, R., Knyphausen-Aufseß, D., & Reding, K. (Ed.), *Zukünftige Geschäftsmodelle* (pp. 249-270). Berlin: Springer.

Corsten, H., Gössinger, R., & Schneider, H. (2006). Grundlagen des Innovationsmanagement. München: Vahlen.

Dahlberg, T., Mallat, N., Ondrus, J., & Zmijewska, A. (2008). Past, present and future of mobile payments research. *Electronic Commerce Research and Applications, 7*(2), 165–181. doi:10.1016/j.elerap.2007.02.001

Davis, F. (1989). Perceived Usefulness, Perceived Ease of Use, and User Acceptance of Information Technology. *MIS Quarterly, 13*(3), 319–340. doi:10.2307/249008

Dholakia, R. R., & Dholakia, N. (2004). Mobility and markets: emerging outlines of m-commerce. *Journal of Business Research, 57*(12), 1391–1396. doi:10.1016/S0148-2963(02)00427-7

Eisenhauer, M., Oppermann, R., & Prinz, W. (2007). Internet der Dinge. In Bullinger, H.-J., & ten Hompel, M. (Ed.), *Internet der Dinge* (pp. 49-62). Berlin: Springer.

Esch, F.-R. (2008). Strategie und Technik der Markenführung. München: Vahlen.

Fleisch, E., Thiesse, F., & Österle, H. (2002). Connected Smart Appliance. In Österle, H., Fleisch, E., & Alt, R. (Ed.), *Business Networking in der Praxis* (pp. 233-248). Berlin: Springer.

Forum, N. F. C. (2008). The NFC Forum. Retrieved October 20, 2008, from http://www.nfc-forum.org/aboutus/

Foscht, T., & Swoboda, B. (2005). Käuferverhalten. Wiesbaden: Gabler.

Fritz, W. (2008). Internet-Marketing und Electronic Commerce. Wiesbaden: Gabler.

Gonzalez, G. R., Organero, M. M., & Kloos, C. D. (2008). Exploring Touching Learning Environments. In Kendall, M., & Samways, B. (Ed.), *Learning to Live in the Knowledge Society* (pp. 93-96). Berlin: Springer.

Gottwald, S. (2007). RFID and Beyond. München: Vahlen.

Hanhart, D. (2008). Mobile Computing und RFID im Facility Management. Berlin: Springer.

Hansen, H. R., & Neumann, G. (2005). [Stuttgart: Lucius & Lucius UTB.]. *Wirtschaftsinformatik*, 2.

Hehl, W. (2008). Trends in der Informationstechnologie. Zürich: vdf Hochschulverlag AG.

Hendry, M. (2007). Multi-application Smart Cards. Cambridge: Cambridge University Press.

Hinrichs, J.-W., Stroborn, K., & van Baal, S. (2004). Mobiles Bezahlen aus der Sicht des Online-Händlers. In Pousttchi, K., & Turowski, K. (Ed.), *Mobile Economy* (pp. 63-78). Bonn: Köllen.

Homburg, C., & Krohmer, H. (2006). Marketingmanagement. Wiesbaden: Gabler.

Hubschneider, M., & Kölmel, B. (2002). Mobile Business. In Bullinger, H.-J., & Berres, A. (Ed.), *E-Business* (pp. 671-686). Berlin: Springer.

Im, I., Kim, Y., & Han, H.-J. (2008). The effects of perceived risk and technology type on users' acceptance of technologies. *Information & Management*, *45*(1), 1–9.

Isomursu, P., Hinman, R., Isomursu, M., & Spasojevic, M. (2007). Metaphors for the Mobile Internet. *Knowledge, Technology, and Policy*, *20*(4), 259–268. doi:10.1007/s12130-007-9033-5

Jaring, P., & Törmänen, V. Siiral E., & Matinmikko, T. (2007). Improving Mobile Solution Workflows and Usability Using Near Field Communication Technology. In Schiele, B., Dey, A. K., Gellersen, H., de Ruyter, B., Tscheligi, M., Wichert, R., Aarts, E., & Buchmann, A. (Ed.), *Ambient Intelligence* (pp. 358-373). Berlin: Springer.

Kagermann, H., & Österle, H. (2006). Geschäftsmodelle 2010. Frankfurt am Main: Frankfurter Allgem. Buch.

Kavassalis, P., Spyropoulou, N., Drossos, D., Mitrokostas, E., Gikas, G., & Hatzistamatiou, A. (2003). Mobile Permission Marketing. *International Journal of Electronic Commerce*, *8*(1), 55–79.

Kern, C. (2006). Anwendung von RFID-Systemen. Berlin: Springer.

Khalifa, M., & Ning Shen, K. (2008). Explaining the adoption of transactional B2C mobile commerce. *Journal of Enterprise Information Management*, *21*(2), 110–124. doi:10.1108/17410390810851372

Kollmann, T. (1998). Akzeptanz innovativer Nutzungsgüter und -systeme. Wiesbaden: Gabler.

Kollmann, T. (2000). Akzeptanzprobleme neuer Technologien – Die Notwendigkeit eines dynamischen Untersuchungsansatzes. In Bliemel, F., Fassot, G., & Theobald, A. (Ed.) *Electronic Commerce* (pp. 27-45). Wiesbaden: Gabler.

Kollmann, T. (2007a). E-Business. Wiesbaden: Gabler.

Kollmann, T. (2007b). Online-Marketing. Stuttgart: Lucius & Lucius UTB.

Komulainen, H., Mainela, T., Tähtinen, J., & Ulkuniemi, P. (2007). Retailers' different value perceptions of mobile advertising service. *International Journal of Service Industry Management*, *18*(4), 368–393. doi:10.1108/09564230710778146

Königstorfer, J., & Gröppel-Klein, A. (2007). Projektive Verfahren zur Ermittlung der Akzeptanz technologischer Innovationen. In Buber, R., & Holzmüller, H. (Ed.) *Qualitative Marktforschung* (pp. 847-862). Wiesbaden: Gabler.

Kotler, P., & Armstrong, G. (2006). Principles of Marketing. Upper Saddle River: Prentice Hall India.

Kotler, P., Keller, K. L., & Bliemel, F. (2007). Marketing-Management. München: Pearson.

Krcmar, H. (2005). Informationsmanagement. Berlin: Springer.

Kreyer, N., Pousttchi, K., & Turowski, K. (2003). Mobile Payment Prodcedures. *eService Journal*, 2 (3), 7-22.

Kroeber-Riel, W., & Weinberg, P. (2003). Konsumentenverhalten. München: Vahlen.

Kuhn, J., & Lehner, F. (2003). Szenarien einer mobilen Zukunft. In Pousttchi, K., & Turowski, K. (Ed.), *Mobile Commerce* (pp.130-142). Bonn: Köllen.

Kuß, A., Tomczak, T., & Reinecke, S. (2007). Marketingplanung. Wiesbaden: Gabler.

Laker, M., Pohl, A., & Dahlhoff, D. (2004). Kundenbindung auf neuen Märkten. In. Hinterhuber, H. H., & Matzler, K. (Ed.), *Kundenorientierte Unternehmensführung* (pp. 147-159). Wiesbaden: Gabler.

Langheinrich, M. (2007). Gibt es in einer total informatisierten Welt noch eine Privatsphäre? In Mattern, F. (Ed.), *Die Informatisierung des Alltags* (pp. 232-264). Berlin: Springer.

Langner, S. (2006). Viral Marketing. Wiesbaden: Vahlen.

Leem, C. S., Suh, H. S., & Kim, D. S. (2004). A classification of mobile business models and its applications. *Industrial Management & Data Systems, 104*(1), 78–87. doi:10.1108/02635570410514115

Lehner, F. (2003). Mobile und drahtlose Informationssysteme. Berlin: Springer.

Link, J. (2003). Die Klärung von Wirtschaftlichkeiten von M-Commerce-Projekten. In Link, J. (Ed.), *Mobile Commerce* (pp. 41-64). Berlin: Springer.

Lu, H. K. (2007). Network smart card review and analysis. *Computer Networks, 51*(9), 2234–2248. doi:10.1016/j.comnet.2007.01.009

Mallat, N., Rossi, M., Tuunainen, V. K., & Öörni, A. (2008). An empirical investigation of mobile ticketing service adoption in public transportation. *Personal and Ubiquitous Computing, 12*(1), 57–65. doi:10.1007/s00779-006-0126-z

Mattern, F. (2005). Die technische Basis für das Internet der Dinge. In Fleisch, E., & Mattern, F. (Ed.), *Das Internet der Dinge* (pp. 39-66). Berlin: Springer.

Mayes, K. E., & Markantonakis, K. (2007). Smart Cards, Tokens, Security and Applications. Berlin: Springer.

Mayes, K. E., & Markantonakis, K. (2008). Mobile communication security controllers. *Information Security Technical Report, 13*(3), 173–192. doi:10.1016/j.istr.2008.09.004

McElroy, D., & Turban, E. (2000). Smart Cards. In Shaw, M., Blanning, R., Strader, T., & Whinston, A. (Éd.), *Handbook on Electronic Commerce* (pp. 289-311). Berlin: Springer.

Messmer, B. T. (2005). Architectures for Mobile Services. In Giordano, M. (Ed.), *Mobile Business* (pp. 201-212). Wiesbaden: Vahlen.

Mobile Marketing Association. (2003). MMA Code for Responsible Mobile Marketing. London: Barnes Graham.

NXP. (2008). MIFARE Classic. Retrieved November 17, 2008, from http://www.nxp.com/#/pip/pip=[pfp=41863]|pp=[t=pfp,i=41863]

Ondrus, J., & Pigneur, Y. (2007). Cross-industry Preferences for Mobile Payments Development in Switzerland. *Electronic Markets, 17*(2), 142–152. doi:10.1080/10196780701296386

Pepels, W. (2005). Käuferverhalten. Berlin: Erich Schmidt.

Petrovic, O., Fallenböck, M., & Kittl, C. (2003). Der Paradigmenwechsel in der vernetzten Wirtschaft. In Petrovic, O., Ksela, M., Fallenböck, M., & Kittl, C. (Ed.), *Trust in the Network Economy* (pp. 3-28). Wien: Springer.

Pils, M. (2004a). Electronic Business und sensible Informations- und Kommunikationssysteme. In Höller, J., Pils, M., & Zlabinger, R. (Ed.), *Internet und Intranet* (pp. 327-359). Berlin: Springer.

Pils, M. (2004b). Mobile Business. In Höller, J., Pils, M., & Zlabinger, R. (Ed.), *Internet und Intranet* (pp. 67-124). Berlin: Springer.

Pleil, T. (2005). Anmerkungen und strategische Ansätze zur Kommunikation von M-Payment. In Hampe, J. F., Lehner, F., Pousttchi, K., Rannenberg, K., & Turowski, K. (Ed.), *Mobile Business* (pp. 74-86). Bonn: Köllen.

Pousttchi, K., & Wiedemann, D. G. (2005). Relativer Vorteil bei mobilen Bezahlsystemen. In Stucky, W., & Schiefer, G. (Ed.), *Perspektiven des Mobile Business* (pp. 35-50). Wiesbaden: Vahlen.

Rannenberg, K., Albers, A., Figge, S., Radmacher, M., & Rossnagel, H. (2005). Mobile Commerce. In: Hampe, J. F., Lehner, F., Pousttchi, K., Rannenberg, K., & Turowski, K. (Ed.), *Mobile Business* (pp. 9-16). Bonn: Köllen.

Rukzio, E., Broll, G., Leichtenstern, K., & Schmidt, A. (2007). Mobile Interaction with the Real World: An Evaluation and Comparison of Physical Mobile Interaction Techniques. In Schiele, B., Dey, A. K., Gellersen, H., de Ruyter, B., Tscheligi, M., Wichert, R., Aarts, E., & Buchmann, A. (Ed.), *Ambient Intelligence* (pp. 1-18). Berlin: Springer.

Rummert, A., Schiefer, G., & Bulander, R. (2005). MVNO - Aussichten des deutschen Mobilfunkmarktes. In Stucky, W., & Schiefer, G. (Ed.), *Perspektiven des Mobile Business* (pp. 97-112). Wiesbaden: Vahlen.

Sander, M. (2004). Marketing-Management. Stuttgart: Lucius & Lucius UTB.

Schwarz, T. (2003). Chancen und Grenzen der elektronischen Kommunikation. In Link, J. (Ed.), *Mobile Commerce* (pp. 145-161). Berlin: Springer.

Seiler, M. (2005). Mobile Marketing. In Giordano, M., & Hummel, J. (Ed.), *Mobile Business* (pp. 373-392). Wiesbaden: Vahlen.

Shneyderman, A., & Casati, A. (2008). Fixed Mobile Convergence. New York: Mcgraw-Hill Professional.

Silberer, G. (2004). Grundlagen und Potentiale der mobilfunkbasierten Kundenbeziehungspflege. In Hippner, H., & Wilde, K. D. (Ed.). *IT-Systeme im CRM* (pp. 453-470). Wiesbaden: Vahlen.

Silberer, G., Magerhans, A., & Wohlfahrt, J. (2002). Kundenzufriedenheit und Kundenbindung im Mobile Commerce. In Silberer, G., Wohlfahrt, J., & Wilhelm T. (Ed.), *Mobile Commerce* (pp. 309-324). Wiesbaden: Vahlen.

Silberer, G., & Wohlfahrt, J. (2001). Akzeptanz und Wirkungen des Mobile Banking. In Nicolai, A. T., & Petersmann, T. (Ed.), *Strategien im M-Commerce* (pp. 161-176). Stuttgart: Schäffer-Poeschel.

Skinner, C. (2007). The Future of Banking in a Globalised World. New York: Wiley.

Stanoevska-Slabeva, K. (2004). Mobile Business. In Stanoevska-Slabeva, K. (Ed.), *The Digital Economy* (pp. 459-475). Berlin: Springer.

StoLPan. (2008). Project Info. Retrieved October 20, 2008, from http://www.stolpan.com/index.php?do=topic&topic=3

Töpfer, A., Silbermann, S., & William, R. (2008). Die Rolle des Web 2.0 im CRM. In Töpfer, A. (Ed.), *Handbuch Kundenmanagement* (pp. 651-675). Berlin: Springer.

van Luijt, A. (2005). Beyond DRM. In Picot, A., & Thielmann, H. (Ed.), *Distribution und Schutz digitaler Medien durch Digital Rights Management* (pp. 61-74). Berlin: Springer.

Venkatesh, V., Morris, M. G., Davis, G. B., & Davis, F. D. (2003). User acceptance of information technology. *MIS Quarterly*, *27*(3), 425–478.

Welfens, P. J. J., Zoche, P., Jungmittag, A., Beckert, B., & Joisten, M. (2005). Internetwirtschaft 2010. Heidelberg: Physica.

Wiedmann, K.-P. (1989). Gesellschaft und Marketing. In Specht, G., Silberer, G., & Engelhardt, W. H. (Ed.), *Marketing-Schnittstellen* (pp. 227-247). Stuttgart: Schäffer-Poeschel.

Wiedmann, K.-P. (2001). Plädoyer für H-Commerce. In Nicolai, A. T., & Petersmann, T. (Ed.), *Strategien im M-Commerce* (pp. 129-146). Stuttgart: Schäffer-Poeschel.

Wiedmann, K.-P., Buckler, F., & Buxel, H. (2000a). Chancenpotenziale und Gestaltungsperspektiven des M-Commerce. *Der markt*, 39 (2), 84-96.

Wiedmann, K.-P., Buckler, F., & Buxel, H. (2000b). Mobile Commerce. *WiSt. Wirtschaftswissenschaftliches Studium*, *29*(12), 684–691.

Wiedmann, K.-P., Keuling, U., & Reeh, M.-O. (2005). Potenziale und Einsatz von RFID/NFC-Systemen im Mobile Commerce. Hannover: Schriftenreihe Marketing und Management.

Wiedmann, K.-P., Ludewig, D., & Reeh, M.-O. (2005). Potenziale des Einsatzes von RFID-Systemen für das Marketing im Einzelhandel. Hannover: Schriftenreihe Marketing und Management.

Willrodt, K. (2005). Marktforschung im Innovationsprozess. In Amelingmeyer, J., & Harland, P.E. (Ed.), *Technologiemanagement & Marketing* (pp. 403-413). Wiesbaden: Vahlen.

Wolfram, G. (2008). Die Metro-Group-Future-Store-Initiative. In Riekhof, H.-C. (Ed.), *Retail Business in Deutschland* (pp. 617-631). Wiesbaden: Vahlen.

Yang, K., & Jolly, L. D. (2008). Age cohort analysis in adoption of mobile data services. *Journal of Consumer Marketing*, *25*(5), 272–280. doi:10.1108/07363760810890507

Yom, M. (2001). Utility und Usability im Mobile Commerce. In Silberer, G., Wohlfahrt, J., & Wilhelm, T. (Ed.), *Mobile Commerce* (pp. 173-184). Wiesbaden: Vahlen.

KEY TERMS AND DEFINITIONS

Implementation Acceptance: A case where a person makes a positive decision in form of repeatedly usage without attempting to change it.

Near Field Communication: NFC is a new short-range wireless communication protocol which enables data exchange between mobile devices.

NFC Payment: The use of an NFC enabled mobile device, e.g., a cell phone for transactional payments in the field of grocery shopping or public transport.

Radio Frequency Identification: RFID is an automatic identification technology, typically using inexpensive RF transponders for process optimization in logistics.

Smart Poster: An advertisement or billboard with an in-build RFID tag for initialling different kinds of mobile commerce applications.

Touchpoint: A touchpoint is any kind of marked object, which is embedded with an RFID tag to start smart applications.

Ubiquitous Computing: A future scenario of human-computer interaction with fully integrated information processing in everyday objects and activities.

Chapter 12
Mobile Customer Acquisition in the Swiss Health Care Industry:
An Empirical Study

Niklas Nohlen
Institute of Research on Information Systems (IRIS), Germany

Stefan Smolnik
European Business School (EBS), Germany

Gerold Riempp
International University Schloss Reichartshausen, Germany

ABSTRACT

Applications based on mobile technologies are predicted to achieve increasing distribution as well as high potential for business processes especially with regard to customer interfaces. This chapter is going to analyze and present perspectives and approaches for initiation efforts by means of mobile technologies of Switzerland's health care industry. On the one hand, acceptance and success factors of such mobile applications are analyzed by means of empirical studies involving customers and companies, on the other hand a prototype of mobile processes for initiation efforts is introduced, which was developed in the course of a case study.

INTRODUCTION

Challenges for Customer Acquisition in Switzerland's Health Care Industry

Switzerland's act on health care insurance (KVG) allows a compulsory basic insurance contract for each obligatorily insured person. Regardless of the chosen insurance company this contract offers the same services for all persons without previous medical examination. Thereby insurance companies have to accept every obligatorily insured person within their regional operating area. Switzerland's health care insurance system comprises a cost sharing for the insured person that consists of a fixed annual fee (so-called Franchise) and a co-payment of 10% which compounds the costs exceeding the franchise. The insured person can reduce the annual fee through an optional higher franchise. In addition to the franchise rate, age, gender, and place of residence also have an influence on the premium

DOI: 10.4018/978-1-60566-074-5.ch012

level. Unlike in Germany, there is no income-dependent premium provided in Switzerland. Due to the newly published annual premiums approved by the Federal Council of Switzerland, the insured have the possibility to switch health insurance providers with a 30 days' notice by December 31 (KVG, 2006). In 2006, for example, about 170.000 people have changed their health care provider (comparis.ch, 2007). Hence, the insurance companies try to acquire via different sales channels and sales promotion activities as many customers as possible who are willing to switch.

One of Switzerland's leading insurance companies – Helsana Versicherungen AG (Helsana, 2007) – can be used as a good example to illustrate the increasing significance of mobile sales channels. In 2003/04, the Helsana Versicherungen AG implemented a mobile offer for the first time. This mobile application, based on the Short Message Service (SMS) technology, enables the interested person to request via SMS the actual insurance fee for the coming year by sending an SMS. The fee is hereby calculated on the basis of age, gender, and the region where the interested person lives, in relation to a standard franchise, whereas no further influence capabilities can be considered concerning the interested person. After receiving the requested information, the prospective customer can, via SMS, ask for a call from the insurance company's call center (Reichold & Schierholz, 2003). Meanwhile, many Swiss insurance companies offer this SMS process of premium calculation. The process as such does no longer offer a characteristic monopoly or innovation potential, and this has led to the loss of users. Nevertheless, further development of mobile technologies provides new possibilities for customer acquisition by means of mobile applications.

In their review of the literature, Ngai and Gnuasekaran provide an overview of mobile commerce (m-commerce). They identify five kinds of publications that cover aspects of m-commerce: M-commerce applications and cases, wireless user infrastructure, mobile middleware, m-commerce theory, and research (Ngai & Gunasekaran, 2007). Our chapter focuses on three of these five aspects: research, m-commerce application, and cases. We specifically wanted to know (1) whether these health care insurance consumers would use m-commerce, and (2) what the key determinants of success are with regard to the use of m-commerce. In section 2, we therefore review the relevant literature and mention some determinants for the successful use of m-commerce.

Relevance and Development of m-Commerce

Mobile commerce can be viewed as a subset of electronic commerce. Devices like cellular phones, personal digital assistants, pagers, notebooks, etc. represent the wireless environment that supports mobile electronic business transactions and gives mobile commerce its name (Coursaris & Hassanein, 2002). Mobile commerce (m-commerce) describes any business transaction, in which transaction partners apply mobile electronic communication techniques (in connection with mobile devices) within the scope of initiation efforts, services agreement, or provision (Turowski & Pousttchi, 2004, p. 1). Thus, the previously described SMS process is a typical example for m-commerce. Considering economic aspects, mobile applications are distinguished according to a generic market model. On the supply side, mobile applications can be employed for optimization and support of added value and services. Within the market transaction, they can help to coordinate and handle benefit transfer. From the consumer perspective, mobile applications as information or service products can turn out to be an economic commodity (Hess, et al., 2005, p. 6). From a technical point of view, mobile technologies form the basis of m-commerce. This includes mobile networks, mobile devices, and available basic services (Scheer, Feld, & Göbl, 2001, p. 8).

The monopoly aspects of m-commerce compared to the static electronic commerce (e-commerce) are location and time independence, identity and context reference, as well as specific interaction possibility and personal reference (Figge & Theysohn, 2006).

Regarding m-commerce, a consistent growth could be observed in the last years. In 2003, for example, there were 94.9 million m-commerce users worldwide. Estimations from 2004 even predicted an increase to 1.64 billion users by 2008, which should bring a return of 554.37 billion USD for the companies via the mobile distribution channel (Bigné, Ruiz, & Sanz, 2007, p. 49). The distribution of mobile devices suitable for Internet use is an important aspect for the success of m-commerce. According to a survey by Detecon & Diebold Consultants, the number of mobile Internet users is expected to increase from 42 million in 2000 to 460 million in 2006 (Selk & Turowski, 2003, p. 9). Contrary to the expectations, the actual development is far behind the predicted distribution (Rannenberg, Schneider, & Figge, 2005, p. 1). Complementary to the drafted market penetration of mobile applications, advancement of mobile technologies could be observed during the last years – especially regarding mobile devices and mobile networks. For example, the network infrastructure was improved by means of wireless transmission techniques with higher fractional bandwidth than UMTS (a 3G technology) (Sinn, 2002). Furthermore, the foundation of the Open Mobile Alliance (OMA) in 2002, which assumed the Wireless Application Protocol (WAP) forum, plays a major role (Holland & Bammel, 2006, p. 34). OMA mainly deals with the development of consistent and open standards orientated to the paradigms of the World Wide Web (WWW) (Turowski & Pousttchi, 2004). WAP 2.0 and XHTML-MP enable, for example, the display of Internet pages on mobile devices according to the traditional WWW (Holland & Bammel, 2006). Modern mobile devices enable direct access to the (mobile) Internet and the WWW via integrated browser. An example is Apple's iPhone, which was recently released on the German market. Such mobile devices enable direct interactions, as known from the (static) WWW, with the interested person. Restrictions exist basically due to smaller sizes of input and output devices.

As a result of the OMA standards, restrictions between mobile and static applications are removed but the necessary investments confront science and industry to find solutions and concepts for safe investments (Figge, 2002). The research regarding economic success of mobile services enjoys increasing popularity, but the research on mobile Internet applications in particular is still not up to date. The description of branch-oriented application possibilities is especially incomplete (Kaspar, 2006). In addition, the investigations of acceptance and reason for utilization of mobile applications from the customers' point of view, as well as of the development of technologies and services benefitting the customers are still not complete. (Hess, et al., 2005; Kaspar, 2006). Particularly against this background, it seems reasonable to employ new ways of sales and performance via mobile applications in the industry.

OBJECTIVES OF THE STUDY AND RESEARCH DESIGN

The context of the empirical studies in this chapter regards the potential of m-commerce applications for the sales of insurance products in the Swiss health care insurance industry. Accordingly, in the first quarter of 2007, an online survey was carried out, involving customers and companies in Switzerland. Motivation of this two-step approach was to gain as many complex perspectives as possible, in the sense of triangulation (Flick, 1995), about the benefits of mobile applications in the Swiss health care insurance industry for science and industry. For representative and evaluation reasons, a quantitative approach was chosen for the customer survey while a qualitative one was used for the expert survey.

The intention of the surveys was to comprise the complexity of the actual and potential utilization of m-commerce applications – particularly those which are based on the use of mobile phones – and to obtain findings about crucial factors for the design of m-commerce applications. The survey therefore focused on the chances of selling insurance products by means of m-commerce applications. The questions, in particular, are if customers in the Swiss health care insurance industry accept m-commerce applications which are based on mobile technologies, how m-commerce application for customer acquisition via the mobile Internet should be designed, and which success factors have to be regarded.

Investigation Design of the Customer Study

To investigate the customer perspective, we developed a questionnaire which considered socio-demographic questions and constructs representing the factors in question. Other surveys on e- and m-commerce include the factors acceptance, trust (Kleinaltenkamp, 1998; Siau & Shen, 2003), as well as individualization and personalization (Kaspar, 2006). In this context, the key determinant of trust could be separated into trust in m-commerce technology or in the vendor of a m-commerce application (Siau & Shen, 2003). Personalization includes personal communication with customers (Schackmann & Schü, 2001) and customized applications development (Müller, 2005).

Lihotzky (Lihotzky, 2003) and Müller (Müller, 2005) have studied different determinants that could improve customer loyalty in the electronic commerce. Müller focused on the determinants of personalization. Lihotzky designed a model with "action parameters" that determine customer loyalty. In their research, Lihotzky and Müller tested the determinants by means of structural equation models. Trumpfheller (Trumpfheller, 2005) explored customer loyalty in insurance companies and Bölscher (Eschke & Bölscher, 2001) provides an overview of electronic commerce in insurance companies. To test their research models, Müller and Trumpfheller surveyed customers directly. Lihotzky surveyed businesses with regard to their customers' views on e-commerce.

Lihotzky, Müller, and Trumpfheller have identified the following additional aspects:

- **Economic change barriers (ECB)** include all the costs related to a business relationship (including non-monetary factors), which would change to sunk costs if the relation were to end. Economic change barriers are theoretically founded in transaction costs theory (Coase, 1937), (Williamson, 1979), Hirschmann's microeconomic theory (Hirschmann, 1974), and Söllner's commitment trust model (Söllner, 1993).
- **Psychological and social change barriers (PSCB)** include psychological and social factors that are responsible for customer loyalty, and are based on Hirschmann's microeconomic theory (Hirschmann, 1974), Söllner's commitment trust model (Söllner, 1993), and socialization theories.
- **Customer satisfaction** is based on the confirmation-disconfirmation model, which describes customer satisfaction or dissatisfaction as the result of a cognitive individual target/actual comparison (Homburg & Bruhn, 2005).
- **Exploratory consumer behavior** has been categorized into curiosity-motivated behavior. This includes the intrinsic desire for knowledge, variety seeking as a means of obtaining stimulation from consumer behavior by alternating between familiar choice objects, and risk tasking as a "change of loss" concept (Steenkamp & Baumgartner, 1992).

Table 1. Operationalization of constructs

Indicator	Question no.	Background of the question
v_124	5.6	Trust
v_125	5.7	Trust
v_123	5.5	Trust
v_68	5.1	Acceptance
v_69	5.2	Acceptance
v_70	5.3	Acceptance
v_71	5.4	Acceptance
v_92	7.2	Individualization and personalization
v_93	7.3	Individualization and personalization
v_94	7.4	Individualization and personalization
v_79	6.1	Probability of utilization in the customer acquisition
v_80	6.2	Probability of utilization in the customer acquisition
v_81	6.3	Probability of utilization in the customer acquisition
v_91	7.1	Probability of utilization in the customer acquisition
v_10	2.3	User type according to utilization frequency

In Lihotzky, Müller and Trumpfheller's research, there are further determinants that can be said to fall under ECB and PSCB. ECB include contract agreements, incentives, technical integration, and free value added services, while PSCB include community, commitment, communication, nearness, emotion, and habit.

Despite the wide range of determinants, we designed a simple model that formed the basis of our survey.

In this model for the Swiss health care insurance companies, the following factors will be examined more closely:

1. Acceptance of mobile applications
2. Trust in mobile technologies
3. Individualization and personalization
4. Probability of utilization in the customer acquisition phase

Furthermore, the actual *degree of utilization* of mobile-based services and technologies will be surveyed as an additional factor, which is named as user type. For the examination of the factors, they will be operationalized through indicators. The classification between variables and indicators from the questionnaire is described in table 1. The items of the questionnaire were constructed by the research team.

This study's data was collected between 19.2.2007 and 19.04.2007 via an online questionnaire using the software "Umfragecenter 5.0" of Globalpark GmbH.

An e-mail campaign was launched in order to reach a representative number of participation in the online survey. 7.400 customers were asked via e-mail to participate in the survey. This covered in total all obligatorily insured persons in Switzerland in the year 2007. The contact data originated from a random choice of addresses provided by marketing databases of several Swiss insurance companies. This of course meant that only Internet experienced customers with Internet connection were interviewed. However, these technical affine customers are considered as the primary addressees of future mobile applications. 18.8%, or 1.396, of the questionnaires were returned. After sorting out incomplete and invalid questionnaires, the data basis was n=1.002. Aside from socio-demographical data, different aspects of user behavior and adjustment data of the above mentioned factors were collected by means of a six-level agreement scale.

Investigation Design of the Expert Study

The survey was conducted in form of a guided interview of the higher management level of Swiss health care insurance companies. The interview outline consisted of two parts. The first part included qualitative questions about sales-oriented m-commerce applications. The focus here was on actual and future application areas, success factors, chances and risks of implementation, and the integration of applications into business processes. The second part included quantitative questions, which regarded the importance of certain factors and the actual state and future development of the implementation of m-commerce applications in the company. Representatives of the largest Swiss health care insurance companies (with a total market share of about 60% in the first six months of 2007) were interviewed personally or by phone for statements relevant to Switzerland's health care insurance industry. The findings of the qualitative data were afterwards transcribed and categorized.

SELECTED STUDY RESULTS

Customer Survey

The survey shows that the distribution of mobile networks among the respondents is very large. The survey participants often had more than one mobile phone. 79% stated that their household has two or more mobile phones. Besides voice telephony, SMS is, at 36%, the most frequently used technology (figure 1). The Multimedia Messaging Service (MMS) technology is not used as frequently, and other technologies like mobile Internet/WWW and e-mail via mobile phone are not or just rarely used. The question about trust in mobile technologies revealed that 64% of the interviewees trust the data transfer of mobile phones via SMS, MMS, WAP, etc.

Figure 1. Utilization of mobile technologies in Switzerland (n=1002)

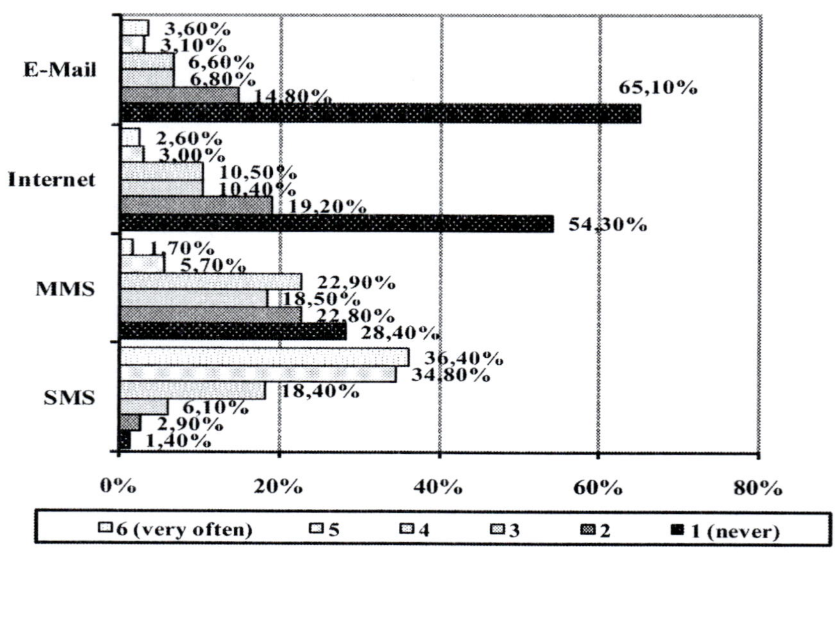

Nevertheless, 76% of the respondents pay attention to the trustworthiness of the provider when transferring data. 73% said that SMS, MMS, WAP, etc. are part of their usual means of communication like, for example, the phone. Personalization and individualization was important to more than 52% of the respondents when deciding about utilization.

On the basis of these factors out of section 2.1, the following structure model (figure 2) will be used to examine the cause-and-effect context between the factors by means of variance-based "Partial Last Squares (PLS)" analysis using the software "SmartPLS 2.0.M3" (Ringle, Wende, & Will, 2007). In this regard, a testing of the measuring model and of the structure equation model will be carried out (see section 3.1.1 and 3.1.2).

Testing of the Measuring Model

In the following section, the quality of the measuring model will be looked at. The reflective construct will be analyzed by means of various quality criterions. The quality degrees used for reflective measuring models are summed up in table 2 and defined in the according sections.

Testing the Content Validity

For testing, an explorative factor analysis is used to check the structure of the indicator factor (Krafft, et al., 2005, p. 72ff.), (Götz & Liehr-Gobbers, 2004, p. 727). The factor analysis is a multivariate analysis process, which reduces m measured units to p theoretical factors, whereby $p < m$. This process correlates the units. Strong correlation is regarded as an index for a joint factor (Eckstein, 2006, p. 306ff.). The correlation matrix can be charted in form of a vector diagram, whereas the respective correlation coefficient describes the angle between two vectors. The vectors are both independent when they come together orthogonally. The connection between angle and correlation coefficient is established by the cosine of the respective angle. In a two-factorial solution, the vectors can be displayed in a coordinate system of the two factors (Backhaus, Erichson, Plinke, & Weiber, 2005, p. 279ff.).

In the case under consideration, a construct is valid if the theoretically associated indicators correlate and therefore are assigned to a factor, which corresponds to the measured construct. Three explorative factor analyses are carried out for testing. Each analysis contains the indicators of the particular construct of the different models for

Figure 2. Structure model

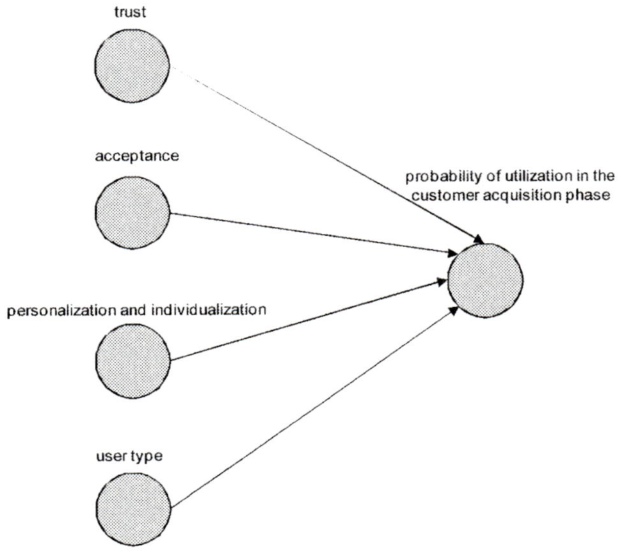

Table 2. Quality criterion of reflective constructs (Krafft, Götz, & Liehr-Gobbers, 2005, p. 75)

Quality degree	Definition
Content validity	Content validity describes the degree to which the indicators belong to the content-semantic domain.
Indicator reliability	The indicator reliability specifies the variance degree of an indicator, which is described by the latent variable of the indicator.
Construct reliability	The construct reliability is responsible for the quality test on the construct level, which is more important than the measuring on the indicator level.
Discriminance validity	The discriminance validity consists of the various measurements of different constructs by one tester.

Table 3. Factor analyses of customer aquisition

	Initial Eigenvalues			Extraction sums of squared loadings			Rotation sums of squared loadings		
Compo-nent	Total	%ofvarian-ce	Cumulative %	Total	%ofvarian-ce	Cumula-tive %	Total	%ofvarian-ce	Cumulative %
1	4,298	30,699	30,699	4,298	30,699	30,699	3,043	21,735	21,735
2	2,099	14,989	45,688	2,099	14,989	45,688	2,604	18,601	40,335
3	1,953	13,950	59,639	1,953	13,950	59,639	2,430	17,354	57,689
4	1,083	7,735	67,373	1,083	7,735	67,373	1,356	9,684	67,373
5	,699	4,996	72,369						
6	,677	4,837	77,206						
7	,558	3,985	81,191						
8	,541	3,865	85,055						
9	,473	3,378	88,434						
10	,453	3,239	91,672						
11	,358	2,554	94,227						
12	,302	2,156	96,383						
13	,261	1,868	98,250						
14	,245	1,750	100,000						

the relation life-cycle phases. The factor calculation includes an automatic orthogonal "Varimax" rotation of the coordinate system. This facilitates the interpretation of the factors. It is mathematically proven that such rotation has no influence on the significance of the factor analysis (Backhaus, et al., 2005, p. 299). Table 3 shows the results of the factor analysis.

SPSS first determines as many factors as indicators were originally integrated into the analysis. Each factor has different key figures:

- Eigenvalue (second column);
- Percentage of total variance of all indicators defined by the respective factor (third column);
- Cumulated share of total variance of all indicators, which are defined by the respective factor (fourth column);
- All factors, whose Eigenvalue is greater than one (fifth column);
- Total variance of all indicators (sixth column);

Table 4. Rotated component matrix

	Component			
	1	2	3	4
v_68	,788			
v_69	,811			
v_70	,755			
v_71	,756			
v_123	,665			
v_124				,725
v_125				,835
v_92			,799	
v_93			,873	
v_94			,867	
v_91		,550		
v_79		,861		
v_80		,873		
v_81		,844		

- Cumulated share of total variance of all indicators, which are defined by the respective factor (seventh column).

The loading of the indicators on the respective factors are illustrated in the rotated component matrix (table 4), based on SPSS. The table shows the indicators in the first column and the respective loading of the indicators on the factors in the subsequent columns, whereas, due to the significance, only factor loadings greater than 0.5 are regarded.

The testing of the first model shows that four factors exist. However, a precise check of the factors and constructs proves that the factor structure of trust and acceptance differs from the theoretical model. It results in two divergent factors, whereas the first can be paraphrased as "trust in provider" (factor 4 in the table) and the second as "trust and acceptance in mobile technologies" (factor 1 in the table). Factor 2 reflects the construct "probability of utilization in the customer acquisition phase" and factor 3 the construct "personalisation and individualisation". As another factor, the formative "user type" is included. This is not regarded for the calculation of the factor analysis; however, the criterion of factor correlation is not essential for formative constructs. This is a construct with only one indicator which was validated by expert opinion (Fassott & Eggert, 2005, p. 38ff.). The outcome is the changed model for the customer acquisition phase illustrated in figure 3.

This adjusted model meets the criterion of content validity and is taken as a reference for further testing of utilization probability of mobile applications in customer acquisition.

Testing of Indicator Reliability and Adjustment of the Model

The indicator reliability specifies the share of total variance of all indicators defined by the respective latent variable. This is the case if more than 50% of the indicator variance results from the latent variable. This criterion is fulfilled if the loadings λ of a latent variable on an indicator variable reach the amount of $\lambda \geq 0.7$. Reflective indicators with a loading smaller than 0.4 should be eliminated (Götz & Liehr-Gobbers, 2004, p. 727). In order to

Figure 3. Adjusted model for utilization probability including assigned indicators

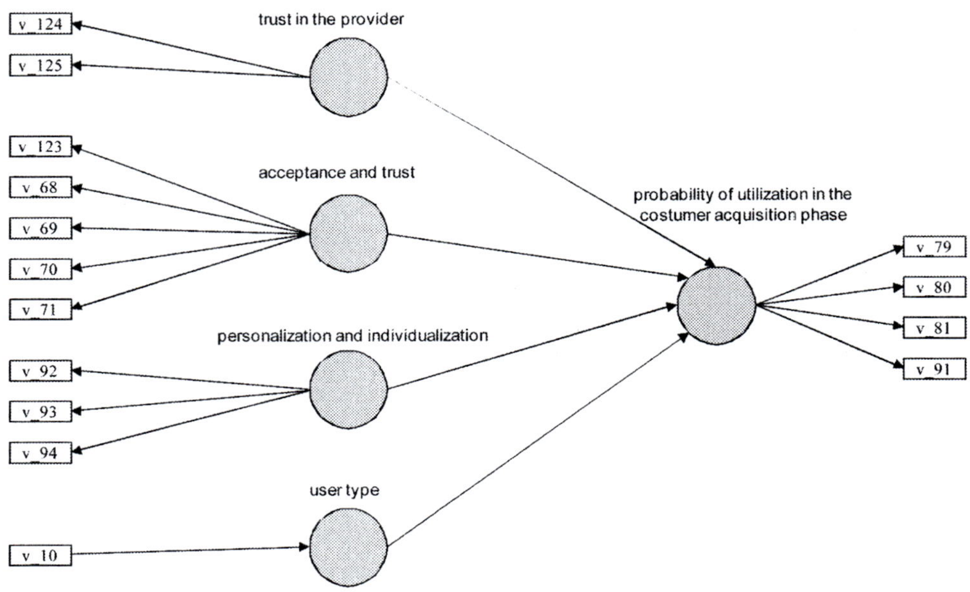

Figure 4. Model including loadings λ on indicators

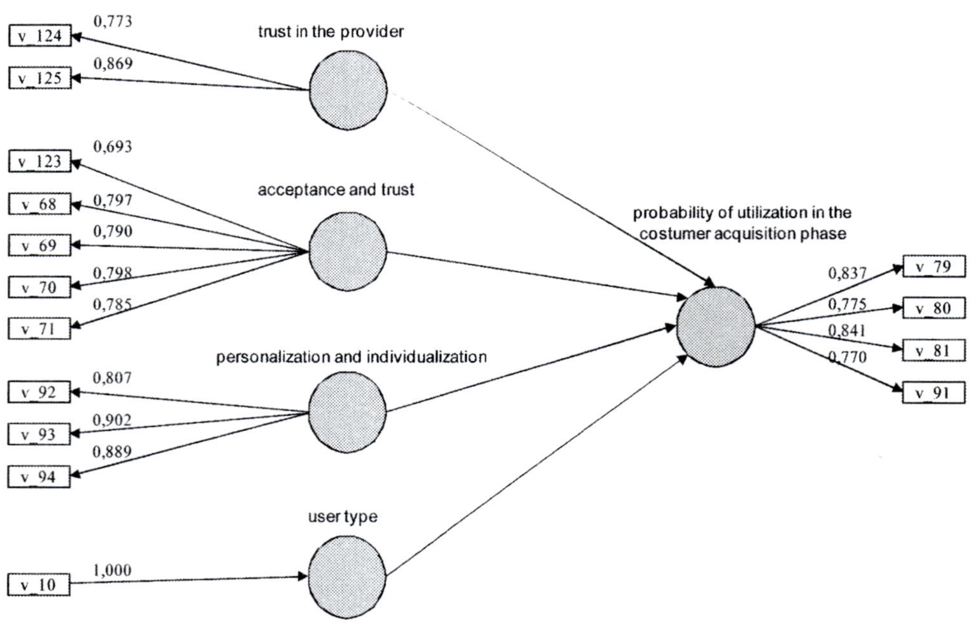

check the indicator reliability, models are assessed with the PLS algorithm by using the software SmartPLS. The loadings λ of latent variables are afterwards read from the path models.

The criterion for indicator reliability is given if, for example, the customer acquisition model regards loadings for the factor "trust in providers" on indicator v_124 with λ=0,773 and on v_125

Table 5. Internal consistency p for constructs

	Trust in the provider	Acceptance and trust	Personalization and individualization	Utilization probability
p	0,800739	0,881365	0,900607	0,881268

Table 6. Average variance extracted (AVE) of the constructs

	Trust in the provider	Acceptance and trust	Personalization and individualization	Utilization probability
AVE	0,668321	0,598373	0,751709	0,650202

with λ=0,869 (figure 4). The indicators' entity of the model fulfills the criterion of indicator reliability. However, the indicator v_123 is just under the limit of 0.7 with a loading λ of 0,693. Since only indicators with loadings λ under 0.4 are eliminated, this one remains in the model.

Testing of Construct Reliability

The construct reliability examines the quality on the construct level which is of higher significance than the measuring on the indicator level. The construct reliability implies that the indicators assigned to the same factor correlate strongly with each other. This internal consistence p can have a value between zero and one (Krafft, et al., 2005, p. 74f.). Construct reliability is given if the value is $p > 0{,}6$ (Ringle, 2004, p. 24), (Götz & Liehr-Gobbers, 2004, p. 727f.). Another quality measure aside from internal consistence is the average variance extracted (AVE). Values greater than 0.5 are basically accepted (Ringle, 2004, p. 24).

As shown in table 5 the criterion $p > 0.6$ is fulfilled for all constructs of the model and construct reliability is given on the part of internal consistency. Another criterion is the value for AVE (table 6).

All constructs show an AVE > 0.5. Therefore, both examined criteria for testing the construct reliability are fulfilled and the constructs can be assumed.

Testing of Discriminance Validity

The discriminance validity is tested in order to complete the evaluation processes. This includes the variety of measures of different constructs with one measuring instrument (Krafft, et al., 2005, p. 74f.). The discriminance validity is given when a manifested variable shows the highest squared correlation coefficient with the assigned latent variable of the measuring model (Ringle, Boysen, Wende, & Will, 2006, p. 87). One evaluation process of discriminance validity is the consideration of correlations between the factor values of latent and other variables contained in the model. If an indicator shows a higher correlation to another latent variable than the assigned one, the model has to be adjusted, as it is not clearly determinable what factor the latent variable reflects. It is essential that a manifested variable shows the highest loading with the assigned variable (Ringle, 2004, p. 25f.). For the control of the discriminance validity, the cross tables with loadings are examined.

The observation of loadings in table 7 shows that all indicators are allocated to the construct to which they show the highest loading. Therefore, discriminance validity is given for the model of customer acquisition.

Assessment of the Structural Equation Model

Quality Criteria for the Evaluation of Structural Models

Now that the measuring models have been assessed in the previous sections, the validity of the entire model will be evaluated. The model evaluation of structural models based on the PLS method is somewhat problematic since there is no global quality criterion for the evaluation of the entire model (Henseler, 2005, p. 75). The problem is explained as such; as a result of missing allocation assumptions, no parametric tests, such as significance tests, could be used. Therefore, a compendium of non-parametric estimate oriented quality criteria will be consulted for the evaluation of the model (Ringle, 2004, p. 16). If deficits are recognized, corresponding modifications will be applied to the model in order to achieve the statistically significant results.

For a structural model based on a PLS determined causal model it is possible to calculate the coefficient of determination R^2, which is known from the statistic method of the regression analysis (Ringle, 2004, p. 18f.). It reflects the determined variance of the latent construct and measures the quality of the adaptation of a regression function on empirically determined manifested indicators. Thereby, it can take values between zero and one. The higher R^2 is, the higher is the portion of the determined dispersion on the gross dispersion (Backhaus, et al., 2005, p. 66ff.). For R^2, values of 0,67 are classified as substantial, 0,33 as moderate, and 0,19 as weak (Chin, 1998, p. 316).

Another quality criterion for the structural model is the evaluation of the path coefficient. Here, single path coefficients are interpreted like the standardized β-coefficient that results from the application of the least squares estimation. Paths that are non significant or that show an algebraic sign that is contrary to the hypothesis formulation, refute the corresponding hypothesis (Krafft, et al., 2005, p. 83f.). Path coefficients should have a minimum value of $|0,1|$ to be accepted in a PLS model (Ringle, 2004, p. 19). Aside from the evaluation of the path coefficients, the Bootstrap based t values provide information about the significance

Table 7. Variable loadings

	Trust in the provider	Acceptance and trust	Personalization and individualization	Utilization probability
v_124	**0,7729**	0,3425	0,2489	0,0987
v_125	**0,8598**	0,1938	0,1772	0,1227
v_123	0,2921	**0,6925**	0,1047	0,2185
v_68	0,2396	**0,7968**	0,1694	0,2737
v_69	0,2256	**0,7904**	0,1379	0,2217
v_70	0,28	**0,7975**	0,1862	0,2836
v_71	0,1969	**0,7853**	0,1258	0,2827
v_92	0,2165	0,1199	**0,8069**	0,2318
v_93	0,2183	0,1851	**0,9024**	0,2722
v_94	0,2288	0,1836	**0,8887**	0,252
v_79	0,1233	0,2948	0,1641	**0,8368**
v_80	0,0544	0,2253	0,0856	**0,7746**
v_81	0,1385	0,2744	0,1918	**0,8414**
v_91	0,1064	0,2689	0,3926	**0,7699**

(Krafft, et al., 2005, p. 83). A t value ≥ 1,648 applies for a 5 percent significance level.

The coefficient of determination R^2 can, on the other hand, be used to determine the substantial explanatory contribution of a variable on the dependent variable. Here, the magnitude of effect f^2 is calculated. The formal statement is as follows:

$$f^2 = \frac{R^2_{incl} - R^2_{excl}}{1 - R^2_{incl}}$$

R^2_{incl} is the estimate of the coefficient of determination of the dependent variable including the non-dependent variable, and R^2_{excl} is the estimate without consideration of the dependent variable. Consequently, the alteration of the coefficient of determination is calculated (Krafft, et al., 2005, p. 84). The calculated values show the effect of an exogenous latent variable on an endogenous latent variable (Krafft, et al., 2005). Thereby, f^2 values of 0,02 stand for a weak, 0,15 for a moderate, and 0,35 for a strong effect of the exogenous on the assigned endogenous latent variable (Ringle, 2004, p. 20).

The non-parametric Stone-Geisser test provides insight on the prognosis relevance Q^2 of the model. The test follows what could be labeled the blindfolding procedure which, during the parameter estimate, systematically assumes part of the raw data material as missing. In the following step, these calculated parameter estimates are consulted in order to calculate the assumed missing raw data. As a result, the blindfolding procedure delivers general cross validation as well as Jackknifing-Standard-Variances of the parameter estimate. The determined value provides information on how well the empirically collected data can be reconstructed with the help of the model and the PLS parameters. It is therefore called the prognosis or estimate relevance. If Q^2 is greater than zero, the model contains a prognosis relevance (Götz & Liehr-Gobbers, 2004, p. 731), (Chin, 1998, p. 318). Thereby, the value can be interpreted as a measure to determine how well empirically collected data can be calculated by the model without the loss of the degree of freedom (Ringle, 2004, p. 21). Table 8 summarizes the quality measurements.

Model Evaluation

To start, the model will be calculated in order to review the model for utilization probability of mobile applications. The coefficient of determination and path coefficient can be evaluated on the basis of the output. Figure 4 shows the complete model of the customer acquisition phase.

As shown in figure 5, the coefficient of determination R^2 has a value of 0,17, hence $R^2 < 0,19$. The effect of the determined dispersion on the gross dispersion can be classified as weak. The analysis of the path coefficient reveals that the constructs "user type" and "trust in the provider" do not fulfill the specification for the path coefficient ≥ |0,1|. For this reason, both constructs are eliminated and the model is recalculated (see figure 6).

Table 8. Quality criteria for the review of the structural model

Quality type	Quality criteria
Coefficient of determination	R^2 of 0,19 = weak R^2 of 0,33 = moderate R^2 of 0,67 = substantial
Path coefficient	Path coefficient ≥ |0,1| T values ≥ 1,648 (5%-level)
Substantial explanatory contribution	Analysis of the f^2 values
Prognosis relevance	$Q^2 > 0$

Figure 5. Model of the customer acquisition phase including R^2

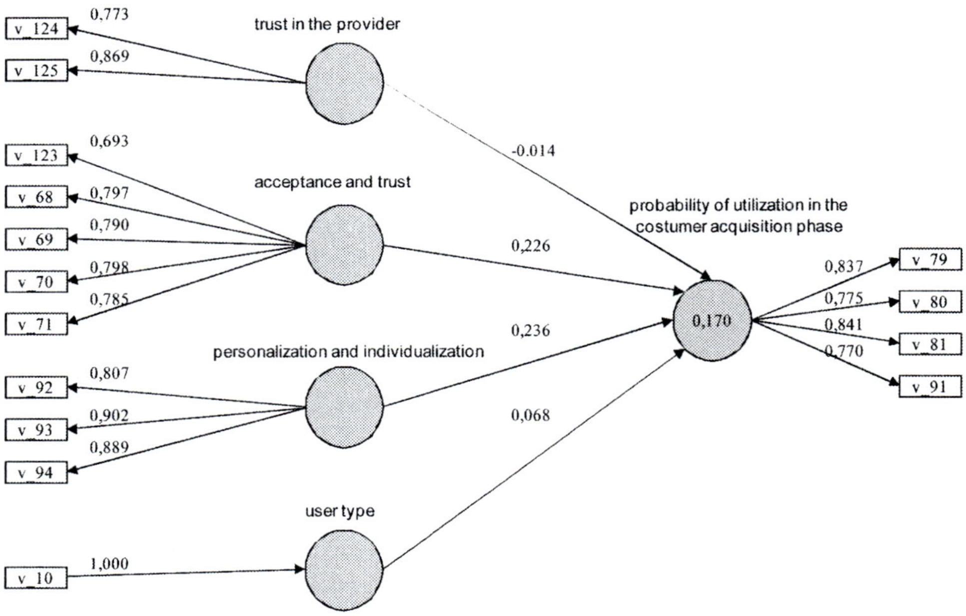

Figure 6. Adjusted model of the customer acquisition phase including R^2

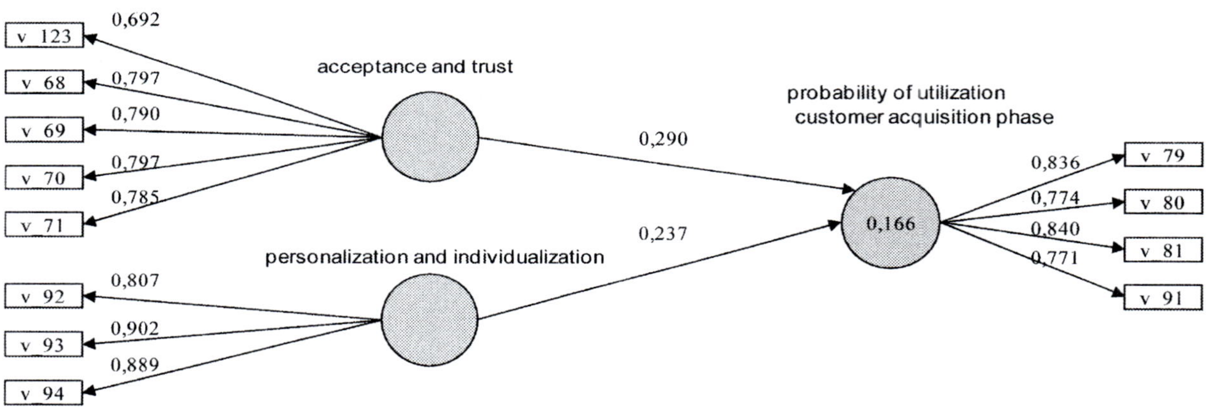

The coefficient of determination R^2 of the adjusted model does not fulfill the quality criteria of $R^2 \geq 0{,}19$. Since the path coefficient clearly fulfills the request for path coefficient $\geq |0{,}1|$, therefore the influence of the determinant is shown, the model is not rejected yet. To start, missing quality criteria are considered. Figure 7 shows the t values of the respective paths.

As shown in Figure 7, the t values with t=3,435 and t = 2,558 are significant on a 5% significance level. The results of the calculation of the f^2 values are shown in table 9.

The f^2 values show that the variables "acceptance and trust" and "personalization and individualization" have a weak or moderate influence on the latent endogenous variable "probability of

Figure 7. Adjusted model of the customer acquisition phase including t values

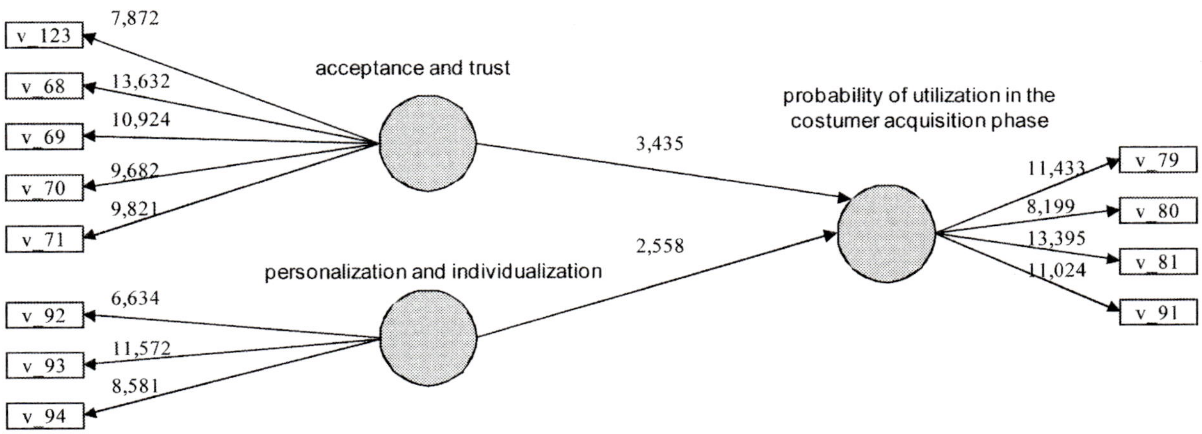

Table 9. f^2 values adjusted model of the customer acquisition

Variable	f^2 values
Acceptance and trust [AV]	0,06594724
Personalization and individualization [PI]	0,06834532

Table 10. Stone-Geisser test criteria of the customer acquisition phase

Variable	Q^2 values
Acceptance and trust	0,397
Personalization and individualization	0,487
Probability of utilization	0,093

utilization in the customer acquisition phase". The prognosis relevance is shown in table 10.

As required, the values of the Stone-Geisser test criteria are higher than zero. Thus, the model has adequate prognosis relevance (see table 10).

Summary of Model Evaluation

Table 11 summarizes the results of examined quality criteria

In summary, the results of the analysis reveal four existing factors. Factor (1), as well as the intensity of utilization, has a negligent low influence on factor (4) (see figure 5). The exogenous factors "acceptance and trust" (2) and "personalization and individualization" (3) have a low or moderate influence on the endogenous factor "probability of utilization in the customer acquisition phase" (4) (see figure 5).

Expert Survey

The Swiss health insurance companies' point of view regarding the current and future role of m-commerce in customer acquisition, customer retrieval, and customer loyalty is summarized in figure 8 It becomes apparent that the experts currently see the highest potential in marketing promotion, with a midpoint of 1,75. The poten-

Table 11. Evaluation of the structural model of the customer acquisition phase

Quality criteria	Result
Coefficient of determination	• R^2 = 0,166 = very weak
Path coefficient	• all path coefficients ≥ |0,1| • all t values ≥ 1,648 (5%-level)
Substantial explanatory contribution	• f^2 values AV = 0,07 • f^2 values PI = 0,07
Prognosis relevance	• all Q^2 > 0

tials for customer retrieval and customer loyalty are estimated to be lower. All in all, the experts see a high development potential within the next three years. Individualization and personalization of contents are assessed as the essential factors of success. On a scale from 1 to 6, the experts evaluate individualization with an importance average of 4,25. Personalization, with an average scale value of 3,88, is lower. As a result, the experts address the personal character of mobile phones, which they see as a risk due to the invasion of privacy.

The expert interviews revealed that companies are currently using applications based on SMS technology. Premium calculators, free value added services, and free SMS transmission are offered. Typical applications in the health insurance branch are information services like reminders for taking pills („pill reminder“) or pollen warnings. Potentials are seen through the new technologies that evolve through WAP 2.0. Aside from the potential in the area of initiation of benefits, the experts see chances for customer loyalty and customer retention in the health insurance branch. These can be realized by means of offline applications like Java technology based coaching applications which can be installed on mobile phones. An example for such an application is the "Brigitte Body Coach". This is a virtual consultant that supports the customer in weight-control issues by providing information on nutrition, sports, etc. (o.V., 2007). This software can be installed

Figure 8. Status quo and development of m-commerce from the experts' view

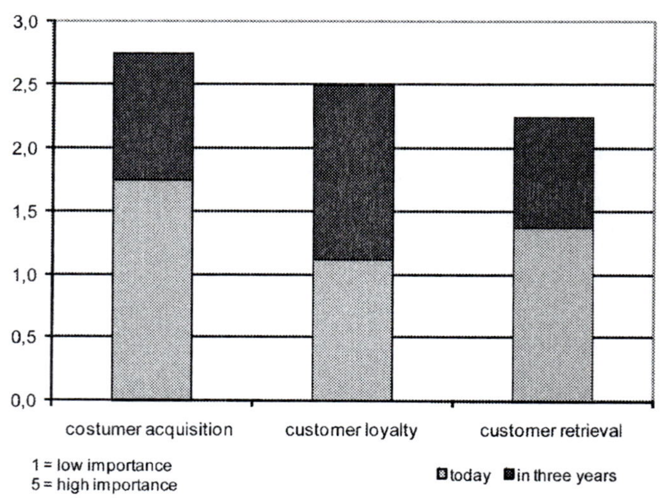

through a link which can be received in a fee-required SMS (labeled as premium SMS) and is available offline for 24 hours.

Discussion of Results

The survey of the Swiss health insurance companies shows that there are currently only few business-to-consumer (B2C) applications in place. SMS technology based applications are assembled for customer acquisition and requests for insurance plan proposals. Few additional applications based on Internet and SMS technology are offered. These, however, do not present purely m-commerce based applications like free SMS transmission through Internet presence of the insurance company (with attached advertisements of the health insurer).

Future potential can be seen in the consolidation of the marketing channels e- and m-commerce, since customer acquisition measures established by the stationary Internet could be carried over to mobile Internet as well as m-commerce. Hereby, new marketing processes by use of mobile phones are made possible and they could eventually take the place the SMS processes currently hold. A prototype offer process will be introduced in the following section 4.

The customer survey makes it clear, that mobile technologies hold a great deal of trust and acceptance from customers. Even today mobile technologies exhibit a high user rate, and thus the potential of m-commerce applications is obvious. Furthermore, the existence of the factors "trust in the provider" and "acceptance and trust" in the technology, as well as "personalization and individualization" and "user willingness" has been confirmed. In addition, the influence of the factors "acceptance and trust" (2) and "personalization and individualization" (3) on the factor "probability of utilization in the customer acquisition phase" (4) has been empirically proven. These should be taken into account in the further development of applications. The essential success factor for mobile applications is the personal and individual customer-oriented design. Both the experts' as well as the customers' opinion confirm this finding. This shows that the customers can handle the technical requirements and accept the technology. With a change volume of approx. 170.000 customers per year and the assumption that only a minimum amount of changers account for m-commerce applications within the benefit initiation process, it is already possible to assume that there is a high attractiveness of the channel. The presumption that customers not planning on changing also request insurance proposals via the mobile channel, and that through accompanying marketing promotions the shift towards the mobile channel will continue to increase, adds to the attractiveness. In addition, experts believe there will be cost savings on the company side because resources will be saved in the insurance proposal creation. They also expect added value due to integral processes and centralized collection of data. These aspects allow m-commerce applications appear to be a reasonable addition in multi-channel management for Swiss health insurance companies.

PROTOTYPE FOR MARKETING RELATED APPLICATIONS

This section proposes an exemplary process as a further development of the proposed SMS processes from Reichold/Schierholz (Reichold & Schierholz, 2003, p. 14) while taking the empirical results into account. This process conduces to the marketing promotion, which according to experts has a high potential, and was developed within the framework of research cooperation together with the Swiss health insurance companies. The mobile Internet (WAP 2.0) was chosen as the base technology for the new m-commerce marketing applications. Advantages are justified in that the mobile Internet allows, just like the stationary Internet, the mapping of complex interactive processes. It is important to point out that it is possible to conduct one-to-one marketing via mobile Internet. This newly proposed process (see figures 6 and 7) is similar to the SMS process of Reichold/Schierholz described in section 1.1, in that it has a cross-media entry and demonstrates, aside from other marketing channels like voice communication, an additional marketing channel within multi-channel marketing.

According to the permission marketing, this results in the entry via pull SMS (see figure 9, picture 1), in which the interested customer sends an SMS containing a code, taken for example from a magazine. The customer then receives an SMS containing a WAP link (see figure 9, picture 2 and following pictures), which enables him/her to interactively create a personalized insurance proposal after providing personal data. Added value for the customer is found in the usability, the fact that his/her personal data is taken into account, as well as the option for the customer to request an insurance proposal according to his product preferences. Therefore, the application is suitable for the customer requirements of personalized applications and the expert opinion on personalization.

Due to the linkage between the WAP promotions and the customer relationship (CRM) systems, we can speak of an integrated approach (see Figure 10). The advantage for the company lies in the generation of a complete customer data record, which is directly fed into the corresponding system and is immediately available for further sales promotions. This way, the entire customer acquisition process, from the offer request to the conclusion of the contract, is displayed. The intervention of an employee is necessary only in special situations, when specifically requested, or in cases of declination of the offer by the customer. A further advantage of this process is that changes can easily be made by means of the used technologies. SMS technology based processes do not offer this possibility, as these work on a text-based transmission in which text fragments in the CRM systems must be translated. This process modification leads to a high amount of maintenance.

CONCLUSION

The presented studies make it apparent that, by using mobile technologies, it is definitely possible to win over new customers in the Swiss health insurance industry. In principle, the technological possibilities and acceptance of these technologies

Figure 9. Prototypical mobile technologies

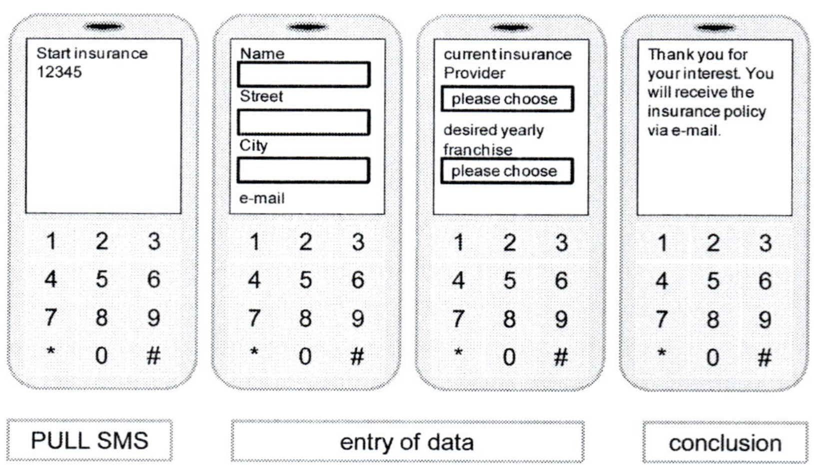

Figure 10. Acquisition process through WAP 2.0

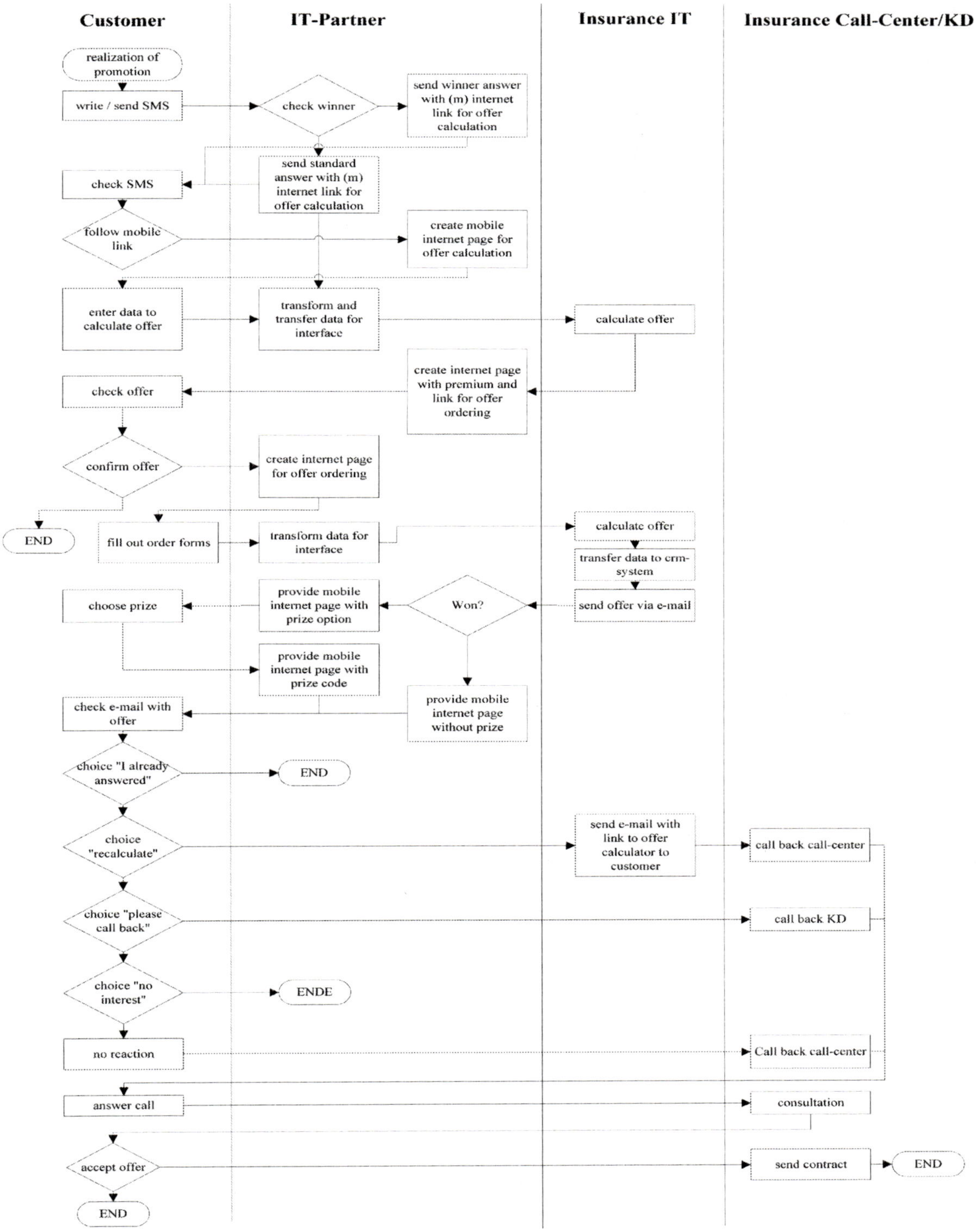

already exist. It is important for the critical success factors that, in the development of m-commerce technologies, special attention is paid to aspects of individualization and personalization. However, the evaluation of the customer perspective also revealed that the willingness to conclude a health insurance policy via m-commerce technologies is still quite low. For this reason, health insurance companies should include mobile channels as a further section in their multi-channel management. The presented prototypical technology is the first step in this direction, whereas to date concrete user numbers do not exist.

By using m-commerce technologies, realized mobile processes provide a cost efficient, integral illustration of marketing processes (Zobel, 2001, p. 158). A central issue in future studies will be the evaluation of the immediate added business value by m-commerce technologies in the Swiss health insurance industry. Here, the connection between receiving an offer through the mobile channel and the preparedness to conclude a contract, respectively the actual conclusion of a contract, should be examined.

REFERENCES

Backhaus, K., Erichson, B., Plinke, W., & Weiber, R. (2005). *Multivariate Analysemethoden* (11. Auflage ed.). Berlin, Heidelberg: Srpinger.

Bigné, E., Ruiz, C., & Sanz, S. (2007). Key Drivers of Mobile Commerce Adoption. An Exploratory Study of Spanish Mobile. *Journal of Theoretical and Applied Electronic Commerce Research, 2*(2), 48–60.

Chin, W. W. (1998). The Partial Least Squares Approach to Structural Equation Modeling. In G. A. Marcoulides (Ed.), *Modern Methods for Business Research* (pp. 295-336). Mahwah: Lawrence Erlbaum Associates.

Coase, R. H. (1937). The Nature of the Firm. *Economica, 4*(16), 386–405. doi:10.1111/j.1468-0335.1937.tb00002.x

comparis.ch. (2007, 09.01.2007). comparis.ch zu den Krankenkassenwechseln 2007 Retrieved 19.04.2007, 2007, from http://www.comparis.ch/comparis/press/communique.aspx?ID=PR_Comm_Communique_070109

Coursaris, C., & Hassanein, K. (2002). UNDERSTANDING M-COMMERCE. *Quarterly Journal of Electronic Commerce, 3*(3), 247.

Eckstein, P. (2006). *Angewandte Statistik mit SPSS. Praktische Einführung für Wirtschaftswissenschaftler* (5. Aufl. ed.). Wiesbaden: Gabler.

Eschke, S., & Bölscher, J. (2001). *Customer Relationship Management in der Versicherungswirtschaft.*

Fassott, G., & Eggert, A. (2005). Zur Verwendung formativer und reflektiver Indikatoren in Strukturgleichungsmodellen: Bestandsaufnahme und Anwendungsempfehlungen. In F. Bliemel, A. Eggert, G. Fassot & J. Henseler (Eds.), *Handbuch PLS-Pfadmodellierung. Methoden Anwendungen Praxisbeispiele*. Stuttgart: Schäffer-Poeschel.

Figge, S. (2002). Die Open Mobile Architecture - Systemumgebung für mobile Dienste der nächsten Generation. *Wirtschaftsinformatik, 44*(4), 375–378.

Figge, S., & Theysohn, S. (2006). Quantifizierung IKS-basierter Marktleistungen - Analyse eines werbefinanzierten Geschäftsmodells für den Mobile Commerce. *Wirtschaftsinformatik, 48*(2), 96–106. doi:10.1007/s11576-006-0024-z

Flick, U. (1995). *Qualitative Forschung. Theorie, Methoden, Anwendungen in Psychologie und Sozialwissenschaften*. Hamburg: rororo.

Götz, O., & Liehr-Gobbers, K. (2004). Analyse von Strukturgleichungsmodellen mit Hilfe der Partial-Least-Squares(PLS)-Methode. *Die Betriebswirtschaft, 64*(6), 714–738.

Helsana (2007). Retrieved 23.08.2007, 2007, from http://www.helsana.ch/?wm=m%287%29&CFID=4809269&CFTOKEN=79035439

Henseler, J. (2005). Einführung in die PLS-Pfadmodellierung. *WiSt. Wirtschaftswissenschaftliches Studium*, (2): 70–75.

Hess, T., Figge, S., Hanekop, H., Hochstatter, I., Hogrefe, D., & Kaspar, C. (2005). Technische Möglichkeiten und Akzeptanz mobiler Anwendungen - Eine interdisziplinäre Betrachtung. *Wirtschaftsinformatik, 47*(1), 6–16.

Hirschmann, A. O. (1974). *Abwanderung und Widerspruch - Reaktionen auf Leistungsabfall bei Unternehmungen, Organisationen und Staaten*. Tübingen: J.C.B. Mohr.

Holland, H., & Bammel, K. (2006). *Mobile Marketing*. München: Vahlen.

Homburg, C., & Bruhn, M. (2005). Kundenbindungsmanagement - Eine Einführung in die theoretischen und praktischen Problemstellungen. In C. Homburg & M. Bruhn (Eds.), *Handbuch Kundenbindungsmanagement* (5. Aufl. ed., pp. 3-40). Wiesbaden: Gabler.

Kaspar, C. (2006). *Individualisierung und mobile Dienste am Beispiel der Medienbranche - Ansätze zum Schaffen von Kundenmehrwert*. Unpublished Dissertation, Institut für Wirtschaftsinformatik 2 Universität Göttingen, Göttingen.

Kleinaltenkamp, M. (1998). *Geschäftsbeziehungsmanagement*. Berlin: Springer.

Krafft, M., Götz, O., & Liehr-Gobbers, K. (2005). Die Validierung von Strukturgleichungsmodellen mit Hilfe des Partial-Least-Squares (PLS)-Ansatzes. In F. Bliemel, A. Eggert, G. Fassott & J. Henseler (Eds.), *Handbuch PLS-Pfadmodellierung* (pp. 71-86). Stuttgart: Schäffer-Poeschel.

KVG. (2006). *Bundesgesetz über die Krankenversicherung (KVG)*.

Lihotzky, N. (2003). *Kundenbindung im Internet - Maßnahmen und Erfolgswirksamkeit im Business-to-Consumer-Bereich*. Wiesbaden.

Müller, U. (2005). *Kundenbindung im E-Commerce - Personalisierung als Instrument des Customer Relationship Marketing*.

Ngai, E. W. T., & Gunasekaran, A. (2007). A review for mobile commerce research and applications. *Decision Support Systems, 43*(1), 3–15. doi:10.1016/j.dss.2005.05.003

Rannenberg, K., Schneider, I., & Figge, S. (2005). Mobile Systeme und Anwendungen - Hammer sucht Nagel. *Wirtschaftsinformatik, 47*(1), 1–2.

Reichold, A., & Schierholz, R. (2003). *Fallstudie Helsana - M-Commerce bei der Helsana: Mobile Prämienerstellung* (No. HSG / IWI / BECS / 30). St. Gallen: Institut für Wirtschaftsinformatik Universität St. Gallen.

Ringle, C. (2004). Gütemaße für den Patrial Last Squares-Ansatz zur Bestimmung von Kausalmodellen. *Hamburg*.

Ringle, C., Boysen, N., Wende, S., & Will, A. (2006). Messung von Kausalmodellen mit dem Partial-Least-Squares-Verfahren. [WiSU]. *Das Wirtschaftsstudium, 35*(1), 81–88.

Ringle, C., Wende, S., & Will, A. (2007). SmartPLS Version: 2.0.M3 Retrieved 01.11.2007, from http://www.smartpls.de

Schackmann, J., & Schü, J. (2001). Personalisierte Portale. *Wirtschaftsinformatik, 43*(6), 623–625.

Scheer, A.-W., Feld, T., & Göbl, M. (2001). Das Mobile Unternehmen. *Information Management & Consulting*, *16*(2), 7–15.

Selk, B., & Turowski, K. (2003). *Mobile Commerce: Stand und Perspektiven.* Unpublished manuscript, Köln.

Siau, K., & Shen, Z. (2003). Building Customer Trust In Mobile Commerce. *Communications of the ACM*, *46*(4), 91–94. doi:10.1145/641205.641211

Sinn, W. (2002). Drahtlose Übertragungssysteme und Zugangsmedien. In R. Reichwald (Ed.), *Mobile Kommunikation* (pp. 113-125). Wiesbaden: Gabler.

Söllner, A. (1993). *Commitment in Geschäftsbeziehungen.* Wiesbaden: nbf.

Steenkamp, J.-B. E. M., & Baumgartner, H. (1992). The Role of Optimum Stimulation Level in Exploratory Consumer Behavior. *The Journal of Consumer Research*, *19*(3), 434–448. doi:10.1086/209313

Trumpfheller, J. (2005). *Kundenbindung in der Versicherungwirtschaft.*

Turowski, K., & Pousttchi, K. (2004). *Mobile Commerce – Grundlagen und Techniken*: Springer.

oV. (2007). Brigitte-Figur-Coach, from http://www.brigitte.de/diaet/figurcoach-sommer2007/index.html

Williamson, O. E. (1979). Transaction-Cost Economics: The Governance of Contractual Relations. *The Journal of Law & Economics*, *22*(2), 233–261. doi:10.1086/466942

Zobel, J. (2001). *Mobile Business und M-Commerce - Die Märkte der Zukunft erobern.* München: Hanser.

ADDITIONAL READING

Blattberg, R. C., Getz, G., & Thomas, J. (2001). Customer Equity: Building and managing relationships as valuable assets. *Mcgraw-Hill-Professional*, 36.

Bliemel, F. W., & Eggert, A. (1998). Kundenbindung die neue Sollstrategie? *Marketing Zeitschrift für Forschung und Praxis*, *20*(1), 37–46.

Bruhn, M., & Georgi, D. (2005). Services Marketing – Managing The Service Value Chain. *Financial Times Prentice Hall*, 443.

Merz, M. (2002). *E-Commerce und E-Business.* Heidelberg: dpunkt Verlag.

Müller, U. (2005). *Kundenbindung im E-Commerce - Personalisierung als Instrument des Customer Relationship Marketing.*

Piller, F., & Schaller, C. (2002). E-Loyalty - Kundenbindung durch Individualisierung im E-Business. In F. Keuper (Ed.), *Electronic Business und Mobile Business* (pp. 439-463). Wiesbaden: Gabler.

Schubert, P., & Hampe, J. F. (2005). *Business Models for Mobile Communities*, Hawaii.

Simon, B. (2001). *Wissensmedien im Bildungssektor: Eine Akzeptanzuntersuchung an Hochschulen.* Unpublished Dissertation, Universität Wien, Wien.

Werani, T. (2004). *Bewertung von Kundenbindungsstrategien in B-to-B Märkten.*

KEY TERMS AND DEFINITIONS

Acceptance: "Acceptance (...) is contradictory to refusal and describes the positive acceptation of an innovation by the user." (Simon, 2001, p. 87)

Confidence: "Customer confidence is a judgmental action to the supplier or to his range of items and services, and results from the customer expectance of being favorably treated by the supplier in the future." (Bliemel & Eggert, 1998, p. 40)

Customer Acquisition: "Customer acquisition includes the first purchase as well as other non-purchase encounters that both precede and follow the purchase, up until the time the customer makes a repeat purchase." (Blattberg, Getz, & Thomas, 2001)

Customer Loyalty: "Customer loyalty is when a customer intentionally realizes or plans repurchases from a certain supplier within an appropriately defined period of time." (Werani, 2004, p. 3)

Customer Recovery: "Customer Recovery is the re-activation of relationships to former customers who have terminated a former relationship with the provider." (Bruhn & Georgi, 2005)

E-Commerce: "Electronic Commerce is the application of communication records, security infrastructures, digital money, electronic shopping malls, electronic data exchange, smart cards, mobile and/or intelligent agents, negotiation records and strategies, electronic notaries, certification authorities, inter-organizational workflow management, electronic contracts and many other technologies for initiation and implementation of business operations on the Internet." (Merz, 2002, p. 20)

Individualization: "…Individualization as extreme orientation to individual requirements and desires of individual consumers. (…) Personalization characterizes (…) the individualization of communication with the customer by using new Internet technologies in terms of a One-to-One-Marketing." (Piller & Schaller, 2002, p. 444)

M-Commerce: "Mobile Commerce refers to all data-driven business transactions and exchanges of value by users of mobile devices via wireless telecommunication networks." (Schubert & Hampe, 2005, p. 3)

Personalization: "Personalization refers to the customer-specific development of Internet offers or Websites or particular elements (e.g. products, prices, layout) on the basis of information about customers." (Müller, 2005, p. 20)

Chapter 13
Opportunistic Networks as an Enabling Technology for Mobile Word-of-Mouth Advertising

Andreas Heinemann
Competence Center for Applied Security Technology, Germany

Tobias Straub
Baden-Württemberg Cooperative State University, Germany

ABSTRACT

The growing share of people using mobile devices, that support wireless peer-to-peer interaction, offers the opportunity to build a ubiquitous infrastructure for electronic word-of-mouth messaging and advertising. This chapter introduces Opportunistic Networks as a layer for one-hop communication that opens up electronic word-of-mouth messaging for mobile devices. The reader will learn about adPASS (short for advertisement passing), a system build on top of Opportunistic Networks for digital advertisement distribution stimulated by an anonymous and tamper-resistant bonus point model. A simulation based on empirical movement patterns indicates the feasibility of our approach. This chapter concludes with a summary and provides an outlook on further research paths starting from our findings.

INTRODUCTION AND MOTIVATION

Mobile marketing as a '*form of marketing communication using mobile communication techniques to promote goods, services and ideas*' (Pousttchi & Wiedemann, 2006) gains more and more attraction with the rapidly increasing number of mobile phones (1.15 billion units were sold in 2007 according to Gartner[1]). In fact each new device generation is equipped with short range wireless communication capabilities. In most cases, either Bluetooth or 802.11 WiFi technology is integrated into the devices. Today's prevalent use of wireless connectivity is to synchronize personal data between a mobile device and a desktop computer (using Bluetooth) or to have easy access to an institution's network (via 802.11 WiFi Wireless Access Points) and further to the Internet (for a general introduction into mobile marketing see Pousttchi & Wiedemann, 2006; Gewei, 2007).

However, with the integration of short range wireless communication technology into mobile devices, a new network type called *Opportunistic Network* (OppNet for short) is possible and especially

DOI: 10.4018/978-1-60566-074-5.ch013

suitable for a new form of digital *word-of-mouth (WOM) recommendation* or *viral marketing.*

Following Bone's (1995) characterisation, WOM communications are '*interpersonal communications in which none of the participants are marketing sources'*. The benefits of WOM recommendation for marketing campaigns have been widely recognized and investigated in the literature. See (Arndt, 1967; Herr, Kardes, & Kim, 1991; Bone, 1995) for research on *face-to-face* WOM or (Helm, 2000; DeBruyn & Lilien, 2004; Hennig-Thurau, Gwinner, Walsh, & Gremler, 2004; Datta, Chowdhury, & Chakraborty, 2005) for Internet-based WOM approaches. As already said, the increasing number of mobile devices now put a research focus on mobile word-of-mouth recommendation (cf. Wiedemann, Haunstetter, & Pousttchi, 2008), in the majority of cases considering cellular network services (SMS, MMS, etc.) for content distribution.

Our approach differs in the sense that we use a *mobile ad hoc peer-to-peer* communication scheme to facilitate the distribution of digital advertisements among interested users. As a starting point, let us first look at word-of-mouth recommendations in everyday live:

Real-World Word-of-Mouth Recommendation: Alice and Bob are co-workers sharing the same office. On her way to work, Alice passes a shop window that announces a digital camera 20% off the regular price. Alice knows that Bob plans to buy such a camera and tells him about the advertisement when she arrives at her office. During lunch break, Bob visits the shop and buys the camera, glad that Alice has directed his attention to the offer.

Word-of-mouth recommendation is a well-known and established way of communication and interaction among humans (cf. Arndt, 1967; Herr, Kardes, & Kim, 1991; Bone, 1995). There are two assumptions here: Alice and Bob know each other's interests in certain aspects of life and they meet on a regular basis (or know how to contact each other) to share information. In addition, on her way to work, Alice needs to be wide awake in order to notice shop window announcements.

OppNets are able to mimic this form of recommendation and to extent it. The adPASS[2] prototype discussed in detail within this chapter makes use of an Opportunistic Network as a communication layer.

Digital Word-of-Mouth Recommendation: Alice carries a mobile device with her. A personal profile, stored on her device, holds information about her interests and knowledge. The device is able to match her profile with other nearby devices by communicating wirelessly and without user interaction.

A shop has put a fixed device next to the shop window announcing digital advertisements to passersby. As Alice passes the shop window, her device learns about the special offer for digital cameras.

Alice physically carries the advertisement with her and passes it further to other users she encounters. All users interested in the advertisement (including herself and her colleague Bob) might take the chance and visit the shop in order to buy the advertised product.

A digital word-of-mouth recommendation differs form its real-world equivalent in several ways: the users who exchange advertisements do not need to know each other personally. A match in their profiles is sufficient to share the information. Next, a user does not need to keep his attention on the device as that the device works autonomously without interaction. This non-intrusive device behaviour allows the exchange of advertisements between mobile users (for example, in a shopping mall or pedestrian zone) without distracting the user from his major occupation.

Chapter Outline

At first the idea of Opportunistic Networks is introduced. We explain their potential as well as their characteristics that give rise to several challenges. In the next section prior art in the area of mobile marketing is reviewed from a technical perspective. In the heart of this chapter adPASS is presented as a prototype that makes use of Opportunistic Networks consisting both of mobile and stationary entities.

It is our purpose to provide guidelines for the implementation of applications on top of Opportunistic Networks by means of an exemplary solution. Although their use is not restricted to mobile marketing, Opportunistic Networks are, as we will see below, very well suited for this scenario. In order to encourage people to participate in the distribution of digital advertisements, a novel incentive scheme has been devised. The reader will learn how a sound business model can be established in a highly volatile and open environment with only few assumptions on infrastructure and trust.

As particular attention is paid to security issues, a brief overview of protection goals and corresponding cryptographic measures with references to implementation details is given. The adPASS bonus point system which is both anonymous and robust against cheating should serve as a practical and privacy-friendly example in order to promote the development and use of such systems in the area of mobile marketing.

The results of simulations, which have been carried out to demonstrate the feasibility of adPASS, are also summarized in this chapter. As a basis, movement patterns that have been gathered in a long-term user observation are combined with synthetic mobility models. This yields a synthetic macro-micro level model which is of interest of its own. Our approach may thus serve as a practical test bed and benchmark for other systems.

OPPORTUNISTIC NETWORKS

Opportunistic Networks consist of mobile users carrying mobile devices, for example, mobile phones or PDAs with short-range, wireless communication capabilities. These devices are able to connect to each other in a spontaneous manner by using the wireless link. Thus, a mobile network is formed.

Origins

OppNets have several roots. They combine ideas from mobile ad hoc networking (MANET) with concepts originating from peer-to-peer (P2P) networks (Datta, 2003; Ding & Bhargava, 2004; Goel, Singh, & Xu, 2002; Hayes & Wilson, 2005; Hu, Das, & Pucha, 2003; Klemm, Lindemann, & Waldhorst, 2004; Lindemann & Waldhorst, 2005). To a certain extent, OppNets are also related to delay tolerant networks (Leguay, Lindgren, Scott, Friedman, & Crowcroft, 2006; Pelusi, Passarella, & Conti, 2006). We will shortly review the most prominent OppNet concepts. A more detailed treatment can be found in (Heinemann, 2008).

Characteristics

Core OppNet features are the following:

- **User vicinity exploitation.** Communication and interaction happens when users are close-by.
- **Profile-based user interest expression** Applications rely on user profiles in order to facilitate non-intrusive and autonomous operation.
- **Data dissemination** Autonomous operation and user profiles suit for arbitrary word-of-mouth schemes in order to disseminate any kind of data.
- **Open and unrelated user group** A priori, there is no relation between OppNet participants and there is also no access restriction to a network. Depending on the application, participants may act completely anonymously.
- **Unpredictable communication pattern** Due to node mobility and autonomous

operation, the network topology and thus communication links and opportunities are unpredictable.

The main use of Opportunistic Networks is the distribution of data by making use of user mobility and so-called *one-hop* forwarding techniques, i.e., information is stored on the device and passed further when an appropriate contact is made.

An OppNet node comes with two core functionalities: *node discovery* and *one-hop message exchange*. Node discovery is needed in order to detect possible communication partners in a node's vicinity. After a successful node discovery, nodes are able to send and receive arbitrary data by direct message exchange. In comparison to MANETs, OppNets have no routing capabilities, but only provide one-hop communication. This is due to the lack of a relation and a common goal among nodes.

Apart from the mobile nodes which we have already discussed above, there is a second type of nodes: stationary nodes or, as we will call them, *Information Sprinklers* since they are placed at dedicated locations in order to provide users with *proximity-based services*. These services offer information or applications that are especially useful at the sprinkler's location, e.g. for advertisement dissemination inside a shop or a shopping mall.

Since OppNets rely on one-hop message exchange, which in turn will only happen if users are close by, any service or application build on top of an OppNet works in a best effort manner. There are no guaranties in delivery time or delivery of an information item at all. This distinguishes OppNets from other mobile networks like GSM or UTMS. However, OppNets do not need a pre-installed infrastructure to work.

Challenges

The use of OppNets for mobile marketing raises a number of challenges to address. We will review them in turn for a better understanding of what follows in the next sections. In the Case Study section, we will see how the design of adPASS takes into account the challenges which are inherent to OppNets.

Privacy OppNets make use of profiles created by individual users. In the case of mobile marketing applications these profiles keep information about product preferences and interests. As said in the last section, OppNet nodes may act autonomously and profile data is revealed without user notice in order to support the mobile marketing application. At this point, user privacy is a risk. We need to find a way to reveal interests in products etc. without revealing a user's identity or other (static) data that can be eventually linked to the user herself.

Incentives OppNets are formed by personal mobile devices carried by individuals. Applications build on top of such a fragile infrastructure need to attract a sufficient user mass in order to be useful and successful. Thus, the question why a user is willing to share her personal device resources (battery power, memory) needs to be answered appropriately.

Data Representation A mobile marketing application needs to find suitable data structures for products, advertisements and user interests. Also, user interests should be matched fast and efficiently against incoming product information and advertisements. Given mobile devices with limited CPU power and memory, complex reasoning algorithms based on ontologies are out of scope. Suitable data structures and matching algorithms working on these data structures need to be fast and flexible to cover a wide rage of marketing domains.

User Interface Design Mobile devices (mobile phones, PDAs) as considered for OppNets come with a limited user interface often without keyboard and a quite small display. Questions arise of how an application should aggregate marketing information in an appealing manner and how a user can express her interests in a user-friendly way. It is reasonable to assume that the mobile device is hooked up to a PC from time to time, e.g. during sync of other pieces of information like dates or phone numbers. Another option is to let the application implicitly learn about user preferences as for instance the SMMART prototype does (see below).

MOBILE MARKETING EXAMPLES AND STATE-OF-THE-ART

In this section we compare approaches for mobile advertising and discuss some prototypes in more detail. Here we are focussing on approaches that make use of ad hoc networking capabilities rather than on commercial services tied to cellular networks. The reason is that, at the time of writing, the latter seem less promising in the light of technical limitations and associated costs arising from their infrastructure requirements. On the other hand, the discussed prototypes provide additional clues concerning the use of social and ad hoc networking.

Solutions Using Mobile Networks as an Infrastructure

Today's commercially available services typically rely on the infrastructures of cellular network providers. In such networks, messages are exchanged via the popular SMS or MMS[3]. Both technologies work in push mode meaning that advertising organizations distribute messages to dedicated users/mobile phone numbers via the network. For two reasons, pull-mode protocols like WAP, GPRS or UMTS[4] are used less often: Obviously advertisements are usually not actively sought for by users. Plus: traffic has to be paid for by the recipient not by the sender as it is the case for SMS/MMS.

Bulk SMS is a standard tool for mobile marketing as it can be combined with auctions, lotteries, or free gimmicks to initiate responses. However the costs for SMS are relatively high especially with relation to the limited message size. Campaigns cost a few cents per SMS while users' responses may be charged for as premium services up to some Euros[5].

Approaches Using Ad Hoc Networking Capabilities

Several authors (see below) have proposed to combine social networking and MANETs for mobile marketing services that require only a decentralized infrastructure or none at all. Their concepts are only sketched in the following since in some regard they are similar to the system adPASS that we discuss in more detail in the Case Study section.

- *Usenet-On-The-Fly* (Becker, Bauer, & Hähner, 2002) allows users to post messages of local interest to "channels" which represent a particular topic. A prototype, based on WiFi node-to-node communication, supports the exchange of recommendations (e.g. concerning shops or restaurants) among users or the distribution of "sensor data" from stationary devices to users (e.g. bus schedules). Users can express their preferences by subscribing to a selection of channels.
- *MobiTip* is a similar system that uses Bluetooth to let users share opinions and comments (Rudström, Svensson, Cöster, & Höök, 2004). Both systems demonstrate the feasibility of disseminating recommendations in MANETs. However, their application scenarios are not primarily commercial.
- A burden of systems like Usenet-On-The-Fly and MobiTip is the fact that they require users to actively express and manage their preferences, a task which can be cumbersome. In contrast, *SMMART* (Kurkovsky & Harihar, 2006) is designed to infer interests from users' actions. By

Table 1. System and Feature Comparison

System	Feature			
	Ad hoc peer to peer recommendation	Information Sprinkler	Incentive scheme	Privacy protection
Usenet-On-The-Fly	√	√		
MobiTip	√	√		
SMMART		√		
MoMa		√		√
eNcentive	√	√	√	
adPASS	√	√	√	√

announcing its preferences a SMMART client retrieves appropriate offers from a server attached to a WiFi access point in a shop. Preferences, which are represented by simple keywords, are automatically updated when purchasing goods or simply browsing through Web pages showing offers and product descriptions.

- Privacy concerns may arise when user preferences get leaked or when user behaviour can be monitored by shops. The *MoMa* concept (Bulander, Decker, Schiefer, & Kolmel, 2005) avoids these threats by means of a trusted party. The system has a level of indirection which disguises the relation between profiles and user identities (which are mobile phone numbers or e-mail addresses in the MoMa context).
- *eNcentive* is another framework that lets customers receive advertisements and discount coupons from WiFi access points in stores (Ratsimor, Finin, Joshi, & Yesha, 2003). eNcentive also encourages users to recommend advertisements and distribute coupons further in an ad-hoc fashion. As an incentive to participate, users are rewarded by bonus points in case the recommendation has been successful.

Table 1 summarizes the discussed systems with respect to major features. The uniqueness found in adPASS is the design of an incentive scheme with an inherent privacy protection mechanism. An evaluation of the concept of digital coupons and a simulation of the effectiveness of their distribution with relation to different user behaviours and strategies is given in (Garyfalos & Almeroth, 2004).

Interestingly enough the idea of distributing advertisements locally without using cellular networks is catching on in the marketing industry. A number of products exists which implement the idea of a stand-alone Information Sprinkler without the supporting ad-hoc communication among users. For instance, *Bloozy*[6] or *BlueBlitz*[7] offer small Bluetooth- and WiFi-enabled appliances.[8]

CASE STUDY: ADPASS

This section presents adPASS, a research prototype for mobile advertising in Opportunistic Networks. For more details see Straub and Heinemann (2004); Heinemann (2007).

System Overview

In the case of adPASS, advertisements are spread and collected according to user preferences. So called *iHave* and *iWish* lists are used as a data structure. While the iHave list keeps advertise-

Figure 1. Incentive scheme – basic idea

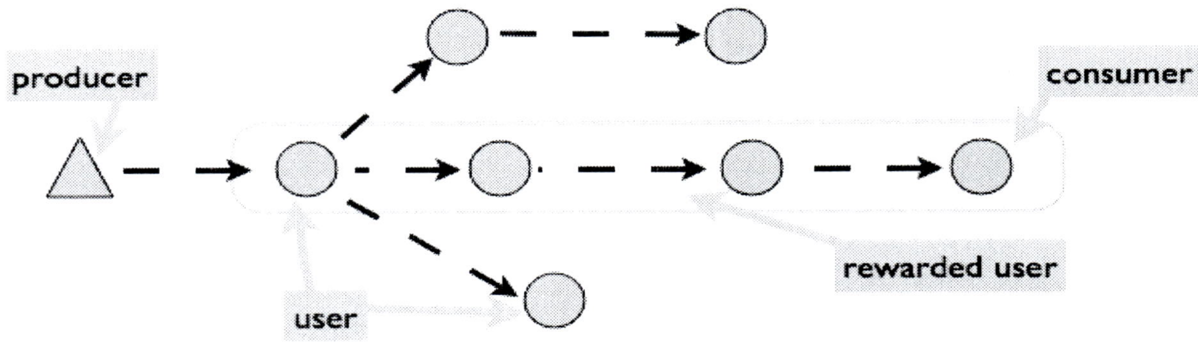

ments the device is willing to share, the iWish lists consists of logical filter expressions that describe desired advertisements. When two devices come into communication range, they exchange their profiles to see if there is a matching between one user's iHave list and the other user's iWish list and vice versa.

Advertisements enter the system via Information Sprinklers mounted at dedicated places, e.g. within a shopping mall or at other public places with high frequency of pedestrians. adPASS rewards users who partly help to carry an advertisement from a producer to a consumer. By carrying around their PDAs, users also transport the advertisements stored on it from one place to another.

Figure 1 illustrates the basic idea of our incentive scheme. A producer passes advertisements to nodes in communication range. The advertisement is passed on from node to node until it reaches a consumer who buys the advertised product. All nodes framed in the box are rewarded by the producer (or consumer), since they helped bringing them together. The set of nodes framed in the box is called *bearer chain*.

We will now distinguish a number of different roles for Opportunistic Network nodes as used in adPASS. The definitions use the more general term *information item* instead of *advertisement* in order to make them applicable to other word-of-mouth recommendation scenarios as well.

Roles. An OppNet node is called *information producer*, if it is the source of an information item and has vital interest in disseminating that information item to a (previously unknown) number of nodes. A node is called *information consumer*, if it is the sink of an information item that was issued by an information producer. The information consumer uses the information item for a personal benefit. A node is called *information bearer*, if it helps to transport an information item from an information producer to an information consumer. As an incentive, a reward is granted by the information producer. We use the term *bonus points* to describe the "currency" of such rewards. Obviously, a mobile node may act as an information bearer and an information consumer alternately or simultaneously.

In order to facilitate bonus point payout, adPASS has a special central component called *mediator*. The mediator acts as a *trusted third party* in the role of an anonymizing proxy between information bearers and an information producer. In this sense, trusted refers to the assumption that the mediator and information producer do not collaborate by exchanging information that would otherwise compromise user privacy. The mediator serves a second purpose as a repository which keeps records of bonus points gained by participating nodes.

An information producer informs the mediator about the users who should be rewarded by bonus

Figure 2. Incentive scheme communication pattern

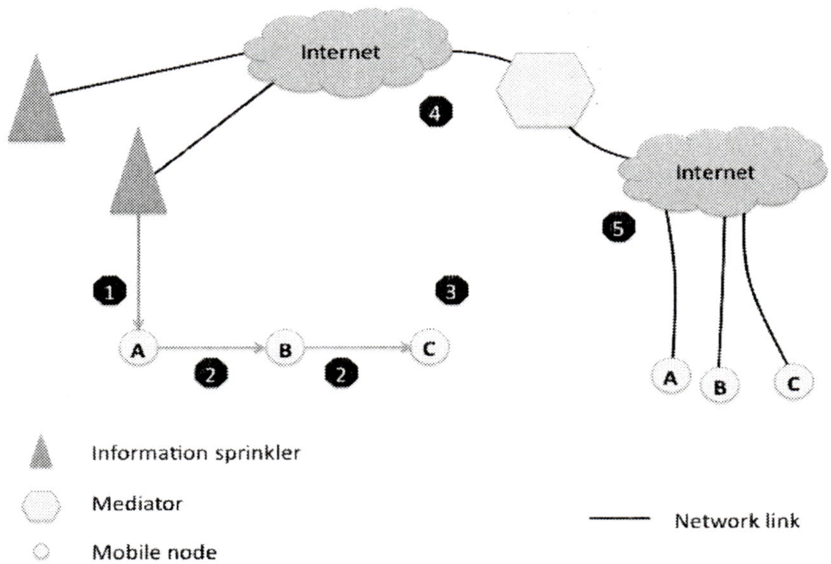

points. An information bearer queries the mediator about the amount of bonus points he gained while participating in the incentive scheme. One might also think of the mediator as a bank which manages the money deposited by information producers in the information bearers' accounts.

Communication Pattern

Figure 2 shows a sample communication of the proposed incentive scheme. It shows two Information Sprinklers that act as information producers. They are connected to a mediator, for example, using an Internet connection. Furthermore, there are three mobile nodes; two of which act as information bearers (*A* and *B*) and one in the role of an information consumer (*C*). The distinct communication steps, labeled with numbers, are:

1 A mobile node *A*, while being in communication range to an Information Sprinkler, learns about an information item. As a prerequisite, it must successfully match against *A*'s iWish-list.
2. Node *A* passes the information item on to *B* and user *B* itself on to *C*.
3. Node *C* takes some action upon receiving the advertisement. This step includes a bearer chain submission (see below) to the information producer. In the adPASS application, node *C* would visit a shop and buy the advertised product and tell the shop owner that the advertisement was received through user *A* and *B*.
4. From the bearer chain, bonus point information is extracted and submitted to the mediator.
5. At a later time, information bearers *A* and *B* query the mediator about gained bonus points using a standard Internet connection.

The illustration in Figure 2 simplifies the setup in two ways. Firstly, Information Sprinklers would be connected to a central hub. Such a server would provide the connection to the mediator as well as manage the items for dissemination via the sprinklers. Secondly, the communication link between adPASS participants and the mediator is not further specified, as there are several possibilities. The connection to the mediator's web site may be established via the cellular network, an Information Sprinkler offering free Internet Access or via a PC at the user's home to which the mobile device is attached to.

Bonus Point Model

An information producer rewards information bearers in the form of bonus points that can be converted to real money or used as a means of payment at a later time. For this, an information producer assigns a maximum number of bonus points to dispense in case an advertisement reaches an information consumer. An intermediate node, i.e., an information bearer, is allowed to claim a certain share of *virtual* bonus points. These virtual bonus points will become *real* bonus points if and only if the advertisement reaches an information consumer and the consumer purchases the advertised product. The distinction between real and virtual bonus points is because of the following: It is a desirable property that advertisements could be duplicated freely in order to propagate them exponentially among participants. However it must not be possible to copy or forge rewards as this would ruin information producers.

Information producers define the total number of bonus point they are willing to grant in case a purchase is made before submitting advertisements to the network. Let the integer $t > 0$ denote the total number of bonus points and let $b_i, i \in I$ denote the participating nodes where I is an index set.

A bearer chain can be expressed by an ordered set $(b_{i_0}, b_{i_1}, \ldots, b_{i_n})$. By v_{i_k} we denote the number of virtual bonus points participant b_{i_k} claims after having received the advertisement from $b_{i_{k-1}}$ and before passing it on to $b_{i_{k+1}}$. The following conditions have to be satisfied:

- All v_{i_k} are positive integers for $0 \le k < n$ and $v_{i_n} \ge 0$.[9]
- The sum $\sum_{k=0}^{n} v_{i_k}$ does not exceed t.

As the last participant in the chain is an information consumer, we can assume that v_{i_n} is maximal, i.e. equality $\sum_{k=0}^{n} v_{i_k} = t$ holds. Furthermore we note that t is an upper bound on the number n of hops. This implies that a copy of an advertisement is removed from the OppNet when there are no more bonus points remaining. This is a mechanism to avoid network traffic quiet similar to the TTL (time to live) parameter in IP packages. The information about the bearer chain as well as the values v_{i_k} are stored in a data structure linked to the advertisement and updated by each participant (see below for details of the chain log).

Information bearers may not only choose their claims at their own discretion, but also individually state criteria whether to accept an advertisement for further dissemination or not. For instance, b_{i_m} may define a lower bound on the absolute or relative number of remaining bonus points, i.e. $r = t - \sum_{k=0}^{m-1} v_{i_k}$ and r / t respectively.

Viejo, Sebé, & Domingo-Ferrer (2007) propose to modify adPASS in a way such that each participant's share of bonus points is computed according to the time the user was storing the advertisement on her device before passing it along. This requires that the chain log be augmented by timestamps. However, a drawback of this proposal is the inverse incentive it creates as participants may chose to keep information to themselves for some time instead of passing it on immediately.

Observe that, although this seems to be the fairest and simplest solution, it is not possible to just reward each participant in the chain with the same number of points. This is due to the desired property of anonymity which in turn allows user to act under several aliases in order to increase their share.

Security Considerations

In this section we will look at the security requirements of adPASS. Security issues are typically examined in terms of *security objectives* and *threats* toward them. This is the first part of our discussion. In the second part, we focus on mechanisms to achieve the mentioned goals (a list of techniques and the corresponding goals are summarized in Table 2).

Security Objectives. The following list enumerates key objectives of IT security and the relevance in our setting. Observe that here we do not consider the objective of *availability*. Additionally we discuss *confidentiality* only in the form of anonymity. As in our setting information should be disseminated freely, there is no need to conceal the content of messages, but merely the source and destination in order to guarantee privacy.

- **Integrity:** *Assure that information cannot be modified without being detected.* Information in the advertisement itself (like price and features of an advertised product) should be kept safe from manipulation. The integrity of the bearer chain, that holds information about bonus point claims, needs to be ensured as well.
- **Data Authentication:** *Ensure that a piece of information originates from the alleged producer.* For all adPASS nodes taking part in the incentive scheme, authentication of an advertisement provides the nodes with assurance that the advertisement was issued by the claimed information producer and not forged. This prevents nodes from sharing their personal resources in vain, since the information producer commits herself to issue a reward .
- **Non-Repudiation**: *Prevent parties from denying having taken some action.* This security objective is relevant to prevent information producers from denying that they have issued certain advertisements with a special price.
- **Anonymity**: *Ensure that an entity remains unidentifiable within a set of parties.* Information bearers should be able to take part in the incentive scheme without revealing their identities to each other. This would hold off an attacker to create user profiles of personal shopping preferences and thus protects user privacy. Also, an information producer should not be able to learn the identities of information bearers by analyzing the bearer chain.
- **Digital Signatures.** The purpose of a digital signature algorithm is to append data (the "payload") with additional information (the "signature") that allows ensuring the integrity and authenticity of the former. Each unauthorized change of the payload and/or the signature can be detected. For this, a signer creates a cryptographic key pair consisting of two bit strings that have to satisfy particular mathematical conditions depending on the algorithm in use. By k^{-1} we denote the private key, which is then used for signing, and by k the corresponding public key, which is used for verification. It is computationally infeasible to calculate k^{-1} when only k is known. This implies that the abilities to verify and sign can be clearly separated from each other.
- **Public Key Certificates.** As key pairs can be generated by everyone, it is necessary for most applications to attribute them to their owner (consider e.g. the case where a public key is used to ensure the authenticity of a web site using the SSL/TLS protocol). This link is provided by a certificate, i.e. a digital statement containing – among others – the public key, a description of its owner (name, e-mail address etc.) and legal information. Certificates are issued by trustworthy authorities. A special case

Table 2. Overview of protection goals and techniques

Goal	Technique
Integrity	Digital signature operation
Data Authentication	Certificates
Non-Repudiation	Qualified signatures and certificates
Anonymity	Multiple key pairs as aliases

of certificates are (according to EU legislation) called *qualified certificates* as they provide strong evidence about the identity of the key owner. Signatures that are created with a private key corresponding to a qualified certificate are the legal equivalent to a handwritten signature.

- **Using Public Keys as Aliases.** Interestingly enough adPASS can nearly do without certificates. Only information producers are required to use certificates to identify themselves and to guarantee that a reward will be paid. In order to provide strong assurance for the information bearers we suggest that qualified signatures be used. Note that information bearers do not need to be provided with a certificate. Instead a public key *k* itself can be used as an alias. In order for the system to work we do not require participants to identify themselves, but to simply prove that they know the corresponding k^{-1}. As a consequence participants can create new aliases as often as they want thus making actions associated with single keys unlinkable.

Implementation

Having explained the security foundations we can now sketch the actual implementation. For the sake of brevity we introduce the following assumptions and notations: As we have seen, an information bearer can be represented by her public key, which we therefore also denote by b_{i_k}. We only look at a single information producer identified by its public key *p* and a single advertisement denoted by *ad*. Advertisements are kept in a data structure that consists of at least a unique identifier called *ad.id* and the description of the product. To indicate that certain payload *data* is being signed with a private key k^{-1} we write $\{data\}_{k^{-1}}$. In case the payload consists of several items, they are written as a comma-separated list.

Let us now look at how the steps in an adPASS communication are implemented.

1. At first, the information producer creates a *signed advertisement object*

$$\{ad, cert(p)\}_{p^{-1}}$$

   to be distributed via Information Sprinklers. The object contains the advertisement as well as a certificate for the information producer's public key.
2. Upon receiving an advertisement from an Information Sprinkler, a mobile node is able to check …
   a) who issued *ad* by retrieving the appropriate information from *cert(p)* and that this issuer is legitimate,
   b) that *ad* was not modified during the dissemination process. (For example, the price of an advertised product has not been altered.)

Both conditions hold as long as signature verification succeeds. It follows from a) that authentication and non-repudiation are achieved and b) implies advertisement integrity.

3. In order to keep track of the bearer chain, some information has to be added to express that a certain advertisement has been copied from the sprinkler to a certain mobile node. This is expressed by a *transfer receipt*

$$\{ad.id, p, b_{i_0}, t\}_{p^{-1}}$$

By inspecting the content and verifying its signature, the information bearer can assure herself that it refers to the advertisement in question and the appropriate number of virtual bonus points is transferred to her.[10]

4. The process of passing on the advertisement from one information bearer to another, say from b_{im-1} to b_{im}, is quite similar to step 3 yielding the following sequence of transfer receipts:

$$\{ad.id, p, b_{i_0}, t\}_{p^{-1}}$$
$$\{ad.id, b_{i_0}, b_{i_1}, t - v_{i_0}\}_{b_{i_0}^{-1}}$$
$$\vdots$$
$$\{ad.id, b_{i_{m-1}}, b_{i_m}, t - \sum_{k=0}^{m-1} v_k\}_{b_{i_{m-1}}^{-1}}$$

Observe that the number of remaining bonus points (the fourth parameter in each line) is strictly monotonically decreasing. Node b_{i_m} is able to check the integrity of the bearer chain step-by-step starting at the topmost entry.

5. Now assume that b_{i_n} wants to purchases the product described in ad for the indicated price. (If an information producer repudiates having disseminated a certain ad and is reluctant to sell the product, b_{i_n} may take legal actions using the qualified signature as strong evidence.) From the bearer chain the pairs of aliases and their corresponding bonus point share, i.e. (b_{i_k}, v_{i_k}), are extracted and send to the mediator. Note that there is no reference to the ad anymore thus providing a high protection against user profiling.

6. Finally, information bearers can turn the bonus points stored at the mediator's site into monetary benefit. Either by having them converted into real money (at an "adPASS-enabled cash machine") or electronic coins. Both alternatives can be implemented in a privacy-friendly way. In order to disguise their identity (namely the IP address) when querying whether there are bonus points available for them, common techniques for online privacy can be used, e.g. *Tor*[11] or *JonDonym*[12].

Simulation

In order to illustrate the effectiveness of advertisement dissemination we present some experimental results. It was the purpose of the simulation to provide figures of how many individual advertisement wishes could be fulfilled within a given time with respect to mobility model and communication range.

Overview. The simulator works in two steps. It combines realistic user mobility data with synthetic mobility models. In the first step, real-world data from the *Reality Mining Project* (see Massachusetts Institute of Technology, 2005) is used in order to get realistic information about the *macro-level* mobility of users. In the second step, synthetic mobility models to simulate *micro-level* user movement are applied. This approach avoids the short-comings of purely synthetic mobility models (see Jardosh, Belding-Royer, Almeroth, and Suri (2003); Bai and Helmy (2006) for a detailed discussion). This combination of coarse-grained user traces with synthetic mobility models on the micro-level allows exploring various communication ranges and user movements while simulations based on pure user traces are bound to the particular technology (traces collected from Bluetooth devices are not suitable to investigate WiFi-like communication behaviour and vice versa).

Macro-Level Mobility. The simulator operates as follows. Its core component extracts a list of user whereabouts from the Reality Mining dataset. Each entry consists of a unique location ID and a list of unique user IDs to determine which users were at a certain time at that location. From this, the *macro level* distance of users can be calculated. The underlying dataset captures communication, proximity, location, and activity information originating from 100 users in Cambridge, Massachusetts (United Sates of America) observed during a 16 month period. Each user was given a mobile phone that runs special software. This software logged communication behaviour of the user as well as location information it learned from its surroundings. We use the cellular network tower ID to which a mobile phone was connected to at a certain time as an approximation to the user's location.

Micro-Level Mobility. When two or more users are co-located according to the macro-level model from step one, the simulator creates an environment of 1000 × 1000 meters for simulating the *micro-level* mobility. Initially, users are placed in the square randomly and then we let them move around for 5 minutes according to the configured micro-level mobility model. If users come in communication range with each other, the simulator executes an advertisement matching algorithm between them. On a match, advertisements between users are exchanged. For the micro-level mobility generation, the *BonnMotion* (University of Bonn, 2002-2005) software package is used which supports different mobility models. Due to space restrictions here we only state the results in the Random Waypoint mobility model (see Heinemann, Kangasharju, and Mühlhäuser (2008) for a discussion of other models).

Results. The simulation results, which are described in the following, were gained from calendar week 45 of the Reality Mining dataset using 10 meters ("Bluetooth") and 100 meters ("WiFi") as a user device's communication range. One Information Sprinkler is located in each cell. In every run of the experiment, a randomly chosen node owns an advertisement at the start and all other nodes are interested in it.

Figure 3. Advertisement dissemination among users over time

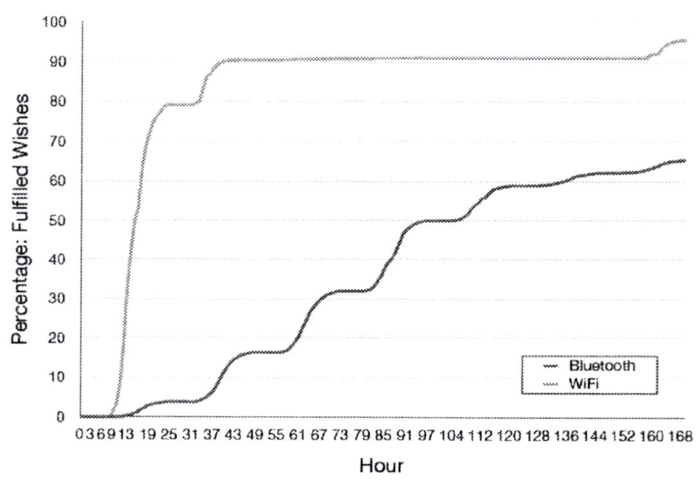

Figure 3 compares results for the Bluetooth and WiFi scenarios. The x-axis represents a whole week starting at Sunday midnight while the percentage of users who received the information is shown on the y-axis. On the micro-level it was modelled that, while they are sleeping, users have their devices switched off.[13] This is clearly recognizable at the Bluetooth curve as it rises during daytimes and remains constant from midnight to 6am. In the WiFi setting mostly all exchange happens within the first two days with an additional increase at the end of the week. Due to the higher communication range, WiFi outperforms Bluetooth by 96% of users reached on average compared to 65% in the latter case.

Figure 4 provides detailed figures of how many hops the information item took to reach users in the Bluetooth scenario (the bars add up to the 65% of users who knew the item at the end of the week). Remarkably most of the users receive the item with relatively small number of hops (about 3–5), but there are also users who receive the item after more than 10 hops.

Coming back to the initial question, i.e. the dissemination coverage in adPASS, we learn that besides an incentive scheme the communication range plays an important role. Thus, a system designer carefully needs to decide on the provided device communication range in order to make a commercial system a success.

SUMMARY

This chapter explained a new form of mobile advertisement based on Opportunistic Networks that puts forward *word-of-mouth* propagation of advertisements to mobile devices. By now the reader should have learnt about major challenges in this area. We have reviewed state of the art research, before we covered one scientific prototype adPASS in greater detail. The adPASS system combines anonymous user action with a simple incentive scheme that stimulates users to participate in this kind of networks. Opportunistic Networks give rise to a number of immanent challenges with respect to business models, usability, computing and storage resource as well as security. adPASS serves as an example of how these challenges can be solved in a coherent way using an adequate combination of technical and non-technical measures.

A selection of simulation results illustrates the effectiveness of adPASS concerning the incentive-

Figure 4. Portion of users reached with a certain hop count (Bluetooth scenario)

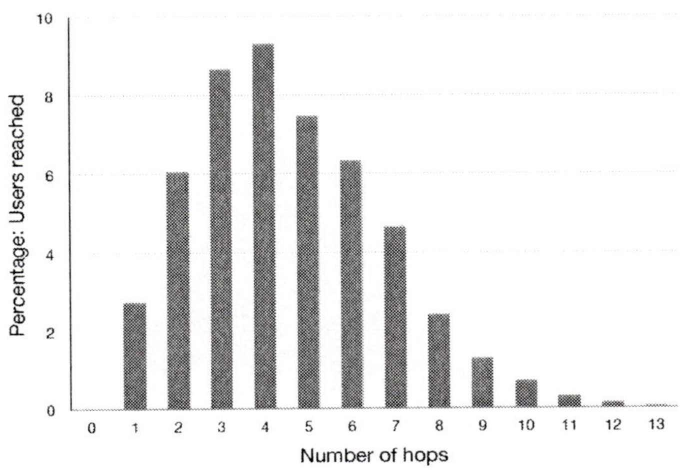

driven dissemination of information. The simulations and a number of realized prototypes (see Heinemann, 2008) clearly show the feasibility of our approach. Current of-the-shelf devices are powerful enough for word-of-mouth propagation facilitating mobile marketing.

For a further treatment on incentive schemes and modelling see the work of Crowcroft, Gibbens, Kelly, and Östring (2004), Garyfalos and Almeroth (2004), Golle, Leyton-Brown, Mironov, and Lilibridge (2001), and Viejo et al. (2007). Further aspects on mobile advertisement in general are covered in (Bulander et al., 2005) and (Decker, Bulander, Högler, & Schiefer, 2006).

Starting from our findings, there are several open questions for future research that we will briefly mention. First, in order to better understand the efficiency of the adPASS communication scheme, more simulations with different settings need to be carried out. The CRAWDAD Project (Dartmouth College, 2005) might be a good starting point to extract user traces for additional simulation runs and to possibly combine them with user field tests in a shopping mall, café, or bookstore. Second, the minimal resource requirements for adPASS-enabled mobile devices is not fully investigated in terms of memory consumption, energy efficiency, processor power, and user interface. With the rapid development of mobile devices, processor power and memory might be neglected, but energy efficiency and a suitable user interface surely not. A final remark since we could not address business implications in depth here: Who are the drivers for adPASS like systems? What are the market opportunities? What are suitable incentives (money, vouchers, free services, etc.) for adPASS and similar systems? These questions should be addressed in cooperation with telecommunication companies, device manufactures, software companies, retailers, and customers.

REFERENCES

Arndt, J. (1967). Role of product-related conversations in the diffusion of a new product. *JMR, Journal of Marketing Research, 4*(3), 291–295. doi:10.2307/3149462

Bai, F., & Helmy, A. (2006). Wireless ad-hoc and sensor networks. In (chap. A Survey of Mobility Models). Springer.

Becker, C., Bauer, M., & Hähner, J. (2002, November 16-20). Usenet-on-the-fly - supporting locality of information in spontaneous networking environments. *Liscano, Ramiro (ed.); Kortuem, Gerd (ed.): Workshop on Ad Hoc Communications and Collaboration in Ubiquitous Computing Environments.*

Bone, P. F. (1995). Word-of-mouth effects on short-term and long-term product judgments. *Journal of Business Research, 32*(3), 213–223. doi:10.1016/0148-2963(94)00047-I

Bulander, R., Decker, M., Schiefer, G., & Kolmel, B. (2005). Comparison of Different Approaches for Mobile Advertising. Mobile Commerce and Services, 2005. WMCS '05. The Second IEEE International Workshop on, 174–182.

Crowcroft, J., Gibbens, R. J., Kelly, F. P., & Östring, S. (2004). Modeling incentives for collaboration in mobile ad hoc networks. *Performance Evaluation - Selected Papers from the First Workshop on Modeling and Optimization in Mobile, Ad Hoc and Wireless Networks (WiOpt'2003), 57*(4), 427–439.

Dartmouth College. (2005). CRAWDAD Project: A Community Resource for Archiving Wireless Data At Dartmouth. http://crawdad.cs.dartmouth.edu/ (seen 11/2008)

Datta, A. (2003). MobiGrid: Peer-to-Peer Overlay and Mobile Ad-Hoc Network Rendezvous – a Data Management Perspective. In *CAiSE 2003 Doctoral Symposium, in conjunction with the 15th Conference On Advanced Information Systems Engineering*. Springer.

Datta, P., Chowdhury, D., & Chakraborty, B. (2005). Viral marketing: New form of word-of-mouth through internet. *Business Review (Federal Reserve Bank of Philadelphia)*, *3*(2), 69–76.

DeBruyn, A., & Lilien, G. (2004). A multi-stage model of word of mouth through electronic referrals. *eBRC Research Paper Series*.

Decker, M., Bulander, R., Högler, T., & Schiefer, G. (2006). m-Advertising: Werbung mit mobilen Endgeräten - ein Überblick. In T. Kirste, B. König-Ries, K. Pousttchi, & K. Turowski (Eds.), *Mms* (Vol. 76, p. 103–114). GI.

Ding, G., & Bhargava, B. (2004). Peer-to-peer File-sharing over Mobile Ad hoc Networks. In *2nd IEEE International Conference on Pervasive Computing and Communications – Workshop on Mobile Peer-to-Peer Computing*. IEEE Computer Society.

Garyfalos, A., & Almeroth, K. C. (2004). Coupon Based Incentive Systems and the Implications of Equilibrium Theory. In *IEEE International Conference on E-Commerce Technology. Proceedings* (pp. 213–220). IEEE Computer Society.

Gewei, Y. (2007). Mobile marketing systems: framework and technology enabler. *International Journal of Mobile Marketing*, *2*(1), 43–49.

Goel, S. K., Singh, M., & Xu, D. (2002). Efficient Peer-to-Peer Data Dissemination in Mobile Ad-Hoc Networks. In *International conference on parallel processing workshops* (pp. 152–158). IEEE Computer Society.

Golle, P., Leyton-Brown, K., Mironov, I., & Lillibridge, M. (2001). Incentives for Sharing in Peer-to-Peer Networks. In *Electronic commerce: Second international workshop, (welcom)* (Vol. 2232, pp. 75–87).

Hayes, A., & Wilson, D. (2005). Peer-to-Peer Information Sharing in a Mobile Ad Hoc Environment. In *Sixth ieee workshop on mobile computing systems and applications (wmcsa '04)* (p. 154–162). IEEE Computer Society.

Heinemann, A. (2007). *Collaboration in Opportunistic Networks, PhD Thesis*. VDM Verlag, Saarbrücken.

Heinemann, A. (2008). Opportunistic Networks. In M. Mühlhäuser & I. Gurevych (Eds.), *Ubiquitous computing technology for real time enterprises* (chap. 9). Information Science Reference.

Heinemann, A., Kangasharju, J., & Mühlhäuser, M. (2008). Opportunistic Data Dissemination Using Real-World User Mobility Traces. In *1st IEEE Intl. Workshop on Opportunistic Networking (WON-08)*.

Helm, S. (2000). Viral marketing – establishing consumer relationships by 'word-of mouse'. *Electronic Markets*, *10*(3), 158–161. doi:10.1080/10196780050177053

Hennig-Thurau, T., Gwinner, K. P., Walsh, G., & Gremler, D. D. (2004). Electronic word-of-mouth via consumer-opinion platforms: What motivates consumers to articulate themselves on the internet? *Journal of Interactive Marketing*, *18*(1), 38–52. doi:10.1002/dir.10073

Herr, P., Kardes, F., & Kim, J. (1991). Effects of word-of-mouth and product-attribute information on persuasion: An accessibility-diagnosticity perspective. *Journal of Consumer Research: An Interdisciplinary Quarterly*, *17*(4), 454–462.

Hu, Y. C., Das, S. M., & Pucha, H. (2003). Exploiting the synergy between peer-to-peer and mobile ad hoc networks. In *Proceedings of HotOS'03: 9th Workshop on Hot Topics in Operating Systems* (p. 37–42). USENIX.

Jardosh, A., Belding-Royer, E. M., Almeroth, K. C., & Suri, S. (2003). Towards Realistic Mobility Models For Mobile Ad Hoc Networks. In *Mobicom '03: Proceedings of the 9th annual international conference on mobile computing and networking* (pp. 217–229). ACM Press.

Klemm, A., Lindemann, C., & Waldhorst, O. P. (2004). Peer-to-Peer Computing in Mobile Ad Hoc Networks. In *Performance tools and applications to networked systems: Revised tutorial lectures* (Vol. 2965, pp. 187–208). Springer.

Kurkovsky, S., & Harihar, K. (2006). Using Ubiquitous Computing in Interactive Mobile Marketing. *Personal and Ubiquitous Computing*, *10*(4), 227–240. doi:10.1007/s00779-005-0044-5

Leguay, J., Lindgren, A., Scott, J., Friedman, T., & Crowcroft, J. (2006). Opportunistic Content Distribution in an Urban Setting. In *CHANTS '06: 2006 SIGCOMM Workshop on Challenged Networks* (pp. 205–212).

Lindemann, C., & Waldhorst, O. P. (2005). Epidemic Data Dissemination for Mobile Peer-to-Peer Lookup Services. In *Peer-to-Peer Systems and Applications* (Vol. 3485, p. 435–455). Springer.

Massachusetts Institute of Technology. (2005). *Reality Mining*. http://reality.media.mit.edu/ (seen 8/2008).

Pelusi, L., Passarella, A., & Conti, M. (2006). Opportunistic Networking: Data Forwarding in Disconnected Mobile Ad Hoc Networks. *IEEE Comm.*, *44*(11), 134–141. doi:10.1109/MCOM.2006.248176

Pousttchi, K., & Wiedemann, D. G. (2006). A Contribution to Theory Building for Mobile Marketing: Categorizing Mobile Marketing Campaigns through Case Study Research. *Proceedings of the International Conference on Mobile Business, Copenhagen, Denmark* (pp. 1–8).

Ratsimor, O., Finin, T., Joshi, A., & Yesha, Y. (2003). eNcentive: A Framework for Intelligent Marketing in Mobile Peer-To-Peer Environments. In *ICEC '03: Proceedings of the 5th international conference on Electronic commerce* (pp. 87–94). ACM Press.

Rudström, Å., Svensson, M., Cöster, R., & Höök, K. (2004). MobiTip: Using Bluetooth as a Mediator of Social Context. In *UbiComp 2004: Ubiquitous Computing: 6th International Conference, Adjunct Proceedings (demo).*

Straub, T., & Heinemann, A. (2004). An Anonymous Bonus Point System For Mobile Commerce Based On Word-Of-Mouth Recommendation. In *Proceedings of the 2004 ACM Symposium on Applied Computing* (pp. 766–773). New York, NY, USA: ACM Press.

University of Bonn. (2002-2005). *BonnMotion – A Mobility Scenario Generation and Analysis Tool.* http://web.informatik.uni-bonn.de/IV/Mitarbeiter/dewaal/BonnMotion/index.html

Viejo, A., Sebé, F., & Domingo-Ferrer, J. (2007). Secure and Private Incentive-Based Advertisement Dissemination in Mobile Ad Hoc Networks. In A. Miyaji, H. Kikuchi, & K. Rannenberg (Eds.), *Iwsec* (Vol. 4752, p. 185–198). Springer.

Wiedemann, D. G., Haunstetter, T., & Pousttchi, K. (2008). Analyzing the basic elements of mobile viral marketing-an empirical study. In *Icmb '08: Proceedings of the 2008 7th international conference on mobile business* (pp. 75–85). Washington, DC, USA: IEEE Computer Society.

KEY TERMS AND DEFINITIONS

adPASS: adPASS (Straub & Heinemann, 2004) is a system for spreading digital advertisements (ads) among interested users using mobile devices. Each user specifies his interests in a profile that is stored on the device. The communication scheme resembles the way information is spread by word of mouth between human beings, e.g., when recommending something to someone else.

Incentive Scheme: A rewarding mechanism or model in order to stimulate user participation in a certain system.

Mobile Marketing: Mobile marketing is a form of marketing communication using mobile communication techniques to promote goods, services, and ideas.

One-Hop Communication: Wireless and short range message exchange between adjacent nodes with no intermediate node on the communication path, i.e. no message routing.

Opportunistic Networks: An opportunistic network is a network of wirelessly connected nodes. Communication range between two connected nodes is not further than walking distance. Nodes are connected only temporarily and the network topology may change due to node mobility or node activation and node deactivation respectively. The network provides at least the following functionalities: *Node discovery.* A network node is able to discover other network nodes in direct communication range. *One-hop message exchange.* A node is able to send and receive arbitrary data to or from any other node in direct communication range.

Viral Marketing: A form of interpersonal (verbal, written) communication in order to promote goods, services and ideas.

Word-of-Mouth: A form of interpersonal verbal communication in which none of the participants are marketing sources.

ENDNOTES

1. http://www.gartner.com/it/page.jsp?id=612207
2. See Section *System Overview* for a summary of the adPASS prototype
3. Short Message Service, Multimedia Messaging Service
4. Wireless Application Protocol, General Packet Radio Service, Universal Mobile Telecommunications System
5. see e.g. http://www.mobile-marketing-system.de/preise/
6. http://www.bloozy.co.uk/bluetooth-marketing.php#Ad-Point
7. http://www.blueblitz.com/mbMagicBeamer.htm
8. A comprehensive list can be found at http://www.advatex.de/BluetoothMarketingFirmenListe.html.
9. In order to model altruistic behaviour as well, one might allow $v_{i_k} = 0$ for some k.
10. To keep things simple, we assumed that the same key pair is used for signing the advertisement and the transfer receipt. This is unrealistic as a key pair associated with a qualified certificate has to be guarded very carefully. For instance, an information producer may create individual key pairs for each sprinkler and equip them with a certificate including a reference to the producer.
11. http://tor.eff.org
12. http://www.jondos.de
13. This is an additional, conservative assumption since the participants in the Reality Mining Dataset did not turn of their devices. Without this restriction the simulation would have indicated information exchange also at night.

Section 3
Research Issues in Mobile Marketing

Chapter 14
Theories behind Mobile Marketing Research

Ramin Vatanparast
Nokia Interactive Advertising, USA

ABSTRACT

The future of mobile communication is expected to rely on mobile services and revenue generated through mobile marketing. Marketing activities supported by mobile devices allow companies to directly communicate with their consumers without location or time barriers. It is becoming vital for today's marketers to understand the processes behind the factors affecting consumers' intention to use and adopt mobile marketing. One can often argue that the mobile marketing adoption is difficult to understand due to a lack of relevant research. However, much research has already been conducted on the adoption of mobile services and technology acceptance that likely can support mobile marketing research. Thus it is essential for mobile marketing researchers to get a thorough understanding of the theory behind mobile service adoption and technology acceptance. Theories in this area have been developed gradually and built up on each other. This chapter covers some of the theories and related models and shows how those could be used in mobile services and mobile marketing research.

INTRODUCTION

The application of technology acceptance models to mobile services originates from an organizational context (Venkatesh et al., 2003). Rogers (1983) defines adoption as "a decision to make a full use of an innovation" which, in our case, means actual use of mobile services. The mobile service market is a consumer market that focuses on enjoyment and new possibilities for leisure use (Rogers 1983). The most popular mobile services are simply designed for better user experience and exploit technological advancement that promotes service use (Teo and Pok, 2003). Mobile marketing can use the acceptance models for a wide variety of mobile services. It can enable mass-market reach through widely used mobile services. To understand the factors affecting mobile marketing, it is

DOI: 10.4018/978-1-60566-074-5.ch014

necessary to get a comprehensive understanding of the theories behind technology acceptance and service adoption.

In 1975 Schultz and Slevin carried out an exploratory factor analysis to measure the typical concerns of ordinary management information system users. They identified seven explanatory dimensions. The performance dimension and "perceived effect of the model on the manager's job performance" were found to be the most highly correlated with the self-predicted use of a decision model. Robey (1979) found that several specific user attitudes or perceptions are positively related to system use. According to his model, "user perceptions or attitudes are formed concerning the value of rewards received from performance, the likelihood that rewards result from performance, and the likelihood that performance results from the use".

The majority of technology acceptance studies are rooted in behavioral intention. The studies contend that a user's choice to adopt a new technology can be sufficiently explained and predicted by their behavioral intention. The challenge is to classify which factors determine an individual's intention to adopt a new technology or service. Venkatesh et al (2003) explains that theoretical models on user acceptance of information technology such as mobile marketing, utilize "intention to use" and "actual use" as the main dependant variables. Research has established that the intention to use information technology is a predictor of behavior (e.g. Sheppard et al. 1988; Ajzen 1991).

Understanding the drivers of adoption can give an idea to service providers on how to shape, tailor or better customize particular services to deliver superior customer experience. Understanding of motives for adoption of technology or services for customer can augment the customer's awareness of his own motives for the use of mobile services, which helps choosing better and more intentionally technology or services. This chapter presents a set of technology acceptance and service adoption models that have been incrementally developed on each other and used in many previous mobile services and mobile marketing research work. It has been suggested that including moderating variables on intention to use mobile services will contribute to important advances in marketing theories (Dabholkar and Bagozzi 2002). In summary, this chapter is organized to first give a background for theories behind mobile services adoption and technology acceptance. The next section provides a review on some recent mobile marketing researches and also application of mobile services adoption and technology acceptance theories in the field. The chapter is closed by discussion and conclusion section and also some proposal for future work.

BACKGROUND

Fishbein's Multi-Attribute Model

The Fishbein multi-attribute model (Ajzen and Fishbein 1980) is a complex and influential conceptualization of the relationships among consumer's belief, attitudes, intentions and behavior. This model recommends that a person's attitude to any service is based on an individual's belief about the service and the evaluation of the service. Fishbein's support for this comes from behavioral learning theory (Kassarjian and Robertson 1991) that says "an attitude toward an object is more or less automatically learned as one learns about a new product, and that learning occurs in the form of beliefs about product attributes." In general, this means that if a person expects a positive outcome from a service, he will have a favorable attitude towards it and if the person expects a negative outcome from a service, he will maintain a negative attitude towards it (Harrison et al 1997). Based on this model, attitude towards a service can change when either evaluative aspect or strength of belief associated with an attribute or consequence changes, or when attributes are added or removed.

The Fishbein model provides a convincing approach to understand how an overall evaluation of a service is derived from a consumer's beliefs. The Fishbein model allows for differences among consumers in the array of service attributes that will be salient or relevant to the overall product evaluation. Additionally, two different consumer's beliefs about the same service may be identical or different. The Fishbein model allows for differences among consumers in strength of beliefs and can readily explain these differences in service evaluation. It suggests that the most positive overall evaluation should result from the integration of many strongly held positive beliefs about the service.

Fishbein's multi-attribute model explains the overall attitude toward an offering as a function of two factors: bi as the belief component that expresses the probability that service has a certain attribute i/consequence i, and ei is the evaluation component associated with the attribute i/consequence i. As a result, attitude toward a service = Σbi x ei. This model can be presented in research settings as either a weighted model or a non-weighted model. It is usual to use non-weighted models in empirical work, which only considers the evaluation component. In other words, researchers focus only on consumers' evaluations of whether they like or dislike the attribute. Fishbein's model is extremely helpful to managers and marketers because it provides reasons for consumer preferences, identifies unfulfilled needs, and provides suggestions for new products.

Theory of Reasoned Action (TRA)

Theory of Reasoned Action (Fishbein and Ajzen 1975) is a tool used to achieve deeper insights into how attitudes and beliefs are correlated with individual intentions to perform. TRA is a refinement of Fishbein's multi-attribute model that clarifies the relationship between attitudes and behavior. In the situation where the customer controls his own behavior and when he is thoughtful about it, TRA can predict consumer's behavior. TRA suggests that the behavioral intention of an individual to perform or not perform certain target behavior is solely and directly responsible for influencing that individual's target behavior. TRA proposes that the most significant determinant of a consumer's actual behavior is the intention to perform a behavior which is a function of attitude towards behavior and subjective norms. The two functions are new contributions to Fisbein's previous model (Figure 1).

Attitude towards behavior is defined as "person's attitude towards performing that behavior" (Kassarjian and Robertson 1991). In addition, based on the expectancy value model of attitude (Fishbein and Ajzen, 1975), attitude towards behavior composed of beliefs about the consequence of the behavior and evaluation of the consequence of the behavior. The intention to behave or to use technology would not be only determined by personal attitude towards behavior but will also be influenced by other's opinions about the behavior. Subjective norms are proposed to account for social influences that the person's behavior is exposed to. Subjective norm can be described as an individual's perception of the social pressure to perform the target behavior (Fishbein and Ajzen 1975, Ajzen and Fishbein 1980). In other words, subjective norm is composed of the user's perception of how others thought about user's behavior, and the user's motivation to comply with the expectations of these referents (Fishbein and Ajzen 1975).

TRA has a proven track-record in explaining "virtually any human behavior" (Ajzen and Fishbein 1980), for example, in the world of marketing applications (Ryan and Bonfield 1980). With respect to mobile marketing, permission-based mobile advertising is usually considered to be a major factor that may affect attitudes (Barwise and Strong 2002; Kavassalis et al. 2003). Incentives, such as free connection time, have an impact on consumer intentions to receive mobile advertising under a given attitude which then affects their

Figure 1. Theory of reasoned action (Adapted from Kassarjian & Robertson, 1991)

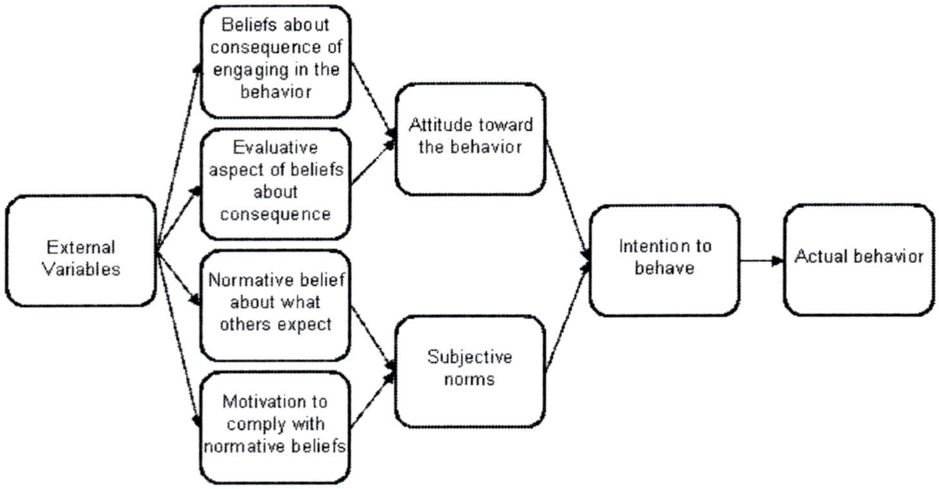

actual advertisement receiving behavior. The non-weighted version of the TRA model is usually used which means it is tested for motivation. The weighted version of the TRA model focuses on the relative strength of the links between attitude and intention and between subjective norms and intention (Kasarjan and Robertson 1991).TRA is one of the most powerful theories for a wide range of human behavior (Sheppard et al 1988; Vekantesh et al 2003).

Theory of Planned Behavior (TPB)

The Theory of Planned Behavior (TPB) broadens TRA to account for conditions where individuals do not have complete control over their behavior (Taylor and Todd 1995). One extension, proposed by Ajzen (1991, is in the form of perceived behavioral control. The inclusion of a third determinant of behavioral intention, perceived behavioral control, is TPB's major point of departure from TRA. Ajzen (1991) shows that perceived behavioral control, attitudes and subjective norms are all positively related to the intentions about the behavior which predicts the actual behavior of a consumer (Figure 2).

Ajzen (1991) defines perceived behavioral control as "the perceived ease or difficulty of performing the behavior". Taylor and Todd (1995) revised the definition of perceived behavioral control in a context of information system research as "perceptions of internal and external constraints on behavior." Perceived behavioral control is further decomposed into control belief "beliefs about resources and opportunities" and perceived power "power to facilitate or obstruct the performance of the behavior" (Ajzen 1988). In the context of system usage, perceived behavioral control relates to the degree to which an individual believes that he has control over personal or external factors that may facilitate or constrain system use (Venkatesh and Brown 2001). For mobile services, attitude towards behavior can then be described as an individual's favorable or unfavorable evaluation of using a specific service while subjective norm can be seen as the perceived social pressure to use or not to use said service.

According to TPB, the behavior can be influenced by factors other than just the personal intention to perform the behavior and subjective norms, "other people's opinion about the behavior", as it was mentioned in TRA model. These

Figure 2. Theory of planned behavior (Adapted from Ajzen, 1991)

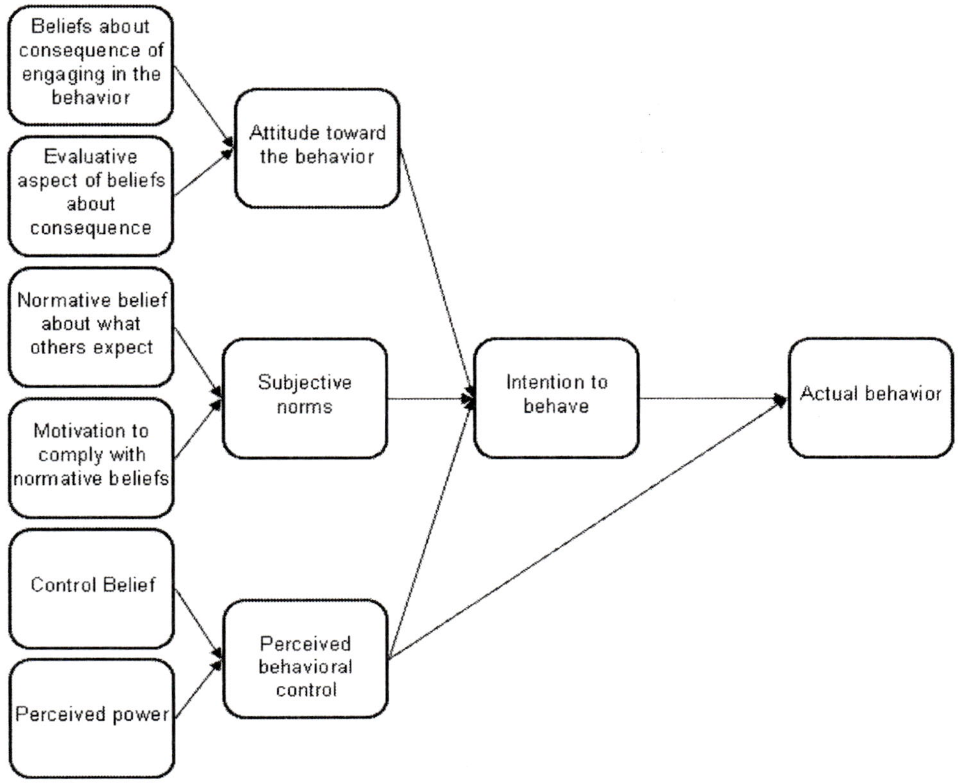

other factors refer to the amount of vital resources and opportunities, such as money, time, skills and cooperation of others (Harrison et al 1997). Hence, it is the intention and the accessibility of resources and opportunities that jointly verify whether or not behavior is performed e.g. use and adopt a service. Similar to other models, if perceived control is not weighted, then it means that there is no emphasis on perceived power element. In the most empirical work, perceived control is considered as a non-weighted factor.

Theory of Planned Behavior has been used to predict intention and behavior in a variety of settings (Ajzen 1991) such as technology (Harrison et al 1997), intention to use a specific information system (Mathieson 1991), weight-loss behavior (Schifter 1984) and student's intention and usage (Ajzen and Madden 1986, Taylor and Todd 1995). The TPB model has not been tested sufficiently in empirical studies because of some limitations such as the ambiguity that surrounds the definition of perceived behavioral control. Furthermore, unconscious motives are not taken into consideration as TPB is grounded on the belief that people think rationally and make logical decisions. Lastly, demographic and personality variables are not taken into account in TPB (Godin and Kok 1996).

Technology Acceptance Model (TAM)

Technology Acceptance Model (Davis et al. 1989), was initially designed to predict a user's acceptance of information technology and usage on the job. TAM has become a well-established, robust, powerful and parsimonious model for predicting user acceptance (Venkatesh and Davis 2000) and become the most broadly applied model of user

Figure 3. Technology acceptance model (Adapted from Ma & Liu, 2004)

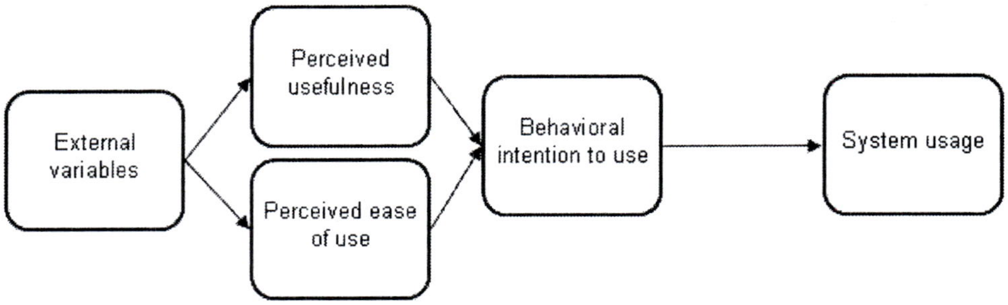

acceptance and usage (Ma and Liu 2004). TAM is grounded in the Theory of Reasoned Action (Fishbein and Ajzen 1975) and Theory of Planned Behavior (Ajzen 1991) and compares favorably with TRA and TPB (Venkates and Davis 2000).

TAM assigns a considerable weight to two key determinants of perceived usefulness: "the degree to which a person believes that using a particular system would enhance his or her job performance" and perceived ease-of-use "the degree to which a person believes that using a particular system would be free of effort" (Davis 1989). These two concepts are fundamental to understanding the core workings of TAM. Perceived usefulness and ease-of-use from user's attitude towards technology leads to the intention to use a technology (Figure 3).

Useful and easy-to-use technologies will have a positive influence on a user's future attitude and intention towards using the technology which leads to increased usage of the technology (Taylor and Todd 1995). Basically, people are more likely to use a system that they believe will help them perform better. But, even if a system is believed to be useful by an individual, if the system is too difficult to use, the potential of enhanced performance benefits to be derived from the system are outweighed by the effort required to use it.

Based on TAM model, the system can be more useful when it is easier to use. Thus, technology acceptance via perceived usefulness is directly influenced by ease-of-use (Davis 1989). Venkatesh and Davis (2000) propose the TAM2 model which includes subjective norms as an additional predictor of intention to use technology, as subjective norms have significant direct effect on usage intentions. According to the TAM2 model, perceived usefulness, ease of use and subjective norms are important drivers of usage intentions. Ma and Liu (2004) confirmed Davis' original findings that the relationship between usefulness and perceived ease-of-use cannot be ignored when the relationships between perceived usefulness and ease-of-use, and perceived usefulness and technology acceptance are strong while a relationship between perceived ease-of-use and technology acceptance is weak.

Many empirical studies have tested the TAM model and confirmed that these factors do correlate with the usage. Therefore, if the technology is useful and easy to use by the consumers, the acceptance of the technology is likely to be positive as well (Nathwani and Eason 2005). Therefore, the perceived usefulness and ease-of-use can influence the consumer's attitude towards a technology (Glassberg et al. 2006). TAM has been used in different fields of research such as skill training (Venkatesh, 1999) and consumer behavior in an online environment (Koufaris, 2002). In the mobile domain, consumers who are more familiar with mobile technology can be considered the consumers who first use mobile services, like mobile advertising. Therefore, consumers who find mobile technology useful and easy to use are consumers who have more positive attitudes towards mobile advertising.

Integrated Model

Different models were integrated together by Nysveen et al (2005) to better predict adoption of mobile services. They proposed new additions that may be relevant for explaining customers' intention to use mobile services. Some of these theories such as information system theories-multi-attribute, TRA, TPB, TAM came from organizational context studies and others such as gratification and domestication theories come from everyday life context (Silverstone and Hirsch 1992; Skog 2002) .

Nysveen et al (2005) claimed that TAM was insufficient in everyday life context as customers themselves are charged for the mobile service and customers' use of mobile service depends on the availability of resources. They argue that the model must be extended by perceived control, TPB (Ajzen 1991) as there are other internal or external constraints on behavior that are not accounted for in TAM. Teo and Pok (2003) recommend subjective norms as a part of information system theories affecting consumers' intentions to use mobile services.

Non-utilitarian motives have their roots in gratification and domestication research (Leung and Wei 2000) which are used to explain the intention to use mobile services. Gratification research suggests that users seek gratification in media and technology use in everyday life based on their individual needs and motivations (Lin 1996). Domestication studies propose that expressiveness is an important driver for intention to use mobile services. Nysveen et al (2005) define perceived expressiveness as "the degree to which users of mobile services perceive the services as suitable for expressing their emotions and social and or personal identity". Leung and Wei (2000) suggest that use of mobile service can be a tool for expressiveness, to display status, personality, fashion, style and publicl image. Stryker and Burke (2000) see expressiveness as people's perception of a mobile service's ability to express both social and personal identity dimensions. Nysveen et al (2005) claim that non-utilitarian motives, such as entertainment, enjoyment and fun seeking, should be used to explain consumers' intentions to use mobile services in everyday life context, whereas the original TAM has been most often used in organizations.

Nysveen et al (2005) could construct an integrated model based on information system theories, gratification and domestic theories, to explain customer intentions to use mobile services (Figure 4). Their study confirmed that; perceived expressiveness and perceived enjoyment show the most promising and notable effect; perceived usefulness, perceived ease-of-use and attitude towards the use are direct or indirect determinants of mobile service usage. Finally subjective norms and perceived control are important antecedents of consumers' intentions to use mobile services.

Innovation Diffusion Theory (IDT)

Another theory that enlightens acceptance and intention to use technology is Innovation Diffusion Theory (Teo and Pok 2003). Rogers (1983) defines innovation diffusion theory as "the process by which innovation is communicated through certain channels over time among the members of a social system." He explains the process of innovation diffusion as one, which is dictated by uncertainty reduction behavior during the introduction of technological innovations. The innovation process that an individual undergoes begins with a basic knowledge of innovation that is formed based on a favorable or unfavorable attitude toward it and transitions through to a decision to either adopt or reject it (Rogers 1983). The uncertainty as to whether the innovations will be superior to existing solutions presents a considerable obstacle in the adoption process. Potential adopters are motivated to seek additional information to counter

Figure 4. Integrated model (Adapted from Nysveen, et al., 2005)

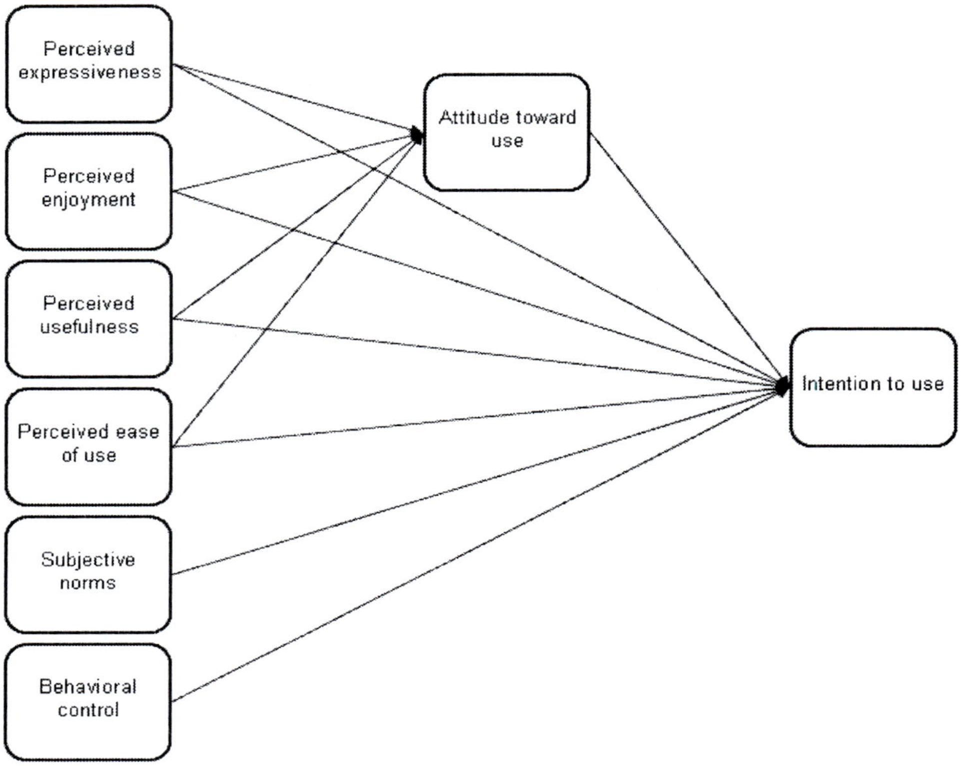

this uncertainty (Brancheau and Wetherbe, 1990). Six major components of IDT are: innovation characteristics, individual user characteristics, adopter distribution over time, diffusion networks, innovativeness and adopter categories, and the individual adoption process (Tornatsky and Klein 1982; Rogers 1983; Brancheau and Wetherbe 1990; Moore and Benbasat 1991; Taylor and Todd 1995). Rogers (1983) claims that the following five attributes of innovation influence the adoption of innovations; relative advantage, compatibility, complexity, trialability and observability (Figure 5). These five attributes are conceptually distinct but empirically interrelated with each other.

Rogers (1983) defines relative advantage as "the degree to which innovation is perceived as being better than the idea it supersedes" such as economic profitability or in status giving. Relative advantage is one of the best predictors of an innovation's rate of adoption as potential adopters need to know the degree to which a new idea is better than an existing one (Rogers 1983). Perceived usefulness from Nysveen et al's (2005) model can be related to relative advantage from Roger's (1983) model. Compatibility is defined by Rogers (1983) as "the degree to which an innovation is perceived as consistent with the existing values, past experiences and needs of potential adopters". Compatibility has a positive direct effect on rate of adoption of innovations as a compatible idea will be less uncertain to the potential adopter. There are no elements from Nysveen et al's (2005) model that can directly overlap with the concept of compatibility. Even though, compatibility could be relatively less important in predicting a rate of adoption than other attributes, but still it makes it an interesting variable to focus on.

Figure 5. Drivers of adoption of innovations (Adapted from Rogers, 1983)

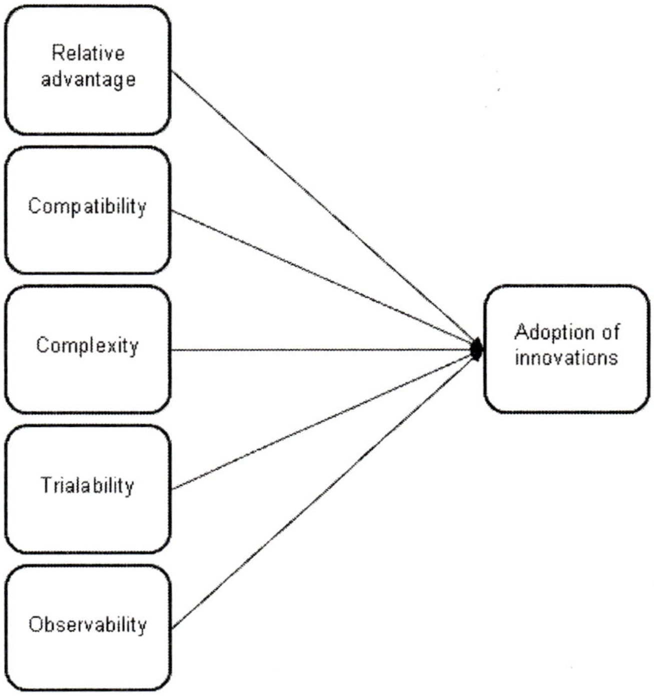

Rogers (1983) defines complexity as "the degree to which an innovation is perceived as relatively difficult to understand and use." The less difficult to understand and use the innovation will be, the less complex it will be perceived and the rate of adoption will be higher. Thus, the complexity of an innovation is negatively related to its rate of adoption (Rogers 1983). Perceived ease of use from Nysveen et al' (2005) model can be related to the complexity from Roger's (1983) model. Trialability is defined by Roger (1983) as "the degree to which an innovation may be experimented with on a limited basis". The trialability on an innovation is directly and positively related to the rate of adoption of innovations (Roger 1983, Singh 1966, Fliegel and Kivlin 1966). Rogers (1983) defines observability as "the degree to which the results of an innovation are visible to others." Rogers suggests a positive relation between observability and adoption rate because the faster the adoption of the innovation will be if the more visible the innovation is to others.

Moore and Benbasat (1991), among other things, built on the work of Rogers and expanded the array of innovation characteristics to seven by adding image, visibility and demonstrability to the model. In their model, three of the seven innovation characteristics; relative advantage, compatibility, and trialability are directly borrowed from Rogers and the fourth characteristic, ease-of-use, is a close relative to Rogers' complexity. They defined image as the degree to which adoption "of an innovation is perceived to enhance one's image or status in a one's social system." Visibility is defined as the "degree to which the innovation is apparent to the sense of sight, visible" (Moore and Benbasat 1991). Demonstrability is defined as the "degree to which the results of using the innovation are observable and communicable to the others." Moore and Benbasat (1991) also pointed out that these definitions are "based on perceptions of the innovation itself, and not on the perceptions of actually using the system". Ajzen and Fishbein (1980) agree that attitudes towards an object and attitudes regarding a particular behavior relating to that object can frequently differ.

Unified Theory of Acceptance and Use of Technology (UTAUT)

Models for technology acceptance and service adoption have been established and tested extensively. In 2003, Venkatash et al. proposed a unified model integrating acceptance determinants across several competing models, such as TAM (Davis 1989), TRA (Fishbein and Ajzen 1975), and IDT (Rogers 1983), through review and empirical testing. Venkatash et al.'s Unified Theory of Acceptance and Use of a Technology (UTAUT) model (see Figure 6) has shown greater explanation power over past models for different empirical settings. According to the UTAUT model, intention to use the information technology (IT) can be determined by three antecedents: performance expectancy, effort expectancy, social influence, and four moderators including age, gender, experience, and voluntaries.

Performance expectancy, measures how much people perceive a system, such as the mobile services, as being useful in achieving their goals in terms of job performance. Regardless of the types of environments, mandatory or voluntarily, the concept of performance expectancy, including perceived usefulness, has been considered the most influential tool for explaining the intention to use the system. On the other hand, effort expectancy explains how much people feel comfortable and find it easy to adopt and employ the system for their jobs. A study by Yan (2003) highlights that effort expectancy has a much more effect on the Chinese consumers, in comparison with the consumers from other developed countries. The Chinese seemed to easily adopt basic and user friendly technology such as Short Message Service (SMS) rather than more advanced ones such as Wireless Application Protocol (WAP). For the European mobile consumers, intention to use mobile technology were explained significantly

Figure 6. The Unified Theory of Acceptance and Use of Technology (UTAUT) (Adapted from Venkatesh, et al., 2003)

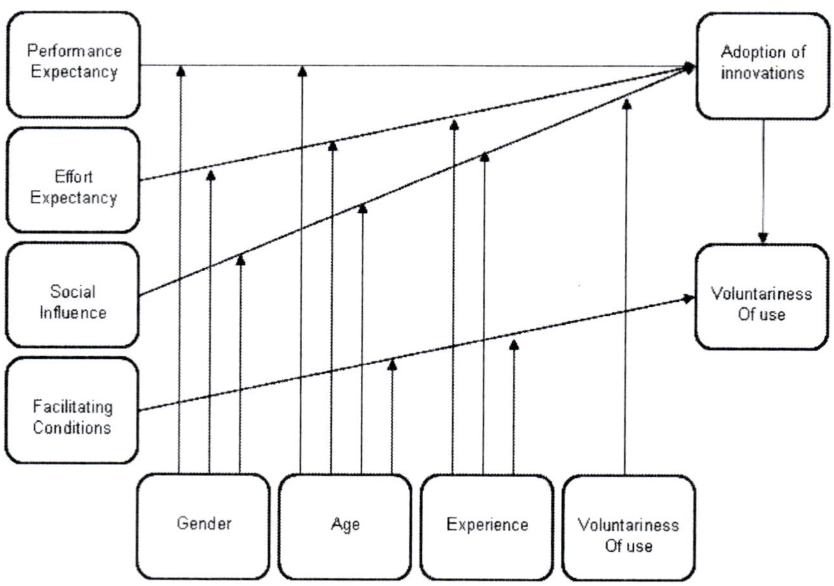

by performance expectancy and effort expectancy (Carlsson et al. 2006).

Social influence described as the influence of others' opinions about a certain system adoption (Venkatesh et al. 2003). The concept of social influence is complex and involves other concepts such as compliance "subjective norm concept" and identification "status gain concept" (Venkatesh and Davis 2000). Facilitating conditions could be defined as people's believe that how existing technical infrastructures could help them to use the system when needed. Facilitating conditions were the only antecedents in UTAUT that were not significant in explaining the behavioral intention (Venkatash et al. (2003).

UTAUT model was tested by Carlsson et al. (2006) on European mobile consumers. The study showed that intention to use mobile technology could be explained significantly by performance expectancy and effort expectancy while a social influence could be used. Chinese consumers experience more diverse economic development and social conditions. As a result, they tend to rely more on social influences where mobile technology use is perceived as conspicuous consumption and mobile phones are regarded as expensive (Zhang and Prybutok 2005). Social influence is a significant, facilitating factor for forming a positive attitude towards adopting mobile technology in China. Even though there are differences across China in building infrastructure for supporting mobile technology (e.g., big cities such as Beijing), the facilitating conditions can be adopted in the same fashion as it is employed in western countries (Tan and Ouyang 2004). As China is in an early stage of adoption in terms of mobile communication technology, behavioral intention can still be used as a meaningful substitute for behavior (Szajna 1996; Agarwal and Prasad 1999).

As mentioned previously, UTAUT has four moderators; age, gender, experience, and voluntaries. Gender would moderate the effect of performance expectancy, effort expectancy, and social influence in UTAUT model. Males with their highly task-oriented nature, expected to rely on performance expectancy when determining to accept a technology. In contrast, effort expectancy and social influence could be the main determinants of a female's technology acceptance as a female has the tendency to measure the amount of effort needed and have concerns about others' opinions and interaction with others. Several studies in the area of technology acceptance have confirmed moderating effect of gender in different set technologies such as e-mail (Gefen and Straub 1997), web-based shopping (Slyke et al. 2002), wireless internet (Lu et al. 2003), Internet banking (Lichtenstein and Williamson 2006), and so on. On the other hand, several recent empirical studies show that the expected gap between genders is diminishing in IT acceptance and usage as the technologies are more widely diffused (Zhou et al. 2007).

With regards to mobile technology acceptance, Bigne et al. (2005) found that Spanish men and women did not show different behaviors in mobile shopping. In agreement, Li and Kishore (2006) found that evaluation of performance expectancy, effort expectancy, and social influence are the same between gender groups in Hong Kong. Mobile technology users in China are mainly educated young generation (Tan and Ouyang 2004). Mao and Palvia's work (2006) on Chinese consumers suggested that with a higher level of experience, people will rely on performance expectancy when forming their attitude towards technology. Effort expectancy seems to be more important to people in earlier stages of adoption (Szajna 1996). People with lower education levels are expected to be more sensitive to effort expectancy factor as technology presents a barrier to them (Venkatesh and Morris 2000).

MOBILE MARKETING RESEARCH

The technological development along with high penetration of mobile phones has created a whole new marketing medium "mobile advertising" for the advertisers. The success of this new marketing medium is largely dependable on user acceptance to receiving ads into their mobile phones (Bauer et al. 2005). While scholars have provided a sizeable body of research on mobile marketing (Leppäniemi et al. 2006), there has been little research about how consumers react to mobile marketing on an international basis. Earlier research on the mobile usage mostly focused on usability issues, technological developments, and mobile telecommunication policies (Kristoffersen and Ljungberg 1999; Gruber 2001). Rarely research has been done to study mobile marketing, utilizing different user acceptance and adoption theories. Since the mobile marketing is a new phenomenon, a study of the pattern of its spread worldwide and within given countries from the viewpoint of user acceptance and adoption literatures could provide valuable insights (Takacs and Freiden 1998). This section provides a review on some recent mobile marketing researches and also application of above mentioned theories in the field.

The results of previous research and studies, e.g. Venkatesh et al. 2003, have proven that behavioral intention (attitude) has a direct effect on the actual behavior (usage). The role of intention as a predictor of behavior (e.g., usage) is critical and has been well-established in IS (Ajzen 1991; Sheppard et al. 1988; Taylor and Todd 1995) which could be applied in moble marketing research. Carlsson et al (2006) have found out that the behavioral intention does have a positive effect on the actual usage of mobile services. The relationship between the attitude and the behavior in regard to mobile marketing can be studied using TRA (Fishbein and Ajzen 1975).

Attitude is about the way individuals perceive their environment and guide the way they respond to it (Aaker et al. 1995). MacKenzie and Lutz (1989) described "attitude towards advertisement" as consumers' learned tendency to respond in a consistently favorable or unfavorable manner towards advertising. As it is known from TRA and TPB (Ajzen 1991), attitudes have a substantial impact on behavior (Churchill and Iacobucci 2002). As the result general attitude towards mobile phones has been major influencing factor on attitude towards mobile advertisement. For example, Japanese consumers show an extraordinarily positive attitude towards mobile phones (Haghirian et al. 2004) and consequently, also towards mobile advertising. Perceived control refers simply to consumers control over when, where, what and how much advertising to receive into the mobile. It is stated that consumers are needy of some control in order to accept advertising into their mobiles (Carroll et al. 2007). Tsang et al. (2004) investigated consumer attitudes towards mobile advertising and the relationship between attitude and behavior. They found out that consumers generally have negative attitudes towards mobile advertising, unless they have specifically consented to it, and there is a direct relationship between consumer attitudes and consumer behavior. Merisavo et al. (2007) studied the drivers of consumer acceptance of SMS-based mobile advertising through a conceptual model. The results show that despite the concerns about privacy, control and trust are not that important to consumers in mobile advertising. As mobile phones are very personal devices, consumer perceptions of controlling that permission as related to the mobile advertising are considered important factors that might affect consumer acceptance of mobile advertising (Nysveen et al. 2005).

In TAM, Davis (1989) defined perceived usefulness as the degree to which a person believes that using a particular system would be advantageous in performing his or her task. TAM theorizes perceived usefulness as one of the predictors of attitude (Kim and Kwahk 2007). If all other things are equal, users in the market will accept whatever marketed product or services they perceive as more useful (Rogers 1983). For this reason, perceived usefulness is an important factor in the initial adoption of information systems (Davis 1989) and also an important post-adoption factor

(Karahanna 1999). Thus perceived usefulness is likely to be a key factor in the initial adoption and the post-adoption behavior of mobile advertising because perceived usefulness is a factor that consistently influences user satisfaction across temporal stages of mobile advertising use. TAM is usually used in consumer acceptance testing. TAM model is tested in many empirical studies, where the results confirm that these factors do correlate with the usage (Nathwani and Eason, 2005). TAM can be used to define how consumers explore the usefulness and the ease of use of mobile advertising. Consumers more familiar to mobile technology can then be considered the consumers first to experience mobile advertising. Therefore, consumers who find mobile technology useful and ease of use, are consumers with more positive attitudes towards mobile advertising.

Nysveen et al (2005) could explain customer intentions to use mobile services, by constructing an integrated model based on information system theories, gratification and domestic theories. Pedersen, Nysveen and Thorbjornsen (Pedersen, 2002, 2005; Nysveen et al., 2003) found that the intention to use is influenced by perceived expressiveness, enjoyment, subjective norm and behavioral control. Nysveen et al (2005) confirmed that perceived usefulness and attitude towards use are direct or indirect determinants of mobile services' usage. Li, Chau, and Lou (2005) suggested that enjoyment shows a significant effect on intention to use instant messaging and will increase when the multimedia communication and entertainment is getting richer e.g. moving from SMS to MMS (Hsu, Lu and Hsu, 2007). Karjaluoto and Alatalo (2007) studied factors that affect consumer attitudes towards mobile marketing by developing a conceptual model. They found out that context, credibility and subjective norm are positively related to the intention to use mobile advertising, whereas perceived behavioral control and financial rewards were not associated with the intention. Consumer attitudes towards mobile advertising are influenced by the explored consumer attitude towards advertising in general and also explored the value on advertising (Bauer et al. 2005). Haghirian et al (2005) studied the relevance of mobile advertising and investigated how perceived advertising value of mobile marketing can be increased based on message content and exposure. The results indicate that the message content is of greatest relevance for the perceived advertising value, while a high frequency of message exposure has a negative impact on it. Karjaluoto et al. (2008) investigated customers' intention to engage in permission-based mobile marketing communications and found that perceived usefulness of mobile communications explains a considerable amount of attitude towards advertising.

Innovations normally offer its adopters novel ways of tackling day-to-day problems, but there is an uncertainty if the new ways will be better than existing solutions. To counter this uncertainty, potential adopters of mobile advertising can be studied using IDT. Such approach is more useful when mobile advertising is perceived as an innovative approach in particular market. Hsu et al. (2007) propose that based on innovation diffusion theory, adopter's intention to use will positively affect by perceived attributes of MMS. The results reveal that image, trial-ability and result demonstrability are not significant predictors of adoption intention but relative advantage is.

UTAUT model is based on studies of eight prominent models in IS adoption research. UTAUT can be used usually to explain possible adoption and acceptance patterns mobile advertising among consumers. The UTAUT provides a useful tool for researchers and managers to assess the likelihood of success for different mobile marketing introductions and helps them understand the drivers of acceptance in order to proactively design interventions targeted at populations of users that may be less inclined to adopt and use mobile advertising. Komulainen et al. (2006) have studied Consumers' perceptions towards mobile advertising by using the determinants of UTAUT as a framework.

Venkatesh et al (2003) analyzed the effect of age as a moderating factor and concluded that the effect of performance expectancy is stronger for the younger workers. The results of the previous studies suggest that the increase on the age should have a negative effect on the intention of the consumers to accept mobile advertising (Haghirian et al 2004). Research on gender differences revealed that men and women perceive mobile phones' usage differently (Ozhan Dedeoglu (2004). Karjaluoto et al. (2008) investigated customers' intention to engage in permission-based mobile marketing communications and found that women have a stronger relationship between mobile marketing communications with both intentions to visit and actual visits in comparison to men. The concept of performance expectancy, including perceived usefulness, has been considered as the most powerful tool for explaining the intention to use the system regardless of the types of environments, be it mandatory or voluntarily (Venkatesh et al 2003). Table 1 illustrates some example of the previous research works which could be utilized in mobile marketing research.

FUTURE TRENDS

The overall mobile marketing use is very moderate and mobile advertising have not really found their way into the everyday lives of people around the world. Academically, in spite of increasing amounts of general empirical studies in mobile marketing, only a few have explored the field of the mobile marketing with a focus on the adoption theories (Okazaki 2005). This chapter attempts to extend the understanding of different theories underlying mobile marketing research and adoption. From a practical perspective, while the entire global economy is keeping its eyes on mobile services and marketing as a good opportunity for growth, the current study may provide a better understanding of theories to help researchers and managers in the field. An understanding of previous theories and research work can provide insights into businesses that confront difficulties in mobile marketing diffusion in different markets. Even though there seems to be a lack of studies applying traditional adoption theories to mobile marketing research, there are some signs of a reorientation in more recent research. Future research can, however, build upon the theories but should focus on bringing additional dimensions and culturally meaningful factors to the field of research. The future work can develop studies that are concerned with differentiating among consumer adoption across levels of complexity in a different market.

CONCLUSION

The technology and service adoption models have been used as conceptual tools to identify central issues in the mobile domain. They highlight factors affecting consumer's intention to use and adopt mobile services. It is necessary to get a thorough understanding of the theories underlying mobile technology and service adoption in order to be successful in mobile marketing. Robey (1979) was among the first researchers who managed to combine earlier work done by Vertinksky et al. (1975) and Schultz and Slevin (1975) and could develop an expectancy model that shows a significant connection between user attitudes (or perceptions) and system use.

Fishbein (1967) multi-attribute model proposes that attitudes are the sum of cognitive beliefs, where attitudes can be determined by relevant cognitive beliefs and the evaluation of those beliefs. Fishbein and Middlestadt, (1995) argue that when using cognitive factors that are inappropriate or measured incorrectly, then only non-cognitive factors influence an attitude. Schwartz (1997) suggests other feelings such as moods are actually cognitive and serve as sources of

Table 1. Some example of categorized theories and previous literatures which could be used in mobile marketing research

Constructs	Authors	Findings
Intention to use	Fishbein and Ajzen (1975)	Theory of Reasoned actions – the relationship between attitude and behavior
	Sheppard et al. 1988; Ajzen 1991; Taylor and Todd 1995	Critical role of intention as a predictor of behavior in information systems(IS)
	Venkatesh, Morris, Davis and Davis (2003)	Direct effect of the behavioral intention (attitude) on the actual behavior (usage)
	Venkatesh et al (2003)	Performance expectancy, including perceived usefulness, has been considered as the most powerful tool for explaining the intention to use the system regardless of the types of environments, be it mandatory or voluntarily
	Pedersen (2002 and 2005); Nysveen et al., 2003	Intention to use is influenced by perceived expressiveness, enjoyment, subjective norm and behavioral control
	Li, Chau, and Lou (2005); Hsu, Lu and Hsu (2007)	Enjoyment shows significant effect on intention to use media and will increase when the media gets richer
	Karjaluoto and Alatalo (2007)	Context, credibility and subjective norm are positively related to intention to use mobile advertising whereas perceived behavioral control and financial rewards were not associated with intention.
	Hsu et al. (2007)	Based on innovation diffusion theory, adopter's intention to use will positively affect by perceived attributes
Perceived usefulness	Davis (1989)	Perceived usefulness is an important factor in the initial adoption of information systems
	Rogers (1983)	Users in the market will adopt whatever product or service they perceive as more useful
	Karahanna 1999	As a post-adoption factor
	Lee et al (2002)	Perceived usefulness as a key factor in the post-adoption behavior as perceived usefulness is a factor that consistently influences user satisfaction across temporal stages of use
	Venkatesh et al 2003	Perceived usefulness, a useful tool for explaining the intention to use
	Nysveen et al (2005)	Perceived usefulness and attitude towards use are direct or indirect determinants of mobile services' usage
	Nysveen et al (2005)	Perceived usefulness and attitude towards use are direct or indirect determinants of mobile services' usage
	Carlsson et al (2006)	The behavioral intention does have a positive effect on the actual usage of mobile services
	Kim and Kwahk 2007	Perceived usefulness as one of the predictor of attitude
	Karjaluoto et al. (2008)	Perceived usefulness of mobile communications explains a considerable amount of attitude towards advertising
Age and intention to use	Venkatesh et al (2003)	Analyzed the effect of age as a moderating factor and concluded that the effect of performance expectancy is stronger for the younger workers.
	Robins (2003)	Mobile devices, more than being as means of communication, are used as fashion accessories
	Haghirian et al (2004)	Increase on the age should have a negative effect on the intention of the consumers to accept mobile advertising
	Ozhan Dedeoglu (2004)	Men and women perceive mobile phones' usage differently
	Karjaluoto et al. (2008)	Women have a stronger relationship between mobile marketing communications with both intentions to visit and actual visits compared to men.

continued on the following page

Table 1. continued

Constructs	Authors	Findings
Education, income and intention to use	Sarker and Wells (2003)	Economic conditions as an influencing factor on adoption and usage of mobile phones
	Dickinger et al (2004)	That young generations are more interested in mobile technologies and services
	Haghirian et al (2004)	The increase on the age should have a negative effect on the intention of the consumers towards mobile marketing
	Velden and Wolbers (2004)	Education is one of the most significant factors affecting socio-economic outcomes
	Ozhan Dedeoglu (2004)	Eeffect of the education on the attitude is found to be negative
Gender differences in their intentions to use	Bem and Allen (1974)	Gender schema theory suggests that gender differences stem from gender roles reinforced from birth rather than biological gender perse
	Minton and Schneider (1980)	Men tend to be highly task-oriented and, therefore, performance expectancies, which focus on task accomplishment, are likely to be especially salient to men
	Rakow and Navarro (1993)	The mobile phone was a device that replicated preexisting gender patterns. Difference in the attitude of the different genders towards technology
	Ling (2001)	Gender differences in the adoption of both voice and other mobile services
	Ozhan Dedeoglu (2004)	Men and women perceive mobile phones' usage differently

information that are evaluated just as beliefs are evaluated in Fishbein's model. Fishbein and Ajzen (1975) developed an adaptive behavioral theory and model called the Theory of Reasoned Action (TRA), It extends Fishbein's multi-attribute model by clarifying the relationship between attitudes and behavior. Two new contributions to Fisbein's model are attitude towards behavior and subjective norms that are affecting the intention to perform a behavior. The model links individual beliefs, attitudes, intentions, and behavior to describe the psychological process that mediates the observed relations between attitudes and behavior (Fishbein and Ajzen 1975). TRA can predict consumer's behavior to perform or not perform in a situation where the customer is solely and directly responsible for his own behavior, and he is considerate. The relationship between the attitude and the behavior in regard to mobile marketing can be studied using Fishbein and Ajzen (1975) theory of reasoned action (TRA).

Ajzen (1991) revises TRA and proposes the inclusion of a third determinant of behavioral intention; perceived behavioral control. According to Theory of Planned Behavior (TPB), behavior is determined by the intention to perform the behavior. The behavior itself is determined by three factors: attitude toward a behavior, subjective norm, and perceived behavioral control (Mathieson 1991). Perceived behavioral control refers to an individual's perceptions of the existence (or non-existence) of the resources, skills, or opportunities required to use a system, or some feature of that system (Ajzen and Madden 1986). Ajzen (1991) shows in his TPB that attitude towards behavior can be described as an individual's favorable or unfavorable evaluation with using a service, while subjective norm can be seen as the perceived social pressure to use or not to use this service. TPB extends TRA to account for conditions where individuals do not have complete volitional control over their behavior (Taylor and Todd 1995; Ajzen and Madden 1986; Ajzen 1991). Based on previous studies, it is clear that general attitude towards mobile phones, consumer attitude towards advertising, consumers control over when, where, what and how much advertising to receive into the mobile, context, messaging content, credibility and subjective norm have major influence on attitude towards mobile advertisement (Haghirian et al. 2004; Bauer et al. 2005; Carroll et al. 2007; Karjaluoto and Alatalo 2007)

Theory of Acceptance Model (TAM) is grounded in the Theory of Reasoned Action (Fishbein and Ajzen 1975) and Theory of Planned Behavior (Ajzen 1991). TAM was initially designed by Davis (1989) to predict user's acceptance of information technology and usage on the job. TAM is based upon intentions to use technology and defines two key constructs; "perceived usefulness" and "perceived ease of use" (Davis 1989). TAM is repeatedly used in studies that focus on users. In studies on the adoption of mobile services, results have complied with central factors in the TAM: perceived ease-of-use and perceived usefulness. In a comparison to TAM and TPB conducted by Mathieson (1991), he found TAM is easier to apply and have a slight empirical advantage over TPB but only supplies very general information on users' opinion about a system. On the other hand, TPB supply more detailed information that could be used to improve the system. Mobile advertising usefulness and ease of use can be studied by TAM. Based on TAM, it is expected that consumers with more experience and positive attetuide toward mobile phone technology should have more positive attitudes towards mobile advertising.

Nysveen, Pedersen, and Thorbjørnsen (2005) developed a new model to better predict adoption of mobile services and suggest several extensions that may be relevant in explaining customers' intention to use mobile services. They integrate different models such as multi-attribute model, TRA, TPB, TAM from information system theories and gratification and domestication theories from everyday life context (Silverstone and Hirsch 1992; Skog 2002). Based on their study, perceived expressiveness and perceived enjoyment show the most promising and notable effect. Direct or indirect determinants of mobile services' usage are perceived usefulness, perceived ease of use and attitude towards use. Finally, subjective norms and perceived control proved to be important antecedents of consumers' intentions to use mobile services. Perceived enjoyment shows a significant effect on intention to use a mobile advertisement which can increase when the multimedia communication and entertainment is getting richer (Li, Chau, and Lou 2005; Hsu, Lu and Hsu, 2007).

Another theory that describes acceptance and intention to use technology is Innovation Diffusion Theory (Teo and Pok 2003). Rogers (1983) explains the process of innovation diffusion as one where the individual passes from an early point of essential knowledge of innovation and creates a decision to either adopt or reject it by forming a favorable or unfavorable attitude towards it. Through the process, the individual reduces uncertainty as to whether the innovation will be superior to existing solutions in a market. Based on IDT, intention to use mobile advertising will positively affect by perceived attributes of mobile advertisement.

The Unified Theory of Acceptance and Use of Technology was formulated by Venkatesh, Morris, Davis and Davis (2003) based on the Davis' Technology Acceptance Model (1989) as well as empirical comparisons, testing and integration of seven other prominent models of information technology acceptance. Based on UTAUT (Venkatash et al. 2003), intention to use can be determined by three antecedents: performance expectancy, effort expectancy and social influence and, as a consequence, intention to use is to exert influence on actual behavior toward a service. Four moderators that contribute to the model are: age, gender, experience, and voluntaries.

Performance expectancy measures how much people perceive a service is useful in achieving their goals. The concept of performance expectancy, including perceived usefulness, has been considered the most powerful tool for explaining the intention to use the service regardless of the types of environments. On the other hand, effort expectancy explains how much people feel comfortable and find it easy to adopt and employ the service. Social influence is the last construct proposed as antecedents of the intention to use. The concept of social influence is likely to be complex while involving compliance related to social pressure such as subjective norms and identification related to self identity standing for social status gains (Venkatesh and Davis 2000). The effect of social influence also depends on environmental characteristics such as mandatory or voluntary (Hartwick and Barki 1994; Karahanna et al. 1999; Venkatesh and Davis 2000).

To reflect individual differences, age, gender, experience, and voluntaries of use moderate the relationship in UTAUT. As an example, based on UTAUT, gender would moderate the effect of performance expectancy, effort expectancy, and social influence. Such moderators have been suggested before by several studies in the area of technology acceptance in a variety of technologies (Gefen and Straub 1997; Slyke et al. 2002; Lichtenstein and Williamson 2006). According to Venkatesh, et al (2003), the UTAUT "provides a useful tool for managers and marketers to assess the likelihood of success for new technology introductions and helps them understand the drivers of acceptance in order to proactively design interventions (including training, marketing, etc.) targeted at populations of users that may be less inclined to adopt and use new systems."

REFERENCES

Aaker, D. A., Kumar, V., & Day, G. (1995). *Marketing Research*, Wiley, New York. p. 254.

Agarwal, R., & Prasad, J. (1999). Are Individual Differences Germane to the Acceptance of New Information Technologies? *Decision Sciences*, *30*(2), 361–391. doi:10.1111/j.1540-5915.1999.tb01614.x

Ajzen, I. (1988). *Attitudes, personality and behavior*. Dorsey Press, Chicago, IL.

Ajzen, I. (1991). Theory of planned behavior. *Organizational Behavior and Human Decision Processes*, *50*, 179–211. doi:10.1016/0749-5978(91)90020-T

Ajzen, I., & Fishbein, M. (1980). *Understanding Attitudes and Predicting Social Behavior*, Englewood Cliffs, New Jersey.

Ajzen, I., & Madden, T. J. (1986). Prediction of goal- directed behavior from attitudinal and normative variables. *Journal of Experimental Social Psychology*, *22*, 453–474. doi:10.1016/0022-1031(86)90045-4

Barwise, P., & Strong, C. (2002). Permission Based Mobile Advertising. *Journal of Interactive Marketing*, *16*(1), 14–24. doi:10.1002/dir.10000

Bauer, H. H., Reichardt, T., Barnes, S. J., & Neumann, M. M. (2005). Driving consumer acceptance of mobile marketing: A theoretical framework and empirical study . *Journal of Electronic Commerce Research*, *6*(3), 181–192.

Bem, D. J., & Allen, A. (1974). On predicting some of the people some of the time: The search for cross-situational consistencies in behavior. [Bigne, E.]. *Psychological Review*, *81*, 506–520. doi:10.1037/h0037130

Brancheau, J. C., & Wetherbe, J. C. (1990). The adoption of spreadsheet software: testing Innovation Diffusion Theory in the context of end-user computing. *Information Systems Research*, *1*(2), 115–143. doi:10.1287/isre.1.2.115

Carlsson, C., Carlsson, J., Hyvönen, K., Puhakainen, J., & Walden, P. (2006). Adoption of Mobile Devices/Services: Searching for Answers with the UTAUT, *Proceedings of the 39th Annual Hawaii International Conference on System Sciences (HICSS'06)*, Track 6: 1-10.

Carroll, A., Barnes, S. J., Scornavacca, E., & Fletcher, K. (2007). Consumer perceptions and attitudes towards SMS advertising: recent evidence from New Zealand . *International Journal of Advertising, 26*(1), 79–98.

Churchill, G. A. J., & Iacobucci, D. (2002). *Marketing Research; Methodological Foundations*. South-Western Publishing, Mason, Ohio.

Dabholkar, P. A., & Bagozzi, R. P. (2002). An attitudinal model of technology- based self- service: Moderating effects of consumer traits and situational factors. *Journal of the Academy of Marketing Science, 30*(3), 184–201.

Davis, F. D. (1989). Perceived Usefulness, Perceived Ease of Use and User Acceptance of Information Technology. *MISQ, 13*, 319–340. doi:10.2307/249008

Davis, F. D., Bagozzi, R. P., and Warshaw, P. R. (1989). User Acceptance of Computer Technology: A Comparison of Two Theoretical Models, *Management Science* (35:8), pp. 982-1002

Dickinger, A., Haghirian, P., Murphy, J., & Scharl, A. (2004). An investigation and conceptual model of SMS marketing. In the Proceedings of 37th Annual Hawaii International Conference on System Sciences, pp. 31–40.

Fishbein, M. (1967). Attitudes and Prediction of Behavior, in: Fishbein M. (ed.), Readings in Attitude Theory and Measurement, New York, John Wiley and Sons, p.477-492.

Fishbein, M., & Ajzen, I. (1975). Belief, Attitude, Intention and Behavior: An Introduction to Theory and Research, Reading, MA: Addison-Wesley.

Fishbein, M., & Middlestadt, S. (1995). Noncognitive effects on attitude formation and change: Fact or artifact? *Journal of Consumer Psychology, 4*(2), 181–202. doi:10.1207/s15327663jcp0402_05

Fliegel, F. C., & Kivlin, J. E. (1966). Attributes of innovations as factors in diffusion. *American Journal of Sociology, 72*, 235–248. doi:10.1086/224292

Gefen, D., & Straub, D. (1997). Gender Differences in the Perception and Use of E-Mail: An Extension to the Technology Acceptance Model . *MISQ, 21*(4), 389–400. doi:10.2307/249720

Glassberg, B.C., Grover, V., and Teng, J.T.C. (2006). *Information System Research with an Attitude*, 76-77

Godin, G., & Kok, G. (1996). The theory of planned behavior: a review of its applications to health-related behaviors. *American Journal of Health Promotion, 11*(2), 87–98.

Gruber, H. (2001). Competition and innovation: The diffusion of mobile telecommunications services in central and Eastern Europe. *Information Economics and Policy, 13*, 19–34. doi:10.1016/S0167-6245(00)00028-7

Haghirian, P., Dickinger, A., & Kohlbacher, F. (2004). Adopting innovative technology – a qualitative study among Japanese mobile consumers. *Proceedings of the 5th International Working with e-Business Conference (WeB-2004)*, November 2004, Perth, Australia.

Haghirian, P., Madlberger, M., & Tanuskova, A. (2005). Increasing advertising value of mobile marketing –An empirical study of antecedents, *Proceedings of the 38th Hawaii International Conference on System Sciences*, Hawaii.

Harrison, D. A., Mykytyn, P. P. Jr, & Riemenschneider, C. K. (1997). Executive decisions about adoption of information technology in small business: Theory and empirical tests. *Information Systems Research, 8*(2), 171–195. doi:10.1287/isre.8.2.171

Hartwick, J., & Barki, H. (1994). Explaining the Role of User Participation in Information System Use. *Management Science*, *40*, 460–465. doi:10.1287/mnsc.40.4.440

Hsu, C.L., Lu, H.P., and Hsu, H.H. (2007). Adoption of the mobile internet: an empirical study of multimedia message service (MMS). *The International Journal of Management Science*, Dec. http://shop.informatm.com/marlin/30000001001/MARKT_EFFORT/marketingid/20001444760?proceed=true&MarEntityId=1167456884244&entHash=100231 1dd57&UType=true Access Jan 2007.

Karahanna, E., Straub, D. W., & Chervany, N. L. (1999). Information technology adoption across time: A cross sectional comparison of pre- adoption and post- adoption beliefs. *MISQ*, *23*(2), 183–213. doi:10.2307/249751

Karjaluoto, H., & Alatalo, T. (2007). Consumers' attitudes towards and intention to participate in mobile marketing . *International Journal of Services Technology and Management*, *8*(2/3), 155–173. doi:10.1504/IJSTM.2007.012866

Karjaluoto, H., Lehto, H., Leppaniemi, M., & Jayawardhena, C. (2008). Exploring Gender Influence on Customer's Intention to Engage Permission-based Mobile Marketing. *Electronic Markets*, *18*(Issue 3), 242–259. doi:10.1080/10196780802265793

Kassarjian, H. H., & Robertson, T. S. (1991). *Perspectives in Consumer Behaviour (4th edition)*. Prentice Hall, New Jersey.

Kavassalis, P., Spyropoulou, N., Drossos, D., Mitrokostas, E., Gikas, G., & Hatzistamatiou, A. (2003). Mobile Permission Marketing: Framing the Market Inquiry. *International Journal of Electronic Commerce*, *8*(1), 55–80.

Kim, H., & Kwahk, K. (2007). Comparing the Usage Behaviour and the Continuance Intention of Mobile Internet Services", *Eight World Congress on Management of eBuisiness* (WCMeB 2007)

Komulainen, H., Ristola, A., & Still, J. (2006). Mobile advertising in the eyes of retailers and consumers – empirical evidence from a real-life experiment, *Proceedings of the International Conference on Mobile Business* (ICMB'06)

Koufaris, M. (2002). Applying the technology acceptance model and flow theory to online consumer behavior. *Information Systems Research*, *13*(2), 205–223. doi:10.1287/isre.13.2.205.83

Kristoffersen, S. and Ljungberg, F. Mobile informatics: Innovation of IT use in mobile settings. *IRIS'21 workshop report. SIGCHI Bulletin* 31, 1 (1999), 29–34.

Lee, Y., Kim, J., & Lee, I. (2002). Kim, h. (2002),"a Cross-cultural Study on the Value Structure of Mobile Internet Usage: Comparison between Korea and Japan. *Journal of Electronic Commerce Research*, *3*(4).

Leppäniemi, M., Sinisalo, J., & Karjaluoto, H. (2006). A review of mobile marketing research'. *International Journal of Mobile Marketing*, *1*(1), 2–12.

Leung, L., & Wei, R. (2000). More than just talk on the move: Uses and gratifications of the cellular phone. *MISQ*, *77*(2), 308–320.

Li, D., Chau, P.Y.K., and Lou, H. (2005). Understanding individual adoption of instant messaging: an empirical investigation. *Journal of the Association for information systems*, vol. 6, 102-129.

Li, J. P., & Kishore, R. (2006). How Robust is the UTAUT Instrument? A multigroup Invariance Analysis in the Context of Acceptance and Use of Online Community Weblog Systems, *Proceedings of the 2006 ACM SIGMIS CPR Conference on Computer Personnel Research*, 183 – 189.

Lichtenstein, S., & Williamson, K. (2006). Understanding Consumer Adoption of Internet Banking: An Interpretive Study in the Australian Banking Context. *Journal of Electronic Commerce Research*, *7*(2), 50–66.

Lin, C. A. (1996). Looking back: The contribution of Blumler and Katz`s uses of mass communication research. *Journal of Broadcasting & Electronic Media*, *40*, 574–582.

Ling, R. (2001) "We Release Them Little by Little" Maturation and Gender Identity as Seen in the Use of Mobile Telephony. Personal and Ubiquitous Computing, 5(123 - 136.

Lu, J., Yu, C.-S., Liu, C., & Yao, J. E. (2003). Technology Acceptance Model for Wireless Internet. *Internet Research*, *13*(3), 206–222. doi:10.1108/10662240310478222

Ma, Q., & Liu, L. (2004). The technology acceptance model: A meta analysis of empirical findings. *Journal of Organizational and End User Computing*, *16*(1), 59–72.

MacKenzie, S. B. and Lutz, R. L. (1989). An empirical examination of the structural antecedents of attitude toward the ad in an advertising pretesting context. *Journal of Marketing*, Vol. 53 (April 1998), 48-65.

Mao, E. and Palvia, P. (2006). Testing an Extended Model of IT Acceptance in the Chinese Cultural Context, *Advances in Information Systems*, Vol. 37, No. 2 and 3: 20-32.

Mathieson, K. (1991). Predicting user intentions: Comparing the technology acceptance model with the theory of planned behaviour. *Information Systems Research*, *2*, 173–191. doi:10.1287/isre.2.3.173

Merisavo, M., Kajalo, S., Karjaluoto, H., Virtanen, V., Salmenkivi, S., & Raulas, M. (2007). An empirical study of the drivers of consumer acceptance of mobile advertising. *Journal of Interactive Advertising*, *7*(2).

Minton, H. L., & Schneider, F. W. *Differential Psychology*, Waveland Press, Prospect Heights, IL, 1980.

Moore, G. C., & Benbasat, I. (1991). Development of an instrument to measure the perceptions of adopting an information technology innovation. *Information Systems Research*, *2*(3), 192–222. doi:10.1287/isre.2.3.192

Nathwani, K., & Eason, K. (2005). Perceptions versus expectations of multimedia messaging service (MMS). *Personal and Ubiquitous Computing*, *9*(4). doi:10.1007/s00779-004-0320-9

Nysveen, H., Pedersen, P. E., & Thorbjørnsen, H. (2003). Using mobile services to strengthen brand relationships: The effects of SMS and MMS channel additions on brand knowledge, satisfaction, loyalty and main channel use. *SNF-report* no. 22/03. Foundation for Research in Economics and Business Administration, Bergen, Norway.

Nysveen, H., Pedersen, P. E., & Thorbjørnsen, H. (2005). Intentions to use mobile services: Antecedents and cross- service comparisons. *Journal of the Academy of Marketing Science*, *33*(3), 330–346. doi:10.1177/0092070305276149

Okazaki, S. (2005). Mobile advertising adoption by multinationals. Senior executives' initial responses. *Internet Research*, *15*(2), 160–180. doi:10.1108/10662240510590342

Ozhan Dedeoglu, A. (2004). The Symbolic Use of Mobile Technology among Turkish Consumers. *Journal of Euromarketing*, *2*(3), 143–162. doi:10.1300/J037v13n02_08

Pedersen, P. E. (2002). The adoption of text messaging services among Norwegian teens: development and test of an extended adoption model. *SNF-report no.* 23/02. Foundation for Research in Economics and Business Administration, Bergen Bergen, Norway.

Pedersen, P. E. (2005). Adoption of mobile Internet services: An exploratory study of mobile commerce early adopters. *Journal of Organizational Computing and Electronic Commerce, 15*, 203–221. doi:10.1207/s15327744joce1503_2

Rakow, L. F., & Navarro, V. (1993). Remote Mothering and the Parallel Shift: Women Meet the Cellular Telephone. *Critical Studies in Mass Communication, 10*, 144–157. doi:10.1080/15295039309366856

Robey, D. (1979). User attitudes and Management Information System use. *Academy of Management Journal*, September, vol.22, no.3, pp.527-538.

Robins, F. (2003). The Marketing of 3G. *Marketing Intelligence & Planning, 21*(6), 370–378. doi:10.1108/02634500310499239

Rogers, E. M. (1983). *Diffusion of Innovations*. The Free Press, London.

Ryan, M.J., and Bonfield, E.H. (1980). Fishbein`s intention model: A test of external and pragmatic validity. *Journal of Marketing*, 44(spring), 82-95.

Sarker, S., & Wells, J. D. (2003). Understanding Mobile . *Communications of the ACM, 46*(12), 35–40. doi:10.1145/953460.953484

Schifter, D. E. (1984). *Attempts, control and weight reduction: An application of a theory of planned behavior.* Master Thesis, University of Massachusetts at Amherst, Amherst, MA.

Schultz, R. L., & Slevin, D. P. (1975). *Implementation and organizational validity: an empirical investigation.* In Implementing Operations Research / Management Science, New York, NY: American Elsevier, pp.153-182.

Schwartz, N. (1997). Moods and attitude judgments: A comment on Fishbein and Middlestadt. *Journal of Consumer Psychology, 6*(1), 93–98. doi:10.1207/s15327663jcp0601_06

Sheppard, B. H., Hartwick, J., & Warshaw, P. R. (1988). The theory of reasoned action: A meta analysis of past research with recommendations for modifications and future research. *The Journal of Consumer Research, 15*(3), 325–343. doi:10.1086/209170

Silverstone, R., & Hirsch, E. (1992). *Consuming technologies*. Routledge, London.

Singh, R. M. (1966). *Characteristics of farm innovations associated with the rate of adoption.* Agricultural Extension Education Report 14. EX (E). Guelph, Ontario.

Skog, B. (2002). *Mobiles and the Norwegian teens: Identity, gender, and class. In perpetual contacts*. Eds. James E. Katz and Mark Aakhus. Cambridge University Press, New York.

Slyke, C. V., Comunale, C. L., & Belanger, F. (2002). Gender Differences in Perceptions of Web-Based Shopping . *Communications of the ACM, 45*(8), 82–86. doi:10.1145/545151.545155

Stryker, S., & Burke, P. J. (2000). The past present and future of identity theory. *Social Psychology Quarterly, 63*, 284–297. doi:10.2307/2695840

Szajna, B. (1996). Empirical Evaluation of the Revised Technology Acceptance Model. *Management Science, 42*(1), 85–92. doi:10.1287/mnsc.42.1.85

Takacs, S. J., & Freiden, J. B. (1998). Changes on the electronic frontier: Growth and opportunity of the World-Wide Web. *Journal of Marketing Theory and Practice, 6*(3), 24–37.

Tan, Z., & Ouyang, W. (2004). Diffusion and Impacts of the Internet and E-Commerce in China. *Electronic Markets, 14*(1), 25–35. doi:10.1080/1019678042000175270

Taylor, S., & Todd, P. A. (1995). Understanding information technology usage: A test of competing models. *Information Systems Research, 6*(2), 144–176. doi:10.1287/isre.6.2.144

Teo, T. S. H., & Pok, S. H. (2003). Adoption of WAP- enabled mobile phones among Internet users. *Omega, 31*, 483–498. doi:10.1016/j.omega.2003.08.005

Tornatsky, L., & Klein, K. (1982). Innovation characteristics and innovation adopting-implementation: A meta-analysis of findings. *IEEE Transactions on Engineering Management, EM-29*, 28–45.

Tsang, M. M., Ho, S. C., & Liang, T. P. (2004). Consumer attitudes toward mobile advertising: An empirical study. *International Journal of Electronic Commerce, 8*(3), 65–78.

Velden, R., & Wolbers, M. (2004). *How much does education matter and why? The effects of education on socio economic outcomes*, ISBN 90-5321-390-2 Sec04.049

Venkatesh, V. (1999, June). Creation of favourable user perceptions: integrating control, intrinsic motivation, and emotion into the technology acceptance model. *MISQ, 23*(2), 239–260. doi:10.2307/249753

Venkatesh, V., & Brown, S. A. (2001, March). A longitudinal investigation of personal computers in homes: adoption determinants and emerging challenges. *MISQ, 25*(1), 71–102. doi:10.2307/3250959

Venkatesh, V., & Davis, F. D. (2000). A theoretical extension of the technology acceptance model: Four longitudinal field studies. *Management Science, 46*(2), 186–204. doi:10.1287/mnsc.46.2.186.11926

Venkatesh, V., & Morris, M. G. (2000). Why don't men ever stop to ask for directions? Gender, social influence, and their role in technology acceptance and usage behavior. *MISQ, 24*, 115–139. doi:10.2307/3250981

Venkatesh, V., Morris, M. G., Davis, G. B., & Davis, F. D. (2003). User Acceptance of Information Technology: Toward a Unified View. *MISQ, 27*(3), 425–479.

Yan, X. (2003). Mobile Data Communications in China. *Communications of the ACM, 46*(12), 81–85. doi:10.1145/953460.953463

Zhang, X., & Prybutok, V. R. (2005). How the Mobile Communication Markets Differ in China, the U.S., and Europe. *Communications of the ACM, 48*(3), 111–114. doi:10.1145/1047671.1047678

Zhou, L., Dai, L., & Zhang, D. (2007). Online Shopping Acceptance Model - A Critical Survey of Consumer Factors in Online Shopping. *Journal of Electronic Commerce Research, 8*(1), 41–62.

KEY TERMS AND DEFINITIONS

Actual Behavior: an individual's intention to perform a behavior which is a function of attitude towards behavior and subjective norms.

Attitude Towards Behavior: an individual's attitude towards performing that behavior which is combination of beliefs and evaluation of the consequence of the behavior.

Intention to Behave: an individual's attitude towards the behavior which is influenced by other's opinions about the behavior..

Perceived Behavioral Control: an individual's believe around control over personal or external factors that may affect usage.

Perceived Ease of Use: an individual's believes that using a particular system would be free of effort.

Perceived Usefulness: an individual's believes that using a particular system would enhance his or her job performance.

Service Adoption: an individual's decision to make a full use of a service, in our case, means actual use of mobile services.

Subjective Norms: an individual's behavior which is exposed to social influences.

Chapter 15
In Search of Successful Mobile Advertising:
Consumer and Business Perspectives

Stuart J. Barnes
University of East Anglia, UK

Eusebio Scornavacca
Victoria University of Wellington, New Zealand

Jonathon McKenzie
Run the Red - Mobile Enables, New Zealand

Amy Carroll
Citigroup, New Zealand

ABSTRACT

The deployment of SMS-based marketing campaigns has been noted as a very successful mean of reaching consumers. With this mind, this chapter reports on three different studies that we have conducted aimed at better understanding what makes SMS marketing campaigns successful. The studies were conducted in the past four years and they explore business as well as consumer perspectives: (i) a consumer scenario perspective, which examines the importance of three factors in SMS advertising acceptance; (ii) a consumer decision perspective, which analyses the opt-in decision for an SMS campaign; and (iii) a business, perspective examining managers' perceptions of the critical success factors in advertising campaigns. The paper rounds off with conclusions and recommendation for future research and practice in the area of mobile advertising.

INTRODUCTION AND BACKGROUND

The tremendous global penetration of digital wireless telephony – estimated at more than 3.3 billion users worldwide - has provided an unprecedented platform for new services delivered via mobile handsets (Scornavacca et al., 2006; Dow Jones Newswire, 2008). Since the average revenue per user from voice services has been declining for several years, operators have been keenly looking to data communications services as the new revenue stream. One key area is that of mobile marketing.

DOI: 10.4018/978-1-60566-074-5.ch015

Mobile or wireless marketing - a subset of electronic marketing - is defined as: "All activities required to communicate with customers through the use of mobile devices in order to promote the selling of products or services and the provision of information about these products and services" (Ververidis and Polyzos, 2002). Mobile advertising has typically been categorised into push- and pull-models (Barnes, 2002). In the pull-model campaign, the marketer sends the information requested by the consumer; whereas in the push-model campaign, the marketer takes the initiative to send messages to the consumer, such as via text messaging. A third type of campaign, as suggested by Jelassi and Enders (2006), revolves around the mobile dialogue model, where the marketer tries to build a long lasting relationship with the consumer.

The most prevalent mode of mobile advertising is SMS (Short Messages Service) to handheld devices, notably mobile phones (Barwise and Strong, 2002; Dickenger et al., 2005). SMS, known as text messaging, is a store-and-forward communication system for the mobile phone. Recent variants, such as MMS (Multimedia Message Service) have added multimedia capabilities. Barwise and Strong (2002) identify six ways of using SMS for advertising: brand building, special offers, timely media 'teasers', competitions, polls/voting, products, services and information requests. Text message ads have been found to boost consumers' inclination to purchase by 36%, which partly explains its growing popularity among marketers (Enpocket, 2005b) and the growing amount of related research (Bamba and Barnes, 2007; Bauer et al., 2005; Carroll et al., 2007; Haghirian et al., 2005; Haghirian and Inoue, 2007; Heinonen and Strandvik, 2007; Karjaluoto and Kautonen, 2006; Kavassalis et al., 2003; Leppäniemi and Karjaluoto, 2005; Merisavo et al., 2006; Nysveen et al., 2005; Scornavacca and McKenzie, 2007; Virtanen et al., 2005). According to Enpocket (2005a), text message campaigns also deliver a 15% response rate, which they estimate is twice as much as direct mail or e-mail campaigns; apparently, text messages are 50% more successful at building brand awareness than TV and 130% more than radio (Enpocket, 2005b). Mobile advertising has clear potential for precise targeting and personalization, recognising social context, individual preferences, time and location.

The wireless channel benefits from the potential for detailed user information and personalization; the message can be tailored for each customer to enable better targeting. Since mobile phones are personal objects marketers can specifically address the person targeted, as well as recognising their social context, individual preferences, time, and location (Barnes, 2002; Barnes and Scornavacca, 2004; Dickenger et al., 2004; Hristova and O'Hare, 2004; Jelassi and Enders, 2006; Okazaki, 2004; Scornavacca and McKenzie, 2007).

The acceptance of SMS advertising has been examined by Bauer et al. (2005). Using a large sample and structural equation modelling they find that the most important factors that affect attitude toward mobile marketing are: consumers' attitudes toward advertising in general; perceived utility (in terms of information, entertainment, and social aspects); perceived risk (in terms of privacy and data security); consumers' knowledge about the technology; and social norms that impact on consumers' behaviour.

The aim of this chapter is to further understanding of the factors underpinning successful SMS advertising campaigns and to explore the different facets of the mobile advertising 'puzzle' that advertisers and operators have been keenly trying to solve in the last few years. In order to achieve these goals, in the following sections, we report three empirical studies that we have undertaken in the past four years. In these studies we explored consumers' as well as businesses' perspectives. The first study takes a consumer scenario perspective and, using qualitative and quantitative research examines the importance of four factors in SMS advertising acceptance (Carroll et al., 2007). The second study provides more depth on consumer

Figure 1. Model for mobile marketing acceptance

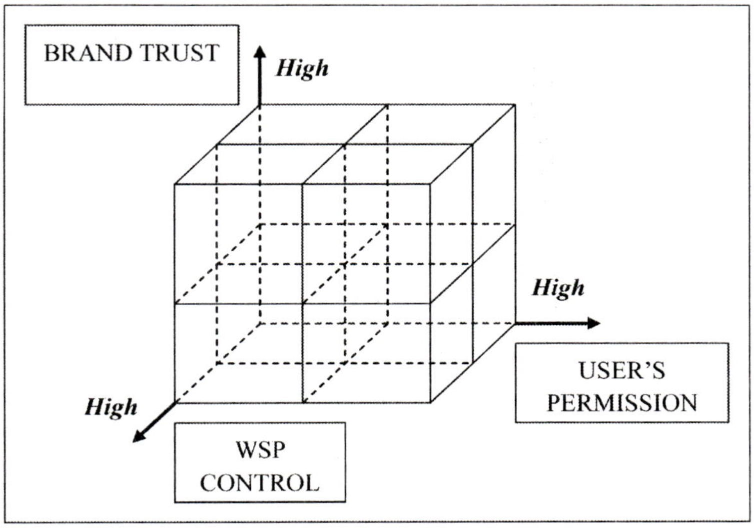

behaviour by investigating individual's decisions to opt-in SMS marketing (Barnes and Scornavacca, 2008). Finally, the last study looks on the other side of the equation by exploring the business perspective and examining managers' perceptions of the critical success factors in mobile advertising campaigns (Scornavacca and McKenzie, 2007). The paper rounds off with conclusions and recommendation for future research and practice in the area of mobile advertising.

Study One: A Consumer Scenario Perspective

In this study we aimed to explore consumer's perceptions and attitudes towards mobile advertising via SMS, and to empirically test Barnes & Scornavacca's (2004) mobile marketing acceptance model.

It has been argued that the acceptance of a mobile marketing message is likely to be influenced by the consumer's acceptance of the mobile medium, the relevance of the content and the context of the marketing message (Enpocket, 2003; Heinonen and Strandvik, 2003; Barnes and Scornavacca, 2004; Dickinger *et al.*, 2004; Bauer *et al.*, 2005). Messages that are short and concise, funny, interactive, entertaining, and relevant to the target group usually achieve higher levels of success (Dickinger *et al.*, 2004; Jelassi and Enders, 2006).

The guiding model used for this research is the conceptual model of permission and acceptance developed by Barnes and Scornavacca (2004). This model (Figure 1) looks at a small subset of factors identified in the literature, which are believed to be the *most* important variables influencing consumer acceptance.

Barnes and Scornavacca (2004) believed that the three most important variables that could influence consumer's acceptance of mobile advertising are: user *permission, wireless service provider (WSP) control* and *brand trust*. User permission was believed to be the most important variable, as consumers were seen to be fearful of SMS mobile advertising and fearful of high levels of spam, as with email marketing. Scornavacca and Barnes (2004) defined permission as the dynamic boundary produced by the combination of one's personal preferences of time, location and information. The concept of permission marketing was coined and popularized by Godin (1999),

Table 1. Scenarios for m-marketing acceptance

User's Permission	WSP Control	Brand Trust	Acceptance
High	*High*	*High*	High Acceptance
High	*High*	*Low*	Acceptable
High	*Low*	*High*	Acceptable
High	*Low*	*Low*	Acceptable
Low	*High*	*High*	Low Acceptance
Low	*High*	*Low*	Low Acceptance
Low	*Low*	*High*	Low Acceptance
Low	*Low*	*Low*	Not Acceptable

permission marketing is the opposite of traditional interruption marketing. Permission marketing is about building an ongoing relationship of increasing depth with customers by obtaining customer consent to receive information from a company. Permitted messages can be tailored to time, location and personal characteristics to get closer to the user.

Barnes and Scornavacca suggested that WSP control is found to increase the probability of user acceptance to mobile advertising. In this situation consumers trust the WSP to monitor and control mobile communications. Trust as a generalised expectation (Rotter, 1967) is defined as a willingness to rely on an exchange partner in whom one has confidence (Moorman *et al.,* 1993). Trust conveys a confidence in the behaviour of the other or an ability to depend on the other or to predict the others' behaviour (Thorelli, 1990; Gundlach and Murphy, 1993). If users are likely to have high levels of trust with their WSP (Ericsson, 2000; Enpocket, 2002a) this transfers into their willingness to accept messages though this WSP.

Brand trust was their third most important variable. It was argued that consumers may be more likely to permit advertising from 'known names' that they trust and align their values and beliefs with. Thus unwarranted use of personal information may leave consumers unwilling to trust not only the offending firm, but mobile advertising activities as a whole.

The model also puts forward eight propositions of varying levels of acceptance according to the different combinations of factors. Table 1 presents Barnes and Scornavacca's (2004) hypothesized acceptability of SMS marketing messages based on high and low levels of permission, WSP control and brand trust. Permission is weighted in the model.

This model is yet to be empirically tested with primary data, which is, of course, a considerable shortcoming. The model is also limited in its scope, with only three variables; these variables, while likely to be important, may not paint the complete picture of advertising acceptance from the consumer's perspective. Notwithstanding, this model was the only one of its kind at the time of the research and these propositions provide a useful starting point in further exploring the factors that could contribute to consumer acceptance of mobile advertising.

The chosen strategy of inquiry for this research is sequential exploratory mixed methods. Sequential procedures are ones in which the researcher uses the findings of one method to elaborate on or expand with another method (Green *et al.,* 1989; Cresswell, 2003). The objectives of the sequential exploratory approach for the purpose of this study is to use two qualitative focus groups to explore perceptions of mobile advertising, focusing on the main variables believed to influence mobile advertising acceptance, and then elaborate on this through experimental research in which the findings of the initial phase will be used. The survey data will hopefully confirm what has been identified from the literature and the findings from the focus groups.

Phase I: Focus Groups

The samples for the focus groups were purposely selected based on convenience sampling, availability and profiling. Participants for both groups were in the age range 20-28 - reflective of one of the major target groups for SMS mobile advertising. Four participants were selected for focus group A and five participants for focus group B. The participants in focus group A had a greater knowledge of mobile commerce technologies and applications than the participants in focus group B, which was purposely achieved in order to canvas a range of experiences and provide differing viewpoints. The participants in this study were students of a University in New Zealand as well as professionals working in the local central business district. This research involved the use of group interviews using open-ended questions and triggers. The results were recorded onto video.

Data analysis for the focus groups involved initially transcribing interviews and sorting the data into groups of information based on various topics. The transcriptions were then read over to look for ideas, depth and credibility of the information from participants; thoughts were noted down in the margins of the transcript (Cresswell, 2003). A coding process was then carried out where the data was organized into clusters before any meaning was derived from it (Rossman and Rallis, 1998). The themes and categories identified from the analysis are the major findings of the qualitative phase, and have been shaped into a general description of the phenomenon of mobile advertising acceptance. Reliability measures were used to check for consistency of themes and patterns, while validity measures (triangulation, member checking, bias discussion and peer debriefing) were used to determine the accuracy of the findings (Cresswell, 2003).

While Focus Group A was more knowledgeable in the area of mobile commerce, mobile technologies and the potential of mobile advertising, both focus groups had only ever experienced mobile advertising through their wireless service providers. To some extent the participants' experience of receiving marketing messages from their service provider influenced their individual perceptions and perceived importance of varying factors contributing to consumer acceptance. The results of both focus groups were consistent with little disparity between the two.

Factors identified in the focus groups as having a significant impact on consumer acceptance of mobile advertising were permission to receive mobile advertising messages, the control of the wireless service provider, relevance of the content, timeliness and frequency of the messages, simplicity and convenience of the messages, the brand or company sending the message, the control of the marketing from the consumer and the privacy of the consumer. Consistent with Barnes and Scornavacca's (2004) model, permission marketing and WSP control were perceived to have a heavy bearing on the acceptance of a mobile advertising message. However brand trust was found to have little or no impact on acceptance.

Brand Trust

The general feeling among both focus groups was that as long as the marketing messages were being filtered by the service provider it wouldn't matter too much who it was from. However if it was from a third party who had not requested permission respondents would be immediately irritated. While the majority of participants argued that it would be communications from the more well-known brands that would be received, some felt they may prefer to receive messages from smaller local shops they were aware of and that the system should make this possible. Again, the WSP control would result in an even higher level of trust for these smaller companies and it was felt consumers should be able to select which companies and brands they receive messages from.

Permission

Prior permission to send messages was raised the most throughout the discussions in each focus group, and was concluded by participants as the most important success factor, which supports the view of Barnes and Scornavacca. Participants stated that consumers should have to 'opt in' before they receive mobile advertising messages of any kind, and have the option to 'opt out' at any stage.

Wireless Service Provider (WSP) Control

Although there was great emphasis on permission, it was also strongly felt that there needed to be a degree of filtering from the service provider. The idea was raised that if participants had just one company to go to which was linked to their service provider's then there would be just one point of contact allowing consumers to easily 'opt in' and 'opt out' rather than tracking down several different companies. Participants agreed that it should be evident in the message that it is being filtered by the service provider and legitimate. However, it was found that another aspect of consumers' acceptance of mobile advertising was the wish to control other aspects of the communication, such as the content, time and frequency of messages. We now discuss these findings.

Control of Personalization and Content

Although once a viable paradigm permission marketing seems to be no longer sufficient as the consumer wants more control. Instead of a company generating messages when it wants to reach its customers, our respondents wanted the company to send only messages that users ask for. This form of marketing is especially suited to the mobile Internet, where intrusive messages are very aggravating.

This control aspect can be seen as one element of privacy (Cespedes & Smith, 1993; Goodwin, 1991; Gardiner Jones, 1991). This argument implies there are two discrete dimensions of control: firstly that of information privacy (control over information disclosure, information dissemination in the consumers' environment, and its inputs, use and control by data users), and secondly physical privacy (control over unwanted physical intrusions into the consumers' environment, the presence of others, unwanted telephone, mail, or personal intrusion in the consumers' environment).

Our respondents agreed that permission regarding content relevance would be critical to the acceptance of a mobile advertising message, with some participants arguing this as the most important factor. It was believed that marketers should make use of the technology and the advantages it provides over traditional forms of marketing and the internet, looking to add value other than just advertising. Other ideas discussed in the focus groups were to tie content in with location and timing, and to ensure the format of the message works with the limitations of the phone.

Frequency

Participants agreed that there would be a limit to the number of mobile advertising messages they wished to receive, and there should be some control over the number of messages which they are receiving depending on what good or service was being marketed or what industry (e.g. food/flowers). Both focus groups agreed that if consumers were to be hounded by marketing messages, it may result in switching providers, or deleting messages without reading them

Time

Participants raised the issue of the time playing an important role in the acceptance of mobile advertising messages. It was believed that it is important for consumers to receive marketing messages at times suitable for them, and consumers are able to not only give permission to receive messages but also choose the time's they wish to receive them.

Technology/Ease of Use

A number of important issues were raised with regards to the mobile technology and convenience of the marketing message, some of which have been pointed out in the previous sections already. The main point raised that falls under this section is that marketing messages shouldn't be a hassle for consumers to receive; they should work with the limitations of the phone and there should be a manageable way to deal with them.

Following these findings the original model was reformulated. The second stage of our research was conducted with the following results.

Phase II: Survey

This phase involved the use of a cross-sectional survey questionnaire to verify the acceptance of mobile advertising messages against sixteen various propositions that were formulated from the results of the focus groups. A total of 78 participants for the quantitative phase of the research were selected using random convenience sampling with eight members of the sample being non-respondents. The sample size was limited due to available resources and the time-consuming nature of the interview survey.

The instrument used in the survey was a modified version of the permission and acceptance model of mobile advertising developed by Barnes and Scornavacca (2004). The three variables included in the Barnes and Scornavacca model - Permission, Control and Brand Trust - were modified, based on the results of the focus groups and brand trust was deleted. It was replaced with two variables related to the content and delivery of the message. Sixteen propositions were formulated around these variables, in a format similar to the original propositions in Table 1. These were tested with a four-point Likert scale ranging from "unacceptable" to "accept enthusiastically". This scale was chosen to focus on the direction of consumer responses, using the justification from Peabody (1967) that intensity is usually poorly represented in Likert scales. Further, Chang (1994) has found the 4-point Likert scale to be more reliable than a 6-point scale.

The sixteen propositions were then placed in a table with the expected and actual levels of acceptance that were found for each proposition. Tabular analysis was conducted in order to analyse the change in SMS mobile advertising acceptance through the various combinations of the set of variables (permission, WSP control, content, and delivery).

Table 2 presents 16 new propositions based on varying combinations of the identified factors, ranked according to the importance of factors in the order of: 1) permission, 2) WSP control, 3) content and 4) delivery, as well as the number of positive factors (0, 1, 2, 3 or 4). As a result, we have a new set of expected acceptance levels.

The survey provides empirical evidence regarding the acceptance of these 16 scenarios. According to the average scores, each scenario was ranked for acceptability. Alongside, scores were categorized according to the level of acceptability:

- Less than 1.5 = unacceptable
- 1.5 to 2.49 = accept reluctantly
- 2.5 to 3.49 = acceptable
- 3.5 or more = accept enthusiastically

The results obtained in the survey demonstrated that propositions 6, 7, 8, 11, 12, 13, 14, 15 and 16 were supported, while propositions 1, 2, 3, 4, 5, 9, and 10 were not found to be supported by the data collected.

Overall, consumer acceptance of mobile advertising messages was much lower than expected. Over 50% of respondents answered unacceptable to more than 10 out of the 16 scenarios put forward to them, with the average number of scenarios answered as unacceptable being 9. On

Table 2. Revised model with the scenarios of marketing acceptance and survey results

Proposition	Permission	WSP Control	Con-tent	Delivery	Expected Acceptance Level	Aver-age	Rank	Actual level
1	***High***	***High***	***High***	***High***	Accept Enthusiastically (4)	3.16	1	Acceptable
2	***High***	***High***	***High***	*Low*	Acceptable (3)	1.60	8	Accept reluctantly
3	***High***	***High***	*Low*	***High***	Acceptable (3)	1.99	3	Accept reluctantly
4	***High***	*Low*	***High***	***High***	Acceptable (3)	2.29	2	Accept reluctantly
5	*Low*	***High***	***High***	***High***	Acceptable (3)	1.91	4	Accept reluctantly
6	***High***	***High***	*Low*	*Low*	Accept reluctantly (2)	1.50	9	Accept reluctantly
7	***High***	*Low*	***High***	*Low*	Accept reluctantly (2)	1.70	5	Accept reluctantly
8	***High***	*Low*	*Low*	***High***	Accept reluctantly (2)	1.63	7	Accept reluctantly
9	*Low*	***High***	***High***	*Low*	Accept reluctantly (2)	1.43	10	Unacceptable
10	*Low*	***High***	*Low*	***High***	Accept reluctantly (2)	1.41	11	Unacceptable
11	*Low*	*Low*	***High***	***High***	Accept reluctantly (2)	1.66	6	Accept reluctantly
12	***High***	*Low*	*Low*	*Low*	Unacceptable (1)	1.41	12	Unacceptable
13	*Low*	***High***	*Low*	*Low*	Unacceptable (1)	1.30	14	Unacceptable
14	*Low*	*Low*	***High***	*Low*	Unacceptable (1)	1.30	15	Unacceptable
15	*Low*	*Low*	*Low*	***High***	Unacceptable (1)	1.39	13	Unacceptable
16	*Low*	*Low*	*Low*	*Low*	Unacceptable (1)	1.19	16	Unacceptable

the other hand, nearly 70% of the respondents did not answer 'accept enthusiastically' to anything, and of the 30% that did give this response for at least one scenario, more than 80% only gave this response for one or two of the questions.

Of all the propositions the highest level of acceptance for mobile advertising was as expected for proposition one, which had the greatest degree of control and privacy trust. However, it can be seen that even where consumers have given permission, the content of the message was relevant, the delivery appropriate, and the message had come through the WSP, a message was on average only acceptable, with just 31% of respondents accepting this message enthusiastically. Alternatively, on average the lowest level of acceptance (unacceptable) was found where there was a low level of all these factors. Only 9 out of the 70 participants answered anything other than unacceptable for this question. This result was expected and consistent in proving proposition 16.

Permission and delivery of the message were the two variables that were found to equally have the most influence on the participant's level of acceptance, while content was found to be the next most important factor with control of the wireless service provider having the least amount of impact on the level of acceptance. Participants were more likely to accept messages that had a lower level of wireless service provider control or irrelevant content than messages which they had not given permission for or that came at an inappropriate time or frequency. This was shown again in Table 2, rows 12-15, where participants found scenarios 13 and 14 more unacceptable, despite having high levels of wireless service provider control and content respectively, than scenarios 12 and 15 where there were higher levels of permission and appropriate delivery respectively.

It is interesting to note that consistent with the propositions, the level of acceptance declined with the number of factors which had low levels, except in the case of proposition two, which was expected to generate the second highest level of acceptance and in actual fact dropped down to position eight. Where all other factors were high, yet the delivery of the message was inappropriate more than 50% of respondents found this message unacceptable, compared to just 26% of respondents who considered a message with low levels of WSP control unacceptable.

Looking at the other rankings of propositions from their expected to actual perceived influence on acceptance, just three propositions stayed in the same ranked position. However of the propositions that did get shuffled in rank, nine of these only moved within one or two ranks, with just three propositions moving three places or more. Participants found all messages which had three or more factors with low levels to be completely unacceptable. This was consistent with the expected results, and supported the propositions 12, 13, 14, 15, and 16. Messages which only had high levels of WSP control or relevant content were found to be 10% less unacceptable than messages with only high levels of permission or appropriate delivery – this supporting the theory that permission and delivery of the messages are perceived to be the most important factors.

DISCUSSION

The findings indicate a number of factors that are critical to the acceptance of mobile advertising by consumers. While the empirical testing showed that some factors are more important than others in influencing the overall level of acceptance, it was found that all factors played a significant role.

Peters and Fletcher (2005) found that where companies understand (and respond appropriately to) consumer concerns regarding interactive communications, then sound customer relationship management can lead to increased customer responsiveness.

Consistent with the literature explicit permission was found to be essential (Godin, 1999; Enpocket, 2003; Barnes and Scornavacca, 2004). It is our position, and one that is shared by other researchers in the field (Gardiner Jones, 1991; Cespedes and Smith, 1993; Dommeyer and Gross, 2003; Peters and Fletcher, 2005) that consumers have different acceptance thresholds depending on the experiences they have had in the past. Consumers are likely to want the same control they have over their web communications, controlling the type of information sent, how it is sent, and (most importantly) who sends it.

Due to the personal nature of the phone, and experiences with unsolicited spam via e-mail, users were weary of receiving marketing to their cell phones, and a number of privacy issues were raised in the focus groups. Literature has suggested that frequency and time is linked to targeting, where users are happy to receive messages at a higher frequency so long as the relevance to them is maintained (Enpocket 2002a). This was supported by the empirical testing where it shows messages with a low level of relevant content yet appropriate delivery were found to be much more acceptable than messages with a low level of relevant content and inappropriate delivery (a higher frequency). While participants in the focus groups made a point of saying that it is useless receiving any messages containing content which is irrelevant, there are a number of possible reasons why the respondents may have found delivery to be more important. If a consumer receives a message which is irrelevant to them occasionally, and it doesn't come at a disturbing time they may not be that bothered by it. On the other hand if they were to receive messages on a topic that was relevant to them, but were receiving these messages continuously and at interruptive times, they are likely to be more unacceptable.

It was interesting to see that the control of the wireless service provider had the least impact on consumer acceptance in the survey results, conflicting with the results of the focus groups where participants expressed their strong opinions towards the importance of WSP control. The results may in fact indicate that where consumers receive messages they find disturbing or intrusive they would rather it hadn't come from the service provider they trust. The focus groups indicated this, stating that they trust their service provider's judgment and would expect them to behave responsibly. Consumer attention seems more likely to divert to the filter when they are receiving unsolicited messages that they find disturbing.

Despite literature showing mobile advertising to be a successful tool in building brand awareness, and an important factor in consumer acceptance (Enpocket 2002b, Dickinger *et al.* 2004), the study revealed that the brand being marketed may have very little impact in acceptance of messages. Consumers are more likely to care whether a brand has been accepted by their service provider and has come through a filter, than about their level of trust between two different brands. Despite having a high trust in a brand, consumers are still doubtful of the bona fide nature of these messages when they have come direct from the company. They are also less likely to care about the brand that is being marketed to them than whether or not the content is relevant. The importance that is placed on brand is likely to increase when all other factors are high, and there is more choice in the market. Currently there are a limited number of brands being marketed through the mobile phone and more attention to brand is likely to arise in the future where consumers receive similar messages, with all other factors being equal, from competing brands.

Study Two: A Multi Criteria Approach to Assessing Customer Opt-In

In this study, we were interested in finding the general basis for an individual's decision to opt-in to a mobile advertising campaign with a view to developing a clearer picture of the decision model (in this case the decision hierarchy) in operation.

As mentioned in the previous study, permission is a core issue for mobile marketing success, and in several countries it is required by law that users opt-in before companies can send them messages via the mobile channel (Barnes and Scornavacca, 2008). In the UK, under Privacy and Electronic Communication Regulations, permission is a requirement of SMS ads, as is opt-out and data protection from misuse and inaccuracy. Similar legislation is found in other parts of the EU; for example, in Scandinavia you cannot approach clientele with SMS in any way before obtaining permission and so other media must be used to attract attention.

Based on a review of the salient research in this area and on concurrent studies by the authors (e.g., Bamba and Barnes, 2007; Bauer et al., 2005; Carroll et al., 2007; Haghirian et al., 2005; Karjaluoto and Kautonen, 2006; Kavassalis et al., 2003; Leppäniemi and Karjaluoto, 2005; Nysveen et al., 2005; Scornavacca and McKenzie, 2007; Virtanen et al., 2005) a broad set of decision criteria were developed. The decision criteria included:

A. The context of receiving a message, e.g., time, location, frequency and social context;
B. The opinion of others, e.g., friends, family and work colleagues;
C. The brand contained in the message (in terms of awareness, trust and loyalty);
D. The characteristics of the message, e.g., content, media and uniqueness;
E. The degree to which messages complement other channels of advertising;
F. The value of the message (information, entertainment or for social interactions); and
G. The control of the advertising channel by the operator to filter content.

A number of second and third order sub-criteria were also included in questionnaire, but not for all first-order categories, since one of these (message context) would be the focus of another, related study. We wanted to make the size of the questionnaire more manageable.

The questionnaire was developed for use in the Expert Choice software package. For each parent node in the hierarchy, starting with the top node, "Goal: The decision to opt-in to a mobile advertising campaign", the respondent was required to rate every permutation of decision criteria options. Thus, the respondent would be asked to rate the decision between criterion A and criterion B on a scale from 1(=equal, the mid-point) through 3(=moderate), 5(=strong), 7(=very strong) and 9(=extreme) preference for A or B. The questionnaire also collected information about gender, age, personal income, mobile replacement plans, average text messages sent and received per day, average advertising messages received per week, and whether mobile advertising has influenced a purchase decision in the past.

The questionnaire was administered in a major New Zealand University in May 2006. Only respondents using a mobile phone that can send and receive text messages were eligible to complete the survey. In all, 111 useable responses were received. The responses were input into Expert Choice for analysis. Thus, the decision priorities of the sample and of various groups were examined, compared and contrasted. Unfortunately, the software does not allow statistical testing of decision priorities between groups, which we were very interested in. To enable this, we printed the global decision hierarchy for each respondent and manually input the data into Excel, along with demographic and other details from Expert Choice. The data was then imported into SPSS for statistical testing. This enabled us to use ANOVA to compare mean global priorities and for additional graphical output such as radar charts.

Findings

The sample collected was 63.1% female, 36% male and with one non-response. Approximately 90% of the sample was under 25 years of age and three-quarters had an income of less than NZ$2500 per month (approximately €1320 or US$1670). Just over half of the sample planned to replace their mobile phone within a year. Some 35.1% of respondents sent and received more than 10 text messages per day, and 79.3% received 0 or 1 advertisement on their mobile phone per week. Overall, 18.9% of respondents stated that they had previously bought a product or service as a result of mobile advertising.

The overall decision hierarchy of the sample (n=111) is given in Figure 2 (where L = local priority and G = global priority), while Figure 3 gives a summary of the decisions between criteria in a matrix format (see the list of criteria above for key to the coding of A to G). Figure 3 shows an acceptably low level of inconsistency (1%).

Figure 2 and Figure 3 together give a picture of how the sample decided between characteristics. Message context is has the highest global priority (G: 0.213) and its preference was expressed over all other characteristics in Figure 3, with preferences over channel complementarity (E), message value (F) and message characteristics (D) being highest. At the other end of the spectrum, message value (F) and operator control (G) had the lowest first-order priorities (G: 0.107 and L: 0.099 respectively) in Figure 1, and all other characteristics were preferred to them in Figure 3, particularly message context (A), message characteristics (D), message brand (C) and social influence (B). Overall, there appears to be little to choose between the second-order priorities of message value (B).

The first-order priorities of other characteristics fall in between, with social influence rating at G: 0.156, followed by message brand (G: 0.153), message characteristics (G: 0.137) and channel complementarity (G: 0.135). Trust and loyalty for

Figure 2. Decision hierarchy for the sample (n=111)

- Goal: The decision to opt-in to a mobile advertisement campaign
  - The context of receiving a message (L: .213 G: .213)
  - The opinion of others, e.g. friends, family and work colleagues (L: .156 G: .156)
  - The brand contained in the message in terms of your awareness, trust and loyalty (L: .153 G: .153)
    - Trust in the brand contained in a message (L: .412 G: .063)
    - Loyalty to the brand contained in a message (L: .359 G: .055)
    - Awareness of the brand contained in a message (L: .228 G: .035)
  - Characteristics of the message, e.g. content, media, uniqueness (L: .137 G: .137)
    - The use of a personalised filter for message content (L: .382 G: .052)
    - Uniqueness of message content (L: .326 G: .045)
    - Use of multimedia in the advertising message (L: .292 G: .040)
      - The use of video content (L: .372 G: .015)
      - The use of graphical content (L: .271 G: .011)
      - The use of audio content (L: .189 G: .008)
      - The use of text content (L: .168 G: .007)
  - Text messages complement other channels of advertising (L: .135 G: .135)
    - Television advertising (L: .389 G: .052)
    - Website advertising (L: .271 G: .037)
    - Print advertising (L: .196 G: .026)
    - Radio advertising (L: .145 G: .019)
  - The value of the message (information, entertainment or for social interactions) (L: .107 G: .107)
    - Information value of the message (L: .358 G: .039)
    - Social value of the message, e.g. for sharing with friends (L: .336 G: .036)
    - Entertainment value of the message, e.g. fun, entertaining (L: .305 G: .033)
  - The control of the advertising channel by the operator, e.g. Vodafone, to filter content (L: .099 G: .099)

a brand had the highest local priorities for C, while television and website advertising were the preferred channels for E. There appears to be a balance in preferred message characteristics, although personal message filtering is slightly ahead in priority (L: 0.382). Video and graphical outputs were by far the most preferred forms of advertising media.

We used some different methods to analyse decision hierarchies than is typically the case. The respondent decision data was compared for various groupings in Excel and SPSS. Figures 4, 5, 6 and 7 show radar charts of four groups compared in terms of their first-order priorities. For each figure the blue line represents the overall sample. Alongside, ANOVA was used to statistically test the differences in priorities for these groups, plus messages sent, messages received, frequency of phone replacement and respondent age. There were no significant differences in the high and low groupings of age and volumes of messages sent and received. There was third-level item that was different ($p=0.014$) between those that planned to replace their phone in under or over a year: the use of video content. Presumably this means people wish to upgrade their phones sooner to receive better video capabilities.

Figure 4 to Figure 7 show the groupings for which the differences were most pronounced. In Figure 4, the biggest differences between high and low income groups appear to be in terms of receiver's context, brand and message characteristics. The first two are significantly different ($p=0.005$ and $p=0.03$ levels respectively), with

Figure 3. Matrix summary of preferences for first order characteristics

	A	B	C	D	E	F	G
A		1.32059	1.36328	1.70175	1.9032	2.18599	1.5595
B			1.07895	1.27231	1.13979	1.26259	1.53669
C				1.25278	(1.06604)	1.59883	1.5427
D					(1.00827)	1.52407	1.65182
E						1.2938	1.27893
F							1.41559
G	Incon: 0.01						

Figure 4. Global priorities – income

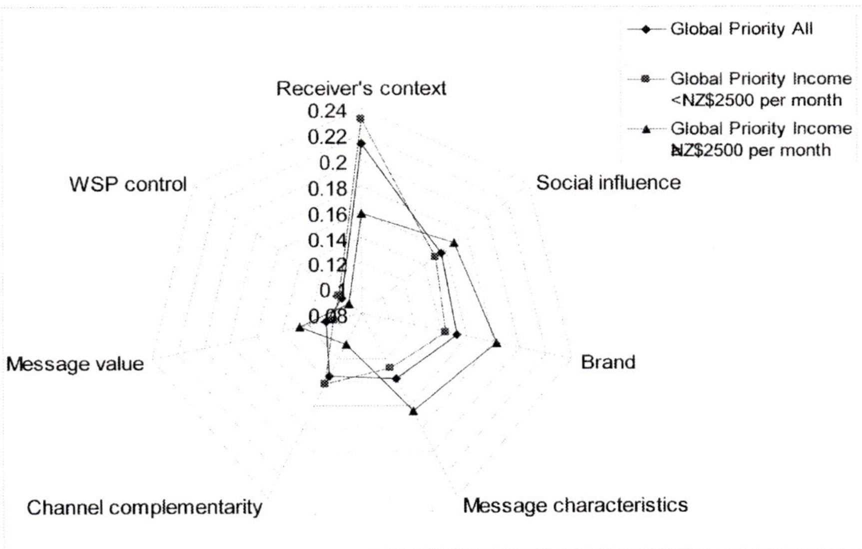

the second-order priority of brand loyalty being significantly different (p=0.02). Other notable differences were second-order, namely message uniqueness (p=0.023) and information value of the message (p=0.038). For gender (see Figure 5), the biggest differences in priority were again in message brand (p=0.045), driven by brand trust (p=0.042), and in message value (p=0.01), driven by information value (p=0.006).

Looking at buying behaviour (see Figure 6), the biggest second-order differences were in brand trust (p=0.023), message uniqueness (p=0.017) and websites as a complementary advertising channel (p=0.027). Finally, Figure 7 looks at differences in advertising messages received. The biggest second-order differences were in brand trust (p=0.038) and websites as a complementary advertising channel (p=0.017).

The results demonstrate significant differences between the decision-making criteria of various groups including by income, gender, volume of advertising messages received, and the influence of mobile advertising on previous purchases. The implications for practice are that every group of consumers for a mobile marketing message must be understood not

Figure 5. Global priorities – gender

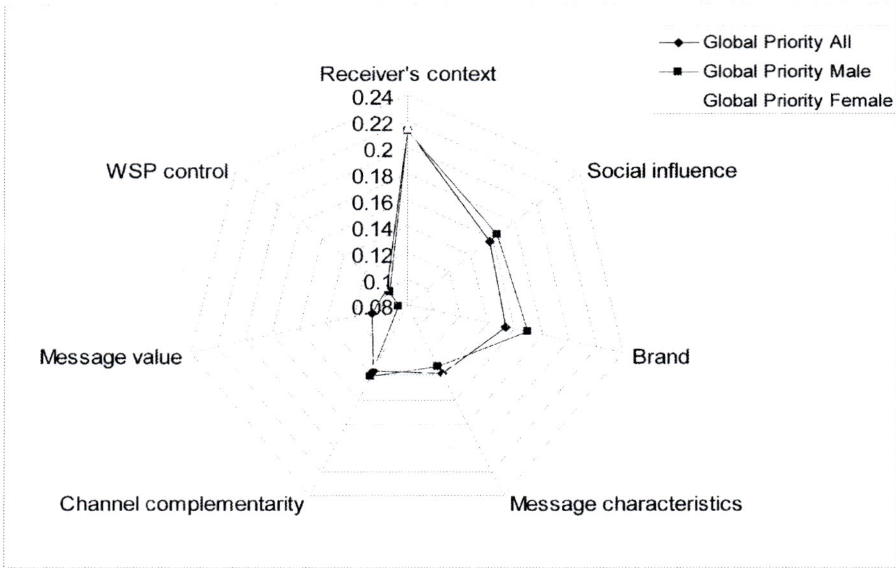

Figure 6. Global priorities – buyers

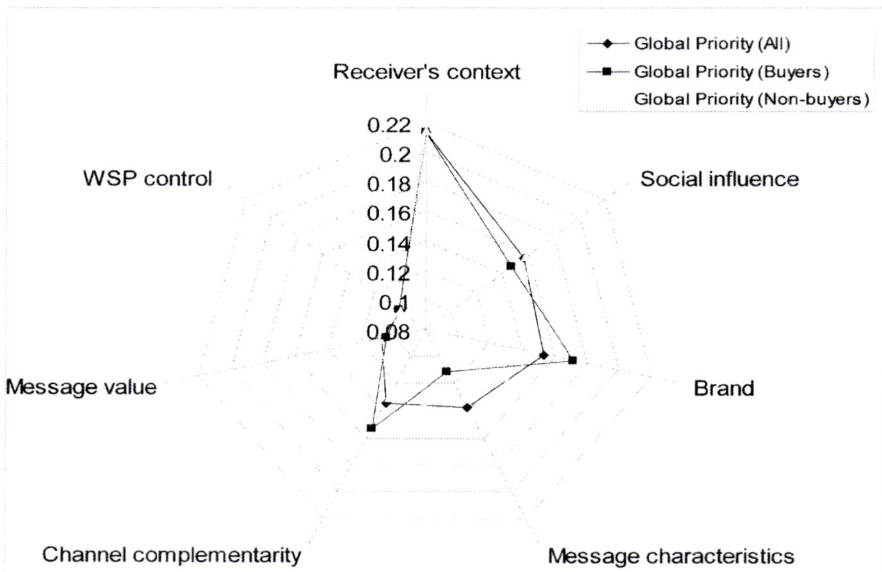

only in terms of consumption patterns, but in terms of decision patterns involved in the initial decision to opt-in to a mobile advertising campaign. Every advertising campaign must be packaged appropriately to achieve opt-in to the right audience. In the present environment no mobile advertising will succeed unless it overcomes the opt-in hurdle; before mobile customers even view advertising messages they must choose to receive them.

Figure 7. Global priorities – ads received

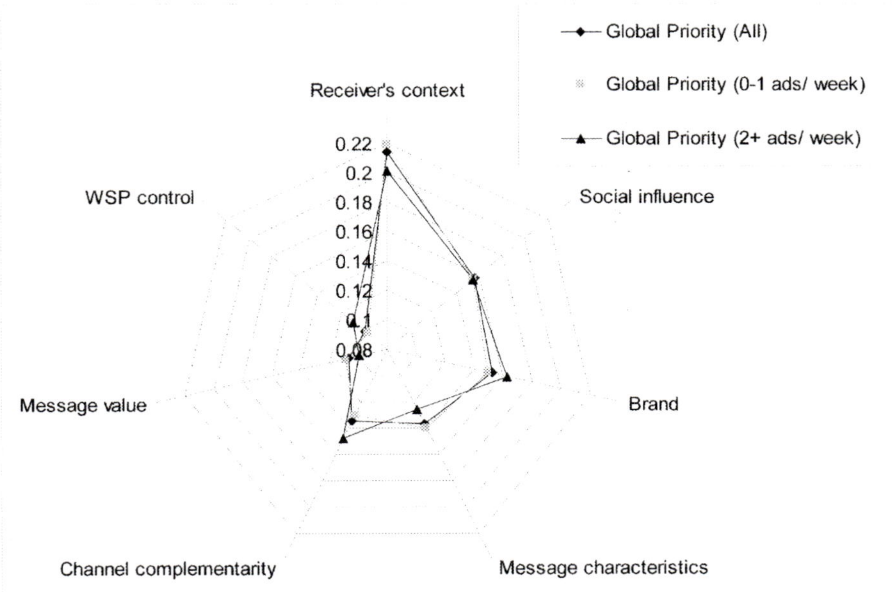

For example, female respondents had significantly less brand trust and more preference for message value, particularly information; opt-in signals for female respondents are likely to be more informative since brand may be less influential. Similarly, those who received more advertising messages trust brand more – probably as a filter – and link more to other electronic channels, namely websites. Thus links to and from branded websites may be more likely to obtain opt-in for this group. Among higher income groups, message context was preferred less and brand loyalty, message uniqueness and information value were more pertinent; those with higher income appear more attuned to branded, unique, valuable content. Similarly, those who have bought from mobile ads before are brand sensitive and value message uniqueness more. They also see websites as a more complementary channel. This last group provides some valuable guidance; those who tend to buy from mobile advertising appear influenced by brand trust, message uniqueness and multiple electronic channels.

Study Three: A Business Perspective on Mobile Advertising

In this last study, our purpose was to investigate the critical success factors of SMS based marketing campaigns from a managerial perspective. In order to achieve this goal we used exploratory multiple case study methodology. Four case studies were conducted with organizations based in New Zealand. Case research is appropriate for researching an area where theories are at formative stages, for emerging technologies, and where little research has been completed (Benbasat et al., 1987). Case research is also appropriate for examining practice based problems, since it allows a researcher to capture the knowledge of practitioners and use it to generate theory (Benbasat et al., 1987).

The investigated organizations were identified opportunistically. The organization selection criteria was the following: (a) should be deploying SMS based campaigns as a marketing tool for the past 12 months; and (b) should be operating in the New Zealand marketplace. Initially six potential companies were identified and contacted via email. Fortunately four of them agreed to participate in this study (names suppressed for confidentiality). In each organization a semi-structured interview with a manager - who has been deeply involved with their SMS based campaigns – was carried out. The cases were as follows:

- **Case 1** is a leading mobile solutions provider that offers a variety of services such as SMS based campaigns. It is located in Wellington and has been involved in some of the most successful SMS based campaigns in the country.
- **Case 2** is a leading international advertising company. It offers a large range of marketing and advertising strategies over multiple communication channels including SMS based campaigns.
- **Case 3** is a Wellington based company offering experiential marketing solutions via non-traditional channels such as SMS.
- **Case 4** is a radio station based in Wellington. The organization combines radio with the

Table 3. CSF of SMS based campaigns found in the mobile business literature

CSF	Definition	Authors
Acceptance	Consumers present a favourable attitude towards mobile marketing.	Barnes and Scornavacca, 2004; Carroll et al., 2007
Complementing	SMS is used in conjunction with other media in order to strengthen advertising campaign.	Dickinger et al., 2004 ; McManus and Scornavacca, 2005
Content	The content of the message (text or multimedia). It should be entertaining, informative and be relevant to the target audience and not irritating.	Dickinger et al., 2004 ; Haghirian et al., 2005 ; Barnes and Scornavacca, 2004
Cost	The cost of running a campaign. It usually can be justified due to campaigns' ability to reach customers.	Dickinger et al., 2004
Credibility	The consumers' perception of truthfulness and believability of advert in general. It also can be understood as brand trust regarding the organization sending the message.	Barnes and Scornavacca, 2004; Carroll et al., 2007; Haghirian et al., 2005
Frequency	The number of advertising messages received per day. It should not cause irritation to customers.	Haghirian et al., 2005; Carroll et al., 2007
Permission	Customers give permission to receive messages from the campaign operator.	Heinonen and Strandvik, 2003; Barnes and Scornavacca, 2004; Haghirian et al., 2004 ; Jelassi and Enders, 2006; Carroll et al., 2007 Haghirian et al., 2005 ; McManus and Scornavacca, 2005
Personalization	Users receive personalised advertising through their unique identifier codes which is customised and targeted correctly.	Barnes and Scornavacca, 2004 Dickinger et al., 2004; Jelassi and Enders, 2006; Haghirian et al., 2005
Reach	The number of consumers that can be reached through a campaign.	Jelassi and Enders, 2006; McManus and Scornavacca, 2005
Response time	The amount of time taken to respond to an event generated by the campaign operator or consumer. In order to generate interactivity, it is imperative the message sent makes the consumer respond quickly.	Dickinger et al., 2004; Haghirian et al., 2005
Richness	Message format, ubiquity, media richness and interactivity.	Jelassi and Enders, 2006; Haghirian et al., 2005 ; McManus and Scornavacca, 2005
Time	The time that the transmission of message occurs. It should not cause irritation towards customer.	Dickinger et al., 2004; Haghirian et al., 2005; Barnes and Scornavacca, 2004

mobile channel (SMS) in order to deliver interactive campaigns.

The data collection was carried out primarily through semi-structured face-to-face interviews (Benbasat et al., 1987; Creswell, 2003). They lasted between 50 minutes to 70 minutes. The interviews were recorded on audiotape and supplementary field notes were made during the interview. Some supplementary data was collected through supporting documents volunteered by the interview participants. All interviews were transcribed as soon as possible after its conclusion. Subsequently, coding procedures and a matrix were used to analyze the data (Creswell, 2003).

In order to avoid mis-interpretation the interviewer presented Rockart's (1979) definition of a critical success factors (CSF). Once a clear understanding of the term was obtained, the following open ended question was presented: "*In your opinion, what are the CSFs of SMS based campaigns in your organization?*" The following step was to present in alphabetical order a list of 12 CSFs that emerged from the literature review (Table 3). For each Critical Success Factor the interviewee was asked to indicate his/her perception regarding: a) what role the CSF plays in SMS based campaigns; and b) how important it is to SMS based campaigns. The final step was to request them to rank the top five CSFs from the provided list. The following section presents the results obtained.

After each manager pointed out and ordered what they believe to be the top five CSFs from Table 3, a point scheme was applied in order to generate an overall ranking. Five points were attributed to CSFs rating "first place", four points to "second place", three points to "third place", two points to "fourth place", and finally one point to the CSFs on "fifth place". CSF that did not make the manager's "top five" list did not obtain any points. Table 4 presents the results of the ranking process.

Findings

During the data collection it was evident that **content** plays a vital role on SMS based campaigns. Each of the managers placed content in their top three of their list - two of them ranked it as the number one CSF. One of them mentioned "*Content is king! Good content makes a differ-*

Table 4. Overall CSF ranking

CSF	Rank (Points)				Final Score	Final Rank
	Case 1	Case 2	Case 3	Case 4		
Content	2 (4)	1 (5)	1 (5)	3 (3)	17	1
Permission	1 (5)	-	3 (3)	1 (5)	13	2
Acceptance	-	3 (3)	-	2 (4)	7	3
Richness	-	2 (4)	4 (2)	-	6	4
Complementing	5 (1)	4 (2)	-	4 (2)	5	5
Credibility	-	-	2 (4)	-	4	6
Personalization	3 (3)	5 (1)	-	-	4	6
Response Time	-	-	5 (1)	5 (1)	2	7
Time	4 (2)	-	-	-	2	7
Cost	-	-	-	-	0	-
Frequency	-	-	-	-	0	-
Reach	-	-	-	-	0	-

ence as it provides a positive user experience". Content mostly refers to whether the message is entertaining, informative and relevant to the audience (Haghirian et al., 2005). Most managers have previously experienced issues with a campaign being unsuccessful when the messages sent to the audience were not clear, entertaining, useful or wanted.

Permission also appeared as one of the most significant CSF. It was perceived as being a key factor to avoid "brand damage" or cause irritation to consumers. Barnes and Scornavacca (2004) point out that permission is the dynamic boundary produced by the combination of one's personal preferences. It is well known among marketers that asking for a customer's permission is better and easier than asking for forgiveness (Peppers and Rogers, 1993). Since a mobile phone is a much more personal environment than a mailbox or an e-mail inbox, undesired messages certainly generate a very negative impact on consumers' perceptions (Barnes and Scornavacca, 2004; Haghirian et al., 2005; McManus and Scornavacca, 2005).

Vodafone and Telecom NZ have installed mechanisms that prevent the occurrence of spam on mobiles phones in New Zealand. Two managers mentioned that as SMS continues to grow; preventing organisations from spamming mobiles may prove to become difficult. They believe that in the future, permission will play a more dominant role in New Zealand's SMS mobile marketing arena. According to them, companies that place an emphasis on permission will be rewarded. In addition, they mentioned that managers of SMS based campaigns should follow strictly a code of ethics in order to be successful.

Acceptance also received a relevant ranking even though most managers did not seem to be concerned about this CSF when answering the open ended question. They perceived that acceptance may be subjective to incentives (e.g. prices) and usually complements other factors such as relevance, promotion and authenticity. Barnes and Scornavacca (2004) understood that consumer's acceptance of mobile marketing derives from user's permission, mobile operator control and brand trust.

Richness was perceived as an important variable on campaigns that requires a high level of interaction. Most managers pointed out that interactivity plays a key role in creating a successful campaign which usually draws interest from a range of target audiences and has the ability to retain users' interest. Jelassi and Enders (2006) observed that richness has three sub dimensions: interactivity, ubiquity and format. Most managers stated that richness will become more important as 3G handsets gain popularity due to the improvement of the technological platform (Evans and Wurster, 1997).

Against the researchers expectations, **complementing** did not receive high a ranking. However, all managers believed that an SMS campaign needs to be complemented by other channel (s). Only one person mentioned that SMS could be used as a "*single channel approach without complementation*", however, he did not regard as a wise strategy. Using a channel such as TV, radio or website to generate a "call-to-action" has proven to be very effective way of initiating interactions with users. In addition, allowing consumers to choose between SMS or a website in order to enter competitions has shown a significant increase in the number of participants in comparison to those campaigns that only used one channel.

Only one manager thought that **credibility** was an important issue. The other three managers perceived that it was not relevant in the New Zealand context due to the high level of spam control and opt-in nature of the campaigns. Surprisingly previous studies with end-users (e.g. Haghirian et al., 2005) found that consumers' strongly relate advertising value with perceived credibility.

Personalization was perceived by most of the managers as a factor that currently is not significant. Perhaps this is due to the fact that most of the campaigns these managers have developed did not offer a high level of personalization. However, they believe that as technology evolves it will certainly play a key role in the future. One mentioned *"people don't like to feel like a number.... personalization will help to avoid this.. "*.

Response time was regarded as an important factor for transaction confirmation and creation of trust. Some managers mentioned that consumers can become concerned if they do not receive an immediate response from the campaign operator right after they send a message. Some managers perceived that many consumers may think that if they do not receive a reply the information they have just sent would be "floating in the air" and did not reach the campaign operator.

Time did not seem to be a relevant issue for most of the managers. However, they mentioned that some recent campaigns are being developed around new concept called "mission critical". In this type of highly interactive campaign consumers opt-in to participate and they must respond to the messages sent to them within a certain amount time in order to be rewarded. Perhaps time was not considered a critical factor because personalized and location dependent campaigns are still in an embryonic stage in New Zealand (Scornavacca and Barnes, 2004). As technology evolves, some factors such as time, location and personalization should play a more significant role in mobile marketing (Carroll et al., 2007).

Another interesting finding was that most managers believed that **cost** was not relevant to the success of an SMS based campaign. They believe that SMS based campaigns (especially in a multi-channel approach) are very cost effective. Concomitantly, **frequency** was also perceived as not relevant due to the fact that all campaigns in New Zealand are based on user's permission. The managers reported as a by-product of permission - *"when a user gives permission to receive messages, they are most likely to be aware of the number and frequency of messages that they will receive"*. It is interesting to note that Haghirian et al. (2005) found that a high frequency of exposure tend to decrease the value of advertising.

Finally **reach** was also not regarded as a key CSF. Most manager perceived it as *"a given"* since New Zealand enjoys a high level of mobile converge – especially in urban areas. It was also perceived as a good selling point when trying to entice a company to invest in an SMS based campaign.

The findings indicate that most campaign managers perceived that having a clear understanding of their clients' needs and goals, good planning, as well as deep knowledge of the main capabilities of the mobile medium are critical factors to ensure the success of SMS based campaigns. In addition, a number of CSFs drawn in the literature were also perceived by the manager as important aspects of SMS marketing.

Content was perceived as the most important CSF of SMS based campaigns. Managers believed that a positive experience for the user relies on the message content being entertaining, interactive informative and relevant to the audience. Permission was also seen as one of the most significant aspects of a SMS based campaign. Since the New Zealand marketplace has strict restrictions on companies' ability to purchase mobile phone numbers and push advertisements without permission, all managers perceive this factor as a key to ensure a sustainable future for this channel. They also pointed out that messages received without permission are likely to cause irritation as well as harm their clients' brand and credibility. In addition, they believed that user acceptance can be increased by providing better incentives. Finally, SMS campaign managers shared a unanimous opinion that SMS should be used in a multi-channel approach, complementing other media.

DISCUSSION AND CONCLUSION

The rapid proliferation of SMS based campaigns is an exciting and growing phenomenon worldwide. The immediacy, interactivity and mobility of wireless devices provide a novel platform for marketing. Through the increasing deployment of mobile technologies as a marketing tool, SMS based campaigns are becoming a more attainable channel for a range of businesses.

The aim of this chapter was to further understanding the critical success factors of SMS advertising campaigns by empirically exploring businesses' and consumers' perspectives on this matter. The three studies reported in this chapter show that most people want protection against unwarranted uses of personal information with minimal damage to their increased choice and flexibility as consumers. Similarly, no company has an interest in spending money on marketing efforts that annoy or alienate potential customers. Most companies want access to consumers so their products and marketing programs can respond to dynamic changes in demand. Unfortunately, this does not always appear to be the case to consumers who often experience concerns regarding their own privacy rights and the responsible use of personal messages by firms. The results in the studies highlight the importance of consumer perceptions and acceptance levels of mobile advertising.

The findings reported throughout the three studies suggest that marketers should be optimistic about choosing to deploy mobile advertising, but should also exercise caution around the factors that will determine consumer acceptance. While consumers can see the potential in the mobile medium, they are weary of receiving messages which they don't want. Quality and richness of content, context and frequency of receiving messages, along with cross-channels linkages and complementarity, particular for the Web and television, are important drivers of successful marketing campaigns.

Further, the findings, especially in the second study, demonstrate significant differences between the decision-making criteria of various groups including by income, gender, volume of advertising messages received, and the influence of mobile advertising on previous purchases. The implications for practice are highlighted in the third study showing that every group of consumers for a mobile marketing message must be understood not only in terms of consumption patterns, but in terms of decision patterns involved in the initial decision to opt-in to a mobile advertising campaign. Every advertising campaign must be packaged appropriately to achieve opt-in to the right audience.

It was noticeable that obtaining user trust and permission will be a key challenge faced by marketers and future research should focus on ways to overcome these challenges. Consumers are more likely to trust messages coming from their service providers than anywhere else and so it is important that service providers give a high level of filtering and protection as reassurance for their users. The findings from the first and third studies emphasize that trust and permission are necessary factors of consumer acceptance; however they should not be seen as the only objectives. Attention needs to be focused around the relevance of the content and the timeliness and frequency of the delivery of marketing messages. Nevertheless, the research showed that simply focusing on contextual, content or permission/control factors in isolation is unlikely to result in a high or even moderate level of acceptance. Any application of these findings suggests that marketers need to take into account all these factors and how varying combinations of these factors will impact on consumer acceptance.

It is expected that popularization of 3G devices will be important in the evolution of mobile marketing campaigns; higher bandwidth allows the deliver a richer and more interactive content. Notwithstanding, it is important to keep the consumer in mind; the key to success is the management of and delivery upon user expectations. In addition, considerable effort is still needed in expanding consumer acceptance, legislation for privacy and data protection, and creating better pricing structures.

REFERENCES

Bamba, F., & Barnes, S. J. (2007). SMS advertising, permission and the consumer: a study. *Business Process Management Journal . Special Issue on Mobile CRM, 13*(6), 815–829.

Barnes, S. (2002). Wireless digital advertising: nature and implications. *International Journal of Advertising, 21*(3), 399–420.

Barnes, S. J., & Scornavacca, E. (2004). Mobile marketing: the role of permission and acceptance. *International Journal of Mobile Communications, 2*(2), 128–139. doi:10.1504/IJMC.2004.004663

Barnes, S. J., & Scornavacca, E. (2008). Uncovering patterns in mobile advertising opt-in behaviour: a decision hierarchy approach. *International Journal of Mobile Communications, 6*(4), 405–416. doi:10.1504/IJMC.2008.018050

Barwise, P., & Strong, C. (2002). Permission-based mobile advertising. *Journal of Interactive Marketing, 16*(1), 14–24. doi:10.1002/dir.10000

Bauer, H., Barnes, S. J., Neumann, M., & Reichardt, T. (2005). Driving consumer acceptance of mobile marketing: A theoretical framework and empirical study. *Journal of Electronic Commerce Research, 6*(4), 181–192.

Benbasat, I., Goldstein, D. K., & Mead, M. (1987). The case research strategy in studies of information systems. *MIS Quarterly, 11*(3), 369–386. doi:10.2307/248684

Carroll, A., Barnes, S. J., Scornavacca, E., & Fletcher, K. (2007). Consumer perceptions and attitudes toward SMS advertising: recent evidence from New Zealand. *International Journal of Advertising, 26*(1), 79–98.

Cespedes, F., & Smith, H. (1993). Database marketing: new rules for policy and practice. *Sloan Management Review, 34*, 7–22.

Chang, L. (1994). A psychometric evaluation of 4-point and 6-point Likert-type scales in relation to reliability and validity. *Applied Psychological Measurement, 18*, 205–216. doi:10.1177/014662169401800302

Creswell, J. (2003) *Research Design Qualitative, Quantitative, and Mixed Methods Approaches*, Second Edition. Thousand Oaks, CA: Sage Publications.

Dickinger, A., Haghirian, P., Murphy, J., & Scharl, A. (2004). An investigation and conceptual model of SMS marketing. *Proceedings of the 37th Hawaii International Conference on System Sciences*, Big Island, HI.

Dickinger, A., Scharl, A., & Murphy, J. (2005). Diffusion and success factors of mobile Marketing. *Electronic Commerce Research and Applications, 4*(2), 159–173. doi:10.1016/j.elerap.2004.10.006

Dommeyer, C., & Gross, B. (2003). What consumers know and what they do: an investigation of consumer knowledge, awareness, and use of privacy protection strategies. *Journal of Interactive Marketing, 17*(2), 34–51. doi:10.1002/dir.10053

Dow Jones Newswire. (2008). Global Mobile Phone Users Top 3.3 Billion By End-2007 – Study. http://www.cellular-news.com/story/31352.php (accessed 8 June 2008).

Enpocket. (2002a). Consumer preferences for SMS marketing in the UK. http://www.enpocket.co.uk (accessed 13 March 2003).

Enpocket. (2002b). The branding performance of SMS advertising. http://www.enpocket.co.uk (accessed 13 March 2003).

Enpocket. (2003). The response performance of SMS advertising. http://www.mda-mobiledata.org (accessed 13 March 2003).

Enpocket. (2005a). *Direct Response Report*. Available http://www.enpocket.co.uk (accessed 10 February 2006).

Enpocket. (2005b). *Consumer Preferences Report*. Available http://www.enpocket.co.uk (accessed 10 February 2006).

Ericsson. (2000). *Wireless Advertising*. Stockholm: Ericsson Ltd.

Evans, P. B., & Wurster, T. S. (1997). Strategy and the new economics of information. *Harvard Business Review*, *75*(5), 70–82.

Gardiner Jones, M. (1991). Privacy: a significant marketing issue for the 1990's. *Journal of Public Policy & Marketing*, *10*(1), 133–148.

Godin, S. (1999). *Permission Marketing: Turning strangers into Friends and Friends into Customers*. New York: Simon and Schuster.

Green, J. C., Caracelli, V. J., & Graham, W. F. (1989). Toward a conceptual framework for mixed method evaluation designs. *Educational Evaluation and Policy Analysis*, *11*(3), 255–274.

Gundlach, G. T., & Murphy, P. E. (1993). Ethical and legal foundations of relational marketing exchanges. *Journal of Marketing*, *57*, 35–46. doi:10.2307/1252217

Haghirian, P., & Inoue, A. (2007). An advanced model of consumer attitudes toward advertising on the mobile Internet. *International Journal of Mobile Communications*, *5*(1), 48–67. doi:10.1504/IJMC.2007.011489

Haghirian, P., Madlberger, M., & Tanuskova, A. (2005). Increasing advertising value of mobile marketing – an empirical study of antecedents. *Proceedings of the 38th Hawaii International Conference on System Sciences*, HI.

Heinonen, K., & Strandvik, T. (2003). Consumer responsiveness to mobile marketing. *Proceedings of the Stockholm Mobility Roundtable*, Stockholm, Sweden.

Heinonen, K., & Strandvik, T. (2007). Consumer responsiveness to mobile marketing. *International Journal of Mobile Communications*, *5*(5), 603–617. doi:10.1504/IJMC.2007.014177

Hristova, N., & O'Hare, G. (2004). Ad-me: Wireless advertising adapted to the user location, device and emotions. *Proceedings of the 37th Hawaii International Conference on System Sciences*, HI.

Jelassi, T., & Enders, A. (2006). Mobile advertising: a European perspective. In S. J. Barnes and E. Scornavacca, E. (Eds.), *Unwired Business: Cases in Mobile Business, (*pp. 82-95). Hershey: IRM Press.

Karjaluoto, H. An&d Kautonen, T. (2006). Trust as a factor affecting consumers' participation in mobile marketing. *Proceedings of the 35th European Marketing Academy Conference*, Athens, Greece.

Kavassalis, P., Spyropoulou, N., Drossos, D., Mitrokostas, E., Gikas, G., & Hatzistamatiou, A. (2003). Mobile permission marketing: framing the market inquiry. *International Journal of Electronic Commerce*, *8*(1), 55–79.

Leppäniemi, M., & Karjaluoto, H. (2005). Factors influencing consumer willingness to accept mobile advertising: A conceptual model. *International Journal of Mobile Communications*, *3*(3), 197–213. doi:10.1504/IJMC.2005.006580

McManus, P., & Scornavacca, E. (2005). Mobile marketing: killer application or new hype? *Proceedings of the 4th International Conference on Mobile Business*, Sydney, Australia.

Merisavo, M., Vesanen, J., Arponen, A., Kajalo, S., & Raulas, M. (2006). The effectiveness of targeted mobile advertising in selling mobile services: an empirical study. *International Journal of Mobile Communications*, *4*(2), 119–127.

Milne, G., & Gordon, M. (1993). Direct mail privacy-efficiency trade-offs within an implied social contract framework. *Journal of Public Policy & Marketing*, *12*(2), 206–216.

Milne, G. M., Boza, M. E., & Rohm, A. (1999). Controlling personal information in marketing databases: A consumer perspective. *Proceedings of the American Marketing Association Winter Meeting*, St. Petersburg, Florida.

Moorman, C., Deshpande, R., & Zaltman, G. (1993). Factors affecting trust in market research relationships. *Journal of Marketing*, *57*, 81–101. doi:10.2307/1252059

Nysveen, H., Pedersen, P. E., & Thorbjornsen, H. (2005). Intentions to use mobile services: antecedents and cross-service comparisons. *Journal of the Academy of Marketing Science*, *33*(3), 330–346. doi:10.1177/0092070305276149

Okazaki, S. (2004). How do Japanese consumers perceive wireless advertising? A multivariate analysis. *International Journal of Advertising*, *23*(4), 429–454.

Peabody, D. (1962). Two components in bipolar scales: Direction and extremeness. *Psychological Review*, *69*, 65–73. doi:10.1037/h0039737

Peppers, D., & Rogers, M. (1993). The new marketing paradigm: One to one. *American Advertising*, *9*(4), 20–22.

Peters, L., & Fletcher, K. (2005). They love me, they love me not: consumer relationship orientation and CRM. *Journal of Customer Behaviour*, *3*(1), 53–80. doi:10.1362/147539204323074600

Rockart, J. F. (1979). Chief executives define their own data needs. *Harvard Business Review*, *57*(2), 15–28.

Rossman, G. B., & Rallis, S. F. (1998). *Learning in the Field: An Introduction to Qualitative Research*. Thousand Oaks, CA: Sage.

Rotter, J. (1967). A new scale for the measurement of interpersonal trust. *Journal of Personality*, *35*(4), 651–665. doi:10.1111/j.1467-6494.1967.tb01454.x

Scornavacca, E., & Barnes, S. J. (2004). Raising the bar: barcode-enabled m-commerce solutions. *Proceedings of the Austin Mobility Roundtable*, Austin, Texas.

Scornavacca, E., Barnes, S. J., & Huff, S. L. (2006). Mobile business research published in 2000-2004: Emergence, current status, and future opportunities. *Communications of the Association for Information Systems*, *17*, 635–646.

Scornavacca, E., & McKenzie, J. (2007). Unveiling managers' perceptions of the critical success factors for SMS-based campaigns. *International Journal of Mobile Communications*, *5*(4), 445–456. doi:10.1504/IJMC.2007.012790

Thorelli, H. (1990). Networks: between markets and hierarchies. In D. Ford (Ed.), *Understanding Business Markets*. London: IMP Group.

Ververidis, C., & Polyzos, G. (2002). Mobile marketing using location based services. *Proceedings of the First International Conference on Mobile Business*, Athens, Greece.

Virtanen, V., Bragge, J., & Tuunanen, T. (2005). Barriers for mobile marketing and how to overcome them. *Proceedings of the Hong Kong Mobility Roundtable*, Hong Kong.

KEY TERMS AND DEFINITIONS

Acceptance (of Mobile Advertising): Consumer have a favourable attitude towards mobile advertising.

Information Privacy: The protection of information concerning a consumer from disclosure and abuse.

Opt-In: The decision of a mobile consumer to receive mobile advertising messages.

Opt-Out: The decision of a mobile consumer who currently receives mobile advertising messages not to receive them in the future.

Personalization (of Mobile Advertising): Mobile advertising messages are tailored to individual recipients using aspects such as preferences, time and location.

Pull Advertising: The marketer takes the initiative to send advertising messages to the consumer.

Push Advertising: The marketer sends information requested by the consumer with advertising embedded.

Chapter 16
The Impact of Gender and Age on Consumer Responsiveness to Permission-Based Mobile Advertising

Heikki Karjaluoto
University of Jyväskylä, Finland

Heikki Lehto
Brandson, Ltd, Finland

Matti Leppäniemi
University of Oulu, Finland

ABSTRACT

This chapter investigates the impact of gender and age on mobile marketing responsiveness among regular customers of Finnish night club chain. The study develops sixteen research propositions which are tested with a sample of 8578 members of the company's permission- based, opt-in mobile advertising list . The results contribute to mobile marketing and technology acceptance literature in various ways. First, the authors find support for four out of eight gender related propositions. Second, five out of the eight age-related hypotheses are supported. The results further show that among opt-in customers, gender and age explain various antecedents of intention to engage in mobile marketing. The study provides several theoretical and managerial contributions and outlines vital avenues for further investigation in the field.

INTRODUCTION

Literature on mobile marketing has grown steadily during the last five years, but still the study of mobile marketing is in its infancy. Even the term "mobile marketing" has not yet gained universally accepted definition (Leppäniemi & Karjaluoto, 2006). In short, mobile marketing refers to the use of a mobile media, as a means of marketing communications (Leppäniemi & Karjaluoto, 2006). Practitioners

DOI: 10.4018/978-1-60566-074-5.ch016

have defined mobile marketing as "*The use of wireless media as an integrated content delivery and direct response vehicle within a crossmedia or stand-alone marketing communications program*" (Mobile Marketing Association, 2008, p. 22). Thus, mobile marketing is a term used to describe marketing on or with a mobile device. By the same token, it is worth noting that the term "mobile advertising" is used to describe advertisements communicated to the consumer/target via a mobile handset. These advertisements include for example mobile web banners, full screen interstitials appearing while mobile web pages are loadings, SMS and MMS ads, and mobile in-game ads (Mobile Marketing Association, 2008). In our study we concentrate on SMS advertising, which is still dominating the mobile marketing landscape in western markets.

A considerable number of the publications on mobile marketing have examined consumer attitudes to mobile marketing (Haghirian & Madlberger, 2004; Leung & Cheung, 2004; Tsang, Ho, & Liang, 2004) and acceptance of mobile marketing (Bauer et al., 2005; Leppäniemi & Karjaluoto, 2005; Barnes & Scornavacca, 2004; Leppäniemi et al., 2005; Karjaluoto et al., 2008), with a growing interest in responsiveness, effectiveness, and branding effects (Leppäniemi & Karjaluoto, 2006). In addition, many new research areas are evolving. For instance, it can be expected that as GPS enabled mobile phones penetrate, location-based mobile marketing will become the next important research area for marketing scholars.

In this chapter our aim is to build on existing conceptualizations of mobile marketing acceptance (Karjaluoto et al., 2008) by looking at how two demographics, namely gender and age, explain the variation in the factors relating to the acceptance of mobile marketing. In other words, our objective is to throw more light on whether males and females differ in their evaluation of the importance of the constructs related to acceptance as well as whether we can find differences between the age groups.

We will build our theoretical background on the discussion of demographic differences in the use of mobile services and mobile marketing (Carlsson et al., 2005; Haghirian, Madlberger & Tanuskova, 2005; Hyvönen & Repo, 2005; Karjaluoto et al., 2006); Nysveen, Pedersen & Thorbjornsen, 2005; Okazaki, 2004). Literature indicates that there are certain differences between males and females in the use of mobile services, but relatively little empirical research has been conducted to date to examine differences in permission-based mobile marketing communications (Karjaluoto et al., 2006).

In a recent study, Karjaluoto et al. (2006) surveyed 4,062 Finnish online users' mobile marketing experience. Their examination covers two areas of mobile communications, namely general mobile communications and the mobile marketing experience. In regard to mobile communications, they report that consumers who are more willing to upgrade their mobile phones earlier are predominantly male, relatively young and with a lower level of education. In addition, the study finds that females and young people are more active in sending and receiving short messages. With regard to mobile marketing, the study shows that there are significant differences between genders and age groups. The age group 21 to 25 receives more mobile marketing messages and is the most active of the different age groups in requesting information via SMS. Interestingly, the study reports that females are more active in engaging in mobile marketing activities such as sweepstakes and interactive TV shows. Hyvönen and Repo (2005) report a similar observation. In a study of mobile service usage they found that females are more active than males in sending and receiving SMS, downloading logos and ring tones and in campaign participation.

BACKGROUND

This paper builds on the earlier work of Karjaluoto et al. (2008). The authors presented the mobile marketing acceptance model which posits that perceived ease of use and perceived usefulness together with trust have a positive interaction with attitude toward advertising in general, which in turn, together with perceived behavioural control and reference group influence, affect intention to participate in mobile marketing. Finally, Karjaluoto et al. found strong support that intention to participate affects intention to behave (in their study "visit night club").

Perceived Ease of Use and Perceived Usefulness of Mobile Communications

Following the original definitions of perceived ease of use and usefulness (Davis, 1989), we define perceived ease of use of mobile communications as the degree to which a person believes that using mobile phones would be free of effort. Perceived usefulness of mobile communications in turn refers to the degree to which a person believes that using mobile phone as a communication vehicle enhances his or her experiences of mobile communications.

With regard to perceived ease of use and perceived usefulness, the technology acceptance literature has a long tradition of acknowledging and investigating the effect of individual differences on acceptance (Burton-Jones & Hubona, 2005). However, partly due to the limited number of studies, there is still a lot of disagreement over whether gender and age have an interaction with ease of use or usefulness. In the seminal article on the topic, Gefen and Straub (1997) state that men are more affected by perceived usefulness while women are more affected by social norms and ease of use. Their study, being rooted in Hofstede's (1980) finding that sex affects perceptions and behaviours, also argues that men and women differ more in their perceptions of technology and less in their actual usage. In addition, Jackson et al. (2001) argue that gender differences also exist in actual usage. In a study of Internet usage in the USA, they found that American females tend to use e-mails more than their male counterparts. Venkatesh and Morris's (2000) study confirms the findings presented by Gefen and Straub (1997) insofar as they propose gender to be an important factor affecting perceived ease of use and perceived usefulness. Morris and Venkatesh (2000) propose that age has a negative direct influence on attitude toward using a new technology (formed by perceived ease of use and usefulness). In other words, younger people will place more emphasis on attitudinal factors than older people will. On this basis, we propose that:

P1a: Women will find it more difficult than men to use mobile services
P1b. Older people will find it more difficult than younger people to use mobile services
P2a: Men will rate the perceived usefulness of mobile communications higher than women will
P2b. Younger people will rate the perceived usefulness of mobile communications higher than older people will

Perceived Behavioural Control

Ajzen (1991) defines perceived behavioural control as a perception of "*the ease or difficulty of performing the behaviour of interest*". (p. 183). Mathieson (1991) argues that perceived behavioural control depends on control beliefs which relate to "*perception of the availability of skills, resources, and opportunities*" and perceived facilitation that refers to "*the individual's assessment of the importance of those resources to the achievement outcomes.*" (p. 176). In the context of mobile marketing and advertising, control over messaging is important insofar that mobile advertising is permission-based in most developed countries. Therefore, consumers can be seen as having control over whether or not to receive communications. However, technology acceptance studies (Venkatesh et al., 2000) have shown that behavioural control is more salient for women than for men in the early stages of adoption. In addition, literature suggests that perceived behavioural control is more salient for older people (Morris & Venkatesh, 2000). Therefore, we propose that

P3a. Women will regard control as being more important than men will

P3b. Older consumers will regard control as being more important than younger people will

Attitude Toward Advertising in General

Bush, Smith and Craig (1999) argue that prior research (e.g., Bauer & Greyser, 1968; Durand & Lambert, 1985) is relatively consistent in suggesting that attitudes toward advertising in general are only moderately related to people's gender and age. However, a review of gender research in advertising (Wolin, 2003) suggests that gender has an effect on the processing of advertisements. Literature also proposes that females process advertisements more elaborately and subjectively than men do (Wolin, 2003).

Although studies have found contradictory evidence that men exhibit more positive attitudes toward advertising than women do (O'Donohue, 1995), most research recognizes the limitations of investigating the direct effects of age and attitude toward advertising (Wolin, 2003). With these limitations in mind, we propose that

P4a. Women and men do not differ in their attitudes toward advertising in general

P4b. Age groups do not differ in their attitudes toward advertising in general

Social Influence

The second predictor of the theory of reasoned action and theory of planned behaviour, social influence, or subjective norm, refers to "*the perceived social pressure to perform or not to perform the behaviour*" (Ajzen & Madden, 1985, p. 454). Literature argues that women are more socially and more network-oriented than men whereas men have been found to be more goal, problem, and task-oriented as well as favouring more hierarchical communications (Gefen & Straub, 1997; Minton & Schneider, 1980). Deaux (1984) furthermore shows that men are more independent and competitive whereas women are more emotional and warm (see Gefen & Straub, 1997). Venkatesh and Morris (2000) support this view by saying that subjective norms are much stronger for women than for men in the initial stages of adoption of new technology. In addition, Venkatesh et al. (2003) find that social influence is more salient for women and older people. With respect to mobile communications, literature indicates that social norms are more important for women than for men (Nysveen et al., 2005). Therefore, we propose that

P5a. Women will be more affected than men by reference groups

P5b. Older people will be more affected than younger people by reference groups

Perceived Trust

Golesorkhi (2006) argues that the area of trust has not received much research attention in gender studies. The limited empirical evidence available indicates that men and women might differ in their level of trust of another party. Wang and Yamagishi (2005) for instance suggest that males are significantly more trusting of unknown partners than females are. Ndubisi (2006) finds that gender plays a strong moderating role between trust and loyalty. Kolsaker and Payne (2002) hypothesized that women are more concerned about trust than men are due to the fact that they are more relationship oriented. However, their study did not find any statistically significant gender differences. Golesorkhi (2006) reports that females regard the trust dimensions of benevolence, integrity and cultural similarity to be significantly more important than men do.

With respect to trust and age, the literature argues that age is a relevant characteristic for trust (Chen & Dillon, 2003; Kenning, 2008). The literature does not explicitly state that age and level of trust could be related. In fact, the suggestion is that level of trust remains relatively constant after childhood (Sutter & Kocher, 2004). Kenning (2008) for example found no relationship between age and general trust. In addition, a study conducted by Okazaki (2007) reveals that Japanese females are more likely to have a higher level of perceived trust when it comes to mobile advertising. On this basis, we propose that

P6a. Women will regard the importance of trust more highly than men will

P6b. Age groups do not differ in their trust perceptions

Intention to Participate and Visit

A growing body of research has shown that gender and age significantly affect a person's intention to participate in online and mobile marketing campaigns. Okazaki (2004) for example argues that in Japan, females are more active users of both e-newsletters and e-mail. Trappey III and Woodside (2005) show that females and younger people are more actively involved with mobile media than men are (see also Karjaluoto et al., 2006). However, research indicates that differences between age groups are also context related as older people in general are adopting SMS and other mobile services at a rapid pace (see Rettie, Grandcolas, & Deakins, 2005). With respect to advertising research, Wolin (2003) states that females need to be exposed to more ads than men. Okazaki (2007) report that Japanese women, regardless of their frequency of usage, are more likely to recall a mobile advertisement. On this basis there seems to be evidence that women and younger people are more likely to participate in mobile marketing campaigns. On the other hand, advertising studies (Wolin, 2003) suggest that men are more likely to respond to advertisements when fewer expositions are used. Based on mobile marketing research (Karjaluoto et al., 2006; Okazaki, 2004; 2007; Trappey III & Woodside, 2005) and advertising research (Wolin, 2003) we propose that

P7a. Females will have a stronger intention than males of participating in mobile marketing campaigns

P7b. Younger people will have a stronger intention than older people of participating in mobile marketing campaigns.

P8a. Men have a stronger intention to visit than women do

P8b. Younger people have a stronger intention to visit than older people do.

A summary of research propositions can be found in Table 1.

METHOD

Questionnaire

The questionnaire was developed on the basis of the research model (Figure 1). Items measuring the study constructs were derived from prior mobile marketing, advertising and technology acceptance research (see Appendix 1). All constructs consist of multiple items, each of which was measured by a five-point Likert scale, from 1 (completely disagree) to 5 (completely agree) with a "do not know" option. After a pre-test of the questionnaire (n=15), the questionnaire was reviewed by a group of mobile marketing professionals and

Table 1. Research propositions

Construct	Proposition	Relevant reference
Perceived ease of use of mobile communications	*P1a: Women will find it more difficult than men to use mobile services* *P1b. Older people will find it more difficult than younger people to use mobile services*	Gefen and Straub (1997); Venkatesh and Morris (2000); Morris and Venkatesh (2000)
Perceived usefulness of mobile communications	*P2a: Men will rate the perceived usefulness of mobile communications higher than women will* *P2b. Younger people will rate perceived usefulness of mobile communications higher than older people will*	Gefen and Straub (1997); Morris and Venkatesh (2000); Venkatesh and Morris (2000)
Behavioural control over messaging	*P3a. Women will regard control as being more important than men will* *P3b. Older consumers will regard control as being more important than younger people will*	Venkatesh et al. (2000); Morris and Venkatesh (2000)
Attitude toward advertising in general	*P4a. Women and men do not differ in their attitudes toward advertising in general* *P4b. Age groups do not differ in their attitudes toward advertising in general*	Bush, Smith and Craig (1999)
Social influence	*P5a. Women will be more affected than men by reference groups* *P5b. Older people will be more affected than younger people by reference groups*	Gefen and Straub (1997); Morris and Venkatesh (2000); Venkatesh et al. (2003)
Perceived trust	*P6a. Women will regard the importance of trust more highly than men will* *P6b. Age groups do not differ in their trust perceptions*	Wang and Yamagishi (2005); Golesorkhi (2006); Sutter and Kocher (2004); Kenning (2008)
Intention to participate	*P7a. Females will have a stronger intention than men of participating in mobile marketing campaigns* *P7b. Younger people will have a stronger intention than older people of participating in mobile marketing campaigns*	Karjaluoto et al. (2006); Okazaki (2004); Trappey III and Woodside (2005); Carlsson et al. (2005)
Intention to visit	*P8a. Men will have a stronger intention than women to visit* *P8b. Younger people will have a stronger intention than older people to visit*	Wolin (2003); Trappey III and Woodside (2005); Karjaluoto et al., 2006)

Figure 1. Mobile marketing acceptance model (Karjaluoto et al., 2008)

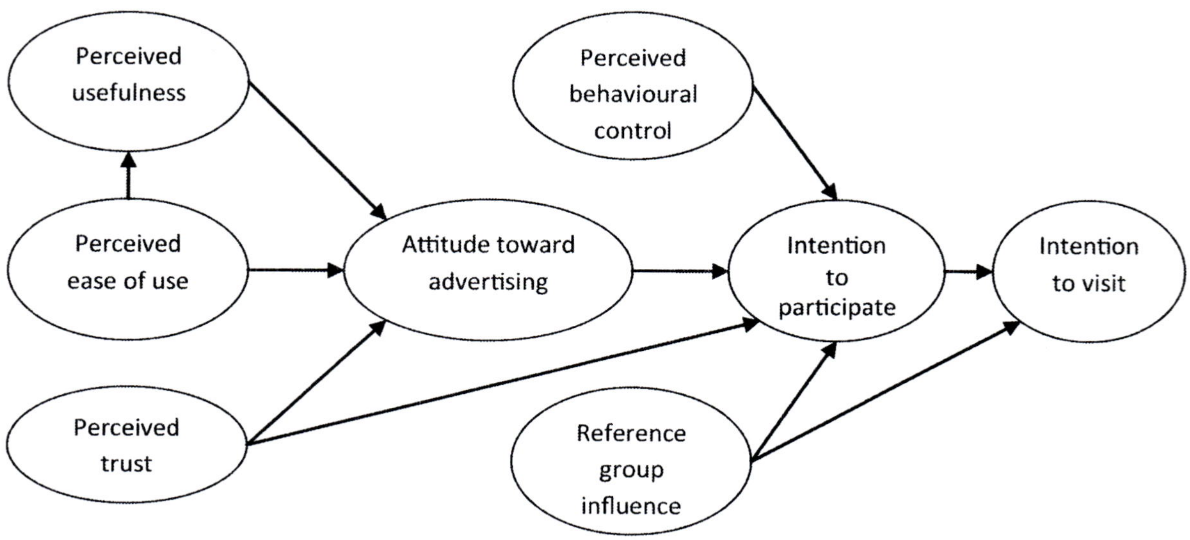

Table 2. t-test for gender and the study constructs

Item	Gender	Construct	s.d.	Item 1	Item 2	Item 3	Item 4	Item 5	Item 6
Perceived ease of use	Female	**4,13**	,70	4,84	4,19	3,82	3,48	4,32	
	Male	**4,37******	,67	4,81	4,50***	4,32***	3,95***	4,29	
Perceived usefulness	Female	**3,82**	,90	4,41	3,79	3,57	3,61	3,71	
	Male	**3,64******	,92	4,24***	3,59***	3,34***	3,43***	3,60***	
Perceived behavioural control	Female	**4,15**	,81	4,05	4,00	4,39			
	Male	**4,08******	,86	3,92***	3,99	4,33			
Attitude toward advertising	Female	**3,14**	,52	3,16	3,06	3,24	3,09		
	Male	**3,00******	,53	2,99***	2,98***	2,99***	3,03		
Social influence	Female	**2,10**	,89	2,30	1,51	1,96	2,62		
	Male	**2,27******	,87	2,41***	1,63***	2,16***	2,88***		
Perceived trust	Female	**3,54**	,73	3,27	3,48	3,43	3,54	3,70	3,85
	Male	**3,53 (ns)**	,74	3,32	3,51	3,44	3,41***	3,64***	3,85
Intention to participate	Female	**2,89**	,86	4,04	3,36	1,99	2,17		
	Male	**3,18******	,88	4,04	3,61***	2,40***	2,69***		
Intention to visit	Female	**2,05**	1,07	1,87	2,63	1,66			
	Male	**2,32******	1,13	2,19***	2,85***	1,93***			

Note: *** at $p< .001$, ** at $p<.01$, * at $p<.05$

academics and, after slight modifications, sent to the respondents.

Data Collection

Respondents for the study were recruited from the participating firm's opt-in e-mail and SMS database. The sample was drawn from five large Finnish cities and represents the firm's overall customer database well, with an emphasis on young adults aged 18 to 35. For the online survey, 8578 acceptable responses were received together resulting in a response rate of 24 percent. No non-response bias was found in the comparison of early and late respondents and their answers in the study constructs. Additionally, no significant differences were found in gender or age between the groups. Therefore, there is enough support to claim that non-response bias is not a problem in the current study.

RESULTS

The sample comprises 5211 females and 3367 males. Approximately 70 percent of the respondents belonged to the 21 to 29 age group. The demographics of the sample are in line with the night club's opt-in customer database.

Gender Differences

Gender differences are investigated with the use of a *t*-test. Table 2 displays the results of the mean difference test regarding gender. In respect of perceived ease of use, we proposed that (P1a) women would find it more difficult to use mobile services than men would. The findings suggest that men regard the use of the mobile phone and its services as being easier than women do. Specifically, there is strong evidence that males have better knowledge of what mobile services they can use on their phone (item 3) and regard downloading mobile services (item 4) on their phones as being easier than females do (both at $p<.001$). Therefore, P1a is supported by the data and declared accepted.

P1b contended that men would rate the perceived usefulness of mobile communications higher than women would. The findings show that females agree more on all the items related to perceived usefulness of mobile communications. In other words, females regard all the various aspects of usefulness as more important than males do ($p<.001$). On this basis, we must reject P1b and state that females rate usefulness higher than men do.

P3a stated that women would regard control as being more important than men would. The results show a statistically significant difference between men (mean 4.08) and women (mean 4.15) at $p<.001$. However, no differences were found in two out of three of the individual items. However, as there is a significant difference in the construct and gender, P3a is declared accepted.

With respect to attitude toward advertising in general, we proposed that (P4a) women and men would not differ in their attitudes toward advertising. Our results suggest that females (mean 3.14) tend to adopt slightly more positive attitudes towards advertising than males do (3.00) (significant at $p<.001$). Therefore, P4a will be rejected.

P5a proposed that women would be more affected by reference groups than men would. Our data does not support this notion as females are less influenced by references (mean 2.10) than men are (mean 2.27). In fact, males have higher mean scores in all the individual items as well and gender difference is in all cases statistically significant ($p<.001$). Therefore, P5a is declared rejected.

With regard to perceived trust, we proposed that (P6a) women would regard the importance of trust more highly than men will. The findings only partially support the hypothesis as there are some statistically significant differences between gender in the individual items, and no difference in the overall trust construct. These findings point to females considering the night club as more customer-centric, in the sense that it takes customers' interests into account in its communications, and believing that the firm is more truthful in its marketing. On this basis, P6a is partially supported by the data.

P7 argued that females would be found to have a stronger intention of participating in mobile marketing than males would. The results show that men (mean 3.18) are more willing to participate in mobile marketing communications with the firm than women (2.89) (at $p<.001$). The mean difference is relatively high as men and women differ on three out of the four items. Only the first item, "I believe I will receive messages from X in the future" did not produce a statistically significant difference. A logical explanation for the rejection of the hypothesis is that items 2-4 are related to more technical aspects of mobile messaging such as browsing a mobile web and downloading material on a mobile phone. With the more advanced mobile messaging solutions, this finding is in line with existing research which confirms that males are more active in participating in mobile campaigns using more developed technologies such as e-mail, JAVA, and browsing the Web (see Carlsson et al., 2005).

Finally, P8 stated that men would have a stronger intention of visiting the club than women would. The intention to visit the club after an SMS advertisement was greater among males (mean 2.32) than females (2.05) (at $p<.001$). There was the same tendency in the individual items as men rated their intention as stronger than women did in all three items. Therefore, P8 receives support from the data and is declared accepted.

Age Differences

Most of the respondents belong to the age category 21-25 (51.6 percent). The rest belong to the groups below 20 (10.1 percent), 26 to 29 (19.9 percent), 30 to 34 (10.8 percent) and 35 or more (7.7 percent). Variance analysis (Table 3) was conducted

Table 3. ANOVA test for age and the study constructs

Item	Age	Construct	s.d.	Item 1	Item 2	Item 3	Item 4	Item 5	Item 6
Perceived ease of use	Under 20	**4.22**	0.67	4.87	4.31	3.94	3.61	4.38	
	21-25	**4.22**	0.68	4.83	4.29	3.98	3.64	4.33	
	26-29	**4.27**	0.67	4.85	4.41	4.08	3.74	4.25	
	30-34	**4.25**	0.75	4.76	4.36	4.14	3.74	4.25	
	35 or over	**4.16 (ns)**	0.77	4.74***	4.14***	4.05***	3.56***	4.31***	
	25 or under	**4.22**	0.68	4.84	4.30	3.97	3.64	4.34	
	26 or over	**4.24 (ns)**	0.72	4.80**	4.34*	4.09***	3.70**	4.26***	
Perceived usefulness	Under 20	**3.96**	0.85	4.45	4.00	3.67	3.71	3.94	
	21-25	**3.73**	0.90	4.35	3.71	3.47	3.48	3.66	
	26-29	**3.64**	0.92	4.28	3.58	3.38	3.44	3.52	
	30-34	**3.70**	0.94	4.30	3.58	3.44	3.58	3.62	
	35 or over	**3.90****	0.97	4.37***	3.83***	3.58***	3.92***	3.80***	
	25 or under	**3.77**	0.90	4.37	3.76	3.51	3.52	3.70	
	26 or over	**3.71****	0.94	4.30**	3.63***	3.44*	3.58*	3.61***	
Perceived behavioural control	Under 20	**4.04**	0.87	4.00	3.89	4.23			
	21-25	**4.10**	0.82	3.95	3.98	4.36			
	26-29	**4.15**	0.82	4.02	4.02	4.40			
	30-34	**4.17**	0.86	4.07	4.05	4.39			
	35 or over	**4.24*****	0.83	4.15***	4.12***	4.45***			
	25 or under	**4.09**	0.83	3.96	3.97	4.34			
	26 or over	**4.18*****	0.83	4.06***	4.05***	4.41**			
Attitude toward advertising	Under 20	**3.28**	0.55	3.24	3.29	3.29	3.30		
	21-25	**3.08**	0.52	3.08	3.01	3.18	3.05		
	26-29	**3.01**	0.51	3.04	2.97	3.05	2.96		
	30-34	**3.03**	0.55	3.08	2.98	3.01	3.05		
	35 or over	**3.09 (ns)**	0.56	3.11***	3.05	3.05***	3.16***		
	25 or under	**3.09**	0.52	3.11	2.94	3.20	3.09		
	26 or over	**3.04*****	0.54	3.07	3.01**	3.04***	3.03**		
Social influence	Under 20	**2.31**	0.91	2.68	1.60	2.21	2.75		
	21-25	**2.18**	0.87	2.35	1.56	2.03	2.77		
	26-29	**2.09**	0.86	2.21	1.52	1.95	2.69		
	30-34	**2.10**	0.88	2.25	1.54	2.02	2.61		
	35 or over	**2.18*****	0.95	2.35***	1.61	2.17***	2.57***		
	25 or under	**2.20**	0.88	2.41	1.57	2.06	2.77		
	26 or over	**2.11*****	0.88	2.25***	1.55	2.01	2.64***		

continued on following page

Table 3. continued

Item	Age	Construct	s.d.	Item 1	Item 2	Item 3	Item 4	Item 5	Item 6
Perceived trust	Under 20	**3.63**	0.75	3.30	3.60	3.54	3.66	3.73	3.94
	21-25	**3.53**	0.73	3.25	3.47	3.42	3.49	3.69	3.86
	26-29	**3.49**	0.71	3.27	3.43	3.37	3.41	3.65	3.81
	30-34	**3.54**	0.72	3.38	3.51	3.45	3.47	3.65	3.81
	35 or over	**3.60*** **	0.81	3.46***	3.60***	3.52***	3.51***	3.65***	3.84***
	25 or under	**3.55**	0.73	3.26	3.49	3.44	3.52	3.70	3.87
	26 or over	**3.53 (ns)**	0.74	3.34***	3.49	3.42	3.45**	3.65*	3.82**
Intention to participate	Under 20	**3.11**	0.86	4.27	3.48	2.23	2.44		
	21-25	**2.97**	0.86	4.06	3.40	2.10	2.33		
	26-29	**2.95**	0.89	3.90	3.43	2.12	2.34		
	30-34	**3.07**	0.91	3.95	3.57	2.25	2.49		
	35 or over	**3.16*** **	0.93	4.07***	3.68***	2.37***	2.51***		
	25 or under	**2.99**	0.86	4.10	3.42	2.12	2.35		
	26 or over	**3.02 (ns)**	0.91	3.95***	3.52***	2.21**	2.42*		
Intention to visit	Under 20	**2.43**	1.12	2.24	3.03	2.01			
	21-25	**2.22**	1.12	2.04	2.79	1.81			
	26-29	**1.99**	1.05	1.84	2.50	1.64			
	30-34	**2.06**	1.07	1.89	2.60	1.68			
	35 or over	**2.03*** **	1.05	1.87***	2.56***	1.66***			
	25 or under	**2.25**	1.12	2.07	2.83	1.84			
	26 or over	**2.02*** **	1.05	1.86***	2.54***	1.65***			

Note: *** at $p< .001$, ** at $p<.01$, * at $p<.05$

to examine age differences in the study constructs. In the analysis, we also recoded the age variable into a dichotomous variable of which '1' is "25 or below" (61.7 percent of respondents) and '2' is "26 and more (38.3 percent of the respondents). The propositions are tested based on both of the categorizations.

P1b contended that older people would find it more difficult to use mobile services than younger people. The results show that there are no statistically significant differences between the age groups and their perceptions of ease of use ($p>.05$). However, when looking at the single items composing the construct perceived ease of use, statistically significant differences in the single items between age categories can be found. Despite the differences found in the single items, we declare P1b rejected as both the ANOVA and *t*-test produce non-significant results.

P2b stated that younger people would rate perceived usefulness of mobile communications higher than older people do. Those aged 25 or below rate perceived usefulness higher (mean 3.77) than those aged 26 or over (mean 3.71). The mean difference is statistically significant ($p<.01$). In addition, the youngest age group (below 20) differs from all the others, except from the group aged 35 or over, significantly in terms of perceived usefulness. There is a tendency among those aged 21 to 34 to regard perceived usefulness as less important than the youngest and oldest groups. No statistically significant differences were found in those aged 21 to 34. It seems that those aged 35 or over are long standing customers of the opt-in database, and thus might represent a more enthusiastic view of usefulness than those (probably newer) customers aged 21 to 34. On this basis, P2b receives enough support from the data to be declared accepted.

P3b asserted that older consumers would regard control as being more important than younger people do. Respondents aged below 20 and between 21 and 25 differ significantly from those aged 35 or over ($p<.001$). At the $p<.01$ level the youngest age group also differs from those aged 30 to 34. As can be seen from Table 2, the category 26 or more rates control higher (mean 4.18) than those 25 or below (mean 4.09). Therefore, we declare P3b accepted.

P4b stated that different age groups would not differ in their attitudes toward advertising in general. Our findings are inconsistent as the t-test provides strong evidence that the two age groups differ ($p<.001$) but in the ANOVA test no statistically significant difference is found. Additionally, there are significant differences between age groups in the individual items. The youngest age group holds more favourable attitudes toward advertising than those aged from 26 to 29 ($p<.001$), 30 to 34 ($p<.01$) and 21 to 25 ($p<.05$). The age group 21 to 25 only differs from the age group 26 to 29 in the sense that those aged 26 to 29 hold a less positive attitude toward advertising ($p<.01$). Interestingly, the oldest age group does not differ from the other age groups in terms of their attitudes ($p>.05$). Therefore, P4b is partially supported by the data and declared partially accepted.

P5b argued that older people would be more affected than younger people by reference groups. The results indicate a statistically significant difference in both analyses, but the data indicates a reversed direction for the proposition, as younger people (mean 2.20) seem to be more influenced by references than older people (mean 2.11). However, as both mean scores are low, it seems that references have a relatively weak impact on intention and behaviour. There is a clear tendency that the youngest age group differs most from the others in relation to social influence. Those below 20 are notably different from all the other age categories ($p<.001$ except from 35 or more $p<.05$). Those aged 21 to 25 also differ from those aged 26 to 29 ($p<.01$). These results indicate that references have more impact on younger customers than on older customers. However, the age group 35 or more is more affected by reference groups than those aged 21 to 34. Thus, although we found some supporting evidence for P5b, we must declare the proposition rejected.

P6b contended that age groups would not be found to differ in their trust perceptions. The mean difference between age groups is statistically insignificant in the t-test but significant in the variance analysis ($p<.001$). Therefore, this proposition is declared partially supported. The group aged under 20 differs from the age group 21 to 25 ($p<.01$) and from the age group 26 to 29 ($p<.001$). Furthermore, the oldest age group differs slightly from the age group 26 to 29 ($p<.05$). No other differences were found. This indicates that trust is of relatively similar significance to all the age groups studied. To explain the differences found, the younger respondents agree more with the statements related to the trust of the night club and its communications. Specifically, the youngest age group rates the night club's responsibility, its ability to take its customers interest into account in its communications, truthfulness and promises more highly than the other groups do. What is somewhat surprising is that the oldest age group rates their level of trust as being higher than the age groups 21 to 34. This said, P6b is accepted.

P7b proposed that younger people would have a stronger intention than older people of participating in mobile marketing campaigns. Our results suggest that in the light of the t-test, age groups do not differ in their intention. However, vari-

Table 4. Summary of the propositions' tests

Proposition	Result
P1a: Women will find it more difficult than men to use mobile services *P1b. Older people will find it more difficult than younger people to use mobile services*	Supported Rejected
P2a: Men will rate the perceived usefulness of mobile communications higher than women will *P2b. Younger people will rate perceived usefulness of mobile communications higher than older people will*	Rejected Supported
P3a. Women will regard control as being more important than men will *P3b. Older consumers will regard control as being more important than younger people will*	Supported Supported
P4a. Women and men do not differ in their attitudes toward advertising *P4b. Age groups do not differ in their attitudes toward advertising*	Rejected Partially Supported
P5a. Women will be more affected than men by reference groups *P5b. Older people will be more affected than younger people by reference groups*	Rejected Rejected
P6a. Women will regard the importance of trust more highly than men will *P6b. Age groups do not differ in their trust perceptions*	Partially supported Partially supported
P7a. Females will have a stronger intention than men of participating in mobile marketing campaigns *P7b. Younger people will have a stronger intention than older people of participating in mobile marketing campaigns*	Rejected Rejected
P8a. Men will have a stronger intention than women to visit *P8b. Younger people will have a stronger intention than older people to visit*	Supported Supported

ance analysis suggests statistically significant age differences between the groups ($p<.001$). Those below 20 and those aged 35 or over are the most willing to participate in digital marketing with the night club. This finding is somewhat surprising as one might have expected that those aged 21 to 29 would be more willing to receive communications than the two oldest age groups. The data further shows that those aged below 20 differ significantly from those aged 21 to 29 ($p<.001$). The age group 21 to 25 differs also slightly from age group 30 to 34 ($p<.05$) and strongly from the age group 35 or more ($p<.001$). The age group 26 to 29 differs from the two oldest age groups slightly in the manner that difference is slight with the age group 30 and 34 ($p<.01$) and strong with the age group 35 or more ($p<.001$). Based on those rather mixed findings, we must declare P7b rejected.

P8b declared that younger people would have a stronger intention to visit the nightclub than the older participants would. The findings show a statistically significant difference between the groups ($p<.001$). The mean tests reveal that the two youngest age groups differ significantly from the three older age groups ($p<.001$). Importantly, those below 20 also differ from those aged 21 to 25. As can be seen from Table 3, the youngest age category is the most willing to visit the night club after receiving an advertisement (Item 2). In addition, the youngest age group also seems to be most receptive to digital marketing (Item 3). The three oldest age groups did not differ significantly from each other ($p>.05$). On this basis, there is enough support to declare P8b accepted.

In summary, of the 16 propositions, only nine are supported by the data, and of the nine, three are partially supported. Table 4 shows the results of the research proposition tests.

DISCUSSION

In terms of gender differences, our results contribute to the existing literature in several ways. First, in line with gender research in technology acceptance literature (Gefen & Straub, 1997; Venkatesh & Morris, 2000), we showed that females see perceived ease of use as more important than men do. Second, our results also pinpoint that females rate perceived usefulness of mobile communications higher than men do. This finding is not directly supported by prior literature, but can be explained by mobile marketing literature that supports the view that females are more interested and involved than men in mobile marketing (Trappey III & Woodside, 2005; Karjaluoto et al., 2006). Third, in line with the literature (Venkatesh et al., 2000), we state that women perceive behavioural control to be more important than men do. Fourth, we showed that women have a more positive attitude toward advertising in general than men do. As advertising literature is not consistent in proposing whether gender differences exist in attitudes toward advertising in general (Bush, Smith, & Craig, 1999), we contribute to the literature by proposing that among active digital marketing respondents, females hold more positive attitudes toward advertising than males. Fifth, we show that men are more influenced by references than women are. This finding contradicts our existing knowledge of the impact of reference groups and gender research. Sixth, we partially confirmed that women have a higher level of trust in mobile marketers than men do. This finding, being in line with mobile advertising research (Okazaki, 2007) as well as propositions posed in e-commerce research (Kolsaker & Payne, 2002), indicates that females are slightly more trusting of the marketer than men are. Seventh, our finding that males are more willing to participate in mobile marketing is in line with Carlsson et al.'s (2005) study that showed men as more active in using developed mobile solutions. Finally, our finding that men have a higher intention to visit the advertiser because of mobile advertising is in line with existing research on advertising effectiveness (Wolin, 2003).

With respect to age, we were able to confirm five of our eight hypotheses. With respect to perceived ease of use, our findings are in contrast with sections of the literature (Morris & Venkatesh, 2000). The data shows that older people do not regard the use of mobile services as being more difficult than younger people do. The underlying reason for this dissimilarity is related to the sample that included only technology-aware people under forty. However, we support the argument that younger people regard perceived usefulness of mobile communications more highly than older people do. This finding confirms the findings presented by Morris and Venkatesh (2000). In line with Morris and Venkatesh (2000), we show that older consumers regard control as being more important than younger people do. Moreover, we partially confirm that age groups do not differ remarkably in their attitudes toward advertising (see Bush, Smith, & Craig, 1999; Wolin, 2003). With respect to social influence, the data shows that younger people are more affected by reference groups. This finding seems to be contrary to the literature (Venkatesh et al., 2003). The data further shows that age groups differ in their trust perceptions but not remarkably. This finding is partly in contrast to the literature that indicates a weak relationship between trust and age (Sutter & Kocher, 2004; Kenning, 2008). It seems that in our sample, trust and experience are related, as the oldest age group rates the importance of trust higher than those aged 21 to 34. In addition, our results contradict existing research (Karjaluoto et al., 2006; Trappey III & Woodside, 2005) in proposing that younger people do not have a stronger intention than older people of participating in mobile marketing campaigns. In fact, we show that those aged 35 or over have the greatest intention of participating in future mobile campaigns with the firm, followed by the age group below 20 and then the 30 to 34 year old group. Finally, our data shows that younger people have a stronger intention than older people to visit the advertiser.

In the light of managerial implications, we can identify the following actions to digital direct marketers. First, although our study focused purely on the effect of gender and age, it is important to use these two important background variables in segmenting opt-in mobile marketing customers. Our results underscore that men and women differ in various aspects of their responsiveness to mobile marketing. Managers should bear in mind that men are more eager to use novel mobile services whereas women place more emphasis on control and trust in mobile marketing. In terms of age, our results underscore that although age is not probably the best segmenting variable when used on its own, younger respondents regard usefulness of mobile communications more highly than older people do, and are more influenced by reference groups than older people are. A specific implication for the study context is the finding that the youngest age group is in many cases the most similar to the oldest age group. Our findings also indicate that these two groups are the most active participants in mobile marketing with the firm, and could thus be used as pilot segments in novel mobile campaigns.

CONCLUSION

The goal of this chapter was to examine gender and age differences in the context of mobile marketing. Building on the conceptualization of Karjaluoto et al. (2008) (see Figure 1), the study formed sixteen propositions that were tested with empirical data (n=8578) collected from opt-in customers drawn from a digital marketing agency's database. The results show that both gender and age are significant factors affecting the various dimensions of responsiveness to permission-based mobile marketing.

The results of our study must be interpreted in the light of certain limitations. As only around half of our hypotheses were confirmed by our data, we will here provide explanations for the somewhat surprising findings. The main reason for the rejection of seven hypotheses might be found in the study context and the sample obtained. The sample is relatively well-aware of digital marketing and also an experienced group in terms of receiving and participating in mobile and e-mail marketing campaigns. The sample also represents only certain age groups with an emphasis on people aged under 35. Therefore, differences in age and gender might be unique to the study context. Also the fact the older customers have more experience of digital marketing with the firm than the newer opt-in customers manipulates our findings. Besides the limitation related to our sample, we acknowledge that in developing the propositions we were forced to make use of literature from other disciplines such as technology acceptance and advertising research. Future studies should validate our findings by using samples obtained from other fields of business. An additional future research avenue that would merit attention is the investigation of gender and age differences in other fields of mobile marketing. As our study only touched the surface of gender and age differences in the perceptions of fields other than SMS advertising, we would encourage future studies to evaluate growing forms of mobile advertising such as mobile banner ads, location-based advertising and key word search in the light of age and gender.

REFERENCES

Ajzen, I. (1991). The Theory of Planned Behavior. *Organizational Behavior and Human Decision Processes*, *50*(2), 179–211. doi:10.1016/0749-5978(91)90020-T

Ajzen, I., & Madden, T. J. (1986). Prediction of goal-directed behavior: attitudes, intentions, and perceived behavioral control. *Journal of Experimental Social Psychology*, *22*, 453–474. doi:10.1016/0022-1031(86)90045-4

Barnes, S. J., & Scornavacca, E. (2004). Mobile marketing: the role of permission and Acceptance. *International Journal of Mobile Communications*, *2*(2), 128–139. doi:10.1504/IJMC.2004.004663

Bauer, H. H., Barnes, S. J., Reichardt, T., & Neumann, M. M. (2005). Driving consumer acceptance of mobile marketing: A theoretical framework and empirical study. *Journal of Electronic Commerce Research, 6*(3), 181–192.

Bauer, R., & Greyser, S. (1968). *Advertising in America: The consumer view*. Boston: Harvard University Press.

Burton-Jones, A., & Hubona, G. S. (2005). Individual differences and usage behavior: Revisiting a technology acceptance model assumption. *The Data Base for Advances in Information Systems, 36*(2), 58–77.

Bush, A., Smith, R., & Craig, M. (1999). The influence of consumer socialization variables on attitude toward advertising: A comparison of African-Americans and Caucasians. *Journal of Advertising, 28*(3), 13–25.

Carlsson, C., Hyvönen, K., Repo, R., & Walden, P. (2005). Asynchronous adoption patterns of mobile services. *Proceedings of the 38th Hawaii International Conference on System Sciences (HICSS-38).* Big Island, HI, January 3-6.

Chen, S. C., & Dhillon, G. S. (2003). Interpreting dimensions of consumer trust in e-commerce. *Information Technology and Management, 4*(2-3), 303–318. doi:10.1023/A:1022962631249

Chiou, J.-S., & Droge, C. (2006). Service quality, trust, specific asset investment, and expertise: Direct and indirect effects in a satisfaction-loyalty framework. *Journal of the Academy of Marketing Science, 34*(4), 613–627. doi:10.1177/0092070306286934

Cox, D., Cox, A. D., & Moschis, G. P. (1990). When consumer behavior goes bad: An investigation of adolescent shoplifting. *The Journal of Consumer Research, 17*(2), 149–159. doi:10.1086/208545

Davis, F. D. (1989). Perceived usefulness, perceived easy of use, and user acceptance of information technology. *MIS Quarterly, 13*(3), 319–340. doi:10.2307/249008

Deaux, K. (1984). From individual differences to social categories. Analysis of a decade's research on gender. *The American Psychologist, 39*(2), 105–116. doi:10.1037/0003-066X.39.2.105

Durand, R., & Lambert, Z. V. (1985). Alienation and criticism of advertising. *Journal of Advertising, 14*(3), 9–17.

Gefen, D., & Straub, D. (1997). Gender difference in the perception and use of e-mail: An extension to the Technology Acceptance Model. *MIS Quarterly, 21*(4), 389–400. doi:10.2307/249720

Golesorkhi, B. (2006). Gender differences and similarities in judgments of trustworthiness. *Women in Management Review, 21*(3), 195–210. doi:10.1108/09649420610657380

Haghirian, P., & Madlberger, M. (2005). Consumer attitude toward advertising via mobile devices - An empirical investigation among Austrian users. *Proceedings of the 13th European Conference on Information Systems*, Regensburg, Germany.

Haghirian, P., Madlberger, M., & Tanuskova, A. (2005). Increasing advertising value of mobile marketing - an empirical study of antecedents. *Proceedings of the 38th Hawaii International Conference on System Sciences (HICSS-38),* Big Island, HI, January 3-6.

Harris, L. C., & Goode, M. M. H. (2004). The four levels of loyalty and the pivotal role of trust: A study of online service dynamics. *Journal of Retailing, 80*(2), 139–158. doi:10.1016/j.jretai.2004.04.002

Hoftstede, G. (1980). *Culture's consequences: International differences in work-related values.* Beverly Hills, CA: Sage.

Hyvönen, K., & Repo, P. (2005). Mobiilipalvelut suomalaisten arjessa [Mobile services in Finnish consumers' everyday life]. In J. Leskinen, H. Hallman, M. Isoniemi et al. (Eds.), *Vox consumptoris –Kuluttajan ääni* [*Consumer's voice*] (pp. 41-59). The National Consumer Research Centre, Kerava, Finland: Savion Kirjapaino Oy.

Jackson, L. A., Ervin, K. S., Gardner, P. D., & Schmitt, N. (2001). Gender and the Internet: Women communicating and men searching. *Sex Roles*, *44*(5), 363–379. doi:10.1023/A:1010937901821

Karjaluoto, H., Leppäniemi, M., Lehto, H., & Jayawardhena, C. (2008). Exploring gender influence on customer's intention to engage permission based mobile marketing. *Electronic Markets*, *18*(3), 242–259. doi:10.1080/10196780802265793

Karjaluoto, H., Leppäniemi, M., Standing, C., Kajalo, S., Merisavo, M., Virtanen, V., & Salmenkivi, S. (2006). Individual differences in the use of mobile services among Finnish consumers. *International Journal of Mobile Marketing*, *1*(2), 4–10.

Kenning, P. (2008). The influence of general trust and specific trust on buying behaviour. *International Journal of Retail & Distribution Management*, *36*(6), 461–476. doi:10.1108/09590550810873938

Kolsaker, A., & Payne, C. (2002). Engendering trust in e-commerce: a study of gender-based concerns. *Marketing Intelligence & Planning*, *20*(4), 206–214. doi:10.1108/02634500210431595

Leppäniemi, M., & Karjaluoto, H. (2005). Factors influencing consumers' willingness to accept mobile advertising: A conceptual model. *International Journal of Mobile Communications*, *3*(3), 197–213. doi:10.1504/IJMC.2005.006580

Leppäniemi, M., & Karjaluoto, H. (2006). A review of mobile marketing research. *International Journal of Mobile Marketing*, *1*(1), 2–12.

Leppäniemi, M., Karjaluoto, H., Salo, J., & Sinisalo, J. (2005). Factors affecting customer's willingness to receive mobile marketing. *Proceedings of the 4th International Conference on Research in Advertising (ICORIA)*, 249-253.

Leung, F. S. K., & Cheung, C. M. K. (2004). Consumer attitude toward mobile advertising. *Proceedings of the Tenth Americas Conference on Information Systems*, New York.

Mathieson, K. (1991). Predicting user intentions: Comparing the Technology Acceptance Model with the Theory of Planned Behavior. *Information Systems Research*, *2*(3), 173–191. doi:10.1287/isre.2.3.173

Merisavo, M., Kajalo, S., Karjaluoto, H., Virtanen, V., Salmenkivi, S., Raulas, M., & Leppäniemi, M. (2007). An empirical study of the drivers of consumer acceptance of mobile advertising. *Journal of Interactive Advertising*, *7*(2). http://www.jiad.org/index.htm.

Minton, H. L., & Schneider, F. W. (1980). *Differential psychology*. Prospect Heights, IL: Waveland Press.

Mobile Marketing Association. (2008). Mobile marketing industry glossary. *MMA Mobile Marketing Association*, Version 068, http://mmaglobal.com/glossary.pdf

Morris, M. G., & Venkatesh, V. (2000). Age differences in technology adoption decisions: Implications for a changing work force. *Personnel Psychology*, *53*(2), 375–403. doi:10.1111/j.1744-6570.2000.tb00206.x

Ndubisi, N. O. (2006). Effect of gender on customer loyalty: a relationship marketing approach. *Marketing Intelligence & Planning*, *24*(1), 48–61. doi:10.1108/02634500610641552

Nysveen, H., Pedersen, P. E., & Thorbjornsen, H. (2005). Intentions to use mobile services: antecedents and cross-service comparisons. *Journal of the Academy of Marketing Science*, *33*(3), 330–346. doi:10.1177/0092070305276149

O'Donohoe, S. (1995). Attitudes to advertising: A review of British and American research. *International Journal of Advertising*, *14*(3), 245–261.

Obermiller, C., & Spangenberg, E. R. (1998). Development of a scale to assess consumer skepticism toward advertising. *Journal of Consumer Psychology*, *7*(2), 187–206. doi:10.1207/s15327663jcp0702_03

Okazaki, S. (2004). How do Japanese consumers perceive wireless ads? A multivariate analysis. *International Journal of Advertising*, *23*(4), 429–454.

Okazaki, S. (2007). Exploring gender effects in a mobile advertising context: On the evaluation of trust, attitudes, and recall. *Sex Roles*, *57*(11), 897–908. doi:10.1007/s11199-007-9300-7

Pollay, R. W., & Mittal, B. (1993). Here's the beef: Factors, determinants, and segments in consumer criticism of advertising. *Journal of Marketing*, *57*(3), 99–114. doi:10.2307/1251857

Rettie, R., Grandcolas, U., & Deakins, B. (2005). Text message advertising: Response rates and branding effects. *Journal of Targeting . Measurement and Analysis for Marketing*, *13*(4), 304–312. doi:10.1057/palgrave.jt.5740158

Sutter, M., & Kocher, M. G. (2004). Age and the development of trust and reciprocity. *Discussion Papers on Strategic Interaction from Max Planck Institute of Economics*, Strategic Interaction Group.

Taylor, S., & Todd, P. (1995). Assessing IT usage: The role of prior experience. *MIS Quarterly*, *19*(4), 561–570. doi:10.2307/249633

Trappey, R. J. III, & Woodside, A. G. (2006). Consumer responses to interactive advertising campaign coupling short-message-service direct marketing and TV commercials. *Journal of Advertising Research*, *45*(4), 382–401. doi:10.1017/S0021849905050476

Tsang, M. M., Ho, S.-C., & Liang, T.-P. (2004). Consumer attitudes toward mobile advertising: An empirical study. *International Journal of Electronic Commerce*, *8*(3), 65–78.

Venkatesh, V., & Davis, F. D. (2000). A theoretical extension of the Technology Acceptance Model: Four longitudinal field studies. *Management Science*, *46*(2), 186–204. doi:10.1287/mnsc.46.2.186.11926

Venkatesh, V., & Morris, M. G. (2000). Why don't men ever stop to ask for directions? Gender, social influence, and their role in technology acceptance and usage behavior. *MIS Quarterly*, *24*(1), 115–139. doi:10.2307/3250981

Venkatesh, V., Morris, M. G., & Ackermann, P. L. (2000). A longitudinal field investigation of gender differences in individual technology adoption decision-making processes. *Organizational Behavior and Human Decision Processes*, *83*(1), 33–60. doi:10.1006/obhd.2000.2896

Venkatesh, V., Morris, M. G., Davis, G. B., & Davis, F. D. (2003). User acceptance of information technology: Toward a unified view. *MIS Quarterly*, *27*(3), 425–478.

Wang, F., & Yamagishi, T. (2005). Group-based trust and gender differences in China. *Asian Journal of Social Psychology*, *8*(2), 199–210. doi:10.1111/j.1467-839x.2005.00167.x

Wolin, L. D. (2003). Gender issues in advertising -An oversight synthesis of research: 1970-2002. *Journal of Advertising Research*, *43*(1), 111–129.

KEY TERMS AND DEFINITIONS

Age: The number of years of life completed in a person's life.

Attitude toward Advertising: A learned predisposition to respond in a consistently favourable or unfavourable manner to advertising in general.

Gender: The social dimension of being male or female.

Mobile Advertising: A form of advertising that is communicated to the consumer/target via a mobile handset. Examples include mobile web banner and mobile web poster, full screen interstitials, SMS and MMS ads, mobile gaming ads and mobile video adds.

Mobile Marketing: The use of mobile handsets as an integrated content delivery and direct response vehicle within a cross-media or stand-alone marketing communications program (Mobile Marketing Association, 2008).

Mobile Communications: Wireless forms of communication using mobile handsets.

Perceived Behavioural Control: People's perceptions of their ability to behave in a given way.

Social Influence: What the consumer believes other people would think of a given behaviour.

Technology Acceptance: How users come to accept and use a technology.

Trust: A belief or expectation that another party can be relied on with confidence to behave in a manner that is beneficial or at least not detrimental to the other party's interests.

APPENDIX 1. LIST OF ITEMS

All items were measured by a five-point Likert scale ranging from 1 (completely disagree) and 5 (completely agree).

PEOU - Perceived ease of use of mobile communications ($\alpha = .791$)
PEOU1. It is easy for me to use a mobile phone
PEOU2. I learn fast to use new mobile services
PEOU3. I know what m-services I can use on my phone
PEOU4. I find it easy to download m-services on my phone
PEOU5. It is easy to receive direct marketing on my phone from the night club

This scale is based on items represented by Davis (1989) and Venkatesh and Davis (2000)

PU - Perceived usefulness of mobile communications ($\alpha = .873$)
PU1. Receiving messages gives me topical information
PU2. Receiving messages saves me time
PU3. Receiving messages gives me monetary benefits
PU4. Receiving messages gives me personal information
PU5. Receiving messages gives me information that interests me

This scale is adapted from Davis (1989) and Venkatesh and Davis (2000)

PBC - Perceived behavioural control ($\alpha = .754$)
PBC1. I find it important that I can choose the form of the messages (e.g. SMS, MMS, e-mail)
PBC2. I find it important that I can easily control the number of messages I receive
PBC3. I find it important that I can easily stop receiving messages

This scale is based on items represented by Mathieson (1991), Taylor and Todd (1995), and Merisavo et al. (2007).

ATT - Attitude toward advertising ($\alpha = .733$)
ATA1. In general I like advertising
ATA2. In general, there is too much advertising around*
ATA3. In general, advertising affects my buying behaviour
ATA4. I receive marketing messages with pleasure on my phone
*reverse coded

This scale is based on items represented by Obermiller and Spangenberg (1998) and Pollay and Mittal (1993)

REF – Social influence ($\alpha = .785$)
REF1. People who are important to me think that I should be a regular customer of the X
REF2. I am receiving messages from X because my friends are receiving them
REF3. Regular customers of X are appreciated among my friends
REF4. I am visiting X because it is popular among my friends

This scale is adapted from Taylor and Todd (1995) and Cox, Cox, and Moschis (1990)

PTR - Perceived trust ($\alpha = .915$)
PTR1. I think X is very honest in its marketing
PTR2. X is very reliable advertiser
PTR3. X is very responsible
PTR4. X takes its customers' interests into account in its communications
PTR5. I think X's marketing is truthful
PTR6. X's promises of various services (e.g. special nights) are mostly true

This scale is adapted from Chiou and Droge (2006) and Harris and Goode (2004)

INR - Intention to participate ($\alpha = .788$)
INR1. I believe I will receive messages from X in the future
INR2. I believe that in the future, I am likely to receive messages other than text messages from X
INR3. I'll browse X's homepage on my mobile phone
INR4. I could download X's material on my mobile in the future

This scale has been developed based on intention measures taken from Ajzen (1991) and Taylor and Todd (1995)

INV - Intention to visit ($\alpha = .813$)
INV1. I had an intention to visit
INV2. After receiving the ad I planned to visit in two weeks
INV3. The ad I received affected my intention to go

This scale has been developed based on intention measures taken from Ajzen (1991) and Taylor and Todd (1995)

Chapter 17
Mobile DM Coupon Promotion in Japan:
A Case Study on Response Behavior Changes in Services Consumption

Fumiyo N. Kondo
The University of Tsukuba, Japan

Yasuhiro Uwadaira
WACCORD Co., Ltd., Japan

Mariko Nakahara
Hitachi Ltd., Japan

Shahriar Akter
East West University, Bangladesh

ABSTRACT

This chapter investigates the changes in customers' responses to mobile direct mail (DM) coupons on the shop visit probability (SVP) of a beauty parlor. Two experiments were carried out to examine the promotional effects of mobile DM coupons. The first experiment, conducted in 2004, compared mobile DM coupons with postcard DM coupons. The mobile DM coupons were found to have no effect on SVP, although positive effects were observed for postcard DM coupons. The second experiment, conducted in 2005 with three types of mobile DM coupons, compared the responses of new customers with those of repeat customers. The results varied according to the settings but demonstrated stability in the same settings. For repeat customers, mail with shop's telephone number for reservation had positive effects on SVP; however, ordinary mail had no effect, while hyperlink mail had negative effects. For new customers, all three types of mobile DM coupons exhibited positive effects on SVP.

INTRODUCTION

Mobile phones will soon play an ideal role as a new, innovative, and one-to-one promotional medium. Experts believe that mobile phones will prevail as the first medium through which advertisers can reach a large audience on an individual basis (Business Wire, 2007, February 28). However, in reality, customer responses to the advertisements

DOI: 10.4018/978-1-60566-074-5.ch017

or promotions delivered through mobile phones have been reported to be extremely complex and little understood. Our research concerns the mixed results that have been reported with regard to the customer responses to such advertisements or promotions. On the one hand, Heun (2005) reported that only 12% of mobile phone customers are willing to receive marketing-related messages based on a consumer investigation, which is not a very promising result. On the other hand, another study revealed that the customers who received mobile advertising exhibited a more positive attitude toward such advertising messages, with only 1% of the consumers refusing to receive future messages (Angell, 2005).

The aim of the research in this chapter is to reveal how such mixed results can be obtained showing a case study in Japan. For this purpose, we conducted two experiments in sequence to study the effects of different types of coupon promotions on the customers of a Japanese beauty parlor. The first experiment involved coupon promotions that were delivered through two types of media: postcards, a form of traditional medium, and mobile phones. In the second experiment, three types of coupons were delivered through mobile platform. It may be noted that mobile phone has provided marketers an unprecedented opportunity of effective promotions to individual customers who can be targeted on a one-to-one basis in *instantaneous*, *interactive*, and *inexpensive* ways.

The next section describes research background and the characteristics of mobile DM coupons as promotional tools for direct marketing. Section 3 describes the experiment design. Section 4 presents the hypotheses of this study and the logistic regression models used in the analyses. Section 5 summarizes the results of the logistic regressions. Section 6 presents the future directions of related research. Finally, section 7 concludes the chapter.

BACKGROUND: MOBILE COUPONS AS A DIRECT MARKETING PROMOTION TOOL

Coupons have frequently been used by services marketers (Hartley & Cross, 1988; Lovelock & Quelch, 1983) with the hope of influencing consumer behavior and enhancing firm profitability. It is predominantly used in restaurants, hair salons, laundry and cleaning services, film processing and travel industries (Narasimhan, 1984; Peattie & Peattie, 1995). Narasimhan (1984) states, "... *the practice of offering a reduced price through such instruments (coupons) is prevalent in the consumer goods industries, more noticeably in the frequently purchased nondurable products and services markets (p. 128).* " For influencing sales and customer counts, coupons have been found effective at the store level in service industries (Matosian, 1982; Varadarajan, 1984; Chapman, 1986).

Research in services marketing clearly indicates that people and organizations buy services differently than they buy goods (George & Meyers, 1981). *Intangibility, inseparability, perishability,* and *variability* are key differences between goods and services that suggest the need for different marketing approaches (Zeithaml, Parasuraman, & Berry, 1985). As a result of these differences, coupon response in a service environment should be examined. This study explores 3 types of coupons (*postcard, hypertext and mobile direct mail*) and their impact on consumer behaviors in the form of visit probabilities through several experiments. Before the explanation of the experiments, we focus our attention to different types of coupons.

Types of Coupons and its Impact

There are different kinds of coupons. A *typical coupon* is a piece of paper or electronic document, which is either distributed widely in a certain geographic area (mass mailing), or mailed to selected shoppers in an area (direct mailing). These coupons typically have an expiry date of a few weeks to a few months associated with them. Coupons are used for "*Loyalty awards*" that are essentially points given on purchase of something and that are redeemed for the purchase later. *Frequent flyer miles* is a good example. Coupons can also be used to carry out '*Two for the price for one*' deals. These deals are partly loyalty award and partly an incentive to increase consumption of the product being promoted. *Post cards coupons* can also be viewed in the framework of coupons. *These* are generally mailed to buyers' home by sellers and redeemed within a specific time period to buy the product or services at a reduced price.

Electronic version of the above mentioned type of coupons can be implemented for use on the Internet which is called *hyperlink direct mail coupons*. These coupons can be issued to shoppers who are vacillating about a purchase in order to induce them to buy the merchandises immediately. These can be delivered through e-mails for distribution of promotional messages which is becoming an important advertising tool. Interestingly, e-mail advertising has grown in recent years. One of the reasons is the inexpensive price compared with that of traditional direct mail with costs ranging from $5 to $7 per thousand consumer addresses, as opposed to $500 to $700 per thousand for direct mail (Gartner, 2002). Secondly, e-mail advertising has been heralded as producing faster response times from consumers (Brown, 2002; Rickman, 2001). Gartner (2002) reports that consumers respond to an e-mail campaign within 10 business days. E-mail advertising also allows for a rapid dissemination of an advertisement to a global target market. In the third, e-mail can encourage interactivity with consumers by including hyperlinks in the e-mail (Brown, 2002; Garden, 2002). These hyperlinks can invite consumers, for example, to visit the company's website by clicking on the hyperlink in the e-mail. Martin et al. (2003) reported that the most important factor on consumer perception of e-mail advertisement is usefulness of e-mail. Useful e-mails appear to influence Finnish consumers to visit the store primarily to buy the product or view the product first hand, rather than visit the company website. On the other hand, Godin (2001) described that new media services, for example, hyperlink direct mail or e-mail, tend to turn users to concerns about unauthorized data access, data manipulation, and unwanted tracking of usage patterns and majority of anonymous mass advertising is despised by consumers leading them to reject the messages. From the users' point of view, invasion of privacy and general security concerns relating to the mobile medium have been identified as one of the main obstacles to the success of mobile advertising (Gohring, 2002).

A mobile coupon is an electronic ticket solicited and/or delivered by mobile phone that can be exchanged for a financial discount or rebate when purchasing a product or service (MMA, 2007). It is generally delivered through mobile e-mails which are perceived as e-mails containing useful information by customers. They are often distributed by SMS, MMS, Bluetooth or other mobile means in the form of QR (quick response) or data matrix barcodes, Universal Product Code (UPC), European Article Number (EAN), Japanese Article Number (JAN), or simply a unique number. The coupons can then be redeemed through special barcode scanners that read them or by inputting the unique number to a relevant website or a point-of-sale (POS) machine that prints out a paper coupon, or by simply showing the coupon at a retail outlet (MMA, 2007). Advanced technologies, such as FeliCa and Bluetooth, enhance the feasibility of applying GPS technology to provide real-time offers to consumers who are shopping in a specific store, or driving in close proximity to a retail outlet (Okazaki and Tylor, 2008). Kondo et al. (2007) studied coupon use probability among three types of DM coupons and obtained the result that hyperlink mail has a significant positive effect on coupon use probability. Mobile coupons services are location based services and consumers treat it as having considerable appeal (Pagani, 2004).

Mobile DM coupons include different types of deals, such as, *cents off or dollars off, free, buy-one-get-one-free, multiple purchase* (consumer is required to buy more than one of a given product), *time release* (coupons with different expiration dates are distributed together to encourage repeat purchase), *crossruffs* (consumer gets a coupon for one product while purchasing another), *sweepstakes- entry or personalized* (redeemable at specific locations), *universal coupons* (manufacturer distributes a high value coupon good on multiple products within manufacturer's product lines) and etc (MMA, 2007).

Direct Marketing in Japan

In Japan, the market toward which direct marketing is being applied is expanding. The Japan Direct Marketing Association reported that the mail-order market in Japan as of 2004 was valued at 3,040 billion yen, an increase of 9.0% from the previous year. In recent years, e-mail has frequently been used as the medium for delivering DM, and its unique properties, as compared with the traditional media used for delivering DM such as postcards, offer new channels for communicating with customers. Various types of DM such as discount coupons or sales vouchers are delivered to the mailboxes or e-mail accounts of customers almost daily. Nowadays, mobile e-mails are being utilized as a new means of delivering DM (mobile DM) because of the rapid penetration of cellular phone messaging services.

A report in Business Wire (2008, May 5) provides detailed insights into the inner workings of mobile marketing and advertising in Japan, where the vast majority of the population uses mobile Internet services. Moreover, mobile marketing in Japan is fast evolving into a standard component of the marketing mix, a scenario soon to become reality in many other parts of the world as well.

In fact, according to the Telecommunications Carriers Association, the number of contracts for cellular phones in Japan at the end of December 2007 exceeded 100 million, with a penetration rate of more than 78% of the Japanese population. Nowadays, the messaging services offered by cellular phone companies have become a more frequently used service than many of the other cellular phone services. Moreover, among the Japanese youth, messaging services are even more frequently used than voice communication services (Hashimoto, 2002). Several companies have noted this phenomenon and begun to implement a new type of promotional tool—namely, mobile DM—to boost sales and persuade customers to visit their outlets. Some examples of such companies are "TSUTAYA online," which is a video rental shop, and "GYUKAKU Sugumeru," which is a barbeque restaurant chain. As a result, mobile mail coupon promotion proliferated in Japan. For example, NTT Docomo makes several hundreds mobile promotions to their users like the customers of gas stations in a year. Now, Japan is at the stage of growth in using mobile coupons and that is why the adoption rate of mobile Internet coupon usage is rapidly expanding. The number of registrants of "beneficial" mobile Internet coupons provided by McDonald's Japan recorded more than 2 million people (1.5%) in April 2007 and increased to 5 million people (4.2%) with an increment of 3 million people within a half year. An example of recent advancements on mobile coupon is Kazasu ("hold up" in Japanese) Coupon that are trialed by McDonald's Japan in 175 stores in May, 2008, based on the Sony Felica RFID technology that is installed on most Japanese phones. Kazasu Coupon is a program that sends digital vouchers to customers' mobile phone which are redeemed at stores by tapping handsets against a contactless POS reader. What customers have to do is to just download an application to their phones and register via mobile web, selecting the coupons you want. McDonald launched mobile coupon campaign to Germany

Table 1. Classification of Sales Promotions in Quelch (1989)

	Immediate	Delayed
Price Cut	Special retail price	Coupon
	Price pack	Rebate
Value-Added	Bonus pack	Mail-in premium
	Gift-with-purchase	Frequent-flier program

and had a successful result. Other leading brands like *Arena Football, Dr. Pepper, Foot Locker, Subway, the U.S. Navy and various movie houses* have embraced mobile direct marketing either to increase traffic or to add value in the world. The growth rate of mobile Internet coupon adoption will vary from country to country, depending on adopted technology on mobile internet and adopted cell phones and psychological obstacles by customers.

CHARACTERISTICS OF MOBILE DM COUPONS AS A DELAYED PRICE CUT PROMOTION

Those employing direct marketing use personal communication to develop and maintain long-term relationships with their customers in order to increase purchase frequencies and customer expenditure levels. For this purpose, it is necessary for companies to invest their efforts into assuring their customers of a good experience that makes a deep and lasting impression while purchasing their products, especially services. There are three major activities that direct marketers should undertake: (1) perform market segmentation in order to take into account their customers' heterogeneity, (2) customize the messages and costs for each segment, and (3) acquire customer responses to marketing promotions (Rudy, 1987). An additional feature of direct marketing is that all transactions are fully recorded in a database (Yokotsuka, 1991). Quelch (1989) classified sales promotions on the basis of the following four attributes: *"immediate," "delayed," "price cut,"* and *"value-added."* DM is often described as a delayed price cut promotion (e.g., coupons or rebates). Bult et al. (1997) noted the interaction between the characteristics of a mail and the characteristics of the target customers who receive the mail, and presented a new target selection strategy. Hanna (2003) constructed a model that calculates the optimal discount rate and time limit for DM promotions. Yuan and Tsao (2003) suggested a mobile advertisement method that would take into consideration each individual customer's preferences.

Table 1 summarizes Quelch's (1989) classification of sales promotion into groups on the basis of the four attributes mentioned above. The attribute of "immediate" indicates that a consumer can gain benefits at the point of purchase, while "delayed" implies that a consumer can gain benefits with a later purchase by accumulating points or coupons. DM is often described as a delayed price cut promotion (e.g., coupons or rebates). A coupon is defined as a sales promotional tool used in marketing communications. Firms and organizations use this tool to deliver their products, services, and philosophies to customers.

For our experiment, a postcard coupon was selected as a representative of the traditional DM coupon. Such types of coupons are often sent to households in Japan and can be redeemed even at the smaller shops. PC-based Internet coupons are also available, but consumers are required to print these coupons in advance in order to avail of them in the shops. On the other hand, mobile coupons are stored directly in mobile phones, which consumers generally carry with them to most places. The "immediate portability" of mobile coupons and postcard coupons is an important feature. Since this research attempts to determine the effectiveness of coupons, the fact that both types of coupons are easily portable restricts the comparison to the differences solely in their promotional effects.

The characteristics of mobile coupons as compared with postcard coupons are summarized below.

Medium: Postcard coupons are printed on paper and mailed through the post office, while mobile coupons are wirelessly delivered and viewed on the display screens of mobile phones.

Permission: The majority of customers who are sent postcard coupons are randomly selected from a database, and permission is not required for sending them promotional material on postcards. Mobile coupons, on the other hand, are sent to all customers who have registered their e-mail addresses and given their consent to receive promotional e-mail.

Cost performance: The costs incurred in creating postcard coupons include postcards and printing costs, and the total cost depends on the number of coupons to be printed. On the other hand, an almost unlimited number of mobile coupons can be delivered to consumers with the running cost primarily consists of Internet communication charges although the initial costs occurs in purchasing a PC and mail software.

Delivery time: Postcard coupons are delivered to a customer's mailbox within a few days. In contrast, mobile coupons reach customers immediately.

***Portability*:** Postcard coupons must be carried around physically by the consumer. However, mobile coupons are already stored in the cell phone, enabling the consumer to use them at any time or place provided he/she is carrying his/her mobile phone at the time.

Extent of information and design complexity: A postcard coupon can have a complex design and a reasonably large amount of product information, including pictures and users' comments. On the other hand, a mobile coupon can only hold a limited amount of information due to the small size of mobile phone screens, which restricts the amount of information that can be viewed at the same time. As a result, mobile coupons typically do not contain pictures, are written in smaller-sized fonts, include a very brief description, and are designed in a simple text format.

Visibility: Recipients tend to put aside or store a postcard coupon if they find it useful. However, a mobile coupon can be buried amidst the daily waves of e-mails unless the customer regularly sorts his/her e-mails into different folders upon receiving them.

Packet communication fee: To receive mobile coupons, consumers are required to pay a packet communication fee of approximately 1 yen per 100 characters (this is based on the charge level prevailing in 2004). On the other hand, customers do not incur any costs for receiving postcard coupons.

EXPERIMENT DESIGN

Customer Bases in 2004 and 2005 Experiments

A new beauty parlor was opened near a university in Japan in May 2003. The number of newly recorded customers had increased from 628 as of the end of March 2004 to 1105 as of the end of March 2005. The number of registrants with a conventional e-mobile mail (hereafter, referred to as mobile mail) address was 307 in March 2004 and increased to 727 in March 2005, more than doubled since the previous year. On the other hand, the number of non-registrants in March 2005 was 378, an increment of 57 from March 2004. Since customers were given the option to register their e-mail addresses, we can presume that they granted the stores permission to send them e-mails.

At the beginning of the study, an initial experiment was conducted between February and October 2004, which targeted the existing customers of the beauty parlor. Coupons were sent to 628 customers who had registered with the beauty parlor before March 31, 2004. Two types of DM were sent to the customers: coupons in the form of postcards and coupons sent directly to their mobile phones. The main objective of this experiment was to compare the behavioral effects of mobile coupons with those of postcard coupons, and specifically, to measure the actual behavioral effects of mobile DM on cellular phone users.

The second experiment was conducted from April 1, 2005 to October 14, 2005. In this experiment, we used only those registrants who provided an e-mail address.

Types of DM Coupons Used in Each Experiment

The postcard coupons and mobile (ordinary mail) coupons for the 2004 experiment were prepared specifically for the experiment. Each coupon was valid for a period of one month. Further, the coupons contained a 25% discount for the following services: haircut, hair coloring, and permanent wave.

In order to increase interactivity with customers, two additional types of mobile DM coupons were introduced in the 2005 experiment. The first contained a hyperlink that could be accessed through the Internet, and the customer could view an image of the discount coupon on his/her computer screen (hereafter, referred to as "hyperlink mail"). The second one contained the telephone number of the shop and attempted to persuade the customer to make a telephone reservation at the beauty parlor (hereafter, referred to as "telephone reservation mail"). The third type of mobile DM coupon was the same as the one used in the 2004 experiment.

The effect of the postcard coupons was not examined in 2005 experiment because our primary concern was customer responses to mobile coupons and we wanted to have enough sample size for three types of mobile coupons. Postcard coupons were widely used in Japan. From the result of 2004 experiment, we could confirm that at least postcard coupon was effective to this particular experiment setting. In addition, we have the same type of ordinary mail coupon, so that we could compare the results of this type (ordinary mail coupon) with the new types.

Three types of mobile coupons were prepared for the 2005 experiment: ordinary mail, hyperlink mail, and telephone reservation mail. Each coupon was valid for a period of one month. In this case, the coupon formed the content of all the mail types, and the discount could be availed of by presenting the mail at the beauty shop. The first type of mobile coupon, ordinary mail, contained a discount coupon identical to the one that was used for the 2004 experiment. The details are explained below.

Hyperlink Mail with Discount Coupon

Customers using mobile Internet could access the discount coupon image through the hyperlink provided in the mail. A hyperlink makes it possible to visit a specific Internet site with a simple click on the appropriate URL, which can be easily included in an e-mail. Customers can access and view the coupon on the Internet site at any time and place with one simple click. They can also save it directly in a data folder on their cell phone rather than in a folder in their e-mail account. This makes it easy for customers to find the coupon and view it quickly; thus, hyperlink mail is useful in improving ease of use for customers who use mobile Internet. Further, the experience of accessing the Internet site and the action of downloading the coupon strengthens their recall of receiving the coupon and creates a more lasting impression. The coupon contained the name of the shop, discount rate, and validity period.

Telephone Reservation Mail

This mail featured the following statement: "The discount is available by making a telephone reservation after the receipt of this mail." The telephone number of the shop was displayed in the mail. This type of mail was directed at customers who may not use mobile Internet and attempted to persuade them to make a reservation at the beauty parlor by calling the number provided in the mail. The act of reservation was introduced in order to induce the customer to remember that the mobile mail was received.

The type of cellular phone determines the number of characters allowed in the text of a mail; thus, there may have been cases in which the sentences were not displayed in their entirety. Therefore, the content included in the additional two mail types was more concise in terms of subject and body text than the content of the ordinary mail. A customer's impression of a mail and whether or not he/she opens it may depend on the subject line; hence, key words comprising the shop name and the words "discount" and "coupon" were included in the subject lines of all the mail types. Further, each individual customer was addressed by name at the beginning of the mail in order to invoke a feeling of "personal communication." Each mail contained the DM content in addition to other details regarding the beauty parlor's name, telephone number, opening hours, and working days.

DELIVERY PLAN FOR DM COUPONS AND DESCRIPTIVE STATISTICS FOR THE SAMPLES

Delivery Plan for Mobile and Postcard Coupons in the 2004 Experiment

The purpose of the first experiment was to measure the effects of postcard DM coupons and mobile DM coupons. Between February and October 2004, postcard coupons and mobile coupons were sent to all of the customers listed in the beauty parlor's records as of the end of January 2004. The primary objective of this experiment was to test the behavioral effects of mobile coupons in comparison with those of postcard coupons. The number of recorded customers at the beauty parlor was 628, of which 307 had registered their e-mail addresses, (hereafter, "e-mail registrants") and 321 had not done so (hereafter, "non-registrants"). More often that not, customers opted to register their e-mail addresses personally; thus, we can presume that they gave their consent to the beauty parlor to send them e-mails.

Table 2 provides a summary of the plan for the first experiment. The design was as follows: (1) Two groups were formed, e-mail registrants (307 customers) and non-registrants (321 customers). The former group was further divided into two groups according to the type of coupon they received, namely, mobile coupons and postcard coupons, denoted as RM and RP, respectively, in Table 2. The non-registrant group (NP) received only postcard coupons. With this design, we avoided a selectivity bias in the estimates for the customer responses, which may have arisen due to the e-mail registration status. (2) The customers within each group were further divided into three period groups, numbered from 1 to 3. (3) Coupons were sent to the three period groups in a manner such that only one group received one particular type of coupon (mobile or postcard) each month. This cycle was repeated every three months such that a DM was sent out each month and each customer received at most one coupon in three months.

On average, the delivery (or mailing) error rate for the coupons was 48.5% for the RM group, 10.5% for the RP group, and 11.3% for the NP group. The non-delivery (error) rates for mobile coupons were much higher than those for postcard coupons due to factors such as changes in mobile phone contracts, changes in e-mail addresses, and

Table 2. DM Delivery Plan for the 2004 Experiment

Group	Mobile Mail	Feb	Mar	Apr	May	Jun	Jul	Aug	Sep	Oct
RM-1	Registrants	MD	×	×	MD	×	×	MD	×	×
RM-2	Registrants	×	MD	×	×	MD	×	×	MD	×
RM-3	Registrants	×	×	MD	×	×	MD	×	×	MD
RP-1	Registrants	PD	×	×	PD	×	×	PD	×	×
RP-2	Registrants	×	PD	×	×	PD	×	×	PD	×
RP-3	Registrants	×	×	PD	×	×	PD	×	×	PD
NP-1	Non-registrants	PD	×	×	PD	×	×	PD	×	×
NP-2	Non-registrants	×	PD	×	×	PD	×	×	PD	×
NP-3	Non-registrants	×	×	PD	×	×	PD	×	×	PD

Note: MD refers to the receipt of a mobile DM coupon, and PD, to a postcard DM coupon

e-mail filters, which are included as a function on mobile phones. Such types of errors could not be fully eliminated.

The data, which was collected from the beauty parlor's customer database, included the following information: customer ID, address, e-mail address, dates of shop visits by customers, services purchased, user or non-user of mobile phones, service purchase frequency, information on the use of DM coupons, and type of mail submitted.

Descriptive Statistics for the Samples of the 2004 Experiment

In a comparison between the samples of e-mail registrants and non-registrants with respect to age group, an obvious relationship between the registrant variable and age variable was found: the younger age group contained a higher percentage of e-mail address registrants than non-registrants. For example, the percentage of customers from the 20–24 age group who were e-mail registrants was 60%, while the percentage for non-registrants was 45%. Additionally, irregular responses by older registrants, who comprised a very small number of the customers, can influence the result to an undesirable extent if such customers are not assigned evenly to each group. Further, research on customer responses to Web sites has revealed differences that are influenced by gender. Therefore, the e-mail registrant groups were formed such that each group contained an approximately equal number of male and female customers from the same age group. However, the assignment of customers to each group was conducted randomly.

The haircut service had been purchased during 97.8% of the previous visits between May and October in 2003. Further, in terms of services purchased, no major differences were found between the postcard DM and mobile DM receivers during the experiment period. Additionally, the coloring and permanent wave services purchased during the period were only 17.3% and 23.3%, respectively, of the previous visits. Minor differences between the groups were found with respect to the purchase of the hair coloring service, but not with respect to the permanent wave service.

Delivery Plan for Mobile Coupons in the 2005 Experiment

The purpose of the second experiment was to measure the effects of three types of mobile DM coupons. All customers received one DM coupon every three months. Table 3 and Table 4 present a summary of the following experimental design.

Table 3. Number of Promotion Plan Customers for Each Period Group for the 2005 Experiment

Round		Apr	May	Jun
	Group No.	**A**	C	E
1st(1)	1	**40**	41	40
	2	**40**	40	41
	3	**40**	40	40
	Group No.	**B**	D	F
15th(2)	1	**41**	41	40
	2	**41**	40	40
	3	**40**	41	41
promoted		**161**	162	161
total		**727**	727	727

(Sample size=727)
Group 1: Ordinary Mail
Group 2: Hyperlink Mail
Group 3: Telephone Reservation Mail
(note: Number of non-delivered e-mails have not considered here.)

(1) All the registrants were divided into three groups, i.e., group 1 (ordinary mail), group 2 (hyperlink mail), group 3 (telephone reservation mail) in Table 3, respectively, through random assignment on the basis of the variables of gender and age. Three mail types were assigned to the three groups, respectively.

(2) The customers in each group were further divided into six period groups, with the alphabetical notation from A through F in Table 3, to create almost equal samples with respect to gender and age.

(3) Mails were delivered on a fortnightly basis to each period group (refer to Table 4). For example, mails were delivered to period groups, A and B in April.

Thus, all customers received one mobile DM mail every three months. For example, on April 1, period group A of each mail type group (group 1, group 2, group 3) received a different type of mobile DM, respectively (promoted group), while the other period groups did not receive any mail within the same period (non-promoted group). In the same manner, on April 15, period group B of

Table 4. Mobile DM Delivery Plan for the 2005 Experiment

Period Group	4/1	4/15	5/1	5/15	6/1	6/15	7/1	...	9/1	9/15	10/1
A	**MD**	×	×	×	×	×	MD		×	×	MD
B	×	**MD**	×	×	×	×	×		×	×	×
C	×	×	MD	×	×	×	×	...	×	×	×
D	×	×	×	MD	×	×	×		×	×	×
E	×	×	×	×	MD	×	×		MD	×	×
F	×	×	×	×	×	MD	×		×	MD	×

Note: MD denotes the receipt of a mobile DM coupon and × denotes the non-receipt of the same in each period

each group received a mobile DM. Period group F received the final mobile DM of its group on June 15. The cycle would then restart on July 1, when each period group A would again receive the same kind of mobile DM. Thus, a mobile DM was submitted twice every month, and one customer received one mobile DM once every three months.

Of the 727 registrants with mobile mail addresses, delivery errors occurred for 214 customers (29.4%), who were excluded from the analysis. This reduced our customer sample size to 513. The delivery errors may have occurred due to frequent changes in e-mail addresses or the relocation of the customers who had graduated from university.

Descriptive Statistics for the Samples in the 2005 Experiment

The total number of mobile DM coupons that were successfully delivered was 513. The breakup for each mail type was as follows: 161 for the ordinary mail group, 173 for the hyperlink mail group, and 179 for the telephone reservation mail group. In order to make these numbers consistent with the number of mobile registrants, a comparison was made between the percentages of each age group from the entire sample of the 2005 experiment, and only mobile registrants from the sample of our 2004 experiment.

The percentage of unavailable birth date increased from 13% in 2004 to 21% in 2005. The reason for this may be due in part to the frequent accidental disclosure of individual information that has occurred recently in Japanese society, which has been broadcasted often in the news. The percentage of 20–24 year-olds was 57%, the highest among the group and not very different from the percentage in 2004, i.e., 60%. This is due to the fact that the store is located near a university, and its core group of customers comprises students. The percentage of males changed from 38% in 2004 to 43% in 2005, resulting in an increase of five points. The characteristics of our data appear to be similar to those of the three groups (high school students, college students, and non-student adults) in DeBaillon and Rockwell (2005), which attempted to determine the differences in cellular phone usage with respect to gender and student-status. According to our data, since the highest frequency of haircuts requested by a customer was once in two weeks, we decided to have two observation periods in one month. Therefore, the maximum number of potential observations was 6669 (513 customers multiplied by 13 time periods). There were a total of 274 visits (4.1%) within the entire period. The number of observations—comprising those customers who both received our DM coupons and visited the shop—used to calculate the SVPs was 120.

Hypotheses and Logistic Regression Models

One of the significant advantages of coupons is its ability to drive traffic to stores or services. It also adds value to the purchase by giving direct discount to the consumer (Kotler & Keller, 2006). Researchers have found that consumers buy more products at one time or buy products at an earlier time than usual in the presence of a coupon (Blattberg, Eppen, & Lieberman, 1981; Shoemaker, 1979). Coupons can also be used to increase consumer inventories (stockpiling) or accelerate purchase timing thereby insulating consumers against competitive activity (Ailloni-Charas, 1984; Blattberg & Neslin, 1990; Krishna & Shoemaker, 1992). In a study of purchase quantity and timing for bathroom tissue and coffee, Neslin, Henderson and Quelch (1985) found that coupons were effective in increasing purchase quantity. So we can hypothesize as follows: (1) there are significant differences between the shop visit probability of customers who received postcard DM coupons and those who did not receive the coupons; (2) there are significant differences between the shop visit probability of customers who received mobile DM coupons and those who did not receive the coupons; (3) under three types of mobile DM coupons, the responses of new customers with those of repeat customers are different. The results varied according to the settings but demonstrated stability in the same settings.

Hypotheses for the 2004 Experiment Involving Postcard and Mobile Promotions

The following three alternative hypotheses, relating to beauty parlor visits and redemption rates, were established for conducting statistical tests for the 2004 experiment.

H1: Shop visits influenced by the availability of PD

There are significant differences between the shop visit probabilities of the customers who received postcard DM coupons and those who did not receive the coupons.

H2: Shop visits influenced by the availability of MD

There are significant differences between the shop visit probabilities of the customers who received mobile DM coupons and those who did not receive the coupons.

H3: Redemption rate

There are significant differences in the redemption rates of the customers who received mobile coupons and those who received postcard coupons.

Hypothesis H3 was established in order to investigate the possibility of statistical differences existing with respect to coupon redemption rates between the mobile coupon and postcard coupon receivers among the e-mail registrant group. Since all the e-mail registrants received either mobile coupons or postcard coupons, it is possible to conduct such a comparison with respect to redemption rates within this group. Furthermore, since postcard coupons are frequently used in Japan, the response rate they generate can serve as a benchmark for printed coupons.

Hypotheses for the 2005 Experiment on Mobile Promotions

The following alternative hypotheses were established for the statistical tests relating to SVP for the 2005 experiment:

H4: Shop visits influenced by the availability of DM.Ord

There are significant differences between the SVP of the customers who received ordinary mobile DM coupons and those who did not receive the coupons.

H5: Shop visits influenced by the availability of DM.Hyper

There are significant differences between the SVP of the customers who received the hyperlink mobile DM coupons and those who did not receive the coupons.

H6: Shop visits influenced by the availability of DM.Tel

There are significant differences between the SVP of the customers who received the telephone reservation mobile DM coupons and those who did not receive the coupons.

H7: Shop visits influenced by the receipt of postcard coupons in 2004

There are significant differences between the SVP of the customers who received the postcard coupon in 2004 and those who did not receive the coupons.

Logistic Regression Model

In order to examine the types of hypotheses for the 2004 and 2005 experiments, logistic regression models were applied to the shop visit data. The data comprised the following information: customer ID, postal address, e-mail address, date of shop visit, purchased services, user or non-user of cellular phone, service purchase frequency, information on the use of DM coupons, and type of mail submitted. The information was obtained from the customer database of the beauty parlor.

SVP Model for the 2004 Experiment

The age and gender data for the customers who did not provide this information were treated as missing values in the analysis. The total number of observations was 5652. Additionally, the customers who were designated recipients of the coupons but did not receive any were treated as customers without coupons. The explanatory variables for coupon types and the control variables associated with customer heterogeneity were as follows:

Age: 15–19, 20–24, 25–29, 30–34, over 35 years, or NA (missing information)
Gender: Male or female
Registration: Whether or not the customer registered his/her e-mail address
Cut.Freq: Total number of times the haircut service was purchased before the experiment
Color.Freq: Total number of times the hair coloring service was purchased before the experiment
Perm.Freq: Total number of times the permanent wave service was purchased before the experiment
Month: Nine time periods from February to October 2004
PD: Postcard coupon or no coupon
MD: Mobile coupon or no coupon

The logistic regression model is given as follows:

$$\log\frac{p}{1-p} = \alpha_0 + \alpha_1 \cdot \mathrm{Age} + \alpha_2 \cdot \mathrm{Gender} + \alpha_3 \cdot \mathrm{Registration} + \alpha_4 \cdot \mathrm{Cut.Freq} + \alpha_5 \cdot \mathrm{Color.Freq} + \alpha_6 \cdot \mathrm{Perm.Freq} + \alpha_7 \cdot \mathrm{Month} + \alpha_8 \cdot \mathrm{Hdm} + \alpha_9 \cdot \mathrm{Mdm}$$

where p represents the SVP and $\alpha_i (i = 0, \ldots, 9)$ are the parameters of the abovementioned explanatory variables. For the categorical variables, a set of linearly independent parameters were estimated through contrasts, and the original effects were recovered by a multiplication of the matrix of contrasts. The statistical software S-PLUS was used, and the missing values (NA) were treated in the S-PLUS command. Polynomial contrasts were used for the ordered categories and Helmert contrasts, for the unordered categories.

The SVP model estimates the probability of customers visiting the shop, corresponding to H1. The dependent variable represents the event of a customer visiting the shop each month (visit = 1, no visit = 0) from February to October 2004. The number of observations was 5652.

SVP Model for the 2005 Experiment

The explanatory variables for the logistic regression models pertaining to the customer purchase data are as follows:

Age: 15–19, 20–24, 25–29, 30–34, over 35, or NA (missing information)
Gender: Male or female
Cut.Freq: Total number of times the haircut service was purchased before the experiment
Color.Freq: Total number of times the hair coloring service was purchased before the experiment
Perm.Freq: Total number of times the permanent wave service was purchased before the experiment

DM.Period: The first 14 days or the remaining days of a month were grouped as one period; mobile DM coupons were sent to customers at the beginning of each period. There was a total of 13 periods

Customer.G: New customers after the end of March 2005 or repeat customers before the end of March 2005

Last.PD: Whether or not postcard coupons were received in the 2004 experiment

DM.Ord: Ordinary mail coupon or no coupon

DM.Hyper: Hyperlink mail coupon or no coupon

DM.Tel: Telephone reservation mail coupon or no coupon

The analyses were conducted by employing the following logistic regression:

$$\log\frac{p}{1-p} = \beta_0 + \beta_1 \cdot \text{Age} + \beta_2 \cdot \text{Gender} + \beta_3 \cdot \text{Cut.Freq} + \beta_4 \cdot \text{Color.Freq} + \beta_5 \cdot \text{Perm.Freq} + \beta_6 \cdot \text{DM. period} + \beta_7 \cdot \text{Customer.G} + \beta_8 \cdot \text{Last. Hdm} + \beta_9 \cdot \text{DM.Ord} + \beta_{10} \cdot \text{DM.Hyper} + \beta_{11} \cdot \text{DM.Tel}$$

where p is the SVP and $\beta_i (i = 0,\ldots,11)$ denote the parameters of the explanatory variables. The explanatory variables represent different types of data such as numeric data, categorical data, and ordered categorical data.

RESULTS

Results of the SVP Model for the 2004 Experiment: Postcard vs. Mobile Coupons

Table 5 provides a summary of the estimated parameters, *t*-statistics, and original effects. For the explanatory variables of Cut.Freq, Month, Age, PD, and Registration, the null hypotheses were rejected at the 1% significance level. The variable Color.Freq was significant at the 5% level; however, the sign of its coefficient was negative. The non-significant variables were Gender, Perm. Freq, and MD. Our findings from Table 5 are summarized below:

1. The SVP of the customers who frequently purchased the haircut service before the experiment is higher than that of the customers who rarely purchased this service before the experiment.
2. The SVP of the customers who received postcard coupons is higher than that of the customers who did not receive any coupon.
3. The SVP of the customers between 30 and 34 years of age is the highest among all the age groups. The customers above 30 years of age comprise only 10% of all customers; they appear to be non-students who live near the parlor and visit the shop frequently. On the other hand, the young customers in their 20s appear to be students and visit the shop infrequently. They also tend to try out many different beauty parlors.
4. The SVP of the customers who registered their e-mail addresses is higher than that of the non-registrants.
5. The SVP of the customers reduces with time; the exception is May, when it increases.

On the basis of the results presented in Table 5, we found that postcard coupons are effective in inducing behavior that leads to increased shop visits, corresponding to H1; this result is consistent with our expectations. However, mobile coupons did not influence SVP. Since the differences between the postcard coupons and mobile coupons lay in the type of medium used to deliver them—namely, on paper (through postal mail) or as e-mails to be viewed on mobile screens (electronically)—it can be inferred from these results that people respond differently to mobile coupons and postcard coupons.

Table 5. SVP Model for the 2004 Experiment

Variable (added sequentially)	DF	Deviance[1]	Pr(Chi)	Estimated Parameters
(Intercept)				–3.471
Cut.Freq	1	173.184	0.000***	0.636
Month	8	110.425	0.000***	–0.109
Age	4	33.210	0.000***	0.804
Registration	1	26.739	0.000***	0.263
PD	1	42.350	0.000***	0.747
Color.Freq	1	6.954	0.009**	–0.189
Perm.Freq	1	2.806	0.094	0.165
Gender	1	1.469	0.226	0.062
MD	1	1.839	0.445	–0.076
Null Deviance: 3682.642 on 5651 degrees of freedom ; Residual Deviance: 3284.913 on 5631 degrees of freedom				

** and *** denote significance at the 0.05 level and 0.001 levels, respectively.

Mobile Coupon Redemption Rate for E-mail Registrants

According to the Theory of Reasoned Action (TRA) by Ajzen and Fishbien (1980), the proximal determinant of an actual behavior is a behavioral intention, which is determined, in turn, by "Attitude" (personal factor) and "Subjective Norm" (social influence). The Theory of Planned Behavior (TPB) by Ajzen (1991, 2002) is an extended version of the TRA with the additional component of "Perceived Behavioral Control," which refers to the extent to which a person has the skills, resources, and other psychological prerequisites required to exhibit a given behavior. Accordingly, e-mail registrants are more likely to hold a favorable impression of the shop they register with and can be considered to have a strong intention to use mobile coupons or other types of coupons. Therefore, we expected that the majority of the e-mail address registrants who visited the shop would also use the coupons.

The coupon redemption rates were calculated only for the e-mail registrants because all the registrants had received either a mobile or postcard coupon; therefore, it is possible to conduct a comparison between these samples. However, the non-registrants received only postcards, and thus, their data could not be used for the comparison.

Table 6 provides a summary of the coupon redemption rates for the postcard coupons and mobile coupons, whose averages were 81.4% and 36.0%, respectively. A one-sided t-test was conducted to test the null hypothesis that the average redemption rates for the mobile and postcard coupons were equal; the alternative hypothesis was that the average redemption rate for postcard coupons was higher than that for mobile coupons. The null hypothesis was rejected at the 1% significance level.

In the case of the postcard coupon receivers, more than 80% of the visitors redeemed their coupon in seven out of the nine periods. Additionally, even for the remaining two periods, the redemption rates exceeded 40%. On the other hand, in the case of the mobile coupon receivers, the rates were extremely unstable. In three out of eight periods, there were instances where the visitors who had received mobile coupons did not redeem them when they visited the parlor. No such instance occurred in any period for the postcard coupon receivers. Further, in two out

Table 6. Movements in the Redemption Rates for Postcard Samples and Mobile Samples; t-test for the Equality of the Average Redemption Rates

	Mobile DM			Postcard DM		
Month	Visit	DM Use	Rates	Visit	DM Use	Rates
Feb	3	3	1.00	12	11	0.92
Mar	4	1	0.25	7	6	0.86
Apr	4	0	0.00	5	3	0.60
May	1	1	1.00	7	3	0.43
Jun	4	0	0.00	6	5	0.83
Jul	3	1	0.33	9	8	0.89
Aug	4	3	0.75	3	3	1.00
Sep	0	0	0	5	5	1.00
Oct	2	0	0	5	4	0.80
Total	25	9	0.360	59	48	0.814

H0: The two redemption rates are equal
H1: Mobile DM rate < Postcard DM rate
Degree of freedom: 1
Chi-square statistics: 14.547
p-value: 0.0001

of eight periods, the redemption rate for mobile coupons was one-third or less. This result indicates that a higher percentage of people are more likely to forget about the receipt of a mobile coupon; consequently, they are more likely to forget to redeem it. If a majority of customers had a habit of redeeming a mobile coupon, the mobile DM approach still has a merit.

Results for the SVP Model in the 2005 Experiment: Three Types of Mobile Coupons

A logistic regression was conducted on three types of data sets comprising the following customer groups: (1) pooled customers, (2) repeat customers, and (3) new customers. For the analyses that treated the old customer and repeat customer data separately, the explanatory variable Customer.G was not used.

SVP Model

The SVP model estimates the probability of a customer visiting the shop, corresponding to H1 and H2. The dependent variable is the event of a customer visiting the shop (visit = 1, no visit = 0) in each period from April 1, 2005 to October 14, 2005. Table 7 provides a summary of the estimated parameters of the pooled data set, repeat customer data set, and new customer data set by giving full information on the most important parameters and brief information on the others (for detailed figures, refer to Kondo and Nakahara, 2007) .

With the pooled data set, the null hypotheses were rejected at the 5% significance level for the explanatory variables of Age, Cut.Freq, Perm.Freq, Color.Freq, Customer.G, DM.Period, DM.Ord, and DM.Tel. The non-significant variables were Gender, Last.PD, and DM.Hyper.

By categorizing the aspects of demographics and periods, past purchase history, and mail types, the results presented in Table 7 can be summarized as follows:

Table 7. SVP Model for the 2005 Experiment

Variable	df	Deviance	Deviance	Deviance	Coeff.	Deviance	Coeff.
		2005	2005	2004		2005	2005
		Pooled Customers	Repeat Customers			New Customers	
Age	5	56.868**	**	***		**	
Gender	1	×	×	×		×	
Cut.Freq	1	347.706**	**	***		**	
Perm.Freq	1	4.231**	×	×		**	
Color.Freq	1	1.673**	×	**(-)		**	
Customer.G	1	6.453**	NA	NA		NA	
DM.Period	1	61.545**	**	***		**	
Last.PD	1	×	×	NA		NA	
DM.Ord	1	5.806**	**0.910**	×	**0.418**	6.508**	**0.969**
DM.Hyper	1	0.683	**(4.034)****	NA	**–1.589**	3.043*	**0.645**
DM.Tel	1	13.904**	3.972**	NA	**0.934**	10.065**	**0.991**

*, **, and *** denote significance at the 0.10, 0.05, and 0.001 levels, respectively; ×, no significance level;
(-), an negative sign; DF, the degree of freedom; and NA, not available.

Demographics and Periods

1. The SVPs by age and by period were significant for all the three data sets, yielding the same results as the 2004 experiment. The variable Age was the second largest factor to affect SVP.
2. The SVP by gender was not significant for any of the three data sets, confirming the result we obtained in the 2004 experiment.
3. For the pooled data set, the SVP of the repeat customers was significantly higher than that of the new customers. This indicates that there was a difference in SVP between the new customers and repeat customers, the latter being more frequent visitors to the shop.

Past Purchase History

4. For all the three data sets, the SVP of the customers who frequently purchased the haircut service before the experiment was higher than that of the customers who rarely purchased this service. This difference in customer behavior was the factor with the largest effect on SVP, consistent with the results obtained in the 2004 experiment.
5. The SVP of the customers who frequently purchased the permanent wave service before the experiment was higher than that of the customers who rarely purchased this service for both the pooled and new customer data sets. However, it was not significant for the repeat customer data set, corresponding to the results obtained in the 2004 experiment. These results indicate that although different customers may behave differently, the behavior of each customer is stable.
6. The SVP of the customers who frequently purchased the hair coloring service before the experiment was higher than that of the customers who rarely purchased this service for the pooled data set; this difference had the smallest effect on SVP among the three services. The result is similar to that of the

permanent wave service in most aspects. However, for the repeat customer data set, it is different from the SVP obtained in the 2004 experiment, the sign of which was negative. This may be caused by random fluctuations that are due to the small number of customers who purchased the hair coloring service.

Mail Types

7. For the pooled data set, the SVP of the customers who received postcard coupons in the 2004 experiment was not significantly higher than that of the customers who did not receive the coupons. Therefore, the event of customers receiving the postcard DM coupon in the 2004 experiment did not influence the SVP in the 2005 experiment.
8. For the pooled data set, there were significant positive effects of the DM.Ord variable, which deviates from the result of the 2004 experiment. For the pooled data set, we also discovered significantly positive effects of the DM.Tel variable.
9. A breakup of the new customer and repeat customer data sets provided us with new findings with respect to customer behavior. For the new customer data set, all three types of mobile coupons had significantly positive effects. Specifically, the SVP of the new customers who received one of the three mobile coupons was higher than that of the customers who did not receive any coupons. Among the three mail types, hyperlink mail yielded the smallest positive effect, since it required customers to use mobile Internet services, which are not necessarily preferred by all customers.
10. For the repeat customer data set, there was no significant effect of ordinary mail, confirming the results of the 2004 experiment. On the other hand, there was a positive significant effect of the telephone reservation mail coupons. This suggests that a new method of coupon delivery, i.e., a combination of mobile mail and telephone reservation, was effective in influencing the SVP of even the repeat customers. This indicates a possibility that the "use of innovation," namely, the combination of "mail + telephone reservation," had occurred. On the other hand, the hyperlink mail was less popular and yielded negative effects. If we regard ordinary mail as being ineffective in increasing the SVP of repeat customers, the customers can be considered to be conservative toward using mobile coupons. Thus, it may also be reasonable to expect the negative sign with respect to the repeat customers.

With respect to the hypotheses, the results of the mail types are summarized as follows. (1) Ordinary mail-mobile coupons in the 2005 experiment were effective in increasing SVP, corresponding to H4. Although this result is different from that obtained in the 2004 experiment, it is consistent for the repeat customer data set, for which ordinary mail mobile coupons did not influence SVP. (2) Hyperlink mail-mobile coupons in the 2005 experiment were found to be effective in increasing SVP for the new customer data set, corresponding to H5; however, they had negative effects on the repeat customer group. (3) Telephone reservation-mail mobile coupons in the 2005 experiment were effective in increasing SVP for all of the three data sets, corresponding to H6. (4) The experience of postcard coupon receipts by customers in the 2004 experiment did not influence the SVP in the 2005 experiment, not supporting H7. In summary, we may be able to conclude that individual behavior generally remains stable within a period of two years, as each type of mobile coupon promotion obtained the same response. However, new customers generally respond more favorably to mobile coupons.

Future Directions

In the 2004 experiment, this research examined the behavioral effects of mobile coupons in comparison with those of postcard coupons. We found that postcard coupons significantly influence SVP and have a high redemption rate, whereas mobile coupons do not display a significant impact on customer behavior. Moreover, the redemption rate of the mobile coupons did not achieve a response level equal to that of the postcard coupons (81.4%). This indicates that the type of medium used to deliver coupons influences their effects on SVP and the redemption rate, even when they contain the same message, discount rate, and validity. A mobile coupon has beneficial characteristics such as real-time delivery, directness, and portability. Further, the nature of mobile DM allows customization of the message or timing according to individual customers. There is a possibility that mobile DM would be effective in inducing store visits under certain conditions. In these cases, the redemption rate of mobile coupons would reach a level close to that of postcard coupons. Our research on the combination of shop visit probability and redemption rate would be useful as a benchmark analysis. Therefore, future research should focus on enhancing the customer's recollection of receiving a coupon by increasing customer involvement and customizing the delivery timing. These will contribute toward increasing the effectiveness of mobile coupon utilization such that it matches the effectiveness of postcard coupon utilization.

By conducting another experiment in 2005, this research examined, on the basis of SVP, the differences in the responses to three types of mobile coupons for two groups of customers. We found that the two aspects of demographics and periods and past purchase history generally remain the same within a period of two years: (1) the SVP by age was significant, (2) seasonal fluctuation by period was significant for all the three data sets, (3) the SVP by gender was not significant, and (4) the SVP of customers who frequently purchased the haircut service before the experiment was higher than that of the customers who rarely purchased the service before the experiment—this behavior was the largest factor to affect SVP.

With regard to mail types, we obtained different results for different combinations of customer data sets and mail types. We found that three types of mobile coupons significantly influenced the SVPs of new customers. Meanwhile, for repeat customers, we obtained mixed responses, as explained in the above sections. These results indicate that the behavior of customers toward the same type of coupon is relatively stable. A new type of mobile coupon, the hyperlink mobile coupon that was enabled by the Internet, yielded the lowest significant effect on new customers and a negative significant effect on repeat customers. These results indicate that the type of medium used to deliver a coupon must be carefully determined by examining whether the targeted customer used a mobile mail service and/or an Internet-enabled hyperlink at the initial stage of registration. This will ensure that the customer will be comfortable with the medium of promotion. Although our research constitutes just one case in which the status on the acceptance of mobile promotions is analyzed, the implication will be useful for other cases as well.

CONCLUSION

The important result of our research from practitioners' point of view is reflected in positive response patterns of customers in adopting mobile DM coupon promotions and corresponding higher visit probability. Most importantly, the result of our research has clearly shown that mobile DM coupon promotion has increased the traffic to the shop, added value to the service purchases by giving direct discounts to the consumers and come up with segmentation and customized targeting possibilities in *instantaneous*, *interactive*, and *inexpensive* manners.

The promotion literature so far focused mostly on consumer packaged goods. The contribution to mobile marketing research is that our study fills a void in mobile marketing for service sectors and sales promotion literature by examining a mobile DM coupon promotion in a service setting previously underserved. In doing so, the study highlights the importance of mobile direct mail coupons and its impact on consumer response.

In conclusion, within a period of two years, an individual will generally behave in the same manner given the same situation; however, he/she may behave differently given a different situation. These results appear to be compatible with the framework of diffusion of innovation, which explains the different rates at which innovation is adopted by different customer groups. Different customers have critically different responses to mobile mail coupons. Such responses are especially exhibited with respect to high-technology related services such as mobile Internet. Therefore, this research implies that segmentation on the basis of the acceptability of mobile Internet should be a primary consideration for mobile marketers. The segment of customers that finds it easier to use mobile Internet will be more favorable in their response to mobile mail. Meanwhile, mobile marketers should take the initiative in designing coupon delivery methods to induce recall to the customer who has a low tendency to respond to mobile mail. This can be accomplished by combining telephone reservation with mobile mail. Our results also highlight the necessity of segmentation strategies because of the uses of different types of media.

REFERENCES

Ailloni-Charas, D. (1984). *Promotion: a guide to effective promotional planning, strategies, and executions*. NewYork: John Wiley & Sons.

Ajzen, I. (1991). The theory of planned behavior. *Organizational Behavior and Human Decision Processes*, *50*, 179–211. doi:10.1016/0749-5978(91)90020-T

Ajzen, I. (2002). Perceived behavioral control, self-efficacy, locus of control, and the theory of planned behavior. *Journal of Applied Social Psychology*, *32*, 665–683. doi:10.1111/j.1559-1816.2002.tb00236.x

Ajzen, I., & Fishbien, M. (1980). *Understanding attitude and predicting social behavior*. Englewood Cliffs, NJ: Prentice-Hall, Inc.

Angell, M. (2005, September 16). Mobile marketers eye ways to connect with wireless users. Investor's Business Daily, A04.

Becker, M. (2005). Effectiveness of Mobile Channel Additions and a Conceptual Model Detailing the Interaction of Influential Variables. Retrieved on November 2, 2006, from http://www.iloopmobile.com/news/mb_research_111705.htm

Blattberg, R., & Neslin, S. (1990). *Sales promotion: concepts, methods, and strategies*. Englewood, NJ: Prentice Hall.

Brown, W. J. (2002). E-mail Marketing: A Better Alternative? *Lodging Hospitality* 58, 5, 20.

Bult, J. R., Scheer, H., & Wansbeek, T. (1997). Interaction between target and mailing characteristics in direct marketing, with an application to health care fund raising . *International Journal of Research in Marketing*, *14*, 301–308. doi:10.1016/S0167-8116(97)00012-8

Business Wire. (2007, February 28). IDC Says Don't Underestimate Full Potential of Mobile Marketing. Retrieved April 10, 2008, from http://www.fiercemobilecontent.com/ node/2941

Business Wire. (2008, May 5). 21st Century Mobile Marketing—Insights into the World's Most Advanced Mobile Society: Japan. Retrieved May 10, 2008, from http://www.researchandmarkets.com/reports/ c90795

Chapman, R. G. (1986). Assessing the profitability of retailer couponing with a low-cost field experiment. *Journal of Retailing*, *62*(1), 19–39.

DeBaillon, L., & Rockwell, P. (2005). Gender and student-status differences in cellular telephone use. *International Journal of Mobile Communications*, *3*(1), 82–98. doi:10.1504/IJMC.2005.005876

Gartner. (n.d.). GartnerG2 Says E-mail Marketing Campaigns Threaten Traditional Direct Mail Promotions. Mar 19, 2002: http://www3.gartner.com/5_about/press_releases.

George, W. R., & Meyers, T. A. (1981). Life underwriters' perceptions of differences in selling goods and services. *CLU Journal,* (April), 44–49.

Godin, S. (2001). *Permission Marketing.* Munich: FinanzBuch-Verlag.

Gohring, N. (2002). And now a word from our sponsors… . *America's Network*, *106*(3), 17.

Hanna, R. C. (2003). The optimization of discount size and time limit for coupon promotions: Application to email offers. *Unpublished doctoral dissertation, Boston University School of Management.*

Hartley, S. W., & Cross, J. (1988). How Sales Promotion Can Work For and Against You. *Journal of Consumer Marketing*, *5*(3), 35–42. doi:10.1108/eb008230

Hashimoto, Y. (2002). The Spread of cellular phones and their influence on young people in Japan, *Review of Media, Information and Society*, Institute of Socio-Information and Communication Studies . *University of Tokyo*, *7*, 97–110.

Heun, C.T. (2005). Ads take aim at cell phones. *Information Week*, 42.

Kondo, F. N., & Nakahara, M. (2007). Differences in Customers' Responsiveness to Mobile Direct Mail Coupon Promotions. *International Journal of Mobile Marketing*, *2*(2), 68–74.

Kondo, F. N., Uwadaira, Y., & Nakahara, M. (2007). Stimulating customer response to promotions: the case of mobile phone coupons. *Journal of Targeting . Measurement and Analysis for Marketing*, *17*(1), 57–67. doi:10.1057/palgrave.jt.5750064

Kotler, P., & Keller, K. L. (2006). *Marketing Management.* Pearson Education.

Krishna, A., & Shoemaker, R. W. (1992). Estimating the effects of higher coupon face values on the timing of redemptions, the mix of coupon redeemers, and purchase quantity. *Psychology and Marketing, 9* (November/ December), 453–467.

Lovelock, C., & Quelch, J. (1983). Consumer Promotions in Services Marketing. *Business Horizons*, *26*(3), 66–75. doi:10.1016/0007-6813(83)90032-0

Martin, B. A. S., Durme, J. V., Raulas, M., & Merisavo, M. (2003). Email advertising: Exploratory insights from Finland. *Journal of Advertising Research*, *43*(3), 293–300.

Matosian, J. (1982). Effectiveness of different coupon delivery methods in building mass transit ridership. *Journal of Advertising Research*, *22*(June-July), 54–59.

Narasimhan, C. (1984). A price discrimination theory of coupons. *Marketing Science, 3*(Spring), 128–147.150 *G.A. Taylor / Journal of Retailing 77 (2001) 139–151.*

Neslin, S. A., Henderson, C., & Quelch, J. (1985). Consumer promotions □ and the acceleration of product purchases. *Marketing Science*, (Spring): 4, 146–165.

Okazaki, S., & Tylor, C. R. (2008). What is SMS advertising and why do multinationals adopt it? Answers from an empirical study in European markets. *Journal of Business Research*, *61*, 4–12. doi:10.1016/j.jbusres.2006.05.003

Pagani, M. (2004). Determinants of adoption of third-generation mobile multimedia services. *Journal of Interactive Marketing*, *18*(3), 46–59. doi:10.1002/dir.20011

Peattie, K., & Peattie, S. (1995). Sales promotion—a missed opportunity for service marketers? *International Journal of Service Industry Management*, *6*, 22–39. doi:10.1108/09564239510078830

Purdy, J.G., & Gandhi (n.d.). When mobile coupons replace paper coupons, everyone wins, http://info.cellfire.com/cellfire/themes/cellfire/downloads/When_Mobile_Coupons_Replace_Paper_Coupons.pdf (accessed on October 30, 2008)

Quelch, J. A. (1989). Sales Promotion Management, Prentice-Hall, Inc., 6–7.

Rickman, C. (n.d.). Harnessing the Killer App: E-mail. February 28, 2001: http://www.digitrends.net/marketing/13640_14649.html

Rudy, K. (1987). Actual Direct Marketing, *Nihonkeizaisinbun*, Inc., (in Japanese). Neslin, S. A., & Shoemaker, R. W. (1989). An alternative explanation for lower repeat rates after promotion purchases. *JMR, Journal of Marketing Research*, *26*(May), 205–213.

Uwadaira, Y., & Kondo, F. N. (2005). 'Direct Mail Promotion Effects on the Responses of Beauty Salon Customers: Mobile DM Coupons vs. Postcards. In the Proceedings in the 5th Annual Hawaii International Conference on Business (pp.3556-3573).

Varadarajan, P. R. (1984). Consumer responses to small business coupon sales promotions. *American Journal of Small Business*, *9*(Fall), 17–26.

Yokotsuka, Y. (1991). Direct Marketing in Era of Multimedia, *Doubunkan Shuppan* Co., Ltd. (in Japanese).

Yuan, S. T., & Tsao, Y. W. (2003). A recommendation mechanism for contextualized mobile advertising. *Expert Systems with Applications*, *24*, 399–414. doi:10.1016/S0957-4174(02)00189-6

Zeithaml, V. A., Parasuraman, A., & Berry, L. L. (1985). Problems and strategies in service marketing. *Journal of Marketing*, *49*(Spring), 33–46. doi:10.2307/1251563

KEY TERMS AND DEFINITIONS

Ease of Use: extent to which something, e.g. mobile DM coupon, is easy to use.

E-Mail Address Registration: When a person agrees to register his/her e-mail address for the purpose of receiving mail, particularly promotional offers.

Hyperlink Mail: An e-mail that contains the URL for a specific Internet site.

Logistic Regression: Regression for dichotomous data.

Mobile DM Coupon: A coupon that is delivered through mobile phones as DM.

Recall of Coupon Receipt: Recalling the fact that the coupon was received.

Responsiveness: Propensity to response quickly.

Segmentation by Promotional Medium: Segmenting a market on the basis of the medium used for a promotion.

ENDNOTE

1 Deviance is a quantity similar to the residual sum of the squares for a Gaussian least squares model, and it is used in the framework of the generalized linear model.

For a logistic model, a transformation from the scale of the mean, μ, to the scale of the linear predictor, η, is performed by using the link function of logit.

Chapter 18
Mobile Store Environment Dynamics:
An Interdisciplinary Approach

Adam P. Vrechopoulos
Athens University of Economics and Business, Greece

Emmanouela E. Manganari
Athens University of Economics and Business, Greece

George J. Siomkos
Athens University of Economics and Business, Greece

ABSTRACT

The development and rapid diffusion of m-commerce has attracted lately a great deal of research interest. Researchers from many different disciplines and domains (e.g. Strategic Marketing, Human Computer Interaction, Consumer Behavior, Environmental Psychology, Information Technology, Retailing, E-Commerce, etc.) attempt to examine and better understand this new medium, following different scientific paths. The current chapter constitutes an interdisciplinary research effort on that field placing particular emphasis on the design qualities of the mobile store environment-atmosphere and its effects on users'/consumers' behavior. To that end, the concept "m-atmospherics" along with a corresponding conceptual model are introduced as the theoretical vehicles that can well support the initiation of future research attempts measuring m-atmospherics effects on consumer behavior.

INTRODUCTION AND OBJECTIVES

The rapid growth of mobile telephony has been the basis for the development of mobile commerce (m-commerce). M-commerce is the buying and selling of goods and services through wireless hand-held devices, such as cellular telephones, personal digital assistants (PDAs) and wireless computers (Michael and Salter, 2006). Proponents of m-commerce estimate that its growth and diffusion will surpass that of electronic commerce (e-commerce) (Lee and Benbasat, 2003). The development and rapid penetration of m-commerce, however, raise considerations about the dynamics of this new alternative shopping channel. Specifically, m-commerce offers new business opportunities, new experiences for users and new challenges for marketers and de-

DOI: 10.4018/978-1-60566-074-5.ch018

signers that should be carefully investigated and strategically planned (Holtzblatt 2005).

Positioned in the mobile retailing (m-tailing) research area, the present chapter investigates the design elements of the mobile store environment. The mobile store environment shares many common characteristics with online (i.e. "traditional" web-based) store environment but it is distinct from it at the same time. This chapter:

1. Presents the objectives of mobile store design.
2. Develops a typology of the principal components/dimensions of the mobile store environment.
3. Highlights the differences between the online store environment and the mobile store environment.
4. Develops a conceptual model for further studying the effects of the mobile store environment on consumer behavior.
5. Overviews the strategic aspects of the mobile design interface.
6. Discusses the implications for marketers and suggests future research ideas for scholars.

The aforementioned issues will be addressed with respect to different, and in some cases conflicting, disciplines. More specifically, principles from strategic marketing planning, consumer behavior, human-computer interaction and environmental psychology will be drawn towards the development of a conceptual model.

MOBILE STORE DESIGN: THE 7C FRAMEWORK OF THE CUSTOMER INTERFACE

Although e-commerce and m-commerce share many common aspects, each also possesses unique characteristics. The most distinctive feature of m-commerce is the facilitation of enhanced information network access (Stafford and Gillenson, 2003). Given the fact that users actually interact through their mobile device interface, the design and development of effective mobile interfaces (i.e. device and interface) is a major determinant for the penetration and growth of m-commerce.

To that end, Lee and Benbasat (2003; 2004) developed a reference framework that facilitates the development of effective m-commerce interfaces based on Rayport and Jaworski's (2001) framework of seven design elements for the customer interface (the 7C framework). The 7C framework was developed primarily for studying and analysing e-commerce interfaces. The seven factors of Rayport and Jaworski's (2001) framework are: ***context***, ***content***, ***community***, ***customization***, ***communication***, ***connection***, and ***commerce***. Given the difference between e-commerce and m-commerce in terms of hardware and software related issues, the vast majority of researchers tend to agree that the design principles from e-commerce cannot be entirely applicable in the context of m-commerce (Chae and Kim, 2003). For this reason, Lee and Benbasat (2004) conducted comparative analysis of m-commerce customer interfaces employing the 7C framework and extended this framework to the mobile context. Similarly, Lee and Benbasat (2003) note that the mobile setting increases the user's cognitive burden and they considered the mobile device constraints that demand careful deliberation on structuring the content appropriate to small screens. Therefore, the 7Cs in the context of m-commerce are formulated as follows:

- The ***Context*** factor refers to the way that the mobile interface is developed with the use of functional and aesthetic cues. Lee and Benbasat (2003) recommend the adoption of a shallow hierarchy (i.e., fewer levels with more choices within each level) rather than a deep structuring hierarchy (i.e., more levels with fewer choices within each level). The shallow hierarchy design is proposed in order to enable users' navigation by decreasing the cognitive burden. Summary and keywords that give a whole picture of information could be applied in order to complement for the comparatively restricted interface in mobile commerce (Lee and Benbasat, 2003)

- ***Content*** focuses on what the mobile interface presents. The content is divided into four subfactors: the offering mix, the appeal mix, the multimedia mix, and the content type. Both visual and audio channels are available in the context of mobile commerce. Non-speech sound (Brewster et al., 1998) and the conversion of visual information to audio format are recommended (Kristoffersen and Ljungberg, 1999).
- ***Community*** refers to the interaction between users which can be accelerated through the use of mobile devices due to the unique characteristics offered (i.e. mobility, high penetration, etc.). Interactive communication can be achieved through the acceleration of interactive information exchange between users (Lee and Benbasat, 2003).
- ***Customization*** refers to the ability of the interface to tailor itself (tailoring) or to be tailored by users based on their preferences (personalization) (Rayport and Jaworski, 2001). Nowadays, mobile devices offer more and more capabilities for tailoring and personalisation. The development of users' profile can act as the vehicle for filtering unnecessary information and retaining only information that is important and constructive (Lee and Benbasat, 2003).
- ***Communication*** is concerned with the dialogue between the interface and the users which can take the form of one-to-one communication. In order to support consumers' limited attention, broadcast messages relevant to consumers' environment and targeted advertising can be applied (Lee and Benbasat, 2003) in the mobile setting.
- ***Connection*** refers to the extent of formal linkages. Interface design is critical in enabling users' connection with other mobile sites without feeling lost due to the constrained mobile interface. The effective and efficient use of icons can support navigability and connectivity.
- ***Commerce*** focuses on mobile interfaces that enable the selling and purchasing of goods and product services. Security and privacy issues along with the minimizations of the mental processing required are the main concerns in the context of m-commerce. Lee and Benbasat (2003) recommended a condensed checkout process or preferably an one-click checkout process.

Why is Design Important in Mobile Commerce?

A well designed store is important in both offline and online shopping environments. Effective store design acts as a contributor in a) applying the retailer's strategy, b) affecting consumer behavior, c) providing flexibility, d) controlling design and maintenance costs and e) meeting legal requirements (Levy and Weitz, 2007). The CEO and co-founder of InContext Entreprises, Karen Holtzblatt, stated that the mobile platform will inspire innovative design (2005). Store design is critical in helping consumers find the products and information they are looking for and in guiding them towards the completion of their shopping tasks. Mobile devices continue to shrink in size and weight. To that end, Holtzblatt (2005) argues that the limited screen space constitutes the greatest challenge in designing the mobile store. In sum, consumers' navigation needs to be enabled despite the restrictive size of the mobile interface.

Mobile store has to be carefully designed in order to be compatible with numerous different kinds of mobile devices. The diversity of devices and the continuous development of new types of mobile devices with different requirements and capabilities put forward another challenge for effective store design in mobile commerce. Security and safety issues also need to be addressed in the context of mobile commerce just like they have been addressed in the context of e-commerce. Another challenge of mobile store design stems from the constant emergence of new sophisticated applications (Holtzblatt, 2005). Mobile store design should enable users in getting familiar with new applications. "These applications are often downloaded over the air (Holtzblatt, 2005)".

Additionally, in the context of m-commerce there can be a significant amount of environmental factors (i.e. other people, objects, activities) that can distract the user's attention from the application itself (Tapesewich, 2003). Indicatively, the mobile nature of m-commerce enables the user to be engaged to other activities while interacting with the mobile device. That is, the user may use a mobile device in order to complete a transaction by putting a minimum effort and attention. Wang et al (2008) found that the Technology Acceptance Model (TAM) serves as a viable model in order to predict consumers' acceptance to the mobile communication innovations. More explicitly, perceived ease of use and perceived usefulness of the mobile interface influence consumers' intention to use the mobile store. Thus, the development and design of mobile applications needs to stand up to this challenge by creating interfaces that require an optimum level of attention. More explicitly, the mobile interface should engage users' attention and interest in interacting with the mobile device but, should not be too demanding in terms of the attention, the information processing or the skills that the user needs to dedicate for the completion of the interaction.

HCI and the Mobile Customer Interface

The user interface design serves as an essential link between the customer and the retail store in Web-based shopping environments (Vrechopoulos et al., 2000). Lohse and Spiller (1998b) predicted that the growth of Internet retail sites would depend on the interface design issues. Similarly, Burke (2002) notes that the promise offered by electronic commerce depends -to a great extent- upon interface design and the way that customers interact with the computer. Along these lines, Gehrke et al. (1999) state that a successful Web storefront actually constitutes a critical success factor to any online store as it influences consumer behavior (e.g. satisfaction).

In the context of m-commerce, users acknowledge the possibilities that were created through the use of new technology. Indicatively, in a focus group research, respondents from Western countries emphasized that mobility offers freedom of choice both at a professional and individual level (Jarvenpaa et al., 2003). Still, respondents also point out its side effects that may be detrimental. Specifically, in some cases, respondents indicated that they felt that the new technology is dominating the user in an undesirable way (Jarvenpaa et al., 2003). That is, mobile technologies are not always aligned with human modes of behavior and social conventions. This could result in users feeling that they have limited - or no – control over the time or the place that they interact with others. Therefore the design and development of mobile environments should be aligned with human modes of behavior. This can be achieved through the use of Human Computer Interaction (HCI) principles.

Human Computer Interaction provides the principles/guidelines for the effective transformation of the existing marketing tools in the virtual marketplace in the sense that it contributes to the effective "casting" of the conventional retail environment determinants within a human computer interface environment. It sets some "presentation constraints" while generating many alternative presentation scenarios within the Web trading environments (Vrechopoulos et al., 2000). Similarly, Dix et al. (1998) note that the appropriate interface layout and design is important in all applications and the issue of presentation has been around for many years, long before computers, interactive systems or HCI. To that end, they report that usability principles are divided into three main categories: (a) learnability, (b) flexibility and (c) robustness.

The Mobile Store Environment Typology

In the previous sections we the seven components of m-commerce interfaces were identified and the importance of effective design in m-commerce was assessed. In the following section the focus is placed in the *context* component. Typologies that have been developed for the conventional and the internet store are presented and then extended to be applied to the mobile store environment. The development of a mobile cue typology is a first step toward understanding the nature and potential implications of this new business format. The mobile cue typology will contribute to the development of an organized and logical structure of the research (Turley and Milliman, 2000) on mobile atmospherics (m-atmospherics) effects. Although the development of a mobile cues' typology and consequently of a conceptual model requires a multidisciplinary approach, the central axis of this chapter is store atmospherics theory developed from a marketing perspective.

Atmospherics in the conventional store attracts a lot of interest from academics who examine the impact of atmospheric cues on consumer behavior (Gardner and Siomkos, 1986; Turley and Milliman, 2000). Atmospherics is defined as "the conscious effort to design buying environments to produce specific emotional effects in the buyer that enhance his purchase probability" (Kotler 1973-4, p. 50). The transition from brick-and-mortar retailing to click-and-mortar or "pure play" environments raised considerations about the dynamics of this new business approach.

Just like retailers provide important information through atmospherics in conventional stores, online retailers also provide an atmosphere via their website, which can affect shoppers' image of and experience with the online store (Eroglu et al., 2000). To that end, (Dailey, 2004, p. 796) defines web-atmospherics as "the conscious designing of web environments to create positive effects in users in order to increase favourable consumer responses"

Although the functional characteristics of mobile store environment are of great importance, companies' focus is shifting from creating mobile interfaces to strategic aspects involving how to best use this new commercial vehicle (e.g. innovative business models and services). The role of atmospherics in the context of mobile commerce was first introduced by Manganari et al. (2007) who claimed that the use of atmospherics cues can make a difference in the overall effectiveness of m-commerce. M-atmospherics can be defined by altering Kotler's (1973-4) definition as "*the conscious design of mobile environments to produce specific emotional internal states in the user that produce favorable consumptive responses*" (Manganari et al., 2007).

In the conventional store environment theory, Lewison (1994) developed a theoretical framework that supports that the retail store environment consists of three basic components:

a) ***Store image***: Store image is about what the consumers see in the store and about the feelings that the store evokes to consumers. In other words, store image is the store's personality. According to Lewison (1994), retail store image and design are among the most powerful tools in attracting and satisfying consumers. It is comprised of the external (e.g., its storefront) and the internal (e.g., the layout of departments) impressions. Aiming at influencing consumers' decision making, retailers focus on store layout and design (Levy and Weitz, 2001). Store design refers to the style or atmosphere of a store that acts as a contributor in projecting a certain image to the market. Store layout involves planning and structuring the internal departments and deciding on the amount of space for each department (Vrechopoulos et al., 2000).

b) ***Store atmospherics***: The store atmosphere refers to the total sensory experience created by the store (Lewison, 1994). Atmospherics have an impact on consumers' perceptual and emotional states and ultimately affect their purchase behavior (Donovan and Rossiter, 1982). Atmospherics are the characteristics or physical elements of a store's interior that establish in the consumer a frame of mind that encourages shopping behavior. The store's atmosphere should constitute a pleasant environment for both the consumer and employees. Such characteristics include, among others: the lighting, elevators, A/C system, carpets, interior design and architecture, decoration, colors, position and width of the aisles, music, flooring, displays, etc.

c) ***Store theatrics:*** Many retailers design their store like a theatre where certain individuals (employees and customers) "hold" specific roles, as if they are part of a theatrical play. Like the theatre, the store design and all its components should work in harmony to support the merchandise rather than competing against it (Levy and Weitz, 2001). A plethora of events and exhibitions that take place within the store are employed by retailers as a tool for the manipulation of the store theatrics variable.

Based on Lewison's framework, Vrechopoulos et al. (2000) developed the Virtual Component Presentation Framework (VCPF) which demonstrates the correspondence between the determinants of the conventional retail store atmosphere with the determinants of the virtual shopping environment (Figure 1). Specifically, the Virtual Component Presentation Framework (VCPF) consists of the following three factors:

a) **Virtual Layout and Design.** Virtual layout includes three layout patterns:
   - The ***tree or hub*** structure (grid in the conventional retailing). It is a rectangular arrangement of the displays and long aisles that generally run parallel to one another. Customers should pass through a hub (e.g., home page) in order to enter another product category either by utilizing the back-forward facility or the existing links (Vrechopoulos et al., 2000). It constitutes a very hierarchical way of displaying and selecting goods. The grid layout provides flexibility and speed in identifying pre-selected products (Levy and Weitz, 2001; Lewison, 1994).
   - The ***pipeline*** structure (free-form in the conventional retailing). Customers' navigation is very flexible within the freeform layout since they are encouraged to flow freely through all the parts of the store and move with no constraint towards any direction within the store (Dunne, Lusch and Griffith, 2002). Customers can access every product category of the store directly by utilizing the multiple links provided at each Web page of the store (Vrechopoulos et al., 2000). It represents a more chaotic but more entertaining structure. It is a free-flowing and asymmetric arrangement of displays and aisles, employing a variety of different sizes, shapes and styles of display.

Figure 1. The Virtual Component Presentation Framework

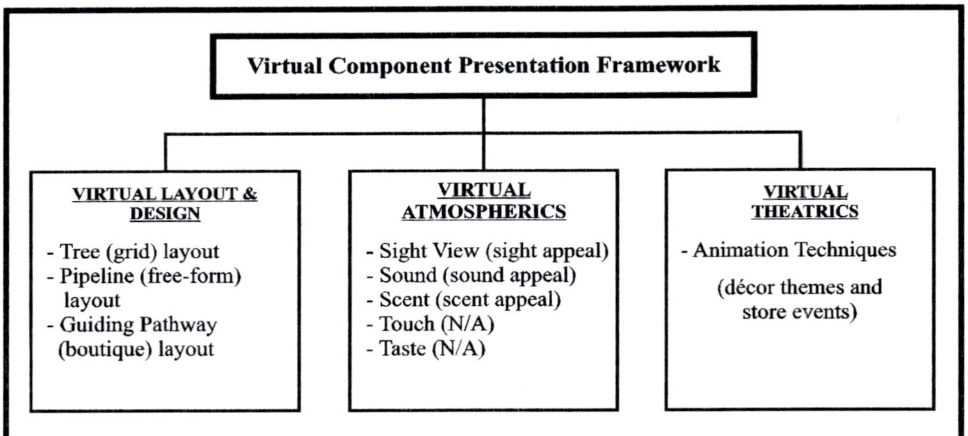

Source: Vrechopoulos et al. (2000).

- The ***guiding pathway*** structure (boutique or racetrack in the conventional retailing). The store is organized into individual, semi-separate areas, each built around a particular shopping theme. Customers are guided by the system to navigate through specific paths of the store in order to reach the products they want (Vrechopoulos et al., 2000).

b) **Virtual Atmospherics.** Virtual atmospherics include the sight view (sight appeal), the sound appeal and the scent appeal. The touch appeal and the taste appeal also appear under the virtual atmospherics, but are not yet applicable online. The touch appeal can nowadays be applied in the context of e-tailing due to technological capabilities but its use is very limited and specific. The taste appeal may also become possible in the future.

c) **Virtual Theatrics.** E-tailers are enabled to make their store look like a "theatre" through the use of images, graphics, animation and icons. Décor themes and store events are the components of virtual theatrics. Virtual theatrics are often manipulated through the use of animation techniques and interactive features.

Based on Lewison's (1994) retail store environment model, on the Virtual Component Presentation Framework (Vrechopoulos et al., 2000) and on studies that claimed the importance of the interface design for mobile commerce (Lee and Benbasat, 2004), Manganari et al. (2007) introduced the M-Commerce Store Environment Framework (MSEF) aiming at providing a comprehensive model for the development of an effective interface design for mobile commerce (Figure 2). Thus, it is proposed that:

P1. The mobile store environment consists of four atmospheric factors: a) store image, b) store atmospherics c) store theatrics and d) social factor.

According to the MSEF, the mobile store environment consists of the following four components:

a) **Store Image.** Store image is the store's personality and comprises the external and internal (e.g., layout) image. The changes

Figure 2. M-Commerce Store Environment Framework (MSEF)

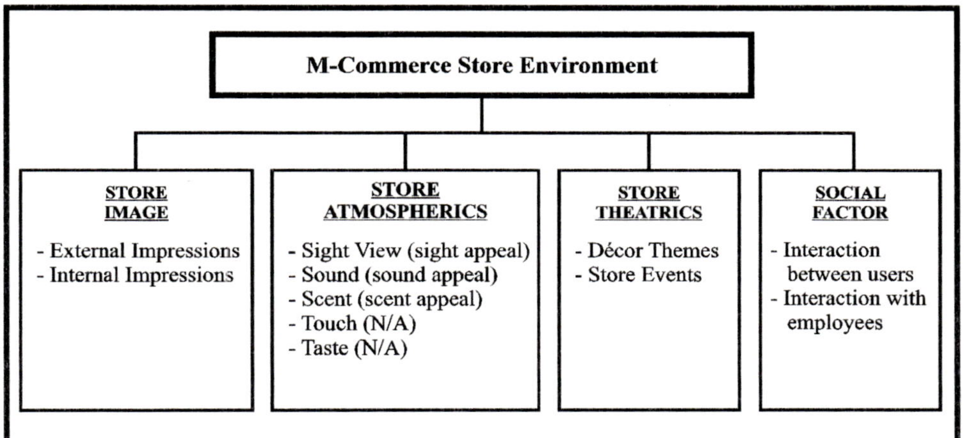

Source: Manganari et al. (2007).

in the design of mobile design interfaces are dynamic and continuous nowadays. The "look and feel" of mobile devices is an important ingredient of the overall store image.

b) **Store Atmospherics.** Store atmospherics in the mobile commerce refers to the physical cues of the store e.g., application of colors, shapes, sounds and music in the environment, that define its atmosphere. It is apparent that the "scent" and "taste" dimensions are not yet applicable in this generation's mobile devices but some of them (e.g. scent) may become available in the future.

c) **Store Theatrics.** Store theatrics refer to the use of animation or décor themes that make the store appear like a theatre. Store theatrics are based on the concept that visiting a store is more than just a shopping navigation procedure. The attainment of the shopping goals is not the only reason for visiting a mobile store. The recreational aspects of consumer behaviour are also important. Store theatrics can influence can enhance consumers' entertainment (Fortin and Dholakia, 2005).

d) **Social Factor.** Manganari et al. (2007) added to this framework a fourth component, the *social factor*. The interaction of users with employees and the presence of other users, all implemented via wireless networks, constitute the social factor. In sum, based on Manganari et al. (2007) mobile shoppers' store environment consists of four atmospheric factors: a) store image, b) store atmospherics and c) store theatrics and d) the social factor.

CONSUMER BEHAVIOR, ENVIRONMENTAL PSYCHOLOGY AND MOBILE STORE ENVIRONMENT

The focal point of this work is that of the atmospheric qualities of the mobile store and their effects on consumer behavior. In the previous section, the components of the mobile store environment were presented. This section focuses on the effects of the mobile store environment on consumer behavior. For this reason a conceptual model is developed in order to motivate research on the effects of m-atmospherics on consumer behavior. No prior research has been conducted in order to empirically measure the effects of m-atmospherics in consumer behavior. The S-O-R Paradigm (Stimulus – Organism – Responses), developed by Mehrabian and Russell (1974), is the conceptual basis for the development of the model.

An adequate S-O-R model possesses the following three components: a stimulus taxonomy, a set of intervening variables and a taxonomy of responses (Donovan and Rossiter, 1982). The *stimulus* part of the S-O-R model refers to the influences that arouse the individual (Eroglu et al., 2001). Due to the myriad of stimuli in any environment, the S-O-R does not elaborate on developing a typology of stimuli. Based on Mehrabian and Russell (1974), there are three basic emotional states that intervene between the stimulus and consumers' responses: a) pleasure – displeasure, b) arousal – non-arousal and c) dominance – submissiveness. Finally, *response* refers to the final outcome, consumers' tendency to approach or to avoid an environmental setting.

Based on the principles of environmental psychology, the S–O–R (Stimulus-Organism-Response) paradigm proposes that, the atmospheric cues of the store influence consumers' emotional states that in turn, influence consumers' overall response to the store (Mehrabian and Russell, 1974). Furthermore, the S-O-R paradigm suggests that consumers' emotional states mediate the relationship between the stimulus and the individual's responses.

In conventional retailing, researchers have extensively used the Stimulus- Organism- Response (S-O-R) paradigm (Mehrabian and Russell, 1974) in order to measure the effects of atmospherics on consumer responses. Mummalaneni (2005) proposed the S-O-R paradigm as a viable model for the investigation of consumer responses in online stores. Despite the relative newness of the research field of web-atmospherics, there is already empirical evidence that web-atmospherics influence various aspects of consumers' online behavior (Eroglu et al., 2003; Mummalaneni, 2005; Menon and Kahn, 2002; Fiore et al., 2005; Vrechopoulos et al., 2004). Specifically, the online store atmosphere has a significant effect on consumers' attitude toward the website (Dailey, 2002; Eroglu et al., 2003; Gorn et al., 2004; Richard, 2005; Fiore et al., 2005) and on satisfaction experienced by the online shopper (Eroglu et al., 2003; McKinney, 2004; Mummalaneni, 2005; De Wulf et al., 2006; Zviran et al., 2006). Similarly, the findings that support that the online atmosphere affects consumers' purchase intention (Van der Heidjen and Verhagen, 2004; Richard, 2005, Fiore et al., 2005; Griffith, 2005; Mummalaneni, 2005) are quite robust and, along with the aforementioned ones, strengthen the contention that "online store atmosphere does indeed make a difference" (Eroglu et al., 2003, p. 148). Finally, some other aspects of consumer behavior that are affected by the induced online atmosphere are: the duration of the online visit (Lee and Benbasat, 2003; Mummalaneni, 2005), approach or avoidance behaviors (Dailey, 2002; Eroglu et al., 2003), the number of product categories or stores – in case of an online mall- examined (Menon and Kahn, 2002), consumers' intention to return to the online store (Rosen and Purinton, 2004) and consumers' intention to recommend the store to others (Gorn et al., 2004).

Manganari et al. (2007) proposed that the S-O-R paradigm also holds for studying m-atmospherics effects on consumer behaviour (Figure 3). In view of applying the S-O-R model in the context of mobile commerce, it is posited that:

P2. The mobile store environment influences shoppers' internal emotional states which then affect their shopping outcomes (approach and avoidance behaviors).

Stimulus: M-Atmospheric Cues

In the conventional store environment theory, Lewison's (1994) theoretical framework supports that the retail store environment consists of three basic components: a) store image, b) store atmospherics and c) store theatrics. As we discussed in the previous section, the MSEF framework sug-

Figure 3. The S-O-R Model for Mobile Retailing

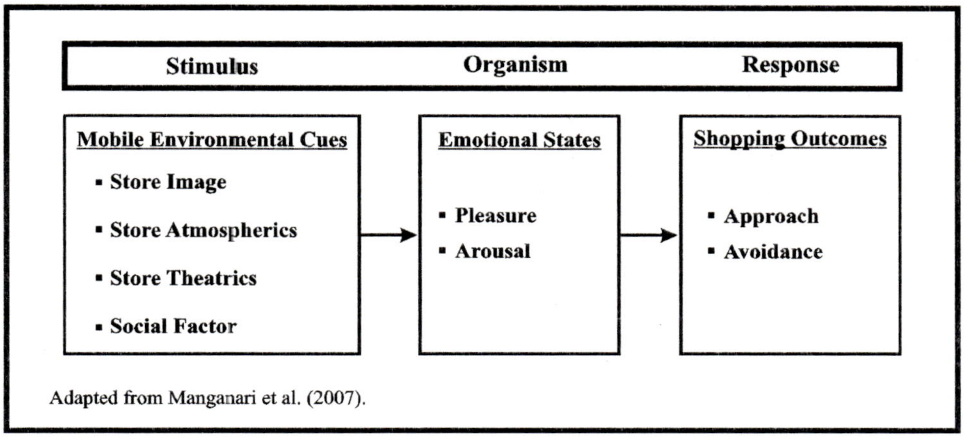

gests that the mobile store environment consists of four atmospheric factors: a) store image, b) store atmospherics c) store theatrics and d) social factor (Manganari et al., 2007).

Organism: Emotional States

According to the S–O–R paradigm, the emotional states mediate the relationship between the stimulus and individual's responses. Three emotional dimensions characterize an individual's feelings in the environment: arousal, pleasure, and dominance (Mehrabian and Russell, 1974). Still, it was proposed that pleasure and arousal alone can adequately represent the range of emotions exhibited in response to environmental stimuli (Russell and Pratt, 1980). In the current approach, the dominance dimension was not included. This is consistent with Russell and Pratt's (1980) proposition of a modification of the Mehrabian-Russell paradigm that deleted the dominance dimension. Consequently, the third proposition posits that:

P3. The emotional internal states (pleasure and arousal) of mobile shoppers mediate the relation between the perceived mobile store environment and shopping outcomes (approach and avoidance behaviors).

Response: Approach/ Avoidance Behaviors

The behavioral outcomes, which stem from the internal states induced by the m-atmospherics, may be positive or negative. Based on Mehrabian and Russell (1974) all responses to an environmental setting can be regarded as either approach or avoidance behaviors. Approach behaviors refer to those positive actions that bring the consumer "closer" to the mobile store. For example, a pleasant mobile atmosphere is proposed to result in approach behaviors e.g., more time spent in the store, intention to revisit or to recommend it to others, unplanned purchases and consumer satisfaction. On the other hand, a mobile atmosphere that is perceived as unpleasant may "create a distance" between the user and the mobile interface, which –in any case- is not desired by marketers. Several studies have empirically found that consumers' perceptions of the store environment affect their approach/avoidance responses in the context of retailing (Donovan and Rossiter, 1982, Eroglu et al., 2005) and e-tailing (Mummalaneni 2005, Eroglu et al. 2003, Davis et al. 2008).

Hence, the fourth proposition is formulated as follows:

P4. Mobile shoppers' positive (negative) emotional states lead to approach (avoidance) behaviors such as mobile store revisit, satisfaction and intention to recommend it to others (no store revisit and limited satisfaction or even dissatisfaction with the mobile store).

Mobile Store Environment Evaluation

Cox and Brittain (1996) support that the right atmosphere follows from a definition of the target customer market and the identification of their requirements. For this reason, information about potential customers of the mobile store is particularly vital for the mobile retailer. Information needs to be collected so that consumers' needs, wishes and preferences are incorporated in the design and development phase of the mobile store environment (Vrechopoulos et al., 2000). To satisfy customer needs, retailers must thoroughly understand how customers make store choices and purchase decisions. To that end, a multiattribute model which indicates what information customers use to decide which store to visit (Levy and Weitz, 2007) could serve as a useful instrument for the effective store environment design. A multiattribute model is based on the notion that customers shape their attitudes toward a retailer or a product through the collection of attributes or characteristics that the retailer or the product possess.

Based on that model, to develop a program for attracting customers, the retailer must conduct market research to collect the following information:

(a) alternative stores that customer consider,
(b) characteristics or benefits that customers consider when making their store evaluation and choice,
(c) customers' ratings of each store's performance on the characteristics, and
(d) the importance weights that customers attach to the characteristics.

The multiattribute model of consumer choice (Mason et. al., 1991) is one which is frequently used to help retailers (Levy and Weitz, 2007) and virtual retailers (Vrechopoulos et al., 2000) understand the importance of various stores features to consumers. The multiattribute model of consumer choice (Mason et. al., 1991) can be applied as follows:

$$As = \sum_{i=1}^{n} Bi\ Wi$$

where:

As = Attitude toward the store
Bi = Belief by a consumer that a store prossesses a particular attribute
Wi = The weight or importance of the attribute to consumers
n = Number of attributes important to consumers in their choice of a store

Mobile retailers could employ this model as a research instrument to understand the importance of various store atmosphere features to consumers. To that end, the components presented in the M-Atmospherics Model (MAM) (i.e. store image, store atmospherics, store theatrics, social factor) could serve as the primary attributes of such kind of surveys. Armed with this information, the mobile retailer can better understand consumers' beliefs. The graphical user interface (GUI) design should be then directed towards creating a pleasant environment that properly addresses the users' wishes, preferences and needs. This would ultimately contribute in building brand loyalty and ensuring frequent shopping and customer satisfaction. Feedback to this multi-attribute model regarding the enrichment of the attribute selection should be provided by a series of extensive consumer surveys focusing on experienced Internet shoppers.

DISCUSSION, RESEARCH IMPLICATIONS AND FUTURE RESEARCH

In the current chapter, we conducted an extensive literature review in order to identify the major determinants for the effective mobile shopping interface design towards the development of an optimum mobile store atmosphere. We conclude with a proposed framework which incorporates all the corresponding factors that determine mobile store atmosphere. This research will serve as a starting point towards the investigation of the effects that each of the proposed virtual atmosphere components has on consumer buying behavior in mobile commerce. The conceptual model presented can be employed as a research instrument in order to measure the impact of various mobile store atmospheric factors on consumer behavior within the mobile shopping interface. Future research should examine the impact of specific atmospheric cues, e.g., color, animation, layout, etc. on consumer behavior through the use of experimental methods.

To that end, an exhaustive list of the available m-atmospheric cues should be developed through the combined use of in depth interviews with experts and consumers. Similarly, a list of the critical consumer behavior variables influenced by m-atmospherics should be also developed. These lists could well serve the purpose of causal research designs where some of the m-atmospherics cues could be manipulated (i.e. treatments) towards measuring their impact on the corresponding consumer behavior variables (dependent ones). In any case, such studies should employ a multidisciplinary research approach (i.e. Human Computer Interaction, Retailing, E-Commerce).

Previous studies showed that users genuinely demand for mobile services only in those cases where mobility really make a difference and that the technical limitations of mobile devices account for the low appreciation of mobile services (Sirkka et al., 2003). These characteristics, however, can be perceived as opportunities and should constitute the true challenge of m-commerce. That is, the uncertain technology standard can be effectively managed through the use of HCI principles and information technology. Just like the computer interface constituted a challenge several years ago, nowadays the restricted mobile store interface (along with the mobile device) is also a challenge. Consumers' perceptions about the effectiveness of this new medium can also be managed through consumer behavior analysis and strategic marketing planning. Marketers need to understand consumers' needs and, in collaborations with the IT department to transform them into attractive and desirable mobile services.

Providing the user with the real usage context is one of each retailer's principal concerns (Kangas and Kinnunen, 2005). The mobile retailer should design the store based not only on the consumer behavior analysis feedback but also combine this feedback with the strategic marketing planning guidelines of the company. Indicatively, Vrechopoulos et al (2000, p. 456) report that "the fact is that "technology enabled", customers could be easily offered the "full-capability" situation of creating the shopping environment based on their own preferences and wishes and on the alternatives provided". This implies that mobile users could easily design a mobile store atmosphere (via their mobile phones) that meets their own preferences and wishes (e.g. select the colors, fonts, product display techniques, etc.). However, such type of service (this is actually a service) may dramatically change the way that store atmosphere manipulation is traditionally perceived by retailers in the conventional retailing world. In other words, since the control of store atmosphere design can be potentially offered to mobile customers (this is not applicable in conventional retailing due to the one-to-many situation) mobile retailers may not be able to directly exploit the influencing power that mobile store atmosphere determinants have on consumer behavior simply because they will offer its manipulation (i.e. control) to their mobile users.

REFERENCES

Brewster, S., Leplatre, C., & Crease, M. (1998). Using non-speech sounds in mobile computing devices. In *Proceedings of the First Workshop on HCI for Mobile Devices, Glasgow,* U.K.

Burke, R. R. (2000). Technology and the Customer Interface: what consumers want in the physical and the virtual store. *Journal of the Academy of Marketing Science, 30*(4), 411–432. doi:10.1177/009207002236914

Chae, M., & Kim, J. (2003). What's so different about the Mobile Internet? *Communications of the ACM, 46*(12), 240–247. doi:10.1145/953460.953506

Cox, R., & Brittain, P. (1996). *Retail Management,* 3rd edition. London: Pitman Publishing.

Dailey, L. C. (2002). Consumers' responses to navigational atmospherics on the Web: An empirical test. *American Marketing Association,* Conference Proceedings, Chicago, 13, 202.

Dailey, L. C. (2004). Navigational web atmospherics: explaining the influence of restrictive navigation cues. *Journal of Business Research, 57*(7), 795–803. doi:10.1016/S0148-2963(02)00364-8

Davis, L., Wang, S., & Lindridge, A. (2008). Culture influences on emotional responses to on-line store atmospheric cues. *Journal of Business Research, 61,* 806–812. doi:10.1016/j.jbusres.2007.08.005

De Wulf, K., Schillewaert, N., Muylle, S., & Rangarajan, D. (2006). The role of pleasure in web site success. *Information & Management, 43*(4), 434–446. doi:10.1016/j.im.2005.10.005

Dix, A., Finlay, J., Abowd, G., & Beale, R. (1998). *Human Computer Interaction.* 2nd edition. Prentice Hall Europe.

Donovan, R. J., & Rossiter, J. R. (1982). Store atmosphere: an environmental psychology approach. *Journal of Retailing, 58*(1), 34–57.

Dunne, P. M., Lusch, R. F., & Griffith, D. A. (2002). *Retailing* (4th ed.) Orlando: Harcourt College.

Eroglu, S. A. A.K., Machleit & L. M., Davis (2000). Online retail atmospherics: empirical test of a cue typology. In J.R. Evans & B. Berman (Eds), Retailing 2000: launching the new millennium (pp. 144-150). *Proccedings of the Sixth Triennial National Retailing Conference presented by the Academy of Marketing Science and the American Collegiate Retailing Association.* Columbus, Ohio, November 2-5.

Eroglu, S. A., Machleit, K. A., & Chebat, J. C. (2005). The interaction of retail density and music tempo: effects on shopper responses. *Journal of Psychology and Marketing, 22*(7), 577–589. doi:10.1002/mar.20074

Eroglu, S. A., Machleit, K. A., & Davis, L. M. (2003). Empirical testing of a model of online store atmospherics and shopper responses. *Psychology and Marketing, 20*(2), 139–150. doi:10.1002/mar.10064

Eroglu, S. A., Machleit, K. A., & Davis Lenita, M. (2001). Atmospheric qualities of online retailing: a conceptual model and implication. *Journal of Business Research, 54*(2), 177–184. doi:10.1016/S0148-2963(99)00087-9

Fiore, A. M., Jin, H., & Kim, J. (2005). For fun and profit: hedonic value from image interactivity and responses toward an online store. *Psychology and Marketing, 22*(8), 669–694. doi:10.1002/mar.20079

Fortin, D. R., & Dholakia, R. R. (2005). Interactivity and vividness effects on social presence and involvement with a web-based advertisement. *Journal of Business Research, 58*(3), 387–396. doi:10.1016/S0148-2963(03)00106-1

Gardner, M. P., & Siomkos, G. J. (1986). Toward a methodology for assessing effects of in-store atmospherics. *Advances in Consumer Research. Association for Consumer Research (U. S.), 13*, 27–31.

Gehrke, D., Turban, E., & Lee, O. K. M. (1999). Consumer and literature survey results of the relative importance of website design categories. *In the Proceedings of the 5th International Conference of the Decision Science Institute,* July 4-7, Athens, Greece.

Gorn, G. J., Chattopadhayay, A., Sengupta, J., & Tripathi, S. (2004). Waiting for the web: how screen color affects time perception. *JMR, Journal of Marketing Research, 41*(2), 215–225. doi:10.1509/jmkr.41.2.215.28668

Griffith, D. A. (2005). An examination of the influences of store layout in online retailing. *Journal of Business Research, 58*(10), 1391–1396. doi:10.1016/j.jbusres.2002.08.001

Holtzblatt, K. (2005). Designing for the mobile device: experiences, challenges and methods. *Communications of the ACM, 48*(7), 33–35. doi:10.1145/1070838.1070862

Jarvenpaa, S. L., & Todd, P. A. (1997). Consumer reactions to electronic shopping on the World Wide Web. *International Journal of Electronic Commerce, 1*(2), 59–88.

Kangas, E., & Kinnunen, T. (2005). Applying user-centered design Development to mobile application. *Communications of the ACM, 48*(7), 55–59. doi:10.1145/1070838.1070866

Kotler, P.(1973-4). Atmospherics as a marketing tool. *Journal of Retailing, 49*, 48–64.

Kristoffersen, S., & Ljungberg, F. (1999). Designing interaction styles for a mobile use context. In *Proceedings of she hiternational Symposium on HanclhirUand Ubiquitous Cotnputing. (HUC99).*

Lee, Y. E., & Benbasat, I. (2003). Interface design for mobile commerce. *Communications of the ACM, 46*(12), 49–52. doi:10.1145/606272.606300

Lee, Y. E., & Benbasat, I. (2004). A framework for the study of customer interface design for mobile commerce. *International Journal of Electronic Commerce, 8*(3), 79–102.

Levy, M., & Weitz, B. A. (2001). *Retailing Management* (4th ed.) McGraw-Hill, IRWIN.

Levy, M., & Weitz, B. A. (2007). *Retailing Management* (6th ed.) McGraw-Hill, IRWIN.

Lewison, D. M. (1994). *Retailing*, 5th edition, New York, NY: Macmillan College Publishing Company.

Lohse, L. G., & Spiller, P. (1998b). Electronic shopping. *Communications of the ACM, 41*(7), 81–87. doi:10.1145/278476.278491

Manganari, E. E., Siomkos, G. J., & Vrechopoulos, A. (2007). Atmospheric qualities in mobile commerce: an initial approach. *Conference Proceedings* European Marketing Academy Conference, EMAC2007, May 22-25, 2007, Reykjavik, Iceland.

Mason, B. J., Mayer, M. L., & Ezell, H. F. (1991). *Retailing.* 4th edition, IRWIN, Homewood, Boston.

McKinney, L. N. (2004). Creating a Satisfying Internet Shopping Experience Via Atmospheric Variables. *International Journal of Consumer Studies*, *28*(3), 268–283. doi:10.1111/j.1470-6431.2004.00368.x

Mehrabian, A., & Russell, J. A. (1974) *An Approach to Environmental Psychology*, Cambridge, MA: MIT Press.

Menon, S., & Kahn, B. (2002). Cross-category effects of induced arousal and pleasure on the internet shopping experience. *Journal of Retailing*, *78*(1), 31–40. doi:10.1016/S0022-4359(01)00064-1

Michael, A., & Slater, B. (2006). *Mobile marketing: achieving competitive advantage through wireless technology*. Oxford, Elsevier Ltd.

Mummalaneni, V. (2005). An empirical investigation of web site Characteristics, consumer emotional states and on-line shopping behaviors. *Journal of Business Research*, *58*, 526–532. doi:10.1016/S0148-2963(03)00143-7

Rayport, J., & Jaworski, B. (2001). *Introduction to E-Commerce*. New York: McGraw- Hill.

Richard, M. O. (2005). Modeling the impact of internet atmospherics on surfer Behavior. *Journal of Business Research*, *58*(12), 1632–1642. doi:10.1016/j.jbusres.2004.07.009

Rosen, D. E., & Purinton, E. (2004). Website design: viewing the web as a cognitive landscape. *Journal of Business Research*, *57*(7), 787–794. doi:10.1016/S0148-2963(02)00353-3

Russell, J. A., & Pratt, G. (1980). A description of the affective quality attributed to environments. *Journal of Personality and Social Psychology*, *38*, 311–322. doi:10.1037/0022-3514.38.2.311

Sirkaa, L., Jarvenpaa, K., Yoko, T., & Virpi, K. (2003). Mobile commerce at crossroads. *Communications of the ACM*, *46*(12), 41–44. doi:10.1145/953460.953485

Stafford, T. F., & Gillenson, M. L. (2003). Mobile commerce: what is and what it could be? *Communications of the ACM*, *56*(12), 33–34. doi:10.1145/953460.953483

Tapesewich, P. (2003). Designing Mobile Commerce Applications. *Communications of the ACM*, *46*(12), 57–50. doi:10.1145/953460.953489

Turley, L. W., & Milliman, R. E. (2000). Atmospheric effects on shopping behavior: a review of the experimental evidence. *Journal of Business Research*, *49*(2), 193–211. doi:10.1016/S0148-2963(99)00010-7

Van der Heijden, H., & Verhagen, T. (2004). Online store image: conceptual foundations and empirical measurement. *Information & Management*, *41*(5), 609–617. doi:10.1016/S0378-7206(03)00100-9

Vrechopoulos, A. P., O'Keefe, R. M., & Doukidis, G. I. (2000). Virtual store atmosphere in internet retailing. In S. Klein, B. O Keefe, J. Gricar, and M. Podlogar, editors, *Proceedings of the 13th Bled Electronic Commerce Conference: The End of the Beginning*, pp. 445-458, June 19-21.

Vrechopoulos, A. P., O'Keefe, R. M., Doukidis, G. I., & Siomkos, G. I. (2004). virtual store layout: an experiment comparison in the context of grocery retail. *Journal of Retailing*, *80*(1), 13–22. doi:10.1016/j.jretai.2004.01.006

Wang, C.-C., Lo, S.-K., & Fang, W. (2008). Extending the technology acceptance model to mobile telecommunication innovation: The existence of network externalities . *Journal of Consumer Behaviour*, *7*, 101–110. doi:10.1002/cb.240

Zviran, M., Glezer, C., & Avni, I. (2006). User satisfaction from commercial web sites: the effect of design and use. *Information & Management*, *43*(2), 157–178. doi:10.1016/j.im.2005.04.002

KEY TERM AND DEFINITION

M-Atmospherics: is the conscious design of mobile environments to produce specific emotional internal states in the user that produce favorable consumptive responses.

Chapter 19
Mobile Loyalty Programs:
Relevance for Relationship Management and Consumer Acceptance

Andreas Mann
SVI-Stiftungslehrstuhl für Dialogmarketing, Germany

Jan Prein
University of Kassel, Germany

ABSTRACT

Mobile loyalty programs hold various advantages to companies compared to traditional card-based programs. Mobile customer cards enable companies to precisely assign advertising reactions and purchase data to individual customers. For this reason customer retention activities can be adjusted to individual preferences and the situational context. However, any successful implementation of mobile technologies in consumer contexts depends primarily on consumer acceptance. In this chapter, we begin by illustrating characteristics of mobile loyalty programs, particularly as it relates to relationship management. Following that we identify and analyse drivers affecting the acceptance of mobile loyalty programs and present our results of an exploratory survey on consumer acceptance by using a fictional mobile customer card offer.

INTRODUCTION

Customer loyalty and customer relationship management (CRM) represent for many businesses important tasks and demands in order to ensure survival in saturated markets with no growth potential and increasing competition. CRM is a matter of retaining customers, especially valuable ones, for a long time and being able to fully satisfy their needs. The successful implementation of a customer loyalty program, however, requires knowledge of the customer. This knowledge serves to assess the value of individual customer relationships and is simultaneously the basis for a more refined customer segmentation and target group formation (Cortiñas, Elorz, Múgica, 2008). In the so-called *Analytical CRM* customer address and reaction data are therefore prepared in Customer Data Warehouses and evaluated for the continuous optimization of all customer-related business transactions. The use of IT-supported CRM-systems makes it possible for companies to precisely x-ray the relationship and to

DOI: 10.4018/978-1-60566-074-5.ch019

purposefully employ resources in the management of business relationships (Payne, Frow, 2006).

The basis for the analytical evaluation of customer data is the targeted generation of the data. The challenges here are particularly large for suppliers carrying out anonymous transactions with consumers because they have practically no relevant information about their customers. This is very often the case with business-to-consumer transactions performed via stationary distribution bases. Card-based loyalty programs represent for them an important opportunity to gain essential customer data, because each customer has to provide his/her basic data, such as name and contact details, upon registration as a program member. At the same time purchase behaviour data is collected with each purchase made by the program member. Only by presenting the customer card is the program member entitled to make use of services provided by the card-based loyalty program (Wübker, Schmidt-Gallas, 2003). Purchases can thus be assigned to individual customers. Problems arise in practice, however, partly as a result of the customer card's often being used collectively by various people known to the customer. In this case the assignment of the "collective" purchasing behaviour to an individual customer card holder can lead to false customer data profiles, which becomes problematic if they are used to derive customer retention measures.

Due to the increasing number of these programs it is becoming increasingly difficult for participants in traditional card-based loyalty programs to carry all of these cards in their wallets. In view of the growing number of plastic cards and the limited capacity of card slots in wallets, a "struggle" is beginning to develop among the various cards for a place in the wallet (Demoulin, Zidda, 2007). The average wallet of the German consumer today already contains between four and five of these cards (e.g. EC-, credit, rail, and medical insurance cards) (TNS Emnid, 2006).

A new possible way of gaining customer data in loyalty programs is offered by mobile marketing. Due to the increasing diffusion of high speed mobile data connections and decreasing prices for mobile data transfers, the cell phone turns out to be a powerful platform for multiple communication and information access applications (Wais, Clemons, 2008, p. 12). This enables companies to use mobile communication technologies not just for push advertising messages but also in the context of customer loyalty programs. For that purpose, cell phones can take over functions of traditional customer loyalty programs and become a kind of mobile customer card holder.

In the following chapter, we will present mobile customer card functions and their importance for relationship management. Afterwards, we develop a model of the consumer acceptance of mobile customer cards, because acceptance among consumers represents a major challenge for the successful implementation of such a program (Bauer et al., 2005a). In addition to the conceptual considerations, results of an exploratory acceptance study will be presented (see also Mann, Prein, 2008). We conclude the chapter with practical implications of the results.

BACKGROUND AND RESEARCH MODEL

Characteristics and Special Features of Mobile Customer Cards

Today, traditional card-based customer loyalty programs have several functions for relationship management. This section points out how the cell phone can assume the main program-functions and how companies can benefit from the implementation of a mobile customer card program (see also Mann, Prein, 2008). Figure 1 depicts the functions of mobile customer cards graphically.

Figure 1. Functions of mobile customer cards

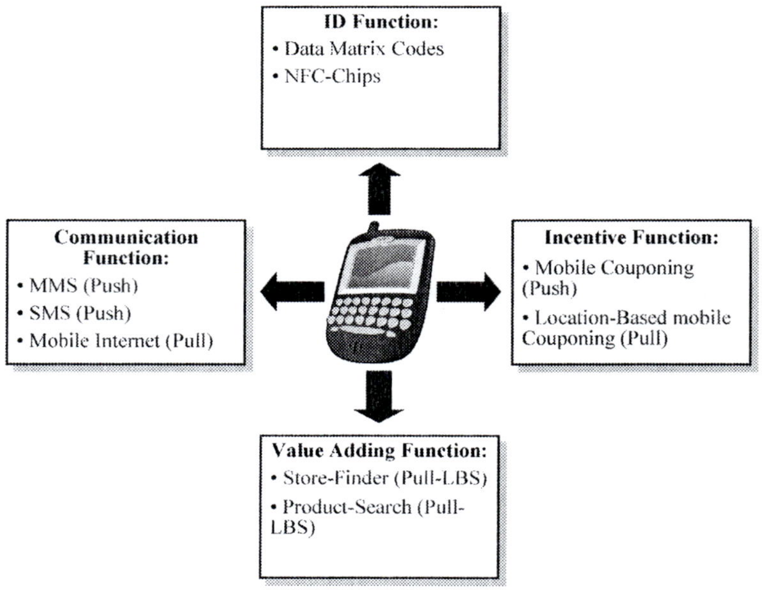

The *identification function* is made possible in mobile loyalty programs by the use of a two-dimensional barcode (data matrix) which can be assigned to customers on an individual basis. They are sent as an image message and can be read with scanner devices even with relatively small cell phone displays. Members registering with the mobile customer card program receive a personalized barcode on their cell phones via SMS or MMS. Alternatively, a code of this sort could be generated on a personalized mobile website. From there, the customer could download the image via a web browser installed on the cell phone. The code, which can be read at the cash desk or on special scanner terminals, then serves to identify the program member at the point of sale and allows for the full use of services provided by the customer card / loyalty program. (cf. Brandstetter, 2006 for a first pilot project). Customers participating in multiple programs must just retain several id-codes on their mobile device.

An alternative to two-dimensional barcodes might be Near Field Communication (NFC) radio chips integrated into cell phones. These too can be assigned to an individual cell phone; they make it possible for individual program members, after registering, to be identified at the point of sale by the provider of the loyalty program. The identification of the cell phone is made via a short-distance (RFID) radio network and is in line with similar security standards, such as closed card-bound ID systems (Wiedmann et al., 2008). This technology is already widespread in Asia. The Japanese provider DoCoMo alone has sold more than 40 million devices, which can also be used to pay for purchases (Balaban, 2008). In Germany cell phones with NFC-technology have hardly been available, and applications such as the purchase of travel tickets in local and long-distance travel are still in the experimental stage (Weidelich, 2008).

Because cell phones are individual products, unlike traditional customer cards, they are not usually given to other people; purchase behaviour can be assigned much more precisely to individual program members by the identification technologies of Data Matrix Codes or NFC-Chips. An additional factor is that cell phones are usually constantly carried around by the users, which mean they can always employ them as a customer card.

The *information and communication function* is a further core task in customer loyalty programs. Among other things, this involves information about the current state of points obtained, product and price information and advertising contacts. These tasks can also be performed effectively by mobile devices. Cell phones of the new generation in particular have displays that enable high resolution graphics to be represented. They are thus able to present multimedia information in a user friendly way. In addition the individual use of cell phones makes it very likely that the advertising messages really get across to (only) the person who is a member of the customer card program. This makes it possible to also send more sensitive information, such as for example access codes or shipment confirmations, on the registered cell phone.

Moreover, with messages that have integrated response requirements the reactions to these (advertising) contacts can be relatively precisely assigned via the Subscriber Identity Module (SIM) to a particular member (Wohlfahrt, 2002). If this information is combined with purchase behaviour data, individualized messages can be created for subsequent contacts, which cater to the special wishes, needs and interests of the individual program member in terms of content and form. In this way the relevance of the messages for the customers is increased (Barnes, Scornavacca, 2004; Carroll et al., 2007; Kurkovsky, Harihar, 2006). In addition, the mobile telephone is usually carried around in stand-by mode, which means that the program member is more or less constantly available. This can be used for the time-specific design of messages.

In this context the sending of individualized coupons on the mobile device is relevant (Wehmeyer, Müller-Lankenau, 2005). It supports the *incentive function* of the mobile customer card program. The coupons are meant – as in traditional customer card systems – to motivate the customers to adopt a certain information and purchasing behaviour. The mobile coupons can in turn be sent via an image message as barcode or in the form of a number code on the device. Like advertising messages, they can be sent time-specifically and tailored in terms of content and form to the individual customer, because they are generated and sent electronically.

However, the sending of mobile push-advertising messages is in some countries limited by legal restrictions. In Germany for example, the federal court of justice (BGH) has decided that the provider of the customer loyalty program Paypack has to receive an opt-in-declaration by the program-member before they can send the program member SMS-advertising messages (BGH, 2008).

In addition to mobile push communication, communicative pull offers, for example in the form of mobile websites, can be reserved for the customer. It is for this reason the provider, Customer Advantage Program GmbH, has created a website for its customer card program specially optimized for mobile telephones. Via this website, program members can call their account information, cash in received coupons and manage personal data.

In addition, product information can be offered via mobile portals and web 2.0 applications, such as forums or other social networks could be transmitted to the mobile internet offer. With the collection and analysis of the content called up and self-generated by the program member, information about individual preferences and interests beyond the purchase behaviour data can be gained, which makes it possible to further optimize customer-specific retention activities (Möhlenbruch et al., 2008).

By localizing cell phones and other mobile devices in the GSM network via COO as well as equipping cell phones via GPS receivers, individual place-dependent services – so-called Location Based Services – can be integrated into the offers available in mobile customer cards. For Example, advertising SMS or mobile coupons, that depend on the geographic position of the customer on a shopping trip can specifically influence his shopping behaviour. Through this situation-related geographic context, the value of *service and value adding function* could increase considerably. Companies thus have an opportunity to align customer retention activities with a geographically limited customer circle (Rudolph, Emerich, 2008).

The first pilot studies with push location-based advertising were carried out in, among other branches, the automotive industry. For example, selected BMW customers received information via cell phone about winter tires, exactly on the day when the first snow flakes started to fall in the place where they were (Anonymous, 2008). Through consistent use of the customer data base the information contained an image of the model being sold together with corresponding wheel rims and detailed product information and the contact data of the nearest local BMW branch office.

The implementation of this kind of push location-based service is actually still limited by the insufficient diffusion of mobile devices with meter-precise GPS receivers because the COO technology, which can be used with every cell phone, guarantees an adequate local transmission only in large conurbations with enough transmitters. Moreover, in some countries, the use of push location-based services is restricted by law. In the European Union for example, the privacy policy for electronic communication allows location-based services only after a declaration of consent by the cell phone user.

Pull location-based services on the other hand can be employed comprehensively through every customer, such as the calling up of special offers from the local partner companies of a collective loyalty program (Bauer et al., 2005b). Localization here is possible through a GPS receiver, if available, or alternatively through manual input of location information by the customer. The Customer Advantage Program GmbH has a mobile application for its customer card program, Happy Digits, in collaboration with the cell phone search service Qiro, which displays Happy-Digits partner branches on a card and can download location-dependent coupons onto the cell phone (Anonymous, 2007).

Finally, the successful use of mobile customer cards depends not only on technically-driven advantages, but also, and above all, on the acceptance on the part of the customer. For this reason the elements and drivers of the acceptance of mobile customer cards are explained in more detail in the following section.

Acceptance of Mobile Customer Cards as Precondition for a Successful Program Implementation

Acceptance in this study means voluntary willingness to adopt patterns of behaviour and to use certain objects which, from the subjective perspective of an individual, are new. In terms of time, three different phases of the acceptance of innovative goods and patterns of behaviour can be distinguished (Kollmann, 1998). The first phase is the attitude to the adoption of the innovation; the second phase consists of actions involved in the adoption of an innovation; the third phase involves the permanent use of the innovation or permanent application of a pattern of behaviour. Because mobile customer card offers are still not common and individual applications still in the pilot phase, the acceptance in the second and third phases can at present be examined only fragmentarily. For this reason the attitude levels to the acceptance of mobile customer cards will examined in more detail here. In this phase of the acceptance process the consumer forms a personal

Figure 2. The structure of the acceptance construct (adapted from Kollmann, 1998)

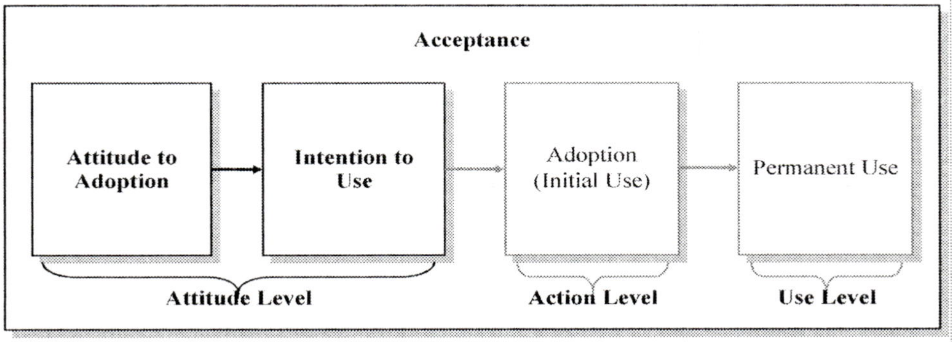

judgement, by means of a subjective perception of the (objective) properties, about their acceptability – the attitude to taking over innovations which forms the basis of the consumer's later adoption behaviour (Regier, 2007). The attitude is to be understood here as, in the words of Eagly and Chaiken (1993), "a psychological tendency that is expressed by evaluating a particular entity with a certain degree of favor or disfavor" (p. 1). The attitude to behaviour, however, does not impact directly on the behaviour, according to the Theory of Planned Behaviour formulated by Fishbein and Ajzen (1975), but rather on the behaviour intention. In addition to the (affective) attitude to adoption the attitude level of acceptance therefore also embraces the intention to use as a cognitive component. Therefore, in our research model, we propose that:

Hypothesis 1: A better (affective) attitude leads to a higher intention to use mobile customer cards.

In Figure 2 the connection between the various phases of acceptance genesis is once again depicted graphically.

Drivers of the Acceptance of Mobile Customer Cards

Expected Benefits for the Card Holders

According to the Technology Acceptance Model (TAM) of Davis (1985, 1989) the perceived usefulness of a technology is decisive for the formation of attitudes and intentions. Mobile customer cards are able, fundamentally and compared with traditional customer cards, to demonstrate numerous advantages for the user.

A major advantage is the convenience of use. Shortage of space in the wallet often means that customers have to decide which cards they need to take with them for specific purchases. If they make the wrong choice, they cannot use the "missing" card when they want to, and the loss of associated benefits can entail a certain amount of frustration in card holders. The program members can normally use the services of the mobile customer card program on a permanent basis because, unlike traditional customer cards, they are used with cell phones, which customers usually have with them. This makes it possible for them to participate in many programs simultaneously.

This advantage of mobile customer cards also applies to the administration and use of mobile coupons, which represent a substantial component of mobile loyalty programs. A time-saving benefit in terms of the Theory of the Allocation of Time (Becker, 1965) is also possible through individualization of coupons. Program members will receive only those coupons really relevant to them (Wehmeyer, Müller-Lankenau, 2005) and thus the time spent on selecting relevant coupons from the entire battery of coupons available is saved. There are also similar advantages with regard to the information and communication measures of mobile loyalty programs because members receive less information from operators and partners of the card program that is of little or no interest to them (Leppäniemi, Karjaluoto, 2005). On mobile portals, consumers can, if necessary, also retrieve information useful to them, independent of time and location. In this case there is in turn a communication benefit for the customer, which may have a positive effect on the acceptance (Bauer et al., 2005a; Merisavo et al., 2007; Rao, Minakakis, 2003).

The constant availability of the mobile customer card and mobile coupons can also mean a financial advantage for the card holder, because the probability of collecting bonus points and using other exclusive price reductions increases. According to the transaction benefit concept proposed by Thaler (1985) the result here is both a higher acquisition and transaction utility over an extended period. The acquisition utility is the result of the improved price-performance ratio, which accrues to the cardholder in the form of discounts and bonuses. From the transaction utility, the customer derives a feeling of having concluded a good deal, because he is able make use of the services provided at a price that is lower than the price quoted and/or reference price.

Hence, we propose that:

Hypothesis 2a: Higher levels of expected benefit (of mobile customer cards) lead to a better (affective) attitude.

Hypothesis 2b: Higher levels of expected benefit (of mobile customer cards) lead to a higher intention to use mobile customer cards.

Expected Ease of Use and Expected Technical Obstacles

According to the Theory of Planned Behaviour (Fishbein, Ajzen, 1975) the perceived behaviour control of an individual also has a substantial influence on the development and character of attitude. It is a matter of estimating whether the behaviour of a person will really lead to the result intended (Ajzen, 1985). Perceived behaviour control is also an important attitude driver in TAM (Davis, 1985; Davis 1989). The model assumes that, in addition to usefulness, the perception of how easy a technology is to use by an individual substantially determines attitude and intention. Perception of the easy use of technology thus represents a major characteristic of perceived behaviour control.

Because of the innovative character of mobile customer loyalty programs and the associated unfamiliarity of the consumers with these applications, they can often not or only with great difficulty, estimate whether they are in a position to really use these offers together with associated benefits. From the theoretical considerations sketched out above, it may be assumed that:

Hypothesis 3: The expected ease of use of a mobile customer card program positively influences the (affective) attitude.

Closely associated with this are possible technical obstacles that make it difficult or impossible to enjoy the benefits promised by mobile customer cards. Not all consumers are equipped with the latest devices and this means limitations on the use of the services offered by the card program, such as location-based services. The handling of mobile coupons or the reception of individual messages can lead to problems of use with displays that have low resolution and/or with relatively small storage capacity for SMS and MMS. This inability to make use of partial performances of mobile customer cards also results, as does the uncertainty of customers about their general competence to use mobile customer cards, in a worse affective attitude.

Hypothesis 4: Expected technical obstacles negatively influence the (affective) attitude.

Reservations about the Use of Personal Data

Because of the geographical context intelligence of mobile customer card offers, providers can collect much more wide-ranging information about individual program members than with traditional customer cards. Although empirical studies, for example on the sensitivity of consumers to data misuse with location-based services, reveal only few reservations about dishonest data use, the increasingly critical perception of customers in this respect can be assumed. (Brakel et al., 2004). Among other things, the growing number of reports in the media about the (more or less) inscrutable collection and use of consumer data for marketing purposes makes this inevitable. Growing scepticism on the part of consumers vis-à-vis the general use of customer cards and mobile services is thus to be expected (cf. Unni, Harmon, 2007 on the subject of mobile services).

If a consumer becomes aware of the (for him) uncontrollable use of his personal, purchase and location data by mobile customer card programs, this can be understood as a restriction of his freedom to decide about how his personal data are used. This restriction of freedom can, according to the Theory of Reactance (Brehm, 1966), produce contrary behaviour in the form of non-use of mobile customer cards as way of re-establishing one's own freedom to decide. It is a precondition of reactant behaviour that the consumer places a certain degree of importance on his personal data. It is therefore to be assumed that:

Hypothesis 5: Reservations about the use of personal data have a negative impact on the (affective) attitude.

Use So Far of Traditional Customer Cards / Customer Card Programs and Mobile Services

If consumers are confronted by the decision whether to accept or reject mobile customer cards, they normally compare the new attitude object with existing attitudes to related or superior attitude objects. Consumers, who have already used traditional customer cards and mobile services and have thereby gained some experience, will probably reveal a positive attitude to mobile loyalty programs. If the experiences with traditional customer cards or mobile services have been negative, mobile customer cards will probably not be used. If the experiences with both areas, however, are quite different, then, according to the Theory of Cognitive Dissonance, there is a tension between the negative attitude to the one and the positive attitude to the other area (Festinger, 1978). Because individuals try to dismantle this psychological tension, they often adjust their affective and conative attitude to a new object to bring it into line with the attitude to the experience object, which, among the competing attitudes, is more relevant. Thus, we expect that

Hypothesis 6a: The use so far of traditional customer cards has a positive impact on the (affective) attitude towards mobile customer cards.

Hypothesis 6b: The use so far of traditional customer cards has a positive impact on the intention to use mobile customer cards.

Hypothesis 6c: The use so far of mobile services has a positive impact on the (affective) attitude towards mobile customer cards.

Figure 3. Research model

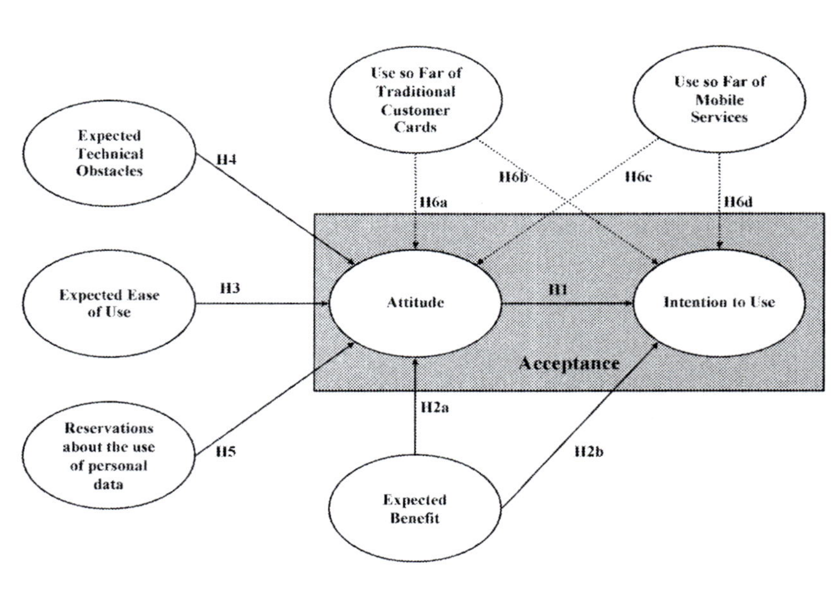

Hypothesis 6d: The use so far of mobile services has a positive impact on the intention to use mobile customer cards.

Based on the conceptual considerations illustrated above, we propose the research model presented in Figure 3.

EMPIRICAL ANALYSIS OF AN ACCEPTANCE MODEL OF MOBILE CUSTOMER CARDS

Data Collection Procedure and Sample Characteristics

The research model was tested in an exploratory empirical study. We employed a written questionnaire on the subject of an imaginary offer via a mobile customer card system which was issued to 118 economics students of the University of Kassel for data collection (cf. also Mann, Prein 2008). 44.8% of the respondents were female and 55.2% were male. The average age was 24 years, whereas about half of the respondents were between 23 and 25 years old. The respondents were shown a fictional advertisement from a mobile multi-partner customer card with a lot of information about the offer. The offer was evaluated within the framework of a standardized questionnaire.

Even though the data sample is a convenience sample with no claim to be representative of larger groups (e. g. older cell phone users), it is still possible to estimate the importance of individual spheres of influence. Young users, because of their mostly intensive use of mobile devices, are not only in a position to competently assess innovative areas of application for mobile communication with respect to possible use and risk variables, but are also at the same time particularly critical of commercial mobile marketing activities (Grant, O'Donohoe, 2007).

This is also revealed in the empirical results of our pilot study. Only 39.2% of the respondents evaluated the offer presented as good and 45.8% of the respondents admitted they found the mobile customer card interesting. As shown in Figure 4, the actual willingness to use it is even lower. Thus,

Figure 4. Empirical findings on the acceptance of mobile loyalty programs

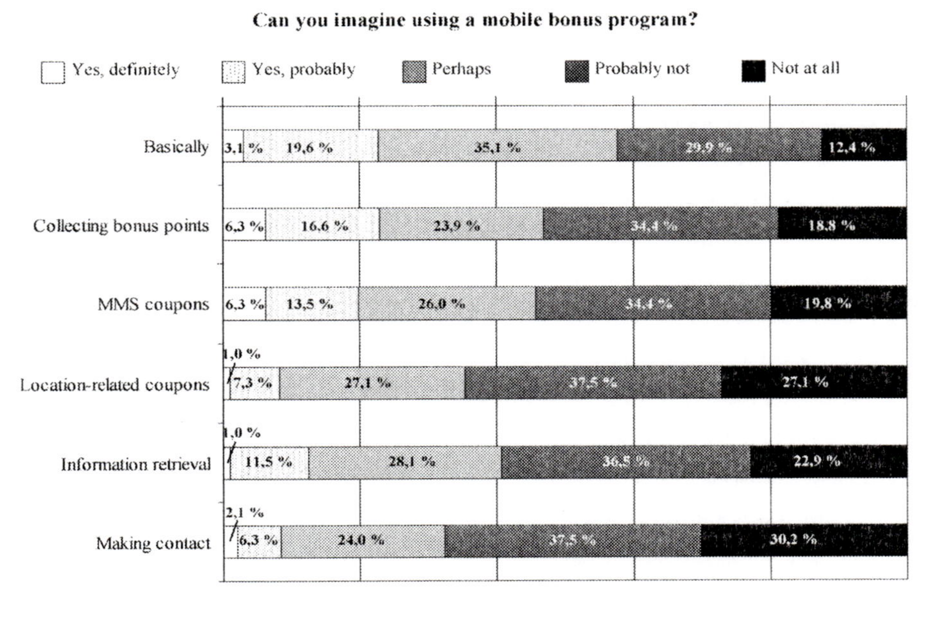

only 22.7% of the respondents could imagine using mobile loyalty programs. 35.1% have no interest in using them, while 42.3% were sceptical about mobile customer cards. Collecting bonus points was the most attractive aspect of mobile customer cards to those surveyed. Thus, 22.9% of the respondents would use bonus points. The willingness to use location-related coupons was lowest. Only 8.3% of the respondents could imagine using them at some time or all the time.

Data Analysis and Results

The above-mentioned acceptance drivers have been tested, apart from the use so far of traditional customer cards and mobile services, using the causal analysis tool of Partial Least Squares procedure (SmartPLS). PLS is preferred to covariance-based structural equation modelling methods such as LISREL because our intention is to examine the validity of using the acceptance determinants to predict the attitude towards and the intention to use mobile customer cards. Covariance-based methods are more appropriate to analyse the covariance of the variables (Moores, Chang 2006, p. 172). The differences between users and non-users of traditional customer cards and mobile services regarding their mobile customer card acceptance were tested with t-tests.

Operationalization and Measurement Model Assessment

The endogenous and exogenous constructs were measured on self-developed five-point Likert scales. The attitude, the intention to use and the reservations about the use of personal data were assessed using multiple reflective indicators, the expected benefit and the expected technical obstacles were operationalized using several formative Items and the expected simplicity of use was assessed with a single-item. Details of all measures are listed in table 3 in the appendix.

To judge the validity and reliability of the measuring models various quality criterions have been tested. Table 1 shows Dillon-Goldstein's Rho, Cronbach's Alpha, average variance extracted (AVE) for the all reflective constructs and loadings and t-values for their indicators. Dillon-Goldstein's Rho and Cronbach's Alpha values of all of the constructs are above the of suggested threshold value of 0.7, indicating sufficient internal consistency. Furthermore, the AVE values for all reflective constructs are over the recommended level of 0.5. This demonstrates adequate convergent validity. The discriminant-validity was tested using the Fornell-Larcker-criterion, which requires the square root of the AVE is greater than all of the inter-construct correlations (Fornell, Larcker 1981). The results suggest that our measurement models demonstrate sufficient discriminant-validity. Moreover all loadings of the reflective indicators are over the 0.7 guideline and significant at the 0.05 level, which indicates satisfactory item reliability.

Because covariance based quality criterions are not applicable for the evaluation of formative measurement models (Chin 1998), the expected benefit and the expected technical obstacle constructs were assessed regarding the relevance of the indicators for the constructs and the multicollinearity of the items. Table 2 shows the weight, variance influence factor (VIF) and t-value for each formative item. The results indicate the expected benefit is significantly formed by EB 1, EB 2 and EB 4, while EB 2 and EB 3 don't significantly contribute to form the construct. Nevertheless, we refrain from eliminating these indicators because this would change the meaning of the variable. The weights for the formative measure of the expected technical obstacles are all significant. Furthermore all values of the VIF are far below the threshold of 10 (Hair et al. 1998, pp. 191 ff.).

Structural Model Assessment and Hypothesis Testing

Figure 5 depict graphically the estimates obtained from the PLS analysis. The R^2 values of 0.558 for the attitude and 0.473 for the intention to use indicate that our model explains a substantial part of the variance of the consumer acceptance of mobile customer cards. Furthermore, the examination of the prognosis relevance by the (non-parametric) Stone-Geisser test (cross validated redundancy) shows that a plausible reconstruction is possible for the attitude ($Q^2 = 0.039$) as well as for the intention to use ($Q^2 = 0.135$) through the exog-

Table 1. Quality criteria of the reflective measurements models

Construct	*Dillon-Goldstein's Rho*	*Cronbach's Alpha*	*AVE*	*Indicator*	*Loading*	*t-Value*
Intention to use	0.888	0.843	0.615	IU 1	0.738	11.978**
				IU 2	0.752	12.289**
				IU 3	0.835	24.864**
				IU 4	0.786	15.826**
				IU 5	0.806	21.095**
Attitude	0.914	0.812	0.841	AT 1	0.903	32.028**
				AT 2	0.930	69.549**
Reservations about the use of personal data	0.922	0.877	0.798	PD 1	0.868	9.596**
				PD 2	0.906	6.322**
				PD 3	0.906	7.491**

enous acceptance determinants. Both endogenous variables show a Q^2 value that is larger than zero (Geisser 1975, p. 101 ff.; Stone 1974).

The positive influence of attitude on the intention to use postulated by TAM (Hypothesis 1) is confirmed in our model (path-coefficient: 0.250; effect size: 0.097). The strongest direct influence factor on the intention to use is the expected benefit, and this is reflected both in the path coefficients (0.498) and in the effect size (0.296). The expected benefit also has an indirect effect on the attitude to the intention to use (path coefficient: 0.531; effect size: 0.559), as a result of which there is a total effect of the path coefficients of 0.631 (0.498 + 0.250 * 0.531) (See Huber et al., 2007 for the calculation of the total effect). Thus, Hypotheses 2a and 2b are also supported.

Further significant influence factors on the attitude and thus indirectly on the intention to

Table 2. Quality criteria of the formative measurement models

Construct	*Indicator*	*Weight*	*VIF*	*t-Value*
Expected benefit	EB 1	0.233	1.875	2.061**
	EB 2	0.536	2.021	4.590**
	EB 3	0.094	1.215	0.901
	EB 4	0.294	2.758	1.676*
	EB 5	0.092	2.828	0.532
Expected technical obstacles	TO 1	0.360	1.065	2.145**
	TO 2	0.848	1.065	8.013**

Figure 5. Structural model of the acceptance of mobile loyalty programs

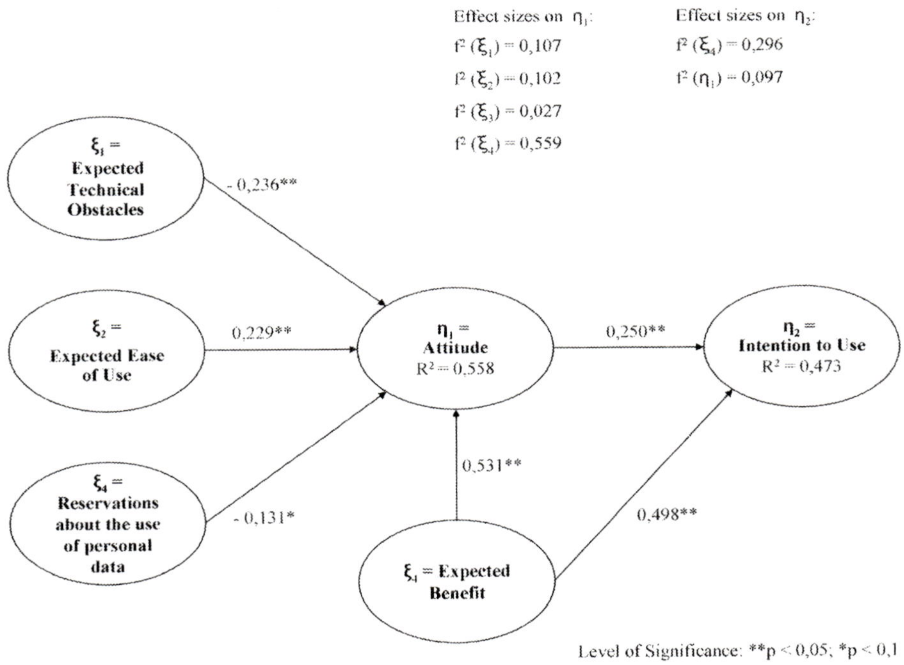

use are the expected simplicity of use (path coefficient: 0.229; effect size: 0.102) and the expected technical obstacles (path coefficient: -0.236; effect size: 0.107). The smallest influence, of relatively weak significance ($p < 0.1$), on the attitude to the mobile customer card program can be seen in the reservations about the use of personal data (path coefficient: -0.131; effect size: 0.027). Therefore, the Hypotheses 3, 4 and 5 could be supported as well.

If one considers the individual determinants a little more closely, one sees the expected benefit of mobile customer cards refers, above all, to financial advantages and use for communication. Thus, 47.9% of the respondents indicate they basically expect to save money by using the mobile customer card presented to them. 54% expect a better price-performance ratio from the use of the mobile customer card when they go shopping. The significance of the financial benefits becomes clear from the fact that receipt of a better price-performance ratio is regarded as important by 85.6% and saving money by as many as 91.6% of the respondents. But this does not differentiate mobile customer card programs from traditional customer card programs which are also used by consumers predominantly because of the financial benefits they bring (Leenheer et al., 2007).

There are also similar user expectations with regard to the information and communication measures of mobile loyalty programs. 29.6% of the respondents assume they will receive less useless advertising from operators and partners of the program, which was regarded as important by 78.1% of all respondents. In this case there is therefore a communication and time-saving benefit for the customer that which tends to have a positive effect on the acceptance attitude.

65.2% of respondents also expect using mobile customer cards will mean they have to carry fewer cards with them. This is particularly relevant because 63.4% of the holders of traditional customer cards have two or more cards. This additional convenience of use, however, does not contribute greatly to the total user expectations and is not significant, because only 37.8% of the respondents consider it important to have to carry fewer cards with them. An equally low and insignificant value is attached to time saving by means of the need-specific and context-dependent sending or use of mobile coupons. Thus, only 33.1% of the respondents expect to have to spend less time sorting out useless coupons. This time savings is seen as important by only 32.2%.

With regard to the perceived simplicity of use, many customers still have reservations about the positive nature of mobile customer card systems. 25.0% of respondents believe accessing mobile websites to redeem mobile coupons on their cell phone would be difficult. For example, the installation of programs on the cell phone is necessary for the storefinder of the customer card program Happy Digits. The installation of such programs would create problems for as many as 30.2% of the respondents.

A similar picture emerges with the expected technical obstacles. For example, 42.7% of respondents were not sure whether the mobile customer card offers were usable on their devices. In addition 69.8% were afraid messages and coupons would clog up the storage capacity of their cell phone and thus impairing its ability to function.

Similar reservations are evident among respondents with regard to the circulation of personal data to the operators and partners of a mobile card system. 66.0% are annoyed loyalty programs are used to spy on consumers. 75.3% of the respondents even indicate they deliberately use various methods to avoid circulating personal data, and 20.5% knowingly give false information in order to protect their personal sphere.

Additional t-tests were performed in order to test the differences between traditional customer card-holders and non-holders with respect to their attitude and intention to use mobile customer cards. As the results show, traditional customer card-holders are significantly more prepared to use the cell phone when shopping in order to collect bonus points than non-holders. Against the expectations, no significant differences could be established with the affective judgement of mobile customer cards and with the willingness to use other additional services of mobile customer cards, such as location-based coupons and offers of information on the cell phone. Thus, Hypothesis 6a could not be supported and Hypothesis 6b just for the intention to use the cell phone in order to collect bonus points.

A comparison between users and non-users of mobile services (e.g. mobile web site-services, MMS newsletter, mobile ticketing) shows users so far cannot be distinguished from non-users in terms of their basic willingness to collect bonus points. However, users of mobile services reveal a fundamentally and significantly better affective attitude to mobile loyalty programs than non-users. In addition, those respondents with mobile service experience are significantly more willing to use new services such as location-based coupons and to make contact with card-issuing companies via the cell phone over those respondents who have so far not made use of mobile services. These results provide support for Hypothesis 6c and partly for Hypothesis 6d.

IMPLICATIONS

The results of our pilot study show it is the expectations of gaining financial advantage and receiving less irrelevant advertising that contributes positively to the creation of acceptance of mobile loyalty programs. Companies that would like to introduce and use mobile technologies within the framework of customer card programs should therefore integrate clear financial incentives into their programs and communicate them to potential program members. Moreover, building on the collection and evaluation of qualified customer data, incentives and advertising information should be tailored in terms of content, form, and if possible, time and place to the interests and needs of the individual customer.

A further factor in the acceptance of cell phones is the perceived ease of use. Negative factors in terms of acceptance however are the expected technical problems involved in using the system and the requirement made of the user to give personal data. Companies wishing to introduce and use mobile customer cards for the purposes of a loyalty program should therefore choose relatively simple solutions to use the cards which will enable the customer to use the technology in a way that is as intuitive as possible. Care must be taken the program functions can be used with ease and trouble-free both on modern touch-screen phones equipped with relatively rapid connections to the mobile internet, such as the Apple iPhone, and other, older devices with traditional keypads, small displays and slow data connections.

Furthermore, issuing card companies should avoid collecting personal data about card-users on a wide and non-transparent scale. Companies should also bear in mind program members who require a particular level of convenience expect special information and services tailored and relevant to them. There is a dilemma here between a limited collection and use of personal customer data and the necessary individualization involved in how customers are addressed, which usually requires customer data, and this dilemma has to be resolved.

CONCLUSION

The starting point for our consideration was the fact traditional card based loyalty programs have different weak points. In this chapter we introduced, how mobile communication technologies can be used in the context of loyalty programs and illustrated which advantages can arise from the transfer of customer card functions on the mobile device.

Purchase behaviour data and reactions to advertising messages can be assigned much more precisely to individual customers. It is also possible via the collection of information and communication behaviour on mobile web portals to generate information about individual preferences and interests and therefore gain more detailed customer data profiles than has been possible with traditional customer card programs. The customer knowledge generated in this way provides an important basis for a differentiated handling and thus a better skimming of the demand potential of their customer.

But the potentials of mobile loyalty programs can only be tapped if there is appropriate acceptance on the part of the customer. For this reason, we explored the phenomenon of the consumer acceptance of mobile loyalty programs more closely. We derived a number of determinants that influence consumer's intention to use mobile loyalty programs and presented and evaluated a conceptual model explaining consumer's acceptance.

The main findings of our exploratory study are threefold. The first is that consumer acceptance of mobile loyalty programs can only be achieved with programs offering a clear benefit for the customer. The second is that all the technical solutions must be easy to use on every mobile device. The third is that mobile loyalty programs should cover the span between the customer's need for privacy protection and its desire for individualization.

REFERENCES

Ajzen, I. (1985). From Intentions to Actions: a Theory of Planned Behavior. In: Kuhl, J., Beckmann, J. (Ed.), Action control: from cognition to behaviour (pp. 11-39). Berlin: Springer.

Anonymous. (2007). HappyDigits Shop Finder. In: *Interaktive Trends 2007/2008, Jahrbuch Deutscher Multimedia Award* (pp. 86-87). Hamburg: J&S Dialog-Medien.

Anonymous. (2008). Gewusst wo: Location Based Services. *direkt marketing*, 8(2008), 18.

Balaban, D. (2008). NFC: When Will The Breakthrough Come? *Cards & Payments*, *21*(1), 30–35.

Barnes, J. A., & Scornavacca, E. (2004). Mobile Marketing: The Role of Permission and Acceptance. *International Journal of Mobile Communications*, *2*(2), 128–139. doi:10.1504/IJMC.2004.004663

Bauer, H. H., Barnes, S. J., Reichardt, T., & Neumann, M. M. (2005a). Driving Consumer Acceptance of Mobile Marketing: A Theoretical Framework and Empirical Study. *Journal of Electronic Commerce Research*, *6*(3), 181–192.

Bauer, H. H., Reichardt, T., & Neumann, M. M. (2005b). Bestimmungsfaktoren der konsumentenseitigen Akzeptanz von Mobile Marketing. *der markt - Zeitschrift für Absatzwirtschaft und Marketing*, 44(2), 59-71.

Becker, G. S. (1965). A Theory of the Allocation of Time. *The Economic Journal*, *75*(299), 493–517. doi:10.2307/2228949

BGH. (2008). Schutz des Verbrauchers vor Werbung durch E-Mail und SMS. Retrieved January 15, 2009, from http://juris.bundesgerichtshof.de/cgi-bin/rechtsprechung/document.py?Gericht=bgh&Art=en&Datum=Aktuell&Sort=12288&anz=627&pos=17&nr=44522&linked=pm&Blank=1.

Brakel, O., Groh, G., Hillebrand, C., & Tasch, A. (2004). *Privacy-Einstellungen bei Lokalisierungsdiensten*. Paper presented at the MKWI04, Essen, Germany.

Brandstetter, E. (2006). Schmucke Erfolge mit mobiler Kundschaft. *Economy Special Innovation, 7*, 14.

Brehm, J. W. (1966). *A Theory of Psychological Reactance.* New York: Academic Press.

Carroll, A., Barnes, S. J., Scornavacca, E., & Fletcher, K. (2007). Consumer Perceptions and Attitudes towards SMS Advertising: Recent Evidence from New Zealand. *International Journal of Advertising, 26*(1), 70–98.

Chin, W. W. (1998). The partial Least Squares Approach to Structural Equation Modeling. In G. A. Marcoulides (ed.), Modern Methods for Business Research (pp. 295-336). Hillsdale, NJ: Lawrence Erlbaum Associates.

Cortiñas, M., Elorz, M., & Múgica, J. M. (2008). The Use of Loyality-Cards Databases: Differences in Regular Price and Discount Sensitivity in the Brand Choice Decision Between Card and Non-Card Holders. *Journal of Retailing and Consumer Services, 15*(1), 52–62. doi:10.1016/j.jretconser.2007.03.006

Davis, F. D. (1985). *A Technology Acceptance Model for Empirically Testing New End-User Information Systems: Theory and Results.* PhD thesis, Massachusetts Institute of Technology, Sloan School of Management.

Davis, F. D. (1989). Perceived Usefulness, Perceived Ease of Use, and User Acceptance of Information Technology. *MIS Quarterly, 13*(3), 318–340. doi:10.2307/249008

Demoulin, N. T. M., & Zidda, P. (2007). On the Impact of Loyalty Cards on Store Loyalty: Does the Customers' Satisfaction with the Reward Scheme Matter? *Journal of Retailing and Consumer Services, 15*(5), 386–398. doi:10.1016/j.jretconser.2007.10.001

Eagly, A. H., & Chaiken, S. (1993). *The Psychology of Attitudes.* Fort Worth: Harcourt Brace Jovanovich.

Emnid, T. N. S. (2006). *Bonusprogramme in Deutschland. Eine Studie des TNS Emnid Medien- und Sozialforschung GmbH.* Bielefeld: TNS Emnid.

Festinger, L. (1978). *Die Theorie der kognitiven Dissonanz.* Bern: Hans Huber.

Fishbein, M., & Ajzen, I. (1975). *Belief, Attitude, Intention and Behavior: An Introduction to Theory and Research.* Reading, MA: Addison-Wesley.

Fornell, C., & Larcker, D. F. (1981). Evaluating Structural Equation Models with Observable Variables and Measurement Errors. *JMR, Journal of Marketing Research, 18*(1), 39–50. doi:10.2307/3151312

Geisser, J. (1975). A Predictive Approach to the Random Effect Model. *Biometrika, 61*(1), 101–107. doi:10.1093/biomet/61.1.101

Grant, I., & O'Donohoe, S. (2007). Why Young Consumers are Not Open to Mobile Marketing Communication. *International Journal of Advertising, 26*(2), 223–246.

Hair, J. F., Anderson, R. E., Tatham, R. L., & Black, W. C. (1998). *Multivariate Data Analysis.* Upper Saddle River, NJ: Prentice Hall.

Huber, F., Herrmann, A., Meyer, F., Vogel, J., & Vollhardt, K. (2007). *Kausalmodellierung mit Partial Least Squares.* Wiesbaden: Gabler.

Kollmann, T. (1998). *Akzeptanz innovativer Nutzungsgüter und -systeme.* Wiesbaden: Gabler.

Kurkovsky, S., & Harihar, K. (2006). Using Ubiquitous Computing in Interactive Mobile Marketing. *Personal and Ubiquitous Computing, 10*(4), 227–240. doi:10.1007/s00779-005-0044-5

Leenheer, J., van Heerde, H. J., Bijmont, T. H. A., & Smidts, A. (2007). Do Loyalty Programms really Enhance Behavioral Loyalty? An Empirical Analysis Accounting for Self-Selecting Members. *International Journal of Research in Marketing*, *24*(1), 31–47. doi:10.1016/j.ijresmar.2006.10.005

Leppäniemi, M., & Karjaluoto, H. (2005). Factors Influencing Consumers' Willingness to Accept Mobile Advertising: A Conceptual Model. *International Journal of Mobile Communications*, *3*(3), 197–213. doi:10.1504/IJMC.2005.006580

Mann, A., & Prein, J. (2008). Akzeptanz mobiler Kundenkarten – konzeptionelle Überlegungen und empirische Befunde. In H. H. Bauer, T. Dirks, & M. D. Bryant (Ed.), *Erfolgsfaktoren des Mobile Marketing* (pp. 241-259). Berlin, Heidelberg: Springer.

Merisavo, M., Kajalo, S., Karjaluoto, H., Virtanen, V., Salmenkivi, S., Raulas, M., & Leppäniemi, M. (2007). An Empirical Study of the Drivers of Consumer Acceptance of Mobile Advertising. *Journal of Interactive Advertising*, 7(2). Retrieved November 25, 2008, from http://www.jiad.org/vol7/no2/merisavo/An%20empirical%20study%20of%20the%20drivers%20of%20consumer%20acceptance.pdf

Möhlenbruch, D. Dölling, S., Ritschel, R. (2008). Web 2.0-Anwendungen im Kundenbindungsmanagement des M-Commerce. In H. H. Bauer, T. Dirks, & M. D. Bryant (Ed.), *Erfolgsfaktoren des Mobile Marketing* (pp. 221-240). Berlin, Heidelberg: Springer.

Moores, T. T., & Chang, J. C.-J. (2006). Ethical Decision Making in Software Piracy: Initial Development and Test of a Four-Component Model. *MIS Quarterly*, *30*(1), 167–180.

Payne, A., & Frow, P. (2006). Customer Relationship Management: from Strategy to Implementation. *Journal of Marketing Management*, *22*(1/2), 135–168. doi:10.1362/026725706776022272

Rao, B., & Minakakis, L. (2003). Evolution of Mobile Location-Based Services. *Communications of the ACM*, *46*(12), 61–65. doi:10.1145/953460.953490

Regier, S. (2007). *Markterfolg radikaler Innovationen. Determinanten des Akzeptanzverhaltens.* Wiesbaden: Gabler.

Rudolph, T., & Emerich, O. (2008). Kundeninteraktion über mobile Services im Handel. In H. H. Bauer, T. Dirks, & M. D. Bryant (Ed.), *Erfolgsfaktoren des Mobile Marketing* (pp. 261-278). Berlin, Heidelberg: Springer.

Stone, M. (1974). Cross-Validatory Choice and Assessment of Statistical Predictions. *Journal of the Royal Statistical Society. Series B. Methodological*, *36*(2), 111–147.

Thaler, R. (1985). Mental Accounting and Consumer Choice. *Marketing Science*, *4*(3), 199–214. doi:10.1287/mksc.4.3.199

Unni, R., & Harmon, R. (2007). Perceived Effectiveness of Push vs. Pull Mobile Location-Based Advertising. *Journal of Interactive Advertising*, 7(2). Retrieved November 25, 2008, from www.jiad.org/vol7/no2/unni/Perceived%20effectiveness%20of%20push%20vs_%20pull.pdf

Wais, J. S., & Clemons, E. K. (2008). Understanding and Implementing Mobile Social Advertising. *International Journal of Mobile Advertising*, *3*(1), 12–18.

Wehmeyer, K., & Müller-Lankenau, C. (2005). *Mobile Couponing - Measuring Consumers' Acceptance and Preferences with a Limit Conjoint Approach.* Paper presented at the 18th Bled eConference, Bled, Slovenia.

Weidelich, F. (2008). CeBIT 2008: RFID-Anwendung Near Field Communication soll 2010 zum Standard bei Handys werden. *VDI Nachrichten*, 14.03.2008, 13.

Wiedmann, K.-P., Reeh, M.-O., & Schuhmacher, H. (2008). Near Field Communication im Mobile Marketing. In H. H. Bauer, T. Dirks, & M. D. Bryant (Ed.), *Erfolgsfaktoren des Mobile Marketing* (pp. 305-325). Berlin, Heidelberg: Springer.

Wohlfahrt, J. (2002). Wireless Advertising. In: G. Silberer, J. Wohlfahrt, & T. Wilhelm (Ed.), *Mobile Commerce* (pp. 245-263). Wiesbaden: Gabler.

Wübker, G., & Schmidt-Gallas, D. (2003). Pricing-Strategien für Banken. In: H. Diller, & A. Herrmann(Ed.). *Handbuch Preispolitik, Strategien – Planung – Organisation – Umsetzung* (pp. 739-761). Wiesbaden: Gabler.

KEY TERMS AND DEFINITIONS

Acceptance: Voluntary willingness to adopt patterns of behaviour and to use certain objects which, from the subjective perspective of an individual, are new.

Data Matrix: Two-dimensional, machine-readable representation of data, consisting of black and white modules arranged in a rectangular grid.

Location Based Service: Individual place-dependent mobile service, accessible with the cell phone.

MMS: A cell phone standard for sending and receiving messages that include multimedia objects such as images-, audio-, video- and text-files.

Mobile Customer Cards: The use of mobile devices, in particular cell phones, that will take over the functions of traditional customer cards and loyalty programs.

Near Field Communication (NFC): Short-distance (RFID) radio network, which enables the exchange of data between mobile devices over about a 10 centimetre distance.

SMS: A cell phone standard for sending and receiving of short text messages of up to 160 characters.

APPENDIX

Table 3. Operationalization of the constructs

Construct	Indicator	Operationalization	Mean value	Standard deviation
Intention to use *Agreement scale from (every case) to 5 (none)*	IU 1	*I could imagine using the cell phone to collect loyalty points.*	3.43	1.16
	IU 2	*I could imagine calling up and cashing in coupons as MMS.*	3.48	1.142
	IU 3	*I could imagine requesting local coupons via the mobile internet.*	3.82	0.951
	IU 4	*I could imagine retrieving information such as the loyalty points status on the mobile internet.*	3.69	0.987
	IU 5	*I could imagine using the cell phone to contact the provider of the loyalty program.*	3.88	0.987
Attitude *Evaluation scale from 1 (good) to 5 (bad)*	AT 1	*How do you basically rank the offer presented? good ... bad*	2.72	0.997
	AT 2	*How do you basically rank the offer presented? interesting ... uninteresting*	2.94	1.366
Reservations about the use of personal data *Agreement scale from 1 (fully and entirely) to 5 (not at all)*	PD 1	*I feel annoyed when companies want personal information from me.*	2.27	1.177
	PD 2	*If companies ask me for personal information, I think twice before giving it.*	1.76	0.899
	PD 3	*It worries me that companies are collecting more and more private information about me.*	2.16	1.038
Expected benefit *Agreement scale from 1 (fully and entirely) to 5 (not at all)*	EB 1	*From using the customer card offer on my cell phone I expect to receive more/better services for the same price.*	2.57	1.020
	EB 2	*From using the customer card offer on my cell phone I expect to save money.*	2.69	1.218
	EB 3	*From using the customer card offer on my cell phone I expect to have to cut out and carry fewer coupons with me.*	2.39	1.139
	EB 4	*From using the customer card offer on my cell phone I expect to receive less advertising that is of no interest to me.*	3.30	1.309
	EB 5	*From using the customer card offer on my cell phone I expect to receive information useful to me each time I need it.*	3.12	1.185
Expected technical obstacles *Agreement scale from 1 (fully and entirely) to 5 (not at all)*	TO 1	*I am not sure if the offers can be used with my cell phone.*	2.91	1.282
	TO 2	*The messages and coupons just clog up my cell phone memory.*	2.01	1.110
Expectedeaseofuse *Rating scale from 1 (I would find it very easy) to 5 (I would find it very difficult)*		*MMS reception, administration and call.*	2.12	1.056

Chapter 20
Analyzing the Forwarding Behavior in Mobile Viral Marketing:
An Empirical Study

Dietmar G. Wiedemann
University of Augsburg, Germany

Tobias Haunstetter
University of Augsburg, Germany

Key Pousttchi
University of Augsburg, Germany

ABSTRACT

As mobile devices are personal communication tools, they are platforms for word-of-mouth marketing. Mobile viral marketing is tremendously attractive for marketers but neglected by academic research. Surprisingly, relatively few studies are directed at its basic elements, i.e., directed at willingness to forward different mobile viral content, understanding characters of those who forward mobile viral content frequently, and characters of the recipients of this content. This chapter presents the findings of an online survey conducted to empirically investigate the consumers' intention to forward different kinds of mobile viral content, to identify the primary target groups for the mobile viral marketing in terms of their forwarding behavior, and to analyze to whom mobile viral content is forwarded.

INTRODUCTION

Word-of-mouth (WOM) has been called the world's most effective marketing strategy (Misner, 1999) and it refers to oral, person-to-person communication between a recipient and a communicator which the recipient perceives as a non-commercial message, regarding a brand, product, or service (Arndt, 1967). With mobile devices being viewed as personal communication tools, they can also be valuable platforms for WOM marketing (Bauer, Barnes, Reichardt & Neumann, 2005; Pousttchi & Wiedemann, 2006). We define *mobile viral marketing* as a concept for distribution or communication that relies on consumers to transmit content via

DOI: 10.4018/978-1-60566-074-5.ch020

mobile communication techniques and mobile devices to other potential consumers in their social sphere and to animate these contacts to also transmit the content. We further define such content as *mobile viral content*; it comprises mobile services or ads. Individuals transmitting mobile viral content are termed *communicators* whereas individuals receiving such content are termed *recipients* (Pousttchi & Wiedemann, 2007). In this chapter, we refer to *mobile viral mavens* as consumers who forward mobile viral content frequently and willingly.

Some studies (Barwise & Strong, 2002; Kümmerlein, 2002; "I-play," 2005; "Intuitive Media Research Services," 2006, see Section 2) have established that mobile viral marketing helps to extend substantially the number of recipients and that it boosts the impact of marketing communication at low company expense (Jelassi & Enders, 2004). For mobile business-to-customer services, mobile viral marketing reduces the cost of distribution as users become part of the firm's salesforce. Therefore, mobile viral marketing is a powerful tool for promotion and distribution in business-to-customer mobile commerce. New entrants are particularly eager to attract customers as quickly as possible in order to reach the break-even point and to turn themselves into established players in mobile commerce. Mobile viral marketing is a suited strategy to cope with these challenges.

Despite mobile viral marketing's usefulness for marketers, relatively few studies are covering its basic elements, i.e., mobile viral content and consumers engaged in mobile viral processes. Whereas the effectiveness of different types of content in mobile marketing campaigns without viral elements is receiving more and more attention in the literature (Barwise & Strong, 2002; Drosos & Giaglis, 2004; Deakins, Grandcolas & Rettie, 2005; Bauer, Barnes, Reichardt & Neumann, 2005; Heinonen & Strandvik, 2006), little is known about the effectiveness of mobile viral content. Such content may consist in e.g., a viral greeting call, a video clip or a recommendation for a Java-based client. A complete list of examples used for the purpose of this chapter is given in Table 2. We assume that consumers' intention to forward mobile viral content varies across content categories because of different characteristics such as the forwarding mechanism and the customer's involvement. Hence, for the marketer's perspective it is vital to know which kind of mobile viral content is most effective in order to understand drivers of successful mobile viral marketing strategies. Targeting the right consumers is the key that triggers viral effects (Phelps, Lewis, Mobilio & Perry, 2004; Helm, 2000). Consequently, it is argued that identifying and addressing mobile viral mavens as initial contacts in the beginning of a mobile viral marketing strategy is an important success factor for mobile viral marketing (Pousttchi & Wiedemann, 2007). Otherwise, marketers will not be able to use the whole potential for exponential growth in the message's exposure and influence (Wilson, 2000). Generally, with the help of one-to-one marketing (Peppers, Rogers & Dorf, 1999), a precise targeting of mobile viral mavens, e.g., by using push SMS is possible. However, in the

Table 1. Mobile viral marketing standard types

Level of network externalities	High	Motivated Evangelism	Signaling Use, Group Membership
	Low	Targeted Recommendation	Awareness Creation, Benefits Signaling
		Active	Passive
		Role of communicator in persuasion	

Table 2. Examples used in the web-based survey

Example	Standard type	Item
Recommendation of an instant messaging service	ME	GIF5
Recommendation of a massive multiplayer mobile game	ME	GIF6
Recommendation of a location based friend finder	ME	GIF9
Recommendation of a mobile community	ME	GIF2
Recommendation of a mobile payment procedure	SUGM	GIF10
Inviting to a contest which requires participation of three people	TR	GIF7
Recommendation of a Java-based cinema portal	TR	GIF1
Forwarding of an SMS-based coupon	TR	GIF1
Forwarding of a video clip via Bluetooth	TR	GIF3
Using an ad-sponsored free SMS service available on a brand website	ACBS	GIF11
Using an IVR-based viral greeting call	ACBS	GIF4

current body of knowledge, limited research is available that focuses specifically on characteristics of mobile viral mavens. Moreover, we are interested in the recipients of the mobile viral content, especially in their relationship to the communicator.

With this background in mind, the present contribution aims to provide a better understanding of mobile viral marketing. Drawing on a simple communication model, this chapter aims to answer the following research question: "Who forwards what to whom in mobile marketing?" Underpinning this aim, the principal objectives of the chapter are: (1) to empirically investigate the consumers' intention to participate in different kinds of mobile viral strategies; (2) to identify the characteristics of mobile viral mavens in terms of their forwarding behavior; (3) to analyze to whom mobile viral content is forwarded. In doing so, we present the findings of an online survey.

The rest of this chapter is organized as follows. In Section 2, we provide broad discussions of mobile viral marketing and some findings from related surveys. In Section 3, we present the theoretical background for analyzing the basic elements of mobile viral marketing. Section 4 specifies details of the method used in this chapter. In Section 5, we present and discuss the results. In Section 6, we present our considerations regarding research and practical implications. Finally, in Section 7, we conclude and summarize our findings.

MOBILE VIRAL MARKETING

In this section we discuss mobile viral marketing, especially its origin, current knowledge from an academic point of view and advantages from a practical point of view.

Mobile viral marketing is based on traditional WOM and on electronic viral marketing (Wiedemann, 2007). Since the early 1950s, researchers have shown that WOM not only influences consumers' choices and purchase decisions (Arndt, 1967; Bansal & Voyer, 2000), but also shapes consumers' expectations (Zeithaml & Bitner, 1996), pre-usage attitudes (Herr, Kardes & Kim, 1991) and even post-usage perceptions of a product or service (Bone, 1995). Therefore, marketers have a vital interest in identifying and capitalizing on consumers who influence others (Kotler & Zaltman, 1976). Consumers' communication environment has been considerably changed and enriched by the development of information and communication technologies such as the stationary

Internet. In this context, WOM has gained new significance (Helm, 2000; Subramani & Rajagopalan, 2003; Dellarocas, 2003; Phelps, Lewis, Mobilio & Perry, 2004). In 1997, the venture capitalist Steve Jurvetson originally used the term "viral marketing" as "network-enhanced word of mouth" with regard to the marketing strategy of the free email service Hotmail (Jurvetson, 1997). By appending an advertising message to every outbound email, "Get your free email at Hotmail", the service signed up over 12 million subscribers in its first one and a half years (Montgomery, 2001). This example illustrates the hybrid nature of viral marketing as a concept of both communication and distribution. The rapid growth of cell phone ownership has opened up a new arena for WOM communication – mobile viral marketing. Being a subset of electronic viral marketing, this marketing instrument relies on mobile communication techniques and mobile devices to spread the word.

As stated in Section 1, scholars have provided a sizeable body of research on mobile marketing, however, research explicitly focusing on mobile viral marketing has received less attention. The below-mentioned surveys allude to the topic only marginally (mainly by one or two survey questions) and focus principally on other aspects of mobile marketing. Notwithstanding, they provide some interesting findings and illustrate the potential of mobile viral marketing for distribution and communication purposes. For example, it has been found out in a trial, that 17 per cent forwarded especially entertaining or informative mobile ads (Barwise & Strong, 2002). In a survey of mobile greeting cards 19 to 65 per cent (dependent on the analyzed campaign) forwarded the mobile viral content once and 22 per cent to 41 per cent repeatedly (Kümmerlein, 2002). According to a survey by Skopos, 30 per cent said a friend's recommendation would convince them to download a mobile application ("I-play," 2005). A study published by Intuitive Media and New Media Age stated that 29 per cent would share and 45 per cent would intend to share music with friends via Bluetooth ("Intuitive Media Research Services," 2006). These are just a few of the readily available statistics in the market that support the idea of mobile viral marketing Analyzing mobile marketing from a theoretical viewpoint, Pousttchi & Wiedemann (2006) defined six mobile marketing objectives, i.e. building brand awareness, changing a brand's image, sales promotion, enhancing brand loyalty, building a customer database and mobile word-of-mouth, with the latter being a key objective. They also developed a conceptual framework that provides insight into achieving these objectives via different forms of campaigns. The exploratory study of Wiedemann (2007) focused exclusively on mobile viral marketing. Using literature review, a conceptualization of mobile viral marketing and a description model were provided. Based on this work, Pousttchi & Wiedemann (2007) derived a typology including four standard types of mobile viral marketing that will be used in Section 3.1 of this chapter. Moreover, they presented a set of eight success factors, i.e. perceived usefulness by recipients, reward for communicators, perceived ease of use, free mobile viral content, initial contacts, first mover's advantage, critical mass and scalability. Other authors have proposed a number of systems in order to stimulate mobile users to share mobile content (Ratsimor, Finin, Yoshi & Yesha, 2003; Garyfalos & Almeroth, 2004; Straub & Heinemann, 2004).

Given the ubiquity and the personal character of mobile devices, mobile viral marketing provides added values for individuals compared to its counterpart on the stationary Internet and WOM. Ubiquity enables transmission of mobile viral content anywhere and at any time and therefore is a prerequisite for immediate forwarding. Moreover, communicators are able to show the content to others to others in their direct surroundings before transmitting mobile viral content. Recipients can gain confidents by testing mobile services on the communicator's device at first. As young people even tend to turn mobile devices into fashion accessories, mobile viral marketing might be better suited for promoting lifestyles and associated products than electronic viral marketing. In a prior study (Marini & Wiedemann, 2006) almost 70 per cent of 44 mobile marketing experts named this innovative marketing instrument as a key issue for mobile marketers. The experts stated that personal messages sent by friends gain more credibility than those coming directly from the self-interested advertiser. Furthermore, people who receive the initial message from a familiar recommender participate more frequently in a campaign as initial contacts. Marketers also have a wider array of seeding supports to choose from at the very start of a mobile viral process. Seeding refers to building up a volume of initial contacts (i.e., the first generation of "infected" people) (Rosen, 2000). In electronic viral marketing, seeding typically involves promoting, placing and endorsing campaign material in the content section of high-traffic connection points on the stationary Internet. In mobile viral marketing, campaign material can be presented in the content section of high-traffic connection points on the mobile and stationary Internet, and, in addition, on any other supporting media serving as triggers for customer requests (Pousttchi & Wiedemann, 2006). As an example, some television programs encourage their audience to answer a quiz via SMS for the opportunity to win prizes. This prize could be a mobile viral content, such as a mobile greeting card.

BACKGROUND

Mobile Viral Content

This section uses the mobile viral marketing typology developed by (Pousttchi & Wiedemann, 2007) to identify different kinds of mobile viral content. Their proposed framework consists of two dimensions. The first dimension, concerning the role of communicator in persuasion, differs depending on whether the attempt of communicators to influence is passive or actively persuasive (Subramanian & Rajagopalan, 2003). Active viral marketing is closely associated with traditional WOM for the communicator is personally involved in the process of acquiring new customers. By contrast, in passive viral marketing, a mechanism to forward the content is integrated into the process that may trigger an additional message sent by the marketer. The second dimension distinguishes high and low levels of network externalities. Network externalities describe the fact that "the value of a unit of the good increases with the expected number of units to be sold" (Economides, 1996, p. 678) or used (Katz & Shapiro, 1985). As shown in Table 1, these two dimensions highlight four quadrants called mobile viral marketing standard types.

Although the standard types represent different major concepts, they are not totally disjoint: any of the current mobile viral marketing strategies can be assigned to (at least) one standard type, but a single mobile viral marketing strategy may use more than one standard type.

Examples for *Motivated Evangelism (ME)* include services like mobile communities, mobile instant messengers, location-based friend finders, or (massive) multi-player mobile games. Mobile communities can be considered as virtual communities to which mobile services are added (Tasch & Brakel, 2004). Social network-based communities or discussion-oriented communities are examples of such mobile communities (Chayko, 2007). An instant messenger's nature is to deliver messages instantaneously and to indicate presence of an individual. The advancement of mobile communication techniques has enabled instant messenger services through cell phones (Hu & Wang, 2007). A location-based friend finder collects location data from mobile users and returns a particular user's location upon request (Tasch & Brakel, 2004; Tatli, 2007). Standalone and single-user games are currently common in the mobile gaming industry. Yet, a new trend for mobile games can be observed, i.e., massive multi-player mobile games addressing a world-wide community simultaneously wandering within virtual scenery (Treu, Martens, Schicker, Breisinger & Küpper, 2007). The viral effect is sustained by the fact that communicators as well as recipients need to use the service to benefit from the usage. This benefit motivates early adopters to actively persuade recipients to also try the service in prospect of future common use. Therefore, this type demands the proactive participation of communicators in acquiring new users.

The standard type *Signaling Use, Group Membership (SUGM)* is illustrated by examples of specific kinds of mobile network products and services, e.g., mobile payment procedures like the LUUPAY solution ("Luupay", n.d.) used in the customer-to-customer scenario. When a user (i.e., the communicator) sends money to an unregistered user (i.e., the recipient), the latter was able to receive the money only by signing up for the mobile payment procedures. The communicators' "recommendation" behavior is passive as the recipients are informed by text messages (Short Message Services, SMS) from the mobile payment service provider. This SMS has the "effect of signaling the user's membership in a group with desirable attributes" (Subramanian & Rajagopalan, 2003, p. 304).

An example of the standard type *Targeted Recommendation (TR)* are send-to-a-friend-options which are functions of Java applications or Wireless Application Protocol (WAP) sites enabling the communicators to directly send an SMS, WAP Push, or email to recipients by at least entering the recipients' mobile telephone number (Mobile Subscriber ISDN, MSISDN) or email address. Commonly, every mobile service can apply send-to-a-friend-options. A Java-based cinema portal served as one example for our study, since high willingness to use such services was detected by Anckar & D'Incau (2002). A less sophisticated solution is adding requests, e.g., "send this to a friend" to mobile ads like mobile coupons. The TR type also includes contests that require the active participation of other people. For example, in the mobile viral marketing campaign for the movie "Mona Lisa's Smile," four people could win a group cinema ticket when the communicator persuaded three recipients to answer an SMS that was sent from the brand website by the communicator (Pousttchi & Wiedemann, 2007). More examples of TR are mobile short films (micro movies) that are downloaded from special portals on the stationary Internet or from the mobile Internet. Typically, Bluetooth is used to transmit the content once it has been loaded from a PC to a mobile device. TR and the next standard type are characterized by low network externalities as there is typically no change of benefit for customers when communicators forward the mobile viral content to other recipients.

Frequent examples of the standard type *Awareness Creation, Benefits Signaling (ACBS)* are mobile voice greeting cards that use Interactive Voice Response (IVR) and are usually combined with an SMS that inform the recipients about the communicator and explain the forwarding mechanism. A further example is the opportunity to send free SMS from a brand website. At the end of the SMS, the brand website's URL is typically indicated (e.g., "free SMS service by XYZ.com").

Mobile Viral Mavens

In this section, we present a conceptual model concerning characteristics of mobile viral mavens. We suppose that mavenism, personality strength, age, gender, and affinity for mobile communication technologies influence the intention to forward mobile viral content and, as a result, characterize mobile viral mavens.

Scholars who explored the phenomenon of WOM designed different profiles to identify possible change agents who influence people and shape opinions in their own social sphere. The concept of market mavens which is well-established in the marketing literature (Gwinner,

Swanson & Walsh, 2004) is used for the purpose of this chapter. According to consumer behavior, communications and social science literatures, *market mavens* refer to "individuals who have information about many kinds of products, places to shop, and other facets of markets, and initiate discussions with consumers and respond to requests from consumers for market information" (Feick & Price, 1987, p. 85). This concept differs fundamentally from previous concepts, notably opinion leaders and early adopters, as the latter are both product-specific (Rodgers & Schneider, 1993). Given that mobile marketers wish to convey various types of mobile viral content and capitalize on a mobile viral maven, "cultured once," for different mobile viral marketing strategies, the market maven concept notion seems particularly well suited to mobile viral marketing. Market mavens show higher levels of marketplace involvement, are engaged in extensive search activities, and demonstrate greater participation in market activities such as couponing, store browsing and reading advertisements (Feick & Price, 1987). Consequently, market mavens might tend to engage in mobile viral strategies similar to the abovementioned activities. In addition, mavens appear to be an accessible market segment for they lend to read magazines and view television more than other consumers (Feick & Price, 1987). Additionally, market mavens demonstrate an earlier awareness of new products and brands than other consumers (Feick, Guskey & Price, 1995). Thus, companies have an active interest in building communication channels with market mavens (Gwinner, Swanson & Walsh, 2004). Yet, these findings raise an interesting issue. As the theoretical underpinnings of the market maven concept depend, to a large extent, on its generalizability across a wide range of product types (Elliott & Warfield, 1993), the question of whether market mavenism can be applied to mobile viral marketing (i.e., forwarding mobile ads and recommending mobile services) is unresolved. Unless, there is some evidence of the generalizability of market mavens' information provision tendency (Slama & Williams, 1990). Hence, we test the following hypothesis:

H1: The higher the degree of market mavenism, the higher the general intention to forward mobile viral content.

The scale of personality strength ("Persönlichkeitsstärke") was created by Noelle-Neumann (1983) to evaluate qualities like leadership and charism. Following the research of Weimann (1991) and Scheufele & Shah (2000), a person displaying these attributes is considered influential and credible by others, making his or her word powerful. Weimann (1991) also stated that personality strength was closely related to opinion leadership as opinion leaders alike convince others because of their perceived expertise concerning a certain product (Rogers, 2003) rather than by gathering information like market mavens would do. To investigate if mobile viral mavens also rely on their reputation as experts to spread the word, it is to be examined whether they display a high degree of personality strength, thus:

H2: The stronger the personality, the higher the general intention to forward mobile viral content.

Young people have been found to be heavy users of mobile services (Haghirian, Madlberger & Tanuskova, 2005). In their early mobile marketing study, Brum & Rettie (2001) established that younger respondents are more likely to forward SMS vouchers. A study published by "Kontraband", "Parys Communications" & "YouGov" (2006) stated that younger age groups have a significantly higher awareness of the viral email concept than older ages. Furthermore, the tendency to forward viral emails on to other people decreases with age. As a mobile viral marketing process consists of the three stages of receipt, usage, and forwarding mobile viral content, we can also refer to studies about adoption of mobile services. For instance, Anckar & D'Incau (2002) examined consumers' willingness to use a number of initial mobile services. The authors concluded that the oldest age group typically shows a much lower willingness to use mobile services than the other groups, and especially the youngest respondents. As we aim to identify the characteristics of mobile viral mavens in terms of their forwarding behavior, we come to presume that older consumers may forward content via mobile devices to a lower extent than younger consumers:

H3: The higher the age of the consumer, the lower the general intention to forward mobile viral content.

Gender is one of the most widespread forms of segmentation in marketing practice as it is easy to identify, easy to access, and large enough to be profitable (Putrevu, 2001). For electronic viral marketing, it was found that women were more ready to pass along viral emails than men by a ratio of 38 per cent to 33 per cent ("Kontraband", "Parys Communications" & "YouGov", 2006). A gender difference is also confirmed by Phelps, Lewis, Mobilio & Perry (2004). This may be explained by the finding of Bakan (1966) that men are more agentic and women are more communal. Agentic orientation is expressed in such traits as being aggressive, achievement-oriented, and self-centered while communal orientation is expressed in being other-oriented, concerned with social acceptability, and being altero-centrist. According to Plant (2001), girls value cell phones as means of social communication, displaying communal use of mobile devices. Brum & Rettie (2001) found that women sent nearly 50 per cent more SMS per week than men. In the light of these past results, women seem to be strongly represented among mobile viral mavens. Put differently, we suppose that mobile viral content is more forwarded by women than by men:

H4: Women's general intention to forward mobile viral content will be higher than men's general intention to forward mobile viral content.

Affinity for mobile communication techniques is defined as the degree to which an individual likes or looks forward to learn about and be involved with mobile communication techniques. We see affinity for mobile communications techniques as a potential indicator of a consumer's intention to engage in different kinds of mobile viral strategies. Some scholars stated that the more familiar a user is with mobile communications in general, the less difficult the use of mobile marketing services will appear to him or her (Bauer, Barnes, Reichardt & Neumann, 2005). Besides, consumers who have high innovativeness traits are able to imagine novel ways to use mobile services and to integrate these services into existing patterns of mobile usage behavior and lifestyle (Drennan & Mort, 2005). Considering the innovativeness of mobile viral marketing, consumers who embrace new mobile communication techniques and applications as well as enjoy the associated learning process and challenges may tend to forward more mobile viral content than others that are uncomfortable with and even fearful of technological change. As a result, the following hypothesis can be formulated:

H5: The higher the affinity for mobile communication techniques of the consumer, the higher the general intention to forward mobile viral content.

Recipients of Mobile Viral Content

We consider one additional factor pertaining to the nature of the relationship between mobile viral mavens and those to whom the mobile viral content is forwarded. Besides the market maven concept, the theory on the strength of weak ties (Granovetter, 1973) provides a promising explanation of WOM (and therefore mobile viral marketing) processes (Brown & Reingen, 1987). Granovetter (1973) defined tie strength as a combination of the amount of time, degree of emotional intensity, level of intimacy, and degree of reciprocity between two individuals. It is assumed that most individuals have a spectrum of social ties ranging from strong (close friends and family members) to weak (mere acquaintances). For WOM communication the strength of the weak ties arises from their important bridging function that allows information to travel from one clump of social structure to another through them. If weak ties did not exist, a system would consist of disjointed subgroups, which would inhibit the widespread diffusion of information (Rogers, 2003). In several studies, e.g., Brown & Reingen (1987) this main premise has been supported. Goldenberg, Libai & Muller (2001) investigated the relative macro-level impact of strong and weak ties and found that the latter may have a bigger impact even though the former are activated more frequently. Moreover, information of a sensitive or personal nature is more likely to be shared by strong ties than by weak ties (Norman & Russell, 2006). Weakly tied pairs engage in fewer, less intimate exchanges, and share fewer types of information (Haythornthwaite, 2001).

Although the theoretical arguments for applying the theory on the strength of weak ties to mobile viral marketing are compelling, its empirical verification is still missing. Given the very personal nature of a mobile device (Turowski & Pousttchi, 2004) we posit the following hypothesis:

H6: The intention to forward mobile viral content to strongly tied consumers is higher than to weakly tied consumers.

METHOD

This study was conducted using an online survey conducted between September and December 2007. Participants were recruited by advertising on a number of German websites and newsletters. The survey instrument was pre-tested with a small sample of business students at the University of Augsburg, Germany, and people with different demographic profiles (n=48). After the pre-test, the wording of some questions was modified.

In order to measure the intention to forward mobile viral content, participants were confronted with eleven examples of viral marketing strategies (Table 2) that were inspired by the work of Pousttchi & Wiedemann (2007) and the mobile viral marketing standard types explained in Section 2.1. In the questionnaire, respondents were instructed to indicate the likelihood (on a 7-point scale: 1 = "yes, definitely" and 7 = "definitely not") that they would forward or recommend the examples. The general forwarding intention was then empirically derived from the total list of these items using principle factor analysis. Results showed that all standard types, i. e. general intention to forward (GIF), personality strength (PS), market mavenism (MM), and affinity for mobile communication techniques (AMCT), related to the forwarding intention loaded on one factor. These abbreviations occur in following tables 2, 4, 6, 7 and 8.

Items determining market mavenism (Feick & Price, 1987), personality strength (Noelle-Neumann, 1983), and affinity for mobile communication techniques (Feierabend & Klingler, 2003) were adapted from previous research. Answers to these items were retrieved using seven-point Likert-type scales, presented in randomized order. The anchors we used were "strongly agree" (1) and "strongly disagree" (7). Finally, the web-based survey included a section on demographics. To analyze the relationship between communicators and recipients, we asked within each example in case of the willingness to forward (i.e. a score below four on a seven-point Likert-type scale) to indicate groups of potential recipients for each kind of content. These groups represented the spectrum of social ties ranging from strong (close friends and family members) to weak (mere acquaintances).

Table 3. Sample demographics

Characteristics	N	Per cent
Age		
≤ 20	85	9.7
21 - 25	278	31.8
26 – 30	229	26.2
31 - 40	121	13.8
41 - 50	98	11.2
> 50	64	7.3
Education		
Have not completed secondary general school	7	0.8
Secondary general school (Hauptschule)	37	4.2
Intermediate school (Realschule)	120	13.7
High school (Gymnasium)	420	48.0
University or university of applied sciences	281	32.1
Other	10	1.1

RESULTS AND DISCUSSION

Sample

We checked plausibility, integrity and completeness of the received questionnaires with the result that 875 could be used for further analysis (Table 3).

The sample consisted of 69.3 per cent males and 30.7 per cent females. The average age was 30.41 years, ranging from 11 to 68 years. We also surveyed previous experience with mobile viral marketing: 43.4 per cent stated that they have already forwarded mobile viral content. Moreover, the majority of respondents used cell phones mainly for making phone calls (98.3%) and SMS (95.8%). Although the usage of multimedia messages (Multimedia Message Service, MMS) (25.3%) and mobile Internet (24.4%) was low, the overall usage pattern – as well as the bigger amount of males – appears representative of the present usage of mobile communication techniques in the German market ("TNS Infratest", 2008). We conclude that our sample does not allow for general statements about the total population but very well of the current mobile marketing target group in Germany.

Reliability and Validity of Constructs

The multi-item scales used in our online survey, i.e., general intention to forward (GIF), personality strength (PS), market mavenism (MM), and affinity for mobile communication techniques (AMCT), proved sufficiently reliable examined with Cronbach's alpha (Table 4).

Using principle factor analysis, the convergent and discriminant validity of four multi-item constructs were supported (GIF, PS, MM and AMCT). After 5 rotations the factors accounted for 48.4 per cent of the total variance. There were no cross construct loadings above 0.5 showing good discriminant validity. All factor loadings were 0.5 or above showing good convergent validity except for items PS9, PS10 and GIF11[1], albeit they did not have a loading above 0.4 on any other factor. These constructs are therefore unidimensional and factorially distinct.

Table 4. Scales of the online survey instrument

Construct	Number of items	Cronbach's alpha
GIF	11	0.821
MM	6	0.869
PS	10	0.838
AMCT	5	0.848

Intention to Participate in Mobile Viral Strategies

Table 5 shows the results regarding consumer intention to participate in mobile viral marketing strategies. The presented frequency represents the cumulative frequency of three affirmative answers on a seven-point Likert scale.

The results indicate a rather low intention in general (mean 4.6), but a relative high intention to forward certain mobile viral content. A remarkably high portion of the respondents indicated an intention to use free SMS provided on a brand website, to transmit videos via Bluetooth, and to forward mobile coupons. Likewise, recommending location based services and mobile instant messengers attracted interest.

Characteristics of Mobile Viral Mavens

Regarding characteristics of mobile viral mavens, we developed a conceptual model and tested five hypotheses on the basis of our online survey. Table 6 presents an overview of the means and standard deviations (SD) of general intention to forward, market mavenism, personality strength, age, and affinity for mobile communication techniques as well as product moment correlations between the factors hypothesized to impact on general intention to forward. A multiple regression analysis was used for testing the hypotheses 1, 2, 3, 5 (Table 7). Multicollinearity was not a concern in the regression model depicted by a low variance inflation factor.

Hypothesis 1 predicts a positive correlation between market mavenism and general intention

Table 5. Means, standard deviations, and cumulative frequency of intention to forward

Example	Means	SD	Cumulative frequency
Using an ad-sponsored free SMS service available on brand website	3.2	2.112	64.6%
Forwarding of a video clip via Bluetooth	3.5	2.066	58.8%
Forwarding of an SMS-based coupon	3.7	2.053	55.1%
Recommendation of a location-based friend finder	3.8	2.135	52.5%
Recommendation of an instant messaging service	3.9	1.957	50.0%
Recommendation of a Java-based cinema portal	4.6	1.926	34.5%
Inviting to a contest which requires participation of three people	5.4	1.966	21.8%
Recommendation of a mobile community	5.1	1.62	17.0%
Recommendation of a massive multiplayer mobile game	5.7	1.657	13.9%
Recommendation of a mobile payment procedure	5.8	1.639	13.5%
Using an IVR-based viral greeting call	5.8	1.599	13.1%

Table 6. Influencing factors on general intention to forward

Construct	Means	SD	Product moment correlations
GIF	4.598	1.135	1
MM	3.262	1.189	0.341
PS	2.829	0.805	0.136
Age	30.420	10.703	0.207
AMCT	3.286	1.422	0.383

to forward mobile viral content. This hypothesis was supported by the data. Earlier research has indicated that mavens are innovative and seek information about many new durable and non-durable products. The present study's findings suggest that mavens' interest and influence may also extend across a wide range of mobile viral content. For stationary Internet, Walsh, Mitchell, Wiedmann, Frenzel & Duvenhorst (2002, as cited by Gwinner, Swanson & Walsh, 2004) identified so-called eMavens, i.e., market mavens on the Internet, and showed that these differ from non-eMavens regarding usage behavior and motives for using web sites. Hence, the notion of a general marketplace influencer with knowledge in a variety of areas appears to be robust and exists not only in a face-to-face context, electronic environment, but, based on our results, also in a mobile environment.

Hypothesis 2 states that personality strength boosts the general intention to forward mobile viral content. However, the comparison between standardized beta coefficients of market mavenism and personality strength shows that personality strength may only have limited impact. Market mavenism's share in explaining the observed general intention to forward mobile viral content appears to be more than eight times higher than that of personality strength. Yet, no final answer can be given to the question whether personality strength plays a role in the process leading to a decision about whether to forward mobile viral content or not. Table 7 shows that the personality strength coefficient is not significantly different from zero, unlike all other coefficients in this regression model. Hence, Hypothesis 2 cannot be supported.

Hypothesis 3 affirms that age is negatively correlated to the general forwarding intention. Surprisingly, this hypothesis could not be supported due to the fact that age has a positive effect on intention. This is in contrast to Brum & Rettie (2001). An explanation may be that these authors only analyzed one specific mobile viral marketing strategy, i.e., forwarding vouchers via SMS, while we used a scale including eleven examples. Likewise, Haghirian, Madlberger & Tanuskova, 2005) could not verify the assumption that mobile

Table 7. Regression analysis summary for influencing factors on general intention to forward

Hypothesis	Factor	B	SEB	Beta	t	p
H1	MM	.205	.034	.215	6.006	.000
H2	PS	-.034	.047	-.024	-.721	.471
H3	Age	.017	.003	.160	5.231	.000
H5	AMCT	.216	.028	.271	7.790	.000
R^2 is 20.7 per cent, $F = 56.75$, $p < 0.001$						

Table 8. Dependent t-test for hypothesis 6

	N	Mean	SD	t	p	Confidence interval boundaries	
GIF1	115	1.035	0.868	12.787	0.000	0.874	1.195
GIF2	515	0.495	0.949	11.843	0.000	0.413	0.578
GIF3	438	0.452	0.898	10.534	0.000	0.368	0.536
GIF4	122	0.475	0.911	5.764	0.000	0.312	0.639
GIF5	302	0.540	0.895	10.482	0.000	0.438	0.641
GIF6	460	0.689	0.961	15.368	0.000	0.601	0.777
GIF7	566	0.435	1.05	9.799	0.000	0.348	0.522
GIF8	118	0.314	1.189	2.865	0.005	0.097	0.530
GIF9	151	0.397	1.020	4.785	0.000	0.233	0.561
GIF10	483	1.122	0.911	27.079	0.000	1.041	1.204
GIF11	243	0.691	0.983	10.963	0.000	0.567	0.816

marketing in general is more effective among younger consumers.

Hypothesis 4 considered gender in relation to general intention to forward. T-test for equality of means was used for testing the hypothesis 4. 606 (69.3 per cent) participants were males (M=4.59; SD=1.14; SEM=.05) and 269 (30.7 per cent) were females (M=4.62; SD=1.13; SEM=.07). Results show no significant differences between men and women (p=0.36). Anckar & D'Incau (2002) found similar results regarding readiness to adopt mobile services.

Hypothesis 5 predicted that affinity for mobile communication techniques is positively related to general intention to forward mobile viral content. This hypothesis was supported by the data. Drennan & Mort (2005) found similar support for the association between mobile self-efficacy (techno-confidence) and intentions to use mobile services.

Hypothesis 6 focused on the role that tie strength plays in people's willingness to share mobile viral content. The hypothesis concerned the primacy of strong ties over weak ties for transmitting mobile viral content. It was tested using a dependent t-test based on subsamples of persons who showed general intention to forward certain mobile viral content. Thus, the sizes of subsamples given in column N of Table 8 vary across examples of mobile viral content. Hence, hypothesis 5 was verified separately for each example.

With all t statistics being significantly different from zero ($p < 0.01$), there is an evident disparity between the activation of strong and weak ties. With the existence of this difference being proven, its direction is yet to be determined. As all eleven confidence intervals are above zero, strong ties are preferred across all eleven examples. Hypothesis 6 is therefore supported. This finding confirms the results of Granovetter (1983).

IMPLICATIONS

The results of this study deliver helpful implications for marketing practitioners and expose some major directions for future research on mobile viral marketing.

With regard to the analyzed intention to participate in mobile viral strategies we can derive a number of implications: Firstly, for a marketer owning a website, the study suggests to incorporate the opportunity to send free SMS (or MMS) from the brand website. This not only involves the sender with the brand, but also the recipient, as the brand can be mentioned in the message.

To control the costs of this marketing instrument, it is advisable to limit the amount of SMS or MMS per user. Secondly, the high interest for the Bluetooth strategies provides a new challenge for mobile marketers. Electronic viral strategies have typically used traceable viruses, i.e., they capture friends of friends' contact dates and are an appropriate means of database building and campaign tracking. Applying this traceable viral technique is providing much more challenges on the mobile channel. For example, with transmission of mobile viral content via Bluetooth, the path of a mobile viral content gets invisible. Hence, users' actions cannot be tracked and databases cannot be built unless they are identified as they purchase the advertised product. Such a scenario has been outlined by Heinemann, Ranke & Straub (2004). Marketers must be aware of this paradigm shift in mobile viral marketing. As an answer to the popularity of transmission via Bluetooth, measuring the success of a campaign may be done by appending contests or other incentives to mobile viral content that lead recipients to respond in a way that they identify themselves to marketers, e. g. on brand websites. Thirdly, due to our results, mobile marketers using coupons should consider adding simple requests to mobile ads like "send this to a friend" to appeal for mobile viral marketing. Finally, the results show that mobile viral marketing should work with mobile services like location based friend finders and mobile instant messaging services.

A second set of implications can be derived from the analyzed theoretical model. Since the market maven concept can be applied to mobile viral marketing, it seems reasonable to use results of past market maven studies for mobile viral marketing. For example, Gwinner, Swanson & Walsh (2004) indicated that market mavens are motivated to a greater extent than other consumers by a sense of obligation to share information, a desire to help others, and feelings of pleasure associated with informing others about products. Mobile marketers could design communication strategies and other tactics to appeal to these three motivations. With the aim of activating mavens' obligation motivation, a mobile marketer can make them part of a special group of consumers that receives new product information, e.g., by email or SMS. Placing them in a privileged group may activate feelings of a special status and, consequently, they may have an obligation to share this product information with others. Concerning leveraging mavens' helping motivation, mobile marketers should make it easy for them to help others. For instance, a mobile marketer could develop a process where the maven could identify to the mobile marketer those individuals that might be most assisted by the mobile viral content. In mobile viral marketing, send-to-a-friend options can perform this task. The pleasure motivation considers the enjoyment that mavens receive from passing information on to others. To take advantage of this motivation, mobile marketers must therefore develop ways to make mobile content sharing a more enjoyable and fun experience. The more difficult it is to share information the more it will be viewed as an undesirable task instead of a pleasant activity. Thus, technological capabilities that make content-sharing easy to do should be developed. In this context, a welcome improvement is the JSR 75 (Java Specification Request) that defines an API (application programming interface) that gives J2ME (Java 2 Micro Edition) enabled devices access to contact data stored in the address book (Pousttchi & Wiedemann (2007). Consequently, future Java-based send-to-a-friend options will operate like forwarding an SMS or MMS as contact data can be easily selected. Besides, Gwinner, Swanson & Walsh (2004) suggested adding on-line chat capability and threaded discussion boards to marketers' web sites where mavens and other consumers can easily and efficiently discuss products.

Some authors (Helm, 2000; Phelps, Lewis, Mobilio & Perry, 2004; Pousttchi & Wiedemann, 2007) highlighted the importance of selecting individuals who will find the marketer's content relevant enough to forward it. Nevertheless, the research failed to deliver empirical evidence for a significant influence of the demographic variables age and gender. As a result, a segmentation of mobile viral mavens based on gender does not appear promising. Furthermore, younger target groups tend to be more reluctant to forward mobile viral content.

The positive association between intention to forward and affinity for mobile communication techniques, though, implicates that marketers can consider targeting communications to mobile viral mavens of technology-complex mobile services. Mobile viral mavens seem to embrace these new technologies and might display characteristics similar to those of early adopters in that field (Rogers, 2003). Typically, mobile network operators are able to identify high mobile service usage of consumers. This usage behavior reveals high affinity for mobile communication techniques, especially innovative ones such as mobile internet, and tendency for mobile viral mavenism. Based on our results, mobile network operators should target such consumers with push messages informing about or transporting mobile viral content if they have previously consented to receiving push messages. This is both effective in terms of ensuring substantial uptake of the mobile viral strategy and also legal, given that mobile network operators have typically built up opt-in databases in the past through customer relationship management efforts and can now use them to spread mobile viral content (Jelassi & Enders, 2004).

As a preference for transmitting mobile viral content to recipients that communicators share strong ties with was detected, mobile viral strategies that are complementary to promotional activities in other media such as print, broadcasting or stationary Internet seem promising. To what extent they may be used to lure entire social networks with less affinity for mobile communication techniques to the mobile channel (Haythornthwaite, 2001) is yet to be explored. Certain examples, such as the recommendation of a Java-based cinema portal, showed a higher correlation between both strong and weak ties, implicating substantial usage of weak ties regardless of the general preference for strong ties. With weak ties serving as bridges between social networks (Granovetter, 1983; Brown & Reingen, 1987), their activation may help spreading this type of mobile viral content at reduced cost if compared to other types of content.

On issues of future research, this chapter offers some potential research avenues. First, one of the major limitations of the study is the bias from the sample in terms of age (average age: 30.41 years) and education (80.1 per cent have graduated from high school, university or university of applied sciences). For that reason, it would be very enriching to answer the research questions with a sample that comprises more consumers aged 14-20 years and is more balanced concerning respondents' education. It also needs to be examined to what extent the results found are also valid outside Germany; in other countries with different mobile phone usage, e.g. Japan, or different price schemes, e.g. Austria. Our research has implicitly assumed that market mavens could be used to enhance dissemination of mobile viral content. Yet, we do not know the extent to which this would occur, or even if mavens would refuse to be utilized by marketers. It is assumed that mavens could show resistance to marketers' attempts to exploit them (Gwinner, Swanson & Walsh, 2004). As a result, the second possibility for future research concerns understanding the manner and degree to which mobile viral mavens will react to a mobile marketer's attempt to more explicitly use them for mobile marketing purposes. Third, as only the personal context was examined as an influencing factor on forwarding intention and factors such as perceived usefulness or ease of use (Pousttchi & Wiedemann, 2007) were neglected, additional research is necessary to sustain our understanding of mobile viral marketing processes. Particularly, characteristics, attitudes, and behavior of those who receive mobile viral content have to be investigated. Our future research will begin to address these challenges. Moreover, we hope that these issues will also be given attention by other researchers in this area.

CONCLUSION

Our consideration started from realizing that mobile viral marketing is very attractive for marketers but has not yet been accorded sufficient attention by academic research in terms of its basic elements.

In a nutshell, the benefit of this chapter is threefold. The first is that it sheds some light on a rather unexplored area in mobile marketing research, i.e., favored mobile viral content and characteristics of mobile viral mavens. The second is that it helps marketers to gain a better understanding of the mobile viral marketing's basic elements and allows predicting whether certain consumers are mobile viral mavens and which kind of mobile viral content seems promising. Hence, based on our findings, they can create mobile viral strategies in a more purposeful way. The third is that this work is a useful stride to help researchers searching to better understand consumers in mobile viral marketing, and makes therefore a contribution to the literature on mobile viral marketing. Nevertheless, it is clear that a great deal of additional research is necessary to obtain a more satisfactory understanding of mobile viral marketing processes.

REFERENCES

Anckar, B., & D'Incau, D. (2002). Value-Added Services in Mobile Commerce: An Analytical Framework and Empirical Findings from a National Consumer Survey. Proceedings of the 35th Hawaii International Conference on System Sciences, Honolulu.

Arndt, J. (1967). Role of product-related conversations in the diffusion of a new product. *JMR, Journal of Marketing Research, 4*(3), 291–295. doi:10.2307/3149462

Bakan, D. (1966). The Duality of Human Existence. Chicago, IL: Rand McNally.

Bansal, H. S., & Voyer, P. A. (2000). Word-of-mouth processes within a services purchase decision context. *Journal of Service Research, 3*(2), 166–177. doi:10.1177/109467050032005

Barwise, P., & Strong, C. (2002). Permission-based Mobile Advertising. *Journal of Interactive Marketing, 16*(1), 14–24. doi:10.1002/dir.10000

Bauer, H. H., Barnes, S. J., Reichardt, T., & Neumann, M. M. (2005). Driving Consumer Acceptance of Mobile Marketing: A Theoretical Framework and Empirical Study. *Journal of Electronic Commerce Research, 6*(3), 181–192.

Bone, P. F. (1995). Word-of-Mouth Effects on Short-term and Long-term Product Judgments. *Journal of Business Research, 32*(3), 213–223. doi:10.1016/0148-2963(94)00047-I

Brown, J., & Reingen, P. (1987). Social Ties and Word-of-Mouth Referral Behavior. *The Journal of Consumer Research, 14*(3), 350–362. doi:10.1086/209118

Brum, M., & Rettie, R. (2001). M-commerce: The Role of SMS text messages. *Proceedings of the 4th biennial International Conference on Telecommunication and Information Markets*, Karlsruhe, Germany.

Chayko, M. (2007). The portable community: envisioning and examining mobile social connectedness. *International Journal of Web Based Communities, 3*(4), 373–385. doi:10.1504/IJWBC.2007.015864

Deakins, B., Grandcolas, U., & Rettie, R. (2005). Text message advertising: Response rates and branding effects. *Journal of Targeting . Measurement and Analysis for Marketing, 13*(4), 304–312. doi:10.1057/palgrave.jt.5740158

Dellarocas, C. (2003). The Digitalization of Word of Mouth: Promise and Challenges of Online Feedback Mechanisms. *Management Science, 49*(10), 1407–1424. doi:10.1287/mnsc.49.10.1407.17308

Drennan, J., & Mort, G. S. (2005). Marketing m-services: Establishing a usage benefit typology related to mobile user characteristics. *Journal of Database Marketing & Customer Strategy Management, 12*(4), 327–341. doi:10.1057/palgrave.dbm.3240269

Drosos, D., & Giaglis, G. M. (2004). *Factors Influencing Mobile Messaging Advertising Effectiveness*. Eltrum Working Paper Series WP 2004-009, Athens University of Economics and Business Department of Management Science and Technology, Athens.

Economides, N. (1996). The Economics of Networks. *International Journal of Industrial Organization, 14*(2), 673–699. doi:10.1016/0167-7187(96)01015-6

Elliott, M. T., & Warfield, A. E. (1993). Do Market Mavens Categorize Brands Differently? *Advances in Consumer Research. Association for Consumer Research (U. S.), 20*(1), 202–208.

Feick, F., & Price, L. (1987). The Market Maven: A Diffuser of Marketplace Information. *Journal of Marketing, 51*, 83–97. doi:10.2307/1251146

Feick, L. F., Guskey, A., & Price, L. (1995). Everyday Market Helping Behavior. *Journal of Public Policy & Marketing, 14*(2), 255–266.

Feierabend, S., & Klingler, W. (2003). Medienverhalten Jugendlicher in Deutschland. *Media Perspektiven, 33*(10), 450–462.

Garyfalos, A., & Almeroth, K. C. (2004). Coupon Based Incentive Systems and the Implications of Equilibrium Theory. *Proceedings of the IEEE International Conference on E-Commerce Technology* (pp. 213-220). Washington, DC: IEEE Computer Society.

Goldenberg, J., Libai, B., & Muller, E. (2001). Talk of the network: A complex systems look at the underlying process of word-of-mouth. *Marketing Letters, 12*(3), 211–223. doi:10.1023/A:1011122126881

Granovetter, J. B. (1973). The Strength of Weak Ties. *American Journal of Sociology, 78*(6), 1360–1380. doi:10.1086/225469

Granovetter, J. B. (1983). The Strength of Weak Ties: A Network Theory Revisited. *Social Structure and Network Analysis, 1*, 105–130.

Gwinner, K., Swanson, K. S., & Walsh, G. (2004). What makes mavens tick? Exploring the motives of market mavens' initiation of information diffusion. *Journal of Consumer Marketing, 21*(2), 109–122. doi:10.1108/07363760410525678

Haghirian, P., Madlberger, M., & Tanuskova, A. (2005) Increasing Advertising Value of Mobile Marketing. An Empirical Study of Antecedents. *Proceedings of the 38th Hawaii International Conference on System Sciences*, Honolulu.

Haythornthwaite, C. (2001). Tie Strength and the Impact of New Media. *Proceedings of the 34th Hawaii International Conference on System Sciences*, Honolulu.

Heinemann, A., Ranke, J., & Straub, T. (2004). Zur rechtsverträglichen Technikgestaltung anhand einer M-Commerce-Anwendung. *Mobile Economy – Transaktionen, Anwendungen, Prozesse und Dienste, Proceedings zum 4. Workshop Mobile Commerce*, Augsburg, Germany.

Heinonen, K., & Strandvik, T. (2006). How Do Consumers React to Mobile Marketing? *Proceedings of the 5th International Conference on Mobile Business*, Copenhagen.

Helm, S. (2000). Viral Marketing – Establishing Customer Relationships by 'Word-of-mouse'. *Electronic Markets, 10*(3), 158–161. doi:10.1080/10196780050177053

Herr, P. M., Kardes, F. R., & Kim, J. (1991). Effects of Word-of-Mouth and Product Attribute Information on Persuasion: An Accessibility-Diagnosticity Perspective. *The Journal of Consumer Research, 17*(4), 454–462. doi:10.1086/208570

Hu, D. H., & Wang, C. L. (2007). GPS-based Location Extraction and Presence Management for Mobile Instant Messenger. *Proceedings of the IFIP International Conference on Embedded and Ubiquitous Computing*, Taipei, Taiwan.

I-play (2005). I-play Outlines Collective Industry Action Required for Mobile Gaming Market to Reach True Potential. Press release.

Infratest, T. N. S. (2008). *GTI Telecoms Report 2007-08 Germany Report*, Munich, Germany.

Intuitive Media Research Services. (2006). Kids Go Mobile. Ownership and Use of Mobile Phones by Children aged 6 to 13 for ICT in Education Sector. Retrieved May 9, 2007 from http://www.nma.co.uk/assets/getAsset.aspx?liAssetID=20934.

Jelassi, T., & Enders, A. (2004). Leveraging Wireless Technology for Mobile Advertising. *Proceedings of the 12th European Conference on Information Systems*, Turku, Finland.

Jurvetson, S. (1997). What is Viral Marketing? *Netscape M-Files*. Retrieved November 11, 2006 from http://www.dfj.com/cgi-bin/artman/publish/steve_may00.shtml.

Katz, M. L., & Shapiro, C. (1985). Network Externalities, Competition, and Compatibility. *The American Economic Review, 75*(3), 424–440.

Kontraband, Parys Communications & YouGov (2006). Viral Marketing Industry Report 2006. Retrieved February 15, 2008 on http://www.the7thchamber.com/pdf/The-Seventh-Chamber_Viral-Report.pdf

Kotler, P., & Zaltman, G. (1976). Targeting prospects for a new product. *Journal of Advertising Research, 16*(1), 7–18.

Kümmerlein, A. (2002). Mobile Marketing Kampagnen. Werbewirkungsanalyse. Düsseldorf.

Luupay. (n.d.). Retrieved December 16, 2008 from https://www.luupay.de/User/De/Help/userguide.aspx

Marini, S., & Wiedemann, D. G. (2006). Entwicklungen im Bereich Mobile Advertising aus der Sicht von Experten. Ergebnisse der Expertenbefragung MM1. In Pousttchi, K. (Ed.), *Studienpapiere der Arbeitsgruppe Mobile Commerce* (pp. 1-49). Augsburg, Germany: Universität Augsburg, Arbeitsgruppe Mobile Commerce.

Misner, I. R. (1999). *The World's Best Known Marketing Secret*. Austin, TX: The Bard Press.

Montgomery, A. L. (2001). Applying Quantitative Marketing Techniques to the Internet. *Interfaces, 31*(2), 90–108. doi:10.1287/inte.31.2.90.10630

Noelle-Neumann, E. (1983). *Persönlichkeitsstärke*. Hamburg, Germany: Spiegel-Dokumentation.

Norman, A. T., & Russell, C. A. (2006). The pass-along effect: Investigating word-of-mouth effects on online survey procedures. *Journal of Computer-Mediated Communication*, 11(4), article 10. Retrieved July 9, 2008, from http://jcmc.indiana.edu/vol11/issue4/norman.html

Peppers, D., Rogers, M., & Dorf, B. (1999). Is Your Company Ready for One-to One Marketing? *Harvard Business Review, 77*(1), 151–160.

Phelps, J. E., Lewis, R., Mobilio, L., & Perry, D. (2004). Viral Marketing or Electronic Word-of-Mouth Advertising: Examining Consumer Responses and Motives to Pass Along Email. *Journal of Advertising Research, 45*(4), 333–348.

Plant, S. (2001). On the mobile - the effects of mobile telephones on social and individual life. Retrieved February 15, 2008 from http://www.motorola.com/mot/doc/0/234_MotDoc.pdf

Pousttchi, K., & Wiedemann, D. G. (2006). A Contribution to Theory Building for Mobile Marketing: Categorizing Mobile Marketing Campaigns through Case Study Research. *Proceedings of the 5th International Conference on Mobile Business*, Copenhagen, Denmark.

Pousttchi, K., & Wiedemann, D. G. (2007). Success Factors in Mobile Viral Marketing: A Multi-Case Study Approach. *Proceedings of the 6th International Conference on Mobile Business*, Toronto, Canada.

Putrevu, S. (2001). Exploring the origins and information processing differences between men and women: Implications for advertisers. *Academy of Marketing Science Review, 2001*, 1–15.

Ratsimor, O., Finin, T., Joshi, A., & Yesha, Y. (2003). eNcentive: A Framework for Intelligent Marketing in Mobile Peer-To-Peer Environments. *ICEC '03: Proceedings* of the 5th International Conference on Electronic Commerce (pp. 87-94). Baltimore, MD: UMBC eBiquity Research Group

Rodgers, W. C., & Schneider, K. C. (1993). A Research Note on Market Mavens and the Dissemination of Positive versus Negative Marketplace Information. *Journal of Marketing Management, 3*(2), 67–72.

Rogers, E. M. (2003). *Diffusion of Innovation*, New York: The Free Press.

Rosen, E. (2000). *The Anatomy of Buzz.* New York: Doubleday.

Scheufele, D. H., & Shah, V. S. (2000). Personality Strength and Social Capital: The Role of Dispositional and Informational Variables in the Production of Civic Participation. *Communication Research, 27*(2), 107–131. doi:10.1177/009365000027002001

Slama, M. E., & Williams, T. G. (1990). Generalization of the market maven's information provision tendency across product categories. *Advances in Consumer Research. Association for Consumer Research (U. S.), 17*(1), 48–52.

Straub, T., & Heinemann, A. (2004). An Anonymous Bonus Point System For Mobile Commerce Based On Word-Of-Mouth Recommendation. In L. M. Liebrock (Ed.), *Applied Computing 2004. Proceedings of the 2004 ACM Symposium on Applied Computing* (pp. 766-773). New York: ACM Press.

Subramani, M. R., & Rajagopalan, B. (2003). Knowledge-Sharing and Influence in Online Social Networks via Viral Marketing. *Communications of the ACM, 46*(12), 300–307. doi:10.1145/953460.953514

Tasch, A., & Brakel, O. (2004). Location Based Community Services. New Services for a new Type of Web Communities. *Proceedings of the IADIS Conference on Web Based Communities*, Lisbon, Portugal.

Tatli, E. I. (2007). Extending P3P/Appel for Friend Finder. *Proceedings of the International Workshop on Privacy-Aware Location-based Mobile Services*, Mannheim, Germany.

Treu, G., Martens, J., Schicker, M., Breisinger, M., & Küpper, A. (2004). Vs Golf — Developing Location-based Multi-Player Games. *Proceedings of the 31st Annual International Computer Software and Applications Conference*, Beijing, China.

Turowski, K., & Pousttchi, K. (2004). *Mobile Commerce. Grundlagen und Techniken*. Heidelberg, Germany: Springer.

Weimann, G. (1991). The Influentials: Back to the Concept of Opinion Leaders? *Public Opinion Quarterly*, *55*(2), 267–279. doi:10.1086/269257

Wiedemann, D. G. (2007). Exploring the Concept of Mobile Viral Marketing through Case Study Research. *Proceedings of the 2nd conference of Mobilität und mobile Informationssysteme* (pp. 49-60). Aachen, Germany.

Wilson, R. F (2000). The Six Simple Principles of Viral Marketing. *Web Marketing Today*, 70.

Zeithaml, V. A., & Bitner, M. J. (1996). *Services Marketing. Integrating Consumer Focus Across the Firm*. New York: McGraw Hill.

KEY TERMS AND DEFINITIONS

Communicators: Are individuals transmitting mobile viral content.

Mobile Viral Content: Is content transmitted from a recipient to a communicator in mobile viral marketing.

Mobile Viral Marketing Standard Types: Are typical mobile viral strategies varying in the role of communicator in persuasion (active or passive) and the level of network externalities (high or low).

Mobile Viral Marketing: Is defined as a concept for distribution or communication that relies on customers to transmit content via mobile communication techniques and mobile devices to other potential customers in their social sphere and to animate these contacts to also transmit the content.

Mobile Viral Mavens: Are consumers who forward mobile viral content frequently and willingly.

Recipients: Are defined as individuals receiving mobile viral content.

Word-of-Mouth (WOM): Refers to oral, person-to-person communication between a recipient and a communicator which the recipient perceives as a non-commercial message, regarding a brand, product or service.

ENDNOTE

1 PS9: I own many things others envy me for; PS10: I often give others advice and suggestions (Weimann, 1991, p. 270); GIF11: Could you imagine Using an ad-sponsored free SMS service available on a brand website?

Chapter 21
Understanding Consumer Recommendation Behavior

Wen-Kuo Chen
Chaoyang University of Technology, Taiwan

Heng-Chiang Huang
National Taiwan University, Taiwan

Seng-Cho T. Chou
National Taiwan University, Taiwan

ABSTRACT

Research shows that the use of mobile phone services is surprisingly low because users would not like to pay money without first using services. We consider that a customer is not only a service user, but also the partner of a firm. Therefore, this chapter investigates how the users recommend mobile phone services to other potential customers based on their experiences. We focus on two determinants of customer recommendation behavior, namely, overall satisfaction and flow experience. We also propose a number of hypotheses related to the constructs of overall satisfaction, and test them using data provided by 3G mobile service subscribers. The results contribute to our understanding of how customers are willing to make recommendations to others. Besides increasing customer satisfaction, we should consider a mediating factor—users' flow experiences. Our findings provide empirical support for most of the hypotheses. The theoretical and practical implications of our findings are also discussed, and suggestions for further research are offered.

INTRODUCTION

Evaluating the adoption and usage of information and communication technology (ICT) has been a critical research issue in MIS field. A large body of research of ICT (e.g., e-mail, Internet, Web, e-services, m-services) has attempted to develop models for understanding and predicting the individuals' technology access using context (e.g., Nysveen et al., 2005). The most important ICT for users in everyday life is the mobile phone because mobile phones are personal communication devices that many people carry with in daily life (Aaltonen et al., 2005). Recently, a mobile phone has equipped with wireless techniques (e.g., 3G, 3.5G, WiFi) that become a new mobile service platform to deliver

DOI: 10.4018/978-1-60566-074-5.ch021

information and services from service providers to receivers.

Mobile telecommunications companies derive important revenues from mobile voice call and obtain additional revenues from data transformation (m-services). Prior research has developed models for examining the m-service acceptance (e.g., Nysveen et al., 2005). Nevertheless, the reason of the low penetration rate of mobile phone services of consumers (Tsai, 2007) is that they would not like to pay money without first using mobile phone services (TNS, 2007).

Recommendation by positive word-of-mouth is increasingly seen as a marketing tool that generates the greatest benefit for supplier firms (Morgan & Rego, 2006; Reichheld, 2006). There is evidence that recommendations have a strong influence on consumer choice, particularly in the pre-purchase stage (East et al., 2005). It stands to reason that in a purchase decision-making context, consumers rely on customer recommendations rather than marketing content or advertising by the provider (Biyalogorsky et al., 2001). Therefore, since present customers can be good sales people, customer recommendations may be considered an important driver for diffusing mobile phone services.

This chapter is important for a variety of reasons. First, it addresses an extremely important topic in the field of IS and marketing. Customer recommendation is an important driver for the diffusion of mobile phone services. Then, how do we urge the customer recommendation to others? Second, we develop and test a more comprehensive model of several key customer perceptions.

Next, we review the relevant literature and develop our research model and hypotheses. Then, we describe the methodology, report the results of our analysis, and consider the implications of the results. We conclude the chapter with a discussion of the contributions and limitations of the present study and make suggestions for future research.

BACKGROUD

Conceptual Framework and Research Hypotheses

The model (see Figure 1) then illustrates a series of relationships that we propose to explain what kind of determinants influence consumers to recommend mobile phone services to others.

Determinants of Customer Recommendations

A number of studies seek to explore the determinants of consumer recommendations (Brown et al., 2005; Shabbir et al., 2007; Pousttchi and Wiedemann 2007). The majority of work suggests that consumer post-purchase responses are the key underlying antecedents of customer recommendations. In particular, most researchers note that satisfied customers are willing to recommend services to others (e.g. Palmatier et al., 2006; Morgan & Rego, 2006). On the other hand, Csikszentimihalyi (1990) proposes that flow experience emphasizes an individual's subjective experience state made by interacting with activities that may also act as a kind of post-purchase responses. Hence, besides customer satisfaction, we posit that flow experience is important to the study of customer recommendation behavior.

Satisfaction, Customer Recommendations, and Flow Experience

The customer recommendations construct derives from word-of-mouth. Word-of-mouth may be categorized as positive, neutral, or negative. Chung and Darke (2006) noted that positive word-of-mouth occurs when consumers are willing to recommend services to others. Moreover, word-of-mouth also impels potential customers to seek advice or actual customers to give advice (East et al., 2005). Thus, using the term

Figure 1. Theoretical Model

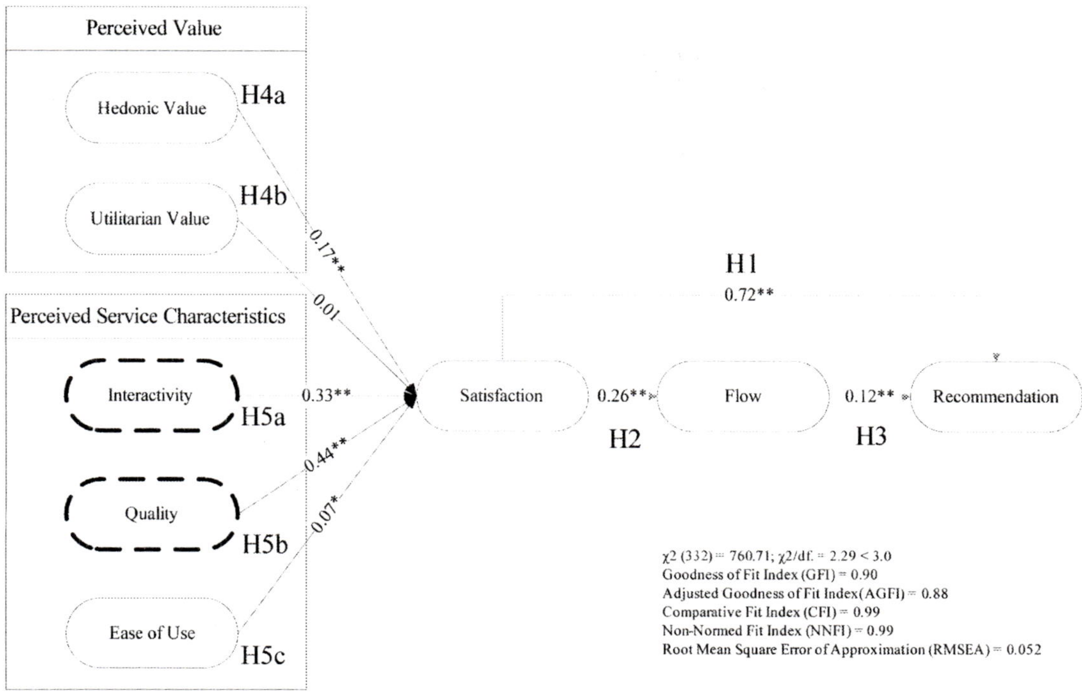

of word-of-mouth may be made ambiguous; we use customer recommendations when customers give positive advice about products or services to others voluntarily. Hence, this chapter defines customer recommendations as the action that occurs when customers voluntarily recommend their experiences with mobile phone services to their relatives and friends.

Overall satisfaction plays a critical role in the study of customer recommendation behavior because it affects individual motivations to recommend products or services to potential customers (e.g. Lam et al., 2004; Morgan et al., 2005). Overall satisfaction includes the total consumer experience after the evaluation of services or products (Anderson et al., 1994; Sweeney & Soutar, 2001) and is the reaction to the 'plausible fulfillment' of needs and desires, (Oliver, 1997).

There was substantial evidence that users' evaluation of service satisfaction had a strong direct impact on customer recommendations. The literature on customer loyalty suggests that satisfaction with products, services, or firms is an important post-purchase response that leads to the spread of customer recommendations to other consumers (Mittal et al., 1999; Del Rio et al., 2001). For example, Morgan et al. (2005) suggested that overall customer satisfaction relates to the post-purchase phenomenon, such as future purchase intentions and the likelihood to recommend. Similarly, Lam et al. (2004) showed that a customer satisfied with a service provider might be willing to recommend the services to other customers. Thus, customer satisfaction plays an important role in facilitating customer recommendations. Therefore, 'willingness to recommend' and 'making recommendations to others' metrics are widely used to assess the impact of customer satisfaction. As a result, we assume that customer satisfaction is a determinant of customer recommendations. Thus, we propose the following hypothesis:

Hypothesis 1 (H1): *Mobile phone service users who experience greater satisfaction are more likely to recommend mobile phone services to others.*

Csikszentmihalyi (1975) proposed Flow Theory to determine why people expend large amounts of energy in the interesting activities that interest them. Csikszentmihalyi (1975, p. 36) defines flow as "holistic sensation that people feel when they act with total involvement." Research suggests that flow experience may be a useful construct for understanding information technology applications, such as mobile services and network navigation (Hoffman & Novak 1996).

A necessary condition of being immersed in the state of flow is to have a clear set of goals that require appropriate responses (Hoffman & Novak 1996). By focusing on those goals, the individual engages in activities and experience flow. According to Anderson et al. (1994), overall satisfaction is "an overall evaluation based on the total purchase and consumption experience with a good or service over time." It stands to reason that the satisfaction of customer needs or desires provides a clear goal for a customer to purchase the service. Therefore, we posit that overall satisfaction is one kind of goal and drives a customer to archive a great level of customer satisfaction: flow experience. Hence, we propose the following hypothesis:

Hypothesis 2 (H2): *Mobile phone service users who experience greater satisfaction are more likely to achieve a greater flow experience.*

Flow Experience and Customer Recommendations

While to the best of our knowledge, no customer recommendations research has confirmed the relationship between flow experience and customer recommendations, there is some indirect evidence to support it. For example, four consequences of flow from Hoffman and Novak's (1996) network navigation model are similar to customer loyalty: namely, increased learning, increased perceived behavioral control, increased exploratory and participatory behavior, and positive subjective experiences. Similarly, Siekpe (2005) confirmed that flow was positively related to the intention to purchase and the intention to return.

In the literature on loyalty, there are two kinds of loyalty: attitude dimension of loyalty and behavior dimension of loyalty (Palmatier et al., 2006). Jones and Sasser (1995) considered that we can measure customer loyalty from intention to repurchase (attitude loyalty), primary behavior (behavior loyalty), and secondary behavior (behavior loyalty). Primary behavior occurs when a consumer actually repurchases the products or services and secondary behavior occurs when a consumer is willing to recommend the products or services to others. It stands to reason that increased exploration and participation is one kind of behavior loyalty. Hence, we believe that it is important to consider the relationship of flow experience to customer recommendations in mobile phone services for theoretical reasons. Thus, we propose the following hypothesis:

Hypothesis 3 (H3): *Mobile phone service users who have better flow experiences are more likely to recommend mobile phone services to others.*

Antecedents to Satisfaction

Perceived Value

Past research has shown that perceived value is an important antecedent for overall satisfaction. Research has conceptualized value simply as the price for which a customer pays to receive a product service (Zeithaml, 1988). However, a number of recent researchers argue that value is not the only price. For example, Babin et al. (1994) and Babin and Attaway (2000) considered two value dimensions: utilitarian value and hedonic value.

Utilitarian value is defined as that value that a customer receives based on a task-related and rational consumption behavior (Babin et al., 1994). Hedonic value is defined as that value a customer receives based on the subject experience of fun and playfulness (Babin et al., 1994). Traditionally, studies of information technology application usage, such as mobile services, believed that users adopt technologies or use information services for utilitarian purposes (e.g., Chiou, 2004). It stands to reason that because technologies or information services are tools for solving customers' problems, past research has focused on technology effectiveness to understand why a customer adopts technology.

However, users do not adopt such applications only for functional benefits but also for entertainment purposes (Holbrook & Hirschman, 1982). Customers may have emotional feelings, which they desire, by using technology. Moreover, utilitarian value and hedonic value do not have a mutually exclusive relationship (Babin et al., 1994) so that customers may not only get hedonic value from using mobile phone services but can satisfy their specific purposes (Babin et al., 1994).

Thus, we consider that the perceived value construct, namely, hedonic value and utilitarian value, are useful for understanding users' evaluations of mobile phone services. Based on the literature, we hypothesize that there is a link between perceived value and user satisfaction as follows:

Hypothesis 4a (H4a): *Perceived Hedonic Value increases user satisfaction with mobile services.*

Hypothesis 4b (H4b): *Perceived Utilitarian Value intensifies user satisfaction with mobile services.*

Perceived Service Characteristics

We suggest that perceived service characteristics that construct interactivity, quality, and ease of use, can help us understand users' evaluations of mobile phone services, particularly customer satisfaction which derives from such services.

Perceived Interactivity: Regarded as the crucial element in a successful human-computer interaction (Hoffman & Novak, 1996; McMillan & Hwang, 2002; Liu & Shrum, 2002; Teo et al., 2003), interactivity is one of the key characteristics of the new media (Hoffman & Novak, 1996). Individuals rate interactivity on the basis of their perceptions of two-way communication, control choices, and time to load (McMillan & Hwang, 2002). It stands to reason that interactivity is the degree to which mobile phone service users have control over items such as video phone, media, objects, and other similar items.

According to Teo et al. (2003), it has been shown that interactivity increases user satisfaction. Similarly, Liu and Shrum (2002) showed that interactivity increases positive attitudes of the human-computer interaction. While consumers use mobile phone services to focus on human-computer interaction, they also interact with other people and information through mobile phones. They experience some feelings about the interaction with services that will help them assess the performance of mobile phone services.

Perceived Quality: High product or service quality is the crucial topic for providers to gain increased market share or become a market leader. High quality is necessary to satisfy user's needs and expectations. Therefore, mobile phone service providers must satisfy customers when they are using services. An information system process perspective from DeLone and McLean (1992; 2003) and Pitt et al. (1995), suggests that a successful system must have high levels of system quality, information quality, and service quality. They believe these factors explain both why consumers use a system and also how they influence subsequent feelings of satisfaction. Moreover, Nelson et al. (2005) also suggested that information quality and system quality may influence user satisfaction.

Thus, we consider that mobile phone service quality include three parts: information quality, system quality, and service quality. The definition is shown below. System quality is defined as that system quality which represents users' perceptions of interaction with mobile phone services over time (Nelson et al., 2005). Information quality is defined as that which provides integration and useful, updated information to satisfy users' needs (Mckinney et al., 2002). Service quality is defined as that with which service providers should satisfy user expectations (Jiang et al., 2000). Thus, we consider that it is perceived quality, namely, system quality, information quality, and service quality, that influences user satisfaction.

Perceived ease of use: Third-generation mobile phones are called interactive medium platforms. The main interaction with services is through a mobile phone. Therefore, an easy-of-use mobile phone is essential to ensure customer satisfaction. For example, Stone et al. (2007) found support for the notion that IT users have more satisfaction when they feel the system is easier to use. In addition, mobile devices that are easy to use are less threatening to an individual (Moon & Kim 2001). Thus, the following hypotheses can be proposed.

Hypothesis 5a (H5a): *Higher Perceived Interactivity results in enhanced user satisfaction when using mobile phone services.*

Hypothesis 5b (H5b): *Higher Usage Quality results in enhanced user satisfaction when using mobile phone services.*

Hypothesis 5c (H5c): *Higher Perceived Ease of Use results in enhanced user satisfaction when using mobile phone services.*

MAIN FOCUS OF THE CHAPTER: RESEARCH METHODS

We designed our study to survey 3G subscribers who used mobile phone services. It is important to note that we are more concerned with service usage in general than usage of a specific service. That is, we focus on users' experience with using multiple services.

Questionnaire Development

We describe below the measures of questionnaire that was used for assessing customer recommendations of mobile phone service The measures of our framework were adapted from the validated measures of prior studies from the literature. We adapted them to suit the context of our study. The respondents assessed all items via a seven-point Likert scale ranging from 1 = "strongly disagree" to 7 = "strongly agree." The measures are listed in full in Appendix A and the descriptive statistics of the individual scales are shown in Table 1.

Sample and Data Collection

Data was collected from subscribers who had used 3G mobile phone services. The data was obtained from three online web-sites that are well known forums for sharing information about mobile phones and mobile phone services.

We posted a message on the websites asking for respondents to complete an online questionnaire. Respondents were limited to website members who had used 3G mobile phone and mobile phone services. We paid US$ 3 for every valid response, and the survey period covered one month. A total of 568 surveys were received, but we eliminated 88 surveys that had incomplete data. The remaining 480 questionnaires represent a useable response rate of 84.51%.

Of these respondents, Sixty-seven percent was male. Eighty-one percent of respondents ranged

Table 1. Means, Standard Deviations, and Intercorrelations Among Latent Variables

	M	SD	1	2	3	4	5	6	7	8
1. Overall Satisfaction	5.02	1.22	1.00							
2. Flow Experience	3.66	1.60	0.26	1.00						
3. Recommendation	4.99	1.37	0.74	0.28	1.00					
4. Hedonic Value	5.07	1.22	0.73	0.19	0.54	1.00				
5. Utilitarian Value	5.43	1.13	0.52	0.14	0.39	0.63	1.00			
6. Perceived Interactivity	4.95	1.14	0.88	0.23	0.65	0.72	0.51	1.00		
7. Perceived Quality	4.83	1.14	0.88	0.23	0.65	0.65	0.47	0.88	1.00	
8. Perceived Ease of Use	5.10	1.24	0.60	0.16	0.45	0.53	0.43	0.58	0.57	1.00

in age from 18 to 35 and fifty-one percent had earned a university degree. Fifty-eight percent of the respondents had used mobile phone services for less than six months and twenty-two percent of the respondents had used the services from six months to one year. Finally, twenty-four percent of the respondents worked in a service business and twenty-three percent of the respondents were students. The services used most frequently were video phones (37.5%), download ring tones or upload/download pictures (18.3%), and information searches (10%).

MAIN FOCUS OF THE CHAPTER: DATA ANALYSIS AND RESULTS

The research model was tested through structural equation modeling (SEM) with LISREL 8.72. Following Anderson and Gerbing (1988) and Hair et al. (1998), we used a two-stage approach to test the model. First, we conducted Confirmatory Factor Analysis (CFA) to assess the construct's reliability and validity in terms of convergent and discriminant validity. Second, we performed structural equation modeling analysis to test the research hypotheses empirically.

Second-Order Factor Estimation

We estimated two second-order factor models of perceived interactivity and perceived quality. The model of perceived interactivity includes three first-order factors: real-time conversation, no delay, and engaging. The posited model fit the data acceptably (χ^2 (32) = 82.84; GFI = .97; AGFI = .94; NFI = .99; CFI = .99; and RMSEA = .058) (Hu & Bentler, 1999). Next, we conceptualized perceived quality as a higher-order factor. The model of perceived quality includes three first-order factors: information quality, system quality, and service quality. The posited model fit the data acceptably (χ^2 (32) = 108.11; GFI = .96; AGFI = .93; NFI = .99; CFI = .99; and RMSEA = .070) (Hu & Bentler, 1999). All composite reliability coefficients greater than .60 demonstrate good reliability of the measures (Bagozzi & Yi, 1988). Thus, we believe that perceived interactivity and perceived quality represents the second-order factors.

Measurement Model Estimation

With respect to the quality of the measurement model, construct reliability and two types of validity were assessed, in terms of convergent and discriminant validity (Boudreau et al., 2001). We examined the composite reliabilities of each of all constructs to provide evidence of construct reliability. The values of the composite reliability range from .80 to .92, exceeding the .7 threshold (Fornell & Larcker, 1981; Hair et al., 1998). Therefore, all the construct measures in the measurement model achieve construct reliability.

In addition, all observable indicators loaded significantly on their intended factors ($p < .01$). The average variance extracted (AVE) from all eight constructs ranges from a low of .50 to a high of .73, well above the critical value of .5 (Fornell & Larcker, 1981). Overall, the analysis demonstrates a good convergent validity of the measures. We provide a summary of the factor loadings, composite reliabilities, average variance extracted, and model-fits in Appendix A. Next, we assessed the discriminant validity by using chi-square difference testes, in which correlations are derived between constrained and unconstrained models (Algesheimer et al., 2005). We ran a series of nested CFA model comparisons (a total of 28 tests in all), and most of pairs were significantly different indicating discriminant validity. Finally, the measurement model indicates an adequate fit of the model to the data ((χ^2 (322) = 622.71; GFI = .92; AGFI = .89; NFI = .98; CFI = .99; and RMSEA = .044). This procedure, along with the CFA results, suggests that all the measurements in this chapter are adequately reliable and valid.

Structural Model: Hypothesis Testing

Figure 1 shows that the fit of this model is excellent: χ^2 (332) = 760.71, GFI = .90, AGFI = .88, CFI = .99, NFI = .99, RMSEA = .052. Thus the model fits well with the data from the study. With the exception of the main effect of the perceived utilitarian value (H4b), all the hypotheses were supported by our research results (see Figure 1).

The purpose of our research is to consider the effects of customer satisfaction and flow experience on customer recommendations. Hypothesis 1 suggests that overall satisfaction has a positive influence on willingness to make recommendations to friends. From the structural equation modeling results and the standardized path coefficient (SPC) of 0.72 ($p < .001$), this contention is supported. As predicted by Hypothesis 2, customer satisfaction exerts a positive influence on a user's flow experience (SPC = .26, $p < .01$). The results support our hypothesis. Regarding Hypothesis 3, there is consistent support that a user's flow experience significantly increases a willingness to make recommendations to friends (SPC = .12, $p < .01$).

The results partially support Hypothesis Set 4. The effect of hedonic value on overall satisfaction was found to be significant (SPC = .17, $p < .01$). However, the effect of utilitarian value on overall satisfaction was not supported (SPC = .01, $p > .05$). Finally, as predicted by Hypothesis Set 5, the effects of perceived service characteristics on overall satisfaction were supported. In terms of relative effect size, perceived quality had the largest effect on overall satisfaction (SPC = .44, $p < .01$). Mobile phone service users' perceptions of interactivity positively related to the overall satisfaction of services (SPC = .33, $p < .01$) while users' perception of ease of use (SPC = .07, $p < .05$) had the lowest effect on overall satisfaction.

MAIN FOCUS OF THE CHAPTER: IMPLICATIONS FOR RESEARCH AND PRACTICE

The primary goal of this chapter was to develop and empirically test a model of user recommendations for mobile services that incorporated both user psychology and the unique features of mobile services. The hypothetical model is characterized by flow experience, overall satisfaction, and customer perceptions. Based on the findings of this chapter, we now consider first the implications of our results for research and practice followed by the limitations of the chapter.

Theoretical Implications

It has been argued that mobile services have unique characteristics that differentiate them from other media, such as the Internet or television. The contribution of this chapter is perhaps the development a new approach to the diffusion of mobile services. In particular, we address the question, "How might a company that provides mobile services encourage higher levels of recommendation behavior among its customers?" The objectives of the chapter are twofold: (1) to verify customer's flow experiences and user satisfaction that strengthen customer recommendations; and (2) to examine the factors that influence user satisfaction. This chapter specifies an explicit structure for direct and indirect influences on customer recommendations and provides a mechanism for marketers to determine external marketing variables relevant to mobile services.

This chapter introduced overall satisfaction and flow experience as potentially important characteristics that induce customer recommendations for mobile services. Our findings show that satisfaction and flow experiences exert significant influence on customer recommendations. Previous studies focused on the idea that high levels of customer satisfaction could increase the effect of customer recommendations (e.g. Mittal et al., 1999; Del Rio et al., 2001). However, they neglected the effect of user's flow experience. When customers use such services, they feel connected and involved in such a way that they lose track of time and their environment. The results of the chapter also show that customers' flow experiences could be a mediating factor that influences satisfied customers to recommend services or products to other potential customers.

Although perceived hedonic and utilitarian values, are believed to be positively linked to user satisfaction (e.g. Cottet et al., 2006), only the effects of hedonic values on satisfaction was supported by this chapter. There is one possible explanation for this. The survey showed that customers use video phones, download ring tones, and upload/download pictures more frequently than they use other mobile services. In other words, consumers use 3G mobile services more for fun rather than utilitarian purposes. This may explain the insignificant utilitarian value. Perceived service characteristics have a direct and positive relationship with satisfaction. This finding shows that a mobile service provider that seeks a mobile service presence must understand the fundamental importance of the following constructs: interactivity, quality, and ease of use.

Practical Implications

The results of this chapter also have some practical implications. Our findings can help mobile service providers design formal programs designed to encourage existing customers to make recommendations. In the past, the development of customer satisfaction was a key objective of marketing managers. However, this chapter has demonstrated that flow experience appears to influence customer recommendations with regard to mobile services. Our results suggest that the concept of flow experience could be used as a driver to encourage customers to recommend services or products to potential customers. Therefore, Mobile service providers should not only focus on developing customer satisfaction, they also need to understand how to help customers enjoy the services.

Moreover, customers' value and service characteristics increase both customer satisfaction and the indirect influence on flow experience. The implication of these results is that mobile service providers should not only focus on service characteristics, such as quality, ease of use, and interactivity, but also provide users' values, especially for hedonic value, when developing mobile services. A good idea to increase users' satisfaction and flow experiences is to enable users to involve in the "atmosphere" of using mobile services. Mobile service providers should particularly pay attention to aspects of quality and interactivity, and develop highly entertainment and fun services to attract users.

FUTURE TRENDS

Limitations and Suggestions for Future Research

We now consider some limitations of our chapter. First, the chapter was conducted in the context of mobile services. Although we drew general findings from mobile service contexts, we should investigate which services allow users to derive more hedonic value or utilitarian value. Possibly the differences in mobile services could moderate the impact of perceived value on overall satisfaction. Second, we tested the construct of flow experience by survey method. Flow experience emphasizes the temporary subjective psychological experience. According to Webster et al. (1993), they found that a measure of the reliability of flow experience by experimental method is better than a measure by survey. Future research should consider a measure of the flow experience by experimental method.

Future research should consider other kinds of information system applications, such as online games and mobile television, because their interaction would produce different degree of flow experiences for users.

CONCLUSION

This chapter was motivated by a broad interest in understanding customer recommendation behavior. The model contributes to our understanding of how customers are willing to make recommendations to others. Besides developing customer satisfaction, we should understand a key mediator—users' flow experiences. Moreover, our model considers that the elements of perceived value and perceived service characteristics could directly influence customer satisfaction and indirectly influence flow experience. Consequently, we believe that customer satisfaction and customer's flow experience induces behavior that can lead to the phenomenon of customer recommendations which, in turn, becomes free advertising for a firm and attracts new customers. Finally, we hope that this chapter will encourage researchers to identify the effects of flow experience in other mobile applications.

REFERENCES

Aaltonen, A., Huuskonen, P., & Lehikoinen, J. (2005). Context Awareness Perspectives for Mobile Personal Media. *Information Systems Management, 22*(4), 43–55. doi:10.1201/1078.10580530/45520.22.4.20050901/90029.5

Algesheimer, R., Dholakia, U. M., & Herrmann, A. (2005). The Social Influence of Brand Community: Evidence from European Car Clubs. *Journal of Marketing, 69*(3), 19–34. doi:10.1509/jmkg.69.3.19.66363

Anderson, E. (1998). Customer Satisfaction and Word of Mouth. *Journal of Service Research, 1*(1), 5–17. doi:10.1177/109467059800100102

Anderson, E. W., Fornell, C., & Lehmann, D. R. (1994). Customer Satisfaction, Market Share and Profitability: Findings from Sweden. *Journal of Marketing, 58*(3), 53–66. doi:10.2307/1252310

Anderson, J. C., & Gerbing, D. W. (1988). Structural equation modeling in practice: A review and recommended two-step approach. *Psychological Bulletin, 103*(3), 411–423. doi:10.1037/0033-2909.103.3.411

Babin, B. J., & Attaway, J. A. (2000). Atmospheric Affect as a Tool for Creating Value and Gaining Share of Customer. *Journal of Business Research, 49*(2), 91–99. doi:10.1016/S0148-2963(99)00011-9

Babin, B. J., Darden, W. R., & Griffin, M. (1994). Work and/or Fun: Measuring Hedonic and Utilitarian Shopping Value. *The Journal of Consumer Research*, *20*(4), 644–656. doi:10.1086/209376

Bagozzi, R. P., & Yi, Y. (1988). On the Evaluation of Structural Equation Models. *Journal of the Academy of Marketing Science*, *16*(1), 74–94. doi:10.1007/BF02723327

Bhattacherjee, A. (2001). Understanding Information Systems Continuance: An Expectation- Confirmation Model. *MIS Quarterly*, *25*(3), 351–370. doi:10.2307/3250921

Biyalogorsky, E., Gerstner, E., & Libai, B. (2001). Customer referral management: optimal reward programs. *Marketing Science*, *20*(1), 82–95. doi:10.1287/mksc.20.1.82.10195

Boudreau, M. C., Gefen, D., & Straub, D. (2001). Validation in IS Research: A State-of-the-Art Assessment. *MIS Quarterly*, *25*(1), 1–16. doi:10.2307/3250956

Brown, T. J., Barry, T. E., Dacin, P. A., & Gunst, R. F. (2005). Spreading the Word: Investigating Antecedents of Consumers' Positive Word-of-Mouth Intentions and Behaviors in a Retailing Context. *Journal of the Academy of Marketing Science*, *33*(2), 123–138. doi:10.1177/0092070304268417

Chiou, J. S. (2004). The antecedents of consumers' loyalty toward Internet Service Providers. *Information & Management*, *41*(6), 685–695. doi:10.1016/j.im.2003.08.006

Chung, C. M. Y., & Darke, P. R. (2006). The Consumer as Advocate: Self-Relevance, Culture, and Word-of-Mouth. *Marketing Letters*, *17*(4), 269–279. doi:10.1007/s11002-006-8426-7

Cottet, P., Lichtle, M. G., & Plichon, V. (2006). The role of value in services: a study in a retail environment. *Journal of Consumer Marketing*, *23*(4), 219–227. doi:10.1108/07363760610674347

Csikszentmihalyi, M. (1975). *Beyond Boredom and Anxiety*. San Francisco: Jossey- Bass.

Csikszentmihalyi, M. (1990). *Flow: The psychology of optimal experience*. New York: Harper Collins.

Del Rio, A. B., Vazquez, R., & Iglesias, V. (2001). The effects of brand associations on the consumer response. *Journal of Consumer Marketing*, *18*(5), 410–425. doi:10.1108/07363760110398808

DeLone, W., & McLean, E. (1992). Information System Success: The Quest for the Dependent Variable. *Information Systems Research*, *3*(1), 60–95. doi:10.1287/isre.3.1.60

DeLone, W., & McLean, E. (2003). The DeLone and McLean Model of Information Systems Success: A Ten-Year Update. *Journal of Management Information Systems*, *19*(4), 9–30.

East, R., Hammond, K., Lomax, W., & Robinson, H. (2005). What is the Effect of a Recommendation? *Marketing Review*, *5*(2), 145–157. doi:10.1362/1469347054426186

Fornell, C., & Larcker, D. (1981). Evaluating structural equation models with unobservable variables and measurement error. *JMR, Journal of Marketing Research*, *18*(1), 39–50. doi:10.2307/3151312

Hair, J. F., Anderson, R. E., Tatham, R. L., & Black, W. C. (1998). *Multivariate Data Analysis (5th ed.)*. New Jersey: Prentice-Hall, Inc.

Hoffman, D. L., & Novak, T. P. (1996). Marketing in Hypermedia Computer-Mediated Environments: conceptual foundations. *Journal of Marketing*, *60*(3), 50–68. doi:10.2307/1251841

Holbrook, M. B., & Hirschman, E. C. (1982). The Experiential Aspects of Consumption: Consumer Fantasies, Feelings and Fun. *The Journal of Consumer Research*, *9*(2), 132–140. doi:10.1086/208906

Hu, L., & Bentler, P. M. (1999). Cutoff Criteria for Fit Indices in Covariance Structure Analysis: Conventional Criteria versus New Alternatives. *Structural Equation Modeling*, *6*(1), 1–56.

Jiang, J. J., Klein, G., & Crampton, S. M. (2000). A note on SERVQUAL reliability and validity in information system service quality measurement. *Decision Sciences*, *31*(3), 725–744. doi:10.1111/j.1540-5915.2000.tb00940.x

Jones, T., & Sasser, W. E. (1995). Why Satisfied Customers Defect. *Harvard Business Review*, *73*(6), 88–99.

Lam, S. Y., Shankar, V., Erramilli, M. K., & Murthy, B. (2004). Customer value, satisfaction, loyalty, and switching costs: an illustration from a business-to-business service context. *Journal of the Academy of Marketing Science*, *32*(3), 293–311. doi:10.1177/0092070304263330

Liu, Y., & Shrum, L. J. (2002). What is interactivity and is it always such a good thing? Implications of definition, person, and situation for the influence of interactivity on advertising effectiveness. *Journal of Advertising*, *31*(4), 53–64.

McKinney, V., Yoon, K., & Zahedi, F. (2002). The measurement of Web-customer satisfaction: An expectation and disconfirmation approach. *Information Systems Research*, *13*(3), 296–315. doi:10.1287/isre.13.3.296.76

McMillan, S. J., & Hwang, J. S. (2002). Measures of Perceived Interactivity: An Exploration of the Role of Direction of Communication, User Control, and Time in shaping Perceptions of Interactivity. *Journal of Advertising*, *31*(3), 29–42.

Mittal, V., Kumar, P., & Tsiros, M. (1999). Attribute-Level Performance, Satisfaction, and Behavioral Intentions over Time: A Consumption-System Approach. *Journal of Marketing*, *63*(2), 88–101. doi:10.2307/1251947

Moon, J. W., & Kim, Y. G. (2001). Extending the TAM for a World-Wide-Web context. *Information & Management*, *38*(4), 217–230. doi:10.1016/S0378-7206(00)00061-6

Morgan, N. A., Anderson, E. W., & Mittal, V. (2005). Understanding Firms' Customer Satisfaction Information Usage. *Journal of Marketing*, *69*(3), 131–151. doi:10.1509/jmkg.69.3.131.66359

Morgan, N. A., & Rego, L. L. (2006). The Value of Different Customer Satisfaction and Loyalty Metrics in Predicting Business Performance. *Marketing Science*, *25*(5), 426–439. doi:10.1287/mksc.1050.0180

Nelson, R. R., Todd, P. A., & Wixom, B. H. (2005). Antecedents of Information and System Quality: An Empirical Examination within the Context of Data Warehousing. *Journal of Management Information Systems*, *21*(4), 199–235.

newsletter%20dec/page2-en.htm.

Novak, T. P., Hoffman, D. L., & Yung, Y. F. (2000). Measuring the customer experience in online environments: A structural modeling approach. *Marketing Science*, *19*(1), 22–42. doi:10.1287/mksc.19.1.22.15184

Nysveen, H., Pedersen, Per E., & Thorbjørnsen, H. (2005). Intentions to Use Mobile Services: Antecedents and Cross-Service Comparisons. *Journal of the Academy of Marketing Science*, *33*(3), 330–346. doi:10.1177/0092070305276149

Oliver, R. L. (1997). *Satisfaction: A Behavioral Perspective on the Consumer.* Boston: McGraw-Hill.

Palmatier, R. W., Dant, R. P., Grewal, D., & Evans, K. R. (2006). Factors Influencing the Effectiveness of Relationship Marketing: A Meta-Analysis. *Journal of Marketing*, *70*(4), 136–153. doi:10.1509/jmkg.70.4.136

Pitt, L. F., Watson, R. T., & Kavan, C. B. (1995). Service quality: a measure of information systems effectiveness. *MIS Quarterly*, *19*(2), 173–187. doi:10.2307/249687

Pousttchi, K., & Wiedemann, D. G. (2007). Success Factors in Mobile Viral Marketing: A Multi-Case Study approach. *Proceedings of the 6th International conference on Mobile Business,* Toronto, Canada.

Reichheld, F. (2006). The Microeconomics of Customer Relationships. *MIT Sloan Management Review*, *47*(2), 73–78.

Shabbir, H., Palihawadana, D., & Thwaites, D. (2007). Determining the Antecedents and Consequences of Donor-Perceived Relationship Quality— A Dimensional Qualitative Research Approach. *Psychology and Marketing*, *24*(3), 271–293. doi:10.1002/mar.20161

Siekpe, J. S. (2005). An examination of the multidimensionality of the flow construct in a computer-mediated environment. *Journal of Electronic Commerce Research*, *6*(1), 31–43.

Stone, R. W., Good, D. J., & Baker-Eveleth, L. (2007). The impact of information technology on individual and firm marketing performance. *Behaviour & Information Technology*, *26*(6), 465–482. doi:10.1080/01449290600571610

Sweeney, J. C., & Soutar, G. N. (2001). Consumer Perceived Value: The Development of a Multiple Item Scale. *Journal of Retailing*, *77*(2), 203–220. doi:10.1016/S0022-4359(01)00041-0

Teo, H. H., Oh, L. B., Liu, C., & Wei, K. K. (2003). An empirical study of the effects of interactivity on web user attitude. *International Journal of Human-Computer Studies*, *58*(3), 281–305. doi:10.1016/S1071-5819(03)00008-9

TNS marketing research company. (2001). Retrieved September 21, 2007, from http://219.232.228.151/infopulse/

Webster, J., Trevino, K. L., & Ryan, L. (1993). The dimensionality and correlates of flow in human-computer interactions. *Computers in Human Behavior*, *9*(4), 411–426. doi:10.1016/0747-5632(93)90032-N

Zeithaml, V. A. (1988). Consumer perceptions of price, quality and value: A means-end model and synthesis of evidence. *Journal of Marketing*, *52*(3), 2–22. doi:10.2307/1251446

KEY TERMS AND DEFINITIONS

Customer Recommendations: Customer recommendations are defined as the action that occurs when customers voluntarily recommend their experiences with mobile phone services to their relatives and friends.

Flow Experience: Csikszentmihalyi (1975, p. 36) defines flow as "holistic sensation that people feel when they act with total involvement."

Hedonic Value: Hedonic value is defined as that value a customer receives based on the subject experience of fun and playfulness (Babin et al. 1994).

Utilitarian Value: Utilitarian value is defined as that value that a customer receives based on a task-related and rational consumption behavior (Babin et al. 1994).

Third Generation (3G) Services: Third generation (3G) services is defined as the services on the third generation of developments in wireless technology.

Interactivity: Interactivity is the degree to which mobile phone service users have control over items such as video phone, media, objects, and other similar items.

Perceived Quality: An information system process perspective from DeLone and McLean (1992, 2003) and Pitt et al. (1995), suggests that a successful system must have high levels of system quality, information quality, and service quality.

NOTE

All t-value are significant at the .001 level. CR = composite reliability and AVE = average variance extracted. Goodness of fit statistics: $\chi^2(322) = 622.71$, GFI = .92, AGFI = .89, NFI = .98, CFI = .99, and RMSEA = .044.

APPENDIX A. ESTIMATED PARAMETERS FOR THE MEASUREMENT MODEL FROM CFA ANALYSIS

Table 2.

Factors and Items	Standardized Factor Loading	Reference source
Overall Satisfaction (CR = .89; AVE = .72)		
I am satisfied with the mobile services I use.	0.83	Bhattacherjee (2001)
I am satisfied with the mobile services offered by the telecommunication company.	0.86	
Overall, the mobile services meet my expectations.	0.86	
Flow Experience (CR = .90; AVE = .70)		
I once experienced flow when using a mobile service.	0.84	Trevino and Webster (1992) and Novak et al. (2000)
I focus on using a mobile service so I do not notice changes in the environment.	0.90	
Using a mobile service makes me curious.	0.81	
When using a mobile service, I feel in control.	0.78	
Customer Recommendations (CR = .84; AVE = .64)		
I have recommended the mobile service to my friends or relatives.	0.76	Brown et al. (2005)
I have often explained the benefit of using the mobile service to my friends.	0.86	
Generally, I have recommended the mobile service to acquaintances.	0.78	
Hedonic Value (CR = .83; AVE = .50)		
While using the mobile service, I felt happy.	0.65	Babin's et al. (1994)
Compared to other things I could have done, the time spent using the mobile service was truly enjoyable.	0.62	
When using the mobile service, I felt excited.	0.74	
I had a very nice time while using the mobile service.	0.81	
While using mobile service, I was able to forget my problems.	0.67	
Utilitarian Value (CR = .80; AVE =, 57)		
While using the mobile service, I used only the service(s) that I must.	0.80	Babin's et al. (1994)
While using the mobile service, I accomplished my goals	0.76	
I have often used mobile services that I really needed.	0.71	
Ease of Use (CR = .92; AVE = .93)		
3G mobile phone is easy to use.	0.78	Staplesa et al. (2002)
Learning how to use a 3G mobile phone is easy.	0.85	
I find 3G mobile phones easy to use.	0.90	
Overall, 3G mobile phones are easy to use.	0.88	
Perceived Interactivity (second-order factor) (CR = .82; AVE = .61)		
Real-time conversation	0.66	McMillan and Hwang (2002)
NO delay	0.80	
Engaging	0.86	
Perceived Quality (second-order factor) (CR = .86; AVE = .68)		
Information Quality	0.80	DeLone and McLean (2003)
System Quality	0.84	
Service Quality	0.83	

Section 4
Managerial Issues in Mobile Marketing

Chapter 22
Integrating Mobile Marketing into the Marketing Communication:
Exemplification of Mobile Marketing Campaigns

Süphan Nasir
Istanbul University, Turkey

ABSTRACT

With the increasing growth and sophistication of mobile communications, companies integrate the mobile channel into their marketing communication and campaigns. Companies are able to execute very different and creative mobile marketing campaigns by using a wide range of mobile platforms such as Wireless Application Protocol (WAP), Short Message Service (SMS), and Multimedia Message Service (MMS). The purpose of this chapter is to review and exemplify the concepts, objectives, strategies and promotional tactics associated with mobile marketing campaigns; so that the review of relevant literature and exemplification of mobile marketing campaigns provide an insight into the nature of mobile marketing campaigns.

INTRODUCTION

Penetration of Internet all over the world, closely followed by mobile technology creates opportunities for companies to reach their potential and current customers. Developments in information and communication technology give rise to new marketing communication channels and companies integrate all these channels to communicate the value propositions to their customers. Mobile communication is a fast-growing medium in terms of penetration and usage. The worldwide mobile subscriber base is still growing, there are more than 3 billion mobile subscribers worldwide (Wireless Intelligence, 2007). Mobile networks and infrastructure make connections better than ever and mobile handsets become better and easier to use. Therefore, mobile communication is becoming a primary means of communication channel. The emergence of high speed wireless network technologies and the increasing market penetration of mobile phones and other mobile devices highlight

DOI: 10.4018/978-1-60566-074-5.ch022

the potential and significance of the mobile channel for marketing.

Marketers have shown a great interest in reaching to and communicating with their customers and prospects through the mobile channel. Mobile communication enables highly targeted and timely marketing communications and it is becoming increasingly popular in the promotion of goods and services. Mobile marketing (m-marketing) campaign provides a company the ability to segment its target group based on specific criteria, to directly and efficiently reach these selected segments at anywhere and anytime (Balasubramanian, Peterson, and Järvenpää 2002; Barwise and Strong, 2002; Barnes and Scornavacca, 2004; Karjaluoto et al., 2007). Since m-marketing campaigns allow directly and efficiently targeting selected segments, it also helps to measure the effectiveness of the campaign at the individual level (Gopal and Tripathi, 2006). M-marketing creates new opportunities for the companies. Marketing has been gradually developing hand in hand with technological changes and the assortment of m-marketing campaign is also constantly changing, developing and transforming; so it is worthwhile to investigate the nature of m-marketing campaigns in a detailed way.

The main objective of this chapter is to provide an insight into the nature of m-marketing campaigns and get a comprehensive understanding about the fundamental concepts of m-marketing campaigns. Towards this end, the main concepts, strategies, objectives and tactics of m-marketing campaigns were reviewed on the basis of the relevant literature and exemplified with the real-life m-marketing campaign cases. Case study is particularly welcome in new situations where only a little is known about the phenomenon and in situations where current theories seem inadequate (Eisenhardt, 1989). The aim of the case study is to attain a comprehensive and detailed a description of the phenomena (Eisenhardt and Graebner, 2007). In this exploratory study, certain cases were chosen from the case database of Mobile Marketing Association (MMA) in order to exemplify m-marketing campaigns. The purpose of exemplifying the m-marketing campaigns with cases is not to make a generalization and attain a generalizable result; but to have a better understanding of the nature of the m-marketing campaigns. Objectives, implementation (specific tactics that are used to reach objective of the campaign), and the results of the campaign were indicated for the cases that were used to exemplification of m-marketing campaigns.

BACKGROUND

Literature Overview

As the mobile communication has been gradually developing, popularity of m-marketing has been also increasing in the academic literature and in the business world. A substantial number of the publications about the topic of m-marketing have dealt with conceptualization of m-marketing (Leppäniemi, Karjaluoto and Sinisalo, 2006), mobile commerce development and its enabling technologies, value added applications, promises and challenges (Siau, Lim, and Shen, 2001), key success factors of m-marketing (Facchetti et al., 2005; Scharl et al. 2005), limitations of mobile communication and services such as networking problems, infrastructure constraints, security concerns and user distrust in mobile applications (Siau and Shen, 2003), consumer adoption of m-marketing channels and key factors that influence consumers' willingness to accept m-marketing (Anckar and D'Incau, 2002b; Barnes and Scornavacca, 2004; Tsang et al., 2004; Bauer et. al., 2005; Leppäniemi and Karjaluoto, 2005), and the role of m-marketing in branding and its effects on brand relationships (Nysveen et al., 2005; Sultan and Rohm, 2005).

The growing body of literature on m-marketing has also focused on the effectiveness of m-marketing (Barwise and Strong, 2002; Rettie,

Grandcolas, and Deakins, 2005; Merisavo et al., 2006), the nature and implications of mobile advertising (Doyle, 2001; Barnes, 2002; Haghirian, Madlberger, and Tanuskova, 2005), challenges and opportunities of mobile advertising (Yunos, Gao, and Shim, 2003), models of mobile advertising (Komulainen et al., 2005), consumers' attitudes toward mobile advertising (Tsang et al., 2004), the developments in the m-marketing (Sullivan and Drennan, 2002), the legal and policy issues raised by mobile advertising (Petty, 2003), the categorization of m-marketing campaigns (Pousttchi and Wiedemann, 2006), the implementation of m-marketing campaigns (Karjaluoto et al., 2007; Leppäniemi and Karjaluoto, 2008), the taxonomy and process of a mobile marketing campaign (Kavassalis et al., 2003), analyzing the basic elements of viral m-marketing (Wiedemann, Haunstetter and Pousttchi, 2008), and creating mobile viral marketing strategies (Wiedemann and Pousttchi, 2008).

Mobile Marketing

Recent years witnessed an increase in the number of publications in the area of mobile marketing (Leppäniemi, Karjaluoto and Sinisalo, 2006) and a considerable number of the publications that have appeared on mobile marketing literature have dealt with defining mobile marketing (e.g., Dickinger et al., 2004; Sultan and Rohm, 2005; Leppäniemi, Karjaluoto and Sinisalo, 2006). Sullivan and Drennan (2002) propose the definition of m-marketing as "*the application of marketing to the mobile environment of smart phones, mobile phones, personal digital assistants (PDA), and telematics*". Kavassalis et al. (2003) define the m-marketing as "*something technologically simple that uses the mobile network as a complementary distribution channel for delivering old-fashioned commercial information and interactive promotional advertising*". Dickinger et al. (2004) define m-marketing as "*using interactive wireless media to provide customers with time and location sensitive, personalized information that promotes goods, services and ideas, thereby generating value for all stakeholders*". Sultan and Rohm (2005) refer m-marketing as "*brand in the hand*" and propose that the combination of the Internet and hand-held mobile devices is making possible marketing communications to be delivered to people in their hands while they are shopping, watching a sporting event, working or doing chores at home. Bauer et al. (2005) identify m-marketing as "*using the mobile phone as a means of conveying commercial content to customers*".

Leppäniemi, Karjaluoto and Sinisalo (2006) review m-marketing literature and their literature review yielded 21 distinct definitions of m-marketing and propose that "*mobile marketing is the use of the mobile medium as a means of marketing communications*". In other words, Leppäniemi, Karjaluoto and Sinisalo (2006) conclude that m-marketing is just a new, different means of communication which should be integrated into the marketing communications mix. Similarly, Pousttchi and Wiedemann (2006) define mobile marketing as "*form of marketing communication using mobile communication techniques to promote goods, services and ideas*". Mobile Marketing Association (MMA) (MMA, 2007b) defines m-marketing as "*the use of wireless media as an integrated content delivery and direct-response vehicle within a cross-media marketing communications program*". This means mobile is one of the many communication channels and the use of mobile phones and other mobile devices to market a brand or message called mobile marketing. Therefore, based on the above definitions, it can be concluded that the focus of mobile marketing is communication.

There is a vast array of possible activities that can be classified as m-marketing. M-marketing is an emerging form of marketing communication that also plays a powerful role in the integrated marketing communication. Marketing communications mix consists of the specific combination of advertising, personal selling, sales promotion, public relations, and direct marketing tools that a marketer uses to pursue its marketing communications objectives (Kotler, 2000; Leppäniemi and Karjaluoto, 2008). Clow and Baack (2007) have expanded these five major components of communication by adding database marketing, sponsorship marketing, and internet marketing as the components of communication (Clow and Baack, 2007). Leppäniemi and Karjaluoto (2008) suggest that due to the fact that personal selling and public relations activities would be extremely difficult to perform in m-marketing context; they only consider three of the main promotion tools: advertising, sales promotion and direct marketing in the context of m-marketing.

Mobile marketing communication channel offer limitless possibilities and opportunities that can not be gotten from any other medium. Mobile devices are always with us and unlike other advertising media; the mobile phone belongs to only one person. Mobile communication allows a company to target consumers with anytime, anywhere marketing (Balasubramanian, Peterson, and Jarvenpaa, 2002). No other form of customer communication or commerce interaction can offer the ability to reach consumers at any time or at any place. Because of the ubiquitous nature of mobile devices, it is the most personalized and customized communication channel (Anchar and D'Incau, 2002a; Barnes, 2002). Mobile phone and mobile devices present the opportunity to interact directly with the consumers and get responses from the consumers. Mobile communication establishes a direct dialogue between the marketer and the potential consumers; so that it is also an interactive media (Barnes, 2002; Dickinger et al., 2004; Bauer et. al., 2005; Haghirian, Madlberger, and Tanuskova, 2005). Karjaluoto et al. (2007) review the existing literature and conclude that mobility and reachability, direct marketing, interactivity and two-way communication, branding, viral-marketing, time, and personalization are special features of the mobile channel. Put it briefly, the personal and ubiquitous nature of the mobile device, the interactivity of the media, and its context-sensitivity make the mobile marketing communication distinctive from other communication channels (Balasubramanian, Peterson, and Jarvenpaa, 2002; Barnes, 2002; Anchar and D'Incau, 2002a; Dickinger et al., 2004, Haghirian, Madlberger, and Tanuskova, 2005; Xu, Liao, and Li, 2008).

M-MARKETING CAMPAIGNS

As markets have fragmented, marketers of today are shifting away from mass marketing to one-to-one marketing, and trying to build closer relationships with well defined micro markets (Kotler and Armstrong, 2008). Developments in mobile telecommunication networks, internet technologies and sophisticated mobile phones enable companies to see mobile communication as an opportunity to do one-to-one marketing and to reach to more consumers. Kavassalis et al., 2003 assess the effectiveness of mobile marketing campaigns through experimental campaigns and from the experimental cases they indicate the relatively high response rates and relatively low costs as the effectiveness indicators for mobile marketing campaigns. As the marketers are realizing the significance and effectiveness of the mobile communication, the use of mobile communication channel in the promotion of products and services is also increasing and companies are now increasingly investing in m-marketing campaigns. Hence, it is meaningful to investigate the topic of m-marketing campaign in a detailed way.

M-Marketing Campaign Objectives

To persuade the customer to make a purchase is one of the primary goals of the sales promotion campaigns. Clow and Baack (2007) indicate that consumer franchise-building and consumer sale-building are the two most general categories of promotion. The former category of promotions is designed with the objective of increasing awareness of and loyalty to a brand. On the other hand, the latter one is designed with the objective of increasing immediate sales through discounts and other enticements (Clow and Baack, 2007). However, in the context of m-marketing, mobile marketing campaigns can also be categorized in terms of the objective of the campaign as in sales promotion campaigns.

Kavassalis et al. (2003) identify four models for a mobile marketing campaign depending on the type of permission given by the customer: one-off push, one-off pull, continued dialogue, and fund raising. In one-off push model, the objective is to generate and maintain a relationship with customers and it is a breath strategy for customer acquisition; in one-off pull m-marketing campaign model, the aim is to make sales promotions to a smaller set of customers who give permission to receive mobile coupons and offerings and it is a depth strategy for customer acquisition/retention; the continued dialogue m-marketing model, which can be described as loyalty creation, is used to interact with customers on a continues basis and it is as a depth strategy for customer retention through mobile membership clubs; however, in the fund raising m-marketing model the objective is to encourage to donate to a charity (Kavassalis et al., 2003). In the same study, Kavassalis et al. (2007) review certain cases and highlight the marketing objectives of these cases, and the findings of the cases indicate creating loyalty, enhancing brand awareness, initiating interactive dialogue, creating database, increasing traffic and sales, and launching a new product are the objectives of the m-marketing campaigns. Karjaluoto et al. (2007) reveal that in many cases the objective of a mobile campaign is to increase sales by creating instant call-to-action. In another study, Pousttchi and Wiedemann (2006) present empirical results about the mobile marketing campaign objectives and they identify building brand awareness (it becomes important for new product launch), changing brand image, sales promotion (to stimulate purchase of a product/service), enhancing brand loyalty, building customer database, and creating mobile word-of-mouth (mobile viral marketing) as the objectives of mobile marketing campaigns. However, MMA (2007c) puts forward the objective of m-marketing campaigns as increasing brand awareness, generating a customer profile opt-in database, boosting up attendance to events or visits to a store, improving customer loyalty, and increasing sales (MMA, 2007c).

As it can be seen from the above literature review, companies have different objectives in creating mobile campaigns. For example, Lipton Ice Tea used m-marketing campaign in order to increase the sales of Lipton Ice Tea in McDonald's Restaurants (Figure 1 provides an overview of the Lipton Ice Tea mobile campaign). The automotive company Hyundai integrated m-marketing campaign to its marketing communication mix to encourage potential customers to visit Hyundai showrooms (to increase store traffic) and to encourage them to participate test drive (to boost up attendance to test drive) (Figure 2 describes the Hyundai Accent Era Campaign).

Akbank, which is one of the three biggest retail banks in Turkey, used SMS and MMS campaigns in order to boost awareness about its new product called SMS based consumer loan. Figure 3 provides an overview about the Akbank's SMS based loan campaign and this case is an example for using m-marketing campaign with the purpose of launching a new product and creating brand awareness for the new product. British Petroleum (BP) implemented mobile campaign with the objective of generating customer database (Figure 4 explains the British Petroleum Mobile CRM Campaign).

Standard Charity Bank in India used mobile led viral campaign in order to launch and create a buzz around its new product Manhattan lifestyle credit card and Figure 5 briefly explains this mobile viral campaign. A modern valentine's fairy tale

Figure 1.

Figure 1 Lipton Ice Tea Campaign

OBJECTIVE: The purpose of this campaign is to increase the sales of Lipton Ice Tea in the McDonald's Restaurants.

IMPLEMENTATION: The customers, who prefer to purchase Litop Ice Tea menu in McDonalds Restaurants, are given scratch cards with a unique code on it. Every participants, who text the unique code to Turkcell 3161 with their names, instantly win 5 free SMS and during the camapign period on each day (between the hours 10.00 am – 7.00 pm) the first participant of the day has the chance to win an iPod.

RESULT: Due to the instant win prizes, this mobile campaign was much more successful and generated twice more sales than the previous two campaigns of Lipton Ice Tea that were implemented in McDonald's Restaurants.

Source: MMA Global Member Case Studies, from http://mmaglobal.com/modules/article/view.article.php/856

Figure 2.

Figure 2 Hyundai Accent Era Campaign

OBJECTIVE: The aim of this mobile marketing is to promote the new Hyundai Accent Era by encouraging potential customers to visit showrooms and to participate in a test drive.

IMPLEMENTATION: The campaign is announced through other communication channels, such as television. Those who take a test drive received a unique code and those who text the code from their mobile phone enter a competition. Participants of the competition have the chance to win 1 Hyundai Accent Era, 10 tickets for the World Cup, and every 50th entrant has the chance to win a soccer ball. Participants also receive a free periodic service.

RESULT: 15,000 potential customers participated to test drive. Participants of the competition filled out a form, this helped Hyundai to increase its customer database for future interaction and loyalty programmes.

Source: MMA Global Member Case Studies, from http://mmaglobal.com/modules/article/view.article.php/324

case, which is described in Figure 6, is another example for creating mobile viral marketing. On the other had, in some instances companies use m-marketing in order to change and create image. The Efes Pilsen case (see Figure 13) is an example of implementing m-marketing to create and strengthen its brand image by supporting of sports in the mobile environment.

Domino's Pizza India used mobile platform for its campaigns in order to create personal and targeted way of communication and interactivity with its customers. Figure 7 illustrates an overview about the Domino's Pizza case and in this interactive mobile campaign Domino's Pizza provided interactivity and consumer involvement through a contest. Moreover, in order to increase customer involvement, Domino's Pizza attached viral element to the same campaign. Viral effects enlarge the reach of mobile marketing campaigns and also enlarge the campaign effectiveness.

M-Marketing Campaign Strategies

Strategies are the statements about how the objectives will be achieved. There is a wide range of strategic approaches that may be adopted in the development of mobile marketing communications (Leppäniemi and Karjaluoto, 2008). After the mobile marketing campaign objective is set, the mobile marketing campaign type, either push or pull, have to be determined (Karjaluoto et al., 2007). Provider-Driven (Push) Campaigns and User-Driven (Pull) Campaigns are the two mobile campaign strategies that are frequently used (Barnes, 2002; Leppäniemi, Karjaluoto and Salo, 2004; Pousttchi and Wiedemann, 2006; Karjaluoto

Figure 3.

Figure 3 AKBANK's SMS-Based Loan Campaign

OBJECTIVE: To launch a new SMS loan product and create brand awareness for the bank's new product.

IMPLEMENTATION: A target consumer profile was selected from the permitted database, which includes approximately 5,000,000 subscribers, according to the billing size, employment info, and location criteria. The bank reached 400,000 consumers through SMS and MMS, and consumers were asked to write their citizenship identity number and send it to the short code 4425 in order to initiate the loan application process. Loan approval is instantly sent to the consumers by the bank through SMS.

RESULT: The response rate for this mobile campaign is over 10% (approximately 40, 000 consumers responded the bank's SMS and MMS about the new loan product).

Source: MMA Global Member Case Studies, from http://mmaglobal.com/modules/article/view.article.php/1741

Figure 4.

Figure 4 British Petroleum – Mobile CRM Campaign

OBJECTIVE: The objective of the BP's mobile CRM campaign is to collect its customer information and develop a customer database in order to develop CRM strategies based on the analysis of this database.

IMPLEMENTATION: During the purchase of gas, 10 digit unique keywords distributed to the customers at the BP gas stations. Then customer asked to send the keyword to a short code via SMS. By sending keywords via SMS, customers win free air time from Turkcell GSM operator and participants also have the chance to participate a weekly lottery of mobile phones and cars.

RESULT: At the end of the campaign, BP collected customer data (name, mobile phone number, purchasing patterns of its customers) from 1,500,000 unique customers.

Source: VeriPark from http://www.veripark.com/Portfolio/SuccessStories/MobileCRM.asp

Figure 5.

Figure 5 Standard Chartered Bank- Manhattan Lifestyle Credit Card

OBJECTIVE: The purpose of the mobile campaign is to create an awareness and build excitement about the bank's new product called Manhattan lifestyle credit card.

IMPLEMENTATION: An interactive wireless quiz about the attributes of the credit card is implemented. Consumers participate the mobile quiz by sending SMS to a short code 8558. Winners of the mobile quiz have the chance to win a ticket for a trip to Manhattan. There is also a viral element in the campaign. If a consumer answered questions in the quiz and sent the promo message to 5 friends, he would get a number of points. If one of his friends in turn answered the questions in the quiz and sent it to his own 5 friends, the initial sender would be credited with additional points.

RESULT: During the 3-month period of the campaign, 350,000 consumer participated the quiz and a total of 60,000 leads were also generated.

Source: MMA Global Member Case Studies, from http://www.mmaglobal.com/studies/standard-chartered-bank-mobile2win

Figure 6.

Figure 6 A Modern Valentine's Fairy Tale

OBJECTIVE: The aim of the Nokia's m-marketing campaign is to encourage people to use the features of mobile phones and to show how a kiss connected people on Valentine's Day in the US market.

IMPLEMENTATION: Nokia invited consumers to send a virtual kiss from their mobile phones to their loved ones' mobile phone. Consumers choose their Valentine's Day kiss, text their name and their Valentine's mobile phone number and then the loved one receives a call with the chosen kiss and a SMS about who has sent this kiss. To increase the viral effect of the kiss campaign, the receiver can forward the kiss by one SMS.

RESULT: On the Valentine's Day more than 6000 participants send the kiss.

Source: MMA Global Member Case Studies, from http://www.mmaglobal.com/studies/modern-valentine%E2%80%99s-fairy-tale-12snap

Figure 7.

Figure 7 Domino's Pizza

OBJECTIVE: The purpose of the Domino's Pizza m-marketing campaign is to create personal and targeted way of communication and interactivity with its customers. Moreover, to provide a richer and more informative experience to its customers.

IMPLEMENTATION: Domino's Pizza launches a WAP based contest and if participants answers 4 consecutive questions correctly, all the participants win discount coupons. The campaign also has a viral element attached; the participtants can refer their friends to the contest through SMS and earn more discount points.

RESULT: During the campaign period, a total of 30,000,000 participtants registered and 65,000 discount coupons were delivered only on the first 6 days of the campaign.

Source: MMA Global Member Case Studies, from http://www.mmaglobal.com/studies/domino%E2%80%99s-mobile1win

et al., 2007; Leppäniemi and Karjaluoto, 2008). In provider-driven (push) campaigns, marketing services are delivered directly to the consumer's the mobile device and the service provider initiates communications (Steinbock, 2006). MMA (2007b) defines the push mobile messaging as any content sent by or on behalf of advertisers and marketers to a wireless mobile device at a time other than when the subscriber requests it (MMA, 2007b). Karjaluoto et al. (2007) define the push campaign as the campaigns aimed at opt-in lists. This means that marketer invites consumers from opt-in list to m-marketing campaign. For example, as it can be seen from the AKBANK's SMS based loan campaign case (see Figure 3), the bank used permitted database and selected its target customer from this database. Then the bank sent SMS and MMS and invited customers to participate SMS based consumer loan campaign. In this case, push strategy is used and the company started the communication. Another example for provider-driven push mobile campaign can be seen in the Figure 8. Renault creates awareness for its new model Symbol, which is going to launch at the Auto Show through a mobile quiz. Renault invites the potential customers to take the mobile quiz via SMS and this invitation SMS is sent to the target group which is chosen from permitted database.

However, MMA (2007b) defines the pull mobile messaging as any content sent to the wireless subscriber upon request, shortly thereafter, on a one time basis (MMA, 2007b). In user-driven (pull) campaigns, the user requests a message (a phone call or a mobile message) from the service provider, typically in response to information on other media (such as television, magazines, newspapers or flyers) (Steinbock, 2006). Leppäniemi, Karjaluoto and Salo (2004) argue that mobile media does not work alone and mobile media needs traditional media in order to increase the effectiveness of mobile media (Leppäniemi, Karjaluoto and Salo, 2004). Kavassalis et al. (2003) also highlights the

Figure 8.

Figure 8 Renault New Symbol Quiz

OBJECTIVE: Renault wants to create awareness for its new model Symbol, which is going to launch at the Auto Show. The aim of the Renault's mobile quiz campaign is to differentiate itself from the other automobile companies at the Auto Show with an attractive mobile marketing campaign.

IMPLEMENTATION: Renault sends a mobile quiz to potential consumers, which are chosen from permitted database before the start of the Auto Show. And participants are asked to text "new symbol" to 3141 and those send SMS to 3141, receive quiz questions about new symbol's features. The first 20 participants that answer all the questions correctly win free entry tickets to Auto Show. Participants are also able to get familiar with new Symbol's features closely.

RESULT: In the first 20th minute of the campaign, all tickets were delivered to the participants who answered the questions correctly.

Source: MMA Global Member Case Studies, from http://www.mmaglobal.com/studies/renault-new-symbol-quiz-electric-agency

Figure 9.

Figure 9 Philips Sonicare Electronic Toothbrush

OBJECTIVE: The objective of Philips is to build awareness about the new toothbrush and increase store traffic and sales in the Netherlands.

IMPLEMENTATION: In order to increase store traffic and create awareness for Sonicare Electronic Toothpaste, Philips develops a mobile marketing campaign and the campaign is also announced from billboards, brochures and the Philips's website. Those interested with the new toothpaste are invited to send their zipcode to a shortcode in order to receive information about the nearest store that carries the product. Based on the postal code and a free SMS is sent about the store address.

RESULT: During the first week of the launch of this mobile service, more than 3000 customers responded this mobile campaign.

Source: MMA Global Member Case Studies, from http://www.mmaglobal.com/studies/philips-sonicare-emexus

significance of mobile media channel integration with other media channels and indicate that in most cases a traditional advertising medium like television, press or flyers is needed in order to invite consumer to participate in a mobile marketing campaign. Mobile campaigns mostly use "pull strategy" because it publicises campaigns via another communication medium such as television, radio and packaging. This means it is the customers who initiate the communication by sending a message to take part in the mobile campaign. Pull campaigns also provide an opportunity build opt-in lists and marketing permissions (Karjaluoto et al., 2007). As it can be seen from the Figure 9, Philips uses pull mobile campaign strategy when launching its new toothbrush. Philips publicises campaigns via another communication medium such as billboards and company's web site. By using these traditional communication channels, Philips invites consumers to send their zipcode to a shortcode in order to receive information about the nearest store that carries the product. Therefore, in this case customer initiates the relationship by sending a message to take part in the mobile campaign.

M-Marketing Campaign Tactics

When the objective of marketing campaign is defined, specific tactics can be developed to reach the objective. Tactics are the activities that are performed to support strategies and tactics include promotional campaigns that are designed based on strategic objectives (Clow and Baack, 2007). Consumer sales promotion includes tools such as samples, coupons, premiums, contests and sweepstakes, refund and rebates, patronage reward, prices off, etc. (Kotler, 2000; Clow and Baack, 2007). However, in the context of m-marketing there are also specific mobile campaign tactics to reach the m-marketing campaign objective.

Leppäniemi and Karjaluoto (2008) propose a wide variety of possible activities that can be classified as sales promotions tactics in m-marketing. Leppäniemi and Karjaluoto (2008) categorize the most common promotional activities in m-marketing context under three groups: branded content category, competition, and other sale promotion activities such as receiving mobile coupons or money off offers to the mobile phone, requesting sampling by sending a text message. The category of branded content includes all kind of downloadable solutions such as branded ring tones, logos wallpapers, and mobile advergames (e.g., the use of mobile games to advertise or promote a product or brand). The category of competition includes all kind of quizzes, voting, and text and win promotions (Leppäniemi and Karjaluoto, 2008).

Pousttchi and Wiedemann (2006) discuss that characteristic of information (e.g., information about the product, information about the nearest point of sale and mobile newsletters), entertainment (e.g., downloadable videos, music, ringtone, and games), raffle (e.g., cash and goods prizes that are won by attending mobile contests, quizzes, and voting) and coupon (e.g., discounts and trial packages) can increase the value of the whole campaign and in turn can affect the consumers' attendance to m-marketing campaigns.

MMA (2007c) develops m-marketing sweepstakes and promotional guide, and in this guide MMA lists sweepstakes, instant win, mobile coupons, contests, polling/voting, mobile web (WAP), and promotional content downloads as m-marketing promotional tactics (MMA, 2007c). Each promotional tactics have unique characteristics, costs, and involves specific channels of communication; hence marketers have to understand the characteristics of m-marketing tactics (Leppäniemi and Karjaluoto, 2008). Therefore it is worthy to further investigate the most common m-marketing campaign tactics (promotional tools) by taking the above literature review as a theoretical base.

Text and Win Campaigns

Mobile messaging campaigns are the most familiar form of m-marketing campaigns and come in two different varieties: SMS-based (Short Message Service) which is text-only messages and MMS-based (Multimedia Messaging Service). MMS is a type of messaging that enables the sending of multimedia objects (e.g., images, audio, and video) as opposed to the simple text messages. Mobile campaigns are usually advertised in other media such as on-pack, in-pack, billboard, or press (Kavassalis et al., 2003; Leppäniemi, Karjaluoto and Salo, 2004; Steinbock, 2006). And with the help of these other media channels, the customer is invited to send a text message to a short code number (typically 4 or 6 digit short numeric numbers to which text messages can be sent from a mobile phone) in order to attend the mobile campaign and have the chance of winning prize. Mobile sweepstakes, instant-win, mobile banking, mobile ticket, mobile coupons, mobile contests are some of the mobile campaign promotional tools that can be used in Text and Win campaigns. The following cases are examples of different types of "text and win" m-marketing campaigns: Lipton Ice Tea (see Figure 1), Hyundai Accent Era (see Figure 2), British Petroleum (see Figure 4), Manhattan Lifestyle Credit Card (see Figure 5), and Renault New Symbol Quiz (see Figure 8).

Mobile Sweepstakes

A sweepstake is a legal game that includes a prize, and a game of chance. No consideration is allowed (MMA, 2007b). Mobile sweepstakes is one of the promotional tactics that can be used in "text and win" campaigns. A mobile sweepstake invites consumers to send a text message to a short code in order to gain a chance to win a prize. By answering a question, voting for something, participating a mobile quiz, or sending a certain code via SMS, participants get the chance to win cash and goods prizes; and the winners are drawn at the end of the campaign (Pousttchi and Wiedemann, 2006; MMA, 2007c). Characteristics of the sweepstakes add value to the m-marketing campaign and also sweepstakes also provides customers a tangible reason to participate the m-marketing campaign (Pousttchi and Wiedemann, 2006). Thus, mobile sweepstakes can be used to attract purchases and awareness. Manhattan Lifestyle Credit Card case is an example for mobile sweepstakes (see Figure 5). In Manhattan Lifestyle Credit Card campaign, consumers take part in the mobile quiz by sending SMS to a short code and winners of the contest have the chance to win a ticket for a trip to Manhattan. During the 3-month period of the campaign, 350,000 consumer attend the sweepstakes to in order to win the ticket. The chance of winning ticket can be considered as a tangible reason for consumers to participate the m-marketing campaign.

Instant-Win

Instant-win can be considered as one of the promotional tactic in the context of "text and win" m-marketing campaigns. In instant-win mobile campaigns consumers have the chance to win something instantly. An instant-win mobile campaign invites consumers to send a text message to a short code in order to have the chance of instantly winning various campaign related prizes. When the consumer sends the message, a text message is returned to the consumer; so that consumers instantly learn if they win a prize. Figure 10 outlines the Ruffles and Lay's m-mobile marketing campaign. As it can be seen from the case, Ruffles and Lay use instant win prizes to reach its campaign objective.

Mobile Banking

In the context of "text and win" m-marketing campaigns, mobile banking services can be used as promotional tactic. The participant of the m-marketing campaign can win cash money prize as an instant win or as a sweepstake. Cash money prize can increase the value of the m-marketing campaign (Pousttchi and Wiedemann, 2006); so that cash money can be considered as desirable and useful promotional tactic. But it was very

Figure 10.

Figure 10 Ruffles and Lay's Gift Flow

OBJECTIVE: The objective is to increase Ruffles and Lay branded snacks' sales during the summer period.

IMPLEMENTATION: Participants are asked to text the unique codes found in the product packages to the campaign's short code. In return, participants have the chance to win 10 YTL (equivalent to 8 USD) of "Mobile Bank" prize instantly and they can draw the money from the ATM by entering the password which is sent it to their mobile phones via SMS. Participtants have also chance to win 8 credits/ 2 minutes airtime from the GSM operator instantly.

RESULT: This mobile campaign generated high volume sales and the redemption rate was reached to 14%.

Source: MMA Global Member Case Studies, from http://www.mmaglobal.com/studies/ruffles-lay%E2%80%99s-gift-flow-aerodeon-turkey

difficult to deliver thousands of cash money prize through mobile phone. Mobile banking allows the company to transfer the cash money prize to the winner by sending a simple SMS. The amount of money prize and a password are sent with the SMS to the winner. When the receiver of the money receives the transaction password through SMS, he goes to the bank's ATM and enters the transaction password, amount of money, and his phone number. So that receiver takes the money easily from the ATM and in order to use this application money receiver does not need to have a bank account or debit card. As it can be seen from the Figure 10, Ruffles snacks use mobile bank prize as a promotional tactic in order to reach its m-marketing campaign objective that is to increase sales.

Mobile Tickets

Mobile tickets can be also used as promotional campaign tool. Participants of the "text and win" m-marketing campaign can win mobile tickets for sweepstakes, events, concerts, cinema, transportation or facilities so on. Winners received their tickets in the form of a 2D code (Data Matrix) in an SMS and holders of mobile tickets are able to verify their entry rights simply by holding their mobile ticket, which is in form of a 2D code and displayed on their mobile phone, to the scanner. Figure 11 provides an overview about how Nokia integrates the mobile tickets to its m-marketing campaign. Nokia executes a m-marketing campaign and delivers the ticket of the sweepstakes through SMS.

Mobile Coupons

Mobile coupon can be also considered as a mobile sales promotional tool (Siau, Lim, and Shen, 2001; Dickinger, et al., 2004; Scharl et al., 2005; Pousttchi and Wiedemann, 2006). A mobile coupon, which is sent and stored on a mobile phone, offers free samples, a rebate, a discount, etc., to a consumer that can be redeemed while making the purchase (Bathwal, 2008; MMA, 2007a). The consumer receives the mobile coupon through pull or push delivery (MMA, 2007a). In pull mobile

Figure 11.

Figure 11 Nokia N-Gage Snow Tour 2004-Mobile Sweepstakes Tickets

OBJECTIVE: In order to generate a strong customer relationship, build a customer database and raise brand awareness, Nokia carries out a mobile marketing campaign during the N-Gage Snow Tour 2004 through many Austrian and Swiss ski resorts in order to promote the new N-Gage game.

IMPLEMENTATION: As a part of the m-marketing campaign all the registered Club Nokia members received a mobile ticket in the form of a 2 D code (Data Matrix) to their mobile phone, which enables the members to participate in the Nokia sweepstakes being held in the N-Gage tour trucks. Those who are not registered with Club Nokia could also receive a mobile sweepstakes ticket by requesting it through texting SMS and sending a special keyword; so that they receive the mobile sweepstake tickets in order to take part in the campaign.

Holders of the mobile sweepstakes tickets can be informed whether they are the winner of the sweepstake by holding their mobile ticket, which is displayed on their mobile phone screen, to the reading system/scanner placed in the N-Gage tour truck at the ski resort.

RESULT: n/a

Source: MMA Global Member Case Studies, from http://www.mmaglobal.com/studies/nokia-n-gage-snow-tour-2004-mobile-sweepstakes-tickets-gavitec

coupon delivery, the consumers must text and send a campaign code, which can be obtained through other media such as on-pack, in-pack, billboard, or press, to a short code number to receive a mobile coupon. And consumer instantly receives the mobile coupons via SMS, MMS, Bluetooth, or other mobile devices. In another form of pull mobile coupon delivery, the consumer registers to an application that can be downloaded on mobile phones by submitting mobile phone number. These downloadable applications let consumers to access mobile coupons which provide great deals and discounts for shopping, restaurants, and entertainment (MMA, 2007a).

However, in push mobile coupon delivery, the company sends the mobile coupons to the consumer without the request of the consumer. There are many types of mobile coupons such as cents off-dollars off, free, buy one get one free, etc. (MMA, 2007a). Mobile coupon campaigns often lead to trial purchases and ideal for when introducing a new product or promoting trial among new target consumers. Mobile coupons are stored in the mobile phone's memory and therefore consumers do not forget to use the mobile coupons (Scharl et al., 2005). Mobile coupons have some advantages such drive traffic to stores, events and entertainment attractions; add value to a purchase; direct discount to the consumer; leverage the distribution; and targeting possibilities (MMA, 2007a; Pousttchi and Wiedemann, 2006).

As it can be seen from the Dunkin' Donuts Hot Lattes m-marketing example (see Figure 12), Dunkin' Donuts uses both push and pull mobile coupon delivery. In pull mobile coupon delivery option, Dunkin' Donuts invites the consumers to receive mobile coupon through other media channel (radio). Those who text and sent the campaign code to a short code number, instantly receives the mobile coupons via SMS. However, Dunkin' Donuts also uses push mobile coupon delivery and sends mobile coupons to 7500 targeted consumer without the request of the consumer.

Mobile Contests

Mobile contest is a promotional mechanism that includes a prize and a game of skill and in the mobile contest there can not be any element of chance (MMA, 2007b). Contests create new ways to encourage creativity, interact with the company's target market and create a buzz around

Figure 12.

Figure 12 Dunkin' Donuts Hot Lattes

OBJECTIVE: The objective of the m-marketing campaign is to increase the trial of Dunkin' Donuts hot lattes among high school and college students in the Boston.

IMPLEMENTATION: As a part of the m-marketing campaign, in-store redemption of $0.99 hot latte mobile coupons are sent to 7500 targeted consumers, which is chosen from the opt-in lists, through SMS. The SMS message promotes the viral element of the coupon because participants have the option of forwarding the message to a friend. Moreover, by using other communication channels such as radio consumers are invited to the participation. By texting to DD-123, they receive their mobile coupons.

RESULT: The targeted SMS messaging to 7500 consumers and coupled with cross media communication create a 21 percent increase in store traffic and redemption of the mobile coupon.

Source: MMA Global Member Case Studies, from http://www.mmaglobal.com/studies/dunkin%E2%80%99-donuts

the company's promotions. Mobile contests are the contests that require consumers to enter via text message. Contests require the participant to perform some type of activity and the winner is selected based on who performs best or provides the most correct answers (Clow and Baack, 2007). This is the difference of between mobile contest and mobile sweepstakes. Mobile quizzes are important element of mobile sweepstakes. Consumers, who answer the mobile quiz correctly, get the chance of entry to the sweepstake and in return have the chance of winning prize. The winner is chosen among those who answer the quiz correctly in the sweepstakes. This means that some of the participants, who answer the quiz correctly, win the prize by chance. However, in the context of contests the participant performs some type of activity and the winner is selected based on who performs best or provides the most correct answers. Domino's Pizza case is an example of mobile contest because all of the participants, who answer the questions correctly, win a prize of discount coupon (see Figure 7). On the other hand, in the case of Renault New Symbol Quiz, which is an example of mobile quiz, only the first 20 participants that answer all the questions correctly win free entry tickets to Auto Show (see Figure 8). The case of Manhattan Lifestyle Credit Card is an example for mobile sweepstakes; because by taking and answering the mobile quiz, consumers get the right to entry to the sweepstake in order to win a ticket to Manhattan (see Figure 5). The role of chance is important in the sweepstakes.

Mobile Web (WAP)

Wireless Application Protocol (WAP) is an open international standard for applications that use wireless communication. Its principal application is to enable access to the Internet from a mobile phone. WAP is a technology platform that helps to create web sites that are easily accessed from handsets (MMA 2007d). A mobile web site is specifically designed and formatted for display on a mobile device. Mobile web (WAP) sites can be viewable on any phone with a mobile Web browser. Mobile Web sites—also known as WAP sites—are the perfect way to reach consumers or give marketers an additional tool for reaching mobile users (MMA 2007d). Mobile web portals can also be incorporated into m-marketing campaigns and in this case mobile web page can provide access to images, information, terms and conditions about the promotion. Moreover, mobile web sites provide an opportunity to gather consumer data during the m-marketing campaigns (MMA,

Figure 13.

Figure 13 Efes Pilsen Sports News WAP Site

OBJECTIVE: Efes Pilsen strengthens the quality, integrity and leadership aspects of its brand image, by sponsoring international cultural, artistic and sports events. The aim of the establishment of WAP site is to strenghten Efes Pilsen's support of sports in the mobile environment.

IMPLEMENTATION: This WAP page provides rich information about all kind of sports activity. Sport lovers able to reach to live football, basketball news , breaking news, score situations, and league tables from their mobile phone any time they want.

RESULT: n/a

Source: MMA Global Member Case Studies, from MMA Global Member Case Studies from http://mmaglobal.com/modules/article/view.article.php/795

2007c). Consumers also get product and service information and download promotional branded contents from the mobile site of a brand.

Efes Pilsen, which is a famous beer brand in Turkey, tries to strengthens the quality, integrity and leadership aspects of its brand image by sponsoring sports events. The Efes Pilsen Sports News WAP Site is designed to strengthen its image of supporter of sports events. This WAP page provides rich content of sports events and establishes connection between the company and consumers.

Promotional Downloadable Content

Promotional downloadable content is usually used to attract consumers to attend m-marketing campaigns (Leppäniemi and Karjaluoto, 2008; Pousttchi and Wiedemann, 2006). The download can be a song or ringtone, wall paper, video, game, or other application, which is usually received on the handset through WAP link in a SMS message. Information about the mobile campaign with downloadable content can reach customers via product packaging, print advertising, bill boards, TV, or online communications (MMA 2007d). Mobile downloadable contents, including logos, wallpapers, or ring tones allow for brand exposure and create buzz about the marketing campaign. Companies can build brand identity by using these software applications. This same content is often shared with a user's friends and family. So that, branded content downloads strengthen consumer loyalty by building communities and also they can be used as an emotional tool that connect brands with consumers emotionally. Figure 14 illustrates how the DOVE (skin, body cleansing and hair care products) uses downloadable content as a m-marketing promotional tactic to make consumers aware of the new product.

Mobile Games

Mobile games have now emerged as the latest tactic to be employed in promoting brand campaigns. Nielsen Mobile (2008) estimates that there were 19 million game downloaders (7% of wireless subscribers) for the first quarter of 2008. Branded mobile game, which is also known as advergame, enables marketers to use gaming as a promotional tool to promote a product or service. Mobile games can also generate brand awareness. One of the benefits of advertising through mobile games is that it is very interactive and it is fun (Mathieson, 2005). Branded mobile game experiences generate buzz and keep consumers coming back. Figure 15 overviews the Seda Teens m-marketing campaign. Seda Teens integrates the mobile game into its

Figure 14.

Figure 14 Dove Fresh Touch Melody

OBJECTIVE: Dove launches a new product and wants to create awareness for the Dove Fresh Touch. Dove includes an interactive mobile element to its marketing mix to create awareness.

IMPLEMENTATION: Consumers are informed through TV advertisements and invited to text DOVE to 3161, and those, who send the SMS, received a WAP link of the polymelody or true tone of Dove jingle. Consumers click on the melody in order to download.

RESULT: Until the beginning of campaign approximately 8.000 consumer downloaded Dove jingle.

Source: MMA Global Member Case Studies from http://www.mmaglobal.com/studies/dove-fresh-touch-melody-aerodeon-turkey

m-marketing campaign in order to communicate with its target customers efficiently.

Location-Based M-Marketing

Location based mobile marketing campaigns deliver promotional information to consumers based on their current location. GPS (Global Position System) is used to determine the position of a consumer's curbetwrent location. In location based m-marketing, promotional information is delivered to the right consumer at the right place and time based on consumers' current location (Fields, 2005). While a consumer is walking near a café, a mobile coupon for a dollar off a latte is pushed to the consumer's mobile device just as the consumer passes the café. Location based services enable companies to offer highly targeted m-marketing campaigns such as mobile coupons (Marriott, 2006).

CONCLUSION

A consumer's mobile device is always on and always connected and that means always available. Mobile communication channel allows a company to target consumers with anytime and anywhere. The mobile campaigns allow organizations to reach out and establish relationships with their current and prospect customers in a more individualized and intimate way. M-marketing campaign tactics are expected to gradually develop, change and transform hand in hand with technological changes in the future. Thus, the objective of this chapter is to provide an insight into the nature of m-marketing campaigns and get a comprehensive understanding about the fundamental concepts of m-marketing campaigns. The main concepts, strategies, objectives and tactics of m-marketing campaigns were reviewed on the basis of the relevant literature and exemplified with the real-life m-marketing campaign cases. There is a wide variety of possible mobile campaign tactics and activities. Hence, the given terms in this chapter about the mobile campaign tactics should not be considered as comprehensive; but rather this study is intended to represent the most common mobile campaign tactics in m-marketing context with cases.

Figure 15.

Figure 15 Seda Teens (Unilever)

OBJECTIVE: Unilever produces a new line of hair products for teenagers called Seda Teens and its target is teen girls from 12 to 17. The purpose of the Unilever is to find a best way to reach its target consumers and communicate with them in an efficient way.

IMPLEMENTATION: Seda Teens Mobile Social Network is designed and from this mobile community consumers can get customized wallpapers, ringtones and games. The mobile game, which is called Dancing Star, is developed based on the hugely popular Dance Revolution.

RESULT: The campaign results rank Seda Teens as one of the most successful mobile strategies in the Brazilian market. The mobile community received a total of 360.000 visits and the game was downloaded 11.000 times.

Source: MMA Global Member Case Studies from http://www.mmaglobal.com/studies/seda-teens-unilever

REFERENCES

Anchar, B., & D'Incau, D. (2002a). Value-Added Services in Mobile Commerce: An Analytical Framework and Empirical Findings from a National Consumer Survey, *35th Hawaii International Conference on System Sciences (HICSS-37),* Hawaii, USA.

Anckar, B., & D'Incau, D. (2002b). Value creation in mobile commerce: Findings from a consumer survey . *Journal of Information Technology Theory and Application*, *4*(1), 43–64.

Balasubramanian, S., Peterson, R. A., & Jarvenpaa, S. L. (2002). Exploring the Implications of M-Commerce for Markets and Marketing . *Journal of the Academy of Marketing Science*, *30*(4), 348–361. doi:10.1177/009207002236910

Barnes, S. C., & Scornavacca, E. (2004). Mobile Marketing: The Role of Permission and Acceptance . *International Journal of Mobile Communications*, *2*(2), 128–139. doi:10.1504/IJMC.2004.004663

Barnes, S. J. (2002). Wireless Digital Advertising: Nature and Implications. *International Journal of Advertising*, *21*, 399–420.

Barwise, P., & Strong, C. (2002). Permission-based Mobile Advertising . *Journal of Interactive Marketing*, *16*(1), 14–24. doi:10.1002/dir.10000

Bathwal, B. (2008). Juniper Research Predicts Growth for ''Mobile Coupons''. Retrieved July 26, 2008, from http://www.tmcnet.com/voip/ip-communications/articles/26836-juniper-research-predicts-growth-mobile-coupons.htm

Bauer, H. H., Reichardt, T., Barnes, S. J., & Neumann, M. M. (2005). Driving Consumer Acceptance of Mobile Marketing: A Theoretical Framework and Empirical Study. *Journal of Electronic Commerce Research*, *6*(3), 181–192.

Carter, E. (2008). M-marketing and Generation Y African-American Mobile Consumers: The Issues and Opportunities. *International Journal of M-marketing*, *3*(1), 62–66.

Clow, K. E., & Baack, D. (2007). *Integrated Advertising, Promotion, and Marketing Communications*. New Jersey: Pearson Education Inc.

Dickinger, A., Haghirian, P., Murphy, J., & Scharl, A. (2004). A conceptual model and investigation of SMS Marketing. *Thirty-Seventh Hawaii International Conference on SystemSciences (HICSS-37)*, Hawaii, USA.

Doyle, S. (2001). Software Review: Using Short Message Services As a Marketing Tool. *Journal of Database Marketing*, *8*(3), 273–277. doi:10.1057/palgrave.jdm.3240043

Eisenhardt, K. M. (1989). Building Theories from Case Study Research. *Academy of Management Review*, *14*(4), 532–550. doi:10.2307/258557

Eisenhardt, K. M., & Graebner, M. E. (2007). Theory Building from Cases. *Academy of Management Journal*, *50*(1), 25–32.

Facchetti, A., Rangone, A., Renga, F. M., & Savoldelli, A. (2005). Mobile Marketing: An Analysis of Key Success Factors and the European Value Chain. *International Journal of Management and Decision Making*, *6*(1), 65–80. doi:10.1504/IJMDM.2005.005966

Fields, L. (2005). Location Based Services – a Marketer's Dream. Retrieved July 25, 2008, from http://mmaglobal.com/modules/article/view.article.php/21

Gopal, R., & Tripathi, A. (2006). Advertising via Wireless Networks. *International Journal of Mobile Communications*, *4*(1), 1–16. doi:10.1504/IJMC.2006.008194

Haghirian, P., Madlberger, M., & Tanuskova, A. (2005). Increasing Advertising Value of Mobile Marketing – An Empirical Study of Antecedents. *38th Hawaii International Conference on System Sciences*, Hawaii, USA.

Karjaluoto, H., Lehto, H., Leppäniemi, M., & Mustonen, T. (2007). Insights into the Implementation of Mobile Marketing Campaigns. *International Journal of Mobile Marketing*, *2*(2), 10–20.

Kavassalis, P., Spyropoulou, N., Drossos, D., Mitrokostas, E., Gikas, G., & Hatzistamatiou, A. (2003). Mobile Permission Marketing: Framing the Market Inquiry. *International Journal of Electronic Commerce*, *8*(1), 55–79.

Komulainen, H., Mainela, T., Sinisalo, J., Tähtinen, J., & Ulkuniemi, P. (2005). Models of Mobile Advertising Network. *E-business Review*, *5*, 95–98.

Kotler, P. (2000). *Marketing Management*. New Jersey: Prentice-Hall, Inc.

Kotler, P., & Armstrong, G. (2008). *Principles of Marketing*. New Jersey: Pearson Education Inc.

Leppäniemi, M., & Karjaluoto, H. (2005). Factors Influencing Consumers' Willingness to Accept Mobile Advertising. A Conceptual Model. *International Journal of Mobile Communications*, *3*(3), 197–213. doi:10.1504/IJMC.2005.006580

Leppäniemi, M., & Karjaluoto, H. (2008). M-marketing: From Marketing Strategy to M-marketing Campaign Implementation. *International Journal of M-marketing*, *3*(1), 50–61.

Leppäniemi, M., Karjaluoto, H., & Salo, J. (2004). The Success Factors of Mobile Advertising Value Chain. *E-Business Review*, *IV*, 93–97.

Leppäniemi, M., Karjaluoto, H., & Sinisalo, J. (2006). A Review of Mobile Marketing Research. *International Journal of Mobile Marketing*, *1*(1), 30–42.

M-marketing Association. (2007a). Introduction to Mobile Coupons. Retrieved July 25, 2008, from http://mmaglobal.com/mobilecoupons.pdf

M-marketing Association. (2007b). *M-marketing Industry Glossary*. Retrieved July 15, 2008, from http://www.mmaglobal.com/glossary.pdf

M-marketing Association. (2007c). M-marketing Sweepstakes and Promotions Guide, July 2007. Retrieved July 15, 2008, from http://www.mmaglobal.com/mobilepromotions.pdf

M-marketing Association. (2007d). Understanding M-marketing: Technology and Reach, May 2007. Retrieved July 25, 2008, from http://www.mmaglobal.com/mobilemarketing102.pdf

Marriott, L. (2006). Location-Based Services: Ready for the Main Stream? Retrieved July 25, 2008, from http://www.clickz.com/showPage.html?page=3623236

Mathieson, R. (2005). Get Your Game On: The Rise of Mobile Advergames. Retrieved July 25, 2008, from http://mmaglobal.com/modules/article/view.article.php/105

Merisavo, M., Vesanen, J., Arponen, A., Kajalo, S., & Raulas, M. (2006). The Effectiveness of Targeted Mobile Advertising in Selling Mobile Services: An Empirical Study. *International Journal of Mobile Communications*, *4*(2), 119–127.

Nielsen Mobile. (2008). Nielsen Provides Mobile Gaming Statistics, Retrieved July 25, 2008, from http://www.nielsenmobile.com/html/press%20releases/MobileGamingStats.html

Nysveen, H., Pedersen, P. E., & Thorbjornsen, H., and Pierre Berthon. (2005). Mobilizing the Brand: The Effects of Mobile Services on Brand Relationships and Main Channel Use. *Journal of Service Research*, *7*(3), 257–276. doi:10.1177/1094670504271151

Petty, R. D. (2003). Wireless Advertising Messaging: Legal Analysis and Public Policy Issues. *Journal of Public Policy & Marketing*, *22*(1), 71–82. doi:10.1509/jppm.22.1.71.17627

Pousttchi, K., & Wiedemann, D. (2006). A Contribution to Theory Building for Mobile Marketing: Categorizing Mobile Marketing Campaigns through Case Study Research. *5th International Conference on Mobile Business (ICMB 2006)*, Denmark, pp. 1-8.

Rettie, R., Grandcolas, U., & Deakins, B. (2005). Text Message Advertising: Response Rates and Branding Effects, *Journal of Targeting. Measurement and Analysis for Marketing*, *13*(4), 304–312. doi:10.1057/palgrave.jt.5740158

Scharl, A., Dickinger, A., & Murphy, J. (2005). Diffusion and Success Factors of Mobile Marketing. *Electronic Commerce Research and Applications*, *4*(2), 159–173. doi:10.1016/j.elerap.2004.10.006

Siau, K., Lim, E.-P., & Shen, Z. (2001). Mobile Commerce: Promises, Challenges, and Research Agenda. *Journal of Database Management*, *12*(3), 4–13.

Siau, K., & Shen, Z. (2003). Mobile Communications and Mobile Services. *International Journal of Mobile Communications*, *1*(1-2), 3–14. doi:10.1504/IJMC.2003.002457

Steinbock, D. (2006). The Missing Link: Why M-marketing Is Different. *International Journal of M-marketing*, *1*(1), 83–95.

Sullivan, M. G., & Drennan, J. (2002). Mobile Digital Technology: Emerging Issues for Marketing. *Journal of Database Marketing*, *10*(1), 9–23. doi:10.1057/palgrave.jdm.3240090

Sultan, F., & Rohm, A. (2005). The Coming Era of "Brand in the Hand" Marketing. *MIT Sloan Management Review*, *47*(1), 83–90.

Tsang, M. M., Ho, S. C., & Liang, T. P. (2004). Consumer Attitudes toward Mobile Advertising: An Empirical Study. *International Journal of Electronic Commerce*, *8*(3), 65–78.

Wiedemann, D. G., Haunstetter, T., & Pousttchi, K. (2008). Analyzing the Basic Elements of Mobile Viral Marketing. An Empirical Study. *7th International Conference on Mobile Business (ICMB 2008)*, Spain, pp. 75-85.

Wiedemann, D. G., & Pousttchi, K. (2008).Creating Successful Mobile Viral Marketing Strategies. *Handbook of Research on Mobile Multimedia*, 2nd Edition. Hershey, PA: Information Science Reference.

Wireless Intelligence. (2007). Subscriber statistics end Q4 2007 Report, Retrieved July 25, 2008, from http://www.gsmworld.com/news/statistics/pdf/gsma_stats_q4_07.pdf

Xu, D. J., Liao, S. L., & Li, Q. (2008). Combining Empirical Experimentation and Modeling Techniques: A Design Research Approach For Personalized Mobile Advertising Applications. *Decision Support Systems*, *44*, 710–724. doi:10.1016/j.dss.2007.10.002

Yunos, H. M., Gao, J. Z., & Shim, S. (2003). Wireless Advertising's Challenges and Opportunities. *Computer*, *36*(5), 30–37. doi:10.1109/MC.2003.1198234

KEY TERMS AND DEFINTIONS

Instant-Win: Instant-win can be considered as one of the promotional tactic in the context of "text and win" m-marketing campaigns. An instant-win mobile campaign invites consumers to send a text message to short code number in order to have the chance of instantly winning various campaign related prizes. When the consumer sends the message, a text message is returned to the consumer, so that consumers instantly learn if they win a prize.

Location-Based Mobile Marketing: In location based m-marketing, promotional information is delivered to the right consumer at the right place and time based on consumers' current location.

Mobile Campaign Tactics: Specific promotional tools (such as sweepstakes, instant win, mobile coupons, mobile web (WAP), and promotional content downloads, etc.) that are used to reach the mobile marketing campaign objective.

Mobile Coupon: A mobile coupon, which is sent and stored on a mobile phone, offers free samples, a rebate, a discount, etc., to a consumer that can be redeemed while making the purchase.

Mobile Marketing: M-marketing is the use of mobile communication channel to market and promote a brand, product, service or a message.

Mobile Sweepstakes: A mobile sweepstakes invites consumers to send a text message to a short code number in order to gain a chance to win various campaign related prizes and winners are drawn at the end of the campaign.

Promotional Downloadable Content: Downloadable content is usually used to attract consumers to attend m-marketing campaigns. The promotional downloadable content, which allows for brand exposure and create buzz about the marketing campaign, can be a song or ringtone, wall paper, video, game, or other application that is usually received on the handset through WAP link in a SMS message.

Provider-Driven (push) Mobile Campaign: Company invites consumers to the m-marketing campaign by delivering the invitation directly to the consumer's mobile device without request of the consumer. In push mobile campaign, the company initiates communication.

Text and Win Campaign: Text and Win campaign is a mobile messaging campaign. The customer is invited to send a text message to a short code number in order to attend the campaign to win campaign related prizes. Mobile sweepstakes, instant-win, mobile banking, mobile ticket, mobile coupons, mobile contests are some of the mobile campaign promotional tools that can be used to encourage attendance to the Text and Win campaigns.

User-Driven (pull) Mobile Campaign: The consumer requests to take part in the m-marketing campaign by sending a mobile text message from the company, typically in response to information on other media (such as television, magazines, newspapers, flyers or packaging).

Chapter 23
Interactive Customer Retention Management for Mobile Commerce

Dirk Möhlenbruch
Martin-Luther-University Halle-Wittenberg, Germany

Steffen Dölling
Martin-Luther-University Halle-Wittenberg, Germany

Falk Ritschel
Martin-Luther-University Halle-Wittenberg, Germany

ABSTRACT

Mobile Commerce is of growing interest for vendors and customers and with that its importance within the mix of marketing and distributing channels increases. This is of particular significance when customer retention and improved service are essential success factors. Because of their focus on customer integration the instruments of Web 2.0 offer new interactive possibilities for customer retention management. With a systematization of the new web-based applications linked to an evaluation of existing possibilities of customer retention this chapter offers a reasonable frame of reference for the utilization of Web 2.0 within the success chain of customer retention management in Mobile Commerce. Combined with existing studies of the acceptance of Web 2.0 recommendations for a successful timely order of introduction are offered to reach optimal diffusion and retention rates.

INTRODUCTION

After the significant upheavals brought by the Internet and all the surrounding technologies, the way we work and communicate is changing again. Within just a couple of years, more people will be accessing the Internet from mobile phones, personal digital assistants (PDAs), pagers, and a wide variety of informational devices than from stationary computers (Sadeh, 2002). That includes the necessity of Mobile Marketing and Mobile Commerce as applications and services which become accessible from Internet-enabled mobile devices. Mobile Commerce is defined by the buying and selling of goods and services through wireless handheld devices such as cellular telephones and personal digital as-

DOI: 10.4018/978-1-60566-074-5.ch023

sistants. Known as next-generation E-Commerce, M-Commerce enables users to access the Internet without needing to find a place to plug in.

The dramatically growing rate of this kind of mobile utensils is foiled by the marginal usage of Mobile Commerce applications and services. The reasons for these circumstances are different technical restrictions as well as the current user behavior. The consensus is that beside the ownership of mobile devices, three additional factors contribute to the establishment of Mobile Commerce. These are changes in the consumer funding model of mobile services, the establishment of fixed electronic commerce as a business phenomenon, and the accelerated service development ushered in by the Internet (May, 2001). With all these factors making strides of progress within the last few years, the breakthrough of Mobile Commerce is coming closer. In this context, the instruments and strategic applications of M-Commerce and marketing settings need to be discussed to serve the customer according to his specific needs. Scientific works on Electronic Commerce show the need of customer specific interaction and for the establishment of longtime relationships in electronic settings because they allow direct contact to each customer. As intensively discussed lately, the instruments of Web 2.0 make it possible for companies to arrange this kind of interactive business relationships. Hence their classification and usage within the specific stages of customer retention management is of growing meaning for the economic success of Electronic and Mobile Commerce. The management of customer retention comprises the analyses, planning, enforcement, and controlling of all arrangements concerning the actual customer, with the aim to keep the individual relationship alive.

The importance of customer retention rose distinctively within the last decade because of the amplification of global competition, ascending dynamics of markets and the change of consumer behavior in many industries (Boehm, 2008; Homburg, Bruhn, 2005). There is a large consensus in the scientific literature that customer retention is able to provide a considerable input to the economic success of a business operation. Customer retention achieves its contribution through the generation of repurchases, cross buying, recommendations and the improvement of tolerance to increased prices (Bruhn, 2001; Diller, Müllner, 1998). Several studies document that the acquisition of new customers is up to ten times more expensive than the retention of existing customers (Stolpmann, 2000; Gummesson, 1999). The consumer is the central authority in the processes of relationship marketing and obtains a strong position in the market, which is linked to a better negotiating point versus distributing companies. Additionally, the market power of customers is particularly strong on the Internet. The electronic market is characterized by high offer transparency, a comprehensive level of information, and a decrease of changing barriers for customers (Förster, Kreuz, 2002; Diller, 2001). The amount of information provided by vendors is increasing at a much faster pace than the volume the customer is demanding (Lihotzky, 2003). The resulting cognitive overload calls for more individualization of information (Kroeber-Riel, Weinberg, 2003). The combination of the rising market power of customers, the cost benefits of customer retention, and the necessity of individualized information search capabilities create an important need for customer retention management. It presents a plethora of possible solutions through the personalization of interaction and buying processes. Compared to the offline environment more opportunities for personalized marketing are offered as well as greater flexibility and convenience to the customer (Srinivasan et al., 2002; Wind, Rangaswamy, 2001).

The objective of this chapter is to analyze and systematize the applications of Web 2.0 within the context of customer retention management in Mobile Commerce considering its main focus functionality. To reach that goal the requirements of customer retention management will

Figure 1. Functional chain of customer retention

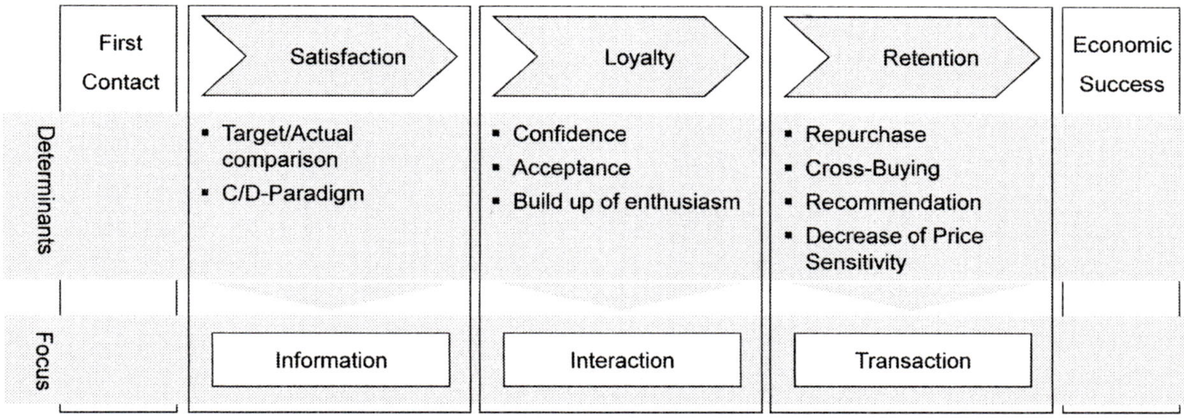

be introduced at first to be followed by the basic findings of Mobile Commerce and the introduction of Web 2.0 applications linked to their role within the instruments of M-Commerce. After this background part the main focus will be on the interfaces between an interactive design of customer relationships, through the instruments of Web 2.0, and the determinants of customer retention management. A specific classification will be displayed and evaluated to document the possible influence of Web 2.0 on customer retention in Mobile Commerce settings. Additionally a diffusion model for Web 2.0 applications is deduced to help M-Commerce providers foster the success of Web 2.0 in the context of customer retention management with an optimized sequence of implementation. The chapter will be concluded by the identification and discussion of future trends as well as further research areas, which are able to exploit the great potential of mobile Web 2.0 applications.

REQUIREMENTS FOR CUSTOMER RETENTION MANAGEMENT

The following segment will display the basic definitions and findings regarding customer retention management and its requirements. This introduction is made to define the major focus of each stage of the functional chain of customer retention management which is necessary for the following analysis of this chapter.

Customer relationships and their stabilization as well as the constitution and utilization of current and prospective potentials of these relations are able to achieve a high economic value in many companies. The importance of customer retention management for the financial success is undisputed and empirical validated (Anderson et al., 1997; Zeithaml et al., 1996). Seen as managerial objectives, links to the customers can be viewed as long-term investments. Customer relationships should be differentiated into dependence and solidarity (Bruhn, 2002; Bliemel, Eggert, 1998). The following comments focus on voluntary customer retention (solidarity), which centers on the development of preparations for increasing satisfaction and reliance. The goal is to maintain the business relationship through a sequence of transactions. Figure 1 shows the functional chain of customer retention with its three main phases, its determinants and the particular focus of every activity.

The first segment of the success chain contains the state of satisfaction. This phase is defined by a comparison (Homburg, Koschate, 2003). The evaluation of the targeted expectations against the actual perceived achievement (Confirmation-/Disconfirmation-Paradigm) results in an individual confidence level (Betz, Krafft, 2002). A pool of information forms the basis for the design of expectations in this concept. It will determine the development of demands and, in turn, the building of customer satisfaction (Capraro et al., 2003). The creation of a subjective information pool will be enhanced through individual communication because of the possibility to provide personalized information to the customer (Diller, 1996). The simplification and individualization of information processes develop a better knowledge level of customers regarding the company and its products and leads to an intensified connection between customer and vendor (Bauer et al., 1999). For these reasons, the constitution of customer satisfaction requires a focus on information.

The second segment of the success chain reflects customer loyalty with its determinants trust and acceptance (Homburg, Bruhn, 2005). Trust is a dimension oriented on the future. It is able to reduce the uncontrollable complexity and risk involved in a decision making process of the customer (Singh, Sirdeshmukh, 2000). Additionally, trust determines the specific expectations on a prospective action and is based primarily on knowledge (Kroeber-Riel, Weinberg, 2003). The development is dominated by the reliability of information and events (Bauer et al., 1999). The consumer willingness to change the purchasing company is smaller on this stage of the success chain. Hence, the exchange of knowledge regarding companies and products causes the customers to deal actively with their vendors. Trust in a relationship affects the commitment of consumers positively (Bauer et al., 1999; Bliemel, Eggert, 1998). The factor acceptance can be seen as consenting value judgment (Wiedmann, Frenzel, 2004; Kroeber-Riel, Weinberg, 2003). The resulting willingness of the customer to excuse minor mistakes of the company and to defend the vendor against recriminations from a third party is based on the exchanges deriving from the relationship of both parties (Bauer et al., 1999). The addressed factors trust and acceptance evolve through the communication. The goal of the second segment should be to integrate dialog supporting activities in the customer retention management of a company. Consumers associate positive attitudes with the vendor and the products as a result of customer loyalty. This might motivate them to offer ideas about product improvements or new products and services (Förster, Kreuz, 2002). The resulting permanent communication between customer and company produces interactivity which enhances trust and leads to commitment (Bauer et al., 1999; Diller, Müllner, 1998). Hence the second stage can be determined through the focus on interaction.

Customer retention is achieved in the third phase of the functional chain if the determinants of repurchase, cross-buying, recommendation and/or decreased price sensitivity are in place (Homburg, Bruhn, 2005). The theory of transaction costs can be used for the analysis of customer retention. It indicates that the inefficiencies of a market cause transaction costs (Williamson, 1991). Repurchases and cross-buying are able to reduce these transaction costs (Wirtz, Lihotzky, 2005; Bruhn, 2001), whereas recommendations are based on the exchange of information. Suggestions and reports about positive experiences result in psychological group memberships and transactions of information. The aspect of the decrease of price sensitivity through customer retention is documented by Homburg and Koschate. They provide evidence that retained customers are less negative about increasing prices than new customers (Homburg, Koschate, 2003). For this reason, all the displayed determinants that are part of this stage are characterized by material and immaterial exchange processes and require a focus on transactions.

WEB 2.0 APPLICATIONS IN MOBILE COMMERCE

The Internet and the electronic realization of business processes through the e-medium have obtained more and more economic relevance for an increasing number of vendors in the last years (Lihotzky, 2003). After the overreaching expectations in the 1990s and the subsequent reassessment of chances and potentials, the development of new business models in E-Commerce and eventually in M-Commerce is starting at the present. The following segment discusses the basic findings of Mobile Commerce identifying its critical areas and introducing the set of marketing instruments. Afterwards Web 2.0 is introduced and its applications will be assigned to the instruments of Mobile Commerce. This will deliver the foundation for the following discussion as possible framework for mobile interactive customer retention management and the subsequent diffusion model for Web 2.0 applications.

Basic Findings for Mobile Commerce

The broad spectrum of applications for mobile technologies possesses numerous possibilities for the electronic initiation and execution of business processes. There are chances for an interactive link between the real and the virtual world through the utilization of computer based mobile applications which allow the access to information from any location (Möhlenbruch, Schmieder, 2001). Mobile Commerce contains every business transaction which is executed with mobile devices (Turowski, Pousttchi, 2004). That makes the concept of wireless Electronic Commerce (Mobile Commerce) a new channel for every E-Commerce business to customer (B2C) and business to business (B2B) transaction (Balasubramanian et al., 2002). But to limit the wireless data transfer for mobile recipients to transaction processes only is, according to current scientific papers, insufficient (Lehner, 2002). The development from a medium focused on information to one with possibilities in the areas of communication and for proposals with strong transactional focus offers extensive possibilities to create economical networks (Möhlenbruch, Schmieder, 2001). Vendors are able to decrease transaction costs, involve customers actively in business processes, and establish personalized interaction. This allows the companies to make new markets accessible and to work more efficiently and effectively on existent business units (Möhlenbruch, Schmieder, 2002b).

The most characteristic attribute of Mobile Commerce is that there is no need of a special location. Mobile technologies allow users to retrieve or receive information and to effect transactions independent of their position (Nysveen et al., 2005; Siau et al., 2003). Additionally, the convenience factor becomes more important because of the simple usability of mobile devices (Kleijnen et al., 2004; Lehner, 2002). The immediate availability is another significant attribute of the medium. Users are not forced to actively establish a connection to the interaction network, but are always present through the mobile data and telephone net. This is possible in chronological compliance between reception and utilization of information as well as for precisely timed data transfers on a well defined occasion (Do, 2003). In 2007 there were about 3 billion mobile phones worldwide in use (EITO, 2008). This remarkable market penetration contains however about 480 million Internet enabled mobile devices which are used for mobile applications (Yu, 2008). That shows the customer behavior of mobile phone users and suggests restrictions as well as the great potential for M-Commerce applications.

The decreasing prices of hardware, the versatile possibilities of communication, and the comprehensive coverage of networks for mobile phones are the fundamentals of mobile data transfer. To exploit the existing potentials, an analysis of the technical restrictions needs to be done as well as an integration of innovative value added services for mobile settings. The highly specialized

Figure 2. Critical Areas of Mobile Internet Applications

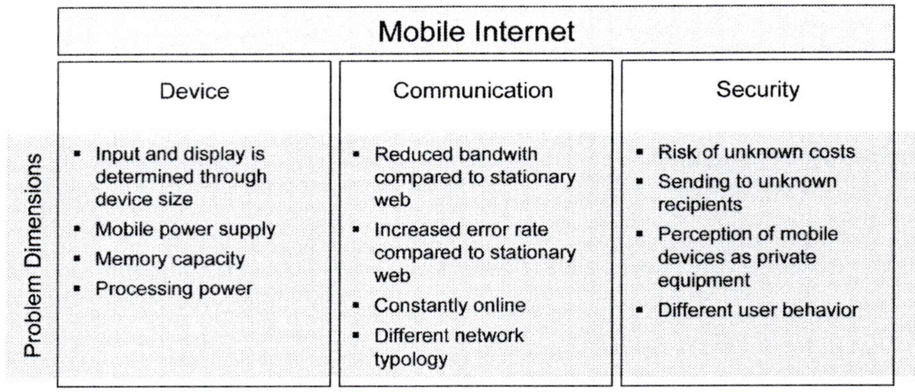

mobile devices can be classified in many ways. The foundations for the analysis in this chapter are the devices of the category handheld such as personal digital assistant (PDA), mobile phones and organizers. The applicability of these utensils is however limited by counter trend success factors. A very good mobile usability represented through good transport ability and long battery lifetime is foiled by slow performance of the central processing unit (CPU), semi optimal displays, and low memory capacity (Lehner, 2003; Nguyen et al., 2003). The different restrictions and critical areas for the success of mobile applications are shown and structured in figure 2.

In particular, the size of the displays is a limiting factor because the amount of viewable information is much smaller than on normal monitors. This makes it difficult for customers to read sites which are created for personal computers and forces them to scroll a lot to get to the required information. 23% of the mobile users criticize the navigation on cellular phones as too complicated and 40% consider the way websites are illustrated on small displays as disturbing (Agosti, Ferro, 2003). Besides this, the input devices are a second important factor. Common users are familiar with handling keyboard and mouse on the stationary web. There is nothing comparable for mobile web devices yet. PDAs, for example, possess only a pen for the communication between user and device, and the input process via character recognition leads to a significant longer handling time (Turowski, Pousttchi, 2004). There are further determining factors for strategic utilization like quality of mobile transfer connection, the likelihood of smaller memory capacity and slower CPU, the convergence of operating systems and the energy consumption (Agosti, Ferro, 2003). In particular, for the energy supply discrepancies between the demands of color displays, the necessary CPU performance and the development of batteries exist.

Scientific studies show that there will be a huge potential for mobile applications to support companies in the areas of customer retention management (Möhlenbruch, Schmieder, 2001). Identified through the explicit assignment of an individual call number, users are able to decide which information they get. This results in a broad security potential for business transactions. Consumers are able to authenticate themselves via subscriber identity module (SIM) or smart cards, thus providing a much better way than via common Internet devices (Siau et al., 2003). Beside this, the easy possibility to locate a user offers potentials in the areas of multichannel marketing. Vendors are able to design target group-specific information and service offers and, with that, build up sustainable customer relationships. Mobile Commerce is defined as mobile channel

Figure 3. Instruments of Mobile Commerce

	Product Line	Price	Communication	Front-End
Characteristics	▪ Dimensions of the product line ▪ Rate of branded goods ▪ Availability	▪ Extent of prices ▪ Amount of goods in different price levels ▪ Special offers ▪ Price differentiation	▪ Message ▪ Communication material ▪ Communication budget	▪ Interface to customers ▪ Linkage ▪ Design of front-end
Customer Retention Management	▪ Combined offerings ▪ Exclusive product offers ▪ Individual mass production	▪ E-Coupons ▪ Discount systems ▪ Bonus systems	▪ Newsletter and Permission Marketing ▪ One-to-One Communication ▪ Viral and cooperative Marketing ▪ Customer cards	▪ Customized websites ▪ Virtual communities ▪ Online-Event Marketing

for Electronic Commerce transactions (Lehner, 2002). Hence, the instruments of demand management of classic retailers can be used for the identification of Mobile Commerce instruments relevant to Electronic Commerce. For example, consider a retail company which is divided into the areas of product, personnel, place, promotion, price, and sales room (Müller-Hagedorn, 2005). The instruments personnel, place, and sales room are less relevant because in an electronic business physical presence is not necessary. The Front-End, the interface to the customer demands, is more important instead (Wilke et al., 2005; Wilke, 2002). The resulting systematization of instruments for Mobile Commerce is shown in figure 3. It is extended with specific attributes of each instrument and the assignable tools of customer retention management.

The instrument product line contains every decision regarding the determination of distributable goods which are going to be purchased by the customer (Möhlenbruch, 2006). It defines the dimensions of the product line and declares the availability as well as the branding of different goods. In this process, the customer retention management is able to manipulate the economic success of a company positively through combined offers, the preparation of exclusive products, or individual mass production. Within the pricing, policy decisions about the actual extent of prices and other different determinants of purchase decisions such as range of special offers and price differentiation have to be made (Müller-Hagedorn, 2005). Principally electronic coupons and discount systems have a special meaning for customer retention management of Mobile Commerce. Communication as an instrument of retail marketing contains decisions about the necessary messages and the reasonable utilization of communication material and channels (Barth et al., 2002). In addition, the budgets for the different communicational arrangements need to be fixed. Activities like newsletter and permission marketing belong to the instrument communication as well as one-to-one communication and viral marketing. The direct interface to the customer in Mobile Commerce settings is defined by the instrument Front-End. Besides, the design of the user interface, the linkage, and the strategic design of the website will be set. Especially relevant for customer retention management are the possible customization of websites and the social effects of virtual communities (Wirtz, Lihotzky, 2005).

A prioritization of the different instruments of Mobile Commerce according to their potential is necessary in order to get a reasonable systematization for the current paper. The instruments promotion and Front-End possess the most significance. Both areas are applied through current Mobile Commerce applications and exhibit huge potentials for interactive technologies. The importance of the instrument promotion results mostly from the way customers use mobile devices (Jaokar, 2006). The scope for design of the Front-End is an essential aspect as well. Vendors have the chance to influence the exposure time, termination rate, and connection costs through the manipulation of the interface between device and customer. Additionally, the convenience factor plays an important role because the design of operational functions is able to increase the acceptance of new applications and media (Lehner, 2002). Both instruments are utilized in current M-Commerce applications. They are able to expose significant potentials in the future in particular when used strategically, for example within the context of customer retention management (Chang, Fang, 2003). Decisions about the product line are not quite possible yet, because of the existing marginal volume of interaction. There are options to monitor page impressions and analyze them statistically. But the technical possibilities for evaluation and recommendation are not used within the context of M-Commerce (Möhlenbruch, Schmieder, 2002a). The same applies for the instrument price which is currently marginal used in mobile applications. This aspect contains potential, particularly within the context of price negotiations, collaborative buying, auctions, and price differentiation which might be developed through the new interactive applications of Web 2.0 (Wamser, Wilfert, 2002).

Web 2.0 Applications in Mobile Commerce

The Internet and the associated technologies go through a rapid development. Consequently, the portfolio of technical resources for an effective and efficient ascertainment, archiving, processing, and transmission of data, information and knowledge is continuously growing. The development over the last years is often summarized under the term "Web 2.0", which is not precisely defined until now. Most of the current attempts to define Web 2.0 go back to a definition of O'Reilly, which understands the term more as an approach than a technology (O'Reilly, 2005b). There are characteristic requirements for applications belonging to the term of Web 2.0 (O'Reilly, 2005a):

- The application is net based; only a web browser is essential.
- The content is not static but is going to be dynamically generated. It will change depending on the input of the user.
- There are possibilities for users to generate content by themselves.
- Role and authorization models enable the differentiation of who is enabled to view which content and who is allowed to edit the content.
- Users own their data and are able to edit it by themselves. There are possibilities for personalized content and layout.
- Users are able to comment on articles by other people and communicate among each other.
- A strong community feeling is going to develop between users.

These requirements are described as the paradigms of Web 2.0. Economic and technical principles are as important for Web 2.0 applications as socio-economic, psychological and legal aspects. The most characteristic attribute is the participation of users (Graham, 2005).

There have been several empirical studies within the last years to validate the theoretical basis regarding the acceptance and user behaviour of Web 2.0. An empirical analysis of the German broadcasting channels ARD and ZDF shows for example that users mainly utilize passive components of Web 2.0 such as Wikis or Pod- and VideoCasts. If they are asked to participate, the vocal point lies on the different kinds of communities which help customers to gather information and to meet new people in private or in their occupational background (Fisch, Gscheidle, 2008). This strong focus on specific Web 2.0 ap-

Figure 4. Applications of Web 2.0 assigned to the Instruments of Mobile Commerce

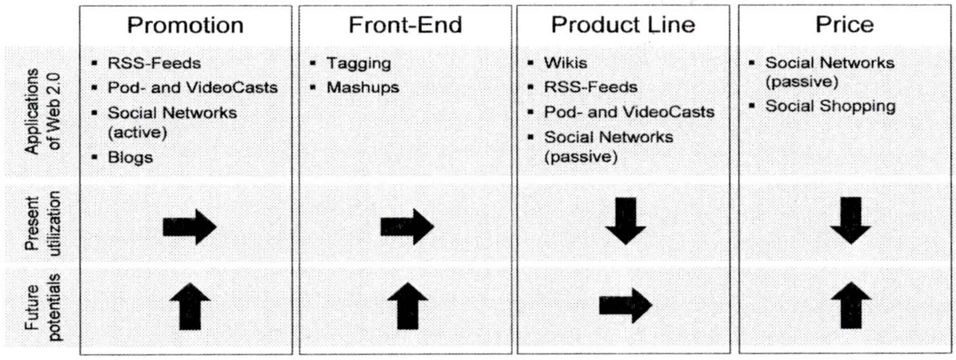

plications is reflected by the investments taken by companies to enhance their interactive profile. A report published by McKinsey shows for instance, that such investments are taken mainly in the areas of web services, collective intelligence, and networking (peer-to-peer and social) which suggests, that the practical understanding of Web 2.0 points in the same direction as the user demands. More interestingly, the retail industry is about to carry out the biggest investments of all industries in Web 2.0 technologies within the next three years (Bughin, Manyika, 2007). This suggests the strong emphasis of the new interactive technologies in this industry area for the channels of electronic and mobile retailing as well as for the stationary part establishing a multichannel strategy.

The applications of Web 2.0 are a result of the previously shown paradigms. It is possible to assign them to the different instruments of M-Commerce according to their main focus and the prioritization argued before. Additionally, the potentials and future perspectives of Web 2.0 applications within the context of Mobile Commerce are exposed under consideration of the empirical findings regarding the existing level of utilization (Figure 4).

The most important applications are displayed within the specific scopes and classified according to their main functional tasks. There is no classification of Web 2.0 applications like Social News, Social Bookmarking, and Internet Based Desktop Applications because they are not directly linked to one of the instruments of Mobile Commerce.

Promotion

Web 2.0 applications get used already within the instruments promotion and front-end of Mobile Commerce. There are enormous potentials especially in the area of communication. Thus, promotion will be of increasing interest within the next years.

The main focus of mobile data applications is on communication. This gives reason for the prominent significance of Web 2.0 applications for this specific instrument and the present utilization within Mobile Marketing Management (Drüner et al., 2007). RSS-Feeds for instance support directly the permission marketing and may be seen as a more goal oriented enhancement of classical short message service (SMS) subscription (Möhlenbruch, Schmieder, 2002b). Feeds are unidirectional and establish an innovative and individual push channel to the customer (Beyer, 2005). The customer requests specific feeds individually unlike classical push channels. The permanent network connection of mobile devices works positive for the timely reception of messages and the following information processed by the customer (Wamser, Wilfert, 2002). The recipient accepts the information actively which makes it possible for companies to develop perfectly assignable customer profiles identified through the SIM card of mobile devices. Vendors will be enabled to collect customer specific information about products and to use them for individual mass production (Förster, Kreuz, 2002). The resulting focus point for the instrument product line will be discussed in the next paragraphs.

Similar customer oriented application areas can be deduced for Pod- and VideoCasts in which different media become important. Pod- or VideoCasts are audio and video files which are provided with RSS-Feeds especially for mobile devices (Faltin, 2006). The files are placed on a web server by the vendors and get replenished at regular intervals. Users are able to retrieve the content singularly or to subscribe for updates on an accepted basis. This enables them to play the files independent of time and place with the requirement of adequate software which is one of the characteristic attributes for mobile business (Lehner, 2002). The media files, in combination with the push channel of RSS-Feeds, give Mobile Commerce vendors the chance to present additional information in an appealing way and to establish user preference profiles simultaneously. The technical possibilities of mobile devices in terms of required data volume and quality of playback of multimedia content are satisfying at present (Do, 2003). Through new and enhanced technologies, there will be additional possibilities of a mobile one-to-one communication enabling a specific transfer of individualized messages.

Social Networks might establish a great potential for the instrument promotion. They allow vendors to influence customers actively if company representatives provide purposeful information within the communities. Such strategies have to be applied very carefully as in case of discovery very negative effects for companies and the community occur (Wiedmann, Langner, 2004). Additionally, companies have to consider that Social Networks via mobile devices offer rather restricted dialog because of the complex possibilities to create input. At this juncture, customers restrict themselves to demonstrations of interests. This has to be considered for the strategic utilization of this application. But the potential is still increasing and the technical possibilities to create input will get better.

Blogs (Weblogs) are written by an author as some kind of online diary which contains contributions to selected topics and is updated on a regular basis (Wolff, 2006). They are basically freely accessible and users are able to comment on the contributions. If a blogger wants to refer to another Blog he or she can place a link. This link will be shown at the end of the article which makes it possible to retrace the origin and the development. Additionally, the linked contributions will hold an evidence of the link. This functionality is known under the name trackback (Groß, Hülsbusch, 2004). The functionalities of comments, the possibility of trackback in connection with tagging blogs, are the ideal stage for fast and broad communication. Vendors are enabled to draw conclusions from the monitoring of Blogs and comments but the technical restrictions of Mobile Commerce define the current utilization of this Web 2.0 application. This is so because a broad range of text is difficult to read on the small monitors of mobile devices and the possibilities to create input are restricted. Nevertheless, the potential for communication of this application should not be underestimated because the social and interactive attributes of Blogs lead to a potential customer oriented communication for Mobile Commerce.

Front End

Beside the applications which belong to the instrument promotion, there is a focus on the front-end as interface to the customer. Web 2.0 applications in this area are able to increase the acceptance of Mobile Commerce business models. The technology Tagging enables users to individually describe content. Customers are able to use individual keywords (tags) for every arbitrary resource. For that a certain amount of keywords is defined and the user is completely free to choose his or her categorization (Smolnik, Riempp, 2006). The content will be described through different subjectively chosen keywords which increases the user friendliness significantly. The resulting cross linking achieves often better and more extensive possibilities for individual information than search engines do (Bächle, 2006). It is obvious that this application is directly assigned to the instrument front-end because with Tagging, the design of the interface between company and customer can be advanced. Given the size of the display and the linked restrictions of a mobile illustration of websites, this technology enables the user to search for information more efficiently especially from a time standpoint. Customers can mark relevant content and are able to get back to the information corresponding to their individual interest and independent of the location. The tags get accumulated to tag-clouds according to the number of clicks. This can be used as a research instrument for companies which are able to monitor trends and preferences.

Mashups are applications which utilize the concept of open interfaces. They are web based services which generate customer value through the combination of different content and data of multiple suppliers. The open interfaces of specific suppliers are used and the essential data gets merged in a new application. Hence the interface of Mobile Commerce vendors to the customer receives an upvaluation and new business models become possible (Göhring et al., 2006). Mashups are uncommon in M-Commerce because of the technical restrictions of mobile devices at present which might change because of the great potential of these kinds of applications.

Product Line

The instruments product line and price are addressed only marginally through Web 2.0 applications in Mobile Commerce at present. This makes it possible to expect high future potentials which will be discussed under estimation of the technical premises in the following text.

Social Networks are enhancements of virtual communities which show the relationships between users as a virtual relationship network (Hippner, 2006). Users are able to design their personal profile and participate in discussion groups which reflect their own areas of interest. There are significant restrictions as a result of the technical limits of mobile devices, regarding in particular the viewable quantity of information. Even so, there are possibilities to optimize the product line through passive monitoring. Companies can learn about new ideas for product development, discover sales arguments and the opinion of the community regarding specific products.

RSS-Feeds provide interesting possibilities for Mobile Commerce companies to optimize the instrument product line. Vendors are able to get an individual view of their customers and use it for strategic decisions. For that, it is possible to present specific information about the availability of products and suggestions for additional and new goods available. There is a potential to decrease inventory, to enable cross selling transactions, and to use the gathered information within multichannel strategies.

Wikis are open cooperative authoring tools for websites which can be edited by every user. Wikis are also collaborative products of many authors with content not directly linked to one specific user. They represent tools for authoring, discussion, and collaboration (Groß, Hülsbusch, 2004) which are able to provide additional value for the instrument product line of Mobile Commerce vendors. With that application, it is possible to point out the importance of products and to offer additional information such as operating manuals and field reports generated by customers. This instrument for E-Commerce vendors plays an inferior role for mobile devices at present. The reasons for that are especially the display sizes, the methods to create input, and the short contact time on mobile sites. Wikis have great potential because user generated content enables companies to integrate value added services into their service offering (Battenberg, 2002).

Price

Beside these application areas, Web 2.0 is of certain importance for the instrument price. Passively monitored Social Networks are able to influence the pricing of vendors because significant pricing information can be raised through their monitoring. This unbiased information has significant consequence for pricing strategies because there is no risk of falsification. The challenge for that strategy is that pricing is not a primary topic for the members of a community. The current situation shows that the consumer adoption of mobile devices is not ready to serve this instrument (Lehner, 2002). But its significance will grow with the increasing utilization of mobile devices for electronic transactions.

Additionally, Social Shopping is of further interest for pricing. This application means that buying transactions get arranged and accomplished by Social Networks. Social Shopping uses group dynamic processes where big groups of users recommend products or buy them together (Komus, 2006). This application of Web 2.0 focuses directly on the field of pricing. Customers are able to discuss prices, realize acquisitions, and use collaborative buying processes as a bargain tool. This application also plays a secondary role in the present field of Mobile Commerce which should change because a sustainable customer orientation is vital for every instrument of Mobile Commerce and pricing is one of the most important instruments in customer oriented strategies. One of the current application areas might be the regional price differentiation due to the localization possibilities of mobile devices. The traditional viewpoints regarding consumer location, retail competition, and pricing decisions might have to be revised when consumers use mobile technologies (Balasubramanian, 2002).

MOBILE INTERACTIVE CUSTOMER RETENTION MANAGEMENT

The findings of the previous chapters feature a reasonable systematization of the applications of Web 2.0 within the context of customer retention management. For that the assignment of the web based applications for the specific instruments of Mobile Commerce will be combined with the requirements of the different stages of the success chain of customer retention management. Additionally a diffusion model for Mobile Web 2.0 applications based on the acceptance of Web 2.0 and the necessity of a sequential view on customer retention management is introduced.

The main focus at the phase of customer satisfaction is information, which is supported mainly through Wikis, RSS-Feeds, Pod- and VideoCasts. RSS-Feeds deliver content on the request of the user and record preferences. Moreover, RSS-Feeds as well as Video- and PodCasts support the increase of customer satisfaction within the promotion because companies can use them as an effective channel for push communication. Wikis are potentially able to generate user driven information. That can be used to cause a positive change of attitude in the customers mind. Requirements would be a further reduction of complexity for data input on mobile devices and the enhancement of quantity regarding the content. Tagging is another information application drawn as a component within the Front-End of M-Commerce solutions. It allows vendors to establish a customer-driven search and hence to achieve a significantly higher transparency and better quality perception from customers.

The second stage of the success chain of customer retention is customer loyalty with its major focus being interaction, supported by numerous instruments of Web 2.0. These applications are not the standard for Mobile Commerce vendors but might have great potential in the future. Social Networks, for example, are interaction oriented applications which are able to enhance the loyalty of customers (Förster, Kreuz, 2002). A passive monitoring of discussion forums will allow the vendors to get valuable information about preferences and sentiments which might be useful for product line design or pricing strategies. The resulting precise activities are able to improve the trust and acceptance of customers and enable an active impact on these communities, which in turn allows the achievement of a positive picture of the company and the products offered. The risk of a reversion of possible positive effects is high because the recognition of an active influence leads to the loss of authenticity in the perception of customers (Göhring et al., 2006). Beyond this, Social Shopping establishes pricing potential through the interaction of users which is able to increase the loyalty of the customers. In particular, this arises during a user driven valuation of prices, which contains direct statements about the willingness to pay. A purposeful utilization of this user driven information creates trust and acceptance. Because of the current marginal utilization of this M-Commerce instrument vendors should use the present technical possibilities of mobile devices and establish multichannel marketing strategies. The integration of the interactive components of Weblogs allows an active communication regarding company related content, which increases customer loyalty. This is even more applicable in Mashups as a combination of services, which in some specifications support interaction and raise the acceptance of websites thus affecting customer loyalty positively.

A direct means of customer retention can be accomplished with Social Shopping, which focuses on the execution of transactions. The tendency to repurchase and cross-buy is increasing due to the collaborative buying processes taking place. Direct recommendations are much easier for the users to trust (Duan et al, 2008). Similar cognitions apply for particular types of Mashups. Value added services for the customer might occur here if the vendor offers relevant information which enables the successful determination of the buying process (Göhring et al., 2006). It can be ascertained that all of the displayed applications of Web 2.0 are able to affect customer retention because of their influence on satisfaction and loyalty. The characteristic of the success chain as a process model of customer retention management also shows the effects of the displayed applications of Web 2.0. Figure 5 documents the possible applications of Web 2.0 as instruments of M-Commerce as well as their connection to the different stages of the success chain of customer retention.

The applications of Web 2.0 have a direct impact on the three main phases of the success chain of customer retention. The integrated consideration of the applications is necessary because the support of the functional chain at every stage achieves an improvement in customer retention management (Garcia, Rennhak, 2006). The consistent adoption of the instruments of Web 2.0 within the framework of customer retention management promises direct positive effects on the relationship between companies and customers. The results of this chapter show that interactive customer retention management is a meaningful complement for existing instruments of Mobile Commerce, but they illustrate as well the technical restrictions. With the decrease of these restrictions, the potentials and the utilization of Web 2.0 within the context of Mobile Commerce will grow significantly.

To help Mobile Commerce providers adopt the success of Web 2.0 in the context of customer retention management the sequence of implementation could be significant. The previously shown differences of the customer acceptance of Web 2.0 applications and the sequential order in

Figure 5. Applications of Web 2.0 in Customer Retention Management of M-Commerce

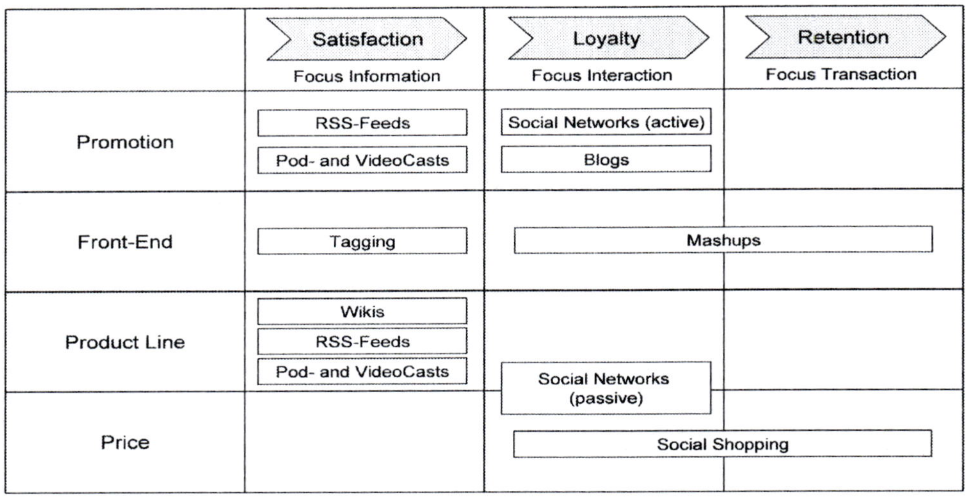

the success chain of customer retention suggest an order of implementation revealed in figure 6. It shows a probable diffusion model for Web 2.0 application in the context of customer retention management. Each phase should bring the focus to specific applications and establish them before moving to the next stage. Technical oriented applications such as Tagging and Mashups are not discussed because they cannot be linked directly to the customer acceptance studies displayed earlier in this chapter.

The research about the acceptance of Web 2.0 instruments showed that users prefer especially the passive utilization of Web 2.0 in the first place. That corresponds with the strong focus on information as first step of customer retention. Consequently, to provide a good start of customer retention management in M-Commerce settings it might be reasonable to begin with accepted instruments which allow the customer the passive retrieval of information. In this range, Wikis can be a good set off followed by RSS-Feeds which allow customers to subscribe relevant content. Customers could get used to the company's environment and the first active contributors would be recruited. This can be followed by applications that are able to foster interaction between the company and customers as well as between customers. The second stage of customer retention management shows, that Blogs and later Social Networks might be able to provide this kind of service for Mobile Commerce. Given the fact that these Web 2.0 applications need a certain amount of users to be relevant, it might be reasonable to start them a certain amount of time after the applications of the first stage. When the interaction is established between the relevant parties, the focus goes on to transaction as the last stage in the success chain of customer retention management. The previous research showed that only Social Shopping is able to create added value in this stage. If the critical amount of users is reached and interaction is established, suppliers might foster the estimated transaction with this application. It is important to emphasize that there should be an ongoing utilization of the instruments of the previous stages. Companies which implement Web 2.0 strategies to their Mobile Marketing should enhance their portfolio from easy adapted instruments like Wikis through Blogs, to the completion with Social Shopping.

Figure 6. Diffusion Model for Mobile Web 2.0 Applications

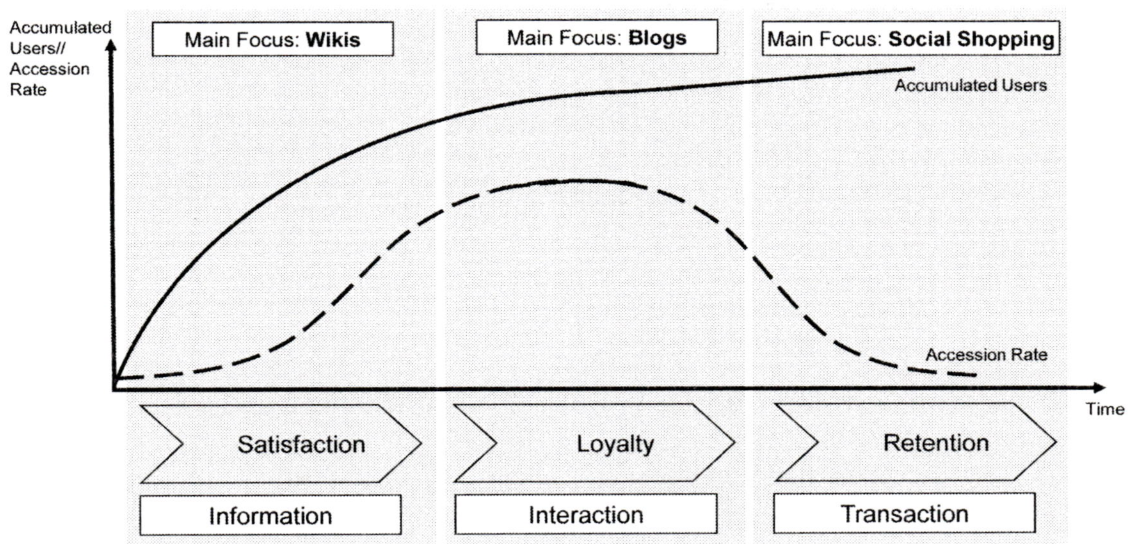

CONCLUSION

Companies intending to use Web 2.0 applications successfully within their customer retention strategies have to deal with the competitive conditions of Web 2.0 and should conduct a systematic and sequential integration in their specific customer retention management. For that, they have to deal with restrictions resulting from mobile devices. That leads to the need of a well-founded analysis of the current user behavior.

The approaches shown here will penetrate the market if they are able to prove their economic value by generating advantages through the increased involvement of customers. The adoption of these new applications promises, besides the attainable emotional commitment, the possibility to effectively differentiate a company from the competition since the customer will most likely use the specific interactive instruments in a limited way. For this reason, vendors have to answer the question of how to generate the critical mass of users which will enable them to produce a competitive advantage through the development of entry barriers. There should be further research to find the right incentives for customers to enhance their participation, which could present an interesting intersection with the field of customer integration. Additionally, research on the order of implementation could be conducted to find specific recommendations for the practical utilization of Web 2.0 instruments in Mobile Commerce.

Because of the present marginal use of mobile devices for economic transactions, the M-Commerce should find its way within multichannel strategies for companies. This shows interesting perspectives for marketing and customer retention management. Vendors might be able to offer their customers different ways to communicate individually. The information which results from this interaction can be used to establish customer and request-oriented transactions that influence the different stages of customer retention positively.

The conclusions displayed create a demand for advanced research. The empirical verification of coherences between the specific focuses of customer retention management and the different applications of Web 2.0 will be of particular interest. Furthermore, the potential of the visualized applications within specific M-Commerce instruments could be firmed up and validated empirically. The possibilities of a utilization of Web 2.0 present a reasonable supplement to existing instruments of customer retention in Mobile Commerce. It remains questionable whether M-Commerce vendors are able to persuade their customers to satisfactory participation during the initiation and processing of business transactions, and whether the decrease of technical restrictions will help to initiate more of the social interactive applications shown in this chapter.

REFERENCES

Agosti, M., & Ferro, N. (2003). Managing the Interactions between Handheld Devices, Mobile Applications, and Users. In Lim, E.-P., & Siau, K. (Eds.), *Advances in Mobile Commerce Technologies* (pp. 205-234). Hershey, PA: Idea Group Publishing.

Anderson, E. W., Fornell, C., & Rust, R. T. (1997). Customer Satisfaction, Productivity, and Profitability: Differences between Goods and Services. *Marketing Science*, *16*(2), 129–145. doi:10.1287/mksc.16.2.129

Bächle, M. (2006). Social Software. *Informatik Spektrum*, *29*(2), 121–124. doi:10.1007/s00287-006-0063-2

Balasubramanian, S., Peterson, R. A., & Jarvenpaa, S. L. (2002). Exploring the Implications of m-Commerce for Markets and Marketing. *Journal of the Academy of Marketing Science*, *30*(4), 348–361. doi:10.1177/009207002236910

Barth, K., Hartmann, M., & Schröder, H. (2007). *Betriebswirtschaftslehre des Handels*. 6th Edition, Wiesbaden, Gabler.

Battenberg, A. (2002). Mobile Business mit Communitys. In Teichmann, R., & Lehner, F. (Eds.), *Mobile Commerce* (pp. 130-143). Berlin, Springer.

Bauer, H., Grether, M., & Leach, M. (1999). Relationship Marketing im Internet. *GfK Jahrbuch der Absatz und Verbrauchsforschung*, *45*(3), 284–302.

Betz, J., & Krafft, M. (2002). Die Wirkung des E-Commerce auf die Kundenzufriedenheit und Kundenbindung. *wiss. Schriftenreihe des Zentrums für Marktorientierte Unternehmensführung*, No. 23, Vallendar.

Beyer, T. (2005). RSS-Feeds im Internet. *CLB Chemie im Labor und Biotechnik*, *56*(5), 144–148.

Bidgoli, H. (2002). *Electronic Commerce – Principles and Practice*. San Diego: Academic Press.

Bliemel, W. F., & Eggert, A. (1998). Kundenbindung – die neue Sollstrategie. *Marketing ZFP*, *20*(1), 37–46.

Boehm, M. (2008). Determining the Impact of Internet Channel Use on a Customer's Lifetime. *Journal of Interactive Marketing*, *22*(3), 2–22. doi:10.1002/dir.20114

Bruhn, M. (2001). *Relationship Marketing*. München: Vahlen.

Bruhn, M. (2002). *Relationship Marketing: Management of Customer Relationships*. Harlow: Prentice Hall.

Capraro, A. J., Broniarczyk, S., & Srivastava, R. K. (2003). Factors Influencing the Likelihood of Customer Defection: The Role of Customer Knowledge. *Journal of the Academy of Marketing Science*, *31*(2), 164–175. doi:10.1177/0092070302250900

Chang, S. S., & Fang, X. (2003). Mobile Commerce and Usability. In Lim, E.-P., & Siau, K. (Eds.), *Advances in Mobile Commerce Technologies* (pp. 235-257). Hershe, PA: Idea Group Publishing.

Diller, H. (1996). Kundenbindung als Marketingziel. *Marketing ZFP, 18*(2), 81–94.

Diller, H. (2001). Die Erfolgsaussichten des Beziehungsmarketing im Internet. In Eggert, A., & Fassott, G. (Eds.), *Electronic Customer Relationship Management - Management der Kundenbeziehungen im Internet-Zeitalter* (pp. 66-85). Stuttgart: Schäffer-Poeschel.

Diller, H., & Müllner, M. (1998). Kundenbindungsmanagement. In Meyer, A. (Ed.), *Handbuch Dienstleistungsmarketing* (pp. 1220-1240). Stuttgart: Schäffer-Poeschel.

Do, V. T. (2003). Mobile E-Commerce on Mobile Phones. In Lim, E.-P., & Siau, K. (Eds.), *Advances in Mobile Commerce Technologies* (pp. 19-43). Hershey, PA: Idea Group Publishing.

Drüner, M., Rattay, R., & Kröger, S. (2007). Web 2.0 - Schneller mehr wissen: Informationsvorsprung durch Nutzung neuer Rückkanäle. *Information Management and Consulting*, 35-42.

Duan, W., Gu, B., & Whinston, A. B. (2008). The dynamics of online word-of-mouth and product sales – An empirical investigation of the movie industry. *Journal of Retailing, 84*(2), 233–242. doi:10.1016/j.jretai.2008.04.005

EITO – European Information Technology Observatory. (2008). *EITO ICT market overview*. Bitkom Research GmbH.

Faltin, C. (2006). Vom Web 2.0 zum Marketing 2.0. *Marketing Journal, 39*(Sonderheft Web 2.0), 10-13.

Fisch, M., & Gscheidle, C. (2008). Mitmachnetz Web 2.0: Rege Beteiligung nur in Communitys. *Media Perspektiven, 12*(7), 356–564.

Förster, A., & Kreuz, P. (2002). *Offensives Marketing im E-Business*. Berlin: Springer.

Funk, J. L. (2004). *Mobile Disruption. The Technologies and Applications Driving the Mobile Internet*. New York: Wiley Publishing.

Garcia, A. G., & Rennhak, C. (2006). Kundenbindung – Grundlagen und Begrifflichkeiten. In Rennhak, C. (Ed.), *Herausforderung Kundenbindung* (pp. 3-14). Wiesbaden, Gabler.

Göhring, M., Happ, S., & Müller, T. (2006). Web 2.0 im Kundenmanagement. In Hildebrand, K., & Hofmann, J. (Eds.), *Social Software, Praxis der Wirtschaftsinformatik, 43*(252), 54-65.

Graham, P. (2005). *Web 2.0*. Retrieved January 30, 2007, from http://www.paulgraham.com/web20.html

Groß, M., & Hülsbusch, W. (2004). Weblogs und Wikis – eine neue Medienrevolution? *Wissensmanagement, 6*(8), 44–48.

Gummesson, E. (1999). *Total Relationship Marketing. Rethinking Marketing Management from 4Ps to 30Rs*. Oxford: Butterworth-Heinemann.

Hippner, H. (2006). Bedeutung, Anwendungen und Ensatzpotentiale von Social Software. In Hildebrand, K., & Hofmann, J. (Eds.), *Social Software, Praxis der Wirtschaftsinformatik, 43*(252), 6-16.

Homburg, C., & Bruhn, M. (2005). Kundenbindungsmanagement – Eine Einführung in die theoretischen und praktischen Problemstellungen. In Homburg, C., & Bruhn, M. (Eds.), *Handbuch Kundenbindungsmanagement* (pp. 3-37). Wiesbaden, Gabler.

Homburg, C., & Koschate, N. (2003). Kann Kundenzufriedenheit negative Reaktionen auf Preiserhöhungen abschwächen? *Die Betriebswirtschaft, 63*(6), 619–634.

Jaokar, A. (2007). *Mobile Web 2.0. Kollektive Intelligenz durch Restricted Devices*, Retrieved June 06, 2007, from http://javamagazin.de/itr/news/psecom,id,30359,nodeid,10.htm

Kleijnen, M. H. P., de Ruyter, J. C., & Wetzels, M. G. M. (2004). Consumer Adoption of Wireless Services: Discovering the Rules, While Playing the Game. *Journal of Interactive Marketing, 18*(2), 51–61. doi:10.1002/dir.20002

Komus, A. (2006). Social Software als organisatorisches Phänomen - Einsatzmöglichkeiten in Unternehmen. In Hildebrand, K., & Hofmann, J. (Eds.), *Social Software, Praxis der Wirtschaftsinformatik, 43*(252), 36-44.

Kroeber-Riel, W., & Weinberg, P. (2003). *Konsumentenverhalten.* 8th Edition. München: Vahlen.

Laudon, K. C., & Traver, C. G. (2002). *E-Commerce*. Boston: Addison Wesley.

Lehner, F. (2002). Einführung und Motivation. In Teichmann, R., & Lehner, F. (Eds.), *Mobile Commerce* (pp. 3-28). Berlin: Springer.

Lehner, F. (2003). *Mobile und drahtlose Informationssysteme. Technologien, Anwendungen und Märkte*. Berlin: Springer.

Lihotzky, N. (2003). *Kundenbindung im Internet – Maßnahmen und Erfolgswirksamkeit im Business-to-Consumer-Bereich.* Wiesbaden, Gabler.

May, P. (2001). *Mobile Commerce: Opportunities, Applications, and Technologies of Wireless Business.* Cambridge: Cambridge University Press.

Möhlenbruch, D. (2006). Sortimentspolitik. In Handelsblatt (Ed.), Wirtschaftslexikon, Band 10 (pp. 5245-5254). Stuttgart: Schäffer-Poeschel.

Möhlenbruch, D., & Schmieder, U.-M. (2001). Gestaltungsmöglichkeiten und Entwicklungspotenziale des Mobile Marketing. In Meier, A. (Ed.), *Mobile Commerce, Praxis der Wirtschaftsinformatik, 39*(220), 15-26.

Möhlenbruch, D., & Schmieder, U.-M. (2002a). Mobile Marketing als Schlüsselgröße für Multichannel-Commerce. In Silberer, G., Wohlfahrt, J., & Wilhelm, T. (Eds.), *Mobile Commerce* (pp. 67-89). Wiesbaden, Verlag.

Möhlenbruch, D., & Schmieder, U.-M. (2002b). Chancen des Mobile Marketing im Rahmen von Multichannel-Strategien. *Thexis – Fachzeitschrift für Marketing, 19*(2), 27-33.

Müller-Hagedorn, L. (2005). *Handelsmarketing.* 4th Edition, Stuttgart, Schäffer-Poeschel.

Nguyen, H. Q., Hackett, M., Johnson, B., & Johnson, R. (2003). *Testing Applications on the Web.* 2nd Edition. New York: Wiley Publishing.

Nysveen, H., Pedersen, P. E., & Thorbjornsen, H. (2005). Intentions to Use Mobile Services: Antecedents and Cross-service Comparisons. *Journal of the Academy of Marketing Science, 33*(3), 330–346. doi:10.1177/0092070305276149

O'Reilly, T. (2005a). *What is the Web 2.0?* Retrieved January 15, 2007, from www.oreilly.de/artikel/web20.html

O'Reilly, T. (2005b). Web 2.0: Compact Definition? Retrieved January 15, 2007, from http://radar.oreilly.com/archives/2005/10/web_20_compact_definition.html

Sadeh, N. (2002). *M-Commerce: Technologies, Services, and Business Models*. New York: Wiley Publishing.

Schütte, C., & Diederich, L. (2006). Neues aus der Blogoshäre – digitale Mundpropaganda via Weblog. *Marketing Journal, 39*(Sonderheft Web 2.0), 26-28.

Shim, J. K., Qureshi, A. A., Siegel, J. G., & Siegel, R. M. (2000). *The International Handbook of Electronic Commerce*. Chicago: Verlag.

Siau, K., Lim, E.-P., & Shen, Z. (2003). Mobile Commerce: Current states and Future Trends. In Lim, E.-P., & Siau, K. (Eds.), *Advances in Mobile Commerce Technologies* (pp. 1-17). Hershey, PA: Idea Group Publishing.

Singh, J., & Sirdeshmukh, D. (2000). Agency and Trust Mechanisms in Customer Satisfaction and Loyalty Judgments. *Journal of the Academy of Marketing Science*, *28*(1), 150–167. doi:10.1177/0092070300281014

Smolnik, S., & Riempp, G. (2006). Nutzenpotentiale, Erfolgsfaktoren und Leistungsindikatoren von Social Software für das organisationale Wissensmanagement. In Hildebrand, K., & Hofmann, J. (Eds.), *Social Software, Praxis der Wirtschaftsinformatik, 43*(252), 17-26.

Srinivasan, S. S., Anderson, R., & Ponnavolu, K. (2002). Customer loyalty in e-commerce: An exploration of its antecedents and consequences. *Journal of Retailing*, *78*(1), 41–50. doi:10.1016/S0022-4359(01)00065-3

Stolpmann, M. (2000). *Kundenbindung im E-Business*. Bonn: Galileo Press.

Turowski, K., & Pousttchi, K. (2004). *Mobile Commerce*. Berlin: Springer.

Wamser, C., & Wilfert, A. (2002). Die wettbewerbsstrategischen Rahmenbedingungen des Mobile Commerce. In Teichmann, R., & Lehner, F. (Eds.), *Mobile Commerce* (pp. 29-50). Berlin: Springer.

Wiedmann, K.-P., & Frenzel, T. (2004). Akzeptanz im E-Commerce – Begriff, Modell, Implikationen. In Wiedmann, K.-P., Buxel, H., Frenzel, T., & Walsh, G. (Eds.), *Konsumentenverhalten im Internet* (pp. 99-117). Wiesbaden, Gabler.

Wiedmann, K.-P., & Langner, S. (2004). Beeinflussung im Internet – Klassische Techniken und aktuelle Phänomene. In Wiedmann, K.-P., Buxel, H., Frenzel, T., & Walsh, G. (Eds.), *Konsumentenverhalten im Internet* (pp. 195-225). Wiesbaden, Gabler.

Wilke, K. (2002). Controlling im E-Commerce. *Handel im Fokus*, *54*(1), 274–294.

Wilke, K., Duscha, A., & Hudetz, K. (2005). Kundenbindung über das Internet. In Hudetz, K. (Ed.), *E-Commerce im Handel – Status Quo und Perspektiven* (pp. 101-133). Gernsbach, Deutscher Betriebswirte-Verlag.

Williamson, O. (1991). Comperative Economic Organizations. The Analysis of Discrete Structural Alternatives . *Administrative Science Quarterly*, *36*, 269–296. doi:10.2307/2393356

Wind, J., & Rangaswamy, A. (2001). Customerization: The next revolution in mass customization. *Journal of Interactive Marketing*, *15*(1), 13–32. doi:10.1002/1520-6653(200124)15:1<13::AID-DIR1001>3.0.CO;2-#

Wirtz, B. W. (2002). Electronic Commerce. In Wirtz, B. W. (Ed.), *Gabler Kompakt-Lexikon eBusiness* (pp. 57-58). Wiesbaden, Gabler.

Wirtz, B. W., & Lihotzky, N. (2005). Kundenbindungsmanagement im Electronic Business: Instrumente und ihre Wirkung . *GfK Jahrbuch der Absatz und Verbrauchsforschung*, *51*(2), 136–155.

Wolff, P. (2006). *Die Macht der Blogs - Chancen und Risiken von Corporate Blogs und Podcasting.* Frechen, Datakontext.

Yu, R. (2008). *Mobile market trends*. London: Ovum Europe Limited.

Zeithaml, V. A., Berry, L. L., & Parasuraman, A. (1996). The Behavioral Consequences of Service Quality. *Journal of Marketing*, *60*(2), 31–46. doi:10.2307/1251929

KEY TERMS AND DEFINITIONS

Web 2.0: Web 2.0 is frequently used as a general term for different web based services and applications which aim to generate added value through user integration. Users are able to generate or comment content which establishes a great dynamic and interactivity.

Wiki: Wikis are open and cooperative authoring systems for websites which can be edited from every user equally. With that, they are a group product of several authors where content is gathered collectively.

Blog: Blogs or weblogs are online diaries written by usually one author. Readers have free access to the content and are able to comment every article. Additionally each blogger can reference his articles to others which is shown at the end of each comment.

Social Networks: Social Networks are advancements of virtual communities where the relationship between users is shown as virtual interaction network. Users of these specific websites are able to post their personal profile and interact with their friends.

Social Shopping: For Social Shopping user build social networks to purchase goods. This application is based on group dynamics where large groups of users recommend products or buy together.

Mobile Commerce: Mobile Commerce is the buying and selling of goods and services through wireless handheld devices such as cellular telephone and personal digital assistants (PDAs). Known as next-generation e-commerce, m-commerce enables users to access the Internet without needing to find a place to plug in.

Product Line: Product Line is an instrument of the mobile commerce marketing mix. The instrument contains every decision regarding the determination of distributable goods which are going to be purchased by the customer. It defines the dimensions of the product line and declares the availability as well as the branding of different goods.

Price: Price is an instrument of the mobile commerce marketing mix. Within the pricing policy decisions about the actual extent of prices and other different determinants of purchase decisions such as range of special offers and price differentiation have to be made.

Communication: Communication is an instrument of the mobile commerce marketing mix. It contains decisions about the necessary messages and the reasonable utilization of communication material and channels. Additionally the budgets for the different communicational arrangements will be defined by this instrument.

Front-End: Front-End is an instrument of the mobile commerce marketing mix. The direct interface to the customer in mobile commerce settings is here defined. Beside this the linkage and the strategic design of the website will be set.

Customer Retention Management: The management of customer retention comprises the analyses, planning, enforcement and controlling of all arrangements concerning the actual customer.

Chapter 24
Mobile Business and Mobile TV:
Available Technologies, Future Opportunities and New Marketing Trends

Marco Garito
Viale Fulvio Testi Milano, Italy

ABSTRACT

The increasing availability of affordable mobile business services (or "m-services") across different telecommunication carriers and platform is now a reality that is changing the way customers perceive products and solutions: mobile TV, one of the latest m-services, can be now seen either as a stand-alone channel or in conjunction with other media, thus providing several levels of engagement and interaction. Marketeers are now talking about "experiential marketing" to explain how customers and businesses are now reacting to market stimuli: Philip Kotler talked about "sense and respond" in his book "Marketing moves". Technology development is now taking a step forward enabling customers to be users and creators of products, services and brand image. This chapter provides and overview about traditional mobile marketing and mobile TV, describing how those interact and how business can be affected in some crucial areas: platform interoperability and revenue models. Some real examples of mobile initiatives are also provided.

MOBILE MARKETING: AN OVERVIEW OF PAST AND CURRENT INITIATIVES CARRIED OUT IN CONJUNCTION WITH OTHER CHANNELS

Mobile marketing is now a very effective communication channel enabling wider and closer relationships with audience for brands, products and services willing to promote themselves. The difficulties of implementation (from both technical and business standpoint) are mostly overtaken due to the improved cooperation between carriers, service providers and companies (Harris and Dennis 2001; May 2001; Settles 2002).

Mobile marketing can be via SMS (limited at 160 characters only) or MMS, browser based via mobile Internet WAP format through banner ads

DOI: 10.4018/978-1-60566-074-5.ch024

and interstitials, voice based using voice lines and as emerging technologies via Java based games, LBS (location based services) and voice mail marketing (Paavilainen 2002).

Some of those formats can be delivered via land-lines to customers, thus enhancing the reach of the message and engaging the recipient in a more comprehensive way.

Structure of a Possible Mobile Marketing Campaign

A simple marketing campaign has this sequence: from the opt-in database, the message is sent to the end customer who reads the message in his/her handset: the message can be set up in different ways. The recipient of the message can purchase the product (or more broadly take action/react upon the message) and the data is received and analyzed. From a merely operation standpoint, the provisioning of a mobile marketing campaign consists of 4 steps:

1) Concept: this phase involves the decision of which the target audience is, the definition of the message and the integration planning where the components are bundled
2) Creative: this step consists of the campaign design, the application design and the build. These three parts are a combination of design and technical work which might involves the use of agency or a third party.
3) Delivery: at this point, the aggregator and the network transmission come into account. The message must be sent across the available networks
4) Analysis: once the campaign has been delivered, analyses regarding response rate, response profile and an overall feedback and improvement constitute the acid test to measure the effectiveness of the marketing initiative and an opportunity to update the CRM data.

Mobile campaign formats can be "push" or "pull" with one or two levels of interaction with the end user/customer who can eventually react via a single or a multiple response.

Examples of pull campaigns are initiatives like text-to-win (where the user receives a 5 digit code - the so called "premium rate"- or a full 11 digit number to dial back. The choice between the two options affects the set up cost being the 5 digit code more expensive especially in those cases where the code is an easy sequence of number to remember, therefore the user can easier call back), ring-tones, games, coupon, vouchers and membership.

The push campaigns add to the above points a crucial issue: the need of compliance to EU regulations about privacy: basically the need of an opt-in database of mobile number to use where the end user agreed to be reached via SMS. This preliminary procedure enables to minimize the risk of spamming, that can lead the offender to be fined, and the negative brand image that can be associated to unsolicited and unauthorized messages to customer.

The mobile marketing proves to be extended into innovative customer communication strategy alongside other channels, mostly voice and browser-based paths, particularly upon the consideration that EU penetration rate of mobile coverage is over 80%.

In addition, companies who adopted mobile marketing as part of their overall strategy, experienced higher response rate, higher brand recall compared to TV and radio and ultimately a decreasing price as many operators are entering the marketplace thus increasing the competition.

The following table suggests a possible view of how mobile market can be a part of an integrated marketing communication alongside other channels (Porter 1980, Plant 1999, Crawford 2007).

A current trend in terms of marketing is the choice between create and use internal resources or rely on agencies/third parties and business partners; the decision is based on: product/solution, market

Table 1. Mobile marketing campaign types as carried out by companies in conjunction with other channels

Title	Description	Considerations	Comments
Messaging			
SMS push examples of companies who used push based campaigns are: Emporio Armani, Reebok, Chupa Chups	Send and SMS to a target audience	Recipient can: call a phone number – send back an SMS – visit a website – receive a voucher	The SMS can be considered as a spam – need of extra care as the matter is being increasingly regulated
SMS pull and push	Recipient opt-in	The message can be specific promotion or value added information	It creates a permission based database – the content is more expensive but increase the perception of value
SMS pull examples of companies who used pull based campaigns are: Cadbury, Coke, Gossard, Channel 5, Mumm Champagne	Recipient see the promotion in other channels and send SMS to respond	Other media can be: TV, radio,, newspapers	This demonstrate how mobile market integrates with other channels – companies can rent a 5 digit code for a short time but this comes to a price
SMS to call back	Recipient send a message for being called back	Can be another feedback opportunity channel in conjunction to other media	Many websites include this function, alongside click-to-chat
Browser based			
Footnote link	The WAP page contains a link to invite the user to take action	WAP pages have limited capabilities and space – there is a need to assess the technology environment	Mobile technology is being developed quickly hence handset manufacturers, platforms and protocols might offer soon better and better results
Interstitials	When a WAP page is loading on a phone, a short ad is presented before the main site	Options increase if graphics, format or amount need to be included	Recipient is charged for the download of this additional ad as quota of data by their carrier, which means that user is paying for something they might not want
Search engine listing	The WAP site is listed in the first places of the search engine	Consistency with the similar situation of search in websites	WAP mobile browser listing is a niche market as it is complex
Portal listing	Brands listed in carrier's portals	Portal positioning has an impact on the number of customers who get a link to the ad	
Voice based			
Register a vote	Users respond to a campaign	Premium rate number – length of the rent of the number	Widely adopted and it is the most common usage of mobile campaign in TV
Follow instruction and submit details	User to phone and receives a code or leaves a message	Premium rate number and callers are kept on the phone as much as possible	Used in high value competition – regulators dictate to make audience aware of the premium rate
Other channels			
MMS	Same as SMS as seen above		Richer format but more expensive than SMS
LBS marketing	Message received upon location	Mandatory opt-in to send such a message – it might be highly intrusive	Used for roaming (for example, when the user is reaching a new country) – privacy issues still open
Downloads	User receives an invitation to download something	Additional cost for this specific item – higher sophistication	Effective for youth market and for those who want to have a brand as their icon

environment and customers (Porter 1980, Plant 1999, Crawford 2007, Nester and Lyall 2003). From a purely monetary and financial point of view, the table below gives an idea of advantages and disadvantages for both the situations: of course, the specific Service Level Agreement might state a different split of tasks and responsibilities.

The table demonstrates how mobile market integrates and enhance the richness and reach enabling companies to potentially stand out of the crowd in a hugely populated world of products and services and interact with customers in a more engaging and comprehensive way, more than ever related to experience and emotion (Porter 1980, Plant 1999, Crawford 2007, Nester and Lyall 2003). It is now time to analyze how companies will be affected by technology developments in mobile.

MOBILE MARKET ECOSYSTEM: AN ANALYSIS

The rising adoption of user-created content and the open distribution platform open a new array of opportunities and the increased number of smaller operators (mostly VAS, value added services companies) who are closer to the end users (Unhelkar 2008) and therefore better placed and able to provide what the customers need implies the redefinition of mobile market players (carriers, content providers and device manufacturers/ OEM and social media platforms).

The table below lists the available business models that are likely to exist for the next few years to come

An alternative way to describe and understand the business models for Mobile TV is given by these figures which are focused on the flow of services and payments/revenues.

Mobile TV Business Model Examples

The following figures explain the available revenue models with the interactions within the stake-holders involved

Mobile TV Value Chain

The figure below describes the value chain for mobile TV. The value chain is not a flat line (Plant 1999, Crawford 2007) therefore the entry gates for new incumbents inter-mediators open considerably

Incumbents and new players can potentially enter the system and dramatically change structure and relationships.

Table 2. Marketing scenario for internal and external deployment

Internal resources		
	Advantages	**Disadvantages**
	Content owner has larger potential Content owner manages the customer relationship, branding and pricing Content owner manages user data and experience	Content owner bear the risk and uncertainty of cash flow Content owner is responsible for set up and capital investment Content owner is responsible for billing and customer service
External resources		
	Partner is responsible for set up and capital investment Partner responsible for billing and customer service Digital right management might make cash flow predictable	Partner owns data and user experience Partner manages pricing and branding Partner take the larger part of revenue for greater risk

Figure 1. The extended value chain in mobile business environment

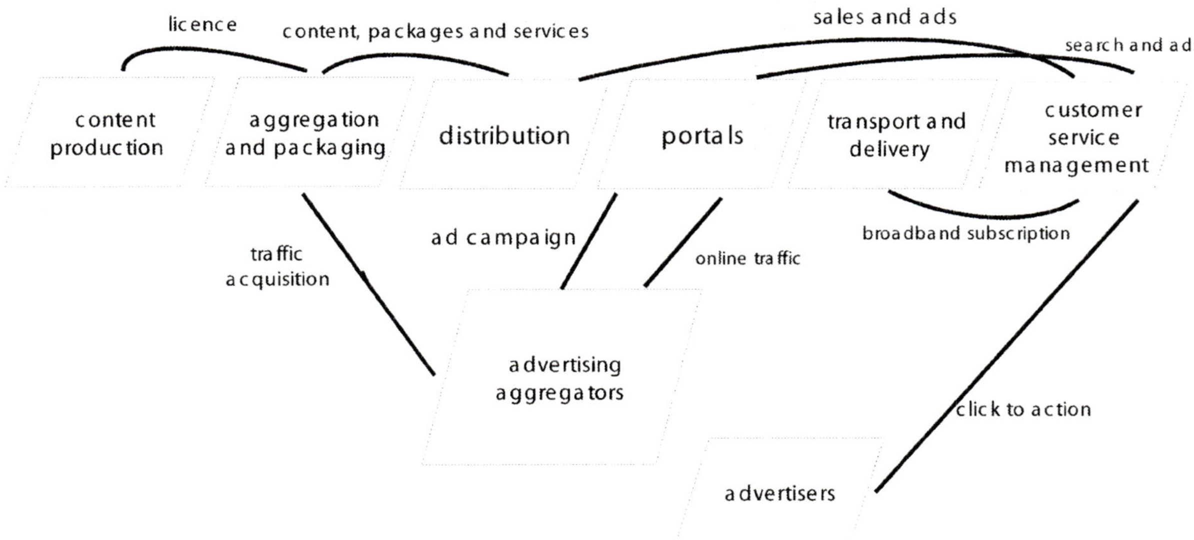

Table 3 New business models overview

Traditional media	Closed communities	New platform aggregations	Content syndication
Professionally made and branded content currently available today. Examples are Columbia Picture or Walt Disney	It is based on niche and user/community generated content. Example: NTT DoCoMo has special interest communities and users interacting via their devices; Comcast announced an agreement with Facebook .com to produce a TV series from user generated videos	User generated content distributed via open platform. Examples are user-driven aggregators as Youtube, Myspace, Second Life, LiveJournal, XING, mixi, dailyLOL and netmusicmakers.	Professionally produced content delivered via open channels. Example: BBC has implemented My BBC interactive media service without a given specific access provider or devices

MOBILE PLATFORMS: A SYNOPSIS OF THE AVAILABLE STANDARDS

Many operators are already offering mobile TV services on a commercial basis over their wireless network: more and more people are watching television on their mobile phones meanwhile subscribers' number is low enough to avoid network congestion (Van der Dam 2006; Wieser 2006). Some western European countries are experiencing overload in their fixed copper based network, due to the increase of users of multimedia applications and social network portals (myspace and youtube to name a few). These portals have been witnessing a dramatic growth in popularity: anyone can produce, share and see videos and rich media files anytime, everywhere, hence the majority of connections to fixed Internet are not suitable to cope with the demanding and increasing request of access from users. The replacement with more powerful network infrastructure (satellite or optic fibers) is lengthy and still expensive, thus putting a serious threat on financial analysis of telecommunication carriers, who are still dealing with the not so successful 3G/UMTS recovery cost.

Mobile TV has the potential to appeal to mass market (Porter 1980, Plant 1999, Crawford 2007, Nester and Lyall 2003) because the coverage of network signals is over 80% in many countries (although some remote areas are not sufficiently

Figure 2. Free to air: (the block arrows describe the service flow – the lines describe the payment flow)

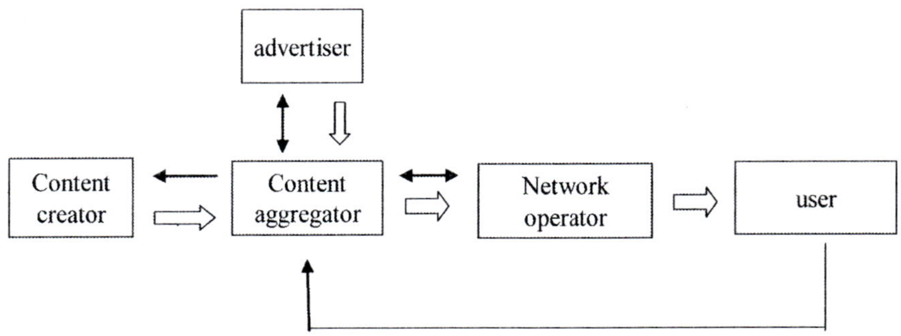

covered). A further segmentation, based on demographic, usage behavior and disposable income can provide a matrix upon which companies can model their strategy (Ahonen 2002). From a purely technological point of view, the situation is far more complicated because of the different standards currently adopted in Mobile TV as shown by the table below (Andersen 2007; Van den Van 2006)

The described situation replicates the so-called "walled garden" (Unhelkar 2006) where users of a platform or customers of a carriers cannot connect and utilize the services made available by others.

It is time now to list a few examples of Mobile TV initiatives and further developments currently carried out by different companies for different purposes, from brand awareness to users' cooperation for content development and advertising.

EXAMPLES OF CURRENT MOBILE TV INITIATIVES AND SOME FUTURE POSSIBLE APPLICATIONS

Some companies have already started to massively use the Mobile TV as a stand-alone channel or in conjunction with other media for many and different reasons: this paragraph lists a few examples. This section also continues with a short description of current researches in IT laboratories to make the mobile business more friendly and integrated into a wider spectrum of situations.

Current TV

Current TV was launched in 2005 as independent media network heavily based on viewer's content: viewers can submit their own clips lasting from a few seconds to 15 minutes and the viewers are those who vote their favorite video. Eventually, Current TV integrates the most popular video and integrates it into programming and broadcasts it.

Figure 3. Mobile operator (block arrows describe the service flow, the tin arrows describe the payment flow)

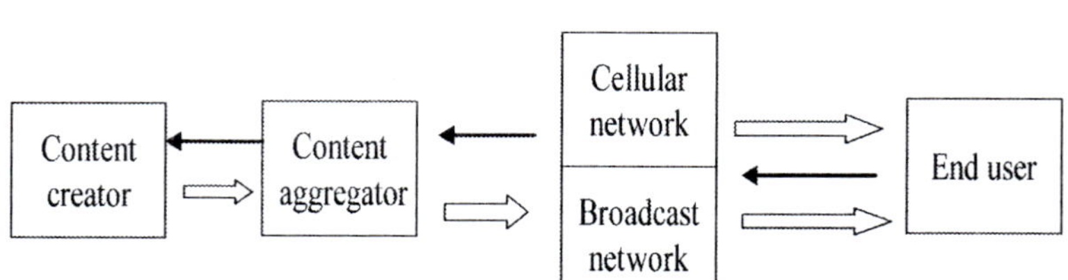

Figure 4. Wholesale

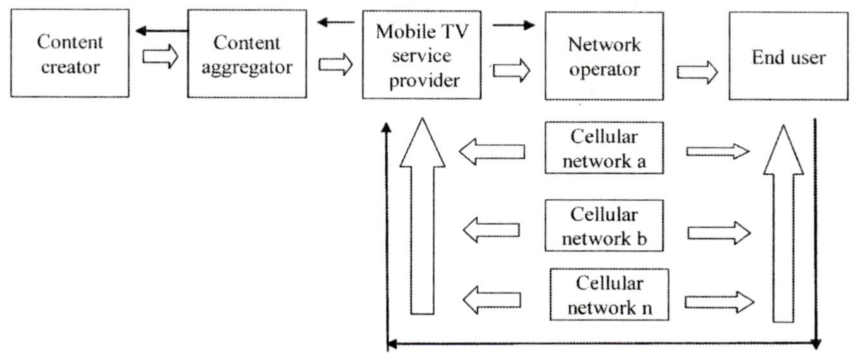

The company also accepts user generated ads that feature any of their corporate sponsors: in this case, the creator of the selected ad receives a sum of money and, if the sponsor wants to use the ad and distribute it, the creator/submitted received additional premium (http://www.currenttv.com)

Bud TV

Bud TV plans to offer advertising programming and branded entertainment content directly to computers, MP3 and other devices; the network also wants to include a loyalty program called "season pass". After sing in, the system download content to user's hard drive and this content is available for view when users want to watch it (Van der Dam 2006; Andersen 2007)

Free Mobile TV from 3 Italia

According to the financial newspaper Il Sole 24 ore (6 June 2008), this is the first trial of free mobile TV, offered by 3 Italia. Some channels like Rai (national broadcaster), Current TV, Sky meteo24 (weather forecast) and the new version of La 3 will be available for free to all user of a 3 Italia SIM card, where signal is geographically available. The new format includes some works from DAMS students (a high profile school of art) in Bologna; Spoiler TV where young people provide subtitles to series episodes like Lost and Dr. House; Myspace videos and 3 Stage dealing with theater productions. Mobile TV 3 Italia has now one million of customers over 8.2 millions of UMTS users.

Figure 5. Mobile TV value chain explained

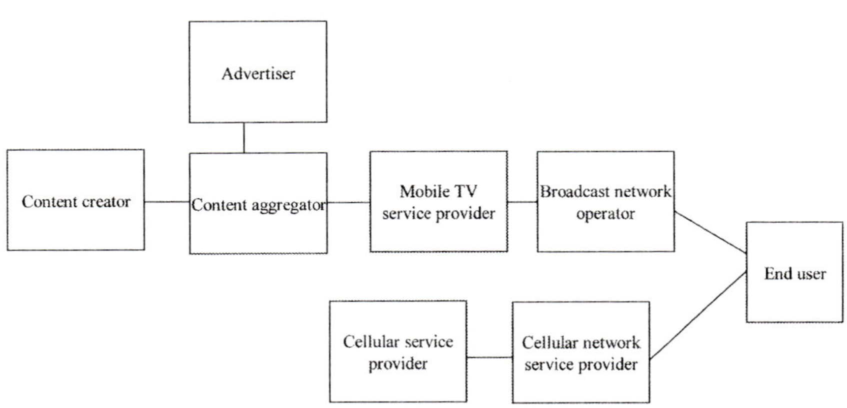

Table 3. standards for Mobile TV

	MBMS (multimedia broadcast and multimedia services over UMTS)	BCMCS (broadcast and multimedia services)	DVB-H	ISDB-T	MEDIAFLO	DMB
standard	OPEN	OPEN	OPEN	OPEN	PROPRIETARY	OPEN
region	EUROPE-ASIA	US-ASIA	US-ASIA-EU-ROPE	JAPAN	US	KOREA
technology	W-CDMA, EDGE/GPRS	CDMA	OFDM (orthogonal frequency division multiplexing interface technology)	OFDM	OFDM	OFDM
availability services	2006/2007	2006-2007	2006	AVAILABLE	2006/7	AVAILABLE
availability handset	2006/2007	2006-2007	2006	AVAILABLE	2006/7	AVAILABLE

Adidas's Brand Reinforcement

In US Adidas's campaign from the autumn 2007 National Basketball Association season stands out for making truly innovative use of the mobile medium; activities related to the campaign ranged from users being able to set up personalised voicemail messages and ringbacks using the voice of Boston Celtics star Kevin Garnett, to posting short codes in athletic footwear stores that users could call to hear leading basketball players explaining which Adidas shoe was their favourite (http://www.adidas.com)

Researches and Developments in Mobile Business

It is now time to take a look at some research labs who are developing the next applications for mobile business: some of these developments are based on existing products or solutions.(Anderson 2007).

The Spoken Web – Voice-enabled mobile commerce to replace or reduce the need of keyboard touches. The Spoken Web aims to transform how people create, build and interact with e-commerce sites on the Web, using the spoken word instead of the written word..

Instant Translation – Real-time communication between multiple languages through mobile phones. Instant Translation aims to facilitate speech among individuals with no common language with the goal to allow free-form dialogue using a PDA (Anderson 2007)

SoulPad – Enabling any portable device to carry your computing application in your pocket . to separate a computer's soul – the programs, settings and data it holds – from its body, the disks, keyboard, screen, processor and other hardware from which it is comprised (Anderson 2007)

Social networking on-the-go (already available). Consumers can communicate with their social network friends regardless of where they are with voice and SMS from either a PC or a mobile phone (Anderson 2007; Berens 2006; Kleinberg 2006)

Good Samaritan – Mobile healthcare information made available in any emergency situation with information on how to aid people in critical medical situations (Anderson et al. 2007; Berens 2006)

OPEN DISCUSSION POINTS FOR MOBILE TV

As discussed previously, the future of Mobile business and of Mobile TV is challenging and at the moment is difficult to clearly predict what it will be: however there are some key points: these same results can be achieved by carrying out a SWOT analysis bearing in mind the industry and product specific features and characteristics.

Many industry reports and carriers (Unhelkar 2006 and 2008; Paavilainen 2002) highlighted decreasing ARPU (average revenue per user) in developing countries due to the maturity and saturated conditions where on one side end-user customers are already owning more mobile devices and business are just starting to take advantage from wireless in some business areas (namely CRM – customer relationship management – and SCM – supply chain management).

The adjacent issue is the network coverage which affect not only the quality of the signal and its availability everywhere (as prescribed by the tender for 3G/UMTS bids) but also the roaming and therefore the split of the costs of signal for people crossing countries (currently under closer review by the European Union Commission as media in Europe are reporting) and ultimately the digital rights management which is even more complex as users can directly create and share content even with companies who can use it for marketing and promotion.

The third point affects the device technical features: power consumption and user friendliness. The use of wireless system requires high availability of energy, particularly for rich media: manufacturers are working on finding a suitable solution, taking into account environmental issues and electronic waste management as required by several regulations in many countries (Ahonen 2002). The quest for a more user friendly technology has been dealt by devices' manufacturers in many and different ways: NTT DoCoMo used the acronym "MAGIC" which stands for Mobile, Anytime, Globally, Integrated and Customized; Ericsson developed the "0-1-2-3" which means 0 written manuals, 1 simple button to connect the Internet, 2 seconds of delay waiting to access the service and 3 keys to gain access to services and features (Ahonen 2002). Lately Apple and other brands, tried to combined the easy of use within their brand characteristics by adopting a touch screen technology which seems to be replacing the keyboard in the next few years for each manufacturers and products.

NEW MEDIA AND APPROACH TOWARD EXPERIENTIAL MARKETING

Every day, each person receives a huge amount of marketing stimuli (Kotler 2004) and companies are competing not only for a share of space on the shelves but also for people's attention: marketing is changing as a consequence of

a) The wider availability of media and channels and
b) More selective consumer's behavior because they are more informed therefore there is shifting power (Porter 1980, Plant 1999) and, as a consequence of social media and new technology capability,
c) The user's capability to create and share content with other, implying that the old marketing structure and theories are moving into a more experiential environment, where customers are experiencing a holistic immersion and fusion with the brands and its products, sensing the products through different channels, each providing a unique feeling. Those three combined factors have determined the increased popularity of social media phenomenon.

The described situation has also an impact on metrics and measurements because in social media marketplace, the calculation of ROI and any other benefit on the bottom line is not a clearly counted against a specific and traceable marketing initiative. As social media is about relationships, new performance measurements must be adopted and marketeers need to take full advantage from social media, refraining from structuring the marketplace as they did in the past.

The new performance measurements in social media has been discussed in 2005 when some companies tried to set up a system for blog's influence in marketing by analyzing the experience of Jeff Jarvis as Dell's customer.

As a consequence, a new marketing theory has born: the so called "experiential marketing" (as opposed to traditional push/pull strategy described a while ago) which can be defined as an attempt to connect consumers with brands in personally relevant and memorable ways. As a marketing strategy, experiential marketing tries to move beyond the usual boundaries of features and benefits (the typical "push approach") where marketing messages and stimuli are designed and sent to wider audience, who can and cannot benefit from a product or service, thus alienating people. Experiential marketing is therefore based on experience that people consciously choose to attend and be part of, after identifying the brand and its offerings as relevant to satisfy their needs (Wolfe 2003).

It is now possible to understand and appreciate how mobile marketing and TV on mobile, in conjunction with the latest technology developments in social network, are enabling a more comprehensive and holistic experience between users/customers and brands

CONCLUSION

Marketing opportunities in mobile environment have been discussed by providing real cases from industry and business who carried out initiative in the wireless world either as a stand alone mode and in conjunction with other channels. The overview of current and available technology infrastructures for mobile marketing has been also provided alongside with an analysis of advantages and disadvantages between open systems and proprietary solution with the intention to demonstrate how the choice between the two approaches impacts the business. It has been also stated that mobile is still an emerging technology hence it is difficult to predict how the industry will be in the near future: however it has been demonstrated that users and customers can enjoy a more complete experience with products and services because the technology development provide additional features and capabilities to companies who want to have a closer interaction with the market.

Websites

http://www.adidas.com
http://currenttv.com/faq
http://marcomele.blog.ilsole24ore.com accessed on 30/9/2008

REFERENCE

Ahonen, T. (2002) *M-profit. Making money from 3G services* New York: John Wiley and Sons.

Anderson, D., Canepa, S., Ekow, N., & Mannel, S. (207). *Navigating the media divide.* USA: IBM Global services.

Crawford, D. (2007). *The mobile TV market.* Colchester, UK: University of Essex.

Graham, H., Saunders, J., Piercy, N., & Nicoulaud, B. (2007). *Marketing strategy and competitive positioning.* Prentice Hall.

Haig, M. (2002). *The b2b ecommerce handbook.* UK: Kogan Page Publisher.

Harris, L., & Charles, D. (2002). *Marketing the e-Business*. UK: Routledge Publisher.

Kotler, P. (2004). *Kotler on marketing.* USA: HBR Publisher.

May, P. (2001). *Mobile commerce.* UK: Cambridge University Press.

Paavilainen, J. (2002). *Mobile business strategy.* USA: Addison Wesley Publishing.

Pircy, N. (2002). *Market-led strategy.* London: B&H Publisher.

Plant, R. (1999). *e-commerce formulation of strategy.* London: Prentice Hall.

Porter, M. (1980). *Competitive Strategy.* New York: Free Press.

Porter, M. (2001). *Strategy and the internet.* HBR Publisher.

Settles, C. (2002). *Wireless Inc.* New York: American Management Association Publishing.

Unhelkar, B. (2006). *Handbook of research in mobile business.* Hershey, PA: Idea Group Reference.

Unhelkar, B. (2008). *Handbook of research in mobile business,* second edition. Hershey, PA: Idea Group Reference.

Van der Dam. (2006, April). *Primetime for mobile television.* IBM global services.

Weber, L. (2008). *Marketing to the Social Web: How Digital Customer Communities Build Your Business.* Wiley.

Wolfe, D. (2003). *Ageless Marketing: strategies for reaching the hearts & minds of the new customer majority.* Chicago, IL: Dearborn Trade Publishing.

KEY TERMS AND DEFINITIONS

ARPU: It means Average Revenue Per User and refers to the profitability of each user of mobile devices for voice, data and multimedia.

Experiential Marketing: It refers to a new marketing approach based on a comprehensive experience between customer and product/service.

Mobile Business Services (M-Services): The expression refers to a wide set of applications and tools through which data, information and files can sent and received via a wireless device.

Mobile Marketing: It refers to a marketing channel and communication where the message is designed and customized to be broadcast to a handset. Mobile marketing can be seen either as a stand-alone initiative and in conjunction with other marketing initiatives.

Mobile TV: It refers to the capability to receive TV signal on a mobile device, assuming that the handset is enabled for this and the geographical area is covered by the network signal.

SMS: Short message (also known as "text") sent and/or received via mobile system but limited to 160 characters. That limitation is a contributing factor for the development and dissemination of new language, practice and habits, also in face-to-face communication.

WAP: It refers to wireless application protocol and is an open international standard to enable access to the Internet via a mobile device.

Chapter 25
Brand Driven Mobile Marketing:
5 Theses for Today and Tomorrow

Christopher Quente
Creative Director, Germany

ABSTRACT

Despite the fact that the mobile phone is a perfectly suited advertising medium also for brand campaigns *a creative perspective is not only desirable but also necessary. Because: Without* creativity, *the latest mass medium would be swiftly rendered desolate. In the creative conceptual design of* brand campaigns *it is important above all to carry out advertising not only on the mobile phone but with it. On the basis of five theses it is illustrated how the mobile phone becomes the essential building block within commercial multi-channel concepts. One key issue being the telling of* 'brand stories', *told across various media or 'stages'. If the stories being told are compelling enough one can convert brand supporters into brand addicts, who in the best case scenario wouldn't want to miss a single performance of their favorite brand.*

INTRODUCTION

To call the market for mobile marketing a large one would be an understatement. It is gigantic – a statement not only maintained by advertising people – a creed known for their notorious exaggerations. Let's compare this to another very popular medium – the Internet: Around 1 billion people worldwide were online at the end of 2007 (ITU, "Trends in Telecommunication Reform 2007 – The road to Next-Generation Networks (NGN) Summary", September 2007, pp 8/9). However, pundits 'Informa' found out that far more people – around 3.3 billion – have contact to mobile services (www.wikipedia.org, November 30, 2008 from http://en.wikipedia.org/wiki/Mobile_phone). Even if one can't necessarily deduce that these people actually own a mobile phone, the fact remains that around half of the global population has access to this latest incarnation of large mass media. This detail alone renders the mobile phone to be "Brand's New Toy" – arguably the most interesting medium for creative advertising campaigns.

DOI: 10.4018/978-1-60566-074-5.ch025

There are many traits that cause the mobile phone to be extraordinarily interesting – let me quote these here for the sake of comprehensiveness: The mobile phone is an always-on, multimedia device offering a feedback channel. Its sheer ubiquity offers advantages through personalization, individualization and localization of advertising messages. These can be designed to be interactive, current, entertaining and highly emotive. It is in particular the latter point – the creation of a touching, emotionally enthralling brand experience through the help of a mobile device – a fact that was continuously doubted in the pioneering days of mobile marketing. This chapter endeavors to prove the opposite. It's all a question of human creativity. And the end of the day, people are laughing and crying in front of a TFT Screen. Nobody said it couldn't be a small one.

To identify opportunities for distinct marketing measures, let's linger another moment on the subject of the market and have a look at how mobile subscribers use their equipment – aside from basic telephony. The following usage pattern develops: 74% of all mobile phone owners use their device to send SMS – and therefore make short messaging to be the most widespread on the mobile phone. Every user writes around 2,6 SMS per day, actively adding to the worldwide turnover reaching one hundred billion US Dollars in 2007 (www.wikipedia.org, November 30, 2008 from http://en.wikipedia.org/wiki/Mobile_phone). The second largest mobile service bringing in around 31 million US Dollars per annum concerns the downloading of mobile music, logos, pictures, games and gambling – not to forget pornography, perceived by many as the 'mother of technological innovation'.

It is comprehensible that advertising to date made use of these successful main applications and services to realize successful campaigns. In the book 'Brand's New Toy – Markenkommunikation mit Handy & Co.' my colleague Bosse Küllenberg and I summarized the most prevalent campaign schemes and processes in mobile marketing and analyzed their application (Küllenberg & Quente, 2006). We were able to categorize the following forms of campaign:

Code-based sweepstake campaigns, their main task being to improve product sales for the trade. This form of campaign is usually developed as an on-pack promotion and lends itself particularly to the consumer goods industry and above all to the 'fast-moving' consumer goods in the lower pricing segment. They actively upgrade the product, add value and directly influence the buying decision at the point of sale. The prize pool mainly consists of the aforementioned downloads. In Germany, scores of Code-based sweepstake campaigns were undertaken by the likes of Coca-Cola and McDonalds (e.g. Coca-Cola AG Campaign, "Festiver 04" Germany/ Switzerland 2004 from (www.festifever.ch, see http://www.spielplatz.cc/presse/wordpress/?cat=7).

Voucher campaigns, the focus of these being the reinforcement of specific distribution channels or the increase of frequency at a certain point of sale. A product introduction or sampling promotion would be valid motivations for this type of campaign. By and large, voucher campaigns are undertaken by the dealership carrying out the promotion. The database for this type of activity is usually bought-in, alternatively the company's own database is utilized. Many current ideas for context-based or location-based services incorporate this campaign method.

Mobile-content campaigns have a very broad range of application. This is mainly due to their potential to easily amalgamate and extend them with other preceding procedures as well as downstream in the campaign flow. The centre of this type of campaign is the dispatched content consisting of various image, sound and video formats. This renders the method ideal for multi-channel campaigns using many communication channels in parallel. Through the mobile phone's omnipresence it provides the erstwhile missing feedback channel for classic media and standardizes the mechanisms of future forms of advertising.

Viral marketing campaigns are a special subset of mobile-content campaigns. However, the conceptual design towards a viral effect – meaning the distribution by the client – makes this form a mechanism in its own right. The benefit: the promoting company does not have to pay for the cost of distribution and media. In turn, the originator relinquishes any means to control the campaign. Content counts – this must be interesting enough to cause the receiver to spend time and effort to forward it on. Chance and risk are at close quarters.

We can see that some campaign schematics and mechanics have already been established by the industry with considerable monetary outlay and huge effort. The consumer has learned how exactly to deal with the new medium – how to enter a code into an SMS and send it off to a particular telephone number, and of course how to claim his digital prize and download it to his mobile device. However, well-practiced behavioral patterns are the natural enemy of any creative solution. Now and again creativity must break up these patterns to make room for astonishing new ideas.

This important insight is not only gained by monitoring worldwide campaigns but in particular a result of own experience in realizing these. While in the early days of mobile marketing it sufficed to play around with innovative mechanics, today the effect of the new and strange has worn off and no longer enthralls the consumer. At some point every medium has to progress into a new phase and offer itself to more advanced tasks. There are many historic examples for the progression from the innovative to the creative phase. In moments like these, innovation hits the masses.

Why innovation usually proves to be breathtaking for a short instance is probably best illustrated by this historical snippet from the early days of cinema. When the brothers Auguste and Louis Lumière invited the public to the first movie premiere in world history it was enough to show a train arriving at the station of La Ciotat to create a huge commotion amongst the masses (Lumière, Auguste & Lumière, Louis (Dir.), 1895 from http://www.imdb.com/title/tt0000012/). As the train speeded towards the auditorium, myth has it that some of the more delicately inclined members of the audience left the cinema in a panic. This well-calculated effect, playing with the naivety of the public managed to secure instant attention to this new medium.

However, such a sensation cannot be repeated perpetually. Filmmakers picked up swiftly on the fact that audiences would get used to new techniques pretty quickly and demanded content instead. In the history to the medium of film this was the point in time where one recognized that from now on stories would have to be told to secure the continuous attention of the public. An insight, which mustn't be disregarded by us members of the creative advertising crowd more than a hundred years later. In the long run we have to use our creativity to tell good 'brand stories'.

But can today's stories actually be compared to those of the days of yore? Yes and no. Of course, the plots and heroes themselves haven't changed much over the years. However the manner of today's storytelling has only developed during the last decades. The audience was used to being told a story, first by roaming wanderers at the market places of the ancient world, later on by modern novelists and scriptwriters. For some time now, the audience actively takes part in the stories. Not only by being emotionally involved, but furthermore through steering its figures interactively through action games.

Truly astonishing: as of recent times, the audience has started to tell its own stories. We can monitor this phenomenon in social networks such as 'Twitter' (http://twitter.com) and 'myspace' (http://www.myspace.com). The twitter community offers the opportunity for senders to let thousands of other, real people, to partake in their life's story – for example by simple text messaging – while an audience of viewers monitors it. The selection of words alludes to the fact that I consider platforms like 'Twitter' as a stage in the classical sense of the word on which a piece is performed. As in the world of theatre, it is possible to take out a subscription to enjoy the performances in continuation.

This behavior can nowadays also be observed in advertising audiences who are used to picking out their preferred pieces, sometimes going as far as actively partaking in a performance as the 'hero'. However to go to such lengths the member of the aforementioned audience will have to at least assume that the efforts will pay off. It is either that the offered incentive (in whatever shape or form) is interesting enough. Alternatively, a strong passion for the brand 'on stage' is felt. However the manner of the contribution: in the best case it is generally regarded as a stand-up ovation for the brand. In the negative case however it is like the ear-deafening chorus of a booing audience.

Taking this insight into consideration there will be hardly any brand wishing to shower their target group with commercial messages on the basis of the 'push principle'. Quite the opposite, it will prefer to apply the 'pull principle', where the consumer autonomously picks out the preferred content and offering. This is why brands have to tell creative 'brand stories' on a stage for brand aficionados. Using the mobile phone, the consumer is immediately involved in the performed piece no matter his location.

Enough for now about the modalities of how a "brand story" can be told and implemented across multiple channels within a "brand campaign". In the following I would like to illustrate to the reader how such a "creative brand campaign" could look like in today's and tomorrow's mobile marketing. My argument rests on five theses, each of them centering around a current trend in mobile communication. Each thesis describes an innovation - and evaluates its opportunities (or lack thereof) with regards to a "brand campaign".

The five theses are:

1. The Mobile Phone Will Be The Universal Feedback Channel.
2. Widgets Are The Mobile Internet's Quality Channel.
3. Mobile Consumers Will Be The Future's Programme Directors.
4. Location-Based Services Allow For Highly Emotive Brand Interaction.
5. Pervasive Gaming Turns Advertising Into An All-Encompassing Game.

For illustratory purposes the individual theses feature real-life examples. These have been carefully selected and described in order to underscore the potential of creative utilization of the mobile phone within brand campaigns. However: What does "creativity" mean in this context?

All examples have three common denominators: they all feature a surprising "change of perspective, a "distortion" and „the breaking of a rule“. Thus, they result in a new insight for both the designer and the consumer. Freely adapted from Holger Jung and Jean-Remy von Matt's statement: Good and creative advertising "[...] is in a manner of speaking a 'light-bulb moment' (Ah!) paired with utter bepuzzlement (Huh?). Or as a slogan: Huh?-Ah!" (Jung & von Matt, 2002, pp. 295, author's own translation)

THESIS 1: THE MOBILE PHONE WILL BE THE UNIVERSAL FEEDBACK CHANNEL

The mobile phone as a mobile device has the capability to connect manifold media channels. It allows us, everywhere and instantly, to react to an impulse. This impulse may be printed in an advert, on a billboard; it may happen on television or communicated via radio. These impulses can be found on the Internet or on the product itself. Only mobile devices such as the mobile phone offer the all-encompassing feature of being able to react to such impulses. They are always available and can at all times establish a connection to the desired channel, be it in the form of speech over a conventional phone line or a continuously growing variety of media formats utilizing data transfer: SMS, MMS, WAP to name but a few.

Currently these possibilities are being used, if at all, in a rudimentary fashion. Time and again, interactive elements were used to prolong the duration of a contact or to increase the depth of information. A sweepstake is a classic feature commonly used to add customer interaction and to insert an activating momentum, usually with a call to participation via mobile phone. In the context of brand communication with the mobile phone or another mobile device it is paramount to deliver a more interesting brand experience, to strongly involve the target group and to gain increased attention. The creative perspective is as follows: give the consumer an active, acting role in promotional measures, allowing him to use the mobile phone like a remote control to interact with the real, molecular world.

To do a new brand experience any justice at all, advertisers have to offer new ways of interaction to the consumer by integrating him right into the centre of action. Nike iD successfully realized this potential in a far-reaching interactive campaign: Back in 2005, passers-by on Times Square in New York could personalize their own Nike iD sneaker using their mobile phone as a remote control and the huge Reuters billboard as a screen (Rodgers, Nike iD Billboard Invites Mobile Users, The ClickZ Network, May 3, 2005 from http://www.clickz.com/3502186). Dialing a toll-free number, the consumer could customize a personal pair of sneakers using the numbers on the telephone keypad. Once happy with the result the stage was then given to the next passer-by. Subsequently the consumer would receive a short message containing a personal access code to Nike's online shop, allowing a hassle-free order of the personal pair of shoes. A mobile phone wallpaper served as a reminder of the event. The beauty of this campaign is based on the public interaction that turns the consumer and brand aficionado into a hero for the time of the interaction. Indeed the described promotion was so unique that the audience reached far beyond New York's city limits. The campaign gained worldwide approval and reflected positively onto the brand.

There are more examples to be found for this new, interactive experience. An installation by Chaos Computer Club in Paris proved equally spectacular. Between September and October 2002, the whole façade of the Bibliothèque Nationale de France was turned into the world's largest computer screen and a stage for the consumer (Chaos Computer Club, Blinkenlights Arcade, Bibliothèque nationale de France, Paris, France, September 25 – October 6, 2002 from http://www.blinkenlights.net/arcade). In preparation for the event, each of the 520 windows of the library's glass front was outfitted with a pair of high-powered lights. Each pair of lights created one of the screen's pixels, allowing for a total resolution of 20x26 pixels – less than a moderately current mobile phone is capable of. Due to its size and location the interactive façade could be viewed by all Paris. Apart from commercial displays the main attraction proved to be classic computer games such as Tetris, Pong and Pac man, which could be played via the mobile phone on the display measuring over 3,000 sq. meters.

The mobile phone is not only ideal to interact with a usually inert environment such as a billboard; Danish agency 'Advance' found another interactive solution for the department store 'Illum' in Copenhagen, combining the auction principle known from eBay with short messaging (Advance A/S (2006), Illum Window Shopping from http://www.flickr.com/photos/angermann/110822340/, http://illum.dk/, http://www.advance.dk/). This idea elevates the branch of the designer fashion chain above the rest of the urban offerings by offering another emotional component even outside conventional business hours.

While on the subject of fashion: That creative mobile marketing is less about gigantism but all about emotions and stories is demonstrated by 'Smart Clothes', garments that offer totally new experiences when combined with mobile devices. Using Bluetooth technology and the mobile phone, these garments are capable of transmitting movement, body heat and pulse picked up by sensitive probes from one wearer to another. Through contraction and warmth the feeling of a comforting or even tender embrace can be induced. The ideal solution for lovers in long-distance relationships, babies separated from their parents, not to forget the elderly who are often alone.

What can we learn from these examples of creative brand communication? Using the mobile phone as a remote control we are able to let the best moments of the digital world materialize in our real environment. This is less about technology but more about the possibility to re-experience the brand in a new context. This aspect is plainly exciting for the consumer. Thus, the potential of brand interaction is not limited to neither specific products nor target groups. The curiosity of the consumers and the instant availability of the mobile phone allow direct interaction in every conceivable context. This is something that nearly every consumer should be capable of.

Enter the emotional potential hidden in every one of these mobile devices: They are well accepted by the consumer, easy-to-use and already an important part of our everyday life. The telephony functionality was only the beginning of another, equally enthralling chapter.

THESIS 2: WIDGETS ARE THE MOBILE INTERNET'S QUALITY CHANNEL

No later than 6 months after Apple's introduction of the iPhone in the USA most communication experts agree: Consumers love the mobile Internet! Not everybody did expect such an unequivocally positive judgment. Slow data transfer technologies and intransparent pricing structures made a pain of using the Internet on the mobile phone. The mobile Internet when viewed from the customer's perspective emoted feelings of caution rather than joy.

The analysis published by M:Metrics in March 2008 paints the following picture: around 85% of iPhone users employ the mobile Internet as a source for information. This compares favorably to 58.2% of Smartphone users and 13.1% when compared to the entire base of mobile phone users in the USA. Most popular seem to be two of the pre-installed Widgets: YouTube (http://www.youtube.com) and Google Maps (http://maps.google.de). Web video, mainly in the shape of the time-killing YouTube, is watched by 30% of iPhone owners. Market average is only 4.5 per cent. Google Maps was used by 36% as of January 2008 (average 2.6%). The analysis further demonstrated a high affinity for social networking, an application being used by nearly 50 per cent of owners. Number one in this sector is Facebook who was the first company to make their service iPhone-compatible. The global market average is a meager 1.5 per cent (:Metrics Press Release, M:Metrics: iPhone Hype Holds Up, Seattle/ London, March 18, 2008 from http://www.mmetrics.com/press/PressRelease.aspx?article=20080318-iphonehype)

What do these figures represent? Is it, as a journalist quibbled in the German 'Handelsblatt', that iPhone users are often bored and looking for orientation? (Postinett, Smartphone-Hype unter der Lupe: iPhone-Nutzer lieben das Web, Handelsblatt, 20.03.2008 from http://www.handelsblatt.com/technologie/mobile-welt/iphone-nutzer-lieben-das-web;1406862)

The iPhone enables its user to browse the Internet in its native layout on a comparatively large display and much more. The possibilities of the mobile Internet are significantly larger than the iPhone itself. And here exactly is the point of entry for brand communication to be found. Why should only services like Google Maps (http://maps.google.de) and YouTube (http://www.youtube.com) open a direct, high-quality channel to the consumer? Why not a popular brand?

The key to this is a combo of 3G mobile telephony, flat rate data plans and widgets, which are small computer programs, which are not individual applications but are integrated into the graphical user interface. Apple introduced widgets in April 2005 with the aid of 'Dashboard'. Of course there are comparable mobile applications running on Nokia's and other manufacturers' mobile devices. Most widgets do not offer complex functionality but simply render dynamic content from other sources (syndication). Examples for this are the embedding of content from video portals or user photographs through 'Flickr' (http://flickr.com). It is these kinds of widgets that played a major role in the proliferation of user-generated content and Web 2.0 applications.

This technology opens up new paths to the creative professional. While up to now there were considerable limitations to the design as well as the functionality of mobile Internet content pages, today's possibilities are countless. User interaction through integrated still or video cameras is conceivable – likewise access to the address book or saved data such as pictures is an option. Of course, widgets also employ the advantages of the mobile phone network, for instance when getting location-based information provided by the phone providers. Widgets offer inconceivable possibilities far beyond the design of a web page. Benefits notwithstanding, these micro applications need not be more expensive, because smaller applications can be released rather cost-effectively due to the simplicity of the programs.

Generally there is a solution to be found for every sector, from the currency calculator using current exchange rates to the full-fledged mobile phone game. It is particularly interesting to watch the multitude of user interfaces being developed. On the one hand attempts are made to involve all buttons on the keypad to allow for quick navigation. On the other hand, in more complex menus, icons are the solution of choice, which can be activated – just like on a PC – through a mouse cursor or in the case of the iPhone, through the touch of a finger. Many applications already make use of the feature to swivel the device by 90 degrees, switching from portrait to landscape format for improved display of pictures and video. Most readers will be familiar with this aspect from their iPhone.

At this point in time, not many companies have embarked on the path to Quality Internet, even though the creation of confidential, high-quality channels would be an ideal base for dialogue marketing to name just one example. It would present itself less than intrusive advertising but as a desired service. There are already some useful applications born out of the efforts of single programmers, creative professionals and the Open-Source community. Large brands have so far acted rather apprehensively when faced with the option of using Widgets on a larger scale; a decision most likely based on the fear that the target group could be too small. A recommendation to creative advertisers: Have a look at some of these small programs – maybe they inspire your own ideas for your target group?

While the author is composing these lines the 'Apple App Store' is being opened, allowing the iPhone user to download free and paid-for Widgets to their device. In this store a new market is being created, a shopping channel just like a stage for every private and commercial developer, a space where every brand large or small can develop and flower. If the idea and its implementation are sound, then there is a good chance to land a worldwide 'hit' in this realm created by Apple, generating ample attention for one's own product or brand. This makes the Widget the equivalent of a hit record. And as a time killer it proves just as popular as a cigarette – without the health implications.

THESIS 3: MOBILE CONSUMERS WILL BE THE FUTURE'S PROGRAMME DIRECTORS

In brand communication, the traditional role of the consumer was to receive and follow orders. The efforts of more conservative marketing strategists focused on combining a visual with a core message with the power to persuade the consumer about the advertised product. It is about compression and to install a well-constructed brand model into the target group. The consumer is not integrated directly into this process – let alone allowing him to design or change this model according to his wishes.

The epitome of the passive consumer however has been invalid for some years now. If a consumer is pondering to buy a product, he will actively seek information on the subject of the product standing up to its claims. The endeavors of the consumer actually go far beyond reading product and campaign reviews on Internet blogs. Since the advent of the Internet and mobile phone, the so-called 'prosumer' originates and designs media content, offering it to the world, sometimes useful, sometimes out of sheer self-interest. This happens in the shape of 'consumer-generated content' or 'Open Source' on the free-to-participate platforms such as 'Wikipedia' (http://wikipedia.org), 'Flickr' (http://flickr.com), 'Technorati' (http://technorati.com), myspace' (http://www.myspace.com) or 'YouTube' (http://www.youtube.com).

Now is the time for brands to change their direction towards attracting this enormous, freely radiating energy and assimilate it. The question remains: How can someone use this passion for a subject or a brand creatively in brand communication? By now, there is an increasing amount of indicators to how campaigns could look like that integrate the brand aficionados as ambassadors and thus use the consumers' productivity for the benefit of the brand. However it is a very broad subject matter. For instance there is still much experimentation going on in the sector of social networking. Especially noteworthy is the election campaign of the Democrat's presidential candidate, Barack Obama (http://my.barackobama.com). He used an enormous network of friends on the Internet to collect millions of Dollars in micro-donations. The power emanating from this movement was strong enough to derail the Clintons' established election machinery. These events are forbearers of a great change being underway.

The mobile phone has long ago been linked to this architecture of join-and-contribute platforms, reaching with extended arms through the handset into our real lives. Of course, this also works in the opposite direction. With the mobile phone as a production tool and new mass medium, we can spontaneously contribute to the architecture of the 'Web 2.0' from everywhere; each to their own inclination. Thanks to a full set of features offered by nearly every modern mobile phone, anyone is capable of producing short movies, photographs, audio snippets or text-based content, and – most importantly – to broadcast it. Unlike the profession PC user, the consumer will limit himself to basic functionalities. Thus, the strength of the mobile phone is more likely to be found in the instantaneous, documentary-style treatment of a story's fabric. The mobile phone is ideally suited to impulsively capture and document emotions in the form of text, image, sound and video.

Thus, the erstwhile passive media consumer turns into a one-man show – for instance in news broadcasting: We were used to be presented with news that was professionally prepared for our consumption by trained journalists. Today we experience more and more talented amateurs entering the business of news dispatch. It the age of the mobile phone, first images of terrorist attacks, catastrophes and VIP sightings are being captured using the phone's built-in camera. The most prominent example in Germany is the 'Volksreporter' appearing in the notorious red-top, 'Bild-Zeitung'.

With comparatively little effort, the reader come journalist send his shot-from-the-hip images to a number via a multi-media message. On successful publication, a fee is awarded. The contributions are furthermore published on a dedicated web site, a 'consumer-generated issue' of 'Bild' (http://www.bild.de/BILD/news/leserreporter/home/leserreporter.html). Hence, a contribution to the 'Bild' project creates insignificantly more effort than an active participation in the aforementioned 'Twitter' (http://twitter.com) community. Important for this argument: it is 'Bild' that benefits most from this promotion. The idea of a people's reporter sharpens the brand image decisively by suggesting to the public: Nobody is closer or more compatible with the masses than 'Bild'!

After showing how to work with brand-related subjects with the aid of mobile concepts and consumer-generated content, I would like to underscore that these kinds of promotions not necessarily have to end up being a brand community. Building on existing communities, the brand aficionado won't even have to create special material at all. This is an approach that makes sense when taking into consideration that most consumers are already active in one or more communities. Due to market saturation, the building from scratch of a new brand community will hardly pay off. While visiting South Korea, land of the 'ubiquitous dream', the author experienced that brands already retire from the public part of the World Wide Web in favor of joining an established community such as 'cyworld' (http://cyworld.com) and opening a branded account with them. At the end of the day, that's the home address for the brand enthusiasts.

The following example depicts how existing communities can be involved in the context of a mobile marketing campaign. How to hook up the mobile channel in a most entertaining fashion was demonstrated by Mediamatic in 2005 (Flickr Peep Show was exhibited at Mediamatic/POST CS in Amsterdam during May/June 2005 from http://www.textually.org/textually/archives/2005/05/008427.htm, http://www.mediamatic.net/page/5617/en). They presented the Flickr-Peep-Show installation in Amsterdam. This comprised an attention-grabbing shop-window installation serving as the virtual stage, allowing viewers to browse the Flickr database with the aid of their mobile phone. The viewer would send an SMS to the server, which in turn would subsequently upload images from the Flickr community relating to the subject, displaying it in different displays across shop-windows. An additional twist was added by covering up the windows save for little peepholes. Thus the consumer turned voyeur in the fashion of peep shows found in the red-light districts across the world's port cities. Via the Internet and mobile phone the communicative tension can be built-up and maintained while an emotional dialogue with the brand takes place. The consumer is being integrated into the communication. The desired effect is a positive customer experience and the brand can enter into an advanced phase of the customer relationship.

THESIS 4: LOCATION-BASED SERVICES ALLOW FOR HIGHLY EMOTIVE BRAND INTERACTION

To receive the right offer, anytime and anywhere – location-based services (LBS) are services delivering information that changes dynamically in relation to the user's position. They are based on the geographical positioning of a person in the provider's radio network. The identification of a mobile phone via Bluetooth belongs to the 1context-based services and does not belong to this type of marketing in a narrower sense.

Location-based services are of particular interest in connection with voucher campaigns and sweepstakes, namely when the customer is in close proximity of the destination. There exist a lot of false expectations about the technical possibilities of LBS. A word of caution to begin with: To the disappointment of many advertisers the continuous positioning of the consumer to facilitate individualized broadcasts is not possible offhand. In order to send out promotional messages on the basis of a localized offer to the consumer's mobile phone, its current position in the mobile network has to be determined using a technique called 'triangulation'. The continuous tracking of a customer is cost-prohibitive. In addition, the consumer has to be identified and contacted, both of which requires prior approval.

The following example for the large potential in location-based marketing demonstrates the fact rather impressively: On May 17, 2004, the world's largest brand surprised its customers with a special idea. Whoever opened his fridge after this point, could expect more than just 'the real thing'. The outlook was even a rather surreal one. As part of the campaign 'Unexpected Summer', the company had 100 cans of Coke outfitted with a GPS (Global Positioning System) transponder and a mobile phone. Despite the high-tech contents the packaging still felt and looked like a standard beverage can. The high-tech cans were then packaged as part of the advertised multi-packs. Through the shrink-wrap, the packs were indiscernible (Coke Sweepstake, Unexpected Summer, USA May 9, 2004 from http://www.usatoday.com/money/industries/food/2004-05-09-coke_x.htm).

Nevertheless, the lucky buyer would know the difference as soon as thirst kicked in: Instead of the usual refreshment however, an exciting and spectacular prize was about to be won. Whoever held one of the 100 sought-after cans in hand, knew that he was a winner – however he was not in the know about what he was about to receive let alone where this would happen. By pressing a prominent button on the can's body the customer would activate it – and only a few moments later be listening to the voice of a Coke representative from inside the fake beverage container. The built-in mobile phone would autonomously speed-dial a call centre in order to record the name and address of the lucky winner and to provide instructions about the further goings-on. From this moment on, the customer would carry the can like a mobile phone with a GPS feature, allowing Coca-Cola to pinpoint their winner at any time with the aid of satellite positioning. The winning customer could anticipate to be surprised by the Coke team at any point in time during a whole week.

Certainly such a promotional sweepstakes could have been undertaken in a totally different manner. It is exactly this slightly different set of mechanics that makes this campaign so special – not the usual simple participation with a quick and easy prize award, but the suspense and following surprise. Coke made perfect use of all available channels to tell an exciting story in all its facets. For instance, the already localized cans were shown with their position on the campaign web site. This way, the consumers always knew how many Coke cans were still up for grabs. His interest was kept alive and he was strung along, continuing to look for winning cans. The campaign was complimented by a TV commercial, which further raised expectations by showing the consumer a glimpse on what would happen on finding the right can. The TV spot showed a 'Super Huey' helicopter unloading the prize, a brand new 'Chevy Equinox' right in front of the consumer's feet.

Even the viral accoutrement – if calculated or not – wasn't missing from the well-oiled campaign machinery. The press openly discussed if Coca Cola would turn out to be a risk to national security. The military was informed about the usage of positioning technology in the cans and urged staff members to control every can of Coca-Cola prior to their entering areas requiring security clearance. After all, personal electronic devices were prohibited in military installations due to their inherent security risks. A possible misuse by third parties could not be ruled out either. However, an expert hit the nail on the head when he commented on CNN: "There are many things that keep you awake at night – a can of Coca-Cola is not one of them." It has to be added that not only the participants had their share of fun with this promotion but all fans of the brand who had learned of this spectacular show. It was this wave of enthusiasm that awarded the actual raison-d'être to Coke's 'Unexpected Summer' campaign. This is how a multi-channel campaign with mobile participation is meant to be: Don't search – because your brand will find you wherever you are!

THESIS 5: PERVASIVE GAMING TURNS ADVERTISING INTO AN ALL-ENCOMPASSING GAME

The attentive viewer has been able to encounter persuasive, highly creative pull-campaigns and game-style promotions since the advent of the dot-com era. Straight at the beginning of the 2001 Steven Spielberg / Stanley Kubrick production 'A.I.' one had to pay very close attention in order not to miss one of the best marketing ideas of the young millennium (A.I. Motion Picture Promotion, The Beast, USA 2001 from http://en.wikipedia.org/wiki/The_Beast_(game)).

A fact that in the first instance sounds rather contradictory to all laws of advertising. Some rather intelligent boffins had cooked up a very tricky scheme to prepare the target group of Sci-Fi fans for the upcoming movie event: a large-scale puzzle that unfolded over hundreds of web sites on the Internet. Step by step a detective story revealed itself that was placed in 2142, the year where the plot of 'A.I.' takes place. Despite the fact that such a quest is not inherently aimed at touching the masses, the success was overwhelming and to this day, marketing experts speak with starry eyes of this promotion that became colloquially as 'The Beast'. It was the first 'alternate reality game' being used for commercial purposes – at the time, it had to make do without the participation of a mobile channel.

Such a hype incentivises brands to think about employing a tricky, interleaved communication in the course of a product promotion. A promotion like this would be a perfect playing ground for all interactive media including the mobile phone. It offers the chance for very innovative positioning and wide-ranging interaction. However these games are very resource-intensive. Thus it comes as no big surprise that further prominent Augmented Reality Games (ARG) were originated by software production companies rather than firms further removed from this sector. For example the ARG 'I Love Bees' was conceived as an accompanying promotion for the Microsoft video game 'Halo 2' – myth has it that game participants subjected themselves to car journeys of hundred of miles in order to receive a hint in the shape of a phone call to be picked up at a particular telephone booth (Halo 2 Game Promotion, I love bees, USA 2004 from http://en.wikipedia.org/wiki/I_Love_Bees, http://ilovebees.com). Sometimes it is incredible to what lengths a brand aficionado will go. Had a mobile channel been integrated, participation would have been a less cumbersome.

The participation in a branded community project such as a so-called 'Big Game' by Area/Code can be implemented a lot easier, especially from the consumer's perspective. Based on the concept of 'Conqwest' from 2004 it becomes obvious how brand communication – in this particular case for US telco 'Qwest' in a number of US cities – works with the aid of 'Pervasive Game' (area/ code, Big Game: ConQwest, USA 2003 from http://www.playareacode.com/work/conqwest/). Five teams consisting of students ready to give everything embarked on a two-hour hunt for treasures consisting of Semacodes. Armed with Qwest mobile phones, a city map and instruction manual, and accompanied by huge inflatable totems it was their task to find around 300 hidden Semacodes and photograph them using the mobile's built-in camera. To make the codes count towards the team's score, it was required to pre-install a Semacode reader on the participants' phones. The task of taking the photograph accomplished, the embedded data was transmitted automatically to the team's HQ and credited to their gaming account. Some of the Semacodes were particularly well hidden and took the shape of small stickers on random items or tucked away in a side alley. In the other extreme, blown-up versions of the Semacodes made for great landmarks on billboards and oversize posters.

These outstanding stagings did not only rouse the interest of gamers but of the whole city who wondered for days on end what the weird signs on billboards were all about. The game was further extended by a strategic component. Enter the large inflatable totems, visible from afar, carried by the participants through the total of eight playing fields. Points could only be gained if one's own totem was in that particular zone. Vice versa, an opponent's totem in a zone would block the collection of the Semacodes. In these cases it had to be mediated via the mobile phone which team would have to leave the sector. Teams could bet a certain number of points to secure a particular sector for a pre-arranged amount of time. It soon becomes apparent that the co-ordination work and organization played key roles during the game. The team to have collected the most 'treasures' in the allotted time span won the jackpot in the shape of a lump sum of money which was subsequently donated to the winning team's school.

But how does one measure the success of a 'Pervasive Game' from a marketing perspective? Is the game a success if as many players as possible participate? Or can success be found in a totally different category? Indeed it would be rather complicated for an agency to present a conversion rate to the client's brand after such a game. Further it would be no mean feat to demonstrate, how many consumers turned into customers after having viewed the game.

An analogy to team sports shows what matters in these 'brand stagings'. When a brand sponsors a soccer team to brand the team shirts it does not do this to get the player's attention. The effort is aimed at getting noticed by the audience. To demonstrate how well the brand suits them. To show whom their heart is beating for. To prove to the audience that the brand has understood what it is all about. Even if at first it appears like the focus is on the game and its players, in reality it is about giving a decisive, inspiring statement to the outside world. A statement that, unlike many other events, has a good chance of being noticed.

In the past a feeling of community and togetherness was created when millions of watchers were in front of their TV sets and enjoyed a movie. The following day the consumed content was subsequently discussed – with any luck one of the commercials became subject of the conversation, too. Such community-engendering moments today are few and far between but still desired by the consumer. Pervasive Games prove the point that it is still possible to create such experiences. It the era of fragmented target groups, these events are being experienced by far smaller clusters of people. Instead, the shared moments are being enjoyed with increasing intensity – be it due to the fact of being a part of the event itself or having actively followed its progress on the Internet.

Basically, these experiences are different from yesterday's conventional TV happenings inasmuch as there is now a direct, interactive element involved by adding mobile and web channels. These technologies are even capable of blurring the frontiers between the real and the virtual, medial world. Brand communication thrives on these moments. Pervasive Games could even make a valuable contribution to a consumer's considerably improved brand experience – over and above what was available to him by consuming the brand's products alone. However you mustn't fail to differentiate Pervasive Games from traditional marketing campaigns with the task to turn a participant into a consumer or at least into a prospect. Pervasive Games send a key signal to the outside world. And this signals says: It's fun to live with this brand.

CONCLUSION

Which direction do the creative solutions of the brand campaigns as described in the 5 theses point us towards? Furthermore, which implications result for makers of campaigns that want to communicate across multiple channels?

The mobile phone as an individual feedback channel for consumer, direct service channels to the consumer via Widgets, consumers that turn into heroes, innovative brand experiences through Pervasive Gaming concepts that can be implemented on a world-wide-scale... hopefully the last sections were able to demonstrate how the mobile phone is suitable not only as a dialogue channel but also as a great tool for exciting brand campaigns. This is not only a desirable perspective but a necessary one. After all, the youngest of all mass media would quickly become desolate if we, designers of the future, would not develop creative ideas which would keep extending the channel again and again over and beyond its former frontiers.

"Don't make creative advertising on the mobile phone, but with it!" – this sentence was put at the centre of our book 'Brand's New Toy' by my colleague Bosse Küllenberg and me (Küllenberg & Quente, 2006). It certainly hasn't lost any of its up-to-dateness after the book's publication. It doesn't actually matter how big a role the mobile phone plays in the context of a multi-channel campaign. Was does matter however is to use the new channel according to its capabilities, for instance everywhere a direct customer response is desired. The mobile phone has unique wherewithal as a crossover channel, as a link between media. If it is to be a big-screen movie transporting the brand message, the mobile phone is ready to control the film, while at the same time interacting with the brand, requesting further information, noting down an address and informing other consumers about the event. We have to change our way of thinking as for the first time we have the chance to tell our stories through the medium that is best suited for them. If using more than one medium adds to the dramaturgy this is possible as well – the mobile phone syncs and controls multiple channels with ease.

One key to good multi-channel campaigns is the telling of exciting 'brand stories', which can be told across many media or 'stages'. If the offered stories are enthralling, they turn brand aficionados into brand addicts, who in the best case don't want to miss a single appearance of their favorite brand. The mobile phone has the chance to play a key role here – just as seen by the author on South Korean TV in a commercial for Broadband Internet Service Provider 'WiBro': In this spot, an old man with a walking cane enters an underground train car in Seoul. Instead of picking one of the available seats, to everyone's surprise the old man suddenly grabs one of the loops suspended from the compartment's ceiling to then start chinning himself (e.g. KT WiBro, South Korea 2008 from http://www.ktwibro.com/ktwibro/main.html).The accidental audience on the train is well impressed, one man even jumps up to film the scene with his mobile phone. With lightning speed, the video propagates throughout the world. – the passengers can watch it on their notebooks in the very same compartment even before the old man has finished with his pull-ups. The mobile phone extends the possibilities of the Internet into the real world and vice versa – and nearly overtakes itself in the process. Especially in those places where a high percentage of broadband internet coverage is established – for instance in South Korea.

Even when taking into consideration the large political and cultural differences between Asia, Europe and the United States – the development towards the mobile phone as an advertising medium is unstoppable. And when stories are, unlike on mobile TV, told with the mobile phone rather than on it, then increasing interactivity raises the question for story management and the organization of channels in advertising campaigns. Especially when one bets on consumer-generated content and the power of Open Source marketing – just as in the aforementioned TV spot. The questions remains: "How do we manage the machines?" It is particularly interesting due to its tendency to polarize. Should brands still be managed in a hierarchical and authoritative manner in the future? Or does one allow the consumer contribute to the brand's image in democratic and interactive way?

Some final words, that may help bringing up a solution path: One viable way could be to offer areas or playing grounds to the consumer, in which he can participate creatively and interactively in the continuation of the 'brand story', while other parts, such as the brand's core or canon remain protected. So happening in the 'Star Wars' universe, where there exists a core of 6 movies, which is continued in installments created by more or less professional fans. This phenomenon has already brought to light its own dynamics consisting of countless novels and fan-produced movies. But there is another – much more radical – example for an uncontrolled but still working model that we all know of and most advertising professionals hardly spend a single thought on, probably because it is too obvious. A story that repeats itself since thousands of years save for small modifications and has even managed to integrate ideas originating in other cultural circles. This story is called 'Christmas'. This story is so enormously huge that it even allows for Christmas trees and Santa Clauses even if those don't event appear in its original version. Christmas works – and with this story in mind one or the other marketing strategist will allow a little more brand democracy and interactivity for his brand. The giant called mobile phone may be of effective assistance.

REFERENCES

Advance A/S. (2006). Illum Window Shopping. Denmark. From http://www.flickr.com/photos/angermann/110822340/ and http://illum.dk/ and http://www.advance.dk/

area/ code, Big Game: ConQwest. (2003). USA. From http://www.playareacode.com/work/conqwest/

Chaos Computer Club. (2002). Blinkenlights Arcade, Bibliothèque nationale de France, Paris, France, September 25 – October 6, 2002 from http://www.blinkenlights.net/arcade

Coca-Cola, A. G. Campaign. (2007). Festiver 04 Germany/ Switzerland. From http://www.spielplatz.cc/presse/wordpress/?cat=7 and www.festifever.ch

Coke Sweepstake. (2004). Unexpected Summer. USA. From http://www.usatoday.com/money/industries/food/2004-05-09-coke_x.htm

Flickr Peep Show was exhibited at Mediamatic/POST CS in Amsterdam during May. (2005, June). From http://www.textually.org/textually/archives/2005/05/008427.htm and http://www.mediamatic.net/page/5617/en

Halo 2 Game Promotion. (2004). I love bees. USA. From http://en.wikipedia.org/wiki/I_Love_Bees), (http://ilovebees.com

ITU. (2007). Trends in Telecommunication Reform 2007–The road to Next-Generation Networks (NGN) Summary.

Jung, Holger, & von Matt (2002). Jean-Remy, Momentum–Die Kraft, die Werbung heute braucht. Berlin, Germany

KT WiBro, South Korea. (2008). From http://www.ktwibro.com/ktwibro/main.html

Küllenberg, B., & Quente, C. (2006). Brand's New Toy–Kreative Markenkommunikation mit Handy & Co. (pp. 211). Munich, Germany.

Lumière, A., & Lumière, L. (Dir.) (1895). Arrivée d'un train à La Ciotat, L'. France. From http://www.imdb.com/title/tt0000012/

M:Metrics Press Release, M:Metrics: iPhone Hype Holds Up, Seattle/ London, March 18, 2008 from http://www.mmetrics.com/press/PressRelease.aspx?article=20080318-iphonehype

Mobile Phone. (2008). www.wikipedia.org, November 30, 2008 from http://en.wikipedia.org/wiki/Mobile_phone)

Motion Picture Promotion, A. I. (2001). The Beast. USA. From http://en.wikipedia.org/wiki/The_Beast_(game)

Postinett, A. (2008). Smartphone-Hype unter der Lupe: iPhone-Nutzer lieben das Web, Handelsblatt, 20.03.2008 from http://www.handelsblatt.com/technologie/mobile-welt/iphone-nutzer-lieben-das-web;1406862

Rodgers, Z. (2005). Nike iD Billboard Invites Mobile Users. The ClickZ Network, May 3, 2005 from http://www.clickz.com/3502186

KEY TERMS AND DEFINITIONS

ARG: An alternate reality game (ARG) is an interactive narrative that uses the real world as a platform, often involving multiple media and game elements, to tell a story that may be affected by participants' ideas or actions. An ARG is a game which deliberately blurs the line between the game and the real world. Players investigate the world of the game using the same tools with which they interact with the real world such as websites, email, telephone conversations and even in-person discussions with actors playing game characters. ARGs are now being recognized by the mainstream entertainment world: e.g. "I love bees" was a promotion for Halo 2 and the "Lost Experience"

Brand Campaign: A brand campaign has the purpose of developing a brand with the aid of communication. It is the part of brand management that is visible to the consumer ('branding"). The main goal of a brand campaign is to differentiate the own product of those proffered by the competition or to differentialte the own company from its competitors. Behind all this is the insight that a brand has a higher value of recognition and that the consumer connects certain attributes, features and achievements with it. Thereby the brand is supposed to aid the consumer's orientation amongst the offerings and radiate trust

Brand Stories: To tell a "brand story" means to dramatize a story suitable to transport and elaborate on the core contents of a brand's message. At the same time, a "brand story" puts the story's components in order across all campaign channels. Ideally, a good "brand story" threads itself through multiple promotional media and completes itself to a whole experience. Alternatively, a "brand story" can be told in time and location-independent climaxes. It is not to be considered a scientific term but rather an auxiliary construct for those involved in the creation of multi-channel campaigns

Consumer Generated Content: "Consumer generated content" (CGC) is also known as "user generated content" (UGC) or "consumer generated media" (CGM) or "user created content (UCC). The term refers to various kinds of media content, publicly available, that are produced by end-users. It entered mainstream usage during 2005 having arisen in new media content production circles and it is used for a wide range of applications including problem processing, news, gossip and research. All digital media technologies are included, such as question-answer databases, digital video, blogging, podcasting, mobile phone photography and wikis

Creativity Creativity is a mental and social process involving the generation of new ideas or concepts, or new associations of the creative mind between existing ideas or concepts. An alternative conception of creativness is that it is simply the act of making something new. From a scientific point of view, the products of creative thought are usually considered to have both originality and appropriateness. It is often useful to explicitly distinguish between creativity and innovation. Creativity is typically used to refer to the act of producing new ideas, approaches or actions, while innovation is the process of both generating and applying such creative ideas in some specific context: Innovation begins with creative ideas

Location Based Services: A location-based service (LBS) is an information and entertainment service, accessible with mobile devices through the mobile network and utilizing the ability to make use of the geographical position of the mobile device. LBS include services to identify a location of a person or object, such as discovering the nearest banking cash machine or the whereabouts of a friend or employee. LBS include personalized weather services and even location-based games. They are an example of telecommunication convergence

Open Source Marketing: Open source is a development methodology, which offers practical accessibility to a product's source (goods and knowledge). Some consider open source as one of various possible design approaches, while others consider it a critical strategic element of their operations. Before open source became widely adopted, developers and producers used a variety of phrases to describe the concept

The Term Open Source Gained Popularity With The Rise Of The Internet: Which provided access to diverse production models, communication paths, and interactive communities. Open Source Marketing allows the consumer the possibility to partake in the brand's design up to a certain point. This participation can range from self-designed adverts to the definition of product names or even the composition of whole products

Pervasive Game: A pervasive games are multi-player games that take place in an urban setting – on city streets and built up urban environments. Pervasive games almost support some kind of localization technology, for example by using satellite positioning like GPS. Current research trends are looking to other embedded mobile technologies such as Bluetooth or mobile tagging

Widget: A widget is a small computer program that does not operate as an independent application but is tied into a graphical user interface (GUI) or a web site. Mainly these are utility or helper applications. In parallel to the development of widgets for operating systems, widgets in the shape of web applications started to evolve. Many of those widgets do not offer complex functionality but simply display information gathered from other sources (syndication) Apple introduced widgets in April 2005 with the integration of "Dashboard" into its latest version of Mac OS X

WiBro: WiBro (Wireless Broadband) is a wireless broadband Internet technology being developed by the South Korean telecoms industry. WiBro is the South Korean service name for IEEE 802.16e (mobile WiMAX) international standard. WiBro base stations will offer an aggregate data throughput of 30 to 50 Mbit/s and cover a radius of 1-5 km allowing for the use of portable internet usage. The technology will also offer Quality of Service that allows for WiBro to stream video content and other loss-sensitive data in a reliable manner. Two South Korean Telco (KT, SKT) launched commercial service in June 2006

Compilation of References

Aaker, D. A., Kumar, V., & Day, G. (1995). *Marketing Research*, Wiley, New York. p. 254.

Aaltonen, A., Huuskonen, P., & Lehikoinen, J. (2005). Context Awareness Perspectives for Mobile Personal Media. *Information Systems Management, 22*(4), 43–55. doi:10.1201/1078.10580530/45520.22.4.20050901/90029.5

Abi Research. (2007). Near Field Communication. Retrieved October 20, 2008, from http://www.abiresearch.com/products/market_research/Near-Field_Communications_(NFC)

Abowd, G. D., & Mynatt, E. D. (2000). Charting past, present, and future research in ubiquitous computing. *ACM Transactions on Computer-Human Interaction, 7*(1), 29–58. doi:10.1145/344949.344988

Ackerman, M., Darrel, T., & Weitzner, D. (2001). Privacy in context. *Human-Computer Interaction, 16*, 167–176. doi:10.1207/S15327051HCI16234_03

Aditham, K. (2007). China's Mobile Charge: Madhouse CEO Joshua Maa Reveals the Framework and Future of the Chinese Mobile Beat. Retrieved August 12, 2007, from Adotas.com

Advance A/S. (2006). Illum Window Shopping. Denmark. From http://www.flickr.com/photos/angermann/110822340/ and http://illum.dk/ and http://www.advance.dk/

Afuah, A., & Tucci, C. (2002). *Internet Business Models and Strategies: Text and Cases 2/e*. McGraw-Hill/Irwin

Agarwal, R., & Prasad, J. (1999). Are Individual Differences Germane to the Acceptance of New Information Technologies? *Decision Sciences, 30*(2), 361–391. doi:10.1111/j.1540-5915.1999.tb01614.x

Agosti, M., & Ferro, N. (2003). Managing the Interactions between Handheld Devices, Mobile Applications, and Users. In Lim, E.-P., & Siau, K. (Eds.), *Advances in Mobile Commerce Technologies* (pp. 205-234). Hershey, PA: Idea Group Publishing.

Ahonen, T. (2002) *M-profit. Making money from 3G services* New York: John Wiley and Sons.

Ailloni-Charas, D. (1984). *Promotion: a guide to effective promotional planning, strategies, and executions*. New York: John Wiley & Sons.

AIS-About US Company Profile. (2007). Retrieved Oct 31, 2008, from http://www.ais.co.th.

Ajzen, I. (1985). From Intentions to Actions: a Theory of Planned Behavior. In: Kuhl, J., Beckmann, J. (Ed.), Action control: from cognition to behaviour (pp. 11-39). Berlin: Springer.

Ajzen, I. (1988). *Attitudes, personality and behavior*. Dorsey Press, Chicago, IL.

Ajzen, I. (1991). The theory of planned behavior. *Organizational Behavior and Human Decision Processes, 50*, 179–211. doi:10.1016/0749-5978(91)90020-T

Ajzen, I. (2002). Perceived behavioral control, self-efficacy, locus of control, and the theory of planned behavior. *Journal of Applied Social Psychology, 32*, 665–683. doi:10.1111/j.1559-1816.2002.tb00236.x

Ajzen, I., & Fishbein, M. (1980). *Understanding Attitudes and Predicting Social Behavior*, Englewood Cliffs, New Jersey.

Ajzen, I., & Madden, T. J. (1986). Prediction of goal- directed behavior from attitudinal and normative variables. *Journal of Experimental Social Psychology, 22*, 453–474. doi:10.1016/0022-1031(86)90045-4

Åkesson, M., & Ihlström Eriksson, C. (2008). From multi channel publishing towards ubiquitous media environment. *Journal of Graphic Technology, 10*, 126–148.

Algesheimer, R., Dholakia, U. M., & Herrmann, A. (2005). The Social Influence of Brand Community: Evidence from European Car Clubs. *Journal of Marketing*, *69*(3), 19–34. doi:10.1509/jmkg.69.3.19.66363

Amit, R., & Zott, C. (2001). Value creation in e-business. *Strategic Management Journal*, *22*(6/7), 493–520. doi:10.1002/smj.187

Anchar, B., & D'Incau, D. (2002a). Value-Added Services in Mobile Commerce: An Analytical Framework and Empirical Findings from a National Consumer Survey, *35th Hawaii International Conference on System Sciences (HICSS-37)*, Hawaii, USA.

Anckar, B., & D'Incau, D. (2002b). Value creation in mobile commerce: Findings from a consumer survey. *Journal of Information Technology Theory and Application*, *4*(1), 43–64.

Anderson, C. (2006). *The long tail: Why the future of business is selling less of more.* New York: Hyperion.

Anderson, D., Canepa, S., Ekow, N., & Mannel, S. (207). *Navigating the media divide.* USA: IBM Global services.

Anderson, E. (1998). Customer Satisfaction and Word of Mouth. *Journal of Service Research*, *1*(1), 5–17. doi:10.1177/109467059800100102

Anderson, E. W., Fornell, C., & Lehmann, D. R. (1994). Customer Satisfaction, Market Share and Profitability: Findings from Sweden. *Journal of Marketing*, *58*(3), 53–66. doi:10.2307/1252310

Anderson, J. C., & Gerbing, D. W. (1988). Structural equation modeling in practice: A review and recommended two-step approach. *Psychological Bulletin*, *103*(3), 411–423. doi:10.1037/0033-2909.103.3.411

Andreasen, A. R. (1966). Geographic mobility and market segmentation. *JMR, Journal of Marketing Research*, *3*(4), 341–348. doi:10.2307/3149849

Angell, M. (2005, September 16). Mobile marketers eye ways to connect with wireless users. Investor's Business Daily, A04.

Anonymous. (2007). HappyDigits Shop Finder. In: *Interaktive Trends 2007/2008, Jahrbuch Deutscher Multimedia Award* (pp. 86-87). Hamburg: J&S Dialog-Medien.

Anonymous. (2008). Gewusst wo: Location Based Services. *direkt marketing*, 8(2008), 18.

area/ code, Big Game: ConQwest. (2003). USA. From http://www.playareacode.com/work/conqwest/

Arndt, J. (1967). Role of product-related conversations in the diffusion of a new product. *JMR, Journal of Marketing Research*, *4*(3), 291–295. doi:10.2307/3149462

Askegaard, S., Arnould, E. J., & Kjeldgaard, D. (2005). Postassimilationist ethnic consumer research: Qualifications and extensions. *The Journal of Consumer Research*, *32*(1), 160–170. doi:10.1086/426625

Assael, H. (1984). *Consumer behavior and marketing action.* 2nd edition. Boston: Kent Publishing Company.

Atkinson, P., Coffey, A., & Delamont, S. (2003). *Key themes in qualitative research: Continuities and change.* Walnut Creek: Altamira Press.

Attwood, R., & Duncan, S. (2000). *Mobile Commerce: Strategies for New Business Paradigm*, London Ft: Telecoms.

Au, Y. A., & Kauffman, R. J. (2008). The economics of mobile payments: Understanding stakeholder issues for an emerging financial technology application. *Electronic Commerce Research and Applications*, *7*(2), 141–164. doi:10.1016/j.elerap.2006.12.004

AxisMobile. (2007). AxisMobile consumer mobile email marketing segmentation White Paper. Retrieved May 5, 2008, from http://www.axismobile.com/ files/ AxisMobile_ Consumer_ Mobile_ Email_ Marketing_ Segmentation_ White_ Paper.pdf

Azuki Systems. (2008). *Today's Mobile Phone Users: Current and Emerging Trends,* September, 2008. Retrieved November 1, 2008 from http://www.azukisystems.com/images/uploads/Azuki_User_Survey_Sept_2008_vfinal.pdf

Babbie, E. (2005). *The basics of social research* (3rd ed.). Belmont, CA: Thomson/Wadsworth.

Babin, B. J., & Attaway, J. A. (2000). Atmospheric Affect as a Tool for Creating Value and Gaining Share of Customer. *Journal of Business Research*, *49*(2), 91–99. doi:10.1016/S0148-2963(99)00011-9

Babin, B. J., Darden, W. R., & Griffin, M. (1994). Work and/or Fun: Measuring Hedonic and Utilitarian Shopping Value. *The Journal of Consumer Research*, *20*(4), 644–656. doi:10.1086/209376

Bächle, M. (2006). Social Software. *Informatik Spektrum*, *29*(2), 121–124. doi:10.1007/s00287-006-0063-2

Backhaus, K., Erichson, B., Plinke, W., & Weiber, R. (2005). *Multivariate Analysemethoden* (11. Auflage ed.). Berlin, Heidelberg: Srpinger.

Bagozzi, R. P., & Yi, Y. (1988). On the Evaluation of Structural Equation Models. *Journal of the Academy of Marketing Science*, *16*(1), 74–94. doi:10.1007/BF02723327

Bai, F., & Helmy, A. (2006). Wireless ad-hoc and sensor networks. In (chap. A Survey of Mobility Models). Springer.

Bakan, D. (1966). The Duality of Human Existence. Chicago, IL: Rand McNally.

Balaban, D. (2008). NFC: When Will The Breakthrough Come? *Cards & Payments*, *21*(1), 30–35.

Balabanis, G., Mitchell, V.-W., & Heinonen-Mavrovouniotis, S. (2007). SMS-based surveys: Strategies to improve participation. *International Journal of Advertising*, *26*(3), 369–385.

Balasubramanian, S., Peterson, R. A., & Jarvenpaa, S. L. (2002). Exploring the Implications of m-Commerce for Markets and Marketing. *Journal of the Academy of Marketing Science*, *30*(4), 348–361. doi:10.1177/009207002236910

Balasubramanian, S., Peterson, R. A., & Jarvenpaa, S. L. (2002). Exploring the Implications of M-Commerce for Markets and Marketing. *Journal of the Academy of Marketing Science*, *30*(4), 348–361. doi:10.1177/009207002236910

Baltas, G. (2003). Determinants of internet advertising effectiveness: An empirical study. *International Journal of Market Research*, *45*(4), 505–513.

Bamba, F., & Barnes, S. J. (2007). SMS advertising, permission and the consumer: a study. *Business Process Management Journal. Special Issue on Mobile CRM*, *13*(6), 815–829.

Banomyoung, A. (2007). Personal Interview at TrueMove. May 2007.

Bansal, H. S., & Voyer, P. A. (2000). Word-of-mouth processes within a services purchase decision context. *Journal of Service Research*, *3*(2), 166–177. doi:10.1177/109467050032005

Barnes, J. A., & Scornavacca, E. (2004). Mobile Marketing: The Role of Permission and Acceptance. *International Journal of Mobile Communications*, *2*(2), 128–139. doi:10.1504/IJMC.2004.004663

Barnes, S. (2002). The mobile commerce value chain: analysis and future developments. *International Journal of Information Management*, *22*, 91–108. doi:10.1016/S0268-4012(01)00047-0

Barnes, S. (2002). Wireless digital advertising: nature and implications. *International Journal of Advertising*, *21*(3), 399–420.

Barnes, S. C., & Scornavacca, E. (2004). Mobile Marketing: The Role of Permission and Acceptance. *International Journal of Mobile Communications*, *2*(2), 128–139. doi:10.1504/IJMC.2004.004663

Barnes, S. J. (2002). Wireless Digital Advertising: Nature and Implications. *International Journal of Advertising*, *21*, 399–420.

Barnes, S. J. (2003). Location-based services: The state-of-the-art. *e-Service Journal*, *2*(3), 59-70.

Barnes, S. J., & Scornavacca, E. (2004). Mobile marketing: the role of permission and Acceptance. *International Journal of Mobile Communications*, *2*(2), 128–139. doi:10.1504/IJMC.2004.004663

Barnes, S. J., & Scornavacca, E. (2004). Mobile marketing: the role of permission and acceptance. *International Journal of Mobile Communications*, *2*(2), 128–139. doi:10.1504/IJMC.2004.004663

Barnes, S. J., & Scornavacca, E. (2008). Uncovering patterns in mobile advertising opt-in behaviour: a decision hierarchy approach. *International Journal of Mobile Communications*, *6*(4), 405–416. doi:10.1504/IJMC.2008.018050

Barnes, S., & Huff, S. (2003). Rising sun: iMode and the wireless Internet. *Communications of the ACM*, *46*(11), 79–84. doi:10.1145/948383.948384

Baron, S., Patterson, A., & Harris, K. (2006). Beyond technology acceptance - understanding consumer practice. *International Journal of Service Industry Management*, *17*(2), 111–135. doi:10.1108/09564230610656962

Barth, K., Hartmann, M., & Schröder, H. (2007). *Betriebswirtschaftslehre des Handels*. 6th Edition, Wiesbaden, Gabler.

Barwise, P., & Strong, C. (2002). Permission Based Mobile Advertising. *Journal of Interactive Marketing*, *16*(1), 14–24. doi:10.1002/dir.10000

Basu, A., Mazumdar, T., & Raj, S. P. (2003). Indirect network effects on product attributes. *Marketing Science*, *22*(2), 209–221. doi:10.1287/mksc.22.2.209.16037

Bathwal, B. (2008). Juniper Research Predicts Growth for "Mobile Coupons". Retrieved July 26, 2008, from http://

www.tmcnet.com/voip/ip-communications/articles/26836-juniper-research-predicts-growth-mobile-coupons.htm

Battenberg, A. (2002). Mobile Business mit Communitys. In Teichmann, R., & Lehner, F. (Eds.), *Mobile Commerce* (pp. 130-143). Berlin, Springer.

Bauer, H. H., Barnes, S. J., Reichardt, T., & Neumann, M. M. (2005). Driving Consumer Acceptance of Mobile Marketing: A Theoretical Framework and Empirical Study. *Journal of Electronic Commerce Research*, *6*(3), 181–192.

Bauer, H. H., Barnes, S., Reichardt, T., & Neumann, M. M. (2005). Driving consumer acceptance of mobile marketing: A theoretical framework and empirical study. *Journal of Electronic Commerce Research*, *6*(3), 181–192.

Bauer, H. H., Haber, T. E., & Reichhardt, T. (2006). Location Based Services. Große Potenziale für Direktmarketing. *Direkt Marketing*, o. Jg., Heft 8, S. 36-38.

Bauer, H. H., Reichardt, T., & Neumann, M. M. (2004). Zu jeder Zeit an jedem Ort – oder über die Potenziale des Mobile Marketing. *Direktmarketing*, *40*(10), 32–37.

Bauer, H. H., Reichardt, T., & Neumann, M. M. (2005b). Bestimmungsfaktoren der konsumentenseitigen Akzeptanz von Mobile Marketing. *der markt - Zeitschrift für Absatzwirtschaft und Marketing*, 44(2), 59-71.

Bauer, H. H., Reichardt, T., Barnes, S. J., & Neumann, M. M. (2005). Driving Consumer Acceptance of Mobile Marketing: A Theoretical Framework and Empirical Study. *Journal of Electronic Commerce Research*, *6*(3), 181–192.

Bauer, H., Grether, M., & Leach, M. (1999). Relationship Marketing im Internet. *GfK Jahrbuch der Absatz und Verbrauchsforschung*, *45*(3), 284–302.

Bauer, H., Reichardt, T., Barnes, S., & Neumann, M. (2005). Driving consumer acceptance of mobile marketing: A theoretical framework and empirical study. *Journal of Electronic Commerce Research*, *6*(3), 181–192.

Bauer, R., & Greyser, S. (1968). *Advertising in America: The consumer view*. Boston: Harvard University Press.

Bayartsaikhan, K., Danielak, P., Dunst, K., Guibert, J., Luxford, L., & Romanossian, R. (2007). Market for third screen: a study of market potential of mobile TV and video across the U.S. and selected European countries. *International Journal of Mobile Marketing*, *2*(1), 12–27.

BBC News. (2005). S Korean consumers spending again. February 2005

BBC. (2007). Ask rolls out search privacy tool. Retrieved December 11, 2007, from http://news.bbc.co.uk/1/hi/technology/7138260.stm

Becker, C., Bauer, M., & Hähner, J. (2002, November 16-20). Usenet-on-the-fly - supporting locality of information in spontaneous networking environments. *Liscano, Ramiro (ed.); Kortuem, Gerd (ed.): Workshop on Ad Hoc Communications and Collaboration in Ubiquitous Computing Environments.*

Becker, G. S. (1965). A Theory of the Allocation of Time. *The Economic Journal*, *75*(299), 493–517. doi:10.2307/2228949

Becker, J., & Winkelmann, A. (2008). Handelscontrolling. Berlin: Gabler.

Becker, M. (2005). Effectiveness of Mobile Channel Additions and a Conceptual Model Detailing the Interaction of Influential Variables. Retrieved on November 2, 2006, from http://www.iloopmobile.com/news/mb_research_111705.htm

Becker, M. (2005). Effectiveness of mobile channel additions and a conceptual model detailing the interaction of influential variables. Access January 2007, from http://www.iloopmobile.com/news/mb_research_111705.htm

Becker, M. (2005). *Effectiveness of mobile channel additions and a conceptual model detailing the interaction of influential variables.* MMA Global. Retrieved December 22, 2008, from http://mmaglobal.com/modules/article/view.article.php/131

Becker, M. (2006). *Academic review: Consumer acceptance of mobile marketing.* MMA Global. Retrieved July 15, 2008, from http://mmaglobal.com/modules/wfsection/article.php?articleid=289

Bell, J. E. (1969). Mobiles-a neglected market segment. *Journal of Marketing*, *33*(2), 37–44. doi:10.2307/1249400

Bem, D. J., & Allen, A. (1974). On predicting some of the people some of the time: The search for cross-situational consistencies in behavior. [Bigne, E.]. *Psychological Review*, *81*, 506–520. doi:10.1037/h0037130

Benbasat, I., Goldstein, D. K., & Mead, M. (1987). The case research strategy in studies of information systems. *MIS Quarterly*, *11*(3), 369–386. doi:10.2307/248684

Bennemann, S., & Möllenberg, A. (2001). *M-Commerce aus der Marketing-Perspektive. Technische Voraussetzungen, State of the Art, Künftige Möglichkeiten.* Norderstedt, Books on Demand.

Berndt, H. (2008). Towards 4G Technologies. New York: Wiley.

Betz, J., & Krafft, M. (2002). Die Wirkung des E-Commerce auf die Kundenzufriedenheit und Kundenbindung. *wiss. Schriftenreihe des Zentrums für Marktorientierte Unternehmensführung*, No. 23, Vallendar.

Beyer, T. (2005). RSS-Feeds im Internet. *CLB Chemie im Labor und Biotechnik, 56*(5), 144–148.

Bezjian-Avery, A., Calder, B., & Iacobucci, D. (1998). Interactive advertising vs. traditional advertising. *Journal of Advertising Research, 38*(4), 23–32.

BGH. (2008). Schutz des Verbrauchers vor Werbung durch E-Mail und SMS. Retrieved January 15, 2009, from http://juris.bundesgerichtshof.de/cgi-bin/rechtsprechung/document.py?Gericht=bgh&Art=en&Datum=Aktuell&Sort=12288&anz=627&pos=17&nr=44522&linked=pm&Blank=1.

Bhattacherjee, A. (2001). Understanding Information Systems Continuance: An Expectation- Confirmation Model. *MIS Quarterly, 25*(3), 351–370. doi:10.2307/3250921

Bidgoli, H. (2002). *Electronic Commerce – Principles and Practice*. San Diego: Academic Press.

Bigné, E., Ruiz, C., & Sanz, S. (2007). Key Drivers of Mobile Commerce Adoption. An Exploratory Study of Spanish Mobile. *Journal of Theoretical and Applied Electronic Commerce Research, 2*(2), 48–60.

Bird, D. (1989). *Commonsense direct marketing.* London: Kogan Page.

BITKOM. (2006). *UMTS - Fünf Länder haben bei UMTS den Durchbruch geschafft*, from http://www.bitkom.org/de/presse/30739_37663.aspx

Biyalogorsky, E., Gerstner, E., & Libai, B. (2001). Customer referral management: optimal reward programs. *Marketing Science, 20*(1), 82–95. doi:10.1287/mksc.20.1.82.10195

Blaber, G., Bouchard, J. P., & Brown, A. (2006). IDC Western European market analysis, mobile phone forecast and analysis 2006-2010.

Blackwell, A., & Green, T. (2003). Notational systems—the cognitive dimensions of notations framework. In J. Carroll (Ed.) *HCI Models,Theories and Frameworks.* Los Altos, CA: Morgan Kaufmann.

Blattberg, R., & Neslin, S. (1990). *Sales promotion: concepts, methods, and strategies.* Englewood, NJ: Prentice Hall.

Bliemel, W. F., & Eggert, A. (1998). Kundenbindung – die neue Sollstrategie. *Marketing ZFP, 20*(1), 37–46.

Blythe, J. (2008). *Consumer behaviour.* London: Thomson.

Bodenstorfer, M., & Hasenauer, R. (2005). Location Based Services. In Holzmüller, H.H., & Schuh, A. (Ed.), *Innovationen im sektoralen Marketing* (pp. 337-350). Heidelberg: Physica.

Boehm, M. (2008). Determining the Impact of Internet Channel Use on a Customer's Lifetime. *Journal of Interactive Marketing, 22*(3), 2–22. doi:10.1002/dir.20114

Bohlin, E., Burgelman, J.-C., & Casal, C. R. (2007). The future of mobile communications in the EU. *Telematics and Informatics, 24*(3), 238–242. doi:10.1016/j.tele.2007.01.011

Bojin, P., & Spencer, I. (2006). Mobile marketing-the Chinese perspective. *International Journal of Mobile Marketing, 1*(2), 50–59.

Bone, P. F. (1995). Word-of-Mouth Effects on Short-term and Long-term Product Judgments. *Journal of Business Research, 32*(3), 213–223. doi:10.1016/0148-2963(94)00047-I

Boudreau, M. C., Gefen, D., & Straub, D. (2001). Validation in IS Research: A State-of-the-Art Assessment. *MIS Quarterly, 25*(1), 1–16. doi:10.2307/3250956

Boulding, W., Staelin, R., Ehret, M., & Johnston, W. J. (2005). A Customer Relationship Management Roadmap: What Is Known, Potential Pitfalls, and Where to Go. *Journal of Marketing, 69*(4), 155–166. doi:10.1509/jmkg.2005.69.4.155

Bourreau, M., & Lethiais, V. (2006). Pricing information goods: free vs. pay content. In Brousseau, E., & Curien, N. (Ed.), *Internet and Digital Economics* (pp. 345-367). Cambridge: Cambridge University Press.

Brakel, O., Groh, G., Hillebrand, C., & Tasch, A. (2004). *Privacy-Einstellungen bei Lokalisierungsdiensten.* Paper presented at the MKWI04, Essen, Germany.

Brancheau, J. C., & Wetherbe, J. C. (1990). The adoption of spreadsheet software: testing Innovation Diffusion Theory in the context of end-user computing. *Information Systems Research, 1*(2), 115–143. doi:10.1287/isre.1.2.115

Brandstetter, E. (2006). Schmucke Erfolge mit mobiler Kundschaft. *Economy Special Innovation, 7,* 14.

Brassington, F., & Pettitt, S. (2006). *Principles of marketing fourth edition.* Hemel Hempstead: Prentice Hall.

Brehm, J. W. (1966). *A Theory of Psychological Reactance.* New York: Academic Press.

Brehm, J. W. (1966). *A theory of psychological reactance*, New York, NY: Academic Press.

Brehm, J. W. (1972). *Response to loss of freedom. A theory of psychological reactance*, Morristown, NJ: General Learning Press.

Brehm, S. S., & Brehm, J. W. (1981). *Psychological reactance: A theory of freedom and control*, New York, NY: Academic Press.

Brewster, S., Leplatre, C., & Crease, M. (1998). Using non-speech sounds in mobile computing devices. In *Proceedings of the First Workshop on HCI for Mobile Devices, Glasgow,* U.K.

Brown, J., & Reingen, P. (1987). Social Ties and Word-of-Mouth Referral Behavior. *The Journal of Consumer Research, 14*(3), 350–362. doi:10.1086/209118

Brown, T. J., Barry, T. E., Dacin, P. A., & Gunst, R. F. (2005). Spreading the Word: Investigating Antecedents of Consumers' Positive Word-of-Mouth Intentions and Behaviors in a Retailing Context. *Journal of the Academy of Marketing Science, 33*(2), 123–138. doi:10.1177/0092070304268417

Brown, W. J. (2002). E-mail Marketing: A Better Alternative? *Lodging Hospitality* 58, 5, 20.

Bruhn, M. (2002). *Relationship Marketing: Management of Customer Relationships.* Harlow: Prentice Hall.

Brum, M., & Rettie, R. (2001). M-commerce: The Role of SMS text messages. *Proceedings of the 4th biennial International Conference on Telecommunication and Information Markets*, Karlsruhe, Germany.

Bruner, G. C. II, & Kumar, A. (2007). Attitude toward location-based advertising. *Journal of Interactive Advertising, 7*(2), 1.

Bruner, G. C., & Kumar, A. (2005). Explaining consumer acceptance of handheld Internet devices. *Journal of Business Research, 58*(5), 553–558. doi:10.1016/j.jbusres.2003.08.002

Büttner, O. B., Schulz, S., & Silberer, G. (2006). Vertrauen, Risiko und Usability bei der Nutzung von Internetapotheken. In Bauer, H. H., Neumann, M.M., & Schüle, A. (Ed.), *Konsumentenvertrauen* (pp. 355-366). München: Vahlen.

Bulander, R., Decker, M., Schiefer, G., & Kolmel, B. (2005). Comparison of different approaches for mobile advertising. Proceedings of the 2nd IEEE International Workshop on Mobile Commerce and Services (WMCS '05), IEEE Computer Society, Munich, Germany, July, 174-182.

Bult, J. R., Scheer, H., & Wansbeek, T. (1997). Interaction between target and mailing characteristics in direct marketing, with an application to health care fund raising. *International Journal of Research in Marketing, 14*, 301–308. doi:10.1016/S0167-8116(97)00012-8

Burke, R. R. (2000). Technology and the Customer Interface: what consumers want in the physical and the virtual store. *Journal of the Academy of Marketing Science, 30*(4), 411–432. doi:10.1177/009207002236914

Burton, D. (2002). Incorporating ethnicity into marketing intelligence and planning. *Marketing Intelligence & Planning, 20*(7), 442–451. doi:10.1108/02634500210450873

Burton-Jones, A., & Hubona, G. S. (2005). Individual differences and usage behavior: Revisiting a technology acceptance model assumption. *The Data Base for Advances in Information Systems, 36*(2), 58–77.

Bush, A., Smith, R., & Craig, M. (1999). The influence of consumer socialization variables on attitude toward advertising: A comparison of African-Americans and Caucasians. *Journal of Advertising, 28*(3), 13–25.

Business Wire. (2007, February 28). IDC Says Don't Underestimate Full Potential of Mobile Marketing. Retrieved April 10, 2008, from http://www.fiercemobilecontent.com/node/2941

Business Wire. (2008, May 5). 21st Century Mobile Marketing—Insights into the World's Most Advanced Mobile Society: Japan. Retrieved May 10, 2008, from http://www.researchandmarkets.com/reports/ c90795

Camponovo, G., & Pigneur, Y. (2003). *Business model analysis applied to mobile business.* Paper presented at the 5th International Conference on Enterprise Information Systems (ICEIS 2003) Angers, France.

Cannon, D. A. (2002). The ethics of database marketing. *Information Management Journal, 36*(3), 42–44.

Capraro, A. J., Broniarczyk, S., & Srivastava, R. K. (2003). Factors Influencing the Likelihood of Customer Defection: The Role of Customer Knowledge. *Journal of the Academy of Marketing Science, 31*(2), 164–175. doi:10.1177/0092070302250900

Carlsson, C., Carlsson, J., Hyvönen, K., Puhakainen, J., & Walden, P. (2006). Adoption of Mobile Devices/Services:

Searching for Answers with the UTAUT, *Proceedings of the 39th Annual Hawaii International Conference on System Sciences (HICSS'06)*, Track 6: 1-10.

Carlsson, C., Hyvönen, K., Repo, R., & Walden, P. (2005). Asynchronous adoption patterns of mobile services. *Proceedings of the 38th Hawaii International Conference on System Sciences (HICSS-38).* Big Island, HI, January 3-6.

Carroll, A., Barnes, S. J., Scornavacca, E., & Fletcher, K. (2007). Consumer Perceptions and Attitudes towards SMS Advertising: Recent Evidence from New Zealand. *International Journal of Advertising, 26*(1), 70–98.

Carter, E. (2008). M-marketing and Generation Y African-American Mobile Consumers: The Issues and Opportunities. *International Journal of M-marketing, 3*(1), 62–66.

Caspers, R. (2002). Neue Geschäftsmodelle der Internet-Ökonomie. In Bieger, T., Bickhoff, N., Caspers, R., Knyphausen-Aufseß, D., & Reding, K. (Ed.), *Zukünftige Geschäftsmodelle* (pp. 249-270). Berlin: Springer.

Cespedes, F., & Smith, H. (1993). Database marketing: new rules for policy and practice. *Sloan Management Review, 34*, 7–22.

Chae, M., & Kim, J. (2003). What's so different about the Mobile Internet? *Communications of the ACM, 46*(12), 240–247. doi:10.1145/953460.953506

Chan, I. (2007). China's Big, Really Big, Mobile Market. *BusinessWeek*, Retrieved August 1, 2007, from Technology Section of BusinessWeek Online located at http://www.businessweek.com/globalbiz /content/jul2007

Chandler, C. (2007). *China's Mobile Maestro.* Retrieved August 12, 2007, from CNNMoney web site, located at http://money.cnn.com/magazines/fortune/fortune_archive/2007/08/06/100156748/index2.htm

Chang, L. (1994). A psychometric evaluation of 4-point and 6-point Likert-type scales in relation to reliability and validity. *Applied Psychological Measurement, 18*, 205–216. doi:10.1177/014662169401800302

Chang, S. S., & Fang, X. (2003). Mobile Commerce and Usability. In Lim, E.-P., & Siau, K. (Eds.), *Advances in Mobile Commerce Technologies* (pp. 235-257). Hershe, PA: Idea Group Publishing.

Chaos Computer Club. (2002). Blinkenlights Arcade, Bibliothèque nationale de France, Paris, France, September 25 – October 6, 2002 from http://www.blinkenlights.net/arcade

Chapin, R (2007), MovieSeer.com interview, May 2007.

Chapman, R. G. (1986). Assessing the profitability of retailer couponing with a low-cost field experiment. *Journal of Retailing, 62*(1), 19–39.

Chayko, M. (2007). The portable community: envisioning and examining mobile social connectedness. *International Journal of Web Based Communities, 3*(4), 373–385. doi:10.1504/IJWBC.2007.015864

Chen, L.-D., & Nath, R. (2004). A Framework for Mobile Business Applications. *International Journal of Mobile Communications, 2*, 368–381. doi:10.1504/IJMC.2004.005857

Chen, S. C., & Dhillon, G. S. (2003). Interpreting dimensions of consumer trust in e-commerce. *Information Technology and Management, 4*(2-3), 303–318. doi:10.1023/A:1022962631249

Chin, W. W. (1998). The partial Least Squares Approach to Structural Equation Modeling. In G. A. Marcoulides (ed.), Modern Methods for Business Research (pp. 295-336). Hillsdale, NJ: Lawrence Erlbaum Associates.

China Mobile Press Release. (2007). Retrieved August 12, 2007, from China Mobilewebsite located at http://www.chinamobileltd.com/doc/ pr/2007/20070321.htm

China Steps into Credit Card Era. CISCA. http://www.cisca.org.hk/english/industry-page.asp?news_id=791, October 8, 2004

Chiou, J. S. (2004). The antecedents of consumers' loyalty toward Internet Service Providers. *Information & Management, 41*(6), 685–695. doi:10.1016/j.im.2003.08.006

Chiou, J.-S., & Droge, C. (2006). Service quality, trust, specific asset investment, and expertise: Direct and indirect effects in a satisfaction-loyalty framework. *Journal of the Academy of Marketing Science, 34*(4), 613–627. doi:10.1177/0092070306286934

Chowdhury, H. K., Parvin, N., Weitenberner, C., & Becker, M. (2006). Consumer attitude toward mobile advertising in an emerging market: An empirical study. *International Journal of Mobile Marketing, 1*(2), 33–42.

Christensen, C. M., & Davis, A. B. (2006). The future of newspapers. Retrieved September 29, 2008, from http://www.forbes.com/home/leadership/2006/10/10/leadership-newspapers-media-lead-innovation-cx_cc_1011clayton.html

Chung, C. M. Y., & Darke, P. R. (2006). The Consumer as Advocate: Self-Relevance, Culture, and Word-of-Mouth.

Marketing Letters, *17*(4), 269–279. doi:10.1007/s11002-006-8426-7

Chung, E. (2000). Navigating the primordial soup: Charting the lived worlds of the migrant consumer. *Journal of Consumer Marketing*, *17*(1), 36–54. doi:10.1108/07363760010309537

Churchill, G. A. J., & Iacobucci, D. (2002). *Marketing Research; Methodological Foundations*. South-Western Publishing, Mason, Ohio.

Claritas. (2008). 66 Prizm marketing segments. Claritas Prizm NE. Retrieved June 3, 2008, from http://www.claritas.com/claritas/ Default.jsp?ci= 3&si= 4&pn= prizmne_segments

Clee, M. A., & Wicklund, R. A. (1980). Consumer behavior and psychological reactance. *The Journal of Consumer Research*, *6*(4), 389–405. doi:10.1086/208782

Clow, K. E., & Baack, D. (2007). *Integrated Advertising, Promotion, and Marketing Communications*. New Jersey: Pearson Education Inc.

CNN, Wireless Society Report. (2003). New switching rules may trash millions of cell phones, cable network news. Access January 2007, from http://www.cnn.com/2003/TECH/ptech/10/31/sprj.ws.cellphone.garbage.ap/

Coase, R. H. (1937). The Nature of the Firm. *Economica*, *4*(16), 386–405. doi:10.1111/j.1468-0335.1937.tb00002.x

Coca-Cola, A. G. Campaign. (2007). Festiver 04 Germany/ Switzerland. From http://www.spielplatz.cc/presse/wordpress/?cat=7 and www.festifever.ch

Coke Sweepstake. (2004). Unexpected Summer. USA. From http://www.usatoday.com/money/industries/food/2004-05-09-coke_x.htm

comparis.ch. (2007, 09.01.2007). comparis.ch zu den Krankenkassenwechseln 2007 Retrieved 19.04.2007, 2007, from http://www.comparis.ch/comparis/press/communique.aspx?ID=PR_Comm_Communique_070109

ComScore M. Metrics. (2008a). *Teens' Mobile Activities by Jen Wu*, August 28, 2008. Retrieved October 25, 2008, from http://www.comscore.com/press/presentations.asp

ComScore M. Metrics. (2008b). *The Mobile Media Authority by Mark Donovan*, August 19, 2008. Retrieved October 25, 2008, from http://www.comscore.com/press/presentations.asp

Cooke, D., Flores, G., Garcia, J., Guzman, E., Lopez, E., Muselli, C., & Seal, K. C. (2004). Influencing Factors for Mobile Values: A Comparative Analysis Between the United States and Iceland, Ireland, the United Kingdom, Belgium and the Netherlands. In the *Proceedings of The Second International Conference of AIMS*, India, 2004.

Corsten, H., Gössinger, R., & Schneider, H. (2006). Grundlagen des Innovationsmanagement. München: Vahlen.

Cortiñas, M., Elorz, M., & Múgica, J. M. (2008). The Use of Loyality-Cards Databases: Differences in Regular Price and Discount Sensitivity in the Brand Choice Decision Between Card and Non-Card Holders. *Journal of Retailing and Consumer Services*, *15*(1), 52–62. doi:10.1016/j.jretconser.2007.03.006

Cottet, P., Lichtle, M. G., & Plichon, V. (2006). The role of value in services: a study in a retail environment. *Journal of Consumer Marketing*, *23*(4), 219–227. doi:10.1108/07363760610674347

Coursaris, C., & Hassanein, K. (2002). UNDERSTANDING M-COMMERCE. *Quarterly Journal of Electronic Commerce*, *3*(3), 247.

Covey, N. (2008). Critical mass, the worldwide state of the mobile Web. Nielsen Mobile. July 2008. Retrieved July 13, 2008, from http://www.nextgenweb.org/ wp-content/uploads/ 2008/07/ nielsen-critical-mass-the-worldwide-state-of-the-mobile-web.pdf

Cox, D., Cox, A. D., & Moschis, G. P. (1990). When consumer behavior goes bad: An investigation of adolescent shoplifting. *The Journal of Consumer Research*, *17*(2), 149–159. doi:10.1086/208545

Cox, R., & Brittain, P. (1996). *Retail Management*, 3rd edition. London: Pitman Publishing.

Crawford, D. (2007). *The mobile TV market*. Colchester, UK: University of Essex.

Creswell, J. (2003) *Research Design Qualitative, Quantitative, and Mixed Methods Approaches*, Second Edition. Thousand Oaks, CA: Sage Publications.

Crowcroft, J., Gibbens, R. J., Kelly, F. P., & Östring, S. (2004). Modeling incentives for collaboration in mobile ad hoc networks. *Performance Evaluation - Selected Papers from the First Workshop on Modeling and Optimization in Mobile, Ad Hoc and Wireless Networks (WiOpt'2003)*, *57*(4), 427–439.

Csikszentmihalyi, M. (1975). *Beyond Boredom and Anxiety.* San Francisco: Jossey- Bass.

Csikszentmihalyi, M. (1990). *Flow: The psychology of optimal experience.* New York: Harper Collins.

Dabholkar, P. A., & Bagozzi, R. P. (2002). An attitudinal model of technology-based self-service: Moderating effects of consumer traits and situational factors. *Journal of the Academy of Marketing Science, 30*(3), 184–201.

Dahlberg, T., Mallat, N., Ondrus, J., & Zmijewska, A. (2008). Past, present and future of mobile payments research. *Electronic Commerce Research and Applications, 7*(2), 165–181. doi:10.1016/j.elerap.2007.02.001

Dailey, L. C. (2002). Consumers' responses to navigational atmospherics on the Web: An empirical test. *American Marketing Association*, Conference Proceedings, Chicago, 13, 202.

Dailey, L. C. (2004). Navigational web atmospherics: explaining the influence of restrictive navigation cues. *Journal of Business Research, 57*(7), 795–803. doi:10.1016/S0148-2963(02)00364-8

Dartmouth College. (2005). CRAWDAD Project: A Community Resource for Archiving Wireless Data At Dartmouth. http://crawdad.cs.dartmouth.edu/ (seen 11/2008)

Datta, A. (2003). MobiGrid: Peer-to-Peer Overlay and Mobile Ad-Hoc Network Rendezvous – a Data Management Perspective. In *CAiSE 2003 Doctoral Symposium, in conjunction with the 15th Conference On Advanced Information Systems Engineering.* Springer.

Datta, P., Chowdhury, D., & Chakraborty, B. (2005). Viral marketing: New form of word-of-mouth through internet. *Business Review (Federal Reserve Bank of Philadelphia), 3*(2), 69–76.

Davenport, T. H., Harris, J. G., & Kohli, A. (2001). How do they know their customers so well? *Sloan Management Review, 42*(2), 63–73.

Davis, F. (1989). Perceived Usefulness, Perceived Ease of Use, and User Acceptance of Information Technology. *MIS Quarterly, 13*(3), 319–340. doi:10.2307/249008

Davis, F. D. (1985). *A Technology Acceptance Model for Empirically Testing New End-User Information Systems: Theory and Results.* PhD thesis, Massachusetts Institute of Technology, Sloan School of Management.

Davis, F. D. (1989). Perceived Usefulness, Perceived Ease of Use, and User Acceptance of Information Technology. *MIS Quarterly, 13*(3), 318–340. doi:10.2307/249008

Davis, F. D. (1989). Perceived usefulness, perceived easy of use, and user acceptance of information technology. *MIS Quarterly, 13*(3), 319–340. doi:10.2307/249008

Davis, F. D., Bagozzi, R. P., and Warshaw, P. R. (1989). User Acceptance of Computer Technology: A Comparison of Two Theoretical Models, *Management Science* (35:8), pp. 982-1002

Davis, L., Wang, S., & Lindridge, A. (2008). Culture influences on emotional responses to on-line store atmospheric cues. *Journal of Business Research, 61*, 806–812. doi:10.1016/j.jbusres.2007.08.005

Day, G. S., & Montgomery, D. B. (1999). Charting new directions for marketing. *Journal of Marketing, 63*(special issue), 3–13. doi:10.2307/1252096

De Reyck, B., & Degraeve, Z. (2003). Broadcast scheduling for mobile advertising. *Operations Research, 51*(4), 509–517. doi:10.1287/opre.51.4.509.16104

De Wulf, K., Schillewaert, N., Muylle, S., & Rangarajan, D. (2006). The role of pleasure in web site success. *Information & Management, 43*(4), 434–446. doi:10.1016/j.im.2005.10.005

Deakins, B., Grandcolas, U., & Rettie, R. (2005). Text message advertising: Response rates and branding effects. *Journal of Targeting. Measurement and Analysis for Marketing, 13*(4), 304–312. doi:10.1057/palgrave.jt.5740158

Deaux, K. (1984). From individual differences to social categories. Analysis of a decade's research on gender. *The American Psychologist, 39*(2), 105–116. doi:10.1037/0003-066X.39.2.105

DeBaillon, L., & Rockwell, P. (2005). Gender and student-status differences in cellular telephone use. *International Journal of Mobile Communications, 3*(1), 82–98. doi:10.1504/IJMC.2005.005876

DeBruyn, A., & Lilien, G. (2004). A multi-stage model of word of mouth through electronic referrals. *eBRC Research Paper Series.*

Decker, M., Bulander, R., Högler, T., & Schiefer, G. (2006). m-Advertising: Werbung mit mobilen Endgeräten - ein Überblick. In T. Kirste, B. König-Ries, K. Pousttchi, & K. Turowski (Eds.), *Mms* (Vol. 76, p. 103–114). GI.

Deighton, J., & Glazer, R. (1998). From the editors. *Journal of Interactive Marketing, 12*(1), 2–4. doi:10.1002/(SICI)1520-6653(199824)12:1<2::AID-DIR1>3.0.CO;2-P

Del Rio, A. B., Vazquez, R., & Iglesias, V. (2001). The effects of brand associations on the consumer response. *Journal of Consumer Marketing*, *18*(5), 410–425. doi:10.1108/07363760110398808

Dellarocas, C. (2003). The Digitalization of Word of Mouth: Promise and Challenges of Online Feedback Mechanisms. *Management Science*, *49*(10), 1407–1424. doi:10.1287/mnsc.49.10.1407.17308

DeLone, W., & McLean, E. (1992). Information System Success: The Quest for the Dependent Variable. *Information Systems Research*, *3*(1), 60–95. doi:10.1287/isre.3.1.60

DeLone, W., & McLean, E. (2003). The DeLone and McLean Model of Information Systems Success: A Ten-Year Update. *Journal of Management Information Systems*, *19*(4), 9–30.

Demoulin, N. T. M., & Zidda, P. (2007). On the Impact of Loyalty Cards on Store Loyalty: Does the Customers' Satisfaction with the Reward Scheme Matter? *Journal of Retailing and Consumer Services*, *15*(5), 386–398. doi:10.1016/j.jretconser.2007.10.001

Dey, A. K. (2001). Understanding and using context. *Personal and Ubiquitous Computing*, *5*, 4–7. doi:10.1007/s007790170019

Dey, A., & Abowd, G. (2000). *The context toolkit: Aiding the development of context-aware applications*. Proceedings of the Workshop on Software Engineering for Wearable and Pervasive Computing, Limerick, Ireland.

DeZoysa, S. (2002). Mobile ads need to get personal. *Telecommunications International*, *36*(2), 8.

DeZoysa, S. (2002). Mobile advertising needs to get personal. *Telecommunications International Ed., 36(2),* 8.

Dholakia, R. R., & Dholakia, N. (2004). Mobility and markets: emerging outlines of m-commerce. *Journal of Business Research*, *57*(12), 1391–1396. doi:10.1016/S0148-2963(02)00427-7

Dholkia, N., Dholkia, R., Lehrer, M., & Kshetri, K. (2004). Global Heterogeneity in the Emerging M-Commerce Landscape. Un Wireless communications and mobile commerce, 2004, accessed at http://cybertech-pub.com/downloads/excerpts/1591401844E.pdf, last accessed on Oct 31, 2008.

Dickenberger, D. (1979). *Ein neues Konzept von Freiheit: Konsequenzen für die Theorie der psychologischen Reaktanz*, Weilheim: Belz .

Dickenberger, D., & Gniech, G. (1982). The theory of psychological reactance. In Irle, M. (Ed.), *Studies in decision making. Social psychological and socio-economic analyses* (pp. 311-341). Berlin, New York: de Gruyter.

Dickenberger, D., Gniech, G., & Grabitz, H.-J. (1993). Die Theorie der psychologischen Reaktanz, In Frey, D., Irle, M. (Ed.), *Theorien der Sozialpsychologie, Band I: Kognitive Theorien* (pp. 243-273). Bern-Göttingen-Toronto-Seattle: Hans Huber Verlag.

Dickinger, A., Haghirian, P., Murphy, J., & Scharl, A. (2004). A conceptual model and investigation of SMS Marketing. *Thirty-Seventh Hawaii International Conference on SystemSciences (HICSS-37)*, Hawaii, USA.

Dickinger, A., Haghirian, P., Murphy, J., & Scharl, A. (2004). An investigation and conceptual model of SMS marketing. *Proceedings of the 37th Hawaii International Conference on System Sciences*, Big Island, HI.

Dickinger, A., Scharl, A., & Murphy, J. (2005). Diffusion and success factors of mobile Marketing. *Electronic Commerce Research and Applications*, *4*(2), 159–173. doi:10.1016/j.elerap.2004.10.006

Diller, H. (1996). Kundenbindung als Marketingziel. *Marketing ZFP*, *18*(2), 81–94.

Diller, H. (2001). Die Erfolgsaussichten des Beziehungsmarketing im Internet. In Eggert, A., & Fassott, G. (Eds.), *Electronic Customer Relationship Management - Management der Kundenbeziehungen im Internet-Zeitalter* (pp. 66-85). Stuttgart: Schäffer-Poeschel.

Diller, H., & Müllner, M. (1998). Kundenbindungsmanagement. In Meyer, A. (Ed.), *Handbuch Dienstleistungsmarketing* (pp. 1220-1240). Stuttgart: Schäffer-Poeschel.

Ding, G., & Bhargava, B. (2004). Peer-to-peer File-sharing over Mobile Ad hoc Networks. In *2nd IEEE International Conference on Pervasive Computing and Communications – Workshop on Mobile Peer-to-Peer Computing*. IEEE Computer Society.

Dix, A., Finlay, J., Abowd, G., & Beale, R. (1998). *Human Computer Interaction.* 2nd edition. Prentice Hall Europe.

Do, V. T. (2003). Mobile E-Commerce on Mobile Phones. In Lim, E.-P., & Siau, K. (Eds.), *Advances in Mobile Commerce Technologies* (pp. 19-43). Hershey, PA: Idea Group Publishing.

Dommeyer, C., & Gross, B. (2003). What consumers know and what they do: an investigation of consumer knowledge, awareness, and use of privacy protection strategies. *Journal of Interactive Marketing*, *17*(2), 34–51. doi:10.1002/dir.10053

Donovan, R. J., & Rossiter, J. R. (1982). Store atmosphere: an environmental psychology approach. *Journal of Retailing*, *58*(1), 34–57.

Dourish, P. (2004). What we talk about when we talk about context. *Personal and Ubiquitous Computing*, *8*(1), 19–30. doi:10.1007/s00779-003-0253-8

Dow Jones Newswire. (2008). Global Mobile Phone Users Top 3.3 Billion By End-2007 – Study. http://www.cellular-news.com/story/31352.php (accessed 8 June 2008).

Doyle, S. (2001). Software Review: Using Short Message Services As a Marketing Tool. *Journal of Database Marketing*, *8*(3), 273–277. doi:10.1057/palgrave.jdm.3240043

Drennan, J., & Mort, G. S. (2005). Marketing m-services: Establishing a usage benefit typology related to mobile user characteristics. *Journal of Database Marketing & Customer Strategy Management*, *12*(4), 327–341. doi:10.1057/palgrave.dbm.3240269

Drosos, D., & Giaglis, G. M. (2004). *Factors Influencing Mobile Messaging Advertising Effectiveness.* Eltrum Working Paper Series WP 2004-009, Athens University of Economics and Business Department of Management Science and Technology, Athens.

Drossos, D., & Giaglis, G. M. (2006). Mobile advertising effectiveness: An exploratory study. *Proceedings of the International Conference on Mobile Business (ICMB'06)*

Drossos, D., Giaglis, G. M., Lekakos, G., Kokkinaki, F., & Stavraki, M. G. (2007). Determinants of effective SMS advertising: An experimental study. *Journal of Interactive Advertising*, *7*(2). Retrieved January 30, 2009, from http://www.jiad.org/article90

Drossos, D., Giaglis, G., Lekakos, G., Kokkinaki, F., & Stavraki, M. (2007). Determinants of effective sms advertising: An experimental study. *Journal of Interactive Advertising*, *7*(2).

Drüner, M., Rattay, R., & Kröger, S. (2007). Web 2.0 - Schneller mehr wissen: Informationsvorsprung durch Nutzung neuer Rückkanäle. *Information Management and Consulting*, 35-42.

Duan, W., Gu, B., & Whinston, A. B. (2008). The dynamics of online word-of-mouth and product sales – An empirical investigation of the movie industry. *Journal of Retailing*, *84*(2), 233–242. doi:10.1016/j.jretai.2008.04.005

Duchnicky, R. L., & Kolers, P. A. (1983). Readability of text scrolled on visual display terminals as a function of window size. *Human Factors*, *25*, 683–692.

Ducoffe, R. H. (1996). Advertising value and advertising on the Web. *Journal of Advertising Research*, *36*(September/October), 21–36.

Duncan, T., & Moriaty, S. E. (1998). A communication-based marketing model for managing relationships. *Journal of Marketing*, *62*(2), 1-13. doi:10.2307/1252157

Dunne, P. M., Lusch, R. F., & Griffith, D. A. (2002). *Retailing* (4th ed.) Orlando: Harcourt College.

Durand, R., & Lambert, Z. V. (1985). Alienation and criticism of advertising. *Journal of Advertising*, *14*(3), 9–17.

Eagly, A. H., & Chaiken, S. (1993). *The Psychology of Attitudes.* Fort Worth: Harcourt Brace Jovanovich.

East, R., Hammond, K., Lomax, W., & Robinson, H. (2005). What is the Effect of a Recommendation? *Marketing Review*, *5*(2), 145–157. doi:10.1362/1469347054426186

Eckstein, P. (2006). *Angewandte Statistik mit SPSS. Praktische Einführung für Wirtschaftswissenschaftler* (5. Aufl. ed.). Wiesbaden: Gabler.

Economides, N. (1996). The Economics of Networks. *International Journal of Industrial Organization*, *14*(2), 673–699. doi:10.1016/0167-7187(96)01015-6

Edens, K. M., & McCormick, C. B. (2000). How do adolescents process advertisements? The influence of ad characteristics, processing objective, and gender. *Contemporary Educational Psychology*, *25*, 450–463. doi:10.1006/ceps.1999.1031

Eisenhardt, K. M. (1989). Building Theories from Case Study Research. *Academy of Management Review*, *14*(4), 532–550. doi:10.2307/258557

Eisenhardt, K. M., & Graebner, M. E. (2007). Theory Building from Cases. *Academy of Management Journal*, *50*(1), 25–32.

Eisenhauer, M., Oppermann, R., & Prinz, W. (2007). Internet der Dinge. In Bullinger, H.-J., & ten Hompel, M. (Ed.), *Internet der Dinge* (pp. 49-62). Berlin: Springer.

EITO – European Information Technology Observatory. (2008). *EITO ICT market overview.* Bitkom Research GmbH.

Elliott, M. T., & Warfield, A. E. (1993). Do Market Mavens Categorize Brands Differently? *Advances in Consumer*

Research. Association for Consumer Research (U. S.), 20(1), 202–208.

eMarketer. (2003). SMS marketing yields strong results. *eMarketer Research Report, UK.*

eMarketer. (2005). Mobile marketing and m-commerce: Global spending and trends. Retrieved January 15, 2006, from http://www.emarketer.com/ Report.aspx?mobile_mktg_feb05

eMarketer. (2008). E-marketer's key predictions for 2009. Available at http://www.emarketer.com/Article.aspx?id=1006818

Emnid, T. N. S. (2006). *Bonusprogramme in Deutschland. Eine Studie des TNS Emnid Medien- und Sozialforschung GmbH.* Bielefeld: TNS Emnid.

Enpocket Research Report 1. (2002). Consumer preferences for SMS marketing in the UK, August–November 12. Access January 2007, from http://www.enpocket.com/news/press-releases/consumers-find-sms-marketing-as-acceptable-as-tv-or-radio-advertising

Enpocket. (2002a). Consumer preferences for SMS marketing in the UK. http://www.enpocket.co.uk (accessed 13 March 2003).

Enpocket. (2002b). The branding performance of SMS advertising. http://www.enpocket.co.uk (accessed 13 March 2003).

Enpocket. (2003). The response performance of SMS advertising. http://www.mda-mobiledata.org (accessed 13 March 2003).

Enpocket. (2005a). *Direct Response Report.* Available http://www.enpocket.co.uk (accessed 10 February 2006).

Enpocket. (2005b). *Consumer Preferences Report.* Available http://www.enpocket.co.uk (accessed 10 February 2006).

Ericsson. (2000). *Wireless Advertising.* Stockholm: Ericsson Ltd.

Eroglu, S. A. A.K., Machleit & L. M., Davis (2000). Online retail atmospherics: empirical test of a cue typology. In J.R. Evans & B. Berman (Eds), Retailing 2000: launching the new millennium (pp. 144-150). *Proccedings of the Sixth Triennial National Retailing Conference presented by the Academy of Marketing Science and the American Collegiate Retailing Association.* Columbus, Ohio, November 2-5.

Eroglu, S. A., Machleit, K. A., & Chebat, J. C. (2005). The interaction of retail density and music tempo: effects on shopper responses. *Journal of Psychology and Marketing, 22*(7), 577–589. doi:10.1002/mar.20074

Eroglu, S. A., Machleit, K. A., & Davis Lenita, M. (2001). Atmospheric qualities of online retailing: a conceptual model and implication. *Journal of Business Research, 54*(2), 177–184. doi:10.1016/S0148-2963(99)00087-9

Eroglu, S. A., Machleit, K. A., & Davis, L. M. (2003). Empirical testing of a model of online store atmospherics and shopper responses. *Psychology and Marketing, 20*(2), 139–150. doi:10.1002/mar.10064

Esch, F.-R. (2008). Strategie und Technik der Markenführung. München: Vahlen.

Eschke, S., & Bölscher, J. (2001). *Customer Relationship Management in der Versicherungswirtschaft.*

Evans, P. B., & Wurster, T. S. (1997). Strategy and the new economics of information. *Harvard Business Review, 75*(5), 70–82.

Eysenck, M. (1982). *Attention and arousal, cognition and performance.* New York: Springer-Verlag.

Faber, E., Ballon, P., Bouwman, H., Haaker, T., Rietkerk, O., & Stern, M. (2003). Designing business models for mobile ICT services. Paper presented at the *16th Bled Electronic Commerce Conference, Bled, June 9-11*, available from http://www.mozaiek.dds.nl/docs/PanelBled03.pdf, last accessed Oct 31, 2008

Facchetti, A., Rangone, A., Renga, F. M., & Savoldelli, A. (2005). Mobile marketing: An analysis of key success factors and the European value chain. *International Journal of Management and Decision Making, 6*(1), 65–80. doi:10.1504/IJMDM.2005.005966

Faltin, C. (2006). Vom Web 2.0 zum Marketing 2.0. *Marketing Journal, 39*(Sonderheft Web 2.0), 10-13.

Fassott, G., & Eggert, A. (2005). Zur Verwendung formativer und reflektiver Indikatoren in Strukturgleichungsmodellen: Bestandsaufnahme und Anwendungsempfehlungen. In F. Bliemel, A. Eggert, G. Fassot & J. Henseler (Eds.), *Handbuch PLS-Pfadmodellierung. Methoden Anwendungen Praxisbeispiele.* Stuttgart: Schäffer-Poeschel.

Faulconbridge, J. R., & Beaverstock, J. V. (forthcoming). Geographies of international business travel in the professional service economy. In D. Hislop (Ed.), *Mobile work/technology: Changing patterns of spatial mobility and mobile technology use in work.* London: Routledge.

Feick, F., & Price, L. (1987). The Market Maven: A Diffuser of Marketplace Information. *Journal of Marketing, 51*, 83–97. doi:10.2307/1251146

Feick, L. F., Guskey, A., & Price, L. (1995). Everyday Market Helping Behavior. *Journal of Public Policy & Marketing, 14*(2), 255–266.

Feierabend, S., & Klingler, W. (2003). Medienverhalten Jugendlicher in Deutschland. *Media Perspektiven, 33*(10), 450–462.

Ferrazzi, K., Chen, J., & Li, Z. (2003). Idea: Playing games with customers. *Harvard Business Review, 81*(4), 21.

Festinger, L. (1978). *Die Theorie der kognitiven Dissonanz*. Bern: Hans Huber.

Fields, L. (2005). Location Based Services – a Marketer's Dream. Retrieved July 25, 2008, from http://mmaglobal.com/modules/article/view.article.php/21

Fife, E., & Pereira, F. (2005). *Global acceptance of technology (GAT) and demand for mobile data services*. Paper presented at the Hong Kong Global Mobility Roundtable, Hong Kong.

Figge, S. (2002). Die Open Mobile Architecture - Systemumgebung für mobile Dienste der nächsten Generation. *Wirtschaftsinformatik, 44*(4), 375–378.

Figge, S. (2007). *Innovatives Mobile Marketing. Kontextabhängige Kundenansprache mit Hilfe mobiler Portale*. Hamburg: Verlag Dr. Kovač.

Figge, S., & Theysohn, S. (2006). Quantifizierung IKS-basierter Marktleistungen - Analyse eines werbefinanzierten Geschäftsmodells für den Mobile Commerce. *Wirtschaftsinformatik, 48*(2), 96–106. doi:10.1007/s11576-006-0024-z

Fiore, A. M., Jin, H., & Kim, J. (2005). For fun and profit: hedonic value from image interactivity and responses toward an online store. *Psychology and Marketing, 22*(8), 669–694. doi:10.1002/mar.20079

Fisch, M., & Gscheidle, C. (2008). Mitmachnetz Web 2.0: Rege Beteiligung nur in Communitys. *Media Perspektiven, 12*(7), 356–564.

Fishbein, M. (1967). Attitudes and Prediction of Behavior, in: Fishbein M. (ed.), Readings in Attitude Theory and Measurement, New York, John Wiley and Sons, p.477-492.

Fishbein, M., & Ajzen, I. (1975). *Belief, attitude, intention, and behaviour: An introduction to theory and research*. Reading, MA: Addison-Wesley.

Fishbein, M., & Middlestadt, S. (1995). Noncognitive effects on attitude formation and change: Fact or artifact? *Journal of Consumer Psychology, 4*(2), 181–202. doi:10.1207/s15327663jcp0402_05

Fitzpatrick, G. (1998). *The locales framework: Understanding and designing for cooperative work*. Unpublished doctoral dissertation, The University of Queensland, Australia.

Fitzpatrick, G. (2003). *The locales framework: Understanding and designing for wicked problems*. Kluwer Academic Publishers.

Fleisch, E., Thiesse, F., & Österle, H. (2002). Connected Smart Appliance. In Österle, H., Fleisch, E., & Alt, R. (Ed.), *Business Networking in der Praxis* (pp. 233-248). Berlin: Springer.

Flick, U. (1995). *Qualitative Forschung. Theorie, Methoden, Anwendungen in Psychologie und Sozialwissenschaften*. Hamburg: rororo.

Flick, U. (2006). *An introduction to qualitative research* (3rd ed.). London: Sage Publications.

Flickr Peep Show was exhibited at Mediamatic/POST CS in Amsterdam during May. (2005, June). From http://www.textually.org/textually/archives/2005/05/008427.htm and http://www.mediamatic.net/page/5617/en

Fliegel, F. C., & Kivlin, J. E. (1966). Attributes of innovations as factors in diffusion. *American Journal of Sociology, 72*, 235–248. doi:10.1086/224292

Fornell, C., & Larcker, D. (1981). Evaluating structural equation models with unobservable variables and measurement error. *JMR, Journal of Marketing Research, 18*(1), 39–50. doi:10.2307/3151312

Forrester Research, Inc. (2001). The marketer's guide to use SMS. *Forrester Research Report*, European Research Center, Amsterdam.

Forrester. (2007). Topic *Overview: Enterprise Mobility. Mobility Becomes an Integral Part of Firms' Network Strategy by Maribel D. Lopez,* July 20, 2007, Retrieved October 29, 2008, from http://www.forrester.com/Research/PDF/0,5110,42404,00.pdf

Förster, A., & Kreuz, P. (2002). *Offensives Marketing im E-Business*. Berlin: Springer.

Fortin, D. R., & Dholakia, R. R. (2005). Interactivity and vividness effects on social presence and involvement with a web-based advertisement. *Journal of Business Research, 58*(3), 387–396. doi:10.1016/S0148-2963(03)00106-1

Fortunati, L. (2001). The mobile phone: An identity on the move. *Personal and Ubiquitous Computing, 5*, 85–98. doi:10.1007/PL00000017

Forum, N. F. C. (2008). The NFC Forum. Retrieved October 20, 2008, from http://www.nfc-forum.org/aboutus/

Forum, U. M. T. S. (2003). UMTS security awareness. *UMTS Forum Report*, November 5, 30. Access January 2007, from http://www.umts-forum.org/servlet/dycon/ztumts/umts/Live/en/umts/Resources_Reports_30_index

Foscht, T., & Swoboda, B. (2005). Käuferverhalten. Wiesbaden: Gabler.

Fritz, W. (2008). Internet-Marketing und Electronic Commerce. Wiesbaden: Gabler.

Frolick, M. N., & Chen, L.-d. (2004). Assesing M-Commerce opportunities. *Information Systems Management, 21*(2), 53–61. doi:10.1201/1078/44118.21.2.20040301/80422.8

Frumkin-Rosengaus, M. (1987). *Increasing transit ridership through a targeted transit marketing approach.* Unpublished doctoral dissertation, State University of New Jersey, New Brunswick, NJ.

Fuller, P. (2003). Why spam doesn't have to happen on mobile devices. Mobile Marketing Association, November 2. Access January 2007, from http://mmaglobal.com/modules/wfsection/article.php?articleid=44.

Funk, J. L. (2004). *Mobile Disruption. The Technologies and Applications Driving the Mobile Internet.* New York: Wiley Publishing.

Future Mobile Handsets, 8th edition. (2006, October). Informa

Gannon, M. (2003). *Understanding Global Cultures: Metaphorical Journeys Through 28 Nations, Clusters of Nations, and Continents.* California: Sage Publications.

Garcia, A. G., & Rennhak, C. (2006). Kundenbindung – Grundlagen und Begrifflichkeiten. In Rennhak, C. (Ed.), *Herausforderung Kundenbindung* (pp. 3-14). Wiesbaden, Gabler.

Gardiner Jones, M. (1991). Privacy: a significant marketing issue for the 1990's. *Journal of Public Policy & Marketing, 10*(1), 133–148.

Gardner, M. P., & Siomkos, G. J. (1986). Toward a methodology for assessing effects of in-store atmospherics. *Advances in Consumer Research. Association for Consumer Research (U. S.), 13*, 27–31.

Gartner. (n.d.). GartnerG2 Says E-mail Marketing Campaigns Threaten Traditional Direct Mail Promotions. Mar 19, 2002: http://www3.gartner.com/5_about/press_releases.

Gartners Forecast: GPS in Mobile Devices, Worldwide, 2004-2010, Dataquest. (2006, November 29). ID Number: G00144746

Garyfalos, A., & Almeroth, K. C. (2004). Coupon Based Incentive Systems and the Implications of Equilibrium Theory. *Proceedings of the IEEE International Conference on E-Commerce Technology* (pp. 213-220). Washington, DC: IEEE Computer Society.

Gebauer, J., & Shaw, M. J. (2004). Success Factors and Impacts of Mobile Business Applications: Results from a Mobile e-Procurement Study. *International Journal of Electronic Commerce, 8*(3), 19–41.

Gefen, D., & Straub, D. (1997). Gender difference in the perception and use of e-mail: An extension to the Technology Acceptance Model. *MIS Quarterly, 21*(4), 389–400. doi:10.2307/249720

Gehrke, D., Turban, E., & Lee, O. K. M. (1999). Consumer and literature survey results of the relative importance of website design categories. *In the Proceedings of the 5th International Conference of the Decision Science Institute,* July 4-7, Athens, Greece.

Geisser, J. (1975). A Predictive Approach to the Random Effect Model. *Biometrika, 61*(1), 101–107. doi:10.1093/biomet/61.1.101

George, W. R., & Meyers, T. A. (1981). Life underwriters' perceptions of differences in selling goods and services. *CLU Journal,* (April), 44–49.

Geser, H. (2004). Towards a sociological theory of the mobile phone. Access January 2007, from www.socio.ch/mobile/t_geser1.htm

Gewei, Y. (2007). Mobile marketing systems: framework and technology enabler. *International Journal of Mobile Marketing, 2*(1), 43–49.

Geysken, I., Gielens, K., & Dekimpe, M. G. (2002). The market valuation of Internet channel addition. *Journal of Marketing, 66*(April), 102–119. doi:10.1509/jmkg.66.2.102.18478

Gibney, M. J. (2008). Asylum and the expansion of deportation in the United Kingdom. *Government and Opposition, 43*(2), 146–167. doi:10.1111/j.1477-7053.2007.00249.x

Glanz, O. (2008). How to get ahead in mobile advertising. *Mobile Marketing Magazine*. Retrieved May 23, 2008, from http://www. mobilemarketingmagazine. co.uk/ 2008/05/ how-to-get-ahea.html

Glassberg, B.C., Grover, V., and Teng, J.T.C. (2006). *Information System Research with an Attitude*, 76-77

Gluck, C. (2003), South Korea's credit card headache. *BBC News*.

Gniech, G., & Dickenberger, D. (1987). Reaktanz. In Frey, D., Greif, S. (Ed.). *Sozialpsychologie (Ein Handbuch in Schlüsselbegriffen)*, 2. Aufl., München-Weinheim: Psychologische Verlagsunion.

Gniech, G., & Grabitz, H.-J. (1978). Freiheitseinengung und psychologische Reaktanz. In Frey, D. (Hg.), *Kognitive Theorien der Sozialpsychologie*. Wien: Huber.

Godin, G., & Kok, G. (1996). The theory of planned behavior: a review of its applications to health-related behaviors. *American Journal of Health Promotion*, *11*(2), 87–98.

Godin, S. (1999). *Permission marketing-turning strangers into friends, and friends into customers*. New York: Simon & Schuster.

Goel, S. K., Singh, M., & Xu, D. (2002). Efficient Peer-to-Peer Data Dissemination in Mobile Ad-Hoc Networks. In *International conference on parallel processing workshops* (pp. 152–158). IEEE Computer Society.

Göhring, M., Happ, S., & Müller, T. (2006). Web 2.0 im Kundenmanagement. In Hildebrand, K., & Hofmann, J. (Eds.), *Social Software, Praxis der Wirtschaftsinformatik, 43*(252), 54-65.

Gohring, N. (2002). And now a word from our sponsors.... *America's Network*, *106*(3), 17.

Goldacre, M. J., Davidson, J. M., & Lambert, T. W. (2004). Country of training and ethnic origin of UK doctors: Database and survey studies. *British Medical Journal*, *329*(7466), 597–600. doi:10.1136/bmj.38202.364271.BE

Goldenberg, J., Libai, B., & Muller, E. (2001). Talk of the network: A complex systems look at the underlying process of word-of-mouth. *Marketing Letters*, *12*(3), 211–223. doi:10.1023/A:1011122126881

Goldsmith, R. E., Lafferty, B. A., & Newell, S. J. (2000). The impact of corporate credibility and celebrity credibility on consumer reaction to advertisements and brands. *Journal of Advertising*, *29*(3), 43–54.

Golesorkhi, B. (2006). Gender differences and similarities in judgments of trustworthiness. *Women in Management Review*, *21*(3), 195–210. doi:10.1108/09649420610657380

Golle, P., Leyton-Brown, K., Mironov, I., & Lillibridge, M. (2001). Incentives for Sharing in Peer-to-Peer Networks. In *Electronic commerce: Second international workshop, (welcom)* (Vol. 2232, pp. 75–87).

Gonzalez, G. R., Organero, M. M., & Kloos, C. D. (2008). Exploring Touching Learning Environments. In Kendall, M., & Samways, B. (Ed.), *Learning to Live in the Knowledge Society* (pp. 93-96). Berlin: Springer.

Gopal, R. D., & Tripathi, A. K. (2006). Advertising via wireless networks. *International Journal of Mobile Communications*, *4*(1), 1–16. doi:10.1504/IJMC.2006.008194

Gorn, G. J., Chattopadhayay, A., Sengupta, J., & Tripathi, S. (2004). Waiting for the web: how screen color affects time perception. *JMR, Journal of Marketing Research*, *41*(2), 215–225. doi:10.1509/jmkr.41.2.215.28668

Gottwald, S. (2007). RFID and Beyond. München: Vahlen.

Götz, O., & Liehr-Gobbers, K. (2004). Analyse von Strukturgleichungsmodellen mit Hilfe der Partial-Least-Squares(PLS)-Methode. *Die Betriebswirtschaft*, *64*(6), 714–738.

Grabitz,-Gniech, G. (1971). Some restrictive conditions for the occurrence of psychological reactance. *Journal of Personality and Social Psychology*, 19(2), 188-196.

Grabitz-Gniech, G., & Grabitz, H.-J. (1973a). Psychologische Reaktanz: Theoretisches Konzept und experimentelle Untersuchungen. *Zeitschrift für Sozialpsychologie*, *4*, 19–35.

Grabitz-Gniech, G., & Grabitz, H.-J. (1973b). Der Einfluß von Freiheitseinengung und Freiheitswiederherstellung auf den Reaktanz-Effekt. *Zeitschrift für Sozialpsychologie*, *4*, 361–365.

Graham, H., Saunders, J., Piercy, N., & Nicoulaud, B. (2007). *Marketing strategy and competitive positioning*. Prentice Hall.

Graham, P. (2005). *Web 2.0*. Retrieved January 30, 2007, from http://www.paulgraham.com/web20.html

Grandjot, T. (2006). *Mobile Scorecard. Entwicklung einer Balanced Scorecard für das mobile Customer Relationship Management*. Hamburg: Rainer Hampp Verlag.

Granovetter, J. B. (1973). The Strength of Weak Ties. *American Journal of Sociology*, *78*(6), 1360–1380. doi:10.1086/225469

Granovetter, J. B. (1983). The Strength of Weak Ties: A Network Theory Revisited. *Social Structure and Network Analysis*, *1*, 105–130.

Grant, I., & O'Donohoe, S. (2007). Why Young Consumers are Not Open to Mobile Marketing Communication. *International Journal of Advertising*, *26*(2), 223–246.

Gray, R. (2007). Immigration, world of opportunity. *The Marketer*, November, 28-31.

Green, B. (2005). *Reflections on the circle of life: Interactions in a pathology environment (medical): A 'wicked problem'*. Unpublished master's thesis, The University of Queensland, Brisbane, Australia.

Green, J. C., Caracelli, V. J., & Graham, W. F. (1989). Toward a conceptual framework for mixed method evaluation designs. *Educational Evaluation and Policy Analysis*, *11*(3), 255–274.

Greengrove, K. (2002). Needs-based segmentation: Principles and practice. *International Journal of Market Research*, *44*(4), 405–422.

Griffith, D. A. (2005). An examination of the influences of store layout in online retailing. *Journal of Business Research*, *58*(10), 1391–1396. doi:10.1016/j.jbusres.2002.08.001

Groß, M., & Hülsbusch, W. (2004). Weblogs und Wikis – eine neue Medienrevolution? *Wissensmanagement*, *6*(8), 44–48.

Gruber, H. (2001). Competition and innovation: The diffusion of mobile telecommunications services in central and Eastern Europe. *Information Economics and Policy*, *13*, 19–34. doi:10.1016/S0167-6245(00)00028-7

Gummesson, E. (1999). *Total Relationship Marketing. Rethinking Marketing Management from 4Ps to 30Rs.* Oxford: Butterworth-Heinemann.

Gundlach, G. T., & Murphy, P. E. (1993). Ethical and legal foundations of relational marketing exchanges. *Journal of Marketing*, *57*, 35–46. doi:10.2307/1252217

Gupta, M. P., Sahu, G. P., & Gauri, S. (2005). Assessing Impact of Mobile Communications on Organizations: A Flexibility Analysis. *Global Journal of Flexible Systems Management*, *6*(3/4), 11–23.

Gupta, S., Jain, D. C., & Sawhney, M. B. (1999). Modeling the evolution of markets with indirect network externalities: An application to digital television. *Marketing Science*, *18*(3), 396–416. doi:10.1287/mksc.18.3.396

Gwinner, K., Swanson, K. S., & Walsh, G. (2004). What makes mavens tick? Exploring the motives of market mavens' initiation of information diffusion. *Journal of Consumer Marketing*, *21*(2), 109–122. doi:10.1108/07363760410525678

Haeckel, S. H. (1998). About the nature and future of interactive marketing. *Journal of Interactive Marketing*, *2*(1), 63–71. doi:10.1002/(SICI)1520-6653(199824)12:1<63::AID-DIR8>3.0.CO;2-C

Haghirian, P., & Inoue, A. (2007). An advanced model of consumer attitudes toward advertising on the mobile Internet. *International Journal of Mobile Communications*, *5*(1), 48–67. doi:10.1504/IJMC.2007.011489

Haghirian, P., & Madlberger, M. (2005). Consumer attitude toward advertising via mobile devices - An empirical investigation among Austrian users. *Proceedings of the 13th European Conference on Information Systems*, Regensburg, Germany.

Haghirian, P., & Madlberger, M. (2006). Cross-cultural consumer perceptions of advertising via mobile devices-some evidence from Europa and Japan. *Proceedings of the ECIS (European Conference on Information Systems)*, Goeteborg, Sweden.

Haghirian, P., Dickinger, A., & Kohlbacher, F. (2004). Adopting innovative technology – a qualitative study among Japanese mobile consumers. *Proceedings of the 5th International Working with e-Business Conference (WeB-2004)*, November 2004, Perth, Australia.

Haghirian, P., Madlberger, M., & Tanuskova, A. (2005). Increasing advertising value of mobile marketing - an empirical study of antecedents. *Proceedings of the 38th Hawaii International Conference on System Sciences (HICSS-38)*, Big Island, HI, January 3-6.

Haghirian, P., Madlberger, M., & Tanuskova, A. (2005). Increasing advertising value of mobile marketing –An empirical study of antecedents, *Proceedings of the 38th Hawaii International Conference on System Sciences*, Hawaii.

Haig, M. (2002). *Mobile marketing: The message revolution.* London: Kogan Page.

Haig, M. (2002). *The b2b ecommerce handbook.* UK: Kogan Page Publisher.

Hair, J. F., Anderson, R. E., Tatham, R. L., & Black, W. C. (1998). *Multivariate Data Analysis (5th ed.)*. New Jersey: Prentice-Hall, Inc.

Halo 2 Game Promotion. (2004). I love bees. USA. From http://en.wikipedia.org/wiki/I_Love_Bees), (http://ilovebees.com

Hanhart, D. (2008). Mobile Computing und RFID im Facility Management. Berlin: Springer.

Hanley, M., Becker, M., & Martinsen, J. (2006). Factors influencing mobile advertising acceptance: Will incentives motivate college students to accept mobile advertisements? *International Journal of Mobile Marketing*, *1*(1), 50–58.

Hanna, R. C. (2003). The optimization of discount size and time limit for coupon promotions: Application to email offers. *Unpublished doctoral dissertation, Boston University School of Management.*

Hansen, H. R., & Neumann, G. (2005). [Stuttgart: Lucius & Lucius UTB.]. *Wirtschaftsinformatik*, 2.

Haque, A. (2004). Mobile commerce: Customer perception and It's prospect on business operation in Malaysia. *Journal of American Academy of Business, Cambridge*, *4*(1/2), 257–262.

Harris Interactive. (2008). Available at http://www.harrisinteractive.com/news/newsletters.asp

Harris, L. C., & Goode, M. M. H. (2004). The four levels of loyalty and the pivotal role of trust: A study of online service dynamics. *Journal of Retailing*, *80*(2), 139–158. doi:10.1016/j.jretai.2004.04.002

Harris, L., & Charles, D. (2002). *Marketing the e-Business*. UK: Routledge Publisher.

Harrison, D. A., Mykytyn, P. P. Jr, & Riemenschneider, C. K. (1997). Executive decisions about adoption of information technology in small business: Theory and empirical tests. *Information Systems Research*, *8*(2), 171–195. doi:10.1287/isre.8.2.171

Hartley, S. W., & Cross, J. (1988). How Sales Promotion Can Work For and Against You. *Journal of Consumer Marketing*, *5*(3), 35–42. doi:10.1108/eb008230

Hartwick, J., & Barki, H. (1994). Explaining the Role of User Participation in Information System Use. *Management Science*, *40*, 460–465. doi:10.1287/mnsc.40.4.440

Hashimoto, Y. (2002). The Spread of cellular phones and their influence on young people in Japan, *Review of Media, Information and Society*, Institute of Socio-Information and Communication Studies. *University of Tokyo*, *7*, 97–110.

Hayes, A., & Wilson, D. (2005). Peer-to-Peer Information Sharing in a Mobile Ad Hoc Environment. In *Sixth ieee workshop on mobile computing systems and applications (wmcsa'04)* (p. 154–162). IEEE Computer Society.

Haythornthwaite, C. (2001). Tie Strength and the Impact of New Media. *Proceedings of the 34th Hawaii International Conference on System Sciences*, Honolulu.

Heflin, J. (2004). *OWL Web ontology language use cases and requirements. W3C Recommendation.* Retrieved December 22, 2008, from http://www.aaai.org/AITopics/html/ontol.html#what

Hehl, W. (2008). Trends in der Informationstechnologie. Zürich: vdf Hochschulverlag AG.

Heinemann, A. (2007). *Collaboration in Opportunistic Networks, PhD Thesis.* VDM Verlag, Saarbrücken.

Heinemann, A. (2008). Opportunistic Networks. In M. Mühlhäuser & I. Gurevych (Eds.), *Ubiquitous computing technology for real time enterprises* (chap. 9). Information Science Reference.

Heinemann, A., Kangasharju, J., & Mühlhäuser, M. (2008). Opportunistic Data Dissemination Using Real-World User Mobility Traces. In *1st IEEE Intl. Workshop on Opportunistic Networking (WON-08).*

Heinemann, A., Ranke, J., & Straub, T. (2004). Zur rechtsverträglichen Technikgestaltung anhand einer M-Commerce-Anwendung. *Mobile Economy – Transaktionen, Anwendungen, Prozesse und Dienste, Proceedings zum 4. Workshop Mobile Commerce*, Augsburg, Germany.

Heinonen, K., & Strandvik, T. (2003). Consumer responsiveness to mobile marketing. *Proceedings of the Stockholm Mobility Roundtable*, Stockholm, Sweden.

Heinonen, K., & Strandvik, T. (2006). How Do Consumers React to Mobile Marketing? *Proceedings of the 5th International Conference on Mobile Business*, Copenhagen.

Heinonen, K., & Strandvik, T. (2007). Consumer responsiveness to mobile marketing. International Journal of Mobile Communications archive, 5(6), 603-617.

Helm, S. (2000). Viral marketing – establishing consumer relationships by 'word-of mouse'. *Electronic Markets*, *10*(3), 158–161. doi:10.1080/10196780050177053

Helsana (2007). Retrieved 23.08.2007, 2007, from http://www.helsana.ch/?wm=m%287%29&CFID=4809269&CFTOKEN=79035439

Hendry, M. (2007). Multi-application Smart Cards. Cambridge: Cambridge University Press.

Hennig-Thurau, T., Gwinner, K. P., Walsh, G., & Gremler, D. D. (2004). Electronic word-of-mouth via consumer-opinion platforms: What motivates consumers to articulate themselves on the internet? *Journal of Interactive Marketing, 18*(1), 38–52. doi:10.1002/dir.10073

Henseler, J. (2005). Einführung in die PLS-Pfadmodellierung. *WiSt. Wirtschaftswissenschaftliches Studium*, (2): 70–75.

Henten, A., Olesen, H., Saugstrup, D., & Tan, S. E. (2003). New mobile systems and services in Europe, Japan and South-Korea. Paper presented at the *Stockholm Mobility Roundtable*, Stockholm, May 22-23.

Herr, P. M., Kardes, F. R., & Kim, J. (1991). Effects of Word-of-Mouth and Product Attribute Information on Persuasion: An Accessibility-Diagnosticity Perspective. *The Journal of Consumer Research, 17*(4), 454–462. doi:10.1086/208570

Herr, P., Kardes, F., & Kim, J. (1991). Effects of word-of-mouth and product-attribute information on persuasion: An accessibility-diagnosticity perspective. *Journal of Consumer Research: An Interdisciplinary Quarterly, 17*(4), 454–462.

Hess, T., Figge, S., Hanekop, H., Hochstatter, I., Hogrefe, D., & Kaspar, C. (2005). Technische Möglichkeiten und Akzeptanz mobiler Anwendungen - Eine interdisziplinäre Betrachtung. *Wirtschaftsinformatik, 47*(1), 6–16.

Heun, C.T. (2005). Ads take aim at cell phones. *Information Week*, 42.

Hill, L., & O'Sullivan, T. (2004). *Foundation marketing.* London: Pearson, Prentice Hall.

Hinde, S. (2003). Spam: The evolution of a nuisance. *Computers & Security, 22*(6), 474–478. doi:10.1016/S0167-4048(03)00603-5

Hinrichs, J.-W., Stroborn, K., & van Baal, S. (2004). Mobiles Bezahlen aus der Sicht des Online-Händlers. In Pousttchi, K., & Turowski, K. (Ed.), *Mobile Economy* (pp. 63-78). Bonn: Köllen.

Hippner, H. (2006). Bedeutung, Anwendungen und Ensatzpotentiale von Social Software. In Hildebrand, K., & Hofmann, J. (Eds.), *Social Software, Praxis der Wirtschaftsinformatik, 43*(252), 6-16.

Hirschmann, A. O. (1974). *Abwanderung und Widerspruch - Reaktionen auf Leistungsabfall bei Unternehmungen, Organisationen und Staaten.* Tübingen: J.C.B. Mohr.

Hitoshi, K. (2005). Personal Interview, VISA, Tokyo Japan, May 2005.

Ho, V. (2007). Celestial Launches Mobile Channel. Retrieved on August 20, 2007 from http://www.varietyasiaonline.com/content/view/1600/

Hoffman, D. L., & Novak, T. P. (1996). Marketing in Hypermedia Computer-Mediated Environments: conceptual foundations. *Journal of Marketing, 60*(3), 50–68. doi:10.2307/1251841

Hofmann, O. (2007). *Young Population Drives Turkey Mobile Market.* Retrieved June 20, 2008, from http://www.euromonitor.com/Young_population_drives_Turkeys_mobile_market

Hoftstede, G. (1980). *Culture's consequences: International differences in work-related values.* Beverly Hills, CA: Sage.

Holbrook, M. B., & Hirschman, E. C. (1982). The Experiential Aspects of Consumption: Consumer Fantasies, Feelings and Fun. *The Journal of Consumer Research, 9*(2), 132–140. doi:10.1086/208906

Holland, H., & Bammel, K. (2006). *Mobile Marketing.* München: Vahlen.

Holtzblatt, K. (2005). Designing for the mobile device: experiences, challenges and methods. *Communications of the ACM, 48*(7), 33–35. doi:10.1145/1070838.1070862

Homburg, C., & Bruhn, M. (2003). Kundenbindungsmanagement – Eine Einführung in die theoretische und praktische Problemstellungen. In Bruhn, M., Homburg, C. (Hg.), *Handbuch Kundenbindungsmanagement Strategien und Instrumente für ein erfolgreiches CRM* (3-37). 4. Aufl., Wiesbaden: Gabler.

Homburg, C., & Bruhn, M. (2005). Kundenbindungsmanagement - Eine Einführung in die theoretischen und praktischen Problemstellungen. In C. Homburg & M. Bruhn (Eds.), *Handbuch Kundenbindungsmanagement* (5. Aufl. ed., pp. 3-40). Wiesbaden: Gabler.

Homburg, C., & Bruhn, M. (2005). Kundenbindungsmanagement – Eine Einführung in die theoretischen und praktischen Problemstellungen. In Homburg, C., & Bruhn, M. (Eds.), *Handbuch Kundenbindungsmanagement* (pp. 3-37). Wiesbaden, Gabler.

Homburg, C., & Koschate, N. (2003). Kann Kundenzufriedenheit negative Reaktionen auf Preiserhöhungen abschwächen? *Die Betriebswirtschaft, 63*(6), 619–634.

Homburg, C., & Krohmer, H. (2006). Marketingmanagement. Wiesbaden: Gabler.

Hong, S., Tam, K. Y., & Kim, J. (2006). Mobile data service fuels the desire for uniqueness. *Communications of the ACM, 49*(9), 89–94. doi:10.1145/1151030.1151034

Hourahine, B., & Howard, M. (2004). Money on the move: Opportunities for financial service providers in the 'third space'. *Journal of Financial Services Marketing, 9*(1), 57–67. doi:10.1057/palgrave.fsm.4770141

Hristova, N., & O'Hare, G. (2004). Ad-me: Wireless advertising adapted to the user location, device and emotions. *Proceedings of the 37th Hawaii International Conference on System Sciences*, HI.

Hsu, C.L., Lu, H.P., and Hsu, H.H. (2007). Adoption of the mobile internet: an empirical study of multimedia message service (MMS). *The International Journal of Management Science*, Dec. http://shop.informatm.com/marlin/30000001001/MARKT_EFFORT/marketingid/20001444760?proceed=true&MarEntityId=1167456884244&entHash=100231ldd57&UType=true Access Jan 2007.

Hsu, T.-h., Wang, Y.-s., & Wen, S.-c. (2006). Using the decomposed theory of planning behavioural to analyse consumer behavioural intention towards mobile text message coupons. *Journal of Targeting. Measurement & Analysis for Marketing, 14*(4), 309–324. doi:10.1057/palgrave.jt.5740191

Hu, D. H., & Wang, C. L. (2007). GPS-based Location Extraction and Presence Management for Mobile Instant Messenger. *Proceedings of the IFIP International Conference on Embedded and Ubiquitous Computing*, Taipei, Taiwan.

Hu, L., & Bentler, P. M. (1999). Cutoff Criteria for Fit Indices in Covariance Structure Analysis: Conventional Criteria versus New Alternatives. *Structural Equation Modeling, 6*(1), 1–56.

Hu, Y. C., Das, S. M., & Pucha, H. (2003). Exploiting the synergy between peer-to-peer and mobile ad hoc networks. In *Proceedings of HotOS'03: 9th Workshop on Hot Topics in Operating Systems* (p. 37–42). USENIX.

Huber, F., Herrmann, A., Meyer, F., Vogel, J., & Vollhardt, K. (2007). *Kausalmodellierung mit Partial Least Squares*. Wiesbaden: Gabler.

Hubschneider, M., & Kölmel, B. (2002). Mobile Business. In Bullinger, H.-J., & Berres, A. (Ed.), *E-Business* (pp. 671-686). Berlin: Springer.

Hyvönen, K., & Repo, P. (2005). Mobiilipalvelut suomalaisten arjessa [Mobile services in Finnish consumers' everyday life]. In J. Leskinen, H. Hallman, M. Isoniemi et al. (Eds.), *Vox consumptoris –Kuluttajan ääni* [*Consumer's voice*] (pp. 41-59). The National Consumer Research Centre, Kerava, Finland: Savion Kirjapaino Oy.

Ihlström Eriksson, C., & Åkesson, M. (2008). Ubiquitous advertising Challenges. In Proceedings of 7th International Conference on Mobile Business, (ICMB´08), Barcelona, Spain.

Im, I., Kim, Y., & Han, H.-J. (2008). The effects of perceived risk and technology type on users' acceptance of technologies. *Information & Management, 45*(1), 1–9.

Infratest, T. N. S. (2008). *GTI Telecoms Report 2007-08 Germany Report*, Munich, Germany.

Intuitive Media Research Services. (2006). Kids Go Mobile. Ownership and Use of Mobile Phones by Children aged 6 to 13 for ICT in Education Sector. Retrieved May 9, 2007 from http://www.nma.co.uk/assets/getAsset.aspx?liAssetID=20934.

I-play (2005). I-play Outlines Collective Industry Action Required for Mobile Gaming Market to Reach True Potential. Press release.

Ireland, C. (2003). Qualitative methods: From boring to brillant. In B. Laurel (Ed.), *Design research: Methods and perspectives*. Cambridge, Massachusetts: The MIT Press.

Isomursu, P., Hinman, R., Isomursu, M., & Spasojevic, M. (2007). Metaphors for the Mobile Internet. *Knowledge, Technology, and Policy, 20*(4), 259–268. doi:10.1007/s12130-007-9033-5

ITU (International Telecommunications Union). (2007). ITU world communication/ICT indicators database. Retrieved January 3, 2009, from http://www.itu.int/ITU-D/ict/statistics/

ITU. (2007). Trends in Telecommunication Reform 2007–The road to Next-Generation Networks (NGN) Summary.

ITU. (2008). International Telecommunications Union statistics. Available at http://www.itu.int/net/home/index.aspx

ITU. (March 2003). Broadband Korea: Internet Case Study.

Jackson, L. A., Ervin, K. S., Gardner, P. D., & Schmitt, N. (2001). Gender and the Internet: Women communicating and men searching. *Sex Roles*, *44*(5), 363–379. doi:10.1023/A:1010937901821

Jaokar, A. (2007). *Mobile Web 2.0. Kollektive Intelligenz durch Restricted Devices*, Retrieved June 06, 2007, from http://javamagazin.de/itr/news/psecom,id,30359,nodeid,10.htm

Jardosh, A., Belding-Royer, E. M., Almeroth, K. C., & Suri, S. (2003). Towards Realistic Mobility Models For Mobile Ad Hoc Networks. In *Mobicom '03: Proceedings of the 9th annual international conference on mobile computing and networking* (pp. 217–229). ACM Press.

Jaring, P., & Törmänen, V. Siiral E., & Matinmikko, T. (2007). Improving Mobile Solution Workflows and Usability Using Near Field Communication Technology. In Schiele, B., Dey, A. K., Gellersen, H., de Ruyter, B., Tscheligi, M., Wichert, R., Aarts, E., & Buchmann, A. (Ed.), *Ambient Intelligence* (pp. 358-373). Berlin: Springer.

Jarvenpaa, S. L., & Todd, P. A. (1997). Consumer reactions to electronic shopping on the World Wide Web. *International Journal of Electronic Commerce*, *1*(2), 59–88.

Jelassi, T., & Enders, A. (2004). Leveraging Wireless Technology for Mobile Advertising. *Proceedings of the 12th European Conference on Information Systems*, Turku, Finland.

Jelassi, T., & Enders, A. (2006). Mobile advertising: a European perspective. In S. J. Barnes and E. Scornavacca, E. (Eds.), *Unwired Business: Cases in Mobile Business, (*pp. 82-95). Hershey: IRM Press.

Jiang, J. J., Klein, G., & Crampton, S. M. (2000). A note on SERVQUAL reliability and validity in information system service quality measurement. *Decision Sciences*, *31*(3), 725–744. doi:10.1111/j.1540-5915.2000.tb00940.x

Jingjun Xu, D. (2006). The influence of personalization in affecting consumer attitudes toward mobile advertising in China. *Journal of Computer Information Systems*, *47*(2), 9–19.

Jones, T., & Sasser, W. E. (1995). Why Satisfied Customers Defect. *Harvard Business Review*, *73*(6), 88–99.

Jung, Holger, & von Matt (2002). Jean-Remy, Momentum–Die Kraft, die Werbung heute braucht. Berlin, Germany

Jurvetson, S. (1997). What is Viral Marketing? *Netscape M-Files*. Retrieved November 11, 2006 from http://www.dfj.com/cgi-bin/artman/publish/ steve_may00.shtml.

Kaeding, N. (2002). Mobile Business und Datenschutz. In R. Reichwald (Ed.), *Mobile Kommunikation Wertschöpfung, Technologien, neue Dienste* (pp. 193-204). Wiesbaden, Gabler Verlag.

Kagermann, H., & Österle, H. (2006). Geschäftsmodelle 2010. Frankfurt am Main: Frankfurter Allgem. Buch.

Kallio, J. (2007). Personal Interview, China, June 2007.

Kangas, E., & Kinnunen, T. (2005). Applying user-centered design Development to mobile application. *Communications of the ACM*, *48*(7), 55–59. doi:10.1145/1070838.1070866

Karahanna, E., Straub, D. W., & Chervany, N. L. (1999). Information technology adoption across time: A cross sectional comparison of pre- adoption and post- adoption beliefs. *MISQ*, *23*(2), 183–213. doi:10.2307/249751

Karjaluoto, H. An&d Kautonen, T. (2006). Trust as a factor affecting consumers' participation in mobile marketing. *Proceedings of the 35th European Marketing Academy Conference*, Athens, Greece.

Karjaluoto, H., & Alatalo, T. (2007). Consumers' attitudes towards and intention to participate in mobile marketing. *International Journal of Services Technology and Management*, *8*(2/3), 155–173. doi:10.1504/IJSTM.2007.012866

Karjaluoto, H., Lehto, H., Leppaniemi, M., & Jayawardhena, C. (2008). Exploring Gender Influence on Customer's Intention to Engage Permission-based Mobile Marketing. *Electronic Markets*, *18*(Issue 3), 242–259. doi:10.1080/10196780802265793

Karjaluoto, H., Lehto, H., Leppäniemi, M., & Mustonen, T. (2007). Insights into the Implementation of Mobile Marketing Campaigns. *International Journal of Mobile Marketing*, *2*(2), 10–20.

Karjaluoto, H., Lehto, H., Leppäniemi, M., & Mustonen, T. (2007)... *International Journal of Mobile Marketing*, *2*(2), 10–20.

Karjaluoto, H., Leppäniemi, M., Lehto, H., & Jayawardhena, C. (2008). Exploring gender influence on customer's intention to engage permission based mobile marketing. *Electronic Markets*, *18*(3), 242–259. doi:10.1080/10196780802265793

Karjaluoto, H., Leppäniemi, M., Standing, C., Kajalo, S., Merisavo, M., Virtanen, V., & Salmenkivi, S. (2006). Individual differences in the use of mobile services among Finnish consumers. *International Journal of Mobile Marketing*, *1*(2), 4–10.

Kaspar, C. (2006). *Individualisierung und mobile Dienste am Beispiel der Medienbranche - Ansätze zum Schaffen von Kundenmehrwert.* Unpublished Dissertation, Institut für Wirtschaftsinformatik 2 Universität Göttingen, Göttingen.

Kaspar, C. (2006). *Individualisierung und mobile Dienste am Beispiel der Medienbranche Ansätze zum schaffen von Kundenmehrwert*, Göttingen: Universitätsverlag.

Kaspar, C., & Hagenhoff, S. (2005). An overview of mobile radio technologies. In Pagani, M. (Ed.), *Encyclopedia of Multimedia Technology and Networking* (pp. 645-651). Hershey, PA: Idea Group Publishing.

Kaspar, C., Diekmann, T., & Hagenhoff, S. (2007). Context-Adaptive Mobile Systems, In Taniar, D. (Ed.), *Encyclopedia of Mobile Computing and Commerce*. Hershey, PA: Idea Group Publishing.

Kassarjian, H. H., & Robertson, T. S. (1991). *Perspectives in Consumer Behaviour (4th edition).* Prentice Hall, New Jersey.

Katz, E., Blummer, J. G., & Gurevitch, M. (1974). Uses and gratifications research. *Public Opinion Quarterly*, *23*, 389–398.

Katz, E., Gurevitch, M., & Haas, H. (1973). On the use of mass media for important things. *American Sociological Review*.

Katz, M. L., & Shapiro, C. (1985). Network Externalities, Competition, and Compatibility. *The American Economic Review*, *75*(3), 424–440.

Kavassalis, P., Spyropoulou, N., Drossos, D., Mitrokostas, E., Gikas, G., & Hatzistamatiou, A. (2003). Mobile permission marketing: Framing the market inquiry. *International Journal of Electronic Commerce*, *8*(1), 55–79.

KDDI. Annual Report. (2007). Retrieved August 12, 2007, from http://www.kddi.com/english/corporate/ir/library/annual_report/2007/pdf/kddi_ar2007_e.pdf

Keller, K. L. (1993). Conceptualizing, measuring, and managing customer-based brand equity. *Journal of Marketing*, *57*(1), 1–22. doi:10.2307/1252054

Kenning, P. (2008). The influence of general trust and specific trust on buying behaviour. *International Journal of Retail & Distribution Management*, *36*(6), 461–476. doi:10.1108/09590550810873938

Kerin, R. A., Hartley, S. W., & Rudelius, W. (2007). *Marketing: The core.* (2nd ed.). McGraw-Hill:Irwin.

Kern, C. (2006). Anwendung von RFID-Systemen. Berlin: Springer.

Khalifa, M., & Ning Shen, K. (2008). Explaining the adoption of transactional B2C mobile commerce. *Journal of Enterprise Information Management*, *21*(2), 110–124. doi:10.1108/17410390810851372

Kim, H., & Kwahk, K. (2007). Comparing the Usage Behaviour and the Continuance Intention of Mobile Internet Services", *Eight World Congress on Management of eBuisiness* (WCMeB 2007)

Kivetz, R. (2005). Promotion reactance: The Role of Effort-Reward Congruity. *The Journal of Consumer Research*, *31*(2), 725–736. doi:10.1086/426606

Kleijnen, M. H. P., de Ruyter, J. C., & Wetzels, M. G. M. (2004). Consumer Adoption of Wireless Services: Discovering the Rules, While Playing the Game. *Journal of Interactive Marketing*, *18*(2), 51–61. doi:10.1002/dir.20002

Kleijnen, M., De Ruyter, K., & Andreassen, T. W. (2005). Image congruence and the adoption of service innovations. *Journal of Service Research*, *7*(4), 343–359. doi:10.1177/1094670504273978

Kleijnen, M., De Ruyter, K., & Wetzels, M. (2004). Consumer adoption of wireless services: discovering the rules, while playing the game. *Journal of Interactive Marketing*, *18*(2), 51–61. doi:10.1002/dir.20002

Kleijnen, M., Wetzels, M., & De Ruyter, K. (2004). Consumer acceptance of wireless finance. *Journal of Financial Services Marketing*, *8*(3), 206–217. doi:10.1057/palgrave.fsm.4770120

Kleinaltenkamp, M. (1998). *Geschäftsbeziehungsmanagement*. Berlin: Springer.

Klemm, A., Lindemann, C., & Waldhorst, O. P. (2004). Peer-to-Peer Computing in Mobile Ad Hoc Networks. In *Performance tools and applications to networked systems: Revised tutorial lectures* (Vol. 2965, pp. 187–208). Springer.

Kodama, M. (2005). Innovation through networked strategic communities: case study on NTT DoCoMo. *Journal of Management Development*, *24*, 169–187. doi:10.1108/02621710510579527

Kohda, Y., & Endo, S. (1996). Ubiquitous advertising on the WWW: Merging advertisement on the browser. *Computer Networks and ISDN Systems*, *28*, 1493–1499. doi:10.1016/0169-7552(96)00070-0

Koike, K. (2005). Personal Interview, NMS Communications, Tokyo, Japan, May 2005.

Koivumäki, T., Ristola, A., & Kesti, M. (2006). Predicting consumer acceptance in mobile services: empirical evidence from an experimental end user environment. *International Journal of Mobile Communications, 4*(4), 418–435.

Kollmann, T. (1998). *Akzeptanz innovativer Nutzungsgüter und -systeme.* Wiesbaden: Gabler.

Kollmann, T. (2000). Akzeptanzprobleme neuer Technologien – Die Notwendigkeit eines dynamischen Untersuchungsansatzes. In Bliemel, F., Fassot, G., & Theobald, A. (Ed.) *Electronic Commerce* (pp. 27-45). Wiesbaden: Gabler.

Kollmann, T. (2007a). E-Business. Wiesbaden: Gabler.

Kollmann, T. (2007b). Online-Marketing. Stuttgart: Lucius & Lucius UTB.

Kolsaker, A., & Payne, C. (2002). Engendering trust in e-commerce: a study of gender-based concerns. *Marketing Intelligence & Planning, 20*(4), 206–214. doi:10.1108/02634500210431595

Komulainen, H., Mainela, T., Sinisalo, J., Tähtinen, J., & Ulkuniemi, P. (2005). Models of Mobile Advertising Network. *E-business Review, 5*, 95–98.

Komulainen, H., Mainela, T., Sinisalo, J., Tähtinen, J., & Ulkuniemi, P. (2004). *Business models in the emerging context of mobile advertising.* Paper presented at the Frontiers of e-Business Research 2004, FeBR 2004, Tampere, Finland.

Komulainen, H., Mainela, T., Tähtinen, J., & Ulkuniemi, P. (2007). Retailers' different value perceptions of mobile advertising service. *International Journal of Service Industry Management, 18*(4), 368–393. doi:10.1108/09564230710778146

Komulainen, H., Ristola, A., & Still, J. (2006). Mobile advertising in the eyes of retailers and consumers – empirical evidence from a real-life experiment, *Proceedings of the International Conference on Mobile Business* (ICMB'06)

Komulainen, H., Ristola, A., & Still, J. (2006). Mobile advertising in the eyes of retailers and consumers–empirical evidence from a real life experience. In Proceedings of the 5th International Conference on Mobile Business, (ICMB'06), Copenhagen, Denmark.

Komus, A. (2006). Social Software als organisatorisches Phänomen - Einsatzmöglichkeiten in Unternehmen. In Hildebrand, K., & Hofmann, J. (Eds.), *Social Software, Praxis der Wirtschaftsinformatik, 43*(252), 36-44.

Kondo, F. N., & Nakahara, M. (2007). Differences in Customers' Responsiveness to Mobile Direct Mail Coupon Promotions. *International Journal of Mobile Marketing, 2*(2), 68–74.

Kondo, F. N., Uwadaira, Y., & Nakahara, M. (2007). Stimulating customer response to promotions: the case of mobile phone coupons. *Journal of Targeting. Measurement and Analysis for Marketing, 17*(1), 57–67. doi:10.1057/palgrave.jt.5750064

Königstorfer, J., & Gröppel-Klein, A. (2007). Experience of Failure and Anger when Using the Mobile and Wired internet – The Interference of Acceptance- and Resistance-driving Factors. *Marketing JRM, 27*(1), 34–47.

Königstorfer, J., & Gröppel-Klein, A. (2007). Projektive Verfahren zur Ermittlung der Akzeptanz technologischer Innovationen. In Buber, R., & Holzmüller, H. (Ed.) *Qualitative Marktforschung* (pp. 847-862). Wiesbaden: Gabler.

Kontraband, Parys Communications & YouGov (2006). Viral Marketing Industry Report 2006. Retrieved February 15, 2008 on http://www.the7thchamber.com/pdf/The-Seventh-Chamber_Viral-Report.pdf

Koranteng, J. (2001). ZAP! There's no escaping the mobile advertisement. *Ad Age Global, January, 1(5),* 9.

Kotler, A. Saunders, & Wong. (2008). *Principles of marketing, fifth European Edition.* Harlow, Pearson Education Limited

Kotler, P. (2000). *Marketing management: The millennium edition.* Upper Saddle River, NJ: Prentice Hall.

Kotler, P. (2004). *Kotler on marketing.* USA: HBR Publisher.

Kotler, P.(1973-4). Atmospherics as a marketing tool. *Journal of Retailing, 49*, 48–64.

Kotler, P., & Armstrong, G. (2008). *Principles of Marketing.* New Jersey: Pearson Education Inc.

Kotler, P., & Keller, K. L. (2006). *Marketing Management.* Pearson Education.

Kotler, P., & Zaltman, G. (1976). Targeting prospects for a new product. *Journal of Advertising Research, 16*(1), 7–18.

Kotler, P., Jain, D., & Maesincee, S. (2002). *Marketing moves: A new approach to profits, growth and renewal.* London: McGraw-Hill.

Kotler, P., Keller, K. L., & Bliemel, F. (2007). Marketing-Management. München: Pearson.

Kotler, P., Wong, V., Saundersa, J., & Armstrong, G. (2005). *Principles of marketing* (fourth European edition). Pearson Education Limited.

Koufaris, M. (2002). Applying the technology acceptance model and flow theory to online consumer behavior. *Information Systems Research, 13*(2), 205–223. doi:10.1287/isre.13.2.205.83

Krafft, M., Götz, O., & Liehr-Gobbers, K. (2005). Die Validierung von Strukturgleichungsmodellen mit Hilfe des Partial-Least-Squares (PLS)-Ansatzes. In F. Bliemel, A. Eggert, G. Fassott & J. Henseler (Eds.), *Handbuch PLS-Pfadmodellierung* (pp. 71-86). Stuttgart: Schäffer-Poeschel.

Krcmar, H. (2005). Informationsmanagement. Berlin: Springer.

Kreyer, N., Pousttchi, K., & Turowski, K. (2003). Mobile Payment Prodcedures. *eService Journal*, 2 (3), 7-22.

Kriewald, M. (2007). *Situationsabhängiges mobiles Customer Relationship Management. Analysen □ Wettbewerbsvorteile □ Beispiele*. Hamburg: Verlag Dr. Kovač.

Krishna, A., & Shoemaker, R. W. (1992). Estimating the effects of higher coupon face values on the timing of redemptions, the mix of coupon redeemers, and purchase quantity. *Psychology and Marketing, 9* (November/ December), 453–467.

Krishnamurthy, S. (2001). A comprehensive analysis of permission marketing. *Journal of Computer Mediated Communication, 6(2).* Retrieved from http://www.ascusc.org/jcmc/vol6/issue2/krishnamurthy.html

Krishnamurthy, S. (2001). A comprehensive analysis of permission marketing. *Journal of Computer-Mediated Communication, 6*(2). Retrieved January 30, 2009, from http://jcmc.indiana.edu/vol6/issue2/krishnamurthy.html

Kristoffersen, S. and Ljungberg, F. Mobile informatics: Innovation of IT use in mobile settings. *IRIS'21 workshop report. SIGCHI Bulletin* 31, 1 (1999), 29–34.

Kristoffersen, S., & Ljungberg, F. (1999). Designing interaction styles for a mobile use context. In *Proceedings of she hiternational Symposium on HanclhirUand Ubiquitous Cotnputing. (HUC99).*

Kroeber-Riel, W., & Weinberg, P. (2003). Konsumentenverhalten. München: Vahlen.

KT WiBro, South Korea. (2008). From http://www.ktwibro.com/ktwibro/main.html

Küllenberg, B., & Quente, C. (2006). Brand's New Toy–Kreative Markenkommunikation mit Handy & Co. (pp. 211). Munich, Germany.

Kümmerlein, A. (2002). Mobile Marketing Kampagnen. Werbewirkungsanalyse. Düsseldorf.

Kuhn, J., & Lehner, F. (2003). Szenarien einer mobilen Zukunft. In Pousttchi, K., & Turowski, K. (Ed.), *Mobile Commerce* (pp.130-142). Bonn: Köllen.

Kurkovsky, S. (2005). Using principles of pervasive computing to design m-commerce applications. *Proceedings of the International Conference on Information Technology: Coding and Computing (ITCC'05),* Volume II, 59-64.

Kurkovsky, S., & Harihar, K. (2006). Using Ubiquitous Computing in Interactive Mobile Marketing. *Personal and Ubiquitous Computing, 10*(4), 227–240. doi:10.1007/s00779-005-0044-5

Kurkovsky, S., & Harihar, K. (2006). Using Ubiquitous Computing in Interactive Mobile Marketing. *Personal and Ubiquitous Computing, 10*(4), 227–240. doi:10.1007/s00779-005-0044-5

Kurkovsky, S., & Harihar, K. (2006). Using ubiquitous computing in interactive mobile marketing. *Personal and Ubiquitous Computing, 10*, 227–240. doi:10.1007/s00779-005-0044-5

Kuß, A., Tomczak, T., & Reinecke, S. (2007). Marketingplanung. Wiesbaden: Gabler.

Laker, M., Pohl, A., & Dahlhoff, D. (2004). Kundenbindung auf neuen Märkten. In. Hinterhuber, H. H., & Matzler, K. (Ed.), *Kundenorientierte Unternehmensführung* (pp. 147-159). Wiesbaden: Gabler.

Lam, S. Y., Shankar, V., Erramilli, M. K., & Murthy, B. (2004). Customer value, satisfaction, loyalty, and switching costs: an illustration from a business-to-business service context. *Journal of the Academy of Marketing Science, 32*(3), 293–311. doi:10.1177/0092070304263330

Lambert, S. C. (2005). Making Sense of Business Models. *The Magnus Journal of Management*, 1.

Langheinrich, M. (2007). Gibt es in einer total informatisierten Welt noch eine Privatsphäre? In Mattern, F. (Ed.), *Die Informatisierung des Alltags* (pp. 232-264). Berlin: Springer.

Langner, S. (2006). Viral Marketing. Wiesbaden: Vahlen.

Laudon, K. C., & Traver, C. G. (2002). *E-Commerce*. Boston: Addison Wesley.

Lee, C., & Shim, J. P. (2006). An Empirical Study on User Satisfaction with Mobile Business Applications Use and Hedonism. [JITTA]. *Journal of Information Technology Theory and Application, 8*(3), 57–74.

Lee, D., & Muhlberger, R. (2006). Mobile marketing as technology intervention: A locales framework analysis. *International Journal of Mobile Marketing, 1*(1), 59–65.

Lee, D., & Muhlberger, R. (2007). A technology introduction view on mobile marketing. In D. Taniar (Ed.), *Encyclopedia of mobile computing & commerce* (Vol. 1). USA: Idea Group Reference.

Lee, R., & Murphy, J. (2006). The consumption of mobile services by Australian university students. *International Journal of Mobile Marketing, 1*(1), 13–20.

Lee, Y. E., & Benbasat, I. (2003). Interface design for mobile commerce. *Communications of the ACM, 46*(12), 49–52. doi:10.1145/606272.606300

Lee, Y. E., & Benbasat, I. (2003). Interface design for mobile commerce. *Communications of the ACM, 46*(12), 49–52. doi:10.1145/606272.606300

Lee, Y. E., & Benbasat, I. (2004). A framework for the study of customer interface design for mobile commerce. *International Journal of Electronic Commerce, 8*(3), 79–102.

Lee, Y., Kim, J., & Lee, I. (2002). Kim, h. (2002),"a Cross-cultural Study on the Value Structure of Mobile Internet Usage: Comparison between Korea and Japan. *Journal of Electronic Commerce Research, 3*(4).

Leem, C. S., Suh, H. S., & Kim, D. S. (2004). A classification of mobile business models and its applications. *Industrial Management & Data Systems, 104*(1), 78–87. doi:10.1108/02635570410514115

Leem, C. S., Suh, S. H., & Kim, D. S. (2004). A Classification of Mobile Business Models and Its Applications. *Industrial Management & Data Systems, 104*(1/2), 78–87. doi:10.1108/02635570410514115

Leenheer, J., van Heerde, H. J., Bijmont, T. H. A., & Smidts, A. (2007). Do Loyalty Programms really Enhance Behavioral Loyalty? An Empirical Analysis Accounting for Self-Selecting Members. *International Journal of Research in Marketing, 24*(1), 31–47. doi:10.1016/j.ijresmar.2006.10.005

Leguay, J., Lindgren, A., Scott, J., Friedman, T., & Crowcroft, J. (2006). Opportunistic Content Distribution in an Urban Setting. In *CHANTS '06: 2006 SIGCOMM Workshop on Challenged Networks* (pp. 205–212).

Lehner, F. (2002). Einführung und Motivation. In Teichmann, R., & Lehner, F. (Eds.), *Mobile Commerce* (pp. 3-28). Berlin: Springer.

Lehner, F. (2003). Mobile und drahtlose Informationssysteme. Berlin: Springer.

Lehner, F. (2003). *Mobile und drahtlose Informationssysteme. Technologien, Anwendungen und Märkte*. Berlin: Springer.

Leighton, D. (2007). In the frame: investigating the use of mobile phone photography in museums. *International Journal of Nonprofit and Voluntary Sector Marketing, 12*(4), 308–319. doi:10.1002/nvsm.320

Leppaniemi, M., & Karjaluoto, H. (2005). Factors influencing consumers' willingness to accept mobile advertising: A conceptual model. *International Journal of Mobile Communications, 3*(3), 197–213. doi:10.1504/IJMC.2005.006580

Leppäniemi, M., & Karjaluoto, H. (2006). A review of mobile marketing research. *International Journal of Mobile Marketing, 1*(1), 2–12.

Leppäniemi, M., & Karjaluoto, H. (2008). M-marketing: From Marketing Strategy to M-marketing Campaign Implementation. *International Journal of M-marketing, 3*(1), 50–61.

Leppäniemi, M., & Karjaluoto, H. (2008). Mobile marketing: From marketing strategy to mobile marketing campaign implementation. *International Journal of Mobile Marketing, 3*(1), 50–61.

Leppaniemi, M., Karjaluoto, H., & Salo, J. (2004). The success factors of mobile advertising value chain. *The E-business Review, 14*(1), 93–97.

Leppäniemi, M., Karjaluoto, H., & Salo, J. (2004). The success factors of mobile advertising value chain. *eBusiness Review IV*, 93-97.

Leppäniemi, M., Karjaluoto, H., & Sinisalo, J. (2006). A Review of Mobile Marketing Research. *International Journal of Mobile Marketing, 1*(1), 30–42.

Leppäniemi, M., Karjaluoto, H., Salo, J., & Sinisalo, J. (2005). Factors affecting customer's willingness to receive mobile marketing. *Proceedings of the 4th International Conference on Research in Advertising (ICORIA)*, 249-253.

Leppaniemi, M., Sinisalo, J., & Karjaluoto, H. (2006). A review of mobile marketing research. *International Journal of Mobile Marketing, 1*(1), 30–40.

Leung, F. S. K., & Cheung, C. M. K. (2004). Consumer attitude toward mobile advertising. *Proceedings of the Tenth Americas Conference on Information Systems*, New York.

Leung, L., & Wei, R. (2000). More than just talk on the move: Uses and gratifications of the cellular phone. *MISQ, 77*(2), 308–320.

Levy, M., & Weitz, B. A. (2001). *Retailing Management* (4th ed.) McGraw-Hill, IRWIN.

Levy, M., & Weitz, B. A. (2007). *Retailing Management* (6th ed.) McGraw-Hill, IRWIN.

Lewis, S. (2001). M-commerce: Ads in the ether. *Asian Business, January, 37(1),* 31.

Lewison, D. M. (1994). *Retailing*, 5th edition, New York, NY: Macmillan College Publishing Company.

Li, D., Chau, P.Y.K., and Lou, H. (2005). Understanding individual adoption of instant messaging: an empirical investigation. *Journal of the Association for information systems*, vol. 6, 102-129.

Li, J. P., & Kishore, R. (2006). How Robust is the UTAUT Instrument? A multigroup Invariance Analysis in the Context of Acceptance and Use of Online Community Weblog Systems, *Proceedings of the 2006 ACM SIGMIS CPR Conference on Computer Personnel Research*, 183 – 189.

Lichtenstein, S., & Williamson, K. (2006). Understanding Consumer Adoption of Internet Banking: An Interpretive Study in the Australian Banking Context. *Journal of Electronic Commerce Research, 7*(2), 50–66.

Liebowitz, S., & Margolis, S. (1998). Network externality. *The New Palgraves Dictionary of Economics and the Law*, Macmillan, London.

Lihotzky, N. (2003). *Kundenbindung im Internet □ Maßnahmen und Erfolgswirksamkeit im Business-to-Consumer-Bereich.* Wiesbaden, Gabler.

Lin, C. A. (1996). Looking back: The contribution of Blumler and Katz's uses of mass communication research. *Journal of Broadcasting & Electronic Media, 40,* 574–582.

Lindemann, C., & Waldhorst, O. P. (2005). Epidemic Data Dissemination for Mobile Peer-to-Peer Lookup Services. In *Peer-to-Peer Systems and Applications* (Vol. 3485, p. 435–455). Springer.

Ling, R. (2001) "We Release Them Little by Little" Maturation and Gender Identity as Seen in the Use of Mobile Telephony. Personal and Ubiquitous Computing, 5(123 - 136.

Link, J. (2003). Die Klärung von Wirtschaftlichkeiten von M-Commerce-Projekten. In Link, J. (Ed.), *Mobile Commerce* (pp. 41-64). Berlin: Springer.

Link, J. (2003). M-Commerce: Die stille Revolution hin zum Electronic Aided Acting. In J. Link (Ed.), *Mobile Commerce. Gewinnpotenziale einer stillen Revolution* (pp. 1-39). Berlin/Heidelberg: Springer Verlag.

Link, J., & Seidl, F. (2008). Der Situationsansatz als Erfolgsfaktor des Mobile Marketing. In H.H. Bauer, T. Dirks, & M.D. Bryant (Ed.), *Erfolgsfaktoren des Mobile Marketing – Strategien, Konzepte und Instrumente* (pp. 51-70). Berlin/Heidelberg: Springer Verlag.

Link, J., & Weiser, C. (2006). *Marketing-Controlling. Systeme für mehr Markt- und Unternehmenserfolg.* München: Vahlen Verlag.

Lippert, I. (2004). Mobile marketing. In R. Gerstner, G. Hunke & H. Sabel (Eds.), *Innovatives marketing* (pp. 109-123). Frankfurt am Main: Fritz Knapp.

Liu, Y., & Shrum, L. J. (2002). What is interactivity and is it always such a good thing? Implications of definition, person, and situation for the influence of interactivity on advertising effectiveness. *Journal of Advertising, 31*(4), 53–64.

Liu, Y., & Shrum, L. J. (2002). What is interactivity and is it always such a good thing? Implications of Definition, Person, and Situation for the Influence of Interactivity on Advertising Effectiveness. *Journal of Marketing, 31*(4), 53–64.

Lohse, L. G., & Spiller, P. (1998b). Electronic shopping. *Communications of the ACM, 41*(7), 81–87. doi:10.1145/278476.278491

Lorenz, R. (2005). Personal Interview, NETS, Singapore, May 2005.

Louvieris, P., & Driver, J. (2001). New frontiers in cybersegmentation: Marketing success in cyberspace depends on IP address. *Qualitative Market Research: An International Journal, 4*(3), 169–181. doi:10.1108/13522750110393071

Lovelock, C., & Quelch, J. (1983). Consumer Promotions in Services Marketing. *Business Horizons, 26*(3), 66–75. doi:10.1016/0007-6813(83)90032-0

Lu, H. K. (2007). Network smart card review and analysis. *Computer Networks, 51*(9), 2234–2248. doi:10.1016/j.comnet.2007.01.009

Lu, J., Yu, C.-S., Liu, C., & Yao, J. E. (2003). Technology Acceptance Model for Wireless Internet. *Internet Research, 13*(3), 206–222. doi:10.1108/10662240310478222

Lu, Y., Dong, Y., & Wang, B. (2007). The Mobile Business Value Chain in China. *Sixth International Conference on the Management of Mobile Business.*

Luarn, P., & Lin, H. H. (2005). Toward an understanding of the behavioral intention to use mobile banking. *Computers in Human Behavior, 21*(6), 873–891. doi:10.1016/j.chb.2004.03.003

Lumière, A., & Lumière, L. (Dir.) (1895). Arrivée d'un train à La Ciotat, L'. France. From http://www.imdb.com/title/tt0000012/

Luupay. (n.d.). Retrieved December 16, 2008 from https://www.luupay.de/User/De/Help/userguide.aspx

M:Metrics Press Release, M:Metrics: iPhone Hype Holds Up, Seattle/ London, March 18, 2008 from http://www.mmetrics.com/press/PressRelease.aspx?article=20080318-iphonehype

M:Metrics. (2006). M:Metrics unveils industrys first definitive mobile marketing metrics. Retrieved January 30, 2009, http://www.mmetrics.com/press/articles/20061003-sms-shorttext.pdf, 2008-08-20.

Ma, Q., & Liu, L. (2004). The technology acceptance model: A meta analysis of empirical findings. *Journal of Organizational and End User Computing, 16*(1), 59–72.

MacDonald, D. (2003). NTT DoCoMo's i-mode: Developing Win-Win Relationships for Mobile Commerce. In B. Mennecke & T. Strader, Mobile Commerce: Technology, Theory, and Applications (1-25). Hershey, PA: IGI Publishing.

MacKenzie, S. B. and Lutz, R. L. (1989). An empirical examination of the structural antecedents of attitude toward the ad in an advertising pretesting context. *Journal of Marketing*, Vol. 53 (April 1998), 48-65.

Mallat, N., Rossi, M., Tuunainen, V. K., & Öörni, A. (2008). An empirical investigation of mobile ticketing service adoption in public transportation. *Personal and Ubiquitous Computing, 12*(1), 57–65. doi:10.1007/s00779-006-0126-z

Maneesoonthorn, C., & Fortin, D. (2006). Texting behaviour and attitudes toward permission mobile advertising: An empirical study of mobile users' acceptance of SMS for marketing purposes. *International Journal of Mobile Marketing, 1*(1), 66–72.

Manganari, E. E., Siomkos, G. J., & Vrechopoulos, A. (2007). Atmospheric qualities in mobile commerce: an initial approach. *Conference Proceedings* European Marketing Academy Conference, EMAC2007, May 22-25, 2007, Reykjavik, Iceland.

Mann, A., & Prein, J. (2008). Akzeptanz mobiler Kundenkarten – konzeptionelle Überlegungen und empirische Befunde. In H. H. Bauer, T. Dirks, & M. D. Bryant (Ed.), *Erfolgsfaktoren des Mobile Marketing* (pp. 241-259). Berlin, Heidelberg: Springer.

Mao, E. and Palvia, P. (2006). Testing an Extended Model of IT Acceptance in the Chinese Cultural Context, *Advances in Information Systems*, Vol. 37, No. 2 and 3: 20-32.

Marini, S., & Wiedemann, D. G. (2006). Entwicklungen im Bereich Mobile Advertising aus der Sicht von Experten. Ergebnisse der Expertenbefragung MM1. In Pousttchi, K. (Ed.), *Studienpapiere der Arbeitsgruppe Mobile Commerce* (pp. 1-49). Augsburg, Germany: Universität Augsburg, Arbeitsgruppe Mobile Commerce.

Marriott, L. (2006). Location-Based Services: Ready for the Main Stream? Retrieved July 25, 2008, from http://www.clickz.com/showPage.html?page=3623236

Martin, B. A. S., Durme, J. V., Raulas, M., & Merisavo, M. (2003). Email advertising: Exploratory insights from Finland. *Journal of Advertising Research, 43*(3), 293–300.

Mason, B. J., Mayer, M. L., & Ezell, H. F. (1991). *Retailing.* 4th edition, IRWIN, Homewood, Boston.

Massachusetts Institute of Technology. (2005). *Reality Mining.* http://reality.media.mit.edu/ (seen 8/2008).

Mathieson, K. (1991). Predicting user intentions: Comparing the Technology Acceptance Model with the Theory of Planned Behavior. *Information Systems Research, 2*(3), 173–191. doi:10.1287/isre.2.3.173

Mathieson, K. (1991). Predicting user intentions: Comparing the technology acceptance model with the theory of planned behaviour. *Information Systems Research, 2*, 173–191. doi:10.1287/isre.2.3.173

Mathieson, R. (2005). Get Your Game On: The Rise of Mobile Advergames. Retrieved July 25, 2008, from http://mmaglobal.com/modules/article/view.article.php/105

Matosian, J. (1982). Effectiveness of different coupon delivery methods in building mass transit ridership. *Journal of Advertising Research, 22*(June-July), 54–59.

Matsuda, M. (2004). International Symposium: The Social Use of Anthropology in the Contemporary World, Session 1: Children and Media. Retrieved August 2, 2007, from http://www10.plala.or.jp/misamatsuda /Mead_Symposium.html

Matsunaga, M. (2000). [*i-mode: An Analogue Account of the Mobile Internet.* Singapore: Chuang-Yi Publishing Pte Ltd.]. *Birth (Berkeley, Calif.)*, i.

Mattern, F. (2005). Die technische Basis für das Internet der Dinge. In Fleisch, E., & Mattern, F. (Ed.), *Das Internet der Dinge* (pp. 39-66). Berlin: Springer.

May, P. (2001). *Mobile commerce.* UK: Cambridge University Press.

May, P. (2001). *Mobile Commerce: Opportunities, Applications, and Technologies of Wireless Business.* Cambridge: Cambridge University Press.

Mayes, K. E., & Markantonakis, K. (2007). Smart Cards, Tokens, Security and Applications. Berlin: Springer.

Mayes, K. E., & Markantonakis, K. (2008). Mobile communication security controllers. *Information Security Technical Report, 13*(3), 173–192. doi:10.1016/j.istr.2008.09.004

McElroy, D., & Turban, E. (2000). Smart Cards. In Shaw, M., Blanning, R., Strader, T., & Whinston, A. (Éd.), *Handbook on Electronic Commerce* (pp. 289-311). Berlin: Springer.

McEwan, G., & Greenberg, S. (2005). *Supporting social worlds with the community bar.* Paper presented at the 2005 international ACM SIGGROUP conference on supporting group work, Sanibel Island, FL.

McGinty, A., & Da Bona, D. (2004). 3G Licensing in China: A Waiting Game. *Computer Law & Security Report, 20*, 480–481. doi:10.1016/S0267-3649(04)00093-7

McKenzie, S. B., & Lutz, R. L. (1989). An empirical examination of the structural antecedents of attitude toward the ad in an advertising pretesting context. *Journal of Marketing, 53*(April), 48–65. doi:10.2307/1251413

McKinney, L. N. (2004). Creating a Satisfying Internet Shopping Experience Via Atmospheric Variables. *International Journal of Consumer Studies, 28*(3), 268–283. doi:10.1111/j.1470-6431.2004.00368.x

McKinney, V., Yoon, K., & Zahedi, F. (2002). The measurement of Web-customer satisfaction: An expectation and disconfirmation approach. *Information Systems Research, 13*(3), 296–315. doi:10.1287/isre.13.3.296.76

McManus, P., & Scornavacca, E. (2005). Mobile marketing: killer application or new hype? *Proceedings of the 4th International Conference on Mobile Business*, Sydney, Australia.

McMillan, S. J., & Hwang, J. S. (2002). Measures of Perceived Interactivity: An Exploration of the Role of Direction of Communication, User Control, and Time in shaping Perceptions of Interactivity. *Journal of Advertising, 31*(3), 29–42.

MDA (Mobile Data Association). (2008). The Q3 2008 UK mobile trends report. Retrieved December 15, 2008, from http://www.themda.org/mda-press-releases/the-q3-2008-uk-mobile-trends-report.php

Meffert, H. (2000). *Marketing Grundlagen marktorientierter Unternehmensführung.* Wiesbaden, Gabler Verlag.

Mehrabian, A., & Russell, J. A. (1974) *An Approach to Environmental Psychology*, Cambridge, MA: MIT Press.

Menon, S., & Kahn, B. (2002). Cross-category effects of induced arousal and pleasure on the internet shopping experience. *Journal of Retailing, 78*(1), 31–40. doi:10.1016/S0022-4359(01)00064-1

Merisavo, M., Kajalo, S., Karjaluoto, H., Virtanen, V., Salmenkivi, S., Raulas, M., & Leppäniemi, M. (2007). An empirical study of the drivers of consumer acceptance of mobile advertising. *Journal of Interactive Advertising, 7*(2). http://www.jiad.org/index.htm.

Merisavo, M., Vesanen, J., Arponen, A., Kajalo, S., & Raulas, M. (2006). The effectiveness of targeted mobile advertising in selling mobile services: an empirical study. *International Journal of Mobile Communications, 4*(2), 119–127.

Messmer, B. T. (2005). Architectures for Mobile Services. In Giordano, M. (Ed.), *Mobile Business* (pp. 201-212). Wiesbaden: Vahlen.

Methlie, L. B., & Pedersen, P. E. (2005). Network effects and intrinsic attributes of mobile services: Mapping perceived service values on business model choices. Presented at the *Hong Kong Mobility Roundtable, 2005,* Hong Kong, June, 1-3.draft pdf version http://ikt.hia.no/perep/hk_mobility.pdf accessed last on Oct 31, 2008

Meyer, M. (2005). *Multidisciplinarity of CRM Integration and its Implications*, in: Proceedings of the 38th Hawaii International Conference on System Sciences, (HICCS 2005).

Michael, A., & Slater, B. (2006). *Mobile marketing: achieving competitive advantage through wireless technology.* Oxford, Elsevier Ltd.

Miles, C. (2007). iPOP interview. May 2007.

Millar, J., & Salt, J. (2008a). Portfolios of mobility: The movement of expertise in transnational corporations in two sectors–aerospace and extractive industries. *Global Networks, 8*(1), 25–50.

Millar, J., & Salt, J. (2008b). *International students and the labour market: Experience in the UK.* Paper presented at Leverhulme Programme Conference on Mobility in International Labour Markets, University College London, UK.

Milne, G. M., Boza, M. E., & Rohm, A. (1999). Controlling personal information in marketing databases: A consumer perspective. *Proceedings of the American Marketing Association Winter Meeting*, St. Petersburg, Florida.

Milne, G., & Gordon, M. (1993). Direct mail privacy-efficiency trade-offs within an implied social contract framework. *Journal of Public Policy & Marketing, 12*(2), 206–216.

Minton, H. L., & Schneider, F. W. (1980). *Differential psychology.* Prospect Heights, IL: Waveland Press.

Minton, H. L., & Schneider, F. W. *Differential Psychology*, Waveland Press, Prospect Heights, IL, 1980.

Miron, A. M., & Brehm, J. W. (2006). Reactance Theory – 40 Years Later. *Zeitschrift für Sozialpsychologie, 37*(1), 9–18. doi:10.1024/0044-3514.37.1.9

Misner, I. R. (1999). *The World's Best Known Marketing Secret.* Austin, TX: The Bard Press.

Missi, F., Alshawi, S., & Fitzgerald, G. (2005). *Why crem efforts fail? A study of the impact of data quality and data integration*, in: Proceedings of the 38th Hawaii International Conference on System Sciences, (HICCS 2005).

Mittal, V., Kumar, P., & Tsiros, M. (1999). Attribute-Level Performance, Satisfaction, and Behavioral Intentions over Time: A Consumption-System Approach. *Journal of Marketing, 63*(2), 88–101. doi:10.2307/1251947

MMA (Mobile Marketing Association). (2008). *Mobile marketing industry glossary.* Version 2008.12. Retrieved December 15, 2008, from http://mmaglobal.com.

MMA. (2008a). MMA's commitment to consumer satisfaction. Available at http://www.mmaglobal.com/consumer

MMA. (2008b). Mobile marketing glossary, version 2008.12. Available at http://mmaglobal.com/uploads/glossary.pdf

M-marketing Association. (2007). Understanding M-marketing: Technology and Reach, May 2007. Retrieved July 25, 2008, from http://www.mmaglobal.com/mobilemarketing102.pdf

M-marketing Association. (2007a). Introduction to Mobile Coupons. Retrieved July 25, 2008, from http://mmaglobal.com/mobilecoupons.pdf

M-marketing Association. (2007b). *M-marketing Industry Glossary.* Retrieved July 15, 2008, from http://www.mmaglobal.com/glossary.pdf

M-marketing Association. (2007c). M-marketing Sweepstakes and Promotions Guide, July 2007. Retrieved July 15, 2008, from http://www.mmaglobal.com/mobilepromotions.pdf

M-marketing Association. (2007d). Understanding M-marketing: Technology and Reach, May 2007. Retrieved July 25, 2008, from http://www.mmaglobal.com/mobilemarketing102.pdf

Mobile Marketing Association. (2003). MMA Code for Responsible Mobile Marketing. London: Barnes Graham.

Mobile Marketing Association. Mobile Marketing Industry Glossary. (2007). Access January 2007, from http://www.mmaglobal.com/modules/content/index.php?id=6

Mobile Phone. (2008). www.wikipedia.org, November 30, 2008 from http://en.wikipedia.org/wiki/Mobile_phone)

Möhlenbruch, D. (2006). Sortimentspolitik. In Handelsblatt (Ed.), Wirtschaftslexikon, Band 10 (pp. 5245-5254). Stuttgart: Schäffer-Poeschel.

Möhlenbruch, D. Dölling, S., Ritschel, R. (2008). Web 2.0-Anwendungen im Kundenbindungsmanagement des M-Commerce. In H. H. Bauer, T. Dirks, & M. D. Bryant (Ed.), *Erfolgsfaktoren des Mobile Marketing* (pp. 221-240). Berlin, Heidelberg: Springer.

Möhlenbruch, D., & Schmieder, U.-M. (2001). Gestaltungsmöglichkeiten und Entwicklungspotenziale des Mobile Marketing. In Meier, A. (Ed.), *Mobile Commerce, Praxis der Wirtschaftsinformatik, 39*(220), 15-26.

Möhlenbruch, D., & Schmieder, U.-M. (2002). Mobile Marketing als Schlüsselgröße für Multichannel-Commerce. In G. Silberer, J. Wohlfahrt, & T. Wilhelm (Ed.), *Mobile Commerce Grundlagen, Geschäftsmodelle, Erfolgsfaktoren* (pp. 67-89). Wiesbaden, Gabler Verlag.

Möhlenbruch, D., & Schmieder, U.-M. (2002a). Mobile Marketing als Schlüsselgröße für Multichannel-Commerce. In Silberer, G., Wohlfahrt, J., &Wilhelm, T. (Eds.), *Mobile Commerce* (pp. 67-89). Wiesbaden, Verlag.

Möhlenbruch, D., & Schmieder, U.-M. (2002b). Chancen des Mobile Marketing im Rahmen von Multichannel-Strategien. *Thexis Fachzeitschrift für Marketing, 19*(2), 27-33.

MOHnz (Ministry of Health New Zealand). (2008). Doctors in New Zealand. Retrieved June 10, 2008, from http://www.moh.govt.nz/moh.nsf/wpg_Index/About-Statistics+about+doctors

Molberg, M. (2006). Success Criteria for Future Mobile Products. Masters thesis, available at http://www.dtu.dk/upload/centre/cict/master_thesis/martin_molberg_success_criteria_for_future_mobile_products.pdf. last accessed, Oct 31, 2008

Montgomery, A. L. (2001). Applying Quantitative Marketing Techniques to the Internet. *Interfaces, 31*(2), 90–108. doi:10.1287/inte.31.2.90.10630

Moon, J. W., & Kim, Y. G. (2001). Extending the TAM for a World-Wide-Web context. *Information & Management, 38*(4), 217–230. doi:10.1016/S0378-7206(00)00061-6

Moore, G. C., & Benbasat, I. (1991). Development of an instrument to measure the perceptions of adopting an information technology innovation. *Information Systems Research, 2*(3), 192–222. doi:10.1287/isre.2.3.192

Moores, T. T., & Chang, J. C.-J. (2006). Ethical Decision Making in Software Piracy: Initial Development and Test of a Four-Component Model. *MIS Quarterly, 30*(1), 167–180.

Moorman, C., Deshpande, R., & Zaltman, G. (1993). Factors affecting trust in market research relationships. *Journal of Marketing, 57*, 81–101. doi:10.2307/1252059

Morgan, N. A., & Rego, L. L. (2006). The Value of Different Customer Satisfaction and Loyalty Metrics in Predicting Business Performance. *Marketing Science, 25*(5), 426–439. doi:10.1287/mksc.1050.0180

Morgan, N. A., Anderson, E. W., & Mittal, V. (2005). Understanding Firms' Customer Satisfaction Information Usage. *Journal of Marketing, 69*(3), 131–151. doi:10.1509/jmkg.69.3.131.66359

Morris, M. G., & Venkatesh, V. (2000). Age differences in technology adoption decisions: Implications for a changing work force. *Personnel Psychology, 53*(2), 375–403. doi:10.1111/j.1744-6570.2000.tb00206.x

Mort, G. S., & Drennan, J. (2002). Mobile digital technology: Emerging issues for marketing. *Journal of Database Marketing, 10*(1), 9. doi:10.1057/palgrave.jdm.3240090

Mort, G. S., & Drennan, J. (2005). Marketing m-services: Establishing a usage benefit typology related to mobile user characteristics. *Journal of Database Marketing & Customer Strategy Management, 12*(4), 327–341. doi:10.1057/palgrave.dbm.3240269

Mossberg, W. (2007). Free My Phone. *Wall Street Journal*, Oct 22, R1-R3.

Motion Picture Promotion, A. I. (2001). The Beast. USA. From http://en.wikipedia.org/wiki/The_Beast_(game)

Müller, U. (2005). *Kundenbindung im E-Commerce - Personalisierung als Instrument des Customer Relationship Marketing.*

Müller-Hagedorn, L. (2005). *Handelsmarketing.* 4th Edition, Stuttgart, Schäffer-Poeschel.

Muk, A. (2007). Cultural influences on adoption of SMS advertising: A study of American and Taiwanese consumers. *Journal of Targeting. Measurement & Analysis for Marketing, 16*(1), 39–47. doi:10.1057/palgrave.jt.5750062

Muk, A., & Babin, J. (2006). U.S. consumers' adoption -- nonadoption of mobile SMS advertising. *International Journal of Mobile Marketing, 1*(1), 21–29.

Mummalaneni, V. (2005). An empirical investigation of web site Characteristics, consumer emotional states and on-line shopping behaviors. *Journal of Business Research, 58*, 526–532. doi:10.1016/S0148-2963(03)00143-7

Myers, P. (2007). ACME Mobile interview, May 2007.

Naccarato, J., & Neuendorf, K. (1998). Content analysis as a predictive methodology: Recall, readership, and evaluations of business-to-business print advertising. *Journal of Advertising Research, 38*(3), 19–33.

Nah, F. F.-H., Siau, K., & Sheng, H. (2005). The VALUE of Mobile Applications: A Utility Company Study. *Communications of the ACM, 48*(2), 85–90. doi:10.1145/1042091.1042095

Narasimhan, C. (1984). A price discrimination theory of coupons. *Marketing Science, 3*(Spring), 128–147.150 *G.A. Taylor / Journal of Retailing 77 (2001) 139–151.*

Nathwani, K., & Eason, K. (2005). Perceptions versus expectations of multimedia messaging service (MMS). *Personal and Ubiquitous Computing, 9*(4). doi:10.1007/s00779-004-0320-9

Natsuno, T. (2003). *The i-mode wireless ecosystem.* Wiley, Yokohama.

Ndubisi, N. O. (2006). Effect of gender on customer loyalty: a relationship marketing approach. *Marketing Intelligence & Planning, 24*(1), 48–61. doi:10.1108/02634500610641552

Nelson, R. (2000). Wireless advertising: What's in it for telcos? *Global Telephony, July, 8(7),* 12.

Nelson, R. R., Todd, P. A., & Wixom, B. H. (2005). Antecedents of Information and System Quality: An Empirical Examination within the Context of Data Warehousing. *Journal of Management Information Systems, 21*(4), 199–235.

Neslin, S. A., Henderson, C., & Quelch, J. (1985). Consumer promotions and the acceleration of product purchases. *Marketing Science*, (Spring): 4, 146–165.

newsletter%20dec/page2-en.htm.

Ngai, E. W. T., & Gunasekaran, A. (2007). A review for mobile commerce research and applications. *Decision Support Systems, 43*(1), 3–15. doi:10.1016/j.dss.2005.05.003

Ngai, W. W. T. (2005). Customer relationship management research (1992 – 2002). An academic literature review and classification. *Marketing Intelligence & Planning, 23*(6), 582–605. doi:10.1108/02634500510624147

Nguyen, H. Q., Hackett, M., Johnson, B., & Johnson, R. (2003). *Testing Applications on the Web.* 2nd Edition. New York: Wiley Publishing.

Nicolai, A. T., & Petersmann, T. (Eds.). (2001). *Strategien im M-Commerce*, Stuttgart: Schäffer-Poeschel.

Nielsen Mobile. (2008). Nielsen Provides Mobile Gaming Statistics, Retrieved July 25, 2008, from http://www.nielsen-mobile.com/html/press%20releases/MobileGamingStats.html

Ning, L. (2007). Personal Interview, China, 2007.

NNA. (2008). Newspaper association of America-advertising expenditure. Available at http://www.naa.org/TrendsandNumbers/Advertising-Expenditures.aspx

Noelle-Neumann, E. (1983). *Persönlichkeitsstärke.* Hamburg, Germany: Spiegel-Dokumentation.

Nokia. (2003). *Make More Profits Now With The Nokia End-To-End EDGE Solution.* Retrieved December 12, 2008, from http://www.nokia.com/BaseProject/Sites/NOKIA_MAIN_18022/CDA/Categories/Operators/MobileOperators&ServiceProviders/MobileNetworks/GSMG-PRSEDGE/_Content/_Static_Files/edge_brochure_final_lo_res.pdf

Norman, A. T., & Russell, C. A. (2006). The pass-along effect: Investigating word-of-mouth effects on online survey procedures. *Journal of Computer-Mediated Communication,* 11(4), article 10. Retrieved July 9, 2008, from http://jcmc.indiana.edu/vol11/issue4/norman.html

Norris, S. (2003). Make the mobile connection. *Revolution Magazine*. Access Jan 2007, from http://www.brandrepublic.com/login/News/182912/

Novak, T. P., Hoffman, D. L., & Yung, Y. F. (2000). Measuring the customer experience in online environments: A structural modeling approach. *Marketing Science, 19*(1), 22–42. doi:10.1287/mksc.19.1.22.15184

NXP. (2008). MIFARE Classic. Retrieved November 17, 2008, from http://www.nxp.com/#/pip/pip=[pfp=41863]|pp=[t=pfp,i=41863]

Nysveen, H., Pedersen, P. E., & Thorbjørnsen, H. (2003). Using mobile services to strengthen brand relationships: The effects of SMS and MMS channel additions on brand knowledge, satisfaction, loyalty and main channel use. *SNF-report* no. 22/03. Foundation for Research in Economics and Business Administration, Bergen, Norway.

Nysveen, H., Pedersen, P. E., & Thorbjornsen, H. (2005). Intentions to use mobile services: antecedents and cross-service comparisons. *Journal of the Academy of Marketing Science, 33*(3), 330–346. doi:10.1177/0092070305276149

Nysveen, H., Pedersen, P. E., & Thorbjornsen, H., and Pierre Berthon. (2005). Mobilizing the Brand: The Effects of Mobile Services on Brand Relationships and Main Channel Use. *Journal of Service Research, 7*(3), 257–276. doi:10.1177/1094670504271151

O'Donohoe, S. (1995). Attitudes to advertising: A review of British and American research. *International Journal of Advertising, 14*(3), 245–261.

O'Reilly, T. (2005a). *What is the Web 2.0?* Retrieved January 15, 2007, from www.oreilly.de/artikel/web20.html

O'Reilly, T. (2005b). Web 2.0: Compact Definition? Retrieved January 15, 2007, from http://radar.oreilly.com/archives/2005/10/web_20_compact_definition.html

Obermiller, C., & Spangenberg, E. R. (1998). Development of a scale to assess consumer skepticism toward advertising. *Journal of Consumer Psychology, 7*(2), 187–206. doi:10.1207/s15327663jcp0702_03

Ogilvy, D. (1963). *Confessions of an advertising man.* NY: Ballantine Books.

Okazaki, S. (2004). How do Japanese consumers perceive wireless ads? A multivariate analysis. *International Journal of Advertising, 23*(4), 429.

Okazaki, S. (2005). Mobile advertising adoption by multinationals. Senior executives' initial responses. *Internet Research, 15*(2), 160–180. doi:10.1108/10662240510590342

Okazaki, S. (2005). New perspectives on m-commerce research. *Journal of Electronic Commerce Research, 6*(3), 160–164.

Okazaki, S. (2006). What do we know about mobile Internet adopters? A cluster analysis. *Information & Management, 43*(2), 127–141. doi:10.1016/j.im.2005.05.001

Okazaki, S. (2007). Exploring gender effects in a mobile advertising context: On the evaluation of trust, attitudes, and recall. *Sex Roles, 57*(11), 897–908. doi:10.1007/s11199-007-9300-7

Okazaki, S., & Tylor, C. R. (2008). What is SMS advertising and why do multinationals adopt it? Answers from an empirical study in European markets. *Journal of Business Research, 61*, 4–12. doi:10.1016/j.jbusres.2006.05.003

Okazaki. (2005). New perspectives on m-commerce research. *Journal of Electronic Commerce Research, 6*(3), 160 - 164.

Oliver, R. L. (1997). *Satisfaction: A Behavioral Perspective on the Consumer.* Boston: McGraw-Hill.

Ondrus, J., & Pigneur, Y. (2007). Cross-industry Preferences for Mobile Payments Development in Switzerland. *Electronic Markets, 17*(2), 142–152. doi:10.1080/10196780701296386

Onoe, S., Nakamura, T., & Higuchi, K. (2007). 3G evolution scenario toward 4G: Super 3G concept. *Wireless Communications and Mobile Computing, 7*, 1013–1019. doi:10.1002/wcm.511

ONS (Office of National Statistics). (2006). Travel trends. Retrieved May 5, 2008, from http://www.statistics.gov.uk

Onwuegbuzie, A. J., & Leech, N. L. (2005). On becoming a pragmatic researcher: The importance of combining quantitative and qualitative research methodologies. *International Journal of Social Research Methodology, 8*(5), 375–387. doi:10.1080/13645570500402447

oV. (2007). Brigitte-Figur-Coach, from http://www.brigitte.de/diaet/figurcoach-sommer2007/index.html

Ovum. (2002). Ovum forecasts: Global wireless markets 2002–2006. Research Report.

Ozhan Dedeoglu, A. (2004). The Symbolic Use of Mobile Technology among Turkish Consumers. *Journal of Euromarketing, 2*(3), 143–162. doi:10.1300/J037v13n02_08

Paavilainen, J. (2002). Mobile business strategies. Understanding the technologies and opportunities. London: Addison Wesley.

Paavilainen, J. (2002). *Mobile business strategy.* USA: Addison Wesley Publishing.

Pagani, M. (2004). Determinants of adoption of third generation mobile multimedia services. *Journal of Interactive Marketing, 18*(3), 46–59. doi:10.1002/dir.20011

Palmatier, R. W., Dant, R. P., Grewal, D., & Evans, K. R. (2006). Factors Influencing the Effectiveness of Relationship Marketing: A Meta-Analysis. *Journal of Marketing, 70*(4), 136–153. doi:10.1509/jmkg.70.4.136

Park, J., & SuJin, Y. (2006). The moderating role of consumer trust and experiences: Value driven usage of mobile technology. *International Journal of Mobile Marketing, 1*(2), 24–32.

Paulissen, K., Milis, K., Brengman, M., Fjermestad, J., & Romano, N. C., Jr. (2007). *Voids in the current CRM Literature: Academic literature review and classification (2000-2005)*, in: Proceedings of the 40th Hawaii International Conference on System Sciences.

Pavlou, P. A., & Stewart, D. W. (2000). Measuring the effects and effectiveness of interactive advertising: A research agenda. *Journal of Interactive Advertising, 1(1).* Access January 2007, from http://jiad.org/vol1/no1/pavlou/

Payne, A., & Frow, P. (2005). A Strategic Framework for Customer Relationship Management. *Journal of Marketing, 69*(4), 167–176. doi:10.1509/jmkg.2005.69.4.167

Payne, A., & Frow, P. (2006). Customer Relationship Management: from Strategy to Implementation. *Journal of Marketing Management, 22*(1/2), 135–168. doi:10.1362/026725706776022272

Peabody, D. (1962). Two components in bipolar scales: Direction and extremeness. *Psychological Review, 69*, 65–73. doi:10.1037/h0039737

Peattie, K., & Peattie, S. (1995). Sales promotion—a missed opportunity for service marketers? *International Journal of Service Industry Management, 6*, 22–39. doi:10.1108/09564239510078830

Pedersen, P. E. (2002). The adoption of text messaging services among Norwegian teens: development and test of an extended adoption model. *SNF-report no.* 23/02. Foundation for Research in Economics and Business Administration, Bergen Bergen, Norway.

Pedersen, P. E. (2005). Adoption of mobile Internet services: An exploratory study of mobile commerce early adopters. *Journal of Organizational Computing and Electronic Commerce, 15*(3), 203–222. doi:10.1207/s15327744joce1503_2

Pedersen, P. E., Methlie, L. B., Gressgård, L., Nysveen, H., & Thorbjörnsen, H. (2005). An exploratory study of the relationships between mobile data services business models and customer value. SNF report no 13/2005, Bergen, March 2005.

Peltier, J. W., Schibrowsky, J. A., & Schultz, D. E. (2003). Interactive integrated marketing communication: Combining the power of IMC. New Media and Database Marketing. *International Journal of Advertising, 22*, 93–115.

Pelusi, L., Passarella, A., & Conti, M. (2006). Opportunistic Networking: Data Forwarding in Disconnected Mobile Ad Hoc Networks. *IEEE Comm., 44*(11), 134–141. doi:10.1109/MCOM.2006.248176

Pepels, W. (2005). Käuferverhalten. Berlin: Erich Schmidt.

Peppers, D., & Rogers, M. (1993). The new marketing paradigm: One to one. *American Advertising, 9*(4), 20–22.

Peppers, D., Rogers, M., & Dorf, B. (1999). Is Your Company Ready for One-to One Marketing? *Harvard Business Review, 77*(1), 151–160.

Perlado, V. R., & Barwise, P. (2004). Mobile advertising: A research agenda. In M. Stafford & R. J. Faber (Eds.), *Advertising and the New Media.* Armonk, NY: M.E. Sharpe.

Peters, L., & Fletcher, K. (2005). They love me, they love me not: consumer relationship orientation and CRM. *Journal of Customer Behaviour, 3*(1), 53–80. doi:10.1362/147539204323074600

Petrovic, O., Fallenböck, M., & Kittl, C. (2003). Der Paradigmenwechsel in der vernetzten Wirtschaft. In Petrovic, O., Ksela, M., Fallenböck, M., & Kittl, C. (Ed.), *Trust in the Network Economy* (pp. 3-28). Wien: Springer.

Petty, R. D. (2003). Wireless Advertising Messaging: Legal Analysis and Public Policy Issues. *Journal of Public Policy & Marketing, 22*(1), 71–82. doi:10.1509/jppm.22.1.71.17627

Phelps, J. E., Lewis, R., Mobilio, L., & Perry, D. (2004). Viral Marketing or Electronic Word-of-Mouth Advertising: Examining Consumer Responses and Motives to Pass Along Email. *Journal of Advertising Research, 45*(4), 333–348.

Phoosuphanusorn, S. (2007). Thailand: Mobile users growing on discounts. Retrieved on August 14, 2007 from http://www.asiamedia.ucla.edu/ article.asp?parentid=74884

Pihlström, M., & Brush, G. J. (2008). Comparing the perceived value of information and entertainment mobile services. *Psychology and Marketing, 25*(8), 732–755. doi:10.1002/mar.20236

Pils, M. (2004a). Electronic Business und sensible Informations- und Kommunikationssysteme. In Höller, J., Pils, M., & Zlabinger, R. (Ed.), *Internet und Intranet* (pp. 327-359). Berlin: Springer.

Pils, M. (2004b). Mobile Business. In Höller, J., Pils, M., & Zlabinger, R. (Ed.), *Internet und Intranet* (pp. 67-124). Berlin: Springer.

Pircy, N. (2002). *Market-led strategy.* London: B&H Publisher.

Pitt, L. F., Watson, R. T., & Kavan, C. B. (1995). Service quality: a measure of information systems effectiveness. *MIS Quarterly, 19*(2), 173–187. doi:10.2307/249687

Plant, R. (1999). *e-commerce formulation of strategy.* London: Prentice Hall.

Plant, S. (2001). On the mobile - the effects of mobile telephones on social and individual life. Retrieved February 15, 2008 from http://www.motorola.com/mot/doc/0/234_MotDoc.pdf

Pleil, T. (2005). Anmerkungen und strategische Ansätze zur Kommunikation von M-Payment. In Hampe, J. F., Lehner, F., Pousttchi, K., Rannenberg, K., & Turowski, K. (Ed.), *Mobile Business* (pp. 74-86). Bonn: Köllen.

Plowman, T. (2003). Ethnography and critical design practice. In B. Laurel (Ed.), *Design research: Methods and perspectives.* Cambridge, MA: The MIT Press.

Pollay, R. W., & Mittal, B. (1993). Here's the beef: Factors, determinants, and segments in consumer criticism of advertising. *Journal of Marketing, 57*(3), 99–114. doi:10.2307/1251857

Porter, M. (1980). *Competitive Strategy.* New York: Free Press.

Porter, M. (2001). *Strategy and the internet.* HBR Publisher.

Postinett, A. (2008). Smartphone-Hype unter der Lupe: iPhone-Nutzer lieben das Web, Handelsblatt, 20.03.2008 from http://www.handelsblatt.com/technologie/mobile-welt/iphone-nutzer-lieben-das-web;1406862

Pousttchi, K., & Wiedemann, D. (2006). A Contribution to Theory Building for Mobile Marketing: Categorizing Mobile Marketing Campaigns through Case Study Research. *5th International Conference on Mobile Business (ICMB 2006)*, Denmark, pp. 1-8.

Pousttchi, K., & Wiedemann, D. G. (2005). Relativer Vorteil bei mobilen Bezahlsystemen. In Stucky, W., & Schiefer, G. (Ed.), *Perspektiven des Mobile Business* (pp. 35-50). Wiesbaden: Vahlen.

Pousttchi, K., & Wiedemann, D. G. (2006). *A contribution to theory building for mobile marketing: Categorizing mobile marketing campaigns through case study research.* Paper presented at the International Conference on Mobile Business (ICMB2006), Copenhagen, Denmark.

Pousttchi, K., & Wiedemann, D. G. (2007). Success Factors in Mobile Viral Marketing: A Multi-Case Study approach. *Proceedings of the 6th International conference on Mobile Business,* Toronto, Canada.

Pousttchi, K., & Wiedemann, D. G. (2007). Success Factors in Mobile Viral Marketing: A Multi-Case Study Approach. *Proceedings of the 6th International Conference on Mobile Business*, Toronto, Canada.

Prete, M. I. (2007). M-politics: Credibility and effectiveness of mobile political communications. *Journal of Targeting. Measurement & Analysis for Marketing, 16*(1), 48–56. doi:10.1057/palgrave.jt.5750063

Promny, M. (2008) YOC AG, integrating the mobile phone into the marketing mix. INSEAD, ECCH case study, Cranfield, UK: ECCH. Retrieved June 15, 2008, from http://www.ecch.com

Punch, K. F. (2005). *Introduction to social research.* Thousand Oaks, California: Sage Publications.

Purdy, J.G., & Gandhi (n.d.). When mobile coupons replace paper coupons, everyone wins, http://info.cellfire.com/cellfire/themes/cellfire/downloads/When_Mobile_Coupons_Replace_Paper_Coupons.pdf (accessed on October 30, 2008)

Putrevu, S. (2001). Exploring the origins and information processing differences between men and women: Implications for advertisers. *Academy of Marketing Science Review, 2001,* 1–15.

Quelch, J. A. (1989). Sales Promotion Management, Prentice-Hall, Inc., 6–7.

Rakow, L. F., & Navarro, V. (1993). Remote Mothering and the Parallel Shift: Women Meet the Cellular Telephone. *Critical Studies in Mass Communication, 10,* 144–157. doi:10.1080/15295039309366856

Ranganathan, A., & Campbell, R. H. (2002). Advertising in a pervasive computing environment. In Proceedings of the 2nd International Workshop on Mobile Commerce. ACM Press.

Rannenberg, K., Albers, A., Figge, S., Radmacher, M., & Rossnagel, H. (2005). Mobile Commerce. In: Hampe, J. F., Lehner, F., Pousttchi, K., Rannenberg, K., & Turowski, K. (Ed.), *Mobile Business* (pp. 9-16). Bonn: Köllen.

Rannenberg, K., Schneider, I., & Figge, S. (2005). Mobile Systeme und Anwendungen - Hammer sucht Nagel. *Wirtschaftsinformatik, 47*(1), 1–2.

Rao, B., & Minakakis, L. (2003). Evolution of Mobile Location-Based Services. *Communications of the ACM, 46*(12), 61–65. doi:10.1145/953460.953490

Rao, B., & Minakakis, L. (2003). Evolution of mobile location-based services. *Communications of the ACM, 46*(12), 61–65. doi:10.1145/953460.953490

Ratsimor, O., Finin, T., Joshi, A., & Yesha, Y. (2003). eNcentive: A Framework for Intelligent Marketing in Mobile Peer-To-Peer Environments. *ICEC '03: Proceedings* of the 5th International Conference on Electronic Commerce (pp. 87-94). Baltimore, MD: UMBC eBiquity Research Group

Ratsimor, O., Finin, T., Joshi, A., & Yesha, Y. (2003). eNcentive: A Framework for Intelligent Marketing in Mobile Peer-To-Peer Environments. In *ICEC '03: Proceedings of the 5th international conference on Electronic commerce* (pp. 87–94). ACM Press.

Rayport, J., & Jaworski, B. (2001). *Introduction to E-Commerce.* New York: McGraw- Hill.

Regier, S. (2007). *Markterfolg radikaler Innovationen. Determinanten des Akzeptanzverhaltens.* Wiesbaden: Gabler.

Reichheld, F. (2006). The Microeconomics of Customer Relationships. *MIT Sloan Management Review, 47*(2), 73–78.

Reichold, A., & Schierholz, R. (2003). *Fallstudie Helsana - M-Commerce bei der Helsana: Mobile Prämienerstellung* (No. HSG / IWI / BECS / 30). St. Gallen: Institut für Wirtschaftsinformatik Universität St. Gallen.

Report, W. (2000). First-to-wireless: Capabilities and benefits of wireless marketing and advertising based on the first national mobile marketing trial. Access January 2007, from www.imapproject.org/imapproject/downloadroot/public3/ftw_report.pdf

Rettie, R., Grandcolas, U., & Deakins, B. (2005). Text message advertising: Response rates and branding effects. *Journal of Targeting. Measurement and Analysis for Marketing, 13*(4), 304–312. doi:10.1057/palgrave.jt.5740158

Rhee, S. (2005). Personal Interview, Clemensen Capital Company, LLC, May 2005

Richard, M. O. (2005). Modeling the impact of internet atmospherics on surfer Behavior. *Journal of Business Research, 58*(12), 1632–1642. doi:10.1016/j.jbusres.2004.07.009

Richards, J., & Curran, C. (2002). Oracles on 'advertising': Searching for a definition. *Journal of Advertising, XXXI*(2), 63–77.

Richtel, M. (2006). *Marketers interested in small screen.* Retrieved December 22, 2008, from http://www.smh.com.au/news/breaking/small-screen-bonanza/2006/01/17/1137466994899.html

Rickman, C. (n.d.). Harnessing the Killer App: E-mail. February 28, 2001: http://www.digitrends.net/marketing/13640_14649.html

Riebel, P. (1994). *Einzelkosten- und Deckungsbeitragsrechnung: Grundfragen einer markt- und entscheidungsorientierten Unternehmensrechnung.* Wiesbaden, Gabler Verlag.

Riecken, D. (2000). Personalized views of personalization. *Communications of the ACM, August, 43(8),* 27-28.

Ringle, C. (2004). Gütemaße für den Patrial Last Squares-Ansatz zur Bestimmung von Kausalmodellen. *Hamburg.*

Ringle, C., Boysen, N., Wende, S., & Will, A. (2006). Messung von Kausalmodellen mit dem Partial-Least-Squares-Verfahren. [WiSU]. *Das Wirtschaftsstudium, 35*(1), 81–88.

Ringle, C., Wende, S., & Will, A. (2007). SmartPLS Version: 2.0.M3 Retrieved 01.11.2007, from http://www.smartpls.de

Robey, D. (1979). User attitudes and Management Information System use. *Academy of Management Journal,* September, vol.22, no.3, pp.527-538.

Robins, F. (2003). The Marketing of 3G. *Marketing Intelligence & Planning, 21*(6), 370–378. doi:10.1108/02634500310499239

Robinson, R. (2007). Dada Asia interview, May 2007.

Rockart, J. F. (1979). Chief executives define their own data needs. *Harvard Business Review, 57*(2), 15–28.

Rodgers, S., & Thorson, E. (2000). The interactive advertising model: How users perceive and process online ads. *Journal of Interactive Advertising, 1(1).* Access January 2007, from www.jiad.org/vol1/no1/rodgers/

Rodgers, W. C., & Schneider, K. C. (1993). A Research Note on Market Mavens and the Dissemination of Positive versus Negative Marketplace Information. *Journal of Marketing Management, 3*(2), 67–72.

Rodgers, Z. (2005). Nike iD Billboard Invites Mobile Users. The ClickZ Network, May 3, 2005 from http://www.clickz.com/3502186

Rogers, E. M. (1983). *Diffusion of Innovations.* The Free Press, London.

Rogers, E. M. (1995). *Diffusion of innovations* (4 ed.). New York, NY: The Free Press.

Rogers, E. M. (1995). *Diffusion of innovations.* New York: Free Press.

Rogers, E. M. (2003). *Diffusion of Innovation*, New York: The Free Press.

Rogers, Y. (1997). Reconfiguring the social scientist: Shifting from prescription to proactive research. In G. C. Bowker, S. L. Star, W. Turner & L. Gasser (Eds.), *Bridging the great divide: Social science, technical systems and cooperative work* (pp. 55-77). Mahwah, NJ: L. Erlbaum.

Rogers, Y., & Muller, H. (2006). A framework for designing sensor-based interactions to promote exploration and reflection in play. *International Journal of Human-Computer Studies, 64,* 1–14. doi:10.1016/j.ijhcs.2005.05.004

Rohm, A. J., & Sultan, F. (2005). Brand in the hand: An exploratory study of mobile marketing communications. *Proceeding of EURAM 2005*, April, Munich.

Rohm, A. J., & Sultan, F. (2006). An exploratory cross-market study of mobile marketing acceptance. *International Journal of Mobile Marketing, 1*(1), 4–12.

Rosen, D. E., & Purinton, E. (2004). Website design: viewing the web as a cognitive landscape. *Journal of Business Research, 57*(7), 787–794. doi:10.1016/S0148-2963(02)00353-3

Rosen, E. (2000). *The Anatomy of Buzz.* New York: Doubleday.

Rossiter, J. R., & Bellman, S. (2005). *Marketing communications: Theory and applications*. Frenchs Forest, NSW: Prentice Hall.

Rossman, G. B., & Rallis, S. F. (1998). *Learning in the Field: An Introduction to Qualitative Research*. Thousand Oaks, CA: Sage.

Rotter, J. (1967). A new scale for the measurement of interpersonal trust. *Journal of Personality*, *35*(4), 651–665. doi:10.1111/j.1467-6494.1967.tb01454.x

Rudolph, T., & Emerich, O. (2008). Kundeninteraktion über mobile Services im Handel. In H. H. Bauer, T. Dirks, & M. D. Bryant (Ed.), *Erfolgsfaktoren des Mobile Marketing* (pp. 261-278). Berlin, Heidelberg: Springer.

Rudström, Å., Svensson, M., Cöster, R., & Höök, K. (2004). MobiTip: Using Bluetooth as a Mediator of Social Context. In *UbiComp 2004: Ubiquitous Computing: 6th International Conference, Adjunct Proceedings (demo)*.

Rudy, K. (1987). Actual Direct Marketing, *Nihonkeizaisinbun*, Inc., (in Japanese). Neslin, S. A., & Shoemaker, R. W. (1989). An alternative explanation for lower repeat rates after promotion purchases. *JMR, Journal of Marketing Research*, *26*(May), 205–213.

Rukzio, E., Broll, G., Leichtenstern, K., & Schmidt, A. (2007). Mobile Interaction with the Real World: An Evaluation and Comparison of Physical Mobile Interaction Techniques. In Schiele, B., Dey, A. K., Gellersen, H., de Ruyter, B., Tscheligi, M., Wichert, R., Aarts, E., & Buchmann, A. (Ed.), *Ambient Intelligence* (pp. 1-18). Berlin: Springer.

Rummert, A., Schiefer, G., & Bulander, R. (2005). MVNO - Aussichten des deutschen Mobilfunkmarktes. In Stucky, W., & Schiefer, G. (Ed.), *Perspektiven des Mobile Business* (pp. 97-112). Wiesbaden: Vahlen.

Russel, S., & Norvig, P. (2003). *Artificial intelligence, a modern approach*, Upper Saddle River, NJ: Pearson Education.

Russell, J. A., & Pratt, G. (1980). A description of the affective quality attributed to environments. *Journal of Personality and Social Psychology*, *38*, 311–322. doi:10.1037/0022-3514.38.2.311

Rust, R. T., Kannan, P. K., & Peng, N. (2002). The customer economics of Internet privacy. *Journal of the Academy of Marketing Science*, *30*(4), 455–464. doi:10.1177/009207002236917

Ryan, M.J., and Bonfield, E.H. (1980). Fishbein's intention model: A test of external and pragmatic validity. *Journal of Marketing*, 44(spring), 82- 95.

Sabat, H. (2002). The evolving mobile wireless value chain and market structure. *Telecommunications Policy*, *26*, 505–535. doi:10.1016/S0308-5961(02)00029-0

Sadeh, N. (2002). *M-Commerce: Technologies, Services, and Business Models*. New York: Wiley Publishing.

Salo, J., & Tähtinen, J. (2005). Retailer use of permission-based mobile advertising. In Clarke III I & Flaherty TB (Eds.), Advances in electronic marketing (pp.139-156). Hershey, PA: Idea Publishing Group.

Sander, M. (2004). Marketing-Management. Stuttgart: Lucius & Lucius UTB.

Sanongphan, S. (2004). Thailand - Telecom and Broadcasting Market Brief. Retrieved July 1, 2007, from http://strategis.ic.gc.ca/epic/site/imr-ri.nsf/en/gr124774e.html

Sarker, S., & Wells, J. D. (2003). Understanding Mobile. *Communications of the ACM*, *46*(12), 35–40. doi:10.1145/953460.953484

Saunders, C. (2003). Studies: Mobile advertisement market to grow, amid risks. *Internet Advertising Report,* November 1. Access January 2007, from http://www.clickz.com/showPage.html?page=915121

Savio, D. C. (2004). Usability aspects in 3G/4G mobile telephony. At http://userpage.fu-berlin.de/~jmueller/its/conf/Madrid02/abstracts/Savio.pdf, last accessed on Oct 31, 2008.

Schackmann, J., & Schü, J. (2001). Personalisierte Portale. *Wirtschaftsinformatik*, *43*(6), 623–625.

Schäfer, J., & Thoma, D. (2008). Trends und Strategien im Mobile Marketing. In H. Bauer, T. Dirks & M. D. Bryant (Ed.). *Erfolgsfaktoren des Mobile Marketing*, Heidelberg: Springer-Verlag.

Scharl, A., Dickinger, A., & Murphy, J. (2005). Diffusion and Success Factors of Mobile Marketing. *Electronic Commerce Research and Applications*, *4*(2), 159–173. doi:10.1016/j.elerap.2004.10.006

Scheer, A.-W., Feld, T., & Göbl, M. (2001). Das Mobile Unternehmen. *Information Management & Consulting*, *16*(2), 7–15.

Scheufele, D. H., & Shah, V. S. (2000). Personality Strength and Social Capital: The Role of Dispositional and Informational Variables in the Production of Civic Participation. *Communication Research, 27*(2), 107–131. doi:10.1177/009365000027002001

Schiffman, L., Kanuk, L., & Hansen, H. (2008). *Consumer behaviour: A European outlook.* London: Pearson, Prentice Hall.

Schifter, D. E. (1984). *Attempts, control and weight reduction: An application of a theory of planned behavior.* Master Thesis, University of Massachusetts at Amherst, Amherst, MA.

Schoenbachler, D. D., & Gordon, G. L. (2002). Trust and customer willingness to provide information in database-driven relationship marketing. *Journal of Interactive Marketing, 16*(3), 2–16. doi:10.1002/dir.10033

Schütte, C., & Diederich, L. (2006). Neues aus der Blogoshäre – digitale Mundpropaganda via Weblog. *Marketing Journal, 39*(Sonderheft Web 2.0), 26-28.

Schultz, R. L., & Slevin, D. P. (1975). *Implementation and organizational validity: an empirical investigation.* In Implementing Operations Research / Management Science, New York, NY: American Elsevier, pp.153-182.

Schwartz, N. (1997). Moods and attitude judgments: A comment on Fishbein and Middlestadt. *Journal of Consumer Psychology, 6*(1), 93–98. doi:10.1207/s15327663jcp0601_06

Schwarz, T. (2003). Chancen und Grenzen der elektronischen Kommunikation. In Link, J. (Ed.), *Mobile Commerce* (pp. 145-161). Berlin: Springer.

Scornavacca, E., & Barnes, S. J. (2004). Raising the bar: barcode-enabled m-commerce solutions. *Proceedings of the Austin Mobility Roundtable*, Austin, Texas.

Scornavacca, E., & McKenzie, J. (2007). Unveiling managers' perceptions of the critical success factors for SMS-based campaigns. *International Journal of Mobile Communications, 5*(4), 445–456. doi:10.1504/IJMC.2007.012790

Scornavacca, E., Barnes, S. J., & Huff, S. L. (2006). Mobile business research published in 2000-2004: Emergence, current status, and future opportunities. *Communications of the Association for Information Systems, 17*, 635–646.

Seal, K. C., Gomez, A., Kong, D., Hernandez, X., Macias, A., Soohoo, K., & Villa, M. (2003). A Comparative Analysis of M-Commerce issues Between Australia, New Zealand, and the United States. In the *Proceedings of The First International Conference of AIMS*, India, Dec 28-30, 2003.

Seiler, M. (2005). Mobile Marketing. In Giordano, M., & Hummel, J. (Ed.), *Mobile Business* (pp. 373-392). Wiesbaden: Vahlen.

Selk, B., & Turowski, K. (2003). *Mobile Commerce: Stand und Perspektiven.*Unpublished manuscript, Köln.

Settles, C. (2002). *Wireless Inc.* New York: American Management Association Publishing.

Sgriccia, M., Nguyen, H., Edra, R., Alworth, A., Brandeis, O., & Escandon, R. (2007). Drivers of Mobile Business Models: Lessons from Four Asian Countries. *International Journal of Mobile Marketing, 2*, 58–67.

Sgriccia, M., Nguyen, H., Edra, R., Alworth, A., Brandeis, O., & Escandon, R. (2007). Drivers of mobile business models: Lessons from four Asian countries. *International Journal of Mobile Marketing, 2*(2), 58–67.

Shabbir, H., Palihawadana, D., & Thwaites, D. (2007). Determining the Antecedents and Consequences of Donor-Perceived Relationship Quality— A Dimensional Qualitative Research Approach. *Psychology and Marketing, 24*(3), 271–293. doi:10.1002/mar.20161

Sharma, A. (2007). What's New in Wireless. Retrieved March 26, 2007, from The Wall Street Journal Online website located at http://online.wsj.com/article/SB117434453204441967.html?mod=todays_us_The_journal_report

Sharma, N., & Ojha, S. (2004). Measuring service performance in mobile communications. *Service Industries Journal, 24*(6), 109–128. doi:10.1080/0264206042000299202

Sharma, R. S., & Wildman, S. S. (2009). The economics of delivering digital content over mobile networks. *Journal of Media Business Studies, 6*(2).

Shavitt, S., Lowrey, P., & Haefner, J. (1998). Public attitudes towards advertising: More favourable than you might think. *Journal of Advertising Research, 38*(4), 7–22.

Shavitt, S., Lowrey, P., & Haefner, J. (1998). Public attitudes toward advertising: More favorable than you might think. *Journal of Advertising Research, 38*(4), 7–22.

Sheng, H., Nah, F. F.-H., & Siau, K. (2005). Strategic implications of mobile technology: A case study using value-focused thinking. *The Journal of Strategic Information Systems, 14*(3), 269–290. doi:10.1016/j.jsis.2005.07.004

Sheppard, B. H., Hartwick, J., & Warshaw, P. R. (1988). The theory of reasoned action: A meta analysis of past research with recommendations for modifications and future research. *The Journal of Consumer Research, 15*(3), 325–343. doi:10.1086/209170

Shiels, M. (2008). The mobile future is calling. BBC News. Retrieved July 14, 2008, from http://news.bbc.co.uk/ go/pr/ fr/-/1/hi/ technology/ 7366403.stm

Shim, J. K., Qureshi, A. A., Siegel, J. G., & Siegel, R. M. (2000). *The International Handbook of Electronic Commerce*. Chicago: Verlag.

Shimokawa, T. (2005). Personal Interview, JCB, Tokyo, Japan. May, 2005.

Shimp, T. A. (2003). *Advertising, promotion and supplemental aspect to integrated marketing communications* (6th ed.). Mason, Ohio: Thomson South-Western.

Shneyderman, A., & Casati, A. (2008). Fixed Mobile Convergence. New York: Mcgraw-Hill Professional.

Siau, K., & Shen, Z. (2003). Building Customer Trust In Mobile Commerce. *Communications of the ACM*, *46*(4), 91–94. doi:10.1145/641205.641211

Siau, K., & Shen, Z. (2003). Mobile communications and mobile services. *International Journal of Mobile Communications*, *1*(1-2), 3–14. doi:10.1504/IJMC.2003.002457

Siekpe, J. S. (2005). An examination of the multidimensionality of the flow construct in a computer-mediated environment. *Journal of Electronic Commerce Research*, *6*(1), 31–43.

Sikora, A. (2001). *802.11: Standard für drahtlose Netze*. In tecChannel, from http://www.techChannel.de/hardware/680/index.html

Silberer, G. (1980). Reaktanz bei Konsumenten. In Graf Hoyos, G., Kroeber-Riel, W., v. Rosenstiel, L., Strümpel, B. (Ed.). *Grundbegriffe der Wirtschaftspsychologie* (386-391), München: Kösel.

Silberer, G. (2004.). Grundlagen und Potentiale der mobilfunkbasierten Kundenbeziehungspflege (mobile eCRM). In Hippner, H., Wilde, K. D. (Ed.). *IT-Systeme im CRM. Aufbau und Potenziale* (453-470), Wiesbaden: Gabler.

Silberer, G., & Wohlfahrt, J. (2001). Akzeptanz und Wirkungen des Mobile Banking. In Nicolai, A. T., & Petersmann, T. (Ed.), *Strategien im M-Commerce* (pp. 161-176). Stuttgart: Schäffer-Poeschel.

Silberer, G., Magerhans, A., & Wohlfahrt, J. (2002). Kundenzufriedenheit und Kundenbindung im Mobile Commerce. In Silberer, G., Wohlfahrt, J., & Wilhelm T. (Ed.), *Mobile Commerce* (pp. 309-324). Wiesbaden: Vahlen.

Silverstone, R., & Hirsch, E. (1992). *Consuming technologies*. Routledge, London.

Singh, J., & Sirdeshmukh, D. (2000). Agency and Trust Mechanisms in Customer Satisfaction and Loyalty Judgments. *Journal of the Academy of Marketing Science*, *28*(1), 150–167. doi:10.1177/0092070300281014

Singh, R. M. (1966). *Characteristics of farm innovations associated with the rate of adoption*. Agricultural Extension Education Report 14. EX (E). Guelph, Ontario.

Sinisalo, J., Salo, J., Karjaluoto, H., & Leppäniemi, M. (2006). *Managing customer relationships through mobile medium-underlying issues and opportunities*. Paper presented at the 39th Hawaii International Conference on System Sciences, Hawaii.

Sinn, W. (2002). Drahtlose Übertragungssysteme und Zugangsmedien. In R. Reichwald (Ed.), *Mobile Kommunikation* (pp. 113-125). Wiesbaden: Gabler.

Sirkaa, L., Jarvenpaa, K., Yoko, T., & Virpi, K. (2003). Mobile commerce at crossroads. *Communications of the ACM*, *46*(12), 41–44. doi:10.1145/953460.953485

Sirkeci, I. (2006). *The environment of insecurity in Turkey and the emigration of Turkish Kurds to Germany*. Lewiston, NY: E. Mellen Press.

Sirkeci, I. (2008). Ethnic marketing potential in England and Wales: New evidence from the 2001 UK census. *Asian Journal of Marketing*, *2*(1), 1–9.

Skinner, C. (2007). The Future of Banking in a Globalised World. New York: Wiley.

Skog, B. (2002). *Mobiles and the Norwegian teens: Identity, gender, and class. In perpetual contacts*. Eds. James E. Katz and Mark Aakhus. Cambridge University Press, New York.

Slama, M. E., & Williams, T. G. (1990). Generalization of the market maven's information provision tendency across product categories. *Advances in Consumer Research. Association for Consumer Research (U. S.)*, *17*(1), 48–52.

Slyke, C. V., Comunale, C. L., & Belanger, F. (2002). Gender Differences in Perceptions of Web-Based Shopping. *Communications of the ACM*, *45*(8), 82–86. doi:10.1145/545151.545155

Smetherham, C., Fenton, S., & Modood, T. (2008). *How global is higher education in the UK?* Paper presented at Leverhulme Programme Conference on Mobility in International Labour Markets, University College London, UK.

Smith-Shomade, B. E. (2002). Narrowcasting in the new world information order, a space for the audience? *Television & New Media*, *5*(1), 69–81. doi:10.1177/1527476403259746

Smolnik, S., & Riempp, G. (2006). Nutzenpotentiale, Erfolgsfaktoren und Leistungsindikatoren von Social Software für das organisationale Wissensmanagement. In Hildebrand, K., & Hofmann, J. (Eds.), *Social Software, Praxis der Wirtschaftsinformatik, 43*(252), 17-26.

Söllner, A. (1993). *Commitment in Geschäftsbeziehungen*. Wiesbaden: nbf.

Sophonthummapharn, K., & Tesar, G. (2007). Technology readiness and propensity of cell phone users to subscribe to commercial messaging services. *Marketing Management Journal, 17*(2), 81–95.

South Korea and Japan agree to launch 4G Forum and Pilot Network. *Asia Pacific Telecom*, January, 2002, http://findarticles.com/p/articles/mi_hb6530/is_200201/ai_n25884820, last accessed Oct 31, 2008

Srinivasan, S. S., Anderson, R., & Ponnavolu, K. (2002). Customer loyalty in e-commerce: An exploration of its antecedents and consequences. *Journal of Retailing, 78*(1), 41–50. doi:10.1016/S0022-4359(01)00065-3

Stabell, C. B., & Fjeldstad, Ø. D. (1998). Configuring value for competitive advantage: on chains, shops, and networks. *Strategic Management Journal, 19*, 413–437. doi:10.1002/(SICI)1097-0266(199805)19:5<413::AID-SMJ946>3.0.CO;2-C

Stafford, T. F., & Gillenson, M. L. (2003). Mobile commerce: what is and what it could be? *Communications of the ACM, 56*(12), 33–34. doi:10.1145/953460.953483

Standing, C., Benson, S., & Karjaluoto, H. (2005). Consumer perspectives on mobile Advertising and marketing. *Proceedings of the Australian & New Zealand Marketing Academy Conference (ANZMAC)*. Perth, Australia, December 5-7, 135-141.

Stanoevska-Slabeva, K. (2004). Mobile Business. In Stanoevska-Slabeva, K. (Ed.), *The Digital Economy* (pp. 459-475). Berlin: Springer.

Steenkamp, J.-B. E. M., & Baumgartner, H. (1992). The Role of Optimum Stimulation Level in Exploratory Consumer Behavior. *The Journal of Consumer Research, 19*(3), 434–448. doi:10.1086/209313

Steinbock, D. (2006). The Missing Link: Why M-marketing Is Different. *International Journal of M-marketing, 1*(1), 83–95.

Stewart, D. W., & Pavlou, P. A. (2002). From consumer response to active consumer: Measuring the effectiveness of interactive media. *Journal of the Academy of Marketing Science, 30*(4), 376–396. doi:10.1177/009207002236912

StoLPan. (2008). Project Info. Retrieved October 20, 2008, from http://www.stolpan.com/index.php?do=topic&topic=3

Stolpmann, M. (2000). *Kundenbindung im E-Business*. Bonn: Galileo Press.

Stone, M. (1974). Cross-Validatory Choice and Assessment of Statistical Predictions. *Journal of the Royal Statistical Society. Series B. Methodological, 36*(2), 111–147.

Stone, R. W., Good, D. J., & Baker-Eveleth, L. (2007). The impact of information technology on individual and firm marketing performance. *Behaviour & Information Technology, 26*(6), 465–482. doi:10.1080/01449290600571610

Strategy Analytics. (2007). *Global Mobile Advertising Update: Outlook Bright as Inventory Expands.*

Straub, T., & Heinemann, A. (2004). An Anonymous Bonus Point System For Mobile Commerce Based On Word-Of-Mouth Recommendation. In L. M. Liebrock (Ed.), *Applied Computing 2004. Proceedings of the 2004 ACM Symposium on Applied Computing* (pp. 766-773). New York: ACM Press.

Straub, T., & Heinemann, A. (2004). An Anonymous Bonus Point System For Mobile Commerce Based On Word-Of-Mouth Recommendation. In *Proceedings of the 2004 ACM Symposium on Applied Computing* (pp. 766–773). New York, NY, USA: ACM Press.

Stryker, S., & Burke, P. J. (2000). The past present and future of identity theory. *Social Psychology Quarterly, 63*, 284–297. doi:10.2307/2695840

Subramani, M. R., & Rajagopalan, B. (2003). Knowledge-Sharing and Influence in Online Social Networks via Viral Marketing. *Communications of the ACM, 46*(12), 300–307. doi:10.1145/953460.953514

Subramani, M. R., & Rajagopalan, B. (2003). Knowledge-sharing and influence in online social networks via viral marketing. *Communications of the ACM, 46*(12), 300–307. doi:10.1145/953460.953514

Sullivan, M. G., & Drennan, J. (2002). Mobile Digital Technology: Emerging Issues for Marketing. *Journal of Database Marketing, 10*(1), 9–23. doi:10.1057/palgrave.jdm.3240090

Sultan, F., & Rohm, A. (2005). The Coming Era of "Brand in the Hand" Marketing. *MIT Sloan Management Review, 47*(1), 83–90.

Sun, S. (2007). KongZhong interview, May 2007.

Suoranta, M., & Mattila, M. (2004). Mobile banking and consumer behaviour: New insights into the diffusion pattern. *Journal of Financial Services Marketing, 8*(4), 354–366. doi:10.1057/palgrave.fsm.4770132

Sutter, M., & Kocher, M. G. (2004). Age and the development of trust and reciprocity. *Discussion Papers on Strategic Interaction from Max Planck Institute of Economics*, Strategic Interaction Group.

Sweeney, J. C., & Soutar, G. N. (2001). Consumer Perceived Value: The Development of a Multiple Item Scale. *Journal of Retailing, 77*(2), 203–220. doi:10.1016/S0022-4359(01)00041-0

Szajna, B. (1996). Empirical Evaluation of the Revised Technology Acceptance Model. *Management Science, 42*(1), 85–92. doi:10.1287/mnsc.42.1.85

Tabachnick, B. G., & Fidell, L. S. (2001). *Using Multivariate Statistics* (4th ed.). Needham Heights, MA: Allyn & Bacon.

Tähtinen, J. (2006). Mobile advertising or mobile marketing. A need for a new concept? Frontiers of e-business Research, 1, 152-164.

Takacs, S. J., & Freiden, J. B. (1998). Changes on the electronic frontier: Growth and opportunity of the World-Wide Web. *Journal of Marketing Theory and Practice, 6*(3), 24–37.

Tamminen, S., Oulasvirta, A., Toiskallio, K., & Kankainen, A. (2004). Understanding mobile contexts. *Personal and Ubiquitous Computing, 8*, 35–143. doi:10.1007/s00779-004-0263-1

Tan, B. (2007). Microsoft interview, May 2007

Tan, Z., & Ouyang, W. (2004). Diffusion and Impacts of the Internet and E-Commerce in China. *Electronic Markets, 14*(1), 25–35. doi:10.1080/1019678042000175270

Tapesewich, P. (2003). Designing Mobile Commerce Applications. *Communications of the ACM, 46*(12), 57–50. doi:10.1145/953460.953489

Tasch, A., & Brakel, O. (2004). Location Based Community Services. New Services for a new Type of Web Communities. *Proceedings of the IADIS Conference on Web Based Communities*, Lisbon, Portugal.

Tatli, E. I. (2007). Extending P3P/Appel for Friend Finder. *Proceedings of the International Workshop on Privacy-Aware Location-based Mobile Services*, Mannheim, Germany.

Taylor, P. (2006). Mobile consumer applications outlook 2007. *Wireless Internet Applications Service, Strategy Analytics,* 24-26

Taylor, S., & Todd, P. (1995). Assessing IT usage: The role of prior experience. *MIS Quarterly, 19*(4), 561–570. doi:10.2307/249633

Taylor, S., & Todd, P. A. (1995). Understanding information technology usage: A test of competing models. *Information Systems Research, 6*(2), 144–176. doi:10.1287/isre.6.2.144

Teo, H. H., Oh, L. B., Liu, C., & Wei, K. K. (2003). An empirical study of the effects of interactivity on web user attitude. *International Journal of Human-Computer Studies, 58*(3), 281–305. doi:10.1016/S1071-5819(03)00008-9

Teo, T. S. H., & Pok, S. H. (2003). Adoption of WAP-enabled mobile phones among Internet users. *Omega, 31*, 483–498. doi:10.1016/j.omega.2003.08.005

Thaler, R. (1985). Mental Accounting and Consumer Choice. *Marketing Science, 4*(3), 199–214. doi:10.1287/mksc.4.3.199

The Economist. (2006). Who killed the newspaper? Retrieved September 29, 2008, from http://www.economist.com/opinion/displaystory.cfm?story_id=7830218

The European Union. (2002). 2002/58/EC. *Official Journal at OJ, L201*(37), 31.

Thorelli, H. (1990). Networks: between markets and hierarchies. In D. Ford (Ed.), *Understanding Business Markets.* London: IMP Group.

Timmers, P. (1998). Business models for e-commerce. *Electronic Markets, 8*, 3–7. doi:10.1080/10196789800000016

TNS marketing research company. (2001). Retrieved September 21, 2007, from http://219.232.228.151/infopulse/

Töpfer, A., Silbermann, S., & William, R. (2008). Die Rolle des Web 2.0 im CRM. In Töpfer, A. (Ed.), *Handbuch Kundenmanagement* (pp. 651-675). Berlin: Springer.

Tornatsky, L., & Klein, K. (1982). Innovation characteristics and innovation adopting-implementation: A meta-analysis of findings. *IEEE Transactions on Engineering Management, EM-29*, 28–45.

Towns, S. (2005). 3 New Entrants Granted Access to Japan's Mobile Phone Market (DCM, NTT, EWJ). http://japan.seekingalpha.com/article/4168, last accessed Oct 31, 2008

Trappey, R. J. III, & Woodside, A. G. (2006). Consumer responses to interactive advertising campaign coupling short-message-service direct marketing and TV commercials. *Journal of Advertising Research, 45*(4), 382–401. doi:10.1017/S0021849905050476

Treu, G., Martens, J., Schicker, M., Breisinger, M., & Küpper, A. (2004). Vs Golf — Developing Location-based Multi-Player Games. *Proceedings of the 31st Annual International Computer Software and Applications Conference*, Beijing, China.

Trumpfheller, J. (2005). *Kundenbindung in der Versicherungwirtschaft*.

Tsai, H. S., & Gururajan, R. (2007). Motivations and Challenges for M-Business Transformation: A Multiple-Case Study. *Journal of Theoretical and Applied Electronic Commerce Research, 2*(2), 19–33.

Tsalgatidou, A., & Pitoura, E. (2001). Business Models and Transactions in Mobile Electronic Commerce: Requirements and Properties. *Computer Networks, 37*, 221–236. doi:10.1016/S1389-1286(01)00216-X

Tsalgatidou, A., & Veijalainen, J. (2000). Mobile Electronic Commerce: Emerging Issues. in the *Proceedings of the International Conference Electronic Commerce and Web Technologies*. Springer Berlin Heidelberg, pp. 477-486.

Tsang, M. M., Ho, S., & Liang, T. (2004). Consumer attitudes toward mobile advertising: An empirical study. *International Journal of Electronic Commerce, 8*(3), 65–78.

Turkcell Communication Services. (2008). First Quarter 2008 Results, Retrieved July 25, 2008, from http://www.turkcell.com.tr/c/docs/presentations/presentations2008Q1.pdf

Turkcell Communication Services. (n.d). *A World Company*, Retrieved July 25, 2008, from http://www.turkcell.com.tr/en/AboutTurkcell/corporateInfo/companyHistory

Turkish Telecommunications Authority. (2007). Annual Report 2007. Retrieved July 25, 2008, from http://www.tk.gov.tr/eng/pdf/2008/fr2007eng.pdf

Turley, L. W., & Milliman, R. E. (2000). Atmospheric effects on shopping behavior: a review of the experimental evidence. *Journal of Business Research, 49*(2), 193–211. doi:10.1016/S0148-2963(99)00010-7

Turowski, K., & Pousttchi, K. (2004). *Mobile Commerce – Grundlagen und Techniken*: Springer.

Turowski, K., & Pousttchi, K. (2004). *Mobile Commerce*. Berlin: Springer.

UNDP (Population Division of the Department of Economic and Social Affairs of the United Nations Secretariat). (2008). Trends in total migrant stock: The 2005 Revision. Retrieved May 5, 2008 from http://esa.un.org/migration

Unhelkar, B. (2006). *Handbook of research in mobile business*. Hershey, PA: Idea Group Reference.

University of Bonn. (2002-2005). *BonnMotion – A Mobility Scenario Generation and Analysis Tool*.http://web.informatik.uni-bonn.de/IV/Mitarbeiter/dewaal/BonnMotion/index.html

Unni, R., & Harmon, R. (2007). Perceived Effectiveness of Push vs. Pull Mobile Location-Based Advertising. *Journal of Interactive Advertising*, 7(2). Retrieved November 25, 2008, from www.jiad.org/vol7/no2/unni/Perceived%20effectiveness%20of%20push%20vs_%20pull.pdf

Ure, J., & Vivorakij, A. (1997). Privatization of Telecoms in Asia. Retrieved July 1, 2007, from Telecom Research Project Hong Kong website located at http://www.trp.hku.hk/publications.php

USA Telecoms and Technology Forecast. (2005). Economist Intelligence Unit.

Usunier, J.-C., & Lee, J. A. (2005). *Marketing across cultures, 4th edition*. London: Pearson, Prentice Hall.

Uwadaira, Y., & Kondo, F. N. (2005). 'Direct Mail Promotion Effects on the Responses of Beauty Salon Customers: Mobile DM Coupons vs. Postcards. In the Proceedings in the 5th Annual Hawaii International Conference on Business (pp.3556-3573).

Valavanis, E., Ververidis, C., Vazirgianis, M., & Polyzos, G. C. (2003). MobiShare: Sharing context-dependent data and services among mobile devices. Retrieved January 3, 2009, from http://www.db-net.aueb.gr

Van Ackeren, R. (2002). Mit dem Handy auf Kundenfang – Potentiale und Erfolgsfaktoren einer mobilen Kundenkommunikation im Einzelhandel. In Ahlert, D., Becker, J., Knackstadt, R., Wunderlich, M. (Ed.). *Customer Relationship Management im Handel – Strategien – Konzepte – Erfahrungen* (343-360), Berlin, New York etc.: Springer.

Van der Dam. (2006, April). *Primetime for mobile television*. IBM global services.

Van der Heijden, H., & Verhagen, T. (2004). Online store image: conceptual foundations and empirical measurement.

Information & Management, *41*(5), 609–617. doi:10.1016/S0378-7206(03)00100-9

van Luijt, A. (2005). Beyond DRM. In Picot, A., & Thielmann, H. (Ed.), *Distribution und Schutz digitaler Medien durch Digital Rights Management* (pp. 61-74). Berlin: Springer.

Varadarajan, P. R. (1984). Consumer responses to small business coupon sales promotions. *American Journal of Small Business*, *9*(Fall), 17–26.

Varanparast, R., & Asil, M. (2007). Factors affecting the use of mobile advertising. *International Journal of Mobile Marketing*, *2*(2), 21–34.

Varshney, U. (2003). Location management for mobile commerce applications in wireless Internet environment. *ACM Transactions on Internet Technology*, *3*(3), 236–255. doi:10.1145/857166.857169

Varshney, U., & Vetter, R. (2002). Mobile commerce: Framework, applications, and networking support. *Mobile Networks and Applications*, *7*, 185–198. doi:10.1023/A:1014570512129

Vatanparast, R. (2007). Piercing the fog of mobile advertising. In Proceedings of the 6th International Conference on Mobile Business (ICMB07′), Toronto, Canada.

Vatanparast, R. (2007). Piercing the fog of mobile advertising. *The Sixth International Conference on Mobile Business ICMB 2007 July 9-11*, 19. Toronto, Ontario, Canada

Vatanparast, R., & Butt, A. H. (2009). Factors affecting use of mobile advertising: A quantitative study. In Proceedings of the 42nd Hawaii International Conference on System Sciences (HICSS42), Hawaii, USA.

Velden, R., & Wolbers, M. (2004). *How much does education matter and why? The effects of education on socio economic outcomes*, ISBN 90-5321-390-2 Sec04.049

Venkatesh, B., Nargundkar, R., Sayed, F. K., & Shahaida, P. (2006). Assessing Indian students' perceptions towards m-learning - some initial conclusions. *International Journal of Mobile Marketing*, *1*(2), 75–79.

Venkatesh, V. (1999, June). Creation of favourable user perceptions: integrating control, intrinsic motivation, and emotion into the technology acceptance model. *MISQ*, *23*(2), 239–260. doi:10.2307/249753

Venkatesh, V., & Brown, S. A. (2001, March). A longitudinal investigation of personal computers in homes: adoption determinants and emerging challenges. *MISQ*, *25*(1), 71–102. doi:10.2307/3250959

Venkatesh, V., & Morris, M. G. (2000). Why don't men ever stop to ask for directions? Gender, social influence, and their role in technology acceptance and usage behavior. *MISQ*, *24*, 115–139. doi:10.2307/3250981

Venkatesh, V., Morris, M. G., & Ackermann, P. L. (2000). A longitudinal field investigation of gender differences in individual technology adoption decision-making processes. *Organizational Behavior and Human Decision Processes*, *83*(1), 33–60. doi:10.1006/obhd.2000.2896

Venkatesh, V., Morris, M. G., Davis, G. B., & Davis, F. D. (2003). User acceptance of information technology: Toward a unified view. *MISQ*, *27*(3), 425–479.

Verkasalo, H. (2007). Handset-based measurement of smartphone service evolution in Finland. *Journal of Targeting. Measurement & Analysis for Marketing*, *16*(1), 7–25. doi:10.1057/palgrave.jt.5750060

Ververidis, C., & Polyzos, G. (2002). Mobile marketing using location based services. *Proceedings of the First International Conference on Mobile Business*, Athens, Greece.

Viejo, A., Sebé, F., & Domingo-Ferrer, J. (2007). Secure and Private Incentive-Based Advertisement Dissemination in Mobile Ad Hoc Networks. In A. Miyaji, H. Kikuchi, & K. Rannenberg (Eds.), *Iwsec* (Vol. 4752, p. 185–198). Springer.

Virtanen, V., Bragge, J., & Tuunanen, T. (2005). Barriers for mobile marketing and how to overcome them. *Proceedings of the Hong Kong Mobility Roundtable*, Hong Kong.

Vrechopoulos, A. P., O' Keefe, R. M., & Doukidis, G. I. (2000). Virtual store atmosphere in internet retailing. In S. Klein, B. O Keefe, J. Gricar, and M. Podlogar, editors, *Proceedings of the 13th Bled Electronic Commerce Conference: The End of the Beginning*, pp. 445-458, June 19-21.

Vrechopoulos, A. P., O'Keefe, R. M., Doukidis, G. I., & Siomkos, G. I. (2004). virtual store layout: an experiment comparison in the context of grocery retail. *Journal of Retailing*, *80*(1), 13–22. doi:10.1016/j.jretai.2004.01.006

Wais, J. S., & Clemons, E. K. (2008). Understanding and Implementing Mobile Social Advertising. *International Journal of Mobile Advertising*, *3*(1), 12–18.

Wamser, C., & Wilfert, A. (2002). Die wettbewerbsstrategischen Rahmenbedingungen des Mobile Commerce. In Teichmann, R., & Lehner, F. (Eds.), *Mobile Commerce* (pp. 29-50). Berlin: Springer.

WAN. (2008). World press trends. Available at http://www.wan-press.org/worldpresstrends/home.php

Wang, A. (2007). Branding over mobile and Internet advertising: The cross-media effect. *International Journal of Mobile Marketing, 2*(1), 34–42.

Wang, C.-C., Lo, S.-K., & Fang, W. (2008). Extending the technology acceptance model to mobile telecommunication innovation: The existence of network externalities. *Journal of Consumer Behaviour, 7*, 101–110. doi:10.1002/cb.240

Wang, F., & Yamagishi, T. (2005). Group-based trust and gender differences in China. *Asian Journal of Social Psychology, 8*(2), 199–210. doi:10.1111/j.1467-839x.2005.00167.x

Wang, Y.-S., Lin, H.-H., & Luarn, P. (2006). Predicting consumer intention to use mobile service. *Information Systems Journal, 16*(2), 157–179. doi:10.1111/j.1365-2575.2006.00213.x

Waterman, D. (1992). "Narrowcasting" and "broadcasting" on nonbroadcast media, a program choice model. *Communication Research, 19*(1), 3–28. doi:10.1177/009365092019001001

Watson, R. T. (2000). Electronic commerce: The strategic perspective. Fort Worth, TX: Dryden.

Webb, W. (2001). *The future of wireless communication*, Boston, MA: Artech House.

Weber, L. (2008). *Marketing to the Social Web: How Digital Customer Communities Build Your Business.* Wiley.

Webster, J., Trevino, K. L., & Ryan, L. (1993). The dimensionality and correlates of flow in human-computer interactions. *Computers in Human Behavior, 9*(4), 411–426. doi:10.1016/0747-5632(93)90032-N

Wedel, M., & Kamakura, W. A. (2000). Market segmentation, conceptual and methodological foundations, 2nd edition. ISQM8. Boston, MA: Kluwer Academic Publishers.

Wehmeyer, K., & Müller-Lankenau, C. (2005). *Mobile Couponing - Measuring Consumers' Acceptance and Preferences with a Limit Conjoint Approach.* Paper presented at the 18th Bled eConference, Bled, Slovenia.

Weidelich, F. (2008). CeBIT 2008: RFID-Anwendung Near Field Communication soll 2010 zum Standard bei Handys werden. *VDI Nachrichten*, 14.03.2008, 13.

Weimann, G. (1991). The Influentials: Back to the Concept of Opinion Leaders? *Public Opinion Quarterly, 55*(2), 267–279. doi:10.1086/269257

Weitenberner, C., Chapman, M., Miranda, G., Tobar, S., Wagoner, E., Akasaki, D., & Seal, K. C. (2006). United States and South East Asian Mobile Markets: A Comparative Analysis of Infrastructure and Cultural Differences. *International Journal of Mobile Marketing, 1*, 73–82.

Weitenberner, C., Chapman, M., Miranda, G., Tobar, S., Wagoner, E., & Akasaki, D. (2006). United States and South East Asian mobile markets: a comparative analysis of infrastructure and cultural differences. *International Journal of Mobile Marketing, 1*(1), 73–82.

Welfens, P. J. J., Zoche, P., Jungmittag, A., Beckert, B., & Joisten, M. (2005). Internetwirtschaft 2010. Heidelberg: Physica.

Wells, W. D. (1997). *Measuring advertising effectiveness.* Mahwah, NJ: Lawrence Erlbaum.

Wicklund, R. A. (1974). *Freedom and reactance*, New York, NY: Wiley.

Wiedemann, D. G. (2007). Exploring the Concept of Mobile Viral Marketing through Case Study Research. *Proceedings of the 2nd conference of Mobilität und mobile Informationssysteme (*pp. 49-60). Aachen, Germany.

Wiedemann, D. G., & Pousttchi, K. (2008).Creating Successful Mobile Viral Marketing Strategies. *Handbook of Research on Mobile Multimedia*, 2nd Edition. Hershey, PA: Information Science Reference.

Wiedemann, D. G., Haunstetter, T., & Pousttchi, K. (2008). Analyzing the Basic Elements of Mobile Viral Marketing. An Empirical Study. *7th International Conference on Mobile Business (ICMB 2008)*, Spain, pp. 75-85.

Wiedmann, K.-P. (1989). Gesellschaft und Marketing. In Specht, G., Silberer, G., & Engelhardt, W. H. (Ed.), *Marketing-Schnittstellen* (pp. 227-247). Stuttgart: Schäffer-Poeschel.

Wiedmann, K.-P. (2001). Plädoyer für H-Commerce. In Nicolai, A. T., & Petersmann, T. (Ed.), *Strategien im M-Commerce* (pp. 129-146). Stuttgart: Schäffer-Poeschel.

Wiedmann, K.-P., & Frenzel, T. (2004). Akzeptanz im E-Commerce – Begriff, Modell, Implikationen. In Wiedmann, K.-P., Buxel, H., Frenzel, T., & Walsh, G. (Eds.), *Konsumentenverhalten im Internet* (pp. 99-117). Wiesbaden, Gabler.

Wiedmann, K.-P., & Langner, S. (2004). Beeinflussung im Internet – Klassische Techniken und aktuelle Phänomene. In Wiedmann, K.-P., Buxel, H., Frenzel, T., & Walsh, G. (Eds.), *Konsumentenverhalten im Internet* (pp. 195-225). Wiesbaden, Gabler.

Wiedmann, K.-P., Buckler, F., & Buxel, H. (2000a). Chancenpotenziale und Gestaltungsperspektiven des M-Commerce. *Der markt*, 39 (2), 84-96.

Wiedmann, K.-P., Buckler, F., & Buxel, H. (2000b). Mobile Commerce. *WiSt. Wirtschaftswissenschaftliches Studium*, *29*(12), 684–691.

Wiedmann, K.-P., Keuling, U., & Reeh, M.-O. (2005). Potenziale und Einsatz von RFID/NFC-Systemen im Mobile Commerce. Hannover: Schriftenreihe Marketing und Management.

Wiedmann, K.-P., Reeh, M.-O., & Schuhmacher, H. (2008). Near Field Communication im Mobile Marketing. In H. H. Bauer, T. Dirks, & M. D. Bryant (Ed.), *Erfolgsfaktoren des Mobile Marketing* (pp. 305-325). Berlin, Heidelberg: Springer.

Wilke, K. (2002). Controlling im E-Commerce. *Handel im Fokus*, *54*(1), 274–294.

Wilke, K., Duscha, A., & Hudetz, K. (2005). Kundenbindung über das Internet. In Hudetz, K. (Ed.), *E-Commerce im Handel – Status Quo und Perspektiven* (pp. 101-133). Gernsbach, Deutscher Betriebswirte-Verlag.

Williamson, O. (1991). Comperative Economic Organizations. The Analysis of Discrete Structural Alternatives. *Administrative Science Quarterly*, *36*, 269–296. doi:10.2307/2393356

Williamson, O. E. (1979). Transaction-Cost Economics: The Governance of Contractual Relations. *The Journal of Law & Economics*, *22*(2), 233–261. doi:10.1086/466942

Willrodt, K. (2005). Marktforschung im Innovationsprozess. In Amelingmeyer, J., & Harland, P.E. (Ed.), *Technologiemanagement & Marketing* (pp. 403-413). Wiesbaden: Vahlen.

Wilson, R. F (2000). The Six Simple Principles of Viral Marketing. *Web Marketing Today*, 70.

Wind, J. a&Mahajan, V. (2000). Digital marketing: Global strategies from the world's leading experts. John Wiley & Sons.

Wind, J., & Rangaswamy, A. (2001). Customerization: The next revolution in mass customization. *Journal of Interactive Marketing*, *15*(1), 13–32. doi:10.1002/1520-6653(200124)15:1<13::AID-DIR1001>3.0.CO;2-#

Wireless Intelligence. (2007). Subscriber statistics end Q4 2007 Report, Retrieved July 25, 2008, from http://www.gsmworld.com/news/statistics/pdf/gsma_stats_q4_07.pdf

Wireless Intelligence. (2007). *Subscriber statistics end Q4 2007 Report*. Retrieved July 25, 2008, from http://www.gsmworld.com/news/statistics/pdf/gsma_stats_q4_07.pdf

Wirtz, B. W. (2002). Electronic Commerce. In Wirtz, B. W. (Ed.), *Gabler Kompakt-Lexikon eBusiness* (pp. 57-58). Wiesbaden, Gabler.

Wirtz, B. W., & Lihotzky, N. (2005). Kundenbindungsmanagement im Electronic Business: Instrumente und ihre Wirkung. *GfK Jahrbuch der Absatz und Verbrauchsforschung*, *51*(2), 136–155.

Wohlfahrt, J. (2002). Wireless Advertising. In: G. Silberer, J. Wohlfahrt, & T. Wilhelm (Ed.), *Mobile Commerce* (pp. 245-263). Wiesbaden: Gabler.

Wohlfahrt, J. (2004). *Akzeptanz und Wirkungen Mobile-Business-Anwendungen*, Hamburg: Dr. Kovac.

Wolfe, D. (2003). *Ageless Marketing: strategies for reaching the hearts & minds of the new customer majority*. Chicago, IL: Dearborn Trade Publishing.

Wolff, P. (2006). *Die Macht der Blogs - Chancen und Risiken von Corporate Blogs und Podcasting*. Frechen, Datakontext.

Wolfram, G. (2008). Die Metro-Group-Future-Store-Initiative. In Riekhof, H.-C. (Ed.), *Retail Business in Deutschland* (pp. 617-631). Wiesbaden: Vahlen.

Wolin, L. D. (2003). Gender issues in advertising -An oversight synthesis of research: 1970-2002. *Journal of Advertising Research*, *43*(1), 111–129.

Woolley, S. (2005). Best Leaders. *Business Week*, 62.

Wray, R. (2008). Cash in hand: Why Africans are banking on the mobile phone. The Guardian, June 17, 2008.

Wright, R. A., & Brehm, S. S. (1982). Reactance as impression management: A critical review. *Journal of Personality and Social Psychology*, *42*(4), 608–618. doi:10.1037/0022-3514.42.4.608

Wu, J.-H., & Wang, S.-C. (2005). What drives mobile commerce? An empirical evaluation of the revised technology acceptance model. *Information & Management*, *42*(5), 719–729. doi:10.1016/j.im.2004.07.001

Wu, P. (2005). Personal Interview, General Manager at Walkgame. May 2005.

Wübker, G., & Schmidt-Gallas, D. (2003). Pricing-Strategien für Banken. In: H. Diller, & A. Herrmann(Ed.). *Handbuch Preispolitik, Strategien – Planung – Organisation – Umsetzung* (pp. 739-761). Wiesbaden: Gabler.

KVG. (2006). *Bundesgesetz über die Krankenversicherung (KVG)*.

Mobile Marketing Association. (2008). Mobile marketing industry glossary. *MMA Mobile Marketing Association*, Version 068, http://mmaglobal.com/glossary.pdf

Xu, D. J., Liao, S. L., & Li, Q. (2008). Combining Empirical Experimentation and Modeling Techniques: A Design Research Approach For Personalized Mobile Advertising Applications. *Decision Support Systems*, *44*, 710–724. doi:10.1016/j.dss.2007.10.002

Xu, M., & Walton, J. (2005). Gaining customer knowledge through analytical CRM. *Industrial Management & Data Systems*, *105*(7), 955–971. doi:10.1108/02635570510616139

Yan, X. (2003). Mobile Data Communications in China. *Communications of the ACM*, *46*(12), 81–85. doi:10.1145/953460.953463

Yang, K. C. C. (2007). Exploring factors affecting consumer intention to use mobile advertising in Taiwan. *Journal of International Consumer Marketing*, *20*(1), 33–49. doi:10.1300/J046v20n01_04

Yang, K., & Jolly, L. D. (2008). Age cohort analysis in adoption of mobile data services. *Journal of Consumer Marketing*, *25*(5), 272–280. doi:10.1108/07363760810890507

Yap, C. (2007). ESPN Mobile Interview, May 2007.

Yokokuchi, S. (2007). Personal Interview, Japan, May 2007.

Yokotsuka, Y. (1991). Direct Marketing in Era of Multimedia, *Doubunkan Shuppan* Co., Ltd. (in Japanese).

Yom, M. (2001). Utility und Usability im Mobile Commerce. In Silberer, G., Wohlfahrt, J., & Wilhelm, T. (Ed.), *Mobile Commerce* (pp. 173-184). Wiesbaden: Vahlen.

Yu, R. (2008). *Mobile market trends*. London: Ovum Europe Limited.

Yuan, S. T., & Tsao, Y. W. (2003). A recommendation mechanism for contextualized mobile advertising. *Expert Systems with Applications*, *24*, 399–414. doi:10.1016/S0957-4174(02)00189-6

Yuan, Y., & Zhang, J. (2003). Towards an Appropriate Business Model for m-commerce. *International Journal of Mobile Communications*, *1*(1-2), 35–56. doi:10.1504/IJMC.2003.002459

Yuan, Y., & Zhang, J. J. (2003). Towards an appropriate business model for m-commerce. *International Journal of Mobile Communications*, *1*(1-2), 35–56. doi:10.1504/IJMC.2003.002459

Yunos, H. M., Gao, J. Z., & Shim, S. (2003). Wireless Advertising's Challenges and Opportunities. *Computer*, *36*(5), 30–37. doi:10.1109/MC.2003.1198234

Yunos, H. M., Gao, J. Z., & Shim, S. (2003). Wireless advertising's challenges and opportunities. *Computer*, *36*(5), 30–37. doi:10.1109/MC.2003.1198234

Yunos, H. M., Gao, J. Z., & Shim, S. (2003). Wireless advertising's challenges and opportunities. *IEEE Computer*, *36*(5), 30–37.

Zeithaml, V. A. (1988). Consumer perceptions of price, quality and value: A means-end model and synthesis of evidence. *Journal of Marketing*, *52*(3), 2–22. doi:10.2307/1251446

Zeithaml, V. A., & Bitner, M. J. (1996). *Services Marketing. Integrating Consumer Focus Across the Firm*. New York: McGraw Hill.

Zeithaml, V. A., Berry, L. L., & Parasuraman, A. (1996). The Behavioral Consequences of Service Quality. *Journal of Marketing*, *60*(2), 31–46. doi:10.2307/1251929

Zeithaml, V. A., Parasuraman, A., & Berry, L. L. (1985). Problems and strategies in service marketing. *Journal of Marketing*, *49*(Spring), 33–46. doi:10.2307/1251563

Zeng, E. Y., Yen, D. C., & Huang, S.-M. (2003). Mobile Commerce: the convergence of e-commerce and wireless technology. *International Journal of Services technology and Management*, 4(3), 302-322.

ZenithOptimedia. (2008). Advertising expenditure forecast December 2008. Available at http://www.zenithoptimedia.com/gff/pdf/Adspend%20forecasts%20March%202008.pdf

Zhang, X., & Prybutok, V. R. (2005). How the Mobile Communication Markets Differ in China, the U.S., and Europe. *Communications of the ACM*, *48*(3), 111–114. doi:10.1145/1047671.1047678

Zhou, L., Dai, L., & Zhang, D. (2007). Online Shopping Acceptance Model - A Critical Survey of Consumer Factors in Online Shopping. *Journal of Electronic Commerce Research*, *8*(1), 41–62.

Zid, L. A.-S. (2005). TV to GO. *Marketing Management*, *14*(1), 5–5.

Zmijewska, A., & Lawrence, E. (2006). *Implementation models in mobile payments*. Paper presented at the International Conference Advances in Computer Science and Technology, Puerto Vallarta, Mexico.

Zobel, J. (2001). *Mobile Business und M-Commerce - Die Märkte der Zukunft erobern.* München: Hanser.

Zviran, M., Glezer, C., & Avni, I. (2006). User satisfaction from commercial web sites: the effect of design and use. *Information & Management*, *43*(2), 157–178. doi:10.1016/j.im.2005.04.002

About the Contributors

Key Pousttchi is an assistant professor and heads the wi-mobile Research Group, University of Augsburg, WI-SE. The focus of his research comprises mobile marketing, mobile financial services, mobile-integrated business processes and the development of mobile markets. The group's research results are applied in strategy consulting for major market players. Prior to his academic career he has been a regular officer of the German army for 12 years, lastly being responsible for integrated information processing and simulation systems in an R&D department.

Dietmar Georg Wiedemann is a researcher and project leader with the wi-mobile Research Group, University of Augsburg, WI-SE. He is an expert in b2c mobile commerce and has executed many qualitative and quantitative studies in Mobile Marketing and Mobile Payment. Since 2007 he is the project leader for the German part of the European Union project on a Secure Mobile Payment System (SEMOPS).

* * *

Maria Åkesson is Lecturer in informatics at Halmstad University and is also doing PhD work within the Media IT research group (media-it.hh.se). She has a professional background in business administration and IT consultancy. Her research interests are related to digital publishing in ubiquitous information environments, digital innovation and value networks. She has been doing research work in several projects such as "DigiNews", "Designing the future e-newspaper", and "Designing Ubiquitous Media Services by Action Research". Currently she is co-leading the project "The Local Newspaper 2.0" with engaged readers. Her work has been published in international conferences and Journal of Electronic Commerce, Journal of Media Business Studies and Journal of Graphic Technology.

Shahriar Akter is currently serving as a Lecturer in the Department of Business Administration of East West University, Bangladesh. He did his BBA (2002) and MBA (2004) from University of Dhaka, Bangladesh. In 2007, he did his Master of Electronic Business Management from International University of Japan. He has already had a good number of local and international publications and presented his research at several international conferences including London Business School and University of Oxford. It is noteworthy that he was awarded Japan Development Scholarship (JDS) in 2006 for doing research on ubiquitous Services.

Stuart J. Barnes is Chair and Professor of Management in the Norwich Business School at the University of East Anglia, United Kingdom. Previously he worked at Victoria University of Wellington,

New Zealand, and the University of Bath. He holds a PhD in Business Administration from Manchester Business School. His primary research interests centre on the successful utilization of new information and communications technologies by businesses, governments and consumers. He has published five books and more than a hundred papers, including those in journals such as Communications of the ACM, the International Journal of Electronic Commerce and Information & Management.

Mark Brown is a Senior Lecturer in Marketing at the UQ Business School, University of Queensland, Australia. His research interests include advertising and communication strategy, sport marketing, customer relationship management, and electronic commerce.

Amy Carroll is Equities Research Associate in the Global Markets division at Citigroup, New Zealand. Her academic background includes a bachelor degree in Commerce and Administration in Information Systems and Marketing, and an honours degree in Information Systems from the School of Information Management at Victoria University of Wellington, New Zealand. Her interests include marketing, finance and mobile technologies.

Wen-Kuo Chen is currently a Ph.D. candidate in the Department of Information Management at National Taiwan University, Taipei, Taiwan. His research focus is e-commerce and m-commerce

Seng-cho T. Chou is Professor in Information Management at National Taiwan University. He received his B.Sc. from the Chinese University of Hong Kong, M.S. from the University of California, and Ph.D. in Computer Science from the University of Illinois at Urbana-Champaign. His current research interests are Web-based technologies and services innovation, e-business and e-commerce, knowledge management, data mining, and mobile & ubiquitous computing.

Steffen Dölling is a research assistant and doctoral student at the chair of Marketing & Retailing, Martin-Luther-University Halle-Wittenberg, since 2006. His research focus lies on relationship marketing, particularly in the retention management for different target groups and on potentials of new interactive media for retail management. He has practical experience as consultant for retailers especially in strategy and implementation for customer integration activities at the point of sale. He is an author of numerous articles on interactive customer retention management.

Dimitris Drossos is Adjunct Lecturer in the Department of Information and Communication Systems Engineering at the University of the Aegean. His research interests are mainly in the areas of e-Business and mobile Commerce, including mobile marketing and advertising effectiveness. He is a member of the International Program Committee of the European Conference on Mobile Government and a reviewer for several IT related journals, including the Journal of Electronic Commerce Research and Information Technology & People. He has published in the International Journal of Electronic Commerce and the Journal of Interactive Advertising.

Carina Ihlström Eriksson (PhD) is Assistant Professor in Informatics at Halmstad University and leads the Media IT research group (media-it.hh.se) within the EIS (Embedded and Intelligent Systems) research environment. Her research interest lies within the fields of digital publishing and she focus on new innovative media and end-user involvement in the innovation process. She is the leader of Halmstad

Living Lab with several research projects and is currently project leader of two research projects "The Local Newspaper 2.0" with engaged readers and "Secure at home" Living lab. Her research is published in international journals such as IT & People, Electronic Markets, Journal of Organizational Computing and Electronic Commerce and International Journal of Services Technology and Management.

Marco Garito is experienced in marketing and business transformation encompassing large enterprise and small medium business: he worked in Italy and in United Kingdom covering roles at for Cisco Systems and IBM. Previously, he worked with a Venture Capital and professional service companies in Italy. His background is within IT and new technology adoption, professional services and consumer goods. He holds a bachelor degree in law from the University of Milan (Italy) and an MBA from the University of Technology in Sydney (Australia).

George M. Giaglis is Associate Professor of eBusiness at the Athens University of Economics and Business. His main research interests lie in the areas of mobile and wireless applications and services; ubiquitous and pervasive information systems; business process modelling and simulation; and information systems evaluation. He has published more than 100 articles in leading journals and international conferences, including the Information Systems Journal, the International Journal of Electronic Commerce, and the International Journal of Information Management. He is a member of the editorial board of seven journals, including the International Journal of Mobile Communications and the Business Process Management Journal. Since 2001, he is the Director of the ISTLab Wireless Research Center, a group pursuing research on mobile business and ubiquitous commerce (http://istlab.dmst.aueb.gr/wrc/, http://www.mobiforum.org/)

Tobias Haunstetter is member of the BAIKA project team at Bayern Innovativ GmbH, Nuremberg, Germany, since 2008. BAIKA is a large international knowledge transfer network of automotive companies. From 2006 to 2008, he was a member of the wi-mobile Research Group, University of Augsburg, Business Informatics/Systems Engineering. He has conducted several studies regarding mobile viral marketing including the development of standard types as well as characteristics and preferences of consumers engaged in mobile viral marketing.

Andreas Heinemann received his diploma degree in computer science in 1999 from the University of Tübingen, Germany. From 2000 to 2001 he worked as a software developer for a financial institute. From 2002 to 2005, he was a PhD student in the PhD program *Enabling Technologies for Electronic Commerce* (DFG Graduiertenkolleg) at Technische Universität Darmstadt. From 2005 to 2007 he has worked as a research assistant and group leader at the Telecooperation Group, led by Prof. Dr. Max. Mühlhäuser at the Technische Universtität Darmstadt, where he received a PhD in computer science (Dr. rer. nat) on Collaboration in Opportunistic Networks. Currently he works for a financial institute that develops online banking solutions for the German market. His research interests are IT security, Opportunistic Networks and mobile computing.

Heng-Chiang Huang is a professor in the Department of International Business at the National Taiwan University, Taiwan. He received his Ph.D. degree in international business from the University of Michigan. His research focuses on international business, international marketing, customer relationship management (CRM), inter-organizational relationship management. His work has been published in Psychology & Marketing, Asia Pacific Journal of Management, Pan Pacific Management Review, Taiwan Academy of Management Journal, NCCU Management Review, NTU Management Review, Sun Yat-Sen Management Review, etc.

Heikki Karjaluoto is a Professor in Marketing at the University of Jyväskylä, Finland. His research interests include electronic business, consumer behavior and marketing communications. Previous publications have appeared in the Electronic Markets, European Journal of Marketing, International Journal of Electronic Business, International Journal of Mobile Marketing, International Journal of Retail & Distribution Management, and Internet Research, among others.

Fumiyo N. Kondo is currently serving as an Assistant Professor in the Department of Social Systems and Management, the University of Tsukuba, Japan. She obtained her Master of Science degree in statistics from Colorado State University, US and a Doctoral Degree in Statistical Science from the Graduate University for Advanced Studies, Japan. She had worked for a marketing research firm, A. C. Nielsen Company in Japan, as an account executive for 5 years and as a statistical officer for 5 years. She has numerous local and international publications including a book on marketing models with a special emphasis on Point-of-Sales data. Her research interest is Mobile Marketing, Marketing Modeling, Forecasting, and Bayesian Statistical Analysis.

Dennis Lee is an Assistant Professor at the School of Business Administration, American University in Dubai, United Arab Emirates. His research interests lie in the areas of Consumer Behavior, Interaction Design, marketing communications and mobile digital communication technologies.

Heikki Lehto is Chief Technology Officer at Brandson, Ltd, Finland. His research interests are in digital marketing communications. His publications have appeared in the *Electronic Markets, International Journal of Electronic Business, International Journal of Mobile Marketing, and Journal of Nonprofit and Public Sector Marketing,* among others.

Matti Leppäniemi is a Postdoctoral Research Fellow in Marketing at the University of Oulu, Finland. His research interests are in the areas of digital marketing and consumer behaviour. His publications have appeared in the Journal of Systems and Information Technology, Electronic Markets, Journal of Interactive Advertising, International Journal of Electronic Business, and International Journal of Mobile Marketing, among others.

Jörg Link is the holder of the Chair of Controlling and Organization at the University of Kassel. His principal topics of research are in the fields of Marketing-Controlling and Customer Relationship Management. He has managed many research and consulting projects in different countries, e.g Germany, Great Britain, the United States of America, and Japan.

Emmanouela E. Manganari, Ph.D., is an Adjunct Lecturer at the Technological Education Institute of Athens. She obtained her Ph.D. from Athens University of Economics and Business (AUEB), while she holds a M.Sc. in Marketing Management from Aston Business School (UK). Her current research interests are in the area of atmospherics in e-tailing and retailing and consumer behaviour. Emmanouela Manganari is the corresponding author and can be contacted at: eem@aueb.gr

Andreas Mann is the holder of the Chair of Dialog Marketing, endowed by the Siegfried Vögele Institute (SVI), and Director of the DMCC (Dialog Marketing Competence Center) at the University of Kassel, Germany. Previously, he was holding the Chair of Management and Marketing at the Technical University Clausthal, Germany. His research subjects include interactive marketing, industrial service policy and sales and distribution management.

Richard Mannix (MBA, University of Hull) is a Senior Lecturer in Marketing at European Business School London, Regent's College. Richard is a member of the Academy of Marketing (MAM), a Member of the Chartered Institute of Marketing (MCIM) and a Fellow of the Higher Education Academy (FHEA). He has previously taught at institutions in both Japan and Sweden, teaching marketing and business communication skills. While at the Folkuniversitetet in Sweden he was in charge of developing and marketing new courses to corporate clients, securing contracts with a number of major listed companies and government agencies. Richard is currently completing a PhD in Marketing at the University of Strathclyde in Glasgow, Scotland. Richard's research interests include the development of social entrepreneurial networks and student profiling and progression in Higher Education. He has recently presented conference papers on these topics in Turkey, Greece and Scotland.

Jonathon McKenzie is business analyst at Run the Red – a New Zealand based mobile marketing company. Before joining Run the Red to is he was a research student at the School of Information Management at Victoria University of Wellington, New Zealand.

Dirk Möhlenbruch is Professor of Business Administration, especially Marketing & Retailing at the Martin-Luther-University Halle-Wittenberg since 1992. He is an authority on retail management, relationship marketing as well as electronic and mobile marketing. He is an author of numerous articles and books on customer retention management, product mix policy and interactive technologies in multi-channel-marketing. He is the managing director of the institute for management economics of the Martin-Luther-University Halle-Wittenberg and member of the supervisory board of Institut für Unternehmensforschung und Unternehmensführung an der Martin-Luther-Universität (ifu) and since 1995 in the board of trustees of the DB Logistics foundation (Deutsche Bahn AG), Berlin. Since 2002 he is the scientific director of Conomic Marketing & Strategy Consultants, Halle (Saale).

Ralf Muhlberger is a Lecturer at the School of Information Technology and Electrical Engineering, University of Queensland, Australia. His key research focus is to investigate and design people's interactive experience with technology in ways that promote theoretical development, stimulate new industry practice, and enhance users' experience. His research falls within the fields of Computer Supported Cooperative Work, Human–Computer Interaction and Interaction Design.

Mariko Nakahara earned a bachelor degree in Management Science and Engineering from College of Policy and Planning Sciences, Third Cluster of Colleges, the University of Tsukuba in 2006. She currently works for Hitachi, Ltd., serving as a sales representative to partner companies in Department of Software Division.

Süphan Nasir is Associate Professor in the Department of Business Administration at Istanbul University, Turkey. She received a Bachelor degree of Sociology with Honors from Boğaziçi University.

She received MBA degree from Işık University and Ph.D. degree in marketing from Istanbul University. Her research interests include customer relationship management, consumer complaint behavior, brand management, and mobile marketing. She has published articles in reputable academic journals and written chapters for well known books. She has also presented her papers at major international conferences.

Niklas Nohlen is a principal consultant at SHS VIVEON AG in the area of customer relationship management. Before joining SHS VIVEON AG in 2008, he worked as a research assistant at the Competence Center of Supplier & Customer Relationship Management at the Institute of Research on Information Systems (IRIS) of the European Business School (EBS), Oestrich-Winkel/Wiesbaden, Germany. He received a degree (diploma) in Business Information Systems from the University of Goettingen, Germany. He also holds a doctoral degree from the European Business School (EBS). Niklas Nohlen has been working within the research field of customer relationship management throughout his studies and while developing his doctoral thesis.

Jan Prein is research assistant and PhD student at the Chair of Dialog Marketing, endowed by the Siegfried Vögele Institute (SVI), at the University of Kassel, Germany. His research subjects include mobile marketing and adoption and diffusion of innovative marketing applications.

Christopher Quente loves to tell stories. After graduation in German studies and philosophy he concentrated his work around the development of cross-channel brand campaigns, encompassing multiple media platforms. His aim is to create unique integrated brand experiences, whose original, powerful ideas don't get lost in single and isolated channels. Working for big international agency networks, his client list includes blue chips like Coca-Cola, Deutsche Telekom, Loewe, Audi, VW, Müller, Payback, UniCredit, AMD, Jägermeister and McDonalds. A whole host of his campaigns have been honored with internationally recognized awards, e.g the Cannes Lion. As an author (he co-authored "Brand's New Toy" Creative Marketing Communication Via Mobile Phones etc. published in 2006), as a speaker on numerous conventions and in his function as a Creative Director, Christopher Quente engages himself for the cause of a more creative handling of mobile media. He lives in the countryside near Hamburg and in Munich.

Marc-Oliver Reeh is PhD student at the Institute of Marketing and Management at the Leibniz Universität Hannover and since 2006 project leader at the Center for Near Field Communication Management. In cooperation with different project partners from science and practice he examines the various ranges of application of RFID/NFC systems. He also collected relevant practical experience by his work at Vodafone Terenci in Ratingen, Germany. Furthermore, he is member of the research cluster *Future Internet - Internet, Information and I*. His main research areas are consumer behavior and acceptance of new technologies like RFID- and NFC systems in electronic- and mobile commerce.

Gerold Riempp is full professor of information systems and executive director of the Institute of Research on Information Systems (IRIS) at European Business School (EBS), Oestrich-Winkel/Wiesbaden, Germany. His main research fields are knowledge management, strategic IT management, and customer & supplier relationship management. In his time as assistant professor, he headed the competence center "customer knowledge management" at the University of St. Gallen's Institute of Information Management in Switzerland from 2000 to 2002. Besides his academic carreer, Prof. Riempp gained more than 14 years of practical experience as manager in several consulting firms and has, among others, introduced IT strategies, CRM, and knowledge management (KM) in many organizations. For example, he was senior manager and headed the KM implementation at PricewaterhouseCoopers in Germany from 1998 to 2000.

Falk Ritschel is a research assistant and doctoral student at the chair of Marketing & Retailing, Martin-Luther-University Halle-Wittenberg, since 2004. His research focus lies on potentials of the new interactive media for retail management and the relationship marketing for retailers, particularly in the area of churn management. He has practical experiences in customer integration and customer retention management by numerous projects with retail partners. He is an author of many articles on interactive customer retention management.

Sebastian Schulz (Dr. rer. pol., University of Göttingen, Germany) worked from 2004 to 2008 as a research assistant at the Institute of Marketing and Retailing at the Georg-August-University Göttingen. Till September 2008 he is Executive Assistant to Chief Sales Officer at the Hannoversche Leben direct insurance. Main research interests include consumer behavior, tracking research in the internet, CRM and online trust. He has chapters in various German marketing readers and his work appeared in proceedings such as Advances in Consumer Research and journals such as transfer werbeforschung und praxis.

Henrik Schumacher is a member of scientific staff and PhD student at the Communications Networks group, which is headed by Prof. Dr.-Ing. Markus Fidler, at the Institute of Communications Technology of the Leibniz Universität Hannover. His main research interests are located in the scope of efficient, reliable and secure protocols for wireless communication networks. This includes NFC as well as other wireless technologies like next generation 802.11 (Wireless LAN) networks, where his work focuses on mobility management protocols for vehicular networks and other challenging scenarios with highly mobile users. Since 2006, he has been responsible for several scientific projects examining these topics in close cooperation with industrial partners.

Eusebio Scornavacca Jr. is Senior Lecturer of Electronic Commerce at SIM. Before moving to Wellington, he spent two years as a researcher at Yokohama National University, Japan. Eusebio has published and presented more than seventy papers in conferences and academic journals. His research interests include mobile business, electronic business, e-surveys, and IS teaching methods. In 2005, he was awarded at the MacDiarmid Young Scientists of the Year awards, and in 2006 he received a VUW Research Excellence Award as well as the Victoria's Award for the best postgraduate supervisor from the PGSA. In 2007, Eusebio received a Teaching Excellence Award from Victoria University of Wellington.

Kala Seal, a computer information systems professor, has been a member of the College of Business Administration faculty since 1990. He received his undergraduate degree in Mechanical Engineering from Indian Institute of Technology in India followed by a Masters in Computer Science and a Ph.D. in Management Science and MIS from University of Texas at Dallas. His teaches information systems and quantitative courses. His primary research interests are in, Spreadsheet Based OR/MS Modeling, Information Systems Frameworks, Mobile-Commerce Success Factors. He has published many articles in journals such as Interfaces, .OR Insight, International Journal of Mobile Marketing and many more.

Franziska Seidl is a research assistant of Prof. Dr. Jörg Link. Her research priorities are in the fields of Marketing-Controlling, Customer Recovery Management and Performance Measurement. Together with Prof. Link she leads a scientific research project called *Customer migration: Early warning, prevention, customer win-back*. Furthermore she works as a lecturer in business studies.

Günter Silberer (Dr. rer. pol., University of Mannheim, Germany) is professor of marketing and director of the Institute of Marketing and Retailing at the Georg-August-University Göttingen. Main research interests include consumer behavior, tracking research, Point of Sale marketing, online marketing and retail management. His work has appeared in journals such as Journal of Consumer Policy, Journal of Business Research and Marketing ZFP. He has chapters in various German marketing readers and edited different publication concerning interactive marketing. He wrote books (in German) on mood as marketing factor, multimedia in sales conversation, on value-oriented business management and on the diffusion and use of comparative product test results.

George J. Siomkos, Ph.D., is Professor of Marketing at the Athens University of Economics and Business (AUEB). He is also the Director of the MSc Graduate Program in Services Management and the Director of the Market and Consumer Behavior Analysis Laboratory at AUEB. He has taught in Stern School of Business, New York University, Long Island University, ALBA, University of Pireaus and University of Macedonia. His research and publishing interests include: consumer behaviour, strategic marketing and planning, product-harm crisis management and electronic marketing. He holds a B.Sc. (Hons), an M.B.A., an M.Sc., an M.Phil. and a Ph.D. in Marketing and Corporate Strategy from the Stern School of Business, New York University. He has published over 80 articles in academic journals including Journal of Retailing, European Journal of Marketing, Journal of Consumer Behaviour, Journal of Business Strategy, Long Range Planning, Advances in Consumer Research. His research work has been presented in many international conferences around the world. He is the author of eight university textbooks on various marketing areas.

Ibrahim Sirkeci (PhD, Sheffield) is a Senior Lecturer teaching international marketing at European Business School London, Regent's College, where he is also an executive team member of the Institute of Contemporary European Studies. He previously worked as Leverhulme Research Fellow at the University of Bristol and Assistant Professor of Management in Ankara, Turkey. His areas of expertise cover transnationalism, demographics, mobility, segmentation, and conflict. He has conducted research in the UK, Germany, Turkey and Iraq. Ibrahim's recent research focuses on ethnic segregation and labour market outcomes in the UK, research reputation in business school choice, conflict model, and mobility of foreign-born. He published several books and more than 40 book chapters, journal articles, and reviews in outlets by international scholarly publishers including ABC-Clio, Blackwell, Edwin Mellen, and Lexington. He is also the managing editor of Migration Letters journal and serves in the editorial boards of several academic journals.

Stefan Smolnik is an Assistant Professor at the Institute of Research on Information Systems (IRIS) at the European Business School (EBS), Oestrich-Winkel/Wiesbaden, Germany. He holds a doctoral degree from the University of Paderborn in Germany. Before joining EBS, he worked as a research and teaching assistant at this university's Groupware Competence Center. Stefan Smolnik has done research on the success and performance measurement of information and knowledge management systems, which has included several benchmarking studies. In addition, he is interested in the successful organizational implementation of social software. His work has been published in the Business & Information Systems Engineering journal, the International Journal of Knowledge Management, the Business Process Management Journal, and the Proceedings of the Annual Hawaii International Conference on System Sciences.

Tobias Straub is a professor of applied computer science at Baden-Württemberg Cooperative State University located at Mannheim, Germany. Prior to that he worked as a researcher at Fraunhofer-Institute for Secure Information Technology in Darmstadt and headed the Fraunhofer Competence Center Public Key Infrastructures (PKI) until mid 2008. From 2002 to 2005 he was enrolled in the PhD program "Enabling Technologies for Electronic Commerce" (DFG Graduiertenkolleg) at Technische Universität Darmstadt. He received a PhD in computer science (Dr. rer. nat.), investigating the usability of PKI-based security mechanisms. Tobias Straub studied mathematics and computer science at Universität Tübingen and ETH Zürich. He holds a diploma degree in mathematics from Universität Tübingen (2001). His research interests cover the areas of IT security and privacy, focusing on applications of cryptography and public key infrastructures.

Yasuhiro Uwadaira currently works for WACCORD Co., Ltd., serving as a sales representative in Department of Marketing and Sales Promotion. He earned a master degree in Master's Program in Management and Public Policy from the Graduate School, the University of Tsukuba, Japan in 2005. He is currently enrolling a doctoral program in Appropriate Technology and Sciences for Sustainable Development, Graduate School of Life and Environmental Sciences, The University of Tsukuba.

Ramin Vatanparast (Ph.D.) has over 16 years of technical and business development experience and has worked in the global telecommunications, consumer electronics, software solutions and mobile marketing industries. Mr. Vatanparast is Senior Strategy and Business Development Manager in Nokia Inc. (Nokia Interactive Advertising). In addition to his role at Nokia, Mr. Vatanparast sits on the Mobile Marketing Association (MMA) Committee and Editorial Review Board of International Journal of Mobile Marketing. He holds two Ph.D. degrees in Technology and in Strategy & International Business. He is also a Faculty Member in University of Phoenix and also a member of Academic Teaching Staff at Asia Pacific International College. He has written more than 30 articles in technology, management and mobile marketing. Mr. Vatanparast is focusing his research on the mobile business, mobile services, user adoption, emerging market strategy and mobile marketing.

Adam P. Vrechopoulos is Assistant Professor at the Athens University of Economics and Business (AUEB), Department of Management Science and Technology and Scientific Coordinator of the Interactive Marketing and Electronic Services (IMES) research group at the ELTRUN Research Center at AEUB. His main research interests are Digital Marketing and Electronic Retailing. He holds a Ph.D. from Brunel University at UK, an M.B.A. from ALBA, and a B.Sc. in Information Systems from AUEB. He has participated in many funded research projects and acted as researcher at the Electronic Business Interaction Research Group at Brunel University. He has published more than 70 papers in peer reviewed journals and academic conferences including Journal of the Academy of Marketing Science, Information Systems Journal, Journal of Retailing, and has acted as a reviewer for several international journals, member of conferences' scientific committees and books' editor. Before starting his academic career he worked in the industry in marketing, sales and project management positions.

Klaus-Peter Wiedmann is a Full Chaired Professor of Marketing and Management and the Director of the Institute of Marketing and Management at the Leibniz Universität Hannover, Germany. He is also the German Representative of the Reputation Institute, New York et al., Deputy Chair of the Academy of Global Business Advancement (AGBA), and as a Visiting Professor at the Henley Business School faculty member of the Henley University of Reading, UK. His main research areas are: Strategic marketing, international marketing, innovation & technology marketing, brand & reputation management, corporate identity, consumer behavior, marketing research & controlling, online & mobile marketing.

Index

D

J

K

L

M

N

R

S

T

U

V

W